PHYSICS for Biology and Pre-Medical Students

SECOND EDITION

PHYSICS for Biology and Pre-Medical Students

SECOND EDITION

DESMOND M. BURNS

SIMON G. G. MACDONALD

University of Dundee

ADDISON-WESLEY PUBLISHERS LIMITED

London · Reading, Massachusetts · Menlo Park, California
Don Mills, Ontario · Amsterdam · Sydney · Manila

WORLD STUDENT SERIES EDITION

 Printed in the United States of America. Published simultaneously in Canada. Library of Congress Catalog Card No. 75-7677.

To Pat and Denise

PREFACE TO THE SECOND EDITION

The first edition has been widely used in a large number of countries and in the preparation of this second edition we have sought advice from all these quarters. Unfortunately, because of the differing backgrounds with which biological and premedical students enter university in the different countries, some of the suggestions for amendments were self-cancelling. The present volume represents what to us seems the best compromise.

There was a widespread desire for many more worked examples in the text and this has been satisfied. In addition, our own experience has accorded with that of some users that the amount of mechanics was excessive, and this section has been severely pruned. In compensation, the properties of liquids have been more extensively treated, and many more examples of relevance to biology and medicine have been given. A number of comments suggested that, tubes being now obsolete except for highly specialized purposes, the chapter on electronics should be rewritten from a solid state point of view, and this has been done. These are the major changes.

In addition, new techniques and new advances have been included where they seemed of major importance, and some omissions have been repaired. A number of new problems have been added. Several additional diagrams are included, and many of the old ones redrawn. We believe that most users will feel that their main suggestions have been attended to.

Our most sincere thanks are due to the many people who have taken the trouble to write to us with helpful suggestions, in particular to Professor K. Maack Bisgård, Dr. B. Buchmann, Professor J. W. Brommert, Dr. M. J. B. Duff, Dr. J. A. Scott, Mr. S. Steenstrup, Dr. D. Tovee, Dr. J. W. Twidell and Dr. E. J. Williams. We should also like to single out Miss P. M. Mitchell for her sterling work in producing an immaculate typescript from almost unreadable copy.

January, 1975

D.M.B.
S.G.G.M.

PREFACE TO THE FIRST EDITION

This book represents an attempt to overcome the difficulties experienced by lecturers in physics when they give courses in their subject to pre-medical, pre-dental, and biology students. They soon discover that there is no textbook they can recommend which adequately covers the material they would like to present. They are faced with old-fashioned approaches, watered-down versions of texts for physical scientists, and nonmathematical and verbose treatments really meant for arts and social science students. None of these copes with the problem.

The recent advances in physical science and their wide application in all fields means that nowadays biologists, doctors, and dentists are dealing with highly sophisticated physical apparatus, and are forced to know something of quite advanced physical concepts. An attempt to give this group of students a physics course suitable to their future needs often founders on the twin difficulties of the lack of time available and the poor mathematical equipment of the students. In the following text an approach has been taken which goes some way toward solving these difficulties.

All biological, medical, and dental departments in the University of Dundee were canvassed as to what physics they would like their students to learn during their first-year course. A long series of discussions with these departments quickly removed the obviously absurd and the clearly impossible, and finally the real essentials were arrived at. Large portions of traditional physics, such as statics, have been eliminated from the course almost completely, and others barely touched on, in order to make room for several topics which are considered of great importance. Some material only treated in a physics honours course has been introduced, simply and relatively nonmathematically, in order to give the students some understanding of matters which loom large in their later courses.

It was found essential to give the students, who attend no mathematics courses in the University, and some of whom have poor school qualifications in that subject, what we would describe as a reading knowledge of mathematics, particularly simple calculus, vector representation, and statistics. The first two of these topics are introduced as painlessly as possible through the physics, in an attempt to show that math is just a convenient and shorthand method of coping with certain problems that must be tackled. Statistics, necessary for a complete understanding of some of the later material and essential in the future careers of these students, has been allotted a chapter early in the book. Some lecturers may

CONTENTS

Chapter 21 Spectrophotometry

Chapter 22 Electrostatics

Chapter 23 Conduction in Solids

Chapter 24 Ionic Conduction

Chapter 25 Magnetism

Chapter 26 Electromagnetic Induction

Chapter 27 Alternating Current

Chapter 28 The Atomic Nature of Matter

Chapter 29 Radioactivity

Chapter 30 The Rutherford-Bohr Atom

CHAPTER 1

PHYSICS AND PHYSICAL MEASUREMENT

1.1 INTRODUCTION

Many biological, pre-medical, and pre-dental students regard the course on physics which they are required to pursue as an unnecessary and unfair obstacle in the path to their qualification. Indeed, we have heard a Dean of Medicine declare that the only purpose the physics and chemistry courses served was to weed out the less able undergraduates, since there were few later courses that could do the job with such efficiency. When wrong-headed ideas of this sort abound, it is worth while starting a textbook of physics aimed at these groups of students by stamping vigorously on such misconceptions of the purpose of a study of the subject.

At one time the avowed aim of general scientific courses was to give a good grounding in scientific method and scientific philosophy to all students. A case could still be made out for doing this; but in the modern climate of opinion ancillary physics courses are unlikely to survive if this is their only purpose. Biologists, doctors, and dentists are quite clear that however worthy the aim of improving the general education of their students may be, this is the least of the purposes which the present course serves. Indeed, their desire is for more physics to be taught, because of the difficulty of dealing with modern biology and medical technology without a good grounding of physical knowledge.

Physics has advanced much more rapidly in the last few hundred years than most other subjects. Its methods have proved so fruitful, and its results so far-reaching, that all other sciences have drawn more and more heavily from it. Let us quote some remarks made in a current textbook of biology in wide use.* "Current study in molecular biology indicates that areas such as genetics, evolution, general physiology, and even the classification of organisms can all be carried to the molecular levels. The molecular approach to biology is based on our present-day understanding of physics and chemistry, and their applicability to biological problems. . . . By using the tools of physics and chemistry, it is possible to treat many biological problems with precision and with rigour."

This represents a typical view of the direction in which biological science is moving, and it is likely that the trend will be even more marked in the future.

---

* *The Study of Biology*, by J. J. W. Baker and G. E. Allen (Addison-Wesley).

The biologist of a few years hence whose physics knowledge is weak will more and more find large areas of biological research closed to him.

A medical student may happily agree that this is true for the biologist but is of no relevance to him, since he is learning to practice a science, or possibly an art, which is different in kind from physics. We would agree with Sir George Pickering, Regius Professor of Medicine at Oxford University, that medicine is neither an art nor a science, but a technology: and technology and science enjoy a symbiotic relationship from which both members benefit. Physics owes much to doctors. Some of the greatest names in physics, such as Copernicus, Bernoulli, Gilbert, Young, and von Helmholtz, were doctors; and Galileo, Foucault, and Davy were medical students who, for one reason or another, never qualified. Much of the modern idea that physics can be useful in instrumentation and in everyday life came from the attempt of doctors to apply scientific principles to the treatment of patients.

Medicine, of course, has benefited in its turn. The last fifty years have seen a spectacular improvement in diagnosis and treatment through the application of scientific discoveries to medicine, and the pace is accelerating all the time. Many of the applications are described in the following pages. It is no longer possible for a doctor to be an expert both in his own field and in the basic sciences, but it is essential that he keep in contact with current ideas in the latter.

It may be asked why the book is not solely concerned with the use and applications of physics in biology and medicine. We would draw attention to the Report of the Working Group on Manpower for Scientific Growth,* which is only the latest of a number of reports to stress the dangers of specialization in subjects which are growing at an ever-accelerating rate. Specialist knowledge learned today may be out of date in five years' time. In this book we have therefore attempted to give a basic understanding of the relevant principles of physics, illustrating these, where possible, with current examples of their use in biology and medicine. The principles will remain the same even if the applications change; and it is a knowledge of the principles which will permit an understanding not only of today's applications but also of tomorrow's.

In the course of preparing and writing this text, the changes that have taken place in some of the fields mentioned have been quite spectacular. The development of artificial pacemakers (cf. Section 24.15) is merely one case in point. At the time of writing the final manuscript there is now a proposal to produce an almost everlasting pacemaker by implanting in the abdomen of the patient an atomic-energy-powered miniature steam-engine which will drive a hydraulic pump to circulate the blood. This is the first step towards a completely artificial heart; yet artificial pacemakers themselves were hardly known when this book was first conceived.

It is because it is certain that the growth of the application of physical ideas to biology and medicine will continue that a book dealing with physical principles

* Cmnd. 3760, H.M.S.O., London.

rather than physical applications is necessary; but we hope that the discussion of current applications will convince the reader that a course of physics not only is of interest but is also essential to a proper understanding of his own field.

1.2 THE METHODS OF PHYSICS

Physics is still in some universities studied under the name of Natural Philosophy. This older title emphasizes the basic nature of the subject and indicates that all other sciences have grown from it; but although a physicist philosophizes about nature, it would be misleading to believe that he uses the same terms of reference as a philosopher. To any scientist truth is determined by the results of experiments only, and any theorizing he does must accord with these results. A scientist is highly successful because he is absolutely precise in the field with which he is dealing, but his field is correspondingly limited in extent.

Pure mathematics by this definition is not a science. No appeal is made to experiment in this case. Axioms are stated and results are deduced from them by logical reasoning, but it is of no interest to the mathematician whether the results he obtains bear any relation to the world around him or not.

Scientific method is something quite different. All scientific enquiry starts with the painstaking collection of facts and a careful examination of the data obtained. Repeatable observations are noted and regularity is observed in certain regions of the data. This suggests that there are underlying laws which control the phenomena under investigation. Inductive reasoning (reasoning from the particular to the general) may indicate the form these laws should take. Rigorous deductions from these laws or axioms are then made and experiments are undertaken to determine whether these deduced phenomena do in fact occur. If they do not, the scientist looks for the flaws in his inductive reasoning and consequently modifies his axioms. If the phenomena do occur, the laws are accepted into the canon of knowledge and used for subsequent deductions and investigations until further data arise which cause him to modify the laws or replace them with others.

The study of dynamics is a good illustration of this process. Various attempts were made from earliest times to explain the motions of bodies, in particular the planets. All of these attempts foundered on the fact that no set of laws postulated could explain all the observed phenomena until Newton's brilliant formulation of the laws of motion and gravitation. Application of these laws was so successful that it appeared for a time that no further basic postulates would ever be required in this subject. However, discrepancies appearing in other branches of physics eventually forced a reappraisal of the whole foundation of dynamics, and it was shown that bodies with velocities close to the velocity of light behaved in unexpected ways. For example, mass, regarded by Newton as an unchanging property of a body, was found to increase markedly as the speed of the body increased. This and other observed phenomena were satisfactorily dealt with in the Special Theory of Relativity.

Further, the very small movement of the perihelion of the planet Mercury was contrary to the results deduced from Newton's laws. Einstein's General Theory of Relativity explained this and predicted the now observed bending of light rays, and the red shift of spectra, in a gravitational field. Newton's laws were seen to be special cases of a more general theory. Now the complete validity of the Theory of Relativity is in question, and in due course a more exact theory is likely to arise.

1.3 THE USE OF MATHEMATICS IN SCIENCE

Although, by the strict definition given earlier, mathematics is not a science, it is increasingly used as a tool in all sciences. No scientist or technologist, in whatever field, can afford to be without a working knowledge of what mathematics can achieve; nor should he be unable to use its results in simple cases. In Appendix A will be found a brief recapitulation of the small amount of mathematics a reader about to start this book is assumed to have already learned. In addition, simple calculus, vector methods, and statistics are introduced in the following pages. The first two of these subjects are brought in as painlessly as possible and at a point which shows their relevance and usefulness. Statistics forms the subject matter of the next chapter. At a later stage of his studies, as well as during the study of physics, the reader will find these simple tools, now so widespread throughout all subjects, quite invaluable.

It is not difficult to see why more and more advanced mathematics appears in any science as time goes by. In the early stages of development of a subject only the simplest facts can be explained. But as more complex ideas are developed to explain more and more complex systems, the underlying laws become more difficult to grasp, to state in precise form, and to argue from. The simplest method of stating laws is in the form of mathematical equations which allow quantitative predictions to be made; and the more complex the system to be described, the more complex the mathematics necessary to explain it. Students often become terrified of mathematics, but mathematics is indeed only a shorthand notation for complex ideas. The student is no doubt correct in believing that he could never have worked out this fearsome notation for himself; but there is no reason why he should not understand what the notation implies in scientific terms, or why he should not acquire a facility, if necessary by rote, in handling the mathematics in the simple cases in which he will use it.

1.4 UNITS AND DIMENSIONS

All sciences rely on experiment to verify the truth of speculations, and thus all sciences have started from the recording of various observed facts. It soon became apparent that quantitative measurements were necessary in order to ensure that all observers agreed on their results, and in order to make predictions precise. The measurement of any physical quantity involves a comparison between the unknown and a standard. Thus lengths are measured against a standard ruler and times against a standard clock. Before measurements can be meaningful, agreement has to be reached between individual scientists as to the standard quantities which will be used in any measuring process.

A physical quantity, in general, will have both a number and a unit attached. The number by itself is insufficient. Thus 1 meter and 1 foot are quite different quantities, as are 1 second and 1 hour. It is also possible to use distinct units to measure the same quantity, the numbers being different in the two cases but obviously related. Over the years many different units have been suggested as standard units against which all further quantities will be measured. A set of units has now been internationally agreed, and this system of units is called the meter-kilogram-second-ampere (MKSA) system.* Systems in common use in the past have been the foot-pound-second (FPS) system, now seldom used except by engineers as more and more countries turn to the metric system, and the centimeter-gram-second (CGS) system. Difficulties arose in electricity and magnetism with the latter system, as is mentioned briefly later (cf. Section 22.5), and the MKSA system has now been universally adopted.

Some quantities have no units attached. Thus relative permittivity is the ratio of two similar quantities, is a pure number, and is said to be dimensionless. Other quantities have dimensions, which is a more basic concept than that of units. A velocity may be described as a number of meters per second or kilometers per hour or miles per hour. All these units are different but they all express the fact that a velocity is defined as the displacement undergone divided by the time of travel. Velocity is thus said to have the dimensions of a length divided by a time; similarly, pressure has the dimensions of a force divided by an area. In shorthand notation one may write the last sentence as

$$[v] = [\mathrm{L/T}]; \qquad [P] = [\mathrm{F/L^2}],$$

$[x]$ meaning the dimensions of x.

The fundamental importance of dimensions is that all the terms of any equation must have the same dimensions. Only quantities with the same dimensions can be added or subtracted or placed equal to one another. Indeed in certain cases this fact is used to determine the form an equation must take (cf. Appendix B).

It is found that the dimensions of any physical quantity are expressible in terms of those of only four others. One therefore chooses four quantities arbitrarily as the basic ones and expresses all others in terms of them. The quantities selected in the MKSA system, the one internationally agreed, are length, time, mass, and electric current. It should be noted that one does not always express a physical quantity in terms of the four basic ones. One may write

$$[P] = [\mathrm{M/LT^2}],$$

but it is often more convenient to write it as it is above when one is more concerned

* The particular variant of the MKSA system adopted is called the Système International d'Unités, or S.I. units for short.

with expressing its relation to force. What the selection of M, L, T, and I as the basic quantities means is that only the units of these four quantities need to be internationally agreed and defined. Once these standard units are established, all other units are automatically fixed.

In practice, two other quantities are defined in the MKSA system: luminous intensity, with which we are not concerned in this book, and temperature. It would be possible to work out a system in which these two quantities were related to the four more basic ones. It is found more convenient to define them separately.

1.5 STANDARDS

At one time there was so much confusion about units that the units of mass, length, and time were internationally agreed in terms of the mass of water, the circumference of the earth, and the length of the day. Since the first two of these three quantities were not known with precision and since, indeed, the values accepted at that time were incorrect, blocks of metal of standard length and standard mass were deposited in Paris to serve as reference standards. Although this was a great advance at the time, it should be clear that it is inherently an unsatisfactory method of definition of standards. Since standards are essentially set up for legal purposes, it is inconvenient to require everyone to compare his units with inaccessible standards in Paris; and, furthermore, it is undoubtedly true that the standards, however carefully preserved, will change with time, as corrosion, crystallization, or any of a number of processes effect alterations. The tendency in recent years has therefore been to define in terms of physical situations which are believed to be unvarying.

The same argument applies to a standard of time tied to the rotation of the earth or the motion of the earth round the sun. The length of the day varies with the time of the year and from one year to the next; the length of the year is no better in constancy. Although it may be essential for many purposes that time should be intimately linked with the motion of the earth, this is not satisfactory for precise scientific measurement.

The current, internationally agreed system of standard units is given below. The meter, abbreviated to m, is the unit of length: it is defined as 1,650,763.73 wavelengths of the red, visible, electromagnetic radiation emitted by an atom of ^{86}Kr when it undergoes a specified atomic transition.

The unit of mass is the kilogram, abbreviated to kg: it is still defined as the mass of a platinum block kept at the International Bureau of Weights and Measures in Sèvres. A secondary definition, which in fact defines atomic masses, is that one kilogram is the mass of 5.0188×10^{25} atoms of ^{12}C. In time, no doubt, the mass of the carbon isotope will be basic and the kilogram will be defined in terms of atomic masses.

The unit of time is the second, abbreviated to s; it is defined as the time taken for the oscillator, which will force atoms of ^{133}Cs to perform a specified transition, to complete 9,192,631,770 oscillations.

The unit of current is the ampere, abbreviated to A: it is defined as that current which exerts a force of repulsion of 2.000×10^{-7} newtons per meter on an identical current when the currents are flowing in opposite directions in long parallel wires one meter apart in vacuum.

The unit of thermodynamic temperature is the kelvin, abbreviated to K: it is defined as 1/273.16 of the thermodynamic temperature of the triple point of water.

The newton, used in the definition of the ampere, is a secondary unit. Secondary units are defined in terms of the basic units from some equation which links them. Thus when a force F is exerted on a body of mass m and produces in it an acceleration a, these quantities are related by the equation

$$F = kma, \tag{1.1}$$

where k is a dimensionless constant. Acceleration is defined as the rate of change of speed, which in turn is defined as the rate of change of distance. The units of speed and acceleration will therefore be meters per second and meters per second per second, respectively (abbreviated to $\mathrm{m\,s^{-1}}$ and $\mathrm{m\,s^{-2}}$). Force must have the units $\mathrm{kg\,m\,s^{-2}}$, which is given the special name of newton (abbreviated to N). The size of the unit is chosen to make the constant k have the value unity. Thus 1 N is that force which produces an acceleration of $1\ \mathrm{m\,s^{-2}}$ on a body of mass 1 kg. The dimensions of F are thus $\mathrm{MLT^{-2}}$.

Note that only those units named after famous scientists have a capital letter in abbreviation. The units used in this book which have special names are shown in Table 1.1. All other units are combinations of those given.

Table 1.1. Standard physical units

Quantity	Unit in basic form	Name of unit	Abbreviation of unit
Length	m	meter	m
Time	s	second	s
Mass	kg	kilogram	kg
Force	$\mathrm{kg\,m\,s^{-2}}$	newton	N
Work or energy	$\mathrm{kg\,m^2\,s^{-2}}$ (Nm)	joule	J
Power	$\mathrm{kg\,m^2\,s^{-3}\ (J\,s^{-1})}$	watt	W
Temperature	K	kelvin	K
Current	A	ampere	A
Charge	As	coulomb	C
Potential difference	$\mathrm{kg\,m^2\,s^{-3}\,A^{-1}\ (JA^{-1}\,s^{-1})}$	volt	V
Resistance	$\mathrm{kg\,m^2\,s^{-3}\,A^{-2}\ (VA^{-1})}$	ohm	Ω
Inductance	$\mathrm{kg\,m^2\,s^{-2}\,A^{-2}\ (\Omega s)}$	henry	H
Capacitance	$\mathrm{A^2\,s^4\,kg^{-1}\,m^{-2}\ (s\Omega^{-1})}$	farad	F
Magnetic flux	$\mathrm{kg\,m^2\,s^{-2}\,A^{-1}\ (Vs)}$	weber	Wb
Magnetic flux density (magnetic induction)	$\mathrm{kg\,s^{-2}\,A^{-1}\ (Wb\,m^{-2})}$	tesla	T
Frequency	$\mathrm{s^{-1}}$	hertz	Hz

In time, all units will no doubt be given in standard form; but there are a number of units which have been for so long in common use that they are accepted for the moment as alternatives to MKSA units. Thus, 10^{-10} m is 1 ångström (Å), 1.6092×10^{-19} J is one electron-volt (eV), 1333 $\mathrm{Nm^{-2}}$ is 1 cm Hg, and so on. A number of such units will be found mentioned throughout the book.

Although the basic units are defined in the manner given above, it is normal practice to create sub-standards for everyday use. Thus length is normally measured by comparison with a ruler, a tape-measure, or some more sophisticated device; time by comparison with a clock; and electric current by use of an instrument known as an ammeter (cf. Section 25.5). Mass is most conveniently measured by the comparison of the weight of an unknown body, i.e. the force exerted on it at a particular place by the gravitational attraction of the earth, with the weight at the same location of a substandard mass. The instrument which does this is the balance.

1.6 POWER-OF-TEN NOTATION: MULTIPLES AND SUBMULTIPLES

Standard units have normally been chosen so that one unit is of convenient magnitude for everyday purposes. Thus 1 m is of the correct order of magnitude for the normal measurement of distances. But other distances will then be of a size which it is inconvenient to write in full. The wavelength of sodium light is 0.0000005893 m, and the distance from the earth to the sun is 149,500,000,000 m. It is ridiculous and wasteful of time and paper to write these quantities in that form. For this reason it is customary to write the magnitudes of all physical quantities in decimal notation as a number between 1 and 10 multiplied by the appropriate power of ten. The two distances quoted above are thus normally written as 5.893×10^{-7} m and 1.495×10^{11} m.

Alternatively, one may write these quantities as 0.5893 μm and 0.1495 Tm. Multiples and submultiples of the meter are being employed and a similar process may be applied to any unit. The internationally agreed abbreviations for multiples and submultiples and the corresponding prefixes added to the name of the unit are given in Table 1.2.

1.7 SIGNIFICANT FIGURES

A student in the laboratory attempting to measure the electronic charge may well give as his result 1.60×10^{-19} C. A reference table gives the accepted value of the same quantity as $(1.60206 \pm 0.00003) \times 10^{-19}$ C. The second value contains more significant figures than the first and, in addition, gives some indication of the accuracy of the result. The quantity $\pm 0.00003 \times 10^{-19}$ C is an estimate of the error in the value given, and implies that the correct value of the constant lies somewhere between 1.60203×10^{-19} C and 1.60209×10^{-19} C. It is this error estimate which determines how many significant figures are given in quoting the value of the physical quantity.

If the likely error in the student's result is $\pm 0.1 \times 10^{-19}$ C (and we shall be explaining in Chapter 2 how errors may be estimated), it is quite incorrect for him to quote the result as 1.60×10^{-19} C. The zero should be left out of the statement of value. Quoting it implies that 1.60×10^{-19} C is a more probable value for the result than 1.61×10^{-19} C or 1.59×10^{-19} C. If the error is likely to be $\pm 0.1 \times 10^{-19}$ C, then this is not true. The number of figures given implies

Table 1.2. Agreed multiples and submultiples of units

Abbreviation	Prefix	Factor
T	tera	10^{12}
G	giga	10^{9}
M	mega	10^{6}
k	kilo	10^{3}
H	hecto	10^{2}
D	deca	10^{1}
d	deci	10^{-1}
c	centi	10^{-2}
m	milli	10^{-3}
μ	micro	10^{-6}
n	nano	10^{-9}
p	pico	10^{-12}
f	femto	10^{-15}
a	atto	10^{-18}

something about the accuracy of the result. Thus 1×10^{-19} C suggests that the 1, though most probable, cannot be relied upon; a figure of 1.6×10^{-19} C states that the 1 is certain but the 6 is not reliable; 1.60×10^{-19} C casts doubt on the value of the zero, and so on. Only the number of figures which are truly significant should therefore be quoted and an error estimate should be given wherever possible.

If, in another experiment, the student measures the mass of a body to be 1.52 kg and measures its acceleration under the action of a force as 2.14 m s^{-2}, he may be tempted to say that the force, being the product of mass and acceleration, is 1.52 kg $\times$ 2.14 m s^{-2} = 3.2528 N. It should be clear, however, that since the mass and the acceleration are both only known to three-figure accuracy, the force cannot be calculated to five-figure accuracy. The greatest number of significant figures that can be quoted is three also. Thus, $F = 3.25$ N. If the errors in the mass and the acceleration estimates are known, it may be found (cf. Section 2.10) that the final 5 is not significant either; without such further evidence, however, it can only be assumed that the force will be known to three-figure accuracy.

1.8 CONVERSION OF UNITS: SYMBOLS

It is often desirable to convert from a particular set of units to another set of units. Thus in a particular problem the speed of a car may be given as 50 kilometres per hour (50 km h^{-1}). If the speed is to be inserted into a formula, it may be necessary to convert the value of the speed given into standard form. This is most conveniently done by inserting conversion factors with units attached and canceling units as well as numbers where possible. Thus

$$50 \text{ km h}^{-1} = \frac{\cancel{50} \, \cancel{\text{km}} \, \cancel{\text{h}^{-1}} \times 10^3 \text{ m } \cancel{\text{km}^{-1}}}{\underset{72}{\cancel{3600}} \text{ s } \cancel{\text{h}^{-1}}}$$

$$= \frac{1000}{72} \text{ m s}^{-1} = 14 \text{ m s}^{-1}.$$

The same unit in a numerator and denominator may be canceled and a unit may be canceled in numerator or denominator with the corresponding unit^{-1} on the same level.

It is often convenient while talking in generalities, or in the solution of problems, to use symbols to denote physical quantities. Indeed, we have done this in discussing Eq. (1.1). Each symbol represents a total quantity, i.e. both the number and the unit. When a symbol in a formula is to be replaced by a particular value in a particular problem, both the number and the unit should be inserted. This ensures that any unknown which is obtained by solution of the equation appears as a number with the correct units attached. A student uses only half the information available if he merely inserts the numbers and guesses the units at the end. Students, in any case, are very poor guessers, and it is ridiculous to trust to luck when success is guaranteed if the correct procedure is adopted.

Once more, units may be canceled out where possible. Thus the formula for the distance s traveled by a body in falling from rest for time t near the earth's surface under the acceleration due to gravity is

$$s = \tfrac{1}{2}gt^2. \tag{1.2}$$

If, in a particular problem, the time of fall is 10 s, since g is a known constant,

$$s = \cancel{\tfrac{1}{2}} \times \overset{4.9}{\cancel{9.8}} \text{ m } \cancel{\text{s}^{-2}} \times 100 \, \cancel{\text{s}^2} = 490 \text{ m}.$$

The result is obtained as a specified number of standard units. The reader will have noted that to avoid confusion between the symbol for distance and the abbreviation for second, symbols are italicized whereas units are given in normal Roman type. The symbols used in this book are those internationally agreed.

PROBLEMS

1.1 What are the dimensions of work and potential difference? (Consult Table 1.1.)

1.2 An atomic particle has an energy of 2560000 eV. Rewrite this in power-of-ten notation and in terms of a multiple of the eV.

1.3 The gravitational constant G has a value of 0.0000000000667 N m^2 kg^{-2}. Rewrite this in terms of power-of-ten notation and in terms of submultiples of the basic units.

1.4 Convert 40 m.p.h. into m s^{-1}.

1.5 Convert 1 lb ft^{-3} into kg m^{-3}.

1.6 Density is mass per unit volume. What are the units of density in the MKSA system?

1.7 The viscous force acting on a sphere of radius r falling through a fluid with speed v is given by the equation $F = 6\pi\eta rv$, where η is called the coefficient of viscosity. What are the units of η in the MKSA system?

CHAPTER 2

STATISTICS

2.1 INTRODUCTION

Since all sciences rely on experiment to verify the truth of speculations, it is necessary to discuss what information can be obtained from the results of experiments and how this information can most easily be extracted and interpreted. An experiment is conducted to obtain facts from which to argue, or to test a particular hypothesis, and a mass of data is the normal result. It is most unusual for a scientist to be able to see immediately from the raw data what the interesting facts obtained from the experiment are, or whether his hypothesis has been proved correct or not. The data are usually in a form too vast to comprehend at a glance, or are at a stage where further calculation has to be done to reduce them to recognizable form. Further, the data are subject to errors due to causes some of which the scientist may be able to eliminate or minimize, but some of which are inherent in the lack of accuracy of the apparatus and are due to factors which the scientist cannot control. These errors make the results more difficult to interpret or to have confidence in.

Statistics is the branch of mathematics most commonly used in this case. The purpose of a statistical analysis of data is first to reduce them from.a bulky form difficult to digest to a form in which they can be easily comprehended, and, second, to assess the importance of the results bearing in mind the inherent errors of the measurements. The first of these aims necessitates some mathematical methods of reduction and presentation of the data—graphically, or by a few parameters. The second implies that, although the results represent a mere small sample of some definite, underlying population,* which we feel will reveal similar results to a future experimenter, it is possible to argue from the particular parameters calculated for the sample to the true parameters which characterize the population; and in the process to assess how near the experimentally obtained and true values are.

As has been stated earlier, a scientist tends to argue inductively from his experimental data, i.e. from the particular results to general laws. The second

* Since a good deal of statistical work has been done in sampling populations, as in public opinion polls, it has now become standard practice in statistical work of all kinds to refer to the situation which is being experimented upon as a population whether a collection of individuals is involved or not.

purpose stated above for a statistical analysis implies that statistics is an inductive branch of mathematics in contrast to most other branches, where deductive reasoning is employed. It would be quite wrong to think that statistics could not be used on occasion to argue deductively as well, and we will start our statistical studies by showing how this can be done. This will show more clearly how theoretical and experimentally obtained values are related.

2.2 PROBABILITY

Any reader who plays bridge or poker and has never heard of percentage plays is almost certain to be a losing card player. Flair plays its part, knowing the psychology of the other players is important, but to play against the odds is foolish. And the mathematical techniques required to establish the best play in any circumstances are extremely simple. It is therefore always a source of wonder that so few card players bother to learn the probabilities of card combinations.

The probability that any event will happen is most simply and most usually calculated experimentally. If m trials of any particular situation are undertaken and n of these yield a particular result, an estimate of the probability of that result occurring is n/m. The value of a probability must always therefore lie between 0 and 1, 0 representing impossibility and 1 certainty. We define the true probability of the result occurring as the limiting value of n/m as m becomes very large, infinite if possible. In shorthand notation we write this as

$$p = \lim_{m \to \infty} (n/m). \tag{2.1}$$

By defining probability in this way, we state our belief that a probability p implies that in m trials on n occasions the desired result will be obtained, where $n = pm$; but that this will only be true if m is very large. In a small number of trials the number of times a desired result is obtained will fluctuate by the random nature of the process about the most probable value. Thus, if we cut a pack of cards, shuffle, cut again, shuffle, and so on, any particular card is as likely to appear in the cut as any other. The probability of the appearance of the ace of spades is thus 1/52. In only 52 trial cuts the ace of spades may not appear, or it may appear once or several times. But we are confident that if the cut were made 520,000 times, the number of times the ace of spades appeared would be very close to 10,000. In this example we have a situation where it is possible by a simple argument (the equal likelihood of any card appearing) to deduce the theoretical value of the probability; and, if the experiment is actually performed, it will be found that the number of times the ace of spades appears divided by the total number of cuts tends more and more to the value 1/52 as the number of cuts increases.

This argument can be extended to slightly more complicated cases. A student is dealt in a poker game four hearts and a spade. If he throws in the spade and gets a further card, what are the chances of its being a heart so that he holds a flush? He has already seen five cards. The remaining 47 contain the 9 other hearts. The probability of his obtaining a flush is therefore 9/47, since the further card has only 9 chances out of 47 of being a heart. This is a good deal worse than the 1 in 4 chance that most people will quote.

It is clear from the last paragraph that alternative probabilities add. The chance of getting a specified heart is 1/47; this is true for each of the hearts remaining. The probability of getting any of the hearts is 9/47, the simple addition of the individual probabilities.

What would be the probability at the beginning of a deal of getting two hearts one after the other? The probability of getting the first one is $13/52 = 1/4$. The probability of getting the second, having already got the first, is clearly 12/51. What is the probability *a priori* of getting both? The probabilities of getting a heart and getting something else with the first card are 1/4 and 3/4, respectively. Having got the heart, the probabilities of the player's getting a heart or something else on the second round are 12/51 and 39/51, respectively; having got a spade, diamond or club on the first card, the probabilities of his getting a heart or something else at the second round are 13/51 and 38/51, respectively. The four possible combinations are HH, HN, NH, NN, where H represents a heart and N not a heart. The sum of the probabilities of these four combinations must add up to 1, since it is certain that one of these hands must be held. But

$$\frac{1}{4}\cdot\frac{12}{51}+\frac{1}{4}\cdot\frac{39}{51}+\frac{3}{4}\cdot\frac{13}{51}+\frac{3}{4}\cdot\frac{38}{51}$$
$$=\frac{12+39+39+114}{4\,.\,51}=\frac{204}{204}=1.$$

It is clear therefore that the probabilities of consecutive events are multiplied to give the combined probability of both events occurring.

2.3 PERMUTATIONS AND COMBINATIONS

A favourite form of competition is one in which a number of TV shows, film stars, fashions, products, etc., have to be arranged in order of merit. In addition, a slogan may have to be invented before the final winner can emerge. Although panels of celebrities are engaged to choose the most apt ordering and the wittiest slogans, this is merely part of the publicity build-up, for, by pure chance alone, any particular order of merit picked at random by the organizers will only be chosen by such a small number of competitors that picking the winner is a relatively easy chore.

Let us consider in how many different ways 12 items can be placed into an order of merit. Into first place can go any of the 12 items, so there are 12 possibilities. The first place having been filled, there are 11 possibilities for second place. There are therefore 12 . 11 ways of filling the first two places. In similar fashion it is easy to show that to fill the whole order of merit table there are

$$12\,.\,11\,.\,10\,.\,9\,.\,8\,.\,7\,.\,6\,.\,5\,.\,4\,.\,3\,.\,2\,.\,1$$

possibilities, i.e. 479,001,600 in all. There cannot be that number of competitors. Any selection chosen at random may not even appear in the entries.

A variation is to give 12 items, say, of which the best 5 are to be placed in an order of merit. By similar arguments to those of the last paragraph, it will be seen that the number of possibilities is 12 . 11 . 10 . 9 . 8 . = 95,040. Now a slogan may be necessary to select a winner from the small number of people who have sent in the correct entry.

In technical language, what has been done in these two examples is to permute 12 items first among 12 places and second among 5 places. From the results it is possible to obtain a general formula. Since it is tedious to write 12 . 11 . 10 . 9 . 8 . 7 . 6 . 5 . 4 . 3 . 2 . 1, or, in general, $n(n - 1)(n - 2) \ldots 3 . 2 . 1$, we write these expressions as $12!$ and $n!$ and call them 12 factorial and n factorial, respectively. The number of permutations of 12 objects among 12 places was $12!$ and of 12 objects among five places was $12!/(12 - 5)!$. In general, in permuting n objects among m places, the number of *permutations* possible is

$$n!/(n - m)!. \tag{2.2}$$

It should be clear from this definition that since permuting n objects among n places gives $n!$ permutations and also $n!/(n - n)!$ permutations, $0! = 1$.

The competition may take a further form. A selection of 12 items being given, the competitor may be asked to choose which 5 are most desirable to have, no order of merit among the 5 being required. How many different entries are now possible? In technical language, one is asking how many *combinations* are possible in selecting 5 objects from 12.

There are 12 . 11 . 10 . 9 . 8 = 95,040 ways of selecting 5 items from 12 if order is important. Once 5 items have been selected, it is clear from our previous discussion that these 5 can be arranged in order of merit in 5! ways.

Therefore $$x \times 5! = 12!/(12 - 5)!$$

Therefore $$x = \frac{12!}{5!(12 - 5)!} = 792.$$

In general, if m objects are to be selected from n objects, order being of no importance, the number of combinations possible is

$$\frac{n!}{m!(n - m)!}. \tag{2.3}$$

It is clear that in the three types of competition mentioned, each entry has a chance of winning of 1/479,001,600, 1/95,040, and 1/792, respectively. Unless a competitor is a lucky guesser, it is obviously a waste of time to spend endless hours on polishing up a slogan in the first two cases.

2.4 PROBABILITY AND COMMON-SENSE

It should be apparent from the example of the poker hand given in Section 2.2 that common-sense is a poor guide to the working out of probabilities. Even experienced players will tell you that the chance of drawing to a flush is 1/4. This is a situation common to many calculations of probability. One should therefore be on one's guard against relying on instinct; probabilities should always be worked out to make sure that common-sense is not leading one astray. A simple example will illustrate this.

If one is in a room which contains 30 people, quite randomly chosen, would it be better to bet for or against there being two people present with the same date of birth, only day and month being considered? Most people would unhesitatingly bet against. The argument would run along the following lines. There are 365 possible dates; there are only 30 people present. The probability of two people having the same birth date must therefore be around 30/365. And that argument is quite false. You are betting with the probabilities if you bet for two people having the same birth date. Indeed the odds are well in your favour.

The correct argument for obtaining the probability of the event is not difficult. The first person in the room has a particular date of birth. The probability that the second has not the same birthday is 364/365. The probabilities that the third, the fourth, the fifth, etc., all have distinct birth dates are 363/365, 362/365, 361/365 etc. The combined probability that n people have all distinct dates of birth is thus

$$p_n = \frac{364}{365} \cdot \frac{363}{365} \cdot \frac{362}{365} \cdot \frac{361}{365} \cdots \frac{365 - (n - 1)}{365} \tag{2.4}$$

When n is small, this value is close to 1. If n is large, the probability is close to zero. For what value of n is the probability 1/2? This occurs when

$$\frac{364 \,.\, 363 \,.\, 362 \,.\, 361 \,\ldots\, 365 - (n - 1)}{365 \,.\, 365 \,.\, 365 \,.\, 365 \,\ldots\ldots\, 365} = 1/2.$$

This equation is not difficult to solve, if necessary by trial and error or graphically. We will merely quote that the result is $n = 23$. With 23 people in the room, it is an even-money bet whether two of the people have the same birthday. The probability that no two of 30 people will have the same birth date is obtained by inserting $n = 30$ in the probability expression of Eq. (2.4). The answer is 0.30. You are therefore betting on a better than 2:1 chance if you bet that there are two people in the room with the same birthday.

In the previous three years to the time of writing, the numbers in the predental and premedical classes at the University of Dundee have been 27, 44; 30, 65; and 27, 70. The corresponding numbers of "twins" have been 0, 1; 1, 4; and 0, 8. In the combined classes of 71, 95, and 97, the numbers with the same birth dates have been 4, 8, and 13.

2.5 PROBABILITY AND GENETICS

In his original experiments on crossing garden peas Mendel realized that the laws of probability were operating in the characteristics displayed by the products of his experiments. This observation laid the foundations of the modern theories of genetics. Let us consider Mendel's original experiments, interpreting them on the basis of modern ideas, and show how the laws of probabilities are involved.

When tall peas are crossed with dwarf peas, all the resulting plants are tall. Each parent has contributed one particular gene to any offspring, the tall pea contributing a tallness factor T and the other a dwarf factor D. The tall factor is dominant, the dwarf factor recessive, and, hence, all offspring are tall. If two offspring are crossed, each contributes a gene to the second-generation pea. The offspring possess T- and D-genes, and it is a matter of pure chance which one is contributed to the next generation. Thus the pea may receive TT, TD, DT, or DD combinations. All these are equally likely. Three contain a T-factor, which is dominant, and therefore grow as tall peas. One contains no T-factor and therefore grows as a dwarf. The probability of a dwarf second-generation plant being produced is thus $1/4$. Mendel actually found that the result of crossing offspring peas was always to produce a crop in the next generation approximately one-quarter of which were dwarf plants.

Another characteristic of peas noticed by Mendel was that some had yellow seeds and some green seeds. Experiments similar to the above showed that yellow was a dominant characteristic and green recessive. Further, the seeds could be smooth or wrinkled, and experiments showed smoothness as dominant and the wrinkled characteristic as recessive. If we hypothesize that each of these characteristics is carried by a gene, what will be the result in the second generation of crossing peas with yellow (Y), smooth (S) characteristics and peas with green (G), wrinkled (W) characteristics?

The first-generation plants will each contain YS characteristics contributed by one parent and GW characteristics contributed by the other. Thus each has a pair of genes YG and another pair SW. The combinations any plant can contribute to the second generation are YS, YW, GS, GW. Each of these combinations can also be contributed by the other parent of the second-generation plant. The latter therefore can have the following combinations of genes: $YYSS$, $YYSW$, $YGSS$, $YGSW$, $YYWS$, $YYWW$, $YGWS$, $YGWW$, $GYSS$, $GYSW$, $GGSS$, $GGSW$, $GYWS$, $GYWW$, $GGWS$, $GGWW$. Each of these combinations is equally likely.

Nine of the combinations contain both the dominant factors Y and S; three, the dominant Y but two recessive W's; three, two recessive G's but the dominant S; and one contains only recessive factors. In any colony of second-generation plants the numbers producing smooth yellow seeds, wrinkled yellow seeds, smooth green seeds, and wrinkled green seeds should therefore be in the ratio $\frac{9}{16} : \frac{3}{16} : \frac{3}{16} : \frac{1}{16}$. This is indeed the ratio found.

This type of result is of enormous importance in assessing the risks of having children when one or other of the parents has a defective gene of some kind. For example, hemophilia is a condition in which the blood will not clot. Until recently it was almost always fatal, and is certainly very disabling. It is carried by an X-chromosome. It is recessive in women, who have two X-chromosomes, but in a man, who has only one, the disease will appear. If a woman is known to have such a defective X-chromosome, none of her daughters will exhibit the disease, assuming her husband is normal, and half of them will not even possess the defective chromosome. Half of her sons will exhibit hemophilia, since they inherit their X-chromosome from her. The probability therefore is that one in every four of her children will exhibit hemophilia.

More important, her first child has a probability of 3/4 of being normal; her first two children have a probability of $3/4 \times 3/4 = 9/16$ of being normal; her first three children a probability of 27/64 of being normal; and so on. It would therefore be highly inadvisable for her to have more than two children. With two she has a better than even chance that they will be normal; with three or more it is more probable than not that one of them at least will be defective.

2.6 THE ARITHMETIC MEAN AND FREQUENCY DISTRIBUTIONS

An even simpler population than a pack of cards is a coin or a collection of coins. If a single coin is tossed, then it may fall either heads (H) or tails (T). Unless the coin has been deliberately biased, the probabilities of occurrence of H and T are equal; and since the coin must come down H or T, each probability will be equal to 1/2.

In an experiment performed by one of the authors 12 coins were tossed simultaneously, and the number of heads appearing was counted. The procedure was repeated 100 times. The numbers of heads appearing in the 100 trials were

6, 9, 6, 10, 6, 4, 4, 8, 7, 7, 5, 4, 7, 7, 5, 7, 5, 4, 7, 4,
6, 5, 7, 9, 8, 6, 8, 6, 8, 8, 4, 7, 2, 7, 5, 7, 2, 4, 6, 8,
8, 5, 4, 5, 7, 9, 5, 5, 6, 6, 8, 4, 7, 6, 5, 8, 6, 8, 5, 5,
4, 11, 5, 4, 9, 7, 7, 5, 6, 4, 9, 3, 6, 3, 6, 4, 7, 8, 7, 7,
5, 7, 5, 7, 7, 6, 6, 4, 8, 6, 7, 3, 2, 6, 5, 9, 4, 8, 6, 5.

The most probable number of heads appearing is 6. Yet it will be noted that had the experiment started at the sixth tossing, it would have been necessary to wait until the sixteenth throw actually to obtain 6 heads. Owing to the random nature of the process, the number of heads appearing fluctuates around the most probable value.

What information can be extracted from this mass of data? The most obvious thing is the number of heads that is most likely to appear when 12 coins are tossed; this piece of information is obtained by taking the arithmetic mean of the 100 numbers given above.

What is the justification for taking the arithmetic mean, a process which the reader will realize is carried out with most experimental data? It is not difficult to show that the arithmetic mean is the best estimate of any quantity in the measurement of which only random errors enter. If x_0 is the true value of the quantity, the n measurements $x_1, x_2, \ldots, x_n$ which have been taken can be written as $x_0 + \varepsilon_1, x_0 + \varepsilon_2, \ldots, x_0 + \varepsilon_n$, where $\varepsilon_1, \varepsilon_2, \ldots, \varepsilon_n$ are the random errors in the measurements. Truly random errors are as likely to be positive as negative, and thus when added will tend to cancel out. Thus

$$x_1 + x_2 + \ldots + x_n = x_0 + \varepsilon_1 + x_0 + \varepsilon_2 + \ldots + x_0 + \varepsilon_n = nx_0 + \varepsilon,$$

when $\varepsilon = \varepsilon_1 + \varepsilon_2 + \ldots + \varepsilon_n$ has a most probable value of zero. Therefore

$$\bar{x} = \frac{x_1 + x_2 + \ldots + x_n}{n} \tag{2.5}$$

is the best estimate of the correct value of x_0.

The mean value of the first five data is $37/5 = 7.40$; for the first 10 data it is $67/10 = 6.70$. If the mean values for the first 20, 40, 60, 80, and 100 data are calculated, the values obtained are 6.10, 6.12, 6.12, 6.11, and 6.06, respectively. Note how the result comes closer to the expected value as the number of trials increases. It is almost at the correct value after 20 trials and, with fluctuations, gradually improves to a final value of 6.06.

With such a large amount of data more information should be obtained than just the mean value. One can, for instance, display the data in the form given in Table 2.1.

Table 2.1. Frequency table for the tossing of 12 coins

0	1	2	3	4	5	6	7	8	9	10	11	12
0	0	0.03	0.03	0.15	0.18	0.19	0.21	0.13	0.06	0.01	0.01	0

The top row tells the number of heads noted in any trial. The corresponding entry in the second row indicates the fraction of times that number of heads appeared. This is called a frequency table, and it is more usual to plot the frequency of occurrence of an event rather than the total number of times of occurrence. In this way any two sets of similar experiments are directly comparable. It is perhaps more revealing to display the data of Table 2.1 in graphical form. Figure 2.1 shows the corresponding frequency *histogram* with a dashed curve superimposed. The latter is a smooth curve drawn through points which represent the theoretical probabilities of obtaining n heads when 12 coins are tossed, n taking all values from 0 to 12. It is clear that the experimental estimates of these probabilities are 0, 0, 0.03, etc., i.e. all the values given in the second row of Table 2.1. But how can theoretical estimates be obtained?

The theoretical probability of getting 0 heads is clearly $\frac{1}{2} \cdot \frac{1}{2} \cdot \frac{1}{2} \cdot \frac{1}{2} \cdot \frac{1}{2} \cdot \frac{1}{2} \cdot \frac{1}{2} \cdot \frac{1}{2}$

$\cdot \frac{1}{2} \cdot \frac{1}{2} \cdot \frac{1}{2} \cdot \frac{1}{2}$ or $1/2^{12}$, which equals zero to two-figure accuracy. But what is the theoretical probability of getting x heads and y tails $(x + y = 12)$? Clearly, from the results of Section 2.3, it must be $1/2^{12}$ multiplied by the number of combinations of x objects selected from 12 objects. Thus

$$p = \frac{1}{2^{12}} \frac{12!}{x!(12 - x)!}. \tag{2.6}$$

If $x = 3$,

$$p = \frac{1}{2^{12}} \frac{12!}{3!(12 - 3)!} = \frac{12 . 11 . 10}{2^{12} . 3 . 2 . 1} = 0.054;$$

if $x = 6$,

$$p = \frac{1}{2^{12}} \frac{12!}{6!(12 - 6)!} = \frac{12 . 11 . 10 . 9 . 8 . 7}{2^{12} . 6 . 5 . 4 . 3 . 2 . 1} = 0.226.$$

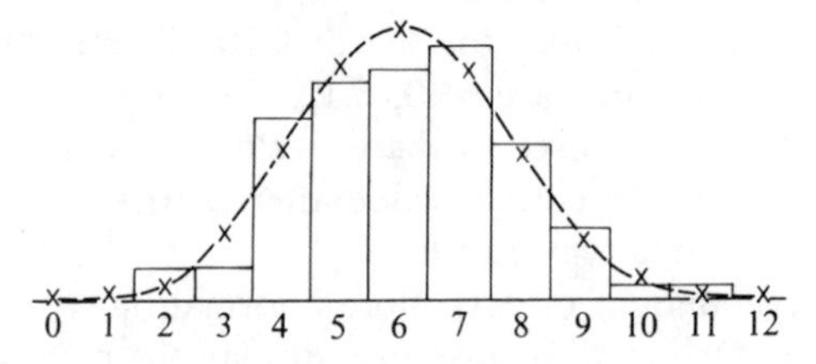

Fig. 2.1. Frequency histogram for the tossing of 12 coins with the theoretical curve superimposed.

In this way the probabilities of obtaining 0, 1, 2, 3, 4, 5, and 6 heads are calculable as 0.000, 0.003, 0.016, 0.054, 0.121, 0.194, 0.226. By symmetry those for 12, 11, 10, 9, 8, 7, and 6 heads are the same. These are the values plotted on the dashed curve.

If the frequency histograms are plotted for 10, 20, 40, 60, and 80 trials, they will be found to approximate less closely to the values on the dashed curve than those for 100 trials. Correspondingly, we believe that, if 1000, 10,000, and 1,000,000 trials were undertaken, the frequency histograms obtained would have values even closer to the theoretical probabilities. This again emphasizes the necessity for a large number of trials before experimental and theoretical probabilities can be expected to have almost the same values.

2.7 STANDARD DEVIATION

Had only five trials been performed in the experiment of the last section, the calculated mean value would have been 7.40; with 10 trials, 6.70. From the whole experiment, the mean was 6.06. The large number of trials brought the experimental value much closer to the theoretical value and therefore reduced the likely error in the final estimate. In all experimental work this will be true. Since it is not always possible to perform a large number of trials in any experimental situation, it would be very helpful to be able to estimate the likely error in the final experimental value quoted. How is this to be done?

A good estimate of the error likely to be present in the final value is the average amount by which the measurements differ from the mean value. The average error will not do, since we have already seen that this has a most probable value of zero whatever the distribution about the mean may be: indeed the mean is taken as the best estimate specifically on this basis. But the average absolute amount* by which the measurements differ from the mean value would be a good measure. Unfortunately, absolute quantities are difficult to handle mathematically. It is therefore usual to measure the sum of the squares of the deviations of all the measurements from the mean value,

$$(\bar{x} - x_1)^2 + (\bar{x} - x_2)^2 + (\bar{x} - x_3)^2 + \cdots (\bar{x} - x_n)^2,$$

which is written in mathematical shorthand as

$$\sum_{i=1}^{n} (\bar{x} - x_i)^2.$$

If this quantity is divided by n to get the average squared deviation, and the square root is then taken, the result is known as the standard deviation, s, and it follows that

$$s = \sqrt{\frac{\sum_{i=1}^{n} (\bar{x} - x_i)^2}{n}}.$$

Each experimenter can then quote his final result as $x \pm s$, giving his best estimate of the quantity measured and a realistic estimate of the likely error involved.

In fact this is not the expression used for s. If only one measurement had been taken, this value would be the mean and s would have the value 0/1 or 0. Thus one measurement would appear to give a value with no error in it, which is obviously absurd. In addition, a mathematician would say that although the n measurements were initially independent, the calculation of the mean has rendered this no longer true. Knowledge of $n - 1$ of the measurements and the mean enables the nth one to be calculated. He would say that there are now only $n - 1$ independent measurements. For these reasons the standard deviation is calculated from the formula

$$s = \sqrt{\frac{\sum_{i=1}^{n} (\bar{x} - x_i)^2}{n - 1}} \qquad (2.7)$$

* The absolute value of x, written $|x|$, is $+x$ if x is positive and $-x$ if x is negative.

For only one measurement, s now has the value 0/0, which is indeterminate. It should be noted that the standard deviation could be calculated with respect to any value but that it is a minimum when taken with respect to the mean. This is another reason why the mean value is taken as the best estimate of a quantity.

For the first five data given in Section 2.6 the mean is 7.40. The standard deviation is thus

$$\sqrt{\left\{\frac{(6-7.40)^2+(9-7.40)^2+(6-7.40)^2+(10-7.40)^2+(6-7.40)^2}{5-1}\right\}}.$$

This can more simply be written as

$$\sqrt{\left\{\frac{3\times(6-7.4)^2+(9-7.4)^2+(10-7.4)^2}{4}\right\}}=\sqrt{3.80}=1.95.$$

The estimate is therefore given as 7.40 ± 1.95.

With 10 data the mean is 6.70. The standard deviation is thus

$$\sqrt{\left\{\frac{3\times(6-6.70)^2+2(7-6.70)^2+(8-6.70)^2+2(4-6.70)^2+(9-6.70)^2+(10-6.70)^2}{10-1}\right\}}=1.95.$$

The estimate is therefore 6.70 ± 1.95.

For all 100 data the standard deviation works out as 1.80. The best estimate of the number of heads appearing in a throw of 12 coins is 6.06 ± 1.80. In accordance with the statement made in Section 1.7 about significant figures, this should be quoted as 6 ± 2.

For an infinite number of trials the standard deviation is ± 1.73.

2.8 NORMAL OR GAUSSIAN DISTRIBUTION

If a large number of students in a physics class measure the time of one oscillation of a simple pendulum of fixed length, by measuring the time of 100 oscillations and dividing by 100, the results obtained may be plotted as a frequency distribution. This will differ from the frequency histogram previously described because not only discrete whole numbers are now possible. Indeed the values obtained may have any magnitude within a small range, and a continuous curve can be drawn through the points plotted. It will be found to be very similar to the dashed curve shown in Fig. 2.1, and the mean $\bar{x}$ and the standard deviation s of the observations may be calculated from it exactly as described in the last section. This type of frequency distribution is by far the most common type of distribution found in a plot of experimental values and is, indeed, one of the only two most people will ever encounter. It is called a normal or Gaussian distribution and its mathematical form is given by

$$f_i = \frac{1}{s\sqrt{2\pi}} \exp\left[-(\bar{x} - x_i)^2/2s^2\right]. \tag{2.8}$$

The $\sqrt{2\pi}$ occurs because the sum of all the individual frequencies of occurrence must add up to unity.

If the normal distribution curve is plotted as the full line in Fig. 2.2 for the frequency of occurrence of the experimentally determined variable x, and the values of $\bar{x} \pm s$, $\bar{x} \pm 2s$, etc., are inserted, ordinates can be drawn at these points. If the equation of the distribution is known, it is not a difficult matter to determine the number of observations which lie inside, or outside, the limits $\bar{x} \pm s$, $\bar{x} \pm 2s$, etc. If you do not know yet how this is done, you will find the mathematical technique for doing so described in Chapter 3. We find that the fractional number of observations lying within the values $\bar{x} \pm s$ is 0.683, within the values $\bar{x} \pm 2s$ is 0.955, within the values $\bar{x} \pm 2.5s$ is 0.988, and within the values $\bar{x} \pm 3s$ is 0.997. Thus when any experimental observation is made, by pure chance it will lie further than s from the mean value of a set of observations, the best estimate of the true value, in 1 in 3 occasions; further than $2s$ in 1 in 20 occasions; further than $2.5s$ in 12 occasions out of 1000; and further than $3s$ in 3 occasions out of 1000. Thus if s is small, as in the dashed curve, it is highly unlikely that any observations will deviate much from the true value; whereas in the dotted curve, where s is large, observations by pure chance may deviate quite markedly from the true value. The standard deviation is thus a good measure of the accuracy of a mean value of a set of observations.

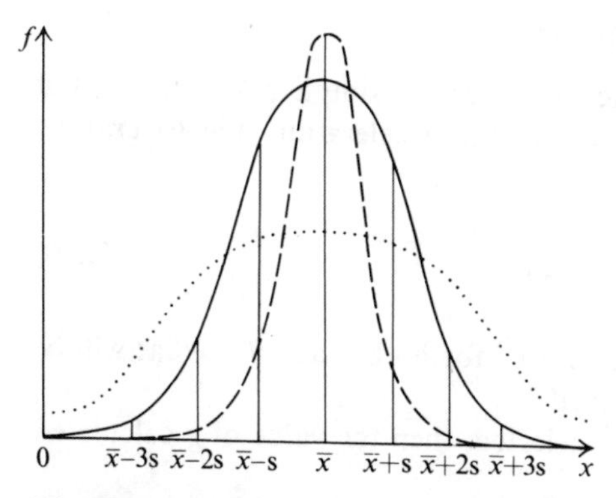

Fig. 2.2. Normal or Gaussian frequency distributions. The full-line curve shows a typical distribution; the dashed curve is a distribution with a smaller standard deviation, and the dotted curve a distribution with a larger standard deviation.

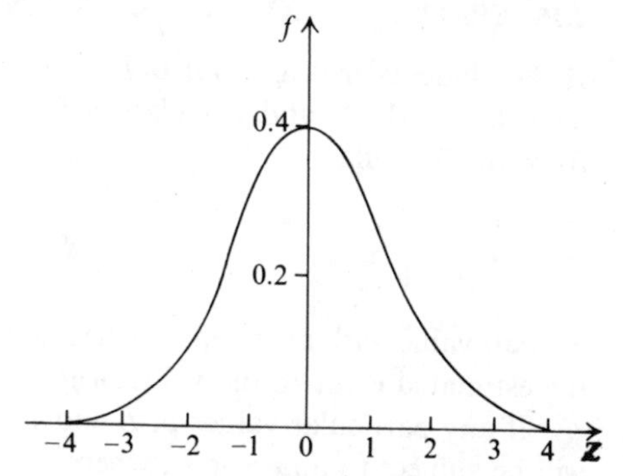

Fig. 2.3. A standardized normal frequency distribution.

In order to compare different normal distributions, it is common to standardize them by moving the origin of coordinates to the mean value and choosing a scale on the x-axis which is in units of s. The variable is now $Z_i = (\bar{x} - x_i)/s$, and the distribution shown in Fig. 2.3 becomes

$$f_i = \frac{1}{\sqrt{2\pi}} \exp\left(-\tfrac{1}{2}Z^2\right). \tag{2.9}$$

2.9 SIGNIFICANCE

If the normal distribution of a particular measured quantity is known and the quantity is measured again under somewhat changed conditions, the value obtained is unlikely to be the mean of the distribution. But whether the value obtained differs from the most likely mean value owing to random errors in the measurement or because the changed circumstances have altered the mean value is not immediately apparent. If it differs little from the mean value, it would be reasonable to assume that no evidence has been presented for other than random errors being involved. If, on the other hand, it is very different from the mean value, it would be just as reasonable to assume that the circumstances have altered the value and that the result is significant. But where is the line to be drawn?

The statistician's criteria are that, if the value obtained deviates from the mean of the distribution by more than $2s$ (i.e. the probability of its occurrence by pure chance is less than 0.05), then the result is significant; if it deviates by more than $2.6s$ (i.e. the probability of its occurrence by pure chance is 0.01), then the result is highly significant.

2.10 ERRORS IN COMPOUND QUANTITIES

If the students measure not only the time period of the simple pendulum T, but also its length L, and calculate a value for g, the acceleration due to gravity, from the formula

$$g = 4\pi^2 \frac{L}{T^2}, \tag{2.10}$$

a mean value with an estimated error will be used for both L and T. What will be the estimated error in the value of g obtained?

If any particular values of L and T are selected, then the value of g calculated will be subject to an error ε_g, where

$$g + \varepsilon_g = 4\pi^2 \frac{L + \varepsilon_L}{(T + \varepsilon_T)^2} = 4\pi^2 \frac{L[1 + (\varepsilon_L/L)]}{T^2[1 + (\varepsilon_T/T)]^2} = 4\pi^2 \frac{L[1 + (\varepsilon_L/L)]}{T^2[1 + (2\varepsilon_T/T)]},$$

since ε_T/T is small and its square may be neglected. If both sides of the equation are divided by g, then, recalling Eq. (2.10),

$$1 + \frac{\varepsilon_g}{g} = \frac{[1 + (\varepsilon_L/L)]}{1 + (2\varepsilon_T/T)} = [1 + (\varepsilon_L/L)][1 + (2\varepsilon_T/T)]^{-1} = [1 + (\varepsilon_L/L)][1 - (2\varepsilon_T/T)],$$

expanding according to the binomial theorem (cf. Appendix A) and ignoring squared terms. Finally, with the same approximation

$$1 + \frac{\varepsilon_g}{g} = \left(1 + \frac{\varepsilon_L}{L} - \frac{2\varepsilon_T}{T}\right);$$

therefore

$$\frac{\varepsilon_g}{g} = \frac{\varepsilon_L}{L} - \frac{2\varepsilon_T}{T};$$

therefore

$$\frac{\varepsilon_g^2}{g^2} = \frac{\varepsilon_L^2}{L^2} + \frac{4\varepsilon_T^2}{T^2} - \frac{4\varepsilon_L\varepsilon_T}{LT}.$$

But because of the random nature of the errors in L and T, the last term has a most probable value of zero if we sum over all values of L and T. Thus

$$\left(\frac{s_g}{g}\right)^2 = \left(\frac{s_L}{L}\right)^2 + \left(\frac{2s_T}{T}\right)^2. \tag{2.11}$$

By similar reasoning we can arrive at the result that if $P = aQ + bR$ or $P = aQ - bR$,

$$s_P = \sqrt{a^2 s_Q^2 + b^2 s_R^2}. \tag{2.12}$$

Thus, in general, if

$$A = \frac{B^\alpha (aC + bD)^\beta}{(cE + dF)^\gamma G^\delta},$$

$$\frac{s_A}{A} = \sqrt{\left\{\alpha^2 \left(\frac{s_B}{B}\right)^2 + \beta^2 \left(\frac{s_{aC+bD}}{aC + bD}\right)^2 + \gamma^2 \left(\frac{s_{cE+dF}}{cE + dF}\right)^2 + \delta^2 \left(\frac{s_G}{G}\right)^2\right\}}, \tag{2.13}$$

where

$$s_{aC+bD} = \sqrt{a^2 s_C^2 + b^2 s_D^2}. \tag{2.14}$$

2.11 ERROR IN AN AVERAGE

All members of the physics class work out from their individual observations values of g, each with an estimated error. One of the students suggests that they average all these results to obtain an even better value for g. What will be the estimated error of the average?

If there are m members of the class, then the final value obtained is

$$\bar{g} = \frac{g_1 + g_2 + \ldots + g_m}{m}.$$

This may be regarded as obtaining a value of g from a number of variables, i.e.

$$\bar{g} = \frac{g_1}{m} + \frac{g_2}{m} + \ldots + \frac{g_m}{m}.$$

Hence,

$$s_g = \sqrt{\frac{s_{g1}^2}{m^2} + \frac{s_{g2}^2}{m^2} + \ldots + \frac{s_{gm}^2}{m^2}} = \frac{1}{m}\sqrt{s_{g_1}^2 + s_{g_2}^2 \ldots + s_{g_m}^2}, \quad (2.15)$$

and, if $s_{g_1} = s_{g_2} \ldots = s_{g_m}$,

$$s_g = \frac{s_{g1}}{\sqrt{m}}. \quad (2.16)$$

2.12 STUDENT'S *t*-TEST

It is unusual to test the significance of a change in circumstances on a particular quantity by a single observation, as was suggested in Section 2.9. It is also somewhat unusual to have an exact knowledge of the normal distribution of the measured quantity. What is much more usual, and this occurs constantly in biology and medicine, is to have a restricted population on which it is possible to experiment, and a similarly restricted population to act as a control group. The effect of a drug, for instance, might be the change in circumstances which is to be investigated. Experiments on the whole population would yield for the quantity in question a mean value of μ with a standard deviation of σ. The control group of n individuals yields a mean m with a standard deviation s; the experimental group of N individuals, a mean M with a standard deviation S. Are these mean values significantly different?

By the results of Section 2.11, m will be distributed normally about μ with standard deviation $\sigma/\sqrt{n}$, and M similarly with standard deviation $\sigma/\sqrt{N}$ if the drug has produced no effect. Thus, according to Section 2.10, $m - M$ will be distributed about the value zero with standard deviation

$$\sqrt{\frac{\sigma^2}{n} + \frac{\sigma^2}{N}} \quad \text{or} \quad \sigma\sqrt{\frac{n+N}{nN}}.$$

The distribution of $m - M$ is standardized as in Section 2.7 by changing to the variable

$$t = \frac{m - M}{\sigma\sqrt{(n + N)/nN}}. \quad (2.17)$$

Unfortunately, σ is not known. It is estimated from $s^2 = \Sigma(x_j - m)^2/(n - 1)$ and $S^2 = \Sigma(X_j - M)^2/(N - 1)$, or, taking the whole population measured, from

$$\bar{s}^2 = \frac{\Sigma(x_j - m)^2 + \Sigma(X_j - M)^2}{(n - 1) + (N - 1)} = \frac{(n - 1)s^2 + (N - 1)S^2}{n + N - 2}. \quad (2.18)$$

The best estimate of t is thus

$$t = \frac{m - M}{\bar{s}} \sqrt{\frac{nN}{n + N}}. \tag{2.19}$$

The way in which t is distributed has been worked out by W. S. Gosset, writing under the pen-name of Student, for various values of $n + N - 2$ in sets of standard tables, and the significance of the deviation of M from m may be worked out in the same way as the deviation of a single observation from the mean value was considered in Section 2.9. The values of t for 5% and 1% significance levels are plotted in Fig. 2.4 as functions of $n + N - 2$.

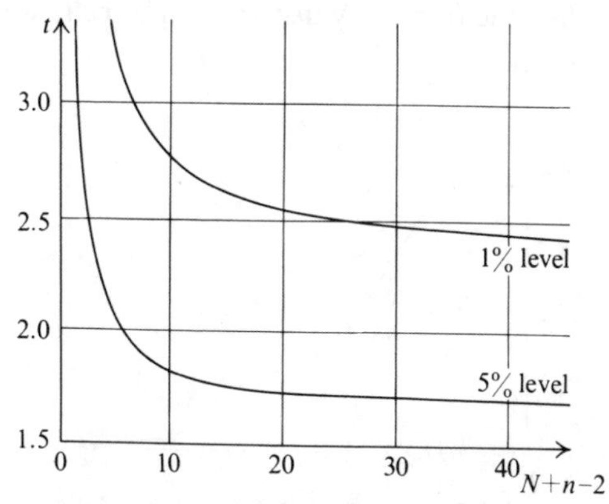

Fig. 2.4. The values of t giving 5% and 1% confidence levels as a function of the number of independent variables $N + n - 2$.

An example may make this clear. In an experiment a drug was fed to 15 nursing mothers in hospital in an attempt to increase lactation. A control group of 12 nursing mothers was fed with an apparently identical but harmless placebo. It is usual to treat both groups identically so that no one learns whether she is the subject of an experiment or not. It has been found that individuals will "improve" if they believe a new drug is being tried on them when they are being fed sugar-pills. The nurses administering the drug are not even told which pill is which in case their attitude is communicated to the patient. The average daily milk supply of the control group was 25.2 oz with a standard deviation of 3.8 oz. For the experimental group the corresponding figures were 29.4 oz and 4.1 oz. Is the difference significant? Can one be confident that the drug produced the improvement?

$$\bar{s} = \sqrt{\frac{(n-1)s^2 + (N-1)S^2}{n + N - 2}} = \sqrt{\frac{(11 \times 3.8^2) + (14 \times 4.1^2)}{11 + 14}} = 4.0.$$

Thus
$$t = \frac{m - M}{\bar{s}} \sqrt{\frac{nN}{n + N}} = \frac{4.2}{4.0} \sqrt{\frac{12 \times 15}{12 + 15}} = 2.7.$$

Figure 2.4 shows that for $n + N - 2 = 25$ independent measurements, or degrees of freedom as they are called, the probability of obtaining a value of t by chance as great as 2.7 is less than 1%. One can therefore assert that the increase in lactation in the experimental group is highly significant, and one may safely assume that the drug has produced the observed effect.

2.13 POISSON DISTRIBUTION

The only other type of frequency curve normally encountered in experimental work is the Poisson distribution, which occurs when the probability of an event's happening is quite low. If a biology class is examining slides only a few of which contain cells at all, then the frequency distribution of cells seen is found to follow

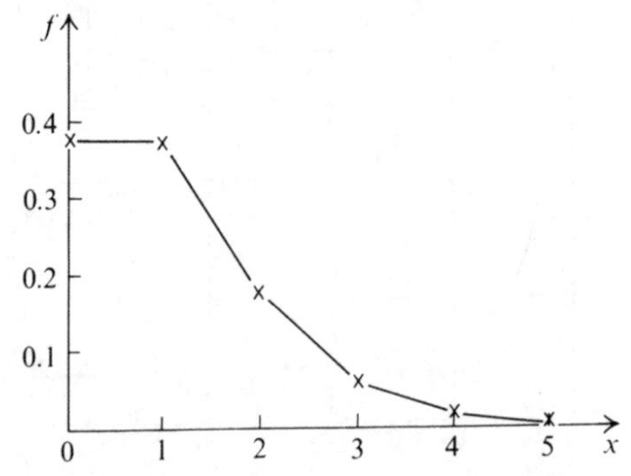

Fig. 2.5. A typical Poisson distribution.

a curve as in Fig. 2.5 rather than a normal distribution. The equation of this Poisson distribution is also known. It is

$$P_n(a) = \frac{e^{-a}a^n}{n!}, \tag{2.20}$$

where $n! = n(n-1)(n-2)\ldots 2.1$ and e is 2.71828 (cf. Section 2.3 and Appendix A). This gives the theoretical frequency or probability with which exactly n events will occur when the average number of events occurring is a. A table such as Table 2.2 can then be constructed for any values of a desired.

Table 2.2 Theoretical frequencies of occurrence of 0, 1, 2, and 3 events in Poisson distributions with various mean values

a	=	0.2	0.6	1	2	3	4	5
$P_0(a)$	=	0.82	0.55	0.37	0.14	0.050	0.018	0.007
$P_1(a)$	=	0.16	0.33	0.37	0.27	0.15	0.073	0.034
$P_2(a)$	=	0.016	0.10	0.18	0.27	0.22	0.14	0.084
$P_3(a)$	=	0.001	0.020	0.061	0.18	0.22	0.19	0.140

An interesting feature of the Poisson distribution is that its standard deviation is equal to the square root of the mean, i.e. $s = \sqrt{a}$. For large values of a, the distribution approximates the normal distribution.

A knowledge of the features of the Poisson distribution helps in the design of some types of experiment. For example, in an experiment to determine the uptake of a radioactive isotope in some part of the human body, it is found from a brief measurement that the number of counts recorded is about 10 per minute and the distribution is Poissonian. If the experiment is only run for one minute, then the standard deviation, which is our measure of the accuracy of the mean number of counts per minute, will be of the order of $\sqrt{10}$, so that our count-rate measurement will be highly inaccurate. If, however, the experiment is run for 1000 minutes, the number of counts recorded will be of the order of 10,000 and the standard deviation will be about $\sqrt{10{,}000} = 100$. Thus the count rate will be 10 ± 0.1 per minute. The count rate, which was only accurate to around $\pm 30\%$ if the counting time occupied one minute, is now accurate to 1%.

In general, in the type of situation where small numbers of events occur in unit time we can calculate how many events must be recorded for a given accuracy. Thus, if the accuracy required is 1%,

$$\frac{s}{a} = \frac{\sqrt{a}}{a} = \frac{1}{\sqrt{a}} = 0.01; \qquad \text{therefore } a = 10{,}000.$$

If the accuracy were 5%,

$$\frac{s}{a} = \frac{\sqrt{a}}{a} = \frac{1}{\sqrt{a}} = 0.05; \qquad \text{therefore } a = 400.$$

Thus 400 events would require to be counted in this case for the accuracy required.

2.14 THE CHI-SQUARED TEST

If a set of n measurements is obtained in an experiment and it is believed that the measurements are samples from a known distribution of mean μ and standard deviation σ, it is possible, as we have seen, to compare the mean m of the measurements with μ to test the truth of the hypothesis. It would also be possible to compare the standard deviation s with σ, although it is more usual to examine chi-squared, defined as

$$\chi^2 = \frac{\sum_{i=1}^{n} (x_i - \mu)^2}{\sigma^2}.$$

If the sample measurements do belong to the expected distribution, then χ^2 should have a value of, or close to, $n - 1$ from the definition of standard deviation.

The χ^2-test is more usually employed when each experimental value is obtained after altering the experimental conditions or deals with a different situation. Thus, in a set of values, one measurement might deal with a control group and each of the others with one of a number of groups which had received different quantities of a given drug. The theoretically expected value μ_j will thus be different in each case. For each single measurement,

$$\chi_j^2 = \frac{(x_j - \mu_j)^2}{\sigma_j^2},$$

or, since we have seen that for a Poisson distribution $\mu_j = \sigma_j^2$, it is more usual to calculate

$$\chi_j^2 = \frac{(x_j - \mu_j)^2}{\mu_j}.$$

If this value is summed over all the experimental measurements, then

$$\chi^2 = \sum_{j=1}^{n} \frac{(x_j - \mu_j)^2}{\mu_j}. \tag{2.21}$$

Tables of the probability of obtaining particular values of χ^2 for differing numbers of independent measurements or variables are listed in many books. Figure 2.6 gives the values of χ^2 for 5% and 1% significance levels as functions of the number of independent variables.

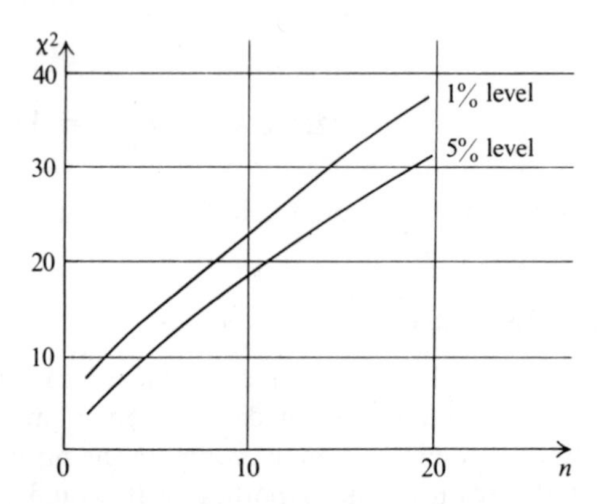

Fig. 2.6. The values of χ^2 giving 5% and 1% confidence levels as a function of the number of independent variables n.

As an example, let us consider a typical biological problem. One hundred fathers and their eldest sons were examined for eye-color. Twenty blue-eyed fathers had blue-eyed sons, and 16 had sons with a different eye-color. For 64 non-blue-eyed fathers the corresponding figures were 14 and 50. We can set this out as in Table 2.3. Thus 36 out of 100 fathers have blue eyes and 34 out of 100 sons. If there is no hereditary factor in the possession of blue eyes, then the corresponding theoretical figures for the incidence of blue eyes in fathers and sons would be those in Table 2.4. These are obtained by distributing the sons with blue eyes and without blue eyes to the two categories of father in the same proportion as for the whole sample population. Thus

$$\chi^2 = \frac{(20 - 12.2)^2}{12.2} + \frac{(16 - 23.8)^2}{23.8} + \frac{(14 - 21.8)^2}{21.8} + \frac{(50 - 42.2)^2}{42.2} = 11.8.$$

Table 2.3. The distribution of 100 sons with respect to their eye-colors and those of their fathers

	S_B	S_N	
F_B	20	16	36
F_N	14	50	64
	34	66	100

Table 2.4. The theoretical distribution of 100 sons with respect to eye-color if the color of the fathers' eyes is of no importance

	S_B	S_N	
F_B	12.2	23.8	36
F_N	21.8	42.2	64
	34	66	100

The number of independent measurements is only one. We started off with four measurements, the numbers of sons in each of four categories; but we fixed the total number at 100, the ratio of blue-eyed to non-blue-eyed fathers as 36:64 and a similar ratio for sons as 34:66, leaving only one independent variable. In other words, having fixed the ratios and the total number, we can construct the whole table knowing only one value.

Consulting Fig. 2.6, we find that for $n = 1$, χ^2 has the values 3.8 and 6.6 for probabilities of occurrence of 0.05 and 0.01. The value of χ^2 obtained is thus highly significant, and it is possible to reject the hypothesis that blue-eyed children are distributed randomly through the population. A hereditary factor is involved. Had the value of χ^2 been small, the opposite would have been true.

2.15 EXPERIMENTAL ERROR

It would be utterly misleading to leave the subject of errors without issuing a word of warning. In a large number of cases which occur in medical and biological work, taking more and more measurements undoubtedly improves the final estimate of the measured quantity and reduces the error inherent in the measurement. In these cases the random error in the measurement is so large that improvement is achieved by defining the distribution better. The same is not always true in more precise measurement and in particular in many situations in physics.

If five students are asked to measure the length of a rod with a meter stick which reads only to the nearest centimeter, the estimates obtained may range from, say, 22.2 cm to 22.6 cm in units of 0.1 cm, the mean being 22.4 cm and the standard deviation 0.1 cm. If now 1000 students make the same measurement, it might appear that the mean can now be given, assuming a normal distribution of values around 22.4 cm, as 22.40 ± 0.01 cm, since the calculated standard deviation will be roughly one-tenth of what it was before. It should, however, be clear that since each student can only estimate to around 0.1 cm at each end of the rod, which we may call the instrumental error, no measurement can be relied upon to more than 0.2 cm. The instrumental error is always the decisive factor and limits the minimum error which can be quoted irrespective of the number of observations made.

It is perhaps easier to see this from the estimate of the length of a room to the nearest foot. If 98 people out of 100 estimate the length as 24 ft and 2 people estimate the length as 25 ft, it should be obvious that to give the length of the room as 24.0 ± 0.1 ft is utterly meaningless. All that can be said is that the length of the room is somewhere between 24 and $24\frac{1}{2}$ ft.

In measurements in physics, if one is dealing with a quantity which is liable to random error, then one can improve the estimate of the quantity and its probable error by increasing the number of observations taken; but where the quantity can only be measured to a given accuracy owing to the scale available for the measurement, this instrumental error is the smallest error that one can ever quote as the probable error in the quantity.

Example 1. In a family the father is normal but the mother is known to have one defective X-chromosome. They have two sons and two daughters. What are the chances that (a) at least one of the sons is hemophilic and (b) at least one of the daughters is a carrier of a defective chromosome?

Solution. a) The chance that either son is hemophilic is $\frac{1}{2}$ because he must have inherited his X-chromosome from his mother. Thus the chance that neither is hemophilic is $\frac{1}{2} \times \frac{1}{2} = \frac{1}{4}$. The chance that at least one is hemophilic is thus $1 - \frac{1}{4} = \frac{3}{4}$.

b) Precisely the same arguments apply to the daughters. They each must have inherited one of their X-chromosomes from their mother and the chance that at least one carries a defective chromosome is thus $\frac{3}{4}$.

Example 2. During an electricity failure, a man fumbles in the dark in a drawer that contains one pair each of white, black, green, and red socks. If he takes out four socks, what are the chances that he has (a) a pair of black socks and (b) at least one pair of socks of any color?

Solution. a) There are 8 socks of which 2 are black. The probability that the first sock taken from the drawer is black is therefore $\frac{2}{8} = \frac{1}{4}$. Similarly, the chance that the second is also black is $\frac{1}{7}$. It is then certain that the third and fourth socks are not black. The total probability of drawing out the socks in the order black, black, not-black, not-black is $\frac{1}{4} \times \frac{1}{7} \times 1 \times 1 = \frac{1}{28}$. The same result is achieved in whatever order these socks are drawn. There are 4! ways of ordering 4 objects and 2! ways of ordering 2 objects. There are thus $\frac{4!}{2!2!} = 6$ distinct ways of ordering the drawing of the two black and the two not-black socks. Thus the probability that the man has two black socks is $6 \times \frac{1}{28} = \frac{3}{14}$.

b) The $\frac{3}{14}$ result of (a) consists of a probability

$$6 \times \tfrac{1}{4} \times \tfrac{1}{7} \times \tfrac{6}{6} \times \tfrac{4}{5} = \tfrac{6}{35}$$

of obtaining one pair of black socks, the other two being of different colors, and a probability

$$6 \times \tfrac{1}{4} \times \tfrac{1}{7} \times \tfrac{6}{6} \times \tfrac{1}{5} = \tfrac{3}{70}$$

of obtaining one pair of black socks and one pair of another color. The same results apply to red, white and green socks. It follows that the chance of getting one pair of socks and two others of different colors is $4 \times \frac{6}{35}$, and the chance of getting two pairs of socks is $2 \times \frac{3}{70}$. We multiply by a factor of 2 and not 4 in the second calculation because the case of, for instance, a red pair + a black pair has already appeared as a black pair + a red pair and multiplying by 4 would mean that every combination of pairs would be counted twice. The probability of getting a pair of socks of any color is thus $\frac{24}{35} + \frac{3}{35} = \frac{27}{35}$.

This result would be obtained more easily by remembering that, if the four contains no pair, each of the socks is of a different color. The probability of this occurring is

$$\tfrac{8}{8} \times \tfrac{6}{7} \times \tfrac{4}{6} \times \tfrac{2}{5} = \tfrac{8}{35}.$$

The probability of at least one pair appearing is thus

$$1 - \tfrac{8}{35} = \tfrac{27}{35}.$$

Example 3. The annual rainfall recorded at St Andrews Old Course over 25 years was

1948, 75 cm; 1949, 77 cm; 1950, 72 cm; 1951, 79 cm; 1952, 68 cm;
1953, 75 cm; 1954, 74 cm; 1955, 71 cm; 1956, 70 cm; 1957, 75 cm;
1958, 63 cm; 1959, 78 cm; 1960, 67 cm; 1961, 78 cm; 1962, 64 cm;
1963, 80 cm; 1964, 71 cm; 1965, 77 cm; 1966, 67 cm; 1967, 76 cm;
1968, 75 cm; 1969, 72 cm; 1970, 81 cm; 1971, 78 cm; 1972, 70 cm.

Calculate the mean and standard deviation of this set of figures.

Solution. The sum of the 25 figures given for annual rainfall is 1833 cm. The mean annual rainfall is thus

$$\frac{1833}{25}\text{ cm} = 73.3\text{ cm}.$$

If 73.3 cm is subtracted from each of the 25 annual rainfall figures and the sum of the squares of the remainders obtained, the value resulting is 585.7 cm². The value of the standard deviation is

$$\sigma = \sqrt{\frac{585.7\text{ cm}^2}{25 - 1}} = 4.9\text{ cm}.$$

In this straightforward way we can calculate the mean and standard deviation required.

However, this method is tedious and cumbersome and in such lengthy calculations mistakes are liable to be made. In the table below the data are given in simpler form and μ and σ are easily calculated. The first column gives the values obtained for the annual rainfall referred to an origin of 70 cm. Thus $x =$ (annual rainfall $-$ 70 cm). The 70 cm is quite arbitrary; any value could have been chosen. The purpose of the shift of origin is to ease the computational burden. The second column gives x^2. The third gives f, the number of times each observation appears. The final two columns give fx and fx^2. The sums of the entries in these columns give $\sum x$ and $\sum x^2$ for all annual rainfalls recorded. $\sum fx/25$ gives μ_x the mean value of x: adding on 70 cm gives the mean annual rainfall.

The standard deviation is

$$\sigma = \sqrt{\frac{\Sigma(x - \mu_x)^2}{n - 1}} = \sqrt{\frac{\{\Sigma x^2 - 2\Sigma x\mu_x + \Sigma\mu_x^2\}}{n - 1}}$$

$$= \sqrt{\frac{\Sigma x^2 - 2\mu_x\Sigma x + n\mu_x^2}{n - 1}} = \sqrt{\frac{\Sigma x^2 - 2n\mu_x^2 + n\mu_x^2}{n - 1}}$$

$$= \sqrt{\frac{\Sigma x^2 - n\mu_x^2}{n - 1}}$$

x (cm)	x^2 (cm^2)	f	fx (cm)	fx^2 (cm^2)
−7	49	1	−7	49
−6	36	1	−6	36
−3	9	2	−6	18
−2	4	1	−2	4
0	0	2	0	0
1	1	2	2	2
2	4	2	4	8
4	16	1	4	16
5	25	4	20	100
6	36	1	6	36
7	49	2	14	98
8	64	3	24	192
9	81	1	9	81
10	100	1	10	100
11	121	1	11	121
		25	83 cm	861 cm^2

$\mu_x = 83\text{ cm}/25 = 3.3\text{ cm}$: $\quad \sigma = \sqrt{\frac{861\text{ cm}^2 - 25 \times (3.3\text{ cm})^2}{24}}$

$\mu = \mu_x + 70\text{ cm} = 73.3\text{ cm}$: $\quad = 4.9\text{ cm}$

The value of Σx^2 is the sum of the final column of the table and the value of σ is thus easily calculable. This is a very convenient expression to use in determining σ since the sum of a collection of numbers and of their squares is normally easily calculable on a modern desk calculating machine.

Example 4. A gas company receives in an average week 101 emergency calls distributed randomly around the clock. Each call lasts on average 1 hour. Calculate the number of staff that should be put on stand-by at any time so that the chances of an emergency call arriving when all the staff are already out is less than 0.5%.

Solution. The average number of emergency calls per hour is $101/(7 \times 24) = 0.6$ and the occurrence of the calls must follow a Poisson distribution. We wish to determine the number of calls per hour whose probability of occurrence is just less than 1/200. But

$$\frac{1}{200} = \frac{e^{-0.6}(0.6)^n}{n!}$$

or, since $e^{-0.6} = 0.55$,

$$\frac{1}{110} = \frac{(0.6)^n}{n!}.$$

For

$$n = 3, \qquad \frac{(0.6)^n}{n!} = \frac{0.216}{6} > \frac{1}{110}.$$

For

$$n = 4, \qquad \frac{(0.6)^n}{n!} = \frac{0.13}{24} < \frac{1}{110}.$$

There must be 4 men detailed for stand-by duty at all times.

PROBLEMS

2.1 A player is dealt three queens in a poker game. If he throws in the other two cards, what are the chances that the replacement cards will contain the fourth queen?

2.2 A player in a poker game is dealt a 3, 4, 5, and 6. If he exchanges his other card, what are the chances that he can complete a straight by obtaining a 2 or a 7?

2.3 A bag contains five black and three white balls. What are the chances that three black balls will be drawn in succession (a) if the ball drawn is replaced each time and (b) if no replacement takes place?

2.4 If two dice are rolled, what are the chances of the total shown adding up to 7?

2.5 Repeat Problem 2.4 with three dice.

2.6 Six balls, all differently colored, and four identical white balls are randomly placed in 10 slots. How many different arrangements are possible?

2.7 Six identical black balls and four identical white balls are randomly placed in 10 slots. How many different arrangements are there?

2.8 A bridge player holds 13 cards made up of 7 spades and 6 hearts. How many different hands are possible with this particular arrangement?

2.9 Black is a dominant color characteristic in cocker spaniels, and red is recessive. If a pure-bred black spaniel is crossed with a pure-bred red, what will be the likely ratio of black to red puppies? If the puppies are crossed, what is likely to be the ratio of black to red in the next generation?

2.10 Red-green color-blindness is transmitted like hemophilia, being carried in the X-chromosome. If the father is normal, and the mother has one defective chromosome, what is the probability that none of their four children will be color-blind? What is the probability if the father is color-blind?

2.11 In addition to the information given in Problem 2.9, it is known that spotting of the coat is a recessive characteristic. If a black spaniel is mated with a spotted red one, and two black, one red, two spotted black, and one spotted red puppies result, what was the gene make-up of the parents?

2.12 Four coins are tossed simultaneously and the experiment repeated 100 times. The number of occasions on which four, three, two, one and zero heads appeared were 5, 28, 35, 24, and 8, respectively. Compare the experimentally determined probability of getting exactly two heads in such a throw with the theoretical probability. Calculate the probability of getting a head in a single toss of a coin from the results and determine whether the theoretical value lies within the expected deviation.

2.13 Ten students in a laboratory class determine the value of g by the use of a compound pendulum. The values obtained are 9.89, 9.70, 10.01, 9.75, 9.78, 9.83, 9.80, 9.69, 9.61, and 9.95 m s^{-2}. Determine the mean and standard deviation of this set of values.

2.14 Equation 17.2 relates Δx, the separation of interference fringes, to the wavelength of the light used λ, the separation of slits d, and the slit-screen distance D. In an experiment the distance of separation of 12 fringes is measured to be 0.80 cm and the slit separation 0.10 cm, both of these measurements being accurate to around 0.05 mm. The distance D is 1.00 m accurate to around 2 mm. Estimate the accuracy to which the wavelength can be calculated.

2.15 The average monthly cost of drugs per member of the population prescribed under the British National Health Service in one region of the country was 56.5 pence in 1967, with a standard deviation of 14 pence. In any practice where the comparable figure for any month is 50% above the regional average, the doctors are asked to justify their high figure. From a statistical point of view, is this a reasonable basis to adopt?

2.16 Customers claimed that the use of a new dye was reducing the strength of a particular yarn. Tests performed on 25 samples of the untreated yarn yielded a breaking strain of 48 units with a standard deviation of 7.1 units, whereas the same test applied to 25 samples of the dyed yarn gave figures of 42 and 6.3, respectively. Is the difference significant? For 48 degrees of freedom the 5% and 1% levels for t are 2.0 and 2.8.

2.17 It was feared that one of the side-effects of a new drug was to reduce the fluid uptake into the body. Clinical trials were therefore performed on a group of 14 patients and a control group of 13 patients. The mean daily fluid output of the experimental group was 63 units, with a standard deviation of 7.5, and for the control group the corresponding figures were 59 and 7.9. Has the experiment confirmed the fears?

2.18 A remote Scottish village has on average 73 births per year. If the incidence of pregnancy occurs randomly throughout the year, on how many days will the village midwife be likely to have to deal with two births?

2.19 A hospital has on average five telephone calls, ingoing and outgoing, occurring at any time during the day. How many external telephone lines are required so that the likelihood of any caller being unable to get a line is less than 1%?

2.20 Because of the unavailability of Mao flu vaccine, a small factory was unable to inoculate its employees. However, 120 of them were inoculated with a different strain of influenza vaccine, and 182 refused the injection. The numbers who contracted Mao flu were 57 and 103, respectively. Is there evidence that the inoculation provided improved immunity?

2.21 Plants infested with a fungus growth were treated by two different methods, A and B. Of 50 plants treated by method A, 36 survived; and of 50 plants treated by method B, 29 survived. Is treatment A significantly better than treatment B?

CHAPTER 3

MOTION IN A STRAIGHT LINE

3.1 INTRODUCTION

That objects in the real world appear to be in a continual state of relative motion is one of the primary data of experience. Traffic moves in various directions in the streets, pedestrians move relative to the buildings, the sun moves over the sky, autumn leaves fall to the ground. Like all really fundamental ideas, the idea of motion is difficult to describe in words, and attempts to do so inevitably reduce to circular statements like "Motion is when things move"—with a shrug to indicate the speaker's contempt for the futility of the question. Yet philosophers of all ages have pondered the problems inherent in the definition of motion. The Greek thinker Zeno, and others following him, came to the conclusion that the popular idea of motion is self-contradictory. Consider Zeno's paradox of Achilles and the tortoise. The assertion was that Achilles ran 10 times as fast as the tortoise, yet, if the tortoise had (say) a start of 100 m, it could never be overtaken. For when Achilles had gone the 100 m, the tortoise would still be 10 m in front of him; by the time he had covered these 10 m it would still be in front of him; and so on for ever. Thus Achilles would get nearer and nearer to the tortoise but would never overtake it. The resolving of the paradox depends on the recognition that the sum of an infinite series of terms need not be infinite, but this explanation rests on the assumption, which Zeno would not have admitted, that space and time are infinitely divisible. Consideration of ancient paradoxes such as this makes clear the need for a new language suitable for describing the motion of objects, and the purpose of the present chapter is to analyze the essential elements of the simplest kind of motion—that in a straight line—and introduce and discuss the language of calculus.

3.2 THE SPACE-TIME DIAGRAM

To take a concrete situation, consider an automobile traveling along a straight stretch of road. At regular intervals of 100 m there are posted observers to record the times at which the automobile passes so that a complete record of the motion can be obtained. The automobile starts opposite one observer who may be called the reference observer and labelled O. All time measurements are synchronized with him, and it is convenient to measure time from the instant the automobile starts. Thus for observer O, $t = 0$, and the times recorded by the other observers are the times that have elapsed since the automobile started. In the same way, the observer O may be taken as the reference point for the positions of the others. Thus successive observers P, Q, R, . . . to the right of O are said to be at positions +100 m, +200 m, +300 m, etc., and observers N, M, L, . . . to the left of O are said to be at positions −100 m, −200 m, etc. These are the coordinates of the various observers along the line of the road, and may be given the symbol x. Thus $x_P = +100$ m, $x_R = +300$ m, $x_M = -200$ m, etc. The displacement of one observer relative to another is the coordinate of the first minus the coordinate of the second. Thus the displacement of observer P relative to observer M is $x_P - x_M = +100 \text{ m} - (-200 \text{ m}) = +300$ m. This means that P is 300 m to the right of M. Conversely, the displacement of M relative to P is −300 m, since M is to the left of P. The coordinate or position of any observer is just his displacement, relative to the reference observer, and frequently the words "coordinate", "position", and "displacement" are used interchangeably, the reference point being understood. If the reference observer is changed, then all the coordinates will change, but the displacement of one point relative to another will not change. If in any interval of time the car changes its position, the displacement of the car is defined as this change of position. Such a displacement may be positive or negative according as the movement is to the right or the left, and the amount of the displacement is independent of the choice of reference observer.

Consider one possible motion of the car when, starting from rest at observer O, it travels from left to right passing observers P, Q, R, etc., in sequence. The times at which is passes these observers may be tabulated as shown in Table 3.1.

Table 3.1. Times of transit of an automobile past a series of observers

Observer	Position, x (m)	Time of transit, t (s)
O	0	0
P	+100	20
Q	+200	30
R	+300	40
S	+400	50

Such a table defines what is known as a *function*, since to every listed position of the car there corresponds a time of transit and to every time of transit there corresponds a position. The position of the car is said to be a function of the time, or the time can be said to be a function of position. In mechanics it is customary to think of position as being a function of the time rather than the other way round, thus according a privileged status to time. Using the symbols x and t for the position and the time, respectively, we may express the functional relation succinctly as $x = x(t)$. When written in this way, t is referred to as the independent variable, and x is referred to as the dependent variable. The data of Table 3.1 may be displayed by plotting them on a piece of graph paper as shown in Fig. 3.1. The actual data are represented only by the points marked by a circled dot, the smooth curve that has been drawn to connect these points representing something new—something that is not contained in the experimental data. Such a curve is called a space-time diagram. This curve represents the belief that if an observer *had* been posted at a position $x = 250$ m, say, then the car would have passed him at a time $t = 35$ s. This process of reading between the lines of a tabulated function by joining up a set of plotted points by means of a smooth curve is called *interpolation*. The validity of the procedure depends on the details of the particular experimental arrangement. Thus in the present example the spacing of the observers at intervals of 100 m would allow a driver to slow up slightly after passing an observer and then accelerate again to reach the next observer at the time shown in Table 3.1. In this case, therefore, the interpolation procedure is plausible but not certain. However, if the spacing of the observers had been 10 m instead of 100 m, then clearly there would have been little uncertainty in the interpolation. In any particular case judgment is needed.

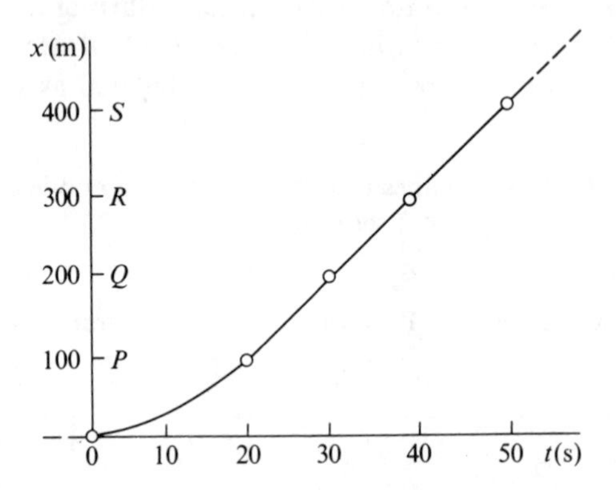

Fig. 3.1. The data of Table 3.1 presented in graphical form.

If an observer had been posted at $x = 450$ m, at what time would the car have passed him? To answer this question the smooth curve must be extended beyond the limit of the actual observations, as shown by the dashed portion of the curve. Such an extension of the curve is called *extrapolation*, and requires even more care than interpolation. The available information about the motion of the car is restricted to the range $t = 0$ to $t = 50$ s, and any statement about what might happen outside this range can only be guesswork. It should be noted, however, that a space-time curve cannot just stop, for this would indicate that the car had ceased to exist. For this reason the space-time curve in Fig. 3.1 is shown extending to time before $t = 0$ as well as to times after $t = 50$ s. Before $t = 0$ the car was stationary at observer O, and so the form of the curve here is known, a straight line along the line $x = 0$.

The space-time curve of Fig. 3.1 is of a particularly simple type in that the displacement of the car from the origin continually increases with time. A more complicated curve is shown in Fig. 3.2. Here the car starts from observer O and moves to the right, passing observers P, Q, and R before stopping at a point just beyond R. It then reverses and proceeds back down the road, passing observers R, Q, P, O, N, and M, and finally slows to a halt at observer L. From this curve it will be seen that although to each value of the time there is only one value of x, there are two possible values of the time for every positive value of x. In other words, a vertical line can cut the curve only once but a horizontal line can cut it more than once. This is because an object can reverse its direction in space but not in time.

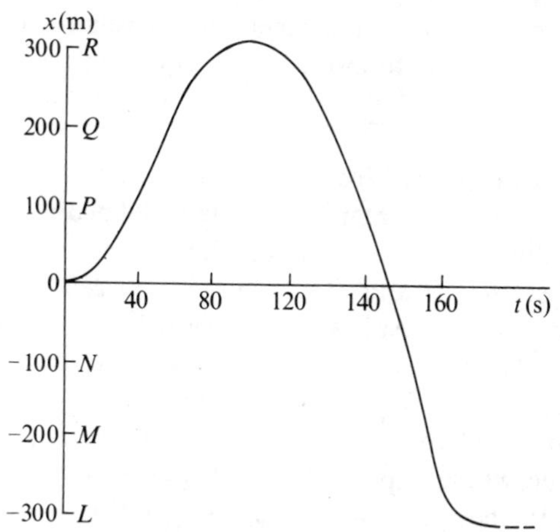

Fig. 3.2. The plot of a more complicated automobile journey than the one shown in Fig. 3.1.

After a time of 146 s from starting, the car has returned to its starting point and its total displacement is zero. In this time it has, of course, traveled a distance of about 640 m. The two concepts of displacement and distance must not be confused. In everyday life it is distance that is important, since the costs of running a car depend on distance and not on displacement. But for the development of mechanics it turns out that displacement is the key concept. The final displacement of the car in this example is -300 m, although the total distance covered is 940 m.

3.3 AVERAGE VELOCITY

Reference to Table 3.1 shows that the car reaches observer P in 20 s from starting, and this corresponds to a displacement from 0 to $+100$ m. The ratio of the displacement of the car in an interval to the length of the interval is defined as the average velocity of the car over that interval. Thus for the first 20 s of its journey its average velocity is $+100$ m/20 s = $+5$ m s^{-1}. For the concept of average velocity the symbol $\bar{v}$ will be used, where the bar over the symbol indicates an average value. By a similar calculation the average velocity after 30 s from the start is $+6.7$ m s^{-1}, after 40 s it is $+7.5$ m s^{-1}, and after 50 s it is $+8$ m s^{-1}. By contrast, consider the average velocity between observers P and Q. The displacement of Q relative to P is $+100$ m, and the time of travel is 10 s: so this average velocity is $+10$ m s^{-1}. This is the same value as is found for the intervals between P and R, between P and S, between Q and R, between Q and S, and between R and S. The average velocities for all these intervals are equal. This equality shows up on the space-time curve, where it will be seen that the four points corresponding to observers P, Q, R, and S lie on a straight line. The common average velocity for any interval between $t = 20$ s and $t = 50$ s is just the slope of this straight line, and is the constant velocity over this interval. If we write v_0 as the symbol for this constant velocity, the displacement relative to observer P of any point between P and S may be found from the equation

$$x - x_P = v_0(t - t_P), \tag{3.1}$$

where $t - t_P$ is the time measured from the instant at which the car passed observer P. The average speed over any interval of time is defined as the distance traveled during that interval divided by the interval. For the motion described by Fig. 3.1 the average speed between O and P is 5 m s^{-1}, between O and Q is 6.7 m s^{-1}, and so on, in every case the result agreeing with that calculated for the average velocity. This is because there is no reversal of direction of the car. If we consider Fig. 3.2, the average velocity for the whole trip from O out to the point past R and then back through O to L is given by -300 m/200 s = -1.5 m s^{-1}. The average speed for the whole trip is 940 m/200 s = 4.7 m s^{-1}. No sign is needed for distance or speed: they are always positive quantities.

Example. A man walks from A to B at an average speed of 1.0 m s^{-1} and immediately returns at an average speed of 1.5 m s^{-1}. What is his average speed for the round trip? What is his average velocity?

Solution. At first sight there would appear to be insufficient data to solve the problem, since two average speeds only are given, with no distance and no times. A naive approach would be to take the mean of the two quoted average speeds to obtain a value of 1.25 m s^{-1} for the overall average speed—and indeed this is the procedure adopted when attempts are made on the world speed record. The two average speeds for two successive runs in opposite directions over a measured mile are recorded and the mean is taken. However, average speed has been precisely defined, and it is not difficult to show that this procedure is incorrect. Let D represent the distance between the points A and B, and let t_1 and t_2 be the times necessary for the two legs of the journey. Then, $D/t_1 = 1.0$ m s^{-1}, and $D/t_2 = 1.5$ m s^{-1}. For the round trip the total distance traveled is $2D$ and the total time taken is $t_1 + t_2$. Consequently, if we apply the definition of average speed, the average speed for the round trip is $2D/(t_1 + t_2)$. The reciprocal of this quantity is then easily found to be

$$\frac{t_1 + t_2}{2D} = \frac{1}{2}\left[\frac{t_1}{D} + \frac{t_2}{D}\right] = \frac{1}{2}\left[\frac{1}{1.0\ \text{m s}^{-1}} + \frac{1}{1.5\ \text{m s}^{-1}}\right] = \frac{1}{1.2\ \text{m s}^{-1}}.$$

Hence, the average speed for the round trip is $1.2\ \text{m s}^{-1}$. This is the so-called *harmonic mean* of the two given average speeds and is the correct mean to take when the two distances involved are the same. The apparently obvious answer of $1.25\ \text{m s}^{-1}$ is the *arithmetic mean*, and would only be correct if the times of the two legs were the same. The word "average" is commonly used instead of arithmetic mean but has been avoided in this example, since the term "average speed" is defined to mean a very specific thing, and great confusion can result from the careless use of words in both a scientifically defined sense and their common, everyday sense. Precision in the definition of concepts is essential for science, but it necessarily involves the loss of any alternative meanings the words used in the definitions may possess.

3.4 INSTANTANEOUS VELOCITY

Again with reference to Fig. 3.1, the average velocity for any interval of time in the range $t = 20$ s to $t = 50$ s corresponding to the straight-line portion of the curve was found to give the same value, this value being labeled as the constant velocity in that range. Consequently, at any instant of time in that range one believes that the velocity has the instantaneous value of $+10\ \text{m s}^{-1}$. For the driver of the car, the needle of his speedometer remains steady over the whole range. For the first 20 s of the trip the average velocity was $+5\ \text{m s}^{-1}$, and clearly this must be the instantaneous velocity at some instant between $t = 0$ and $t = 20$ s. The driver sees the needle of the speedometer move steadily over the dial, and at some instant it must be at a reading corresponding to $5\ \text{m s}^{-1}$. At every instant of time, t, during the trip there corresponds one and only one position of the car. In other words, $x = x(t)$ is a single-valued function of the time. Correspondingly, at the same instant of time the car will have one, and only one, velocity, which may be written as $v = v(t)$. The problem is to correlate the function $x(t)$ with the function $v(t)$, and consideration of this problem leads ineluctably to the branch of mathematics known as differential calculus.

Consider a portion of a space-time curve as shown in Fig. 3.3(a). P_1 and P_2 are two points on the curve corresponding to values x_1, x_2 of the coordinates and t_1, t_2 of the time. From the definition of the previous section, the average velocity over this interval is $(x_2 - x_1)/(t_2 - t_1)$, and this is just the slope P_2Q/P_1Q of the chord P_1P_2. This value of the average velocity is not the same thing as the instantaneous velocity at P, but, if the interval $t_2 - t_1$ is not too large, it will give a reasonable estimate of that instantaneous velocity. To get a better estimate, a smaller interval can be chosen (Fig. 3.3b), this smaller interval still including the time t corresponding to the point P. Once again the slope of the chord P_1P_2 gives the average velocity over this shorter interval. To get an even better estimate, the interval may be shortened even more, as shown in Fig. 3.3(c), where P_1 and P_2 are now so close together that one can be very confident that the average velocity so calculated will be very close to the required velocity at P. At this point a conceptual leap of fundamental importance has to be made. The process of taking the average velocity over progressively decreasing intervals surrounding

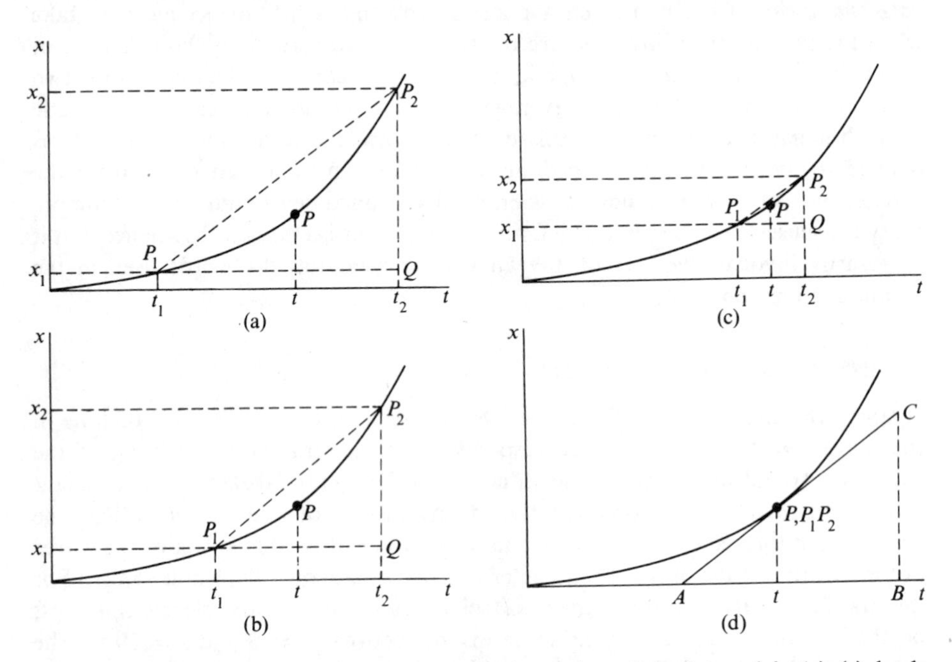

Fig. 3.3. Space-time curves showing how the shortening of the interval in (a)–(c) leads to the average velocity in that interval being a better and better approximation to the instantaneous velocity at a point in the interval: in (d), where the interval has become zero, the instantaneous velocity is the slope of the tangent line drawn at the point in question.

the point P gives increasingly accurate estimates of the instantaneous velocity at the instant t. Hence, if this process could be continued until the interval became zero, the successive values of the average velocity would tend to the precise value of the instantaneous velocity at P. Accordingly, the instantaneous velocity v at any instant t is *defined* as the limiting value of the average velocity over an interval that includes the instant t as this interval is reduced to zero. The process of proceeding to the limit may be interpreted geometrically, since, as the points P_1 and P_2 approach P from each side, the chord P_1P_2 becomes progressively closer to the tangent line at P, and in the limit it actually becomes the tangent line, as shown in Fig. 3.3(d). Indeed, the tangent line to a curve is defined as the limit of chords in this way. Hence, the instantaneous velocity at any instant t is given by the slope of the tangent line to the space-time curve at the instant.

It is most important that students of modest mathematical attainments, who tend to have an exaggerated fear of mathematical arguments and symbolism, should appreciate that the determining of the instantaneous velocity at any time is no more complicated than obtaining the slope of the tangent line to the space-time curve. A graphical method will yield a satisfactory result even when the functional relation between x and t is extremely complicated. In Fig. 3.3(d) the tangent line at P slopes from bottom left to top right, corresponding to a position coordinate x that is positive and increasing with time. In other words, the velocity at P is positive. Its value is given by the ratio BC/AB, where BC is measured in the units chosen along the x-axis, and AB in the units chosen along the t-axis. A positive velocity also corresponds to a position coordinate x that is negative and whose magnitude is decreasing with the time. Both of these situations correspond to motion from left to right. Conversely, a negative velocity is from right to left, and here the tangent line slopes from bottom right to top left. The position coordinate x is either positive and decreasing, or negative and of increasing magnitude. When the tangent line is horizontal, its slope is zero. This corresponds to a position of rest. A vertical tangent line is impossible in space–time diagrams since it would correspond to an infinite velocity. Of course curves that are not space–time curves may have vertical tangents. Figure 3.2 should be studied with all these points in mind, and the various portions of the curve should be interpreted in terms of the motion of the car.

3.5 DIRECT CALCULATION OF THE INSTANTANEOUS VELOCITY

In those cases in which the functional form of the space-time curve is known, the instantaneous velocity may be calculated. Before this is done in a simple case, it is necessary to introduce and explain the meaning of the symbol Δ, which will occur frequently throughout this book. This is the Greek *delta* and is used as a shorthand for the words "a very small increment of". Thus $x + \Delta x$ means a point very slightly to the right of the point x; $t + \Delta t$ means a time very slightly later than the instant t. Note that Δx does not mean Δ multiplied by x. The two symbols are to be read as one to mean "a small increment of x".

Consider a specific problem in which the displacement from the origin of an object traveling along a straight line increases as the square of the time from starting. This specifies the functional form $x = x(t)$ as $x = At^2$, where A is a constant. The graph of this relation is shown in Fig. 3.4. A general point P of the curve corresponding to values x, t is shown, and very close to it a neighboring point P_2, its distance from P being greatly exaggerated in the diagram for clarity. The point P_2 corresponds to a displacement from the origin of $x + \Delta x$ and a time $t + \Delta t$ in accordance with the meaning of the symbol explained above. The instantaneous velocity at the point P is obtained by calculating the average velocity for an interval that includes P, and then shrinking the interval to zero. In Fig. 3.3

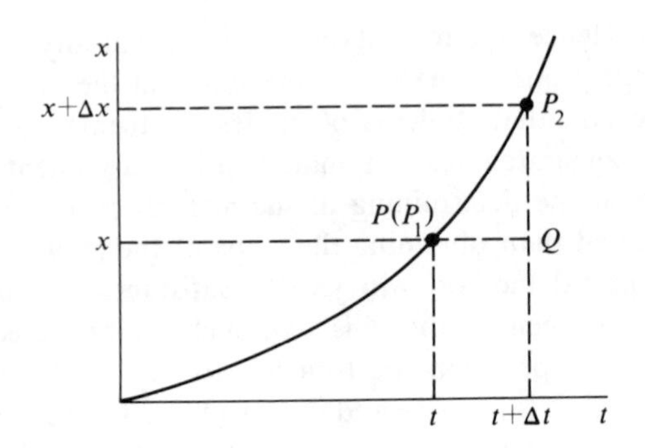

Fig. 3.4. The space-time graph for a body whose displacement increases as the square of the time.

the point P was taken inside the interval, with P_1 to the left of P and P_2 to the right. Clearly, because of the shrinking process, it cannot make any difference where exactly in the interval the point P lies, and here it is more convenient to take it at the left-hand extremity, to coincide with P_1 as shown. The shrinking process now involves moving only P_2 to the left until it coincides with P, which implies shrinking both Δx and Δt to zero. The average velocity is the slope of the chord PP_2, which is $P_2Q/PQ = \Delta x/\Delta t$, and it is the limiting value of this ratio that is required. Since the Point P_2 lies on the curve, its coordinates satisfy the equation

$$x + \Delta x = A(t + \Delta t)^2. \tag{3.2}$$

Squaring out the bracket on the right gives

$$\begin{aligned} x + \Delta x &= A\{t^2 + 2t(\Delta t) + (\Delta t)^2\} \\ &= At^2 + 2At(\Delta t) + A(\Delta t)^2. \end{aligned} \tag{3.3}$$

Since $x = At^2$, subtraction gives

$$\Delta x = 2At(\Delta t) + A(\Delta t)^2,$$

and so

$$\frac{\Delta x}{\Delta t} = 2At + A(\Delta t). \tag{3.4}$$

As the interval is shrunk to zero, the second term on the right-hand side of Eq. (3.4) becomes zero, but the first term does not. Hence, the instantaneous velocity at any time t is found to be

$$v \equiv \lim_{\Delta t \to 0} \frac{\Delta x}{\Delta t} = 2At. \tag{3.5}$$

This equation contains the definition of instantaneous velocity in an obvious shorthand. The velocity $2At$ is called the *derivative* of the displacement At^2. The sign $\equiv$ is to be read as "is defined to be", and must be distinguished from the usual equality sign used in equations. A simple example of the distinction between these signs is the following. One would write $1 ≡ 100 cents, because this merely defines the symbol for dollar in terms of the word "cent" (or vice versa); but one would write £1 = $2.40, because this is an equation expressing a truth about the world of international finance.

Finding the instantaneous velocity when the space-time diagram is specified as a particular function, as in the example above, is a branch of mathematics known as differential calculus. Those of modest mathematical attainments may well feel frightened at this idea, but there is no real need for this. The important point is to grasp the basic idea of taking successive average velocities and proceeding in this way to a limit. The manipulative processes of actually doing this for a given space-time equation are not too important in themselves for medical or biological scientists, especially since they have been worked out once and for all for any of the functions likely to be met with. A short list is given in Table 3.2. Column I contains the most commonly occurring functions and column II contains the derivatives of these. This simply means that if the first column represents the displacement at any time t, then the second column represents the velocity at the same instant. The third column gives the derivatives of column II so that the pattern of obtaining such derivatives may be made clearer. Note that the general rules for

Table 3.2. A short list of derivatives. Column I gives the function; column II, the derivative of the function in column I; and column III, the derivative of the function in column II

I	II	III	
A	0	0	
At	A	0	
At^2	$2At$	$2A$	
At^3	$3At^2$	$6At$	
.	.	.	
.	.	.	
.	.	.	
At^n	nAt^{n-1}	$n(n-1)At^{n-2}$	*
At^{-1}	$-At^{-2}$	$+2At^{-3}$	
.	.	.	
.	.	.	
.	.	.	
At^{-n}	$-nAt^{-n-1}$	$+n(n+1)At^{-n-2}$	*
$A \sin \omega t$	$\omega A \cos \omega t$	$-\omega^2 A \sin \omega t$	
$A \cos \omega t$	$-\omega A \sin \omega t$	$-\omega^2 A \cos \omega t$	
$Ae^{+\omega t}$	$\omega Ae^{+\omega t}$	$+\omega^2 Ae^{+\omega t}$	
$Ae^{-\omega t}$	$-\omega Ae^{-\omega t}$	$+\omega^2 Ae^{-\omega t}$	
$\log_e x$ ($\ln x$)	$1/x$	$-1/x^2$	

positive and negative integral powers of t which are marked by asterisks are really one and the same rule. This holds also when the index n is a positive or negative fraction. A, n, and ω in the above table are constants.

3.6 ACCELERATION

Just as velocity is the rate at which displacement changes with time, so acceleration is the rate at which velocity changes with time. Thus suppose that the relation between displacement and time is

$$x = At^3,$$

where A is a constant. Then from Table 3.2 we find the value of the instantaneous velocity to be

$$v = 3At^2,$$

and we may plot a graph of this relation as shown in Fig. 3.5. The average acceleration over the interval between the points $P(P_1)$ and P_2 is given by the slope of the chord PP_2; or, in symbols,

$$\bar{a} \equiv \frac{v_2 - v_1}{t_2 - t_1} \equiv \frac{\Delta v}{\Delta t}\,.$$

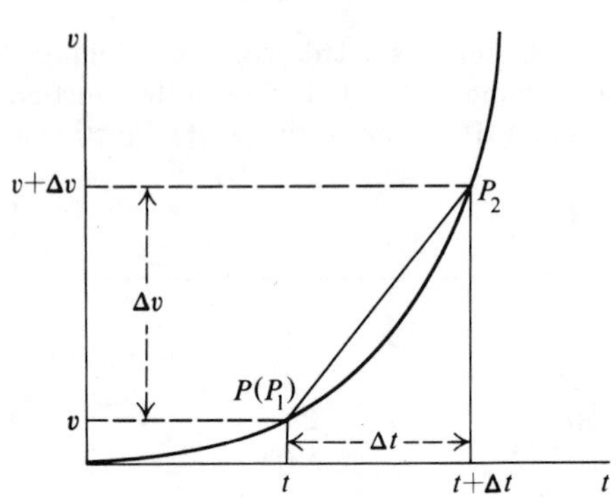

Fig. 3.5. The velocity-time graph for a body whose velocity increases as the square of the time.

The problem of finding the instantaneous acceleration is precisely the same as the previous problem of obtaining the instantaneous velocity, and it is solved in precisely the same way by taking a set of average accelerations for progressively diminishing intervals until the limit of zero interval is reached. The instantaneous acceleration a at any instant t is defined as the limiting value of the average acceleration over an interval that includes the instant t as this interval is reduced to zero. In symbols,

$$a \equiv \lim_{\Delta t \to 0} \frac{\Delta v}{\Delta t}\,. \tag{3.6}$$

Geometrically the acceleration at any instant is the slope of the tangent line to the velocity-time curve at that instant.

Clearly one need not stop here, but could go on and take the rate of change of acceleration with time, and then the rate of change of this new quantity with time, and so on. This is not done, in fact, since it emerges from an experimental study that acceleration is the key dynamical concept, and there is little need for the higher rates of change.

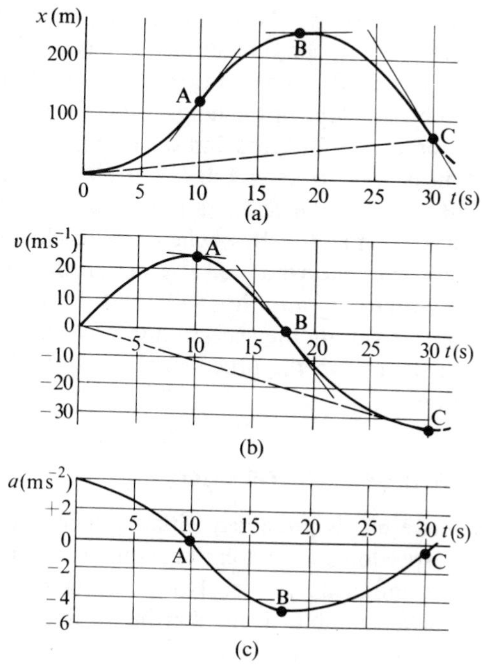

Fig. 3.6. (a) A general space-time diagram: (b) the velocity–time diagram calculated from (a): (c) the acceleration-time diagram calculated from (b).

Consider as an example the motion of an object along the x-axis, the space-time diagram being as shown in Fig. 3.6(a). Inspection of this shows that the object starting from the origin with zero velocity accelerates to the right until it is about 150 m away, then gradually reduces speed until it comes to rest 250 m from the origin. It immediately reverses its direction and proceeds back along the negative direction, increasing its speed all the time. Its last recorded position corresponding to the point C on the curve is at a displacement of +80 m from its starting point. The continuation of the curve is shown dashed, the implication being that we have no information about this part of the motion.

The velocity of the object at any instant can be found by measuring the slope of the tangent line at that instant. Thus at A the tangent line has a slope of +25 m s^{-1}, and at B the tangent line is horizontal, corresponding to the point where the direction of motion is reversed and the instantaneous velocity is zero. The average velocity for the whole interval of 30 s for which data are available is given by the slope of the chord OC, or 2.7 m s^{-1}, whereas the average speed over the 30 s is 14 m s^{-1}, the total distance traveled being 420 m.

If the velocity is measured from the slope of the tangent line at a series of times, a velocity-time curve can be constructed; this is shown in Fig. 3.6(b). The velocity is seen to increase to a maximum value of 25 m s^{-1} at $t = 10$ s corresponding to the point A on the space-time curve. It then decreases to zero at $t = 18$ s, when the change of direction occurs. Thereafter the velocity is negative, since the motion is now from right to left. Its magnitude increases until at the point C it reaches a value of -33.3 m s^{-1}. Beyond that there is no information. The acceleration at any time is given by the slope of the tangent line to this velocity-time curve. This may be measured for a series of times and the result may be plotted as in Fig. 3.6(c). The acceleration is seen to be $+4$ m s^{-2} at $t = 0$, and then to decrease to zero at $t = 10$ s. Thereafter the acceleration is negative until the end. Its magnitude reaches a maximum at $t = 18$ s, where the acceleration is -4.8 m s^{-2}. The average acceleration over the whole 30 s is obtained from the slope of the chord OC of the velocity-time diagram: it is -1.1 m s^{-2}. The important thing to realize is that the process of getting the acceleration-time curve from the velocity-time curve is exactly analogous to the process of getting the velocity-time curve from the space-time curve.

3.7 WORKING BACKWARDS: INTEGRATION

Consider a situation where one is presented with a velocity-time curve, obtained, perhaps, by reading the speedometer of a car at set intervals during a trip. Figure 3.7 shows a possible form of such a curve. How can one obtain from this curve the displacement of the automobile from starting? Consider two instants t_A

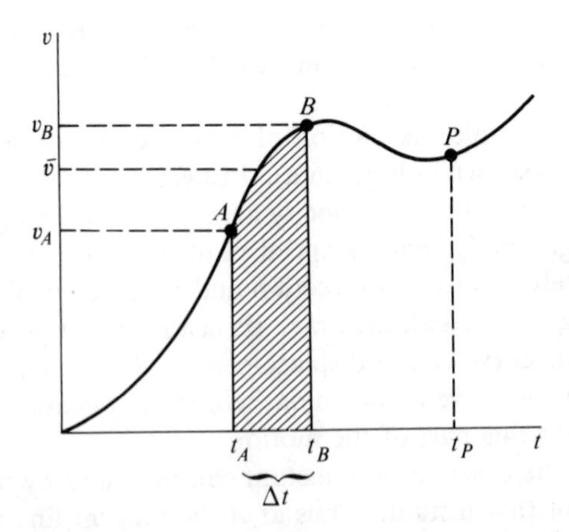

Fig. 3.7. A general velocity-time diagram with the construction for obtaining the displacement.

and t_B, where the interval $t_B - t_A$ may be denoted by Δt. If $\bar{v}$ is the average velocity over the interval Δt, then the displacement in that interval is just $\bar{v} \,.\, \Delta t$. But what value are we to take for $\bar{v}$? If v_A and v_B are the velocities corresponding to the instants t_A and t_B, then clearly v_A is less than $\bar{v}$ and v_B is greater than $\bar{v}$ for the particular situation illustrated. But if the interval Δt is made smaller and smaller, the difference between v_A and v_B also becomes less and less, and either of them may be taken as representing $\bar{v}$ without appreciable error. All three products $v_A \,.\, \Delta t$, $v_B \,.\, \Delta t$, and $\bar{v} \,.\, \Delta t$ are indistinguishable in the limit as Δt tends to zero, and they are all equal to the area of the vertical strip shown shaded in Fig. 3.7.

Finally the total displacement of the automobile up to the point P corresponding to a total time t_P from starting is obtained by adding up all these elementary strips. The procedure is as follows. On the velocity–time graph, a whole series of closely spaced vertical lines are drawn which divide the area under the curve into a large number of areas like that shown in Fig 3.7, but with very much smaller bases. The value of the velocity at the top of each little strip is read off (and it does not matter in the least which point on the curved portion is used) and this velocity is multiplied into the base to give that little area. All these little areas are then added up, and the sum clearly gives just the whole area under the curve between $t = 0$ and $t = t_P$. The conclusion is that the total displacement of an object in time t from starting is given by the area under the velocity-time curve between the origin and time t. If the area under the curve lies below the time axis, the corresponding displacement is negative since this corresponds to an object traveling from right to left. A velocity-time curve for such a situation is shown in Fig. 3.8. The area under the positive portion of the curve would be found first, and then the area under the negative portion. The total displacement in time t_P is then just the algebraic sum of these. If the areas are considered both to be positive, then the sum would give the total *distance* traveled instead of the displacement.

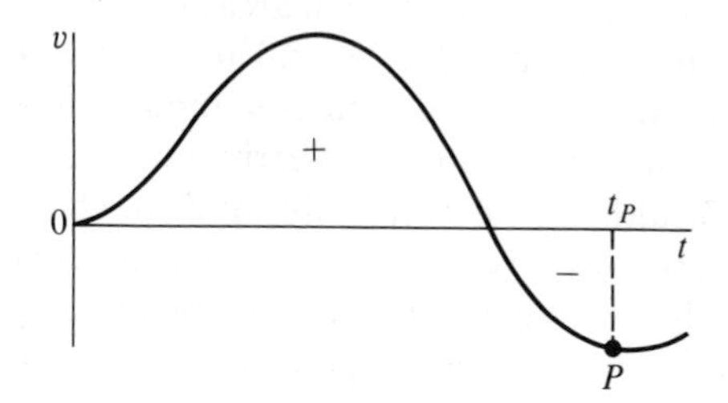

Fig. 3.8. A velocity-time graph showing positive and negative areas.

The process of finding the displacement from the velocity-time curve is known as integration and the displacement is said to be the *integral* of the velocity. It should be clear that the velocity is the integral of the acceleration in exactly the same way. Considering Fig. 3.6 again: the process of differentiation takes us from diagram (a) to (b) and then to (c); the process of integration takes us from (c) to (b) and then to (a).

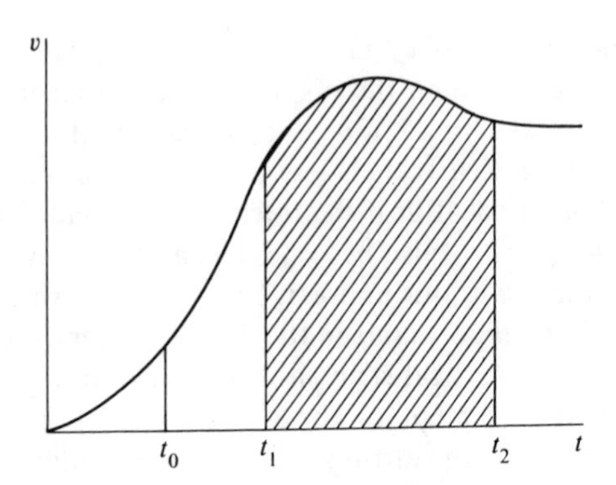

Fig. 3.9. A velocity-time graph in which the area between two arbitrary times is to be determined.

3.8 DEPENDENCE OF THE INTEGRAL ON THE END-POINTS

In the above discussion the displacement in the whole time from starting has been found. Frequently one is more interested in the displacement between two times t_1 and t_2, and this will obviously be given by the area under the velocity-time curve between those two times, as shown in Fig. 3.9. The value of this area will depend on both t_1 and t_2, since varying either of them will alter the area. It is a serious inconvenience to have to deal with quantities that depend in this way on two variables, since it means that we cannot write down, for any particular functional form of the v-t curve, the displacement as a function of one variable as we should like to do. There is, fortunately, a very easy way out of the dilemma. For the area under the curve between t_1 and t_2 is just the area between t_0 and t_2 minus the area between t_0 and t_1, where t_0 is any point to the left of t_1. If the point t_0 is taken as an agreed standard point, then areas between this standard point and any other point t can be tabulated once and for all for any form of v-t curve. Then the area between any two points such as t_1 and t_2 can be found by subtraction of the appropriate tabulated values. Nearly always, the standard point is chosen to be $t_0 = 0$, although occasionally another choice is made for special functional forms of the v-t curve. The area under the curve between two points such as t_1 and t_2 is called the *definite integral* between these points; the area under the curve between the agreed standard point t_0 and any other general point t is called the *indefinite integral*. The indefinite integral gives the shape of the displacement-time curve, but an arbitrary constant can always be added on to it. This can readily be seen by considering the two displacement curves shown in Fig. 3.10 The two curves are identical in shape, but curve (b) has been drawn so that every point of it is higher than curve (a) by an amount x_0. This shift of the curve has no effect on the slope of the tangent line at any point, and so both of these curves will give the same velocity-time curve. Obviously, then, if we are given this common v-t curve to start with, it will not be possible to determine which of the two displacement-time curves it was derived from. Integration gives the

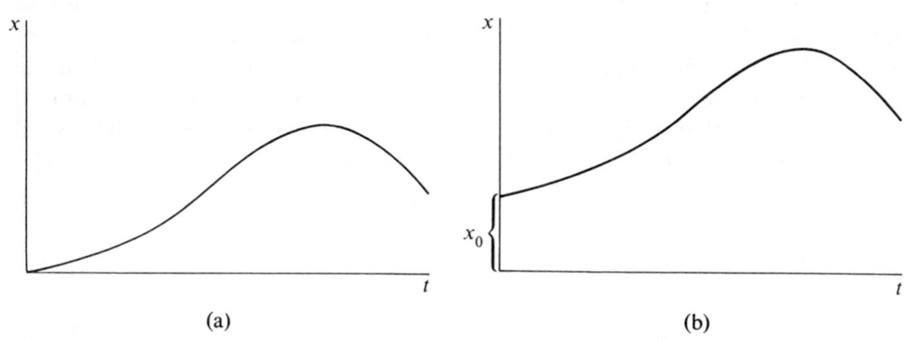

Fig. 3.10. Two different space-time graphs which will give the same velocity-time graph.

shape of the displacement-time curve but not the value of x_0. In order to fix the value of the arbitrary *constant of integration*, as it is called, it is necessary to have more information about the motion. Specifically, one must know the value of the displacement at time $t = 0$. In exactly the same way the indefinite integral of the acceleration will give the shape of the velocity-time curve, but to solve the problem completely one must be told the value of the velocity at $t = 0$. Since the definite integral is found by subtracting two indefinite integrals, this indeterminacy does not affect it, the arbitrary constant canceling out of the subtraction.

3.9 NUMERICAL INTEGRATION

In discussing how the area under a curve is to be found, we must remember that only rarely will an actual curve be given. Instead one can expect to be presented with a table of corresponding values of velocity and time. The basic problem, therefore, is how to reconstruct from such a table another table of corresponding values of displacement and time. One very simple method which is always valid is actually to plot the points on graph paper, and then find the required area or areas by counting the number of squares under the curve between the appropriate end-points. This number of squares is then reduced to the displacement by multiplying by a scale factor that will depend on the scales chosen along the two axes. A very little experience with this process will soon convince anyone that it is exceedingly tiresome and that it is not easy to feel confident of the result. Fortunately, there is a short-cut method that is much easier. Suppose that the data are given in a table as follows:

t	t_0	t_1	t_2	t_3	t_4	. . .	t_n
v	v_0	v_1	v_2	v_3	v_4	. . .	v_n

where it is assumed that the values of the time are equally spaced, e.g. $t = 0, 1, 2, 3, \ldots, n$ s. The common interval of time $t_1 - t_0 = t_2 - t_1 =$ etc., will be denoted by the symbol h. One might ask why the symbol Δt is not used. The answer is that the interval is not now assumed to be very small, and consequently the use of Δt would not be appropriate. Another, and better, reason is that the formulae to be given are not at all restricted to velocity-time curves, but can be applied to any curve, whether the horizontal axis denote the time or not. Hence, it is useful to have a more general symbol for the interval along the axis.

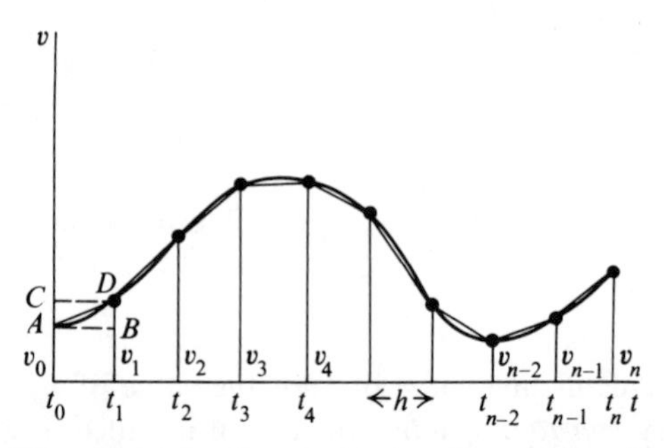

Fig. 3.11. The approximation to an area by a series of trapezoids.

The simplest approximation to the true area is obtained by joining all the plotted points by straight lines and calculating the area as the sum of all the trapezoids so formed. This is shown in Fig. 3.11. The calculation of the area of a trapezoid is demonstrated for the first one in this figure. The straight line at the top divides the little rectangle *ABCD* into two halves, so the area of the trapezoid is the mean of the areas of the rectangles whose tops are *AB* and *CD*. This gives the area of the trapezoid as $\frac{1}{2}(v_0h + v_1h) = \frac{1}{2}(v_0 + v_1)h$. The area is one-half the sum of the parallel sides times the perpendicular distance between them. Using this expression for the areas of all the trapezoids gives the total area under the curve as:

$$\begin{aligned}\text{Area} &= h[\tfrac{1}{2}(v_0 + v_1) + \tfrac{1}{2}(v_1 + v_2) + \tfrac{1}{2}(v_2 + v_3) + \ldots \\ &\qquad + \tfrac{1}{2}(v_{n-2} + v_{n-1}) + \tfrac{1}{2}(v_{n-1} + v_n)] \\ &= h[\tfrac{1}{2}v_0 + v_1 + v_2 + v_3 + \ldots + v_{n-2} + v_{n-1} + \tfrac{1}{2}v_n]. \qquad (3.7)\end{aligned}$$

This very simple formula is called the *trapezoidal rule*. It can be used for any value of n.

A second rule, better than the trapezoidal rule, is *Simpson's rule*, which gives the area as follows:

$$\begin{aligned}\text{Area} &= \frac{h}{3}[v_0 + 4(v_1 + v_3 + v_5 + \ldots + v_{n-1}) \\ &\qquad + 2(v_2 + v_4 + \ldots + v_{n-2}) + v_n]. \qquad (3.8)\end{aligned}$$

This rule can be applied only when n is an even number, that is for an odd number of data points. In practical cases Simpson's rule should be used where possible, the trapezoidal rule only being resorted to if n is odd. If is often possible to arrange things—for example, in collecting experimental data—so that n is even. It is not important to know how Simpson arrived at his rule; it is only important to appreciate its use. Either rule is applied to the tabular data, no actual curve being needed.

Example. An automobile starts from rest and its velocity at quarter-second intervals from starting is given in the following table:

t (s)	0	0.25	0.50	0.75	1.00	1.25	1.50
v (m s^{-1})	0	0.055	0.193	0.378	0.582	0.784	0.969

How far has it traveled after 1.5 s?

Solution. a) By the trapezoidal rule, since $h = 0.25$ s,

$$x = 0.25\text{ s}\left[\frac{0}{2} + 0.055 + 0.193 + 0.378 + 0.582 + 0.784 + \frac{0.969}{2}\right]\text{m s}^{-1}$$
$$= 0.25\text{ s} \times 2.477\text{ m s}^{-1} = 0.6193\text{ m}.$$

b) By Simpson's rule, which is possible since $n = 6$,

$$x = \frac{0.25\text{ s}}{3}[0 + 4(0.055 + 0.378 + 0.784) + 2(0.193 + 0.582) + 0.969]\text{ m s}^{-1}$$
$$= \frac{0.25\text{ s}}{3} \times 7.387\text{ m s}^{-1} = 0.6156\text{ m}.$$

The correct answer is 0.6155 m, showing clearly the superiority of Simpson's rule.

If the velocity-time diagram is given as a graph and not as a table of corresponding values, then the procedure is to draw a series of equally spaced vertical lines to divide the area under the graph into an *even* number of parts. The values of the velocity at the top of these vertical lines are read off from the graph, and then used as $v_0, v_1, \ldots, v_n$ in Simpson's rule. Other formulae for numerical integration exist and can be looked up if needed, but the above two rules will suffice for all ordinary cases met with.

3.10 THE NOTATION OF CALCULUS

The whole of the chapter to this point has been concerned with the basic notions of calculus, since it is impossible to understand simple motion in a straight line without these notions. Indeed, calculus was invented specifically for the description of motion. Thus the process of obtaining the instantaneous velocity of an object at a given time t by finding the limiting value of the average velocity for successively smaller intervals surrounding t must be clearly understood, although it is not necessary to be familiar with the technical notation used by mathematicians. However, increasingly frequently, biomedical scientists make use of calculus and it would be absurd not to give students the rudiments of the notation used. This is being done in a separate section because experience has shown that it is possible to be so frightened by an unfamiliar notation as to fail completely to grasp the underlying ideas. The really important thing is the idea, not the notation.

There are two aspects of mathematical notation that should be borne in mind. The first is its use merely as a shorthand for a concept, or a relation between concepts, that is already known and understood. The use of the Greek symbol Δ provides an example of this. The notation "Δx" is merely a translation into symbols of the words "a small increment of the variable x"; nothing more. As such it serves precisely the same purpose as the shorthand symbols used so effortlessly by stenographers. This comparison is intended to stress the absurdity of being afraid of notation by itself when used in this way. The second aspect of notation is its use as a language of logical deduction. This is where difficulties arise, and where specialized training is necessary. A certain familiarity with mathematical analysis of this type is necessary for all scientists, whether biologists or physicists, but on the former the demands are relatively light. The comparison between the two aspects of mathematics can be made more cogent by considering language studies. The specialist in a foreign language such as French can not only read the road-signs and the magazine captions in Paris, but can also write papers and books in French. He is the equivalent of a mathematical physicist who uses the language of mathematics quite freely. But a mere reading knowledge of French is also useful, although it does not require a detailed knowledge of syntax, or even a large vocabulary if a dictionary is available. All serious biomedical scientists should have at least a reading knowledge of mathematics, and that is the purpose of the present section.

In obtaining the instantaneous velocity from the displacement-time curve a certain limiting process is gone through, defined by the shorthand symbolism

$$v \equiv \lim_{\Delta t \to 0} \frac{\Delta x}{\Delta t}.$$

This expression is an excellent example of mathematical notation in its purely shorthand aspect. It is merely a translation of the words "the instantaneous velocity at any instant t is defined as the limiting value of the average velocity over a short interval of time that includes t as this short interval is reduced to zero". This meaning can be read without difficulty from the expression once the meaning of the shorthand symbols $\equiv$ and Δ are known. However, even this shorthand notation is not short enough, and a new symbol d/dt is defined as follows:

$$\lim_{\Delta t \to 0} \frac{\Delta x}{\Delta t} \equiv \left(\frac{d}{dt}\right) x \equiv \frac{dx}{dt}. \tag{3.9}$$

This is a very expressive notation, since it looks like the ratio of dx to dt which reminds us that v is obtained by taking the ratio of Δx to Δt and then proceeding to the limit. However, this is not really so, and the first way of writing it, with the symbol (d/dt) and the displacement x clearly separated, is really correct. The whole expression dx/dt is not the ratio of an infinitely small displacement to an infinitely small time, although in truth it may nearly always be manipulated as if it were. The symbol (d/dt) stands for the whole limiting process of obtaining the slope of the tangent line to the curve. Using this symbol, we have the relations

$$v \equiv \frac{d}{dt} x$$

and

$$a \equiv \frac{d}{dt} v,$$

so clearly

$$a \equiv \frac{d}{dt} \frac{d}{dt} x,$$

which is written $(d^2/dt^2)x$. dx/dt is called the *derivative* of x with respect to t. This notation is due to Leibnitz and is the most common for general calculus use. An alternative notation due to Newton is restricted to mechanics, in which the independent variable is nearly always the time. In this notation the time rate of change of a variable is denoted by putting a dot over the variable. This single dot stands for the whole process of proceeding to the limit as discussed above. Thus in Newton's notation

$$v \equiv \dot{x}, \qquad a \equiv \dot{v}, \qquad \text{and} \qquad a \equiv \ddot{x}.$$

In mechanics this is a most convenient notation and it will be used freely in ensuing chapters. It must not be used if the independent variable is anything other than the time. Thus if the pressure of a gas is plotted along the vertical axis as dependent variable and the volume is plotted along the horizontal axis as independent variable, then the slope of this graph at any value of the volume is written as dP/dV and never as $\dot{P}$. The meaning of dP/dV is expressable in words as follows: the derivative dP/dV at any value of V is defined as the limiting value of the average rate of change of pressure with volume, $\Delta P/\Delta V$, over an increment of volume ΔV that includes the volume V as this increment of volume is reduced to zero. This is in complete accord with the definition of velocity: only the variables have been changed.

Example 1. What is the slope of the tangent line to the curve $y = 4x^5$ at the point $x = 2$?

Solution. This is an example of a pure mathematical curve, where y and x represent dimensionless numbers. In such cases it is conventional to draw the axis of x horizontal and the axis of y vertical, thus choosing x and y to be the independent and dependent variables, respectively. The slope of the tangent line is then given by dy/dx, and it is required to find the value of this derivative at $x = 2$. Table 3.2 gives the derivative of At^n to be nAt^{n-1}, and, consequently, the derivative of

Ax^n is nAx^{n-1}. Hence, the derivative of $4x^5$ is just $20x^4$, and the value of this at $x = 2$ is 320. This, then, is the desired slope.

Example 2. When a body falls through air it suffers a retardation that is proportional to its speed in addition to the constant acceleration of gravity. Write down the equation of motion in mathematical form.

Solution. The *free* fall of objects under gravity is discussed in the next section, where the constant acceleration of gravity is given the symbol g. Here there is an additional negative acceleration whose magnitude is proportional to the speed, say bv, where b is a constant. Hence, writing the acceleration as $\dot{v}$, the mathematical statement of the problem is

$$\dot{v} = g - bv,$$

or

$$\dot{v} + bv = g. \tag{3.10}$$

An alternative mathematical formulation can be made in terms of the distance fallen from rest. For if y is used to denote the depth below the point of release at any time, then $v = \dot{y}$ and $\dot{v} = \ddot{y}$, and the equation may be written as

$$\ddot{y} + b\dot{y} = g. \tag{3.11}$$

It is left as an exercise to show that Eq. (3.10) is satisfied by the expression

$$v = \frac{g}{b}(1 - e^{-bt}),$$

which shows that the velocity approaches a constant value exponentially.

Having discussed the notation used in the differential calculus, let us now consider the symbolic representation of the process of obtaining the area under a curve by dividing it up into a large number of very narrow trapezoids, adding up the areas of all the trapezoids, and then proceeding to the limit by shrinking the widths of the trapezoids to zero. For this whole process of integration the symbol is $\int^t \ldots dt$. This is to be read as one symbol, with the actual functional form of the curve still to be written in on the dotted line. Thus the displacement of any time is $x \equiv \int^t vdt$, and the velocity at any time is $v \equiv \int^t adt$. The dt written at the end of the symbol serves to remind us of which variable is plotted along the horizontal axis. The product vdt, for example, may be thought of as the area of an infinitely narrow trapezoid of height v and width dt. The curly sign at the beginning is an old form of the letter S, and stands for "summation". The symbol $\int^t \ldots dt$ means the indefinite integral, where the area is calculated from some agreed reference point t_0 as discussed above. Usually, but not always, $t_0 = 0$. If the area is wanted between two definite times t_1 and t_2, then the appropriate symbol is $\int_{t_1}^{t_2} \ldots dt$. This is the definite integral. The relation between the integrals was explained geometrically in Section 3.8, and this relation may be symbolically written

$$\int_{t_1}^{t_2} \ldots dt = \int^{t_2} \ldots dt - \int^{t_1} \ldots dt.$$

The indefinite integral when evaluated appears as a function of the time. To evaluate the definite integral, first the value of $t = t_1$ is substituted into the general expression and then the value $t = t_2$ is substituted in. The first result is subtracted from the second to give the required value of the definite integral. The definite integral is *not* a function of the time but only of the two specific times t_1 and t_2.

Consider now the differential relation and the integral relation between the displacement x and the velocity v. They are:

$$v \equiv \frac{dx}{dt} \quad \text{and} \quad x \equiv \int^t v\,dt.$$

Substituting the second into the first gives

$$v \equiv \frac{dx}{dt} \equiv \frac{d}{dt}x \equiv \frac{d}{dt}\int^t v\,dt. \tag{3.12}$$

This tells us how to set about finding indefinite integrals. We need to find that function which, when we take the derivative, will give us the function we wish to integrate.

Example 1. Find the indefinite integral of At^2, where A is a constant.

Solution. From Table 3.2 a value of At^3 in column I is seen to correspond to a value $3At^2$ in column II, which means that

$$\frac{d}{dt}(At^3) = 3At^2.$$

Hence, dividing both sides of this equation by 3,

$$\frac{d}{dt}(\tfrac{1}{3}At^3) = At^2.$$

The required indefinite integral is therefore $\frac{1}{3}At^3$. However, $(d/dt)(\frac{1}{3}At^3 + B)$ is also equal to At^2, since the rate of change of a constant such as B is zero. The most general expression for the indefinite integral is thus $\frac{1}{3}At^3 + B$, where B is quite arbitrary. This matter was discussed in Section 3.8. The value of the arbitrary constant can be found only if additional information is given.

Example 2. A particle is traveling with a velocity $v = At^2$. What is its displacement at any time, and what is its displacement between $t = 2$ s and $t = 8$ s?

Solution. The indefinite integral $\int^t At^2 dt = \frac{1}{3}At^3 + B$ as discussed in the previous example, and this is the displacement at any time. To find the definite integral between $t = 2$ s, and $t = 8$ s, we must substitute those values of t in turn into the indefinite integral. For $t = 8$ s, the value is $\frac{1}{3}A(8\text{ s})^3 + B$; for $t = 2$ s, the value is $\frac{1}{3}A(2\text{ s})^3 + B$. Subtracting the second from the first gives for the definite integral the value $168A\text{ s}^3$, which is the answer required. Notice that the arbitrary constant B cancels out. The definite integral does not involve an arbitrary constant.

Example 3. Find the general formula for the integral of t^n, where n is any positive, negative, or fractional number except the one value $n = -1$.

Solution. The general formula for the derivative of t^n is given by $(d/dt)t^n = nt^{n-1}$. (See Table 3.2.) It follows that

$$\frac{d}{dt}t^{n+1} = (n + 1)t^n,$$

and, hence, that

$$\frac{1}{(n + 1)}\frac{d}{dt}t^{n+1} = t^n.$$

So the required formula is

$$\int^t t^n\,dt = t^{n+1}/(n + 1). \tag{3.13}$$

As division by zero is not permitted, the single value $n = -1$ must be excluded, since it makes the denominator of the result zero.

Considering Table 3.2, it should by now be clear that to find the derivative of any function (i.e. the slope of the graph of that function) we look up the function in the table and read off the derivative from the adjacent entry in the column immediately to the right. The second derivative d^2/dt^2—for example, acceleration from displacement—is found in the next but one column to the right. Adjustment of the constants may be necessary. To find the indefinite integral of a function we find the function in the table and the integral will be in the next column to the left. In this case adjustment of the constants as in the examples above will certainly be necessary, but is not difficult.

In terms of Fig. 3.6, differentiation means moving downwards from (a) to (b) to (c), and integration means moving upwards from (c) to (b) to (a). It must be remembered that any constant factors are lost in differentiation and so cannot be regained in the reverse process of integration.

3.11 UNIFORMLY ACCELERATED MOTION

The special case of motion in a straight line with constant acceleration is very important and affords an excellent illustration of the methods discussed in this chapter. To make the problem as general as possible, it will be assumed that an object is moving along the positive direction of the axis of x with a constant acceleration a, and that at $t = 0$ it is at a point of the axis of coordinate $x = x_0$ and

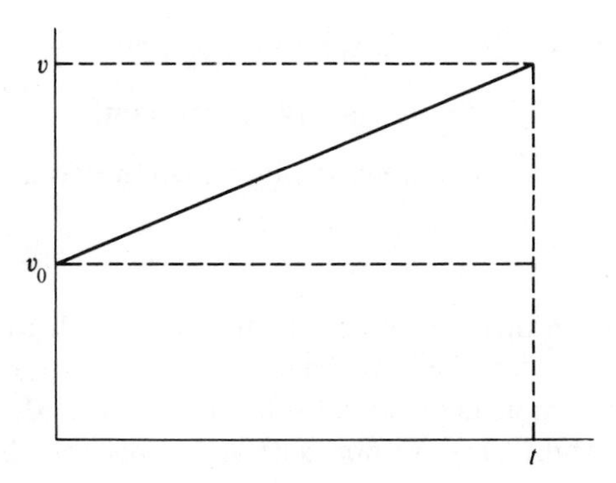

Fig. 3.12. The velocity-time graph for an object with uniform acceleration.

has velocity v_0. We first solve the problem graphically as follows. Since the acceleration is constant, the velocity-time curve is a straight line of slope a, as shown in Fig. 3.12. By elementary geometry, $a = (v - v_0)/t$, which may be rewritten

$$v = v_0 + at. \tag{3.14}$$

The area under the velocity-time curve between $t = 0$ and any other time gives the displacement in that time. Since the object is at $x = x_0$ at time $t = 0$, the displacement after time t will be $x - x_0$, where x is the position coordinate of the object at time t. The area under the curve consists of a rectangle of area v_0t and a triangle of area $\frac{1}{2}t(v - v_0)$. Hence, the displacement in time t is given by

$$x - x_0 = v_0t + \tfrac{1}{2}t(v - v_0).$$

From Eq. (3.14) we get $(v - v_0) = at$, and substituting this into the expression for the displacement gives us the value of the coordinate x at any time,

$$x = x_0 + v_0t + \tfrac{1}{2}at^2. \tag{3.15}$$

A third equation connecting v and x can be obtained by eliminating the time between Eqs. (3.14) and (3.15). This leads to

$$v^2 = v_0^2 + 2a(x - x_0). \tag{3.16}$$

In many applications $x_0 = 0$, which simplifies Eqs. (3.15) and (3.16). This problem may alternatively be solved as an exercise in calculus. Here Newton's notation will be used. The statement of the problem is

$$\ddot{x} = \text{constant} = a,$$

and the indefinite integral of this, complete with arbitrary constant, is simply

$$\dot{x} = at + \text{constant}.$$

To find the constant, we require to know the value of $\dot{x}$ at $t = 0$. This we are told is v_0. Hence, $\dot{x} = v_0 + at$. Integrating again gives

$$x = v_0 t + \tfrac{1}{2}at^2 + \text{constant},$$

where, to fix the constant, we require the information that at $t = 0$ the coordinate is $x = x_0$. Thus

$$x = x_0 + v_0 t + \tfrac{1}{2}at^2.$$

The third equation can then be found by elimination as before.

It is to be emphasized that this calculus method is no better and no more convincing than the simple geometrical solution. It must also be emphasized that Eqs. (3.14), (3.15), and (3.16) are not generally applicable, but apply only in the special circumstance of constant acceleration.

By far the most important problem to which these equations apply is the problem of the free fall of objects near the earth. Experimentally it is found that, when due allowance is made for air resistance, all objects in the same location fall to earth with the same acceleration. The value of this acceleration varies slightly with geographical location, but anywhere on earth it may be assigned the value 9.8 m s^{-2}. The acceleration is downwards towards the earth, of course, and this raises the question of the sign to be given to the acceleration. This is often a real stumbling block for students. The accepted convention is to use the symbol g for the *magnitude* of the acceleration. That is, $g = +9.8$ m s^{-2}. It is then necessary to choose a set of coordinate axes, and here there are two possibilities. The axis of x is always chosen to be the horizontal, and the axis of y is chosen to be the vertical, but which direction, up or down, is to be the positive direction of y? If the upward vertical is chosen as positive, then the acceleration is $-g = -9.8$ m s^{-2}, since it is downwards. If the downward vertical is chosen as positive, then the acceleration is $+g = +9.8$ m s^{-2}, since its direction coincides now with the positive direction of the y-axis. Which choice is made is a matter of convenience only, but errors are likely to creep in unless the above convention is adhered to. All the equations derived above can be used in problems of free fall, with the slight modification that the position and velocity will now be written y and $\dot{y}$ instead of x and $\dot{x}$.

Example. A ball is thrown upwards with a velocity of 14.7 m s^{-1} from a boy's hand which is 1.5 m above the ground at the instant the ball leaves it. (a) How long does the ball take to reach the highest point of its flight? (b) How high above the ground is it then? (c) At what times will it be 11.3 m above the ground? (d) What will be its velocities at these times?

Solution. We shall choose the positive direction of the y-axis vertically upwards. Then the equations corresponding to (3.14), (3.15) and (3.16) are:

$$v = v_0 - gt, \tag{3.17}$$

$$y = y_0 + v_0 t - \tfrac{1}{2}gt^2, \tag{3.18}$$

$$v^2 = v_0^2 - 2g(y - y_0), \tag{3.19}$$

since the acceleration is negative with this choice of y-axis.

a) At the highest point of the flight the velocity of the ball is instantaneously zero, and so from Eq. (3.17) we get $0 = v_0 - gt$, or

$$t = \frac{v_0}{g} = \frac{14.7 \text{ m s}^{-1}}{9.8 \text{ m s}^{-2}} = 1.5 \text{ s}.$$

b) To find the height reached we may use Eq. (3.18) to give

$$\begin{aligned} y &= 1.5 \text{ m} + (14.7 \text{ m s}^{-1})(1.5 \text{ s}) - \tfrac{1}{2}(9.8 \text{ m s}^{-2})(1.5 \text{ s})^2 \\ &= 1.5 \text{ m} + 22.05 \text{ m} - 11.025 \text{ m} \\ &= 12.525 \text{ m} = 12.5 \text{ m}. \end{aligned}$$

Notice the round-off to one decimal place, which is all the quoted accuracy of the data warrants.

The height may alternatively be found from Eq. (3.19) by setting $v = 0$. This gives

$$y - y_0 = \frac{v_0^2}{2g} = \frac{(14.7 \text{ m s}^{-1})^2}{2 \times 9.8 \text{ m s}^{-2}} = 11.025 \text{ m},$$

which gives $y = 12.5$ m as before.

c) The formula for the height at any time is Eq. (3.18) which may be rearranged to read

$$t^2 - \frac{2v_0}{g} t + \frac{2(y - y_0)}{g} = 0.$$

This is a quadratic equation in the variable t, which means that for any value of the height y there are two possible values of the time, one corresponding to the motion upwards, the other to the motion downwards. Putting in the numerical values, the equation becomes $t^2 - (3 \text{ s})t + 2 \text{ s}^2 = 0$. This is a very simple equation, which factorizes quite easily into $(t - 2 \text{ s})(t - 1 \text{ s}) = 0$. The product of the two brackets is zero when either of them is zero. Equating the brackets to zero in turn gives the two roots as $t = 2$ s or $t = 1$ s.

d) The velocities at these times are given by Eq. (3.17) as follows:

$$\begin{aligned} \text{at } t = 1 \text{ s}, \quad v &= 14.7 \text{ m s}^{-1} - (9.8 \text{ m s}^{-2})(1 \text{ s}) \\ &= +4.9 \text{ m s}^{-1}; \\ \text{at } t = 2 \text{ s}, \quad v &= 14.7 \text{ m s}^{-1} - (9.8 \text{ m s}^{-2})(2 \text{ s}) \\ &= -4.9 \text{ ms}^{-1}. \end{aligned}$$

The two velocities are equal in magnitude but opposite in direction.

PROBLEMS

3.1 The position of a particle as a function of time is given in the following table:

t (s)	0	1	2	3	4	5	6	7	8	9	10
x (m)	0	2.3	8.4	17.1	27.2	37.5	46.8	53.9	57.6	56.7	50.0

a) What is the average velocity of the particle over the first 5 s of its trip? Over the first 8 s? Over the whole trip?

b) What is the average velocity over the interval between $t = 3$ s and $t = 7$ s? Over the interval between $t = 4$ s and $t = 6$ s?

c) Plot the data on a sheet of graph paper, and join the points up with a smooth curve. Use the curve to obtain estimates of the position of the particle at $t = 4.5$ s and $t = 5.5$ s. Hence estimate the average velocity over the interval between $t = 4.5$ s and $t = 5.5$ s.

d) Draw the tangent to the curve at $t = 5$ s, and hence determine the instantaneous velocity at $t = 5$ s. Compare this with the average velocities obtained in (b) and (c) above.

e) Determine the instantaneous velocity at times $t = 1$ s and $t = 9$ s.

f) Estimate from the space-time curve the position of the particle when it comes instantaneously to rest and the time at which this occurs.

3.2 The position of a particle as a function of time is given in the following table:

t (s)	0	1	2	3	4	5	6	7	8	9	10
x (m)	0	2.2	6.9	13.9	23.1	34.3	47.2	61.6	77.1	93.4	110.0

Answer (a)–(e) as in Problem 3.1.

3.3 The position of a point traveling along the axis of x is given as a function of the time by the relation $x = At^3$, where A is a constant. What are the units of A in the MKSA system? Derive from first principles the velocity of the particle as a function of time.

3.4 The position of a particle is given by $x = A + Bt + Ct^2$, where A, B, and C are constants. What are the units of A, B, and C? Find the velocity of the particle as a function of time from first principles.

3.5 Use Table 3.2 to find the velocity of a particle whose position is given by $x = A \sin \omega t + B \cos \omega t$, where A, B are constants. What are the units of A and B? What are the units of ω?

3.6 Use Table 3.2 to find the velocity of a particle whose position is given by $x = X(1 - e^{-kt})$, where X and k are both constant. What are the units of X and k?

3.7 The velocity of a point traveling along the axis of x is given as a function of the time by the following table:

t (s)	0	1	2	3	4	5	6	7	8	9	10
v (m s^{-1})	0	8.8	15.2	19.2	20.8	20.0	16.8	11.2	3.2	−7.2	−20.0

a) What is the average acceleration of the point over the first 5 s of the motion? Over the first 8 s? Over the whole 10 s?

b) What is the average acceleration over the interval between $t = 1$ s and $t = 5$ s? Over the interval between $t = 2$ s and $t = 4$ s?

c) Plot the data carefully and join the plotted points by a smooth curve. Estimate the velocity of the point at $t = 2.5$ s and $t = 3.5$ s. Hence estimate the average acceleration over this interval.

d) Draw the tangent to the curve at $t = 3$ s and hence determine the instantaneous acceleration at $t = 3$ s. Compare this with the average accelerations obtained in (b) and (c) above. Contrast this case with what you found in (b), (c), and (d) of Problem 3.1.

e) Find the average acceleration over successive 1-s intervals by subtracting every entry in the table from the one immediately following it. Taking these as the accelerations at $t = 0.5, 1.5, 2.5, \ldots, 9.5$ s, plot a graph of acceleration against time. Attempt to find a mathematical relation between the acceleration and the time.

f) By counting squares, estimate the displacement of the point in the first 8 s. Estimate also the displacement of the point between $t = 2$ s and $t = 5$ s.

3.8 Using the data of the previous Problem, use the trapezoidal rule to calculate the displacement of the point at $t = 1, 2, 3, \ldots, 9$, and 10 s. Using Simpson's rule, calculate the displacement of the point at $t = 2, 4, 6, 8$, and 10 s.

3.9 A particle travels along the x-axis, its velocity at 0.5 s intervals from starting being as follows:

t (s)	0	0.5	1.0	1.5	2.0	2.5	3.0	3.5	4.0	4.5	5.0
v (m s^{-1})	0	1.3	5.0	11.4	20.6	32.8	48.3	67.5	91.1	119.8	154.7

Use Simpson's rule to calculate the displacement of the particle after the whole 5s.

3.10 Express the following laws in the symbolism of calculus:

a) The rate of change of the area of a circle with respect to its radius is proportional to its radius.

b) The velocity of a body that has traveled a distance x in a time t is proportional to x.

c) The rate at which a head of liquid, H, decreases owing to the liquid escaping from a hole in the bottom of the container is proportional to $H^{\frac{1}{2}}$.

d) The acceleration of a point moving in a straight line with velocity v is proportional to v^2.

e) The rate at which the number of bacteria, N, in an unlimited culture increases is proportional to the number N.

f) The rate of decrease of temperature T of a body standing in a strong draft is proportional to the excess of temperature of the body over its surroundings.

3.11 From the following table show as nearly as possible that if $y = e^x$, then $dy/dx = y$.

x	1	1.01	2	2.01	3	3.01
e^x	2.718	2.746	7.389	7.463	20.09	20.29

3.12 A body is falling through the air, and its acceleration is given by the expression $\dot{v} = a - bv^2$, where $a = 9.8 \text{ m s}^{-2}$ and $b = 0.016 \text{ m}^{-1}$. Calculate the maximum velocity acquired by the body.

3.13 The position of a body is given as a function of the time by the relation $x = (2 \text{ m s}^{-3})t^3 - (9 \text{ m s}^{-2})t^2 + (12 \text{ m s}^{-1})t + 6 \text{ m}$. When is the acceleration of the body zero, and what is its velocity at that time?

3.14 The position of a particle as a function of time is given by the relation $x = at + bt^3$. The particle has an initial velocity of 4 m s^{-1} and after 1 s this has been reduced to 1 m s^{-1}. Calculate its velocity after 2 s.

3.15 Integrate with respect to x the following expressions:

(i) $x^2 + 3$ (ii) $8x^2$ (iii) $6x^3$ (iv) $2x^5$ (v) $18x^{11}$ (vi) $4x^3 + 5x^2 - x + 7$ (vii) $(x + 2)^2$ (viii) $(x - 2)(2x + 1)$ (ix) x^{-3} (x) $9x^{-8}$ (xi) $x^2 + x^{-2}$ (xii) $x^{-1/2}$ (xiii) $x^{-0.4}$ (xiv) $3x^{-2} - 4x^{1/2}$ (xv) $(2x - 1)/x^3$ (xvi) $(x - 1)^4$.

3.16 Evaluate the following definite integrals:

(i) $\int_0^3 x^2\, dx$ (ii) $\int_{-3}^{-2} (3x^2 + 2x + 1)\, dx$ (iii) $\int_1^2 4x^3\, dx$.

3.17 A train of length 69 m starts from rest with an acceleration of 2 m s^{-2}. A railway worker is standing by the side of the track 100 m from the front of the train as it starts. Calculate the speeds with which the front and the back of the train pass this man, and the time the train takes to pass him.

3.18 An automobile traveling with constant acceleration moves from point A to a point B 150 m away in 10 s, reaching B with a speed of 20 m s^{-1}. What was its speed as it passed A? What is its acceleration? How far from A did the automobile start?

CHAPTER 4

MOTION IN TWO DIMENSIONS

4.1 INTRODUCTION

So far in our discussion of the motion of objects we have restricted ourselves to motion along a line. The position (x), the displacement ($x_2 - x_1$), the velocity (v) and the acceleration (a) of an object have all been defined, and we have seen that these quantities must be treated *algebraically*. This means that we ascribe to them either a positive or a negative sign, according as they are directed to the right or to the left. In this chapter we shall extend the discussion to motion in a plane, and we shall find that these quantities are *not* simply algebraic. Only in the case of motion along a line may we so treat them.

Consider a fly crawling over a table, and imagine that we are recording its movement. At $t = 0$ we observe the fly to be at a point O, at $t = t_1$ we observe it at another point P, and at $t = t_2$ we observe it at Q. The track of the fly is erratic, and is shown by the dashed line of Fig. 4.1.

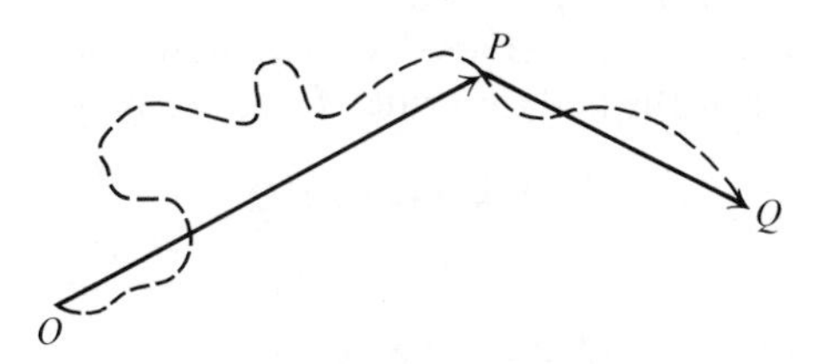

Fig. 4.1. A fly crawling on a table covers the path shown by the dashed line. In an interval t_1 it has moved from O to P, and the straight line OP is the displacement of the fly from O in that interval. Similarly the straight line PQ is the displacement in the interval $t_2 - t_1$. The total distance covered by the fly between O and Q is much greater than the sum of the lengths of the two straight lines OP and PQ.

The straight-line OP is the shortest distance between the starting point O and the finishing point P, and this straight line is defined as the *displacement* of the fly during the time interval t_1. Conventionally it is drawn as an arrow, with the tip at the point corresponding to the later instant, as shown in Fig. 4.1. It will be seen that the actual distance covered by the fly in moving from O to P is very much greater than the length of the line OP. This length is known as the *magnitude* of the displacement. In order to specify the displacement of the fly it is not enough to specify the magnitude: we must also give the *direction* of the line OP. A physical quantity like displacement that requires for its specification both a magnitude and a direction is known as a *vector*. A physical quantity like distance that has magnitude only, with no directional properties, is known as a *scalar*.

4.2 ADDITION AND SUBTRACTION OF VECTORS

In the same way as the arrow OP represents the displacement of the fly between $t = 0$ and $t = t_1$, so the arrow PQ represents the displacement between $t = t_1$ and $t = t_2$. These two displacements are carried out in succession, and they are equivalent to a single displacement represented by the arrow OQ as shown in Fig. 4.2. This diagram illustrates the general triangle rule for adding vector quantities: the vectors are laid tip to tail, and the sum is defined as the single vector from the free tail to the free tip. This rule is obvious in the case of displacements, since it merely states that a displacement from O to P followed by a displacement from P to Q brings the fly to a point which it could have reached by moving directly from O to Q. With this rule of addition, we may now formally define a vector as follows:

> *A vector is defined as a physical quantity that has both magnitude and direction, and that obeys the triangle law of addition illustrated in Fig. 4.2*

The law of addition may readily be extended to the sum of any number of vectors. The first two vectors are added, and then their sum is added to the third vector, and then their sum is added to the fourth, . . . and so on. The order in which the vectors are added is of no importance.

With the addition of vectors defined, we may now understand what is meant by vector subtraction. For the addition rule of Fig. 4.2 may be written as

$$OP + PQ = OQ$$

which implies that

$$PQ = OQ - OP.$$

The two vectors are drawn from the same point, and their difference is the vector joining the heads of the two vectors. In Fig. 4.2, since OP is being subtracted from OQ, the vector difference points from P to Q.

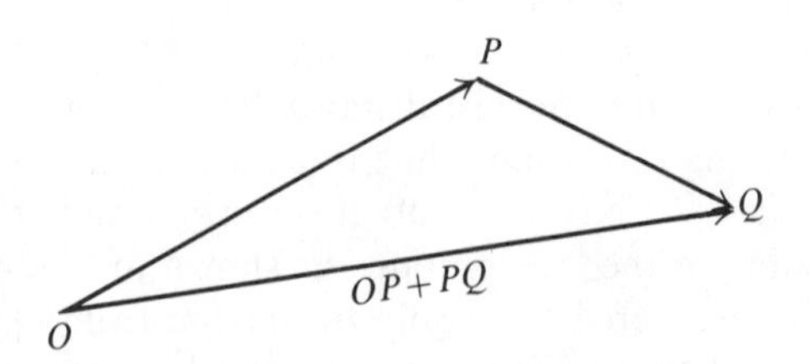

Fig. 4.2. Illustrating the rule for adding two vector quantities. The vectors are laid tip to tail, and sum is the single vector from the free tail to the free tip.

There is another way of looking at the process of vector subtraction. For by the negative of a vector we mean a vector of the same magnitude and the same direction but the opposite sense. Thus the vector $-OP$ is just the vector PO, as shown in Fig. 4.3, where the arrow now points from P to O instead of from O to P. Then the rule for subtraction may be written

$$PQ = OQ - OP = OQ + (-OP) = OQ + PO = PO + OQ.$$

The difference PQ is now seen to be the sum, calculated by the general rule for addition, of the vector OQ with the negative of OP. This way of looking at subtraction is useful, since it means that only one rule, namely that of addition, has to be remembered.

By the introduction of vectors, we have enlarged the meanings of the words addition and subtraction: they now have a geometrical significance and not simply an algebraic one. Only in one dimension, i.e. along a straight line, do vectors have simple algebraic properties, since direction along the line reduces to a simple choice between the two opposite senses of traversing the line.

In addition to knowing how to add and subtract vectors, we also need to know how to multiply or divide a vector by a scalar. The result of this is to give a new vector which has the same direction as the original vector, but which has a magnitude equal to the original magnitude multiplied or divided by the scalar. If the scalar is not a mere number, then the new vector will be a physical quantity of a kind different from the original vector. For example, if we divide a displacement of 10 m towards the East by a time interval of 5 s, then (since time is a scalar quantity) we obtain a velocity of 2 m s^{-1} towards the East. The direction has not changed, but the magnitude has changed from 10 m to 2 m s^{-1}. Clearly, therefore, velocity is a vector, and similarly acceleration is also a vector.

It is convenient to have a special notation for vector quantities that will tell us what kind of vector we are dealing with. To indicate both the vector nature and the type of physical quantity, we use in printing a bold typeface. For example, we represent a velocity by the symbol $\mathbf{v}$ and an acceleration by the symbol $\mathbf{a}$. The symbol tells us what kind of physical quantity it is, and the special typeface tells us it is a vector. Symbols in ordinary italic type, v and a, are then used to represent the magnitudes of the vectors $\mathbf{v}$ and $\mathbf{a}$.

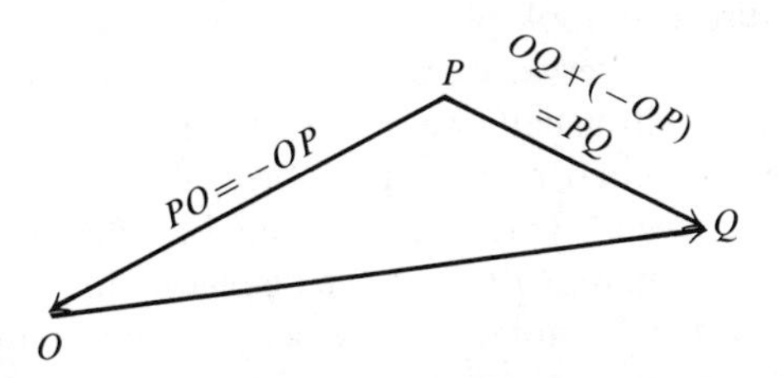

Fig. 4.3. The vector PQ is the difference between the vectors OQ and OP of Fig. 4.2, and it may be regarded as the sum of the vector $-OP = PO$ and the vector OQ.

4.3 POSITION AND VELOCITY VECTORS

Consider a particle moving along the dotted line of Fig. 4.4. The particle is at the point P_1 at the instant t_1, and at the point P_2 at the later instant t_2. In order to specify the positions of P_1 and P_2 in the plane, it is necessary to choose a fixed origin O. Then the vectors OP_1 and OP_2 define completely these two positions. We call them the *position vectors* of the points P_1 and P_2 and represent them by the symbols $\mathbf{r}_1$ and $\mathbf{r}_2$. The origin O may be chosen quite arbitrarily, since only changes of position are of importance. We may choose O to lie on the path of the particle as we did for the motion of the fly (Fig. 4.1), or not (as here): it makes no difference.

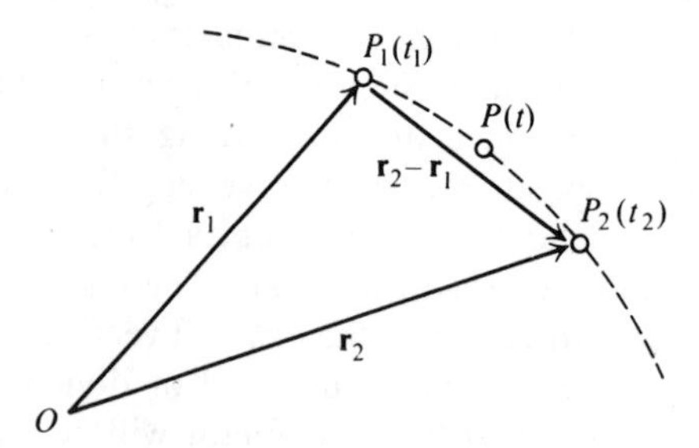

Fig. 4.4. The vector displacement of an object referred to an arbitrary origin.

The displacement of the particle during the interval $t_2 - t_1$ is the vector P_1P_2, and this is simply $\mathbf{r}_2 - \mathbf{r}_1$. The ratio of this displacement to the time interval is defined as the average velocity $\bar{\mathbf{v}}$ for the interval, the bar over the symbol again being used to indicate an average.

Thus

$$\bar{\mathbf{v}} \equiv (\mathbf{r}_2 - \mathbf{r}_1)/(t_2 - t_1).$$

This definition of average velocity is completely consistent with that given for straight-line motion in the previous chapter, and, indeed, reduces to that previous definition when $\mathbf{r}_1$ and $\mathbf{r}_2$ are in the same direction. In the straight-line case, except where a reversal of direction took place, the average velocity over an interval and the average speed over an interval were the same. In two dimensions the average speed over any interval will usually be greater than the average velocity over the same interval, since the actual path of the object between the end-points of the displacement will be curved. This is clear from Fig. 4.4, where the actual path of the object is along the curve P_1PP_2 and the displacement is the straight line P_1P_2.

In order to define the instantaneous velocity at the point P of the path, the two points P_1 and P_2 are made to approach P from either side. The limiting value of the average velocity when P_1, P, and P_2 coincide is defined as the instantaneous velocity vector at the instant t corresponding to the position P on the path. Geometrically, it should be obvious that the direction of the instantaneous velocity at P is along the tangent line to the path at P, since the chord P_1P_2 becomes this tangent line in the limit.

If an object is traveling along a curve in a plane, the velocity at any point of the path has the direction of the tangent line at that point.

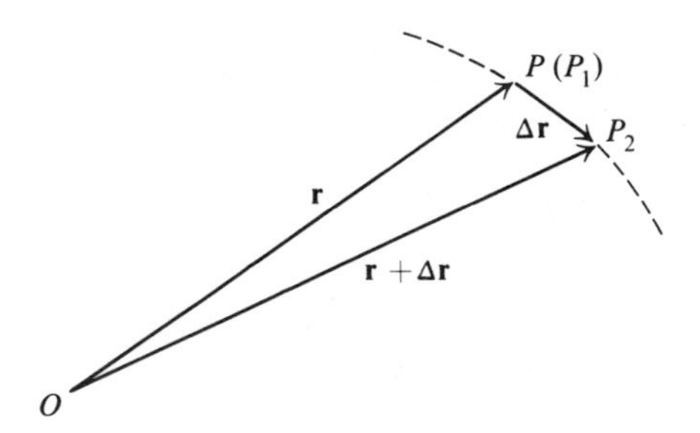

Fig. 4.5. Figure 4.4 for the case of an infinitesimal displacement.

Figure 4.5 shows essentially the same situation as Fig. 4.4, but the displacement P_1P_2 has been much reduced. Since this displacement is shrunk to zero as the interval is shrunk to zero, it is convenient to use the symbol Δ again, as in Chapter 3. The change in position vector $\mathbf{r}_2 - \mathbf{r}_1$ that gives the displacement is written as $\Delta\mathbf{r}$ and the interval $t_2 - t_1$ as Δt. The point P at which the velocity is wanted is made to coincide with P_1 and the limiting process is simplified to moving P_2 backwards along the path until it coincides with P. With these conventions the position vector $\mathbf{r}_1$ may be written simply as $\mathbf{r}$, and the position vector $\mathbf{r}_2 = \mathbf{r} + \Delta\mathbf{r}$ by the triangle law. The resemblance between this and Fig. 3.3 of Chapter 3 will be seen. The latter figure, however, was a space-time diagram describing motion along a straight line: Fig. 4.5 is a diagram of the actual path of the object. The two different kinds of diagram must not be confused. With the notation of Chapter 3, the above argument defining the instantaneous velocity at P as the limiting value of the average velocity over the interval Δt may be summarized in the shorthand notation

$$\lim_{\Delta t \to 0} \frac{\Delta\mathbf{r}}{\Delta t} \equiv \frac{d\mathbf{r}}{dt} \equiv \dot{\mathbf{r}} \equiv \mathbf{v}.$$

The three different notations for velocity have been written into this definition. The Leibniz notation $d\mathbf{r}/dt$ will not be used further, but both the Newton notation $\dot{\mathbf{r}}$ and the single-symbol notation $\mathbf{v}$ will be used in what follows.

A general result of the utmost importance is apparent from the geometry of Fig. 4.5: the vectors $\mathbf{r}$ and $\dot{\mathbf{r}}$ are not in the same direction. Of course, the origin

O is arbitrary, and so for the particular point P it would be possible to choose an origin for which $\mathbf{r}$ and $\dot{\mathbf{r}}$ were parallel; but this parallelism would be destroyed as the object moved on from P along the path. Clearly, the only path for which $\mathbf{r}$ and $\dot{\mathbf{r}}$ are always parallel is a straight line. This general result holds for any vector.

A vector $\mathbf{A}$ *and its time rate of change* $\dot{\mathbf{A}}$ *are not, in general, in the same direction.*

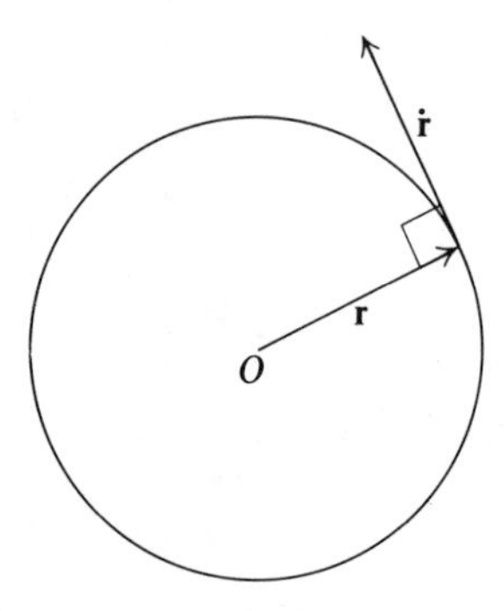

Fig. 4.6. The relation of the position vector and the velocity vector for a particle moving in a circle.

Example. An object travels along a path and, relative to a fixed point as origin, the position vector $\mathbf{r}$ and the velocity vector $\dot{\mathbf{r}}$ are always mutually perpendicular. What is the form of the path?

Solution. Since a circle has the property that the tangent at any point is perpendicular to the radius at the same point, it follows that the required path is a circle, the center of the circle being the required origin. For motion in a circle the velocity is always perpendicular to the corresponding position vector, as shown in Fig. 4.6.

4.4 ACCELERATION

The acceleration is defined as the instantaneous rate of change of velocity, and is found by first determining the average acceleration over an interval and then shrinking this interval to zero. In order to understand the process, consider Fig. 4.7 which shows the velocity vectors $\mathbf{v}_1$ and $\mathbf{v}_2$ corresponding to the instants t_1 and t_2 when the particle of Fig. 4.4 is at positions $\mathbf{r}_1$ and $\mathbf{r}_2$. These velocity vectors have been drawn parallel to the tangent lines at P_1 and P_2, and with their tails at the points P_1 and P_2. Since we wish to subtract $\mathbf{v}_1$ from $\mathbf{v}_2$, it is convenient to move both vectors parallel to themselves to a common origin, as shown in Fig. 4.8. Vectors may always be moved about parallel to themselves in this way in order to simplify calculations. The change in velocity over the interval $t_2 - t_1$ is $\mathbf{v}_2 - \mathbf{v}_1$, and this is

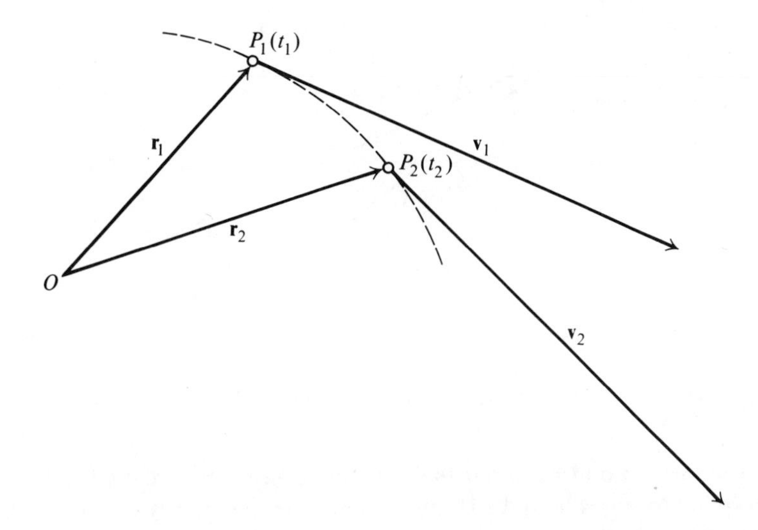

Fig. 4.7. The velocity vectors $\mathbf{v}_1$ and $\mathbf{v}_2$ for the particle of Fig. 4.4 at the instants t_1 and t_2.

the vector P_1P_2 as shown. The average acceleration over the interval is defined as

$$\bar{\mathbf{a}} \equiv (\mathbf{v}_2 - \mathbf{v}_1)/(t_2 - t_1).$$

As the interval is reduced, the situation approaches that shown in Fig. 4.9, where $\mathbf{v}_1$ has been written simply as $\mathbf{v}$, $\mathbf{v}_2 - \mathbf{v}_1$ as $\Delta\mathbf{v}$, and $t_2 - t_1$ as Δt. The average acceleration is now written $\bar{\mathbf{a}} = \Delta\mathbf{v}/\Delta t$, and the limiting value of this ratio as the interval Δt is shrunk to zero is defined as the instantaneous acceleration at the instant t. In the usual shorthand symbols

$$\lim_{\Delta t \to 0} \frac{\Delta \mathbf{v}}{\Delta t} \equiv \dot{\mathbf{v}} \equiv \mathbf{a} \equiv \ddot{\mathbf{r}}.$$

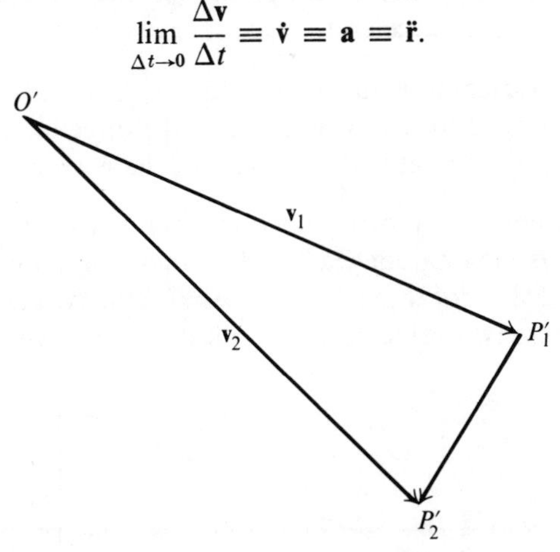

Fig. 4.8. The velocity vectors $\mathbf{v}_1$ and $\mathbf{v}_2$ of Fig. 4.7 drawn from a common origin.

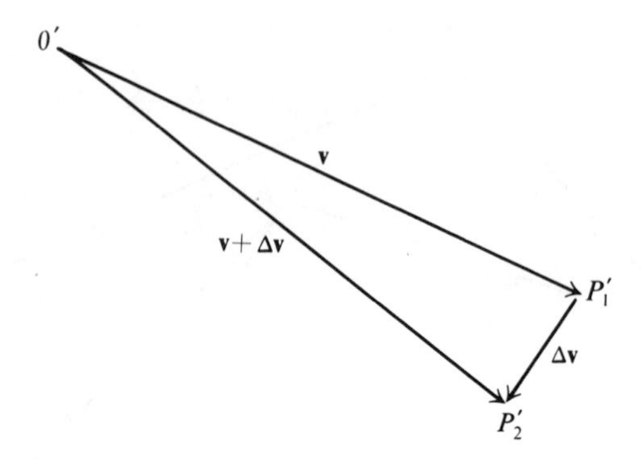

Fig. 4.9. Figure 4.8 for the case of an infinitesimal change of velocity.

The directions of $\mathbf{v}$ and $\dot{\mathbf{v}}$ are not the same in general. Figure 4.9 has been drawn to correspond to Fig. 4.5, and a study of these two diagrams will convince us of the following general rule.

The acceleration vector always points towards the concave side of the path of the particle.

This is a purely geometrical result, and it will prove of great utility when we come to study the dynamics of motion in a plane. It must be carefully noted that a particle can have an acceleration even when its speed is constant. Any particle that is moving along a curved path has a velocity that is constantly changing its direction. Even if its speed along the path is constant, its velocity (which is a vector quantity) is not constant, and therefore there must be an acceleration. There is unfortunately no word in physics for the rate of change of speed, and the vector nature of acceleration must always be borne in mind. Only when the motion is confined to a straight line is the magnitude of the acceleration the same as the rate of change of speed.

Example 1. A ship is steaming due East at 12 m s^{-1}. A passenger runs across the deck at 5 m s^{-1} in a direction perpendicular to the direction of motion of the ship and towards the North. What is the velocity of the passenger relative to the sea?

Solution. The passenger has two velocities simultaneously. Because he is on board the ship, he is moving due East at 12 m s^{-1}. In addition he has a velocity of 5 m s^{-1} across the deck of the ship. His actual velocity, therefore, relative to the sea is the sum of these two velocities. The addition is shown in Fig. 4.10.

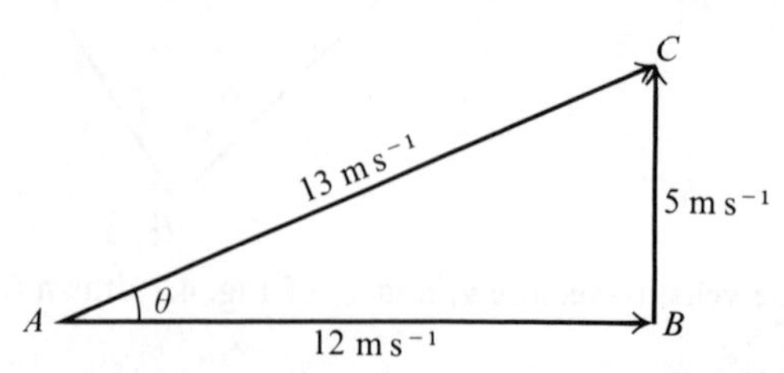

Fig. 4.10.

The vector sum of the ship's velocity AB and the passenger's velocity relative to the ship BC is the vector AC drawn from the free tail to the free head. In this case the triangle is right-angled, and we get from Pythagoras's theorem:

$$AC^2 = AB^2 + BC^2 = (12^2 + 5^2)\ \text{m}^2\ \text{s}^{-2} = 13^2\ \text{m}^2\ \text{s}^{-2}.$$

Hence the velocity of the passenger relative to the sea has a magnitude of 13 m s^{-1}. To find the direction of this velocity, we have

$$\tan\theta = 5/12 = 0.4167,$$

and therefore

$$\theta = 22°37' \text{ North of East.}$$

Example 2. An automobile, traveling at 20 m s^{-1} due North along the highway makes a right turn on to a side road that heads due East. It takes 50 s for the automobile to make the turn, and at the end of this period it has a speed of 15 m s^{-1} along the side road. Calculate the average acceleration over the 50-s interval.

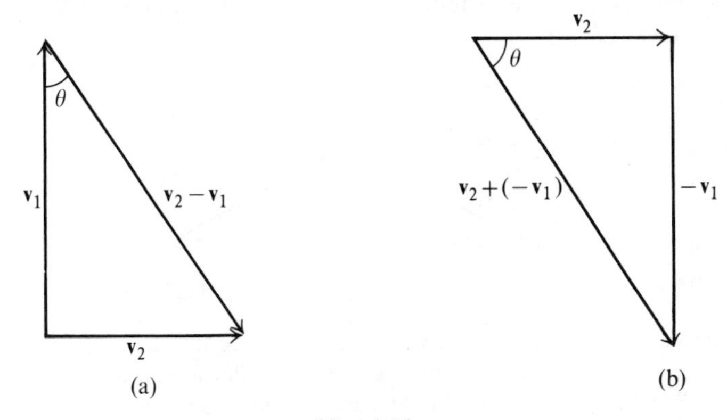

Fig. 4.11.

Solution. The change in velocity of the automobile is $\mathbf{v}_2 - \mathbf{v}_1$, where $\mathbf{v}_2 = 15$ m s^{-1} due East, and $\mathbf{v}_1 = 20$ m s^{-1} due North. This vector subtraction is illustrated in Fig. 4.11(a). Since we are subtracting $\mathbf{v}_1$ from $\mathbf{v}_2$ the difference is the vector from the tip of $\mathbf{v}_1$ to the tip of $\mathbf{v}_2$. The alternative way of subtracting the vectors, by writing $\mathbf{v}_2 - \mathbf{v}_1 = \mathbf{v}_2 + (-\mathbf{v}_1)$ where $-\mathbf{v}_1$ is 20 m s^{-1} due South, is shown in Fig. 4.11(b). Clearly both methods lead to identical right-angled triangles for the determination of the vector difference. The magnitude of this difference is given by the theorem of Pythagoras as

$$\sqrt{(15\ \text{m s}^{-1})^2 + (20\ \text{m s}^{-1})^2} = 25\ \text{m s}^{-1}$$

and the direction is at an angle θ East of South as shown, where

$$\tan\theta = 15/20 = 0.75$$

giving

$$\theta = 37°.$$

Hence, since the interval is 50 s, the average acceleration during the turn is

$$\bar{\mathbf{a}} = \frac{25 \text{ m s}^{-1}}{50 \text{ s}} = 0.50 \text{ m s}^{-2}$$

in a direction 37° East of South.

4.5 THE RESOLUTION OF VECTORS

In discussing the concept of vector velocity, we took the point O as a fixed origin in the plane. Without such a reference point the word "position" has no meaning. It does not matter what point we choose as origin, since we are interested only in changes of position, but we must choose *some* point if we wish to say where the particle is. In the same way we usually want to specify directions with reference to some fixed direction, and this essentially means choosing a set of coordinate axes. Fig. 4.12 shows the two points P and Q of Fig. 4.1 located with respect to a suitable coordinate system, with O as origin. Point P has coordinates (x_1, y_1) and point Q has coordinates (x_2, y_2).

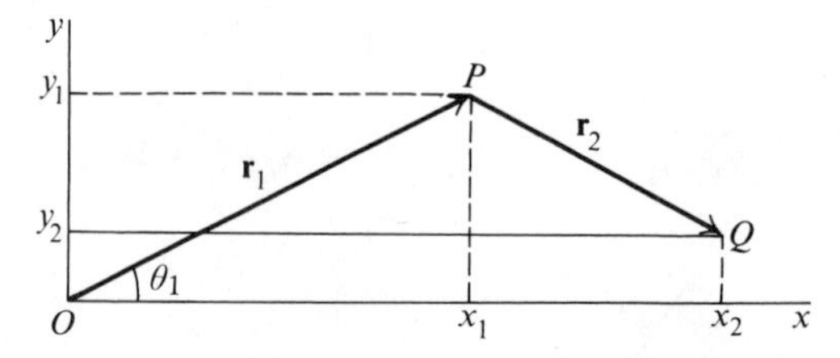

Fig. 4.12. The vectors of Fig. 4.1 shown relative to a coordinate system. The tail of vector OP has been chosen as the origin, and perpendicular axes drawn through O. The projection of OP and PQ on the coordinate axes are known as components.

Consider first the vector OP. Let its magnitude be denoted by r_1 and its direction by the angle θ_1 it makes with the positive direction of the x-axis. By elementary trigonometry, the following relations are seen to hold:

$$\begin{aligned} x_1 &= r_1 \cos \theta_1; \qquad y_1 = r_1 \sin \theta_1 \\ r_1^2 &= x_1^2 + y_1^2; \qquad \tan \theta_1 = y_1/x_1. \end{aligned} \tag{4.1}$$

The quantities x_1, y_1 are referred to as the *components* of the vector OP along the chosen axes. By using Eqs (4.1) we may specify both the magnitude and the direction of the vector by using *either* r_1, θ_1 *or* x_1, y_1. The components of a vector are the projections of the vector on the two coordinate axes. Thus the vector PQ is seen to have components $(x_2 - x_1)$ and $(y_2 - y_1)$ respectively, where $(y_2 - y_1)$ is negative corresponding to the fact that the vector PQ has a direction from upper left to lower right.

If we now add the components of OP to the components of PQ we get:

$$x_1 + (x_2 - x_1) = x_2; \qquad y_1 + (y_2 - y_1) = y_2,$$

and the two answers are the components of the vector OQ. We see, therefore, that the rule for adding vectors can be broken down into two ordinary algebraic additions, one along each of the chosen axes. What this means is that motion along a curve in a plane can be regarded as the sum of two independent linear motions, one along the x-axis and the other along the y-axis. All vector quantities are expressed in terms of their components along the chosen axes: this process is known as *resolution*, and we speak of *resolving* a vector into its components. When this has been done, we may treat the two linear motions separately, and combine the results at the end if we wish by using Eq. (4.1). Figure 4.13 shows the resolution of a velocity vector **v**. The components of velocity may be written either as $\dot{x}$, $\dot{y}$ or as v_x, v_y, although this latter form is decidely more cumbersome.

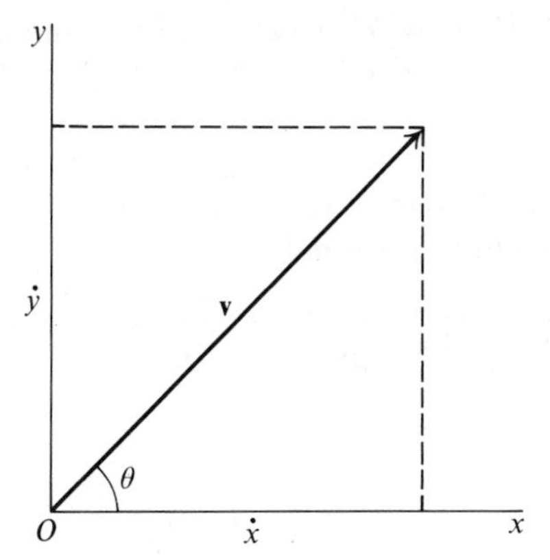

Fig. 4.13. A velocity **v** may be resolved into components $\dot{x}$ and $\dot{y}$ along the coordinate axes.

Equations of the same form as Eqs. (4.1) hold for any vector. For a velocity vector they are

$$\begin{aligned} \dot{x} &= v_x = v \cos\theta; \qquad & \dot{y} &= v_y = v \sin\theta; \\ v^2 &= \dot{x}^2 + \dot{y}^2; & \tan\theta &= \dot{y}/\dot{x}. \end{aligned} \tag{4.2}$$

In choosing coordinate axes there is no need for the x- and y-axes to be horizontal and vertical: any two perpendicular directions will serve the purpose equally well. As an example, Fig. 4.14 shows a horizontal acceleration **a** resolved parallel and perpendicular to a line that slopes downwards at an angle θ to the horizontal. The components may be written as $\ddot{x}$, $\ddot{y}$ or as a_x, a_y.

The values of the components are

$$\begin{aligned} \ddot{x} &= a_x = a \cos\theta; \qquad & \ddot{y} &= a_y = a \sin\theta \\ a^2 &= \ddot{x}^2 + \ddot{y}^2; & \tan\theta &= \ddot{y}/\ddot{x}. \end{aligned} \tag{4.3}$$

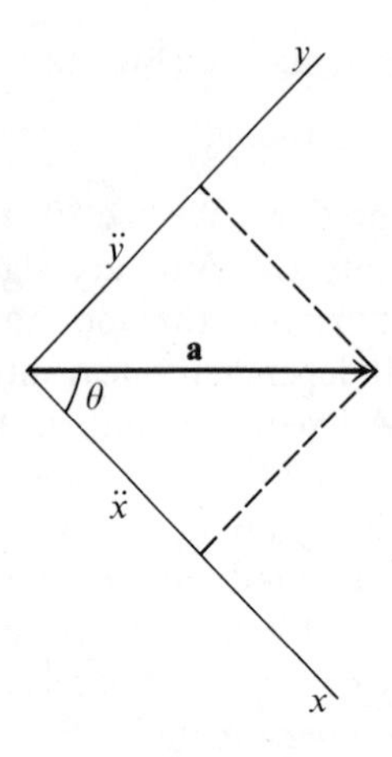

Fig. 4.14. A horizontal acceleration is resolved into components $\ddot{x}$ and $\ddot{y}$ parallel and perpendicular to a line making an angle θ with the horizontal.

It will be appreciated that, apart from a change of orientation, Fig. 4.14 contains nothing essentially different from Figs. 4.12 and 4.13, and Eqs (4.3) are of precisely the same form as before.

In actually solving problems involving vectors, it is nearly always more convenient to resolve the vectors along two mutually perpendicular directions, and then deal with each of the directions separately.

Example 1. What is the component of the gravitational acceleration parallel to a hill that makes an angle of 15° with the horizontal?

Solution

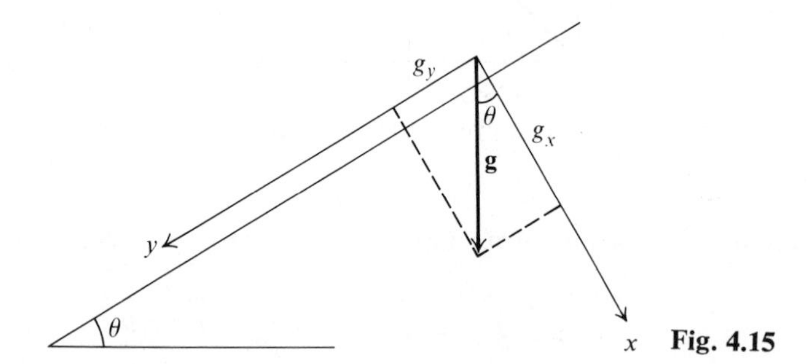

Fig. 4.15

The gravitational acceleration is a vector **g** of magnitude g that acts vertically downwards. If we choose axes x and y as shown in Fig. 4.15, then the angle between **g** and the x-axis is just 15°, the angle the hill makes with the horizontal. In this case it is more appropriate to denote the components by g_x and g_y, and it will be seen that the component down the hill is

$$\begin{aligned} g_x &= g \sin \theta = g \sin 15^\circ \\ &= 9.8 \text{ m s}^{-2} \times 0.259 \\ &= 2.55 \text{ m s}^{-2} \end{aligned}$$

Example 2. Three successive displacement vectors have components referred to rectangular axes of +2.4 m, +0.5 m; −4.6 m, +3.3 m; and −2.8 m, −15.8 m. What are the components of the resultant displacement? What is the magnitude of the resultant?

Solution. The x-component of the resultant displacement is

$$+2.4 \text{ m} - 4.6 \text{ m} - 2.8 \text{ m} = -5.0 \text{ m},$$

and the y-component is

$$+0.5 \text{ m} + 3.3 \text{ m} - 15.8 \text{ m} = -12.0 \text{ m}.$$

The magnitude of the resultant displacement is therefore

$$\sqrt{(-5.0 \text{ m})^2 + (-12.0 \text{ m})^2} = 13.0 \text{ m}$$

Notice that, although positive signs have been written for the components where appropriate, we do not need to write the magnitude as +13 m since magnitude is an *essentially* positive quantity.

4.6 MOTION WITH CONSTANT ACCELERATION

In Chapter 3 motion in a straight line with constant acceleration was discussed and the equation for the position of an object at any time t derived as

$$x = x_0 + v_0 t + \tfrac{1}{2}at^2,$$

where x_0 and v_0 are the position and velocity of the object at $t = 0$ and a is the constant acceleration. The vector form of this equation is

$$\mathbf{r} = \mathbf{r}_0 + \mathbf{v}_0 t + \tfrac{1}{2}\mathbf{a}t^2, \tag{4.4}$$

where $\mathbf{r}_0$ and $\mathbf{v}_0$ are the position and velocity vectors at $t = 0$, and $\mathbf{a}$ is the constant vector acceleration. The important application of this equation is to the motion of an object near the earth. In this case the constant acceleration is written $\mathbf{g}$, the magnitude of which is $g = +9.8 \text{ m s}^{-2}$. The axis of x is always chosen along the horizontal, with the axis of y vertical. As discussed in Chapter 3, there is a choice involved in the positive direction of the y-axis. If the positive direction is chosen upwards, then the y-component of the magnitude of $\mathbf{g}$ is $-9.8 \text{ m s}^{-2} = -g$. In what follows, the positive direction will always be chosen upwards.

The implications of Eq. (4.4) must be accepted, namely that $\mathbf{r}_0$, $\mathbf{v}_0$, and $\mathbf{g}$ are not in general in the same direction. Hence, the sum of the three terms must be obtained by the triangle rule applied twice. The vector $\mathbf{v}_0 t$ is in the same direction as $\mathbf{v}_0$ itself but its magnitude increases linearly with the time. The vector $\frac{1}{2}\mathbf{g}t^2$ is in the same direction as $\mathbf{g}$ but its magnitude increases as the square of the time. The process of compounding the vectors is most easily grasped by actually drawing the vectors in a particular case. Consider a projectile fired from the earth with an initial velocity whose magnitude is 30 m s^{-1} and whose direction makes an angle of 37° with the horizontal. The magnitudes of the two vectors

Table 4.1. The magnitudes of the vector components of the displacement from the origin of a projectile at various times

t (s)	v_0t (m)	$\frac{1}{2}gt^2$ (m)
0	0	0
1	30	4.9
2	60	19.6
3	90	44.1

$\mathbf{v}_0t$ and $\frac{1}{2}\mathbf{g}t^2$ at different times are listed in Table 4.1. The origin of coordinates is chosen at the point of projection so that $\mathbf{r}_0 = 0$. Figure 4.16 shows the successive displacement vectors $\mathbf{r}_1$, $\mathbf{r}_2$, and $\mathbf{r}_3$ obtained by the triangle rule. The figure is drawn to scale, and the student should repeat the drawing of it himself. The line $OA_1A_2A_3$ is at 37° to the horizontal, which is a very easy angle to draw, since $\sin 37° = \frac{3}{5}$, $\cos 37° = \frac{4}{5}$, and $\tan 37° = \frac{3}{4}$ to a good approximation. The vector OA_1 has a length proportional to 30 m, the magnitude of $\mathbf{v}_0t$ after 1 s, and A_1B_1 has a length proportional to 4.9 m, the magnitude of $\frac{1}{2}\mathbf{g}t^2$ after 1 s. Similarly, OA_2,A_2B_2 and OA_3,A_3B_3 have lengths proportional to the magnitudes of $\mathbf{v}_0t$ and $\frac{1}{2}\mathbf{g}t^2$ at $t = 2$ s and $t = 3$ s, respectively. By the triangle rule the sum $\mathbf{r} = \mathbf{v}_0t + \frac{1}{2}\mathbf{g}t^2$ is OB_1 at $t = 1$ s, OB_2 at $t = 2$ s, and OB_3 at $t = 3$ s. Hence, the points B_1, B_2, and B_3 are points on the trajectory of the projectile. The smooth curve drawn through these points, shown dashed in Fig. 4.16, is a parabola. It will be seen that the projectile rises to a height of about 16.5 m and returns to earth at a distance of about 88 m from its point of projection. Note that the acceleration vector $\mathbf{g}$ points inwards toward the concave side of the parabola. This is a general result of great importance.

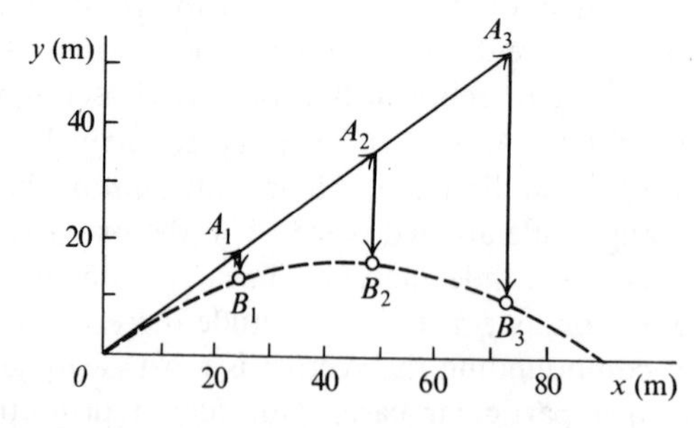

Fig. 4.16. Successive displacement vectors, calculated from the data of Table 4.1.

In an intriguing class-room demonstration an air-gun is aimed accurately at a target suspended near the roof by sighting along the barrel. As the bullet leaves the barrel, it breaks a fine wire that in turn breaks an electrical circuit, and the target begins to fall. The bullet always hits the target, even when the air pressure is adjusted so that the collision takes place near the ground. Inspection of Fig. 4.16 provides the explanation. For the gun is aimed along the line $OA_1A_2A_3$, and the initial position of the target is therefore along this line. Let us write the initial position of the target as $\mathbf{r}_0$, its initial velocity being zero, since it falls from rest. The position of the target at any time after release is therefore given by

$$\mathbf{r}' = \mathbf{r}_0 + \tfrac{1}{2}\mathbf{g}t^2, \tag{4.5}$$

and the position of the bullet is

$$\mathbf{r} = \mathbf{v}_0 t + \tfrac{1}{2}\mathbf{g}t^2, \tag{4.6}$$

the origin being taken at the muzzle of the gun. Subtracting these two expressions gives

$$\mathbf{r} - \mathbf{r}' = \mathbf{v}_0 t - \mathbf{r}_0. \tag{4.7}$$

Since $\mathbf{r}_0$ is along the line $OA_1A_2A_3$, it is parallel to $\mathbf{v}_0$, and so, at some time after firing, the vector $\mathbf{v}_0 t$ will be equal to $\mathbf{r}_0$, and at that time $\mathbf{r} - \mathbf{r}'$ will be zero. This means that the target and the bullet are in the same position, which means they collide. In practice, air resistance has some effect but usually not enough to prevent the success of the experiment. The physical explanation, implicit in the above argument, is that gravity affects both the target and the bullet equally. Since, in the absence of gravity, the bullet would certainly hit the target if properly aimed, it will still hit it when gravity is effective.

The above treatment of projectile motion near the earth shows both the power of general vector methods and their limitations. If one wishes to find the exact values of the maximum height reached by the projectile and the distance along the horizontal at which its strikes the ground (called the *range*), then the very posing of these questions implies a privileged set of coordinate axes, and working in components along these axes will always yield such results much more

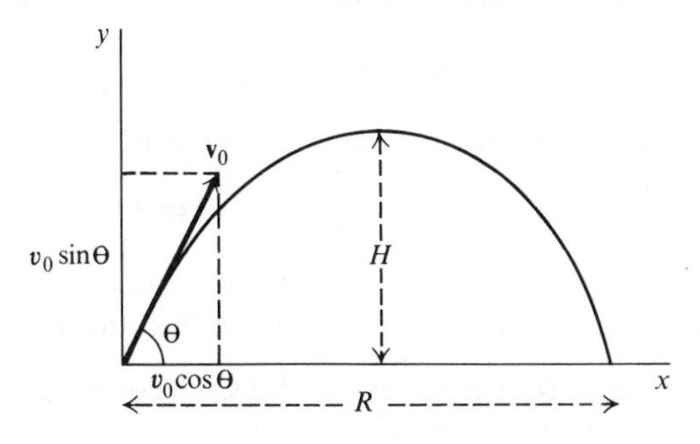

Fig. 4.17. The path of a projectile in the absence of air resistance.

easily. It is usual in problems of projectile motion to specify the direction of the initial velocity by the angle θ, called the *elevation*. The maximum height H and the range R are shown in Fig. 4.17. In terms of θ, the initial velocity $\mathbf{v}_0$ has components $v_0 \cos\theta$ and $v_0 \sin\theta$ along the axes of x and y, respectively, as shown. The equations of motion for the horizontal and vertical directions are simply the equations of uniform motion in a straight line, as discussed in Section 3.11. They may be written down side by side as follows.

x-direction	y-direction
$\ddot{x} = 0$	$\ddot{y} = -g$ (constant)
$\dot{x} = \text{constant} = v_0 \cos\theta$	$\dot{y} = (v_0 \sin\theta) - gt$
$x = (v_0 \cos\theta)t$	$y = (v_0 \sin\theta)t - \frac{1}{2}gt^2$

The motion in the x-direction is very simple, since there is no acceleration in this direction and the projectile moves along x with constant speed. In the y-direction the equations are identical to Eqs. (3.17), (3.18), and (3.19), already discussed. The projectile increases its height up to a maximum, at which point $\dot{y} = 0$. Then it reverses its vertical direction of motion and returns to earth, striking the ground with a speed v_0, the same as its initial speed. The equation of the trajectory is obtained by substituting $t = x/(v_0 \cos\theta)$ into the equation for y, giving

$$y = (v_0 \sin\theta)\left(\frac{x}{v_0 \cos\theta}\right) - \tfrac{1}{2}g\left(\frac{x}{v_0 \cos\theta}\right)^2$$

$$= x \tan\theta - \left(\frac{g}{2v_0^2 \cos^2\theta}\right)x^2, \qquad (4.8)$$

which is the equation of a parabola.

To find the maximum height H, one uses the fact that the velocity $\dot{y}$ is zero there, which means that the maximum height is reached after a time

$$t_1 = v_0 \sin\theta/g.$$

The maximum height is then just the value of y at that time, or

$$H = (v_0 \sin\theta)t_1 - \tfrac{1}{2}gt_1^2 = (v_0^2 \sin^2\theta)/2g. \qquad (4.9)$$

The trajectory is a symmetrical curve, so that the total time of flight is just $2t_1$. The range is obtained by substituting this value of the time into the equation for x:

$$R = (v_0 \cos\theta)(2t_1) = v_0^2(2 \sin\theta \cos\theta)/g,$$

which is

$$R = v_0^2 \sin 2\theta/g \qquad (4.10)$$

Since the maximum value of the sine of an angle is unity, it is seen that the maximum possible range is

$$R_{\text{max}} = v_0^2/g, \tag{4.11}$$

and this occurs when $\sin 2\theta = +1$, which means $2\theta = 90°$, and therefore $\theta = 45°$. This result may be roughly verified by experimenting with a garden hose, the stream of water from which follows a parabolic path as described here.

Example 1. From the top of a wall of height 20 m a ball is thrown horizontally with a speed of 6.0 m s^{-1}. How far from the wall will the ball land?

Solution. It would be possible to solve this problem by recognizing that the path of the ball as it falls is just the second half of Fig. 4.17, and to apply the equations already derived. But it is preferable to tackle problems like this one from scratch, and to derive the particular forms of the equations of motion that apply to the given situation. We choose the origin at the top of the wall and take the x-axis and y-axis as shown in Fig. 4.18. If v_0 denotes the initial speed of the ball, then v_0 is also the x-component of the initial velocity, the y-component of the initial velocity being zero since the motion is initially entirely horizontal. Consequently the equations of motion appropriate to this situation are:

$$\dot{x} = v_0; \qquad \dot{y} = -gt;$$

$$x = v_0 t; \qquad y = -\tfrac{1}{2}gt^2.$$

The ball reaches the ground when $y = -H$, where H is the height of the wall. Hence the time taken to reach the ground is given by $-H = -\frac{1}{2}gt_1^2$, or

$$t_1 = \sqrt{2H/g} = \sqrt{2 \times 20 \text{ m}/9.8 \text{ m s}^{-2}} = 2.0 \text{ s}.$$

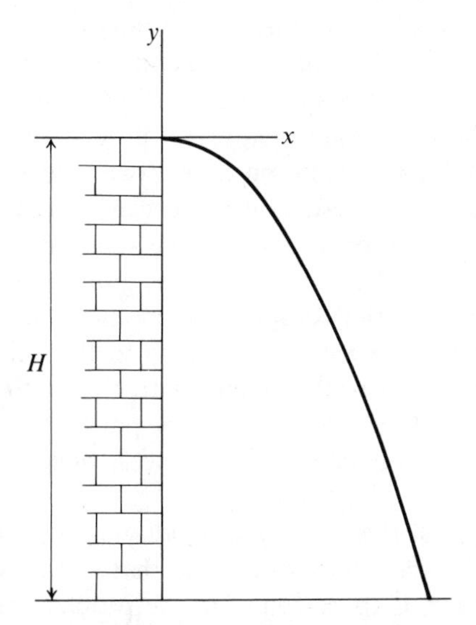

Fig. 4.18. Illustrating Example 1.

It will be noticed that this is the same time as would be required for the ball to fall straight down from rest: the horizontal motion has no effect on the vertical motion. The value of x at $t = t_1$ is therefore

$$v_0 t_1 = 6.0 \text{ m s}^{-1} \times 2.0 \text{ s} = 12 \text{ m},$$

and this is the distance from the bottom of the wall at which the ball lands.

Example 2. A projectile has a range of 50 m and reaches a maximum height of 10 m. What is the elevation of the projectile?

Solution. If we divide Eq. (4.9) by (4.10) we get

$$\frac{H}{R} = \frac{v_0^2 \sin^2\theta/2g}{v_0^2\, 2 \sin\theta \cos\theta/g} = \tfrac{1}{4}\tan\theta.$$

Therefore

$$\tan\theta = 4H/R = 4 \times 10 \text{ m}/50 \text{ m} = 0.8,$$

which yields

$$\theta = 38^\circ\, 40'.$$

4.7 THE PHYSICS OF PROJECTILE MOTION

The interaction of mathematics and physics in this simple example of two-dimensional motion is worthy of attention. The physical assumption is that objects near the earth have a constant acceleration downwards. The other derived results, such as the equation of the trajectory and the expressions for the maximum height and the range, are mathematical deductions from this simple physical assumption. They are aspects of the physical situation not immediately obvious in the bald statement $\ddot{\mathbf{r}} = \mathbf{g}$, but they are, nevertheless, logically equivalent to it. In making the statement about the constancy of the acceleration, the physicist does so in the belief that he is making a statement about reality—a belief based, perhaps, on observation of falling bodies. When the logical process of mathematical analysis is applied, the derived results can be no more true than the statement itself. If some of these derived results are tested by experiment and found not to correspond with reality, then the conclusion is inescapable that the original statement did not correspond with reality either. In the process of deducing the consequences of any physical assumption, all the resources of mathematics can be utilized, and sometimes the complexity involved is frightening to the uninitiated. So much so that they tend to confuse the truth content with the mathematical detail. But, no matter how sophisticated the mathematics, the physical content never increases; if one wishes to alter the predicted consequences, one must alter the physical assumptions.

An excellent example of this process of modifying the physical statement of a problem, in the light of experimental evidence that conflicts with results derived from a previous statement, is to be found in the problem of the flight of a golf-ball. The assumption that the acceleration of the ball is constant and equal to

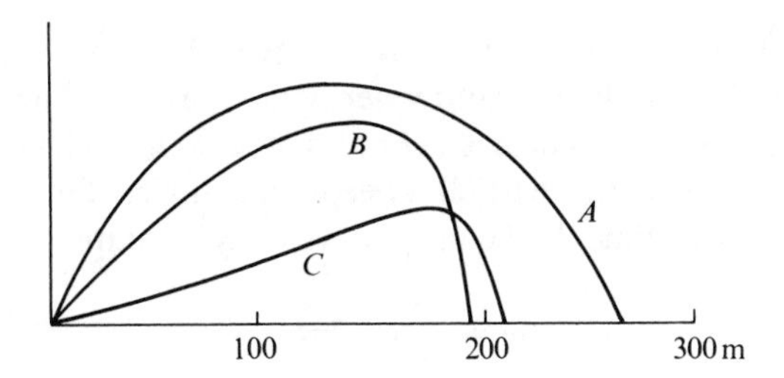

Fig. 4.19. The trajectory of a golf-ball. Curve A shows the path if only gravity acts on the ball; curve B the path if air resistance is also considered; and curve C the path when the effects of backspin are taken into account.

9.8 m s^{-2} downwards is not compatible with the observed facts, for it leads to predicted times of flight that are very much shorter than those actually observed and to a terminal speed that is equal to the speed on leaving the tee. This would make a golf-ball a lethal weapon indeed. Changing the physical assumptions to include the effect of air resistance modifies these results considerably. As shown in Fig. 4.19, the trajectory is no longer symmetrical, which is in agreement with observed trajectories, and the calculated terminal speed comes out to be much more reasonable. But the times of flight for angles of elevation comparable to those observed ($< 15°$) all turn out to be much too short, only two or three seconds. No amount of juggling with the parameters of the problem can give a satisfactory fit with the observed facts. In particular, it is not unknown for good drives to bend upwards for a good part of their total carry, and this implies, as we have seen, an upward acceleration toward the concave side of the curve. Hence, the conclusion is forced on the physicist that there is another influence at work, and this can be none other than the spin of the ball. When this is taken into account, good agreement can be obtained with observation. The lift on a ball with backspin is quite analogous to the lift on an aerofoil. The importance of backspin in the golf drive was first realised by P. G. Tait in 1896. His approach, briefly indicated above, was that of the true physicist, and his sequence of assumption, mathematical deduction of consequences, checking with observation, reformulation of assumption, . . . , and so on, was a perfect example of scientific method, even if the application was to what heretics might regard as only a game. As a moral tailpiece, it might be added that, although he knew more about the physics of golf than any man of his day, Tait himself was only a mediocre player of the game.

4.8 UNIFORM CIRCULAR MOTION

Consider a particle traveling round a circular path with constant speed v. The time to go once round is called the *period* of the motion and is written T. Since the circumference of a circle is $2\pi r$, the relation between speed, radius, and period is

$$T = 2\pi r/v. \tag{4.12}$$

The reciprocal of the period is called the frequency and will be written in this book as f. Other common symbols for the frequency are n and ν. The angular velocity or angular frequency of the particle is defined as the rate at which the radius from the center of the circle to the particle sweeps out angle. Since by definition it sweeps out 2π radians in time T, the angular velocity is simply

$$\omega = \frac{2\pi}{T} = 2\pi f. \tag{4.13}$$

Since the period is a time, it has units s, therefore both f and ω have units s^{-1}. It is conventional to write the units of ω as rad s^{-1}, and the units of f as simply s^{-1}, or c/s, standing for "cycles per second". The cycle per second is also known as the hertz (Hz). The radian unit of angle is discussed in Appendix A. It has no dimensions.

Substituting the value of T from Eq. (4.12) into (4.13) gives

$$\omega = v/r, \qquad \text{or} \qquad v = \omega r. \tag{4.14}$$

Consider the position vectors $\mathbf{r}_1$ and $\mathbf{r}_2$ for two successive instants of time, as shown in Fig. 4.20. These vectors are, of course, both of length r, equal to the radius of the circle, and so there is no change in the magnitude of $\mathbf{r}$ with time. But there is a change of direction, and this means that the vector $\mathbf{r}$ is changing with time. The displacement of the particle between the two instants is the difference

$$\Delta\mathbf{r} = \mathbf{r}_2 - \mathbf{r}_1 = P_1P_2.$$

It will be apparent that as the interval of time is shrunk to zero in the usual way, then this vector displacement will tend to become perpendicular to the radius. Since the velocity vector is defined by the relation

$$\mathbf{v} \equiv \dot{\mathbf{r}} \equiv \lim_{\Delta t \to 0} \frac{\Delta\mathbf{r}}{\Delta t},$$

the conclusion is that, for circular motion, the velocity vector is perpendicular to the position vector. The velocity vector is *always* along the tangent line to the path, but it is only perpendicular to the position vector for a circular path, since this is a unique geometrical property of a circle. This is true whether the speed is constant or not. The rule for obtaining the velocity from the position vector may be formulated as follows.

a) *The magnitude of the rate of change of position in uniform circular motion is obtained by multiplying the magnitude of the position vector by the angular velocity* ω *(Eq.* 4.14).

b) *The direction of the rate of change of position is perpendicular to the position vector, the sense being given by a rotation of the position vector in the direction of motion.*

This rule is illustrated in Fig. 4.21. The velocity vector and the position vector have been drawn from a common point, which may be taken as the center of the

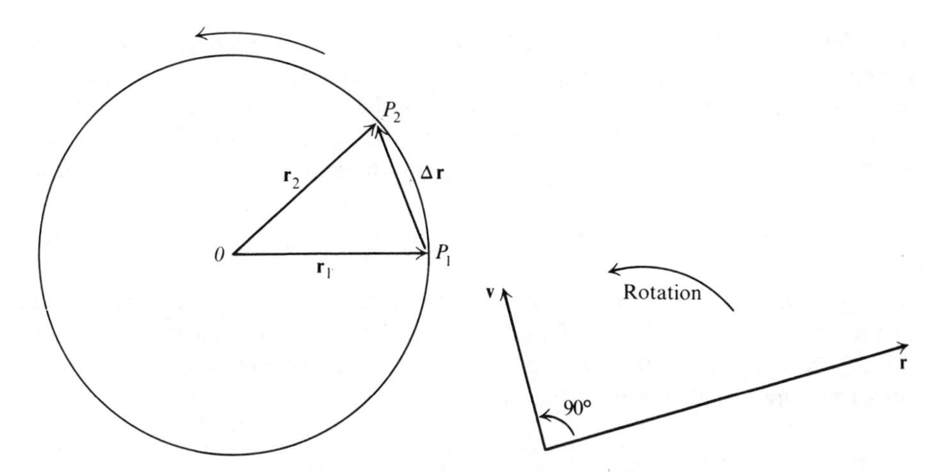

Fig. 4.20. The displacement of a particle which is undergoing uniform circular motion.

Fig. 4.21. The relation of the position vector and the velocity vector in uniform circular motion.

circle. Vectors may always be slid about parallel to themselves on the page: all that matters is that the magnitude and the direction be right.

As the particle moves with constant speed round the circle, the position vector (which is simply the directed line from the center to the particle) rotates with constant angular velocity ω, and the velocity vector also rotates with constant angular velocity ω but is 90° ahead of the position vector by the rule above. In Fig. 4.22 the position vectors at two instants of time have been drawn, as in Fig. 4.20, and the velocity vectors $\mathbf{v}_1$ and $\mathbf{v}_2$ for the same two instants have also been drawn from the center of the circle. The angles P_1OP_1' and P_2OP_2' are both 90°.

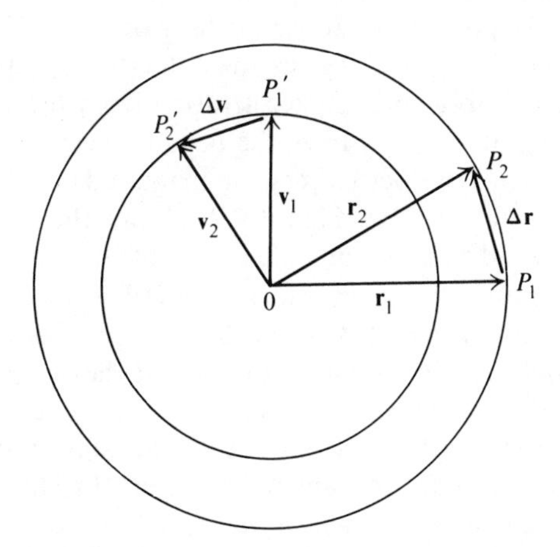

Fig. 4.22. The displacement and the change of velocity of a particle undergoing uniform circular motion.

The change in velocity is $\Delta\mathbf{v} = P_1'P_2' = \mathbf{v}_2 - \mathbf{v}_1$, and so the average acceleration over the interval is $\Delta\mathbf{v}/\Delta t = (\mathbf{v}_2 - \mathbf{v}_1)/(t_2 - t_1)$. As the interval Δt is shrunk to zero, $\Delta\mathbf{v}$ becomes perpendicular to $\mathbf{v}$ in exactly the same way as $\Delta\mathbf{r}$ becomes perpendicular to $\mathbf{r}$. Note that the magnitude of the velocity does not change with time, only its direction. It follows that the instantaneous acceleration defined by

$$\mathbf{a} \equiv \dot{\mathbf{v}} \equiv \ddot{\mathbf{r}} \equiv \lim_{\Delta t \to 0} \frac{\Delta\mathbf{v}}{\Delta t}$$

is perpendicular to the velocity, just as the instantaneous velocity is perpendicular to the position vector. Study of Fig. 4.22 will show that the triangles OP_1P_2 and $OP_1'P_2'$ are similar, and therefore, by the properties of similar triangles, the ratios of corresponding sides are equal. That is,

$$\Delta r/r = \Delta v/v,$$

since $r_1 = r_2 = r$, and $v_1 = v_2 = v$. Dividing both sides of this equation by Δt gives

$$\frac{1}{r}\frac{\Delta r}{\Delta t} = \frac{1}{v}\frac{\Delta v}{\Delta t}.$$

Proceeding to the limit by shrinking the interval Δt to zero leads directly to

$$v/r = a/v.$$

This gives the following alternative expressions for the magnitude of the acceleration, by the use of Eq. (4.14):

$$a = v^2/r = \omega^2 r = \omega v. \tag{4.15}$$

The direction of the acceleration is perpendicular to the velocity just as the velocity is perpendicular to the position vector, and study of Fig. 4.22 shows that the acceleration leads the velocity by 90° just as the velocity leads the position vector by 90°. Thus the rule for obtaining the acceleration from the velocity is the same as the rule for obtaining the velocity from the position vector. The magnitude of the acceleration is the angular velocity times the magnitude of the velocity; the direction is 90° ahead of the velocity. Figure 4.23 shows the three vectors $\mathbf{r}$, $\mathbf{v}$, and $\mathbf{a}$ for the same instant, all drawn from the center of the circle. The acceleration is 180° ahead of the position vector, which is another way of saying that it acts toward the center of the circle. It is called the *centripetal* acceleration. The magnitudes of the three vectors are constant in time, but the vectors themselves are not, since their directions are constantly changing. The general rule for finding the rate of change of any vector in uniform circular motion should now be clear from the above discussion; and if, for example, one wanted to find the rate of change of the acceleration, then, clearly, one would find the magnitude by the product ωa, and the direction by rotating once again through 90° in the sense of the rotation of the particle. This would give the vector shown dotted in Fig. 4.23.

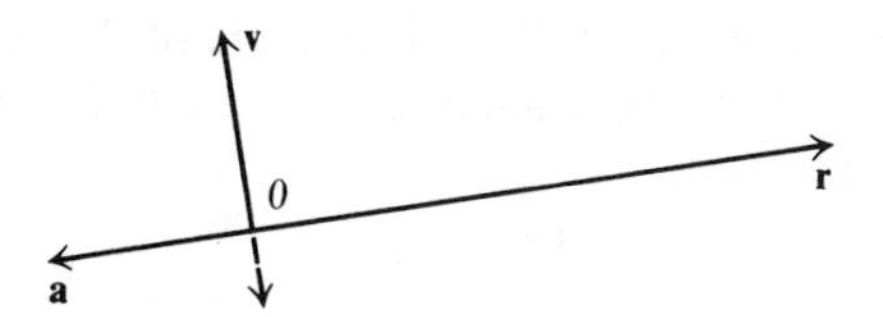

Fig. 4.23. The relation of the position vector, the velocity vector, and the acceleration vector in uniform circular motion.

Example 1. What is the period of an earth satellite traveling in a circular orbit 10^5 m above the earth where the acceleration of gravity is 9.5 m s^{-2}? The radius of the earth is 6.4×10^6 m.

Solution. A satellite coasting round the earth is in free fall, since it is under the influence of gravity alone. Hence, the centripetal acceleration of the satellite is just the local gravitational acceleration, or $\omega^2 r = g$. The period of the satellite is given in terms of ω by Eq. (4.13), whence we get

$$T = 2\pi\sqrt{r/g}.$$

From the data given, $g = 9.5 \text{ m s}^{-2}$ and $r = (6.4 \times 10^6 + 10^5) \text{ m} = 6.5 \times 10^6$ m. Substituting these values gives

$$T = 2\pi\sqrt{\frac{6.5 \times 10^6 \text{ m}}{9.5 \text{ m s}^{-2}}} = 5196 \text{ s} = 1.44 \text{ h}.$$

Example 2. What is the acceleration of a point on the edge of an LP disk?

Solution. Long-playing disks rotate at $33\frac{1}{3}$ r.p.m. or rev min^{-1}, and their standard radius is very nearly 0.15 m. To convert rev min^{-1} to units of rad s^{-1} we utilize conversion factors in the form

$$\left(\frac{1 \text{ min}}{60 \text{ s}}\right) \quad \text{and} \quad \left(\frac{2\pi \text{ rad}}{1 \text{ rev}}\right).$$

Both of these expressions are simply the number +1, which means that, if required, they may be squared or cubed and so on without affecting their value. This strategy is strongly recommended whenever a change of units has to be made. Thus

$$33.3 \text{ rev min}^{-1} \times \left(\frac{1 \text{ min}}{60 \text{ s}}\right) \times \left(\frac{2\pi \text{ rad}}{1 \text{ rev}}\right) = 3.49 \text{ rad s}^{-1}.$$

The acceleration required is therefore

$$\begin{aligned} a = \omega^2 r &= (3.49 \text{ rad s}^{-1})^2 \times 0.15 \text{ m} \\ &= 1.83 \text{ m s}^{-2} \end{aligned}$$

towards the center spindle.

Note that we do not write "rad" into the units of the centripetal acceleration. It is convenient to retain it in the angular velocity, but it is never kept once the reference to angle is dropped.

PROBLEMS

4.1 A man walks 8.5 m due east, then 6.0 m due north. What is his resultant displacement?

4.2 A particle is subjected to three consecutive displacements as follows: 30 m to the north-east; 40 m to the north-west; 50 m in a direction 8° east of south. What is the resultant displacement?

4.3 What single velocity is equivalent to velocities of 20 m s^{-1} along the positive x-axis and 30 m s^{-1} along the positive y-axis?

4.4 An object slides down an inclined plane which makes an angle of 37° with the horizontal at a constant speed of 0.5 m s^{-1}. Resolve its velocity into a component along the horizontal direction and a component along the vertical direction.

4.5 A boy throws a baseball vertically upwards while standing on an inclined plane of angle 30°. If the initial velocity of the ball is 10 m s^{-1}, resolve this into components parallel and perpendicular to the slope.

4.6 A motor launch travels at 6 m s^{-1} relative to the water. In a river flowing at 2 m s^{-1}, what is the velocity of the launch relative to the banks when it is traveling (a) upstream, (b) downstream, and (c) across the river?

4.7 A motor launch takes 30 s to travel 120 m upstream, and 20 s to travel the same distance downstream. Calculate the speed of the current, and the velocity of the launch relative to the water.

4.8 An aeroplane has a velocity of 250 m s^{-1} relative to the air. The pilot wishes to fly due south, and there is a wind blowing due west of speed 25 m s^{-1}. In what direction should he steer the plane, and what will be its velocity relative to the earth?

4.9 Ship A, which is steaming south-east at 6 m s^{-1}, is 10 km due north of ship B which is steaming north-east at 8 m s^{-1}. What is the velocity of ship A relative to a frame of reference moving with ship B? What is the velocity of ship B relative to a frame of reference moving with ship A? What is the shortest distance between the two ships?

4.10 Three displacement vectors have component magnitudes referred to rectangular axes of 3.2 m, 2.1 m; −1.6 m, 6.2 m; and −0.5 m, −1.8 m. Use a sheet of graph paper to find the resultant displacement.

4.11 Find graphically the result of subtracting a velocity with component magnitudes 8.8 m s^{-1}, 6.4 m s^{-1} from a velocity with component magnitudes −5.7 m s^{-1}, 9.2 m s^{-1}.

4.12 A point is moving in a plane, and its position with respect to rectangular coordinate axes is noted for a period of 10 s at intervals of 1 s. The results are as follows:

t (s)	0	1	2	3	4	5	6	7	8	9	10
x (m)	2.08	4.24	6.36	8.32	10.00	11.28	12.04	12.16	11.52	10.00	7.48
y (m)	1.05	2.20	3.45	4.80	6.25	7.80	9.45	11.20	13.05	15.00	17.05

Plot the curved path of the point on a sheet of graph paper and, using a piece of thread, estimate the total distance covered by the point in the whole 10 s. What is the average speed of the point over the 10 s?

4.13 Using the data of the previous problem, find the displacement of the point during the 10 s, and hence calculate the average velocity over the whole period of the motion.

4.14 Using the same data, estimate the average speeds over successive 1-s intervals using a piece of thread. Assuming that these are a reasonable estimate of the instantaneous speeds at the instants $t = 0.5, 1.5, 2.5, \ldots, 8.5$, and 9.5 s, plot a graph of instantaneous speed against time for the motion of the point.

4.15 Draw displacement-time graphs $x - t$ and $y - t$, and from these find the values of $\dot{x}$ and $\dot{y}$ at $t = 0.5, 1.5, \ldots, 8.5$, and 9.5 s. Hence determine the magnitude of the velocity at these times and compare with your previous results.

4.16 On the curved path of the point draw the velocity vectors at $t = 4.5$ s and $t = 8.5$ s, making the length of the vectors proportional to the instantaneous speeds calculated in Problem 4.14. Verify that the components of these two vectors agree with the $\dot{x}$- and $\dot{y}$-values found from the displacement-time graphs of the previous problem.

4.17 Draw graphs of $\dot{x}$ against time and of $\dot{y}$ against time, using the values of $\dot{x}$ and $\dot{y}$ obtained previously. From these graphs determine the components of acceleration $\ddot{x}$ and $\ddot{y}$ at $t = 9$ s and use these to draw in the vector acceleration on the path of the point at that time. Verify that this vector points to the concave side of the path, although it does not lie along the normal to the path. Why not?

4.18 A projectile is fired with an initial speed of 500 m s^{-1} horizontally from the top of a cliff of height 19.6 m. At what distance from the foot of the cliff does it strike the ground?

4.19 A bullet is fired from a gun with a muzzle velocity whose horizontal component is 500 m s^{-1} and whose vertical component is 200 m s^{-1}. How long does the bullet take to reach the highest point in its flight? What is its velocity at that point? What is its acceleration at that point? What is the total time of flight of the bullet? What is the total horizontal range of the bullet? Do the calculated answers seem reasonable? If not, explain why.

4.20 A golf-ball falls from a height of 10 m on to a steel plate at ground level from which it rebounds without change of speed at an angle of 15° to the horizontal. How far will it travel horizontally before it strikes the ground? How long will it take to cover this horizontal distance?

4.21 A tank fires its gun while traveling at 12 m s^{-1}. The muzzle velocity of the gun is 400 m s^{-1} and the elevation of the barrel is 15° to the horizontal. Calculate the range of the shell when the gun is fired (a) forwards, (b) backwards, and (c) perpendicular to the direction of motion of the tank.

4.22 Prove that a boy who can throw a stone to a maximum distance of 64 m on the level can also throw it so as to clear a wall 24 m high at a distance of 32 m, the angle of elevation being 63°26′.

4.23 The same boy as in the previous problem wishes to throw a stone over a wall that is 40 m away from him. What is the maximum height of the wall?

4.24 A rifle bullet is fired at and hits a target 100 m distant. If the bullet rises to a maximum height of 2 cm above the horizontal line between the muzzle of the rifle and the point of impact on the target, calculate the muzzle velocity.

4.25 An automobile travels round a curve of radius 150 m at a constant speed of 20 m s^{-1}. What is the acceleration of the car in magnitude and in direction?

4.26 To a good approximation the earth travels round the sun in a circular orbit of radius 1.50×10^{11} m once every 365.3 days. Calculate (a) the orbital velocity of the earth in meters per second, and (b) the radial acceleration of the earth toward the sun.

CHAPTER 5

THE LAWS OF TRANSLATIONAL MOTION

5.1 INTRODUCTION

In the previous chapters the motion of objects along a line and in a plane has been considered, and the concepts of displacement, velocity, and acceleration have been introduced and precisely defined. All these quantities are essentially geometrical in character, provided that time is allowed as a coordinate on the same basis as the usual space coordinates. Since the basic units of length and time are known, all of these concepts are properly defined. In discussing displacement in a plane, it has been tacitly assumed that, when the object is not of negligible size, all points of the body undergo the same displacement. Such a displacement is known as a *translation*, and the present chapter will be confined to discussion of this type of motion. Consideration of rotational motion is deferred to Chapter 6.

In discussing the laws governing the translational motion of objects, the third basic mechanical dimension of mass must be introduced. The mass of an object is a scalar quantity to be determined ultimately by comparison with a legally accepted standard, the international prototype kilogram kept at Sèvres, near Paris. All other standardizing laboratories in the world have copies of this prototype, and these in turn are used to provide the necessary standardization down to the level of those kept in ordinary laboratories. An unknown mass is compared with such sub-standards by means of the equal-arm balance, and this permits the determination of mass over a range of about 10^{-9} kg up to several hundreds of kilograms with great accuracy. For objects whose mass lies outside this range other indirect methods must be used.

The importance of mass in the motion of an object is a matter of common experience. To be struck on the head by a baseball is not at all the same thing as being struck on the head by a tennis-ball traveling at the same speed. It is the combined effect of the mass of the ball and its speed that is important, and this leads to the introduction of a new concept, *momentum*. The momentum of an object is defined as the product of the mass and the velocity. In symbols,

$$\mathbf{p} \equiv m\mathbf{v}. \tag{5.1}$$

Momentum, being defined as the product of a scalar and a vector, is itself a vector quantity. Its units are kg m s^{-1}, for which there is no special name.

5.2 FORCE

We ask the question: what is natural motion? In other words, what motion is self-explanatory? When we see a stone fall to the ground, do we seek a cause for this? This is an important question, because the followers of Aristotle saw no need to seek a cause for this motion; it was to them a natural motion that needed no further explanation. On the other hand, they needed a cause to explain motion in a straight line with constant speed, since this was for them an unnatural motion. There is nothing contrary to reason or observation in this attitude of the Aristotelians; it is merely that the physics built on their assumptions was sterile and led to no significant advance. The alternative point of view of Galileo and Newton—that the only natural motion is that in a straight line with constant speed—in fact appears at first sight to be less plausible than that of Aristotle, but this starting point has proved to be enormously fertile, and is in fact the key to the whole development of science since the seventeenth century. It may be expressed as follows.

The momentum of an object does not change unless the object is acted on by an unbalanced force.

This is the first law of motion. A change of momentum can occur in two ways. The velocity can remain unchanged but the mass can change. This possibility will be ignored in this book, since it is of little importance in biological and medical contexts. Or, for constant mass, the velocity can change. A change of velocity can be a change in the magnitude, or a change in the direction, or both. Thus any motion in a curved path, whether the speed be constant or not, requires an explanation in terms of some unbalanced force. Where a deviation from natural motion with constant momentum is observed, a cause must be sought.

Consider a very ordinary example, a tractor pulling a log along level ground. Initially the log is at rest on the ground and the tractor must exert a large force on the chain connecting it to the log in order to start the log moving. Thereafter the log moves along the ground with constant velocity. In Newton's view, both the state of rest and the final state of constant velocity are states of mechanical equilibrium in which no unbalanced force acts on the log. There are forces acting, of course, but these balance out. At rest the weight of the log manifests itself as a force which the log exerts on the ground, and this is just balanced by the force which the ground exerts upwards on the log. These two forces balance throughout the motion so long as the ground remains level. In motion with constant velocity the forward force exerted by the tractor is much smaller than the force initially needed to start the log moving, and this forward force is balanced by the backward drag between the log and the ground. Such forces as this, resulting from the relative motion of surfaces in contact, are called frictional forces. Between the state of rest and the state of constant velocity, the log must be accelerated forwards, and during this time the forward force exerted by the tractor must be greater than the retarding force of friction, the difference between them being the unbalanced force that causes the acceleration.

The force exerted by the tractor in this example is of a direct and intuitively obvious kind, and so is the frictional force between the log and the ground. All pushes and pulls, whether by muscular effort on the part of an animal or some other means, involve a mechanism of transmission, as in the chain connecting the tractor to the log or in the direct contact between the log and the ground. That other kinds of force exists is apparent from consideration of the weight of the log, which manifests itself as a downward force exerted on the ground. But what agency is exerting this force? To conceive of the log as pushing downwards is to attribute animal-like qualities to it which it does not possess. The weight of the log in any case does not depend on its being in contact with the ground, for we do not believe that the weight will vanish if the ground suddenly opens up beneath it. Weight is not a force exerted *by* the log at all, but a force exerted *on* the log—by the earth as we know—with no apparent mechanism of transmission. And what of the force that the sun exerts on the earth? That a force exists follows from the fact that the earth is not traveling with constant velocity, but no apparent mechanism of transmission of such a force exists. Today we simply accept the existence of such forces as a fact, forced upon us by our acceptance of Newton's first law of motion, and leave speculation about the mechanism of transmission to the philosophers. The force is to be regarded as understood when its mathematical form is known.

The quantitative measure of force is provided by Newton's second law of motion.

The unbalanced force acting on an object is proportional to the rate of change of momentum of the object.

In symbols,

$$\mathbf{F} = k\dot{\mathbf{p}}, \tag{5.2}$$

where k is a constant. In circumstances where the mass may be regarded as constant, this may be rewritten in the form

$$\mathbf{F} = km\dot{\mathbf{v}} = km\mathbf{a}, \tag{5.3}$$

since acceleration is rate of change of velocity. These equations have been written in vector form, and this implies that the unbalanced force and the acceleration are in the same direction. In this book we follow Newton and accept force as a concept with an intuitive meaning formed ultimately by analogy with human muscular force. On this view, Eqs. (5.2) and (5.3) are to be regarded as genuine equations and not as mere definitions. Suitable units of force may be defined by specifying the constant k, and the simplest and best choice is to take k as the number $+1$. The units of force are then the units of acceleration times the unit of mass, and the name *newton* (abbreviated N) is given to that unbalanced force that will cause a mass of one kilogram to have an acceleration of one meter per second: $1\ \mathrm{N} \equiv 1\ \mathrm{kg\ m\ s^{-2}}$. Some idea of the magnitude of the newton can be obtained by remembering that a medium-sized apple held in the hand exerts a force of roughly one newton on the hand. Eqs. (5.2) and (5.3) may now be rewritten as

$$\mathbf{F} = \dot{\mathbf{p}} = m\mathbf{a}, \tag{5.4}$$

with the understanding that MKS units are being used.

5.3 THE THIRD LAW OF MOTION

Newton's principal contribution to mechanics lay in his clear understanding that forces always occur in pairs. Thus, when a baseball is struck by a bat, there is not only the force that the bat exerts on the ball but also the force that the ball exerts on the bat: The interaction of ball and bat necessarily involves both these forces. The third law of motion expresses the fundamental equality of these two forces.

> *When an object A exerts a force on an object B, then the object B exerts an equal but opposite force on A.*

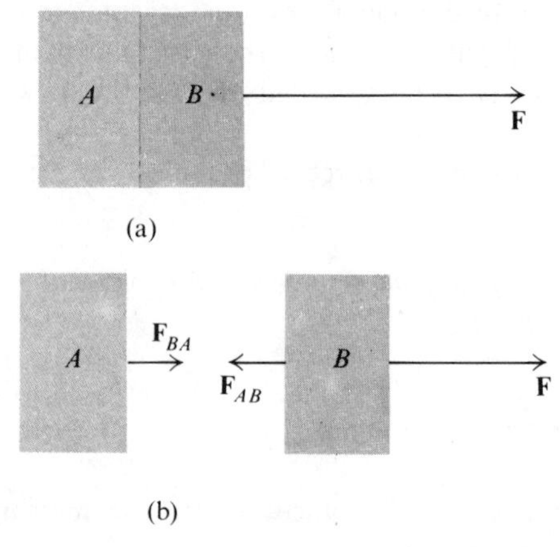

Fig. 5.1. (a) A force acting on a composite body; (b) the forces acting on the individual bodies.

If the force exerted by A on B is written as $\mathbf{F}_{AB}$ and the force exerted by B on A as $\mathbf{F}_{BA}$, then the third law may be written

$$\mathbf{F}_{AB} = -\mathbf{F}_{BA}. \tag{5.5}$$

These two forces are referred to as an action-reaction pair. This law of motion provides us with the justification for treating the motion of objects of finite size directly, without introducing the idea of a particle, as is usually done. For consider an object being accelerated to the right by an unbalanced force **F**, and imagine it as two separate objects, A and B, glued together as shown in Fig. 5.1. The second law applied to the whole object gives simply $\mathbf{F} = m\mathbf{a}$, where $m = m_A + m_B$ is the total mass of the whole object. The two parts A and B are held together by forces, and these are shown in Fig. 5.1(b), where for convenience the parts are shown separated. The third law gives

$$\mathbf{F}_{AB} = -\mathbf{F}_{BA}.$$

The equations of motion for the whole object and for the separate parts of the object can be written down separately, either in vector form or in component form by use of the magnitudes of the vectors along the line of the motion. These two ways of writing the equations are displayed side by side below.

Whole object:

$$\mathbf{F} = m\mathbf{a}; \qquad F = ma \tag{5.6}$$

Part B:

$$\mathbf{F} + \mathbf{F}_{AB} = m_B\mathbf{a}; \qquad F - F_{AB} = m_B a \tag{5.7}$$

Part A:

$$\mathbf{F}_{BA} = m_A\mathbf{a}; \qquad F_{BA} = m_A a \tag{5.8}$$

The acceleration is the same in all these equations, since the two parts of the object remain together at all times. Observe that in Eq. (5.7) a plus sign is used when the forces are written vectorially, whereas the difference in direction between the two forces has to be explicitly recognized by the use of a minus sign when components are used. In either case the sum of Eqs. (5.7) and (5.8) is just Eq. (5.6).

In other words, when the constituent parts A and B of the object are considered as a whole, the forces of interaction cancel out, and the motion of the whole is determined only by the external force **F**. The same result will obviously be true no matter how the object is divided up. We may even conceive it to be split up into its constituent atoms and all these atoms to be exerting internal forces on one another. Nevertheless, when considered as a whole, all these interactions cancel out, and the motion of the object depends only on the external forces acting on it. If there is no external unbalanced force acting, then the momentum of the object is constant. This, the so-called *principle of conservation of momentum*, is seen to be a direct consequence of the laws of motion.

Example 1. A rifle bullet has a mass of 12 g, its muzzle velocity is 800 m s^{-1} and the length of the barrel is 0.8 m. Calculate the force accelerating the bullet, assuming it to be constant.

Solution. If the force is constant, then the acceleration is constant, and the equations of uniformly accelerated motion may be applied. In particular, for a particle that starts from rest and reaches a speed v in a distance x we have $a = v^2/2x$. Hence, by Newton's second law,

$$\begin{aligned} F = ma &= mv^2/2x \\ &= \frac{0.012 \text{ kg} \times (800 \text{ m s}^{-1})^2}{2 \times 0.8 \text{ m}} \\ &= 4800 \text{ N}. \end{aligned}$$

Example 2. An object of mass 10 kg is subjected simultaneously to two constant forces: a force $\mathbf{F}_1$ of magnitude 5 N towards the North, and a force $\mathbf{F}_2$ of magnitude 12 N towards the East. What is the acceleration of the object?

Solution. The unbalanced force acting on the object is $\mathbf{F}_1 + \mathbf{F}_2$, and this vector sum is illustrated in Fig. 5.2. The magnitude of the sum is given by

$$F^2 = (5 \text{ N})^2 + (12 \text{ N})^2 = (13 \text{ N})^2,$$

$$\therefore F = 13 \text{ N}.$$

The force $\mathbf{F}$ acts along the line OP at an angle θ North of East as shown, where

$$\tan \theta = 5/12 = 0.417.$$

Therefore $\theta = 22° 37'$. The acceleration is also directed along OP, and is of magnitude

$$a = F/m = 13 \text{ N}/10 \text{ kg} = 1.3 \text{ m s}^{-2}.$$

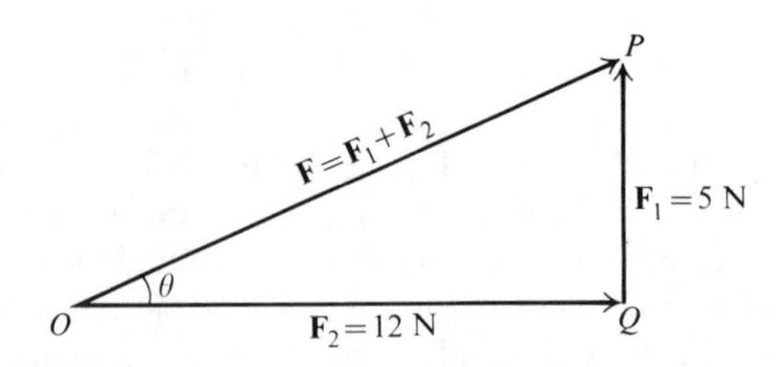

Fig. 5.2. Illustrating Example 2.

Example 3. Two sliders on an air track are fitted with magnets arranged in such a way that the sliders repel one another. The sliders are pushed together until they are nearly in contact and are then released. One slider, of mass 0.8 kg, is observed to move away with an initial acceleration of 3 m s^{-2}. What is the initial acceleration of the other slider if its mass is 0.6 kg?

Solution. We do not need to know anything about magnetism to solve this problem: all that is necessary is to realize that with magnets present the sliders will interact with one another, and that Newton's third law will apply to this interaction. From Eq. (5.5) we get

$$\mathbf{F}_1 = -\mathbf{F}_2, \qquad \text{and so} \qquad m_1\mathbf{a}_1 = -m_2\mathbf{a}_2.$$

The negative sign tells us that the second slider moves off with an acceleration in a direction opposite to the acceleration of the first slider, and the magnitude of the acceleration is

$$a_2 = \frac{0.8 \text{ kg} \times 3 \text{ m s}^{-2}}{0.6 \text{ kg}} = 4 \text{ m s}^{-2}.$$

5.4 THE FORCE OF GRAVITY

All of Newton's efforts in mechanics were directed toward the aim of explaining the motions of the planets round the sun and of the moon round the earth. His supreme achievement lay in the recognition that the force that caused an apple to fall to the ground and the force that kept the moon in place in its orbit round the earth were only different manifestations of one universal force. The form of this force necessary to explain the known facts about the motion of the planets and the moon was worked out logically, and his law of universal gravitation was formulated as follows.

Every particle in the universe attracts every other particle with a force that varies directly with the product of their masses and inversely with the square of their distance apart.

To be of much use this formulation, which is in terms of particles, must be supplemented by a second law so that it can be applied to the sun, the earth, and the moon. This second law may be stated as follows.

In its external gravitational action, a spherical body with a spherically symmetric distribution of mass acts as if its mass were concentrated at its center.

This second law does not have the same status, of course, as the law of gravitation itself, since it can be shown to follow from it. Nevertheless, it is convenient to state it overtly, since the proof is too difficult for an introductory book such as this one. Note that it is not enough for the body to be spherical. Thus a hemisphere of lead and a hemisphere of wood of the same radius can be joined face to face to form a sphere, but in such a case the sphere would not act as if all the mass were concentrated at the center, since the distribution of mass would clearly not have spherical symmetry. On the other hand, the earth has a much higher density in its core than near the surface, but the variation of density only depends on the distance from the center of the earth and so the mass distribution has spherical

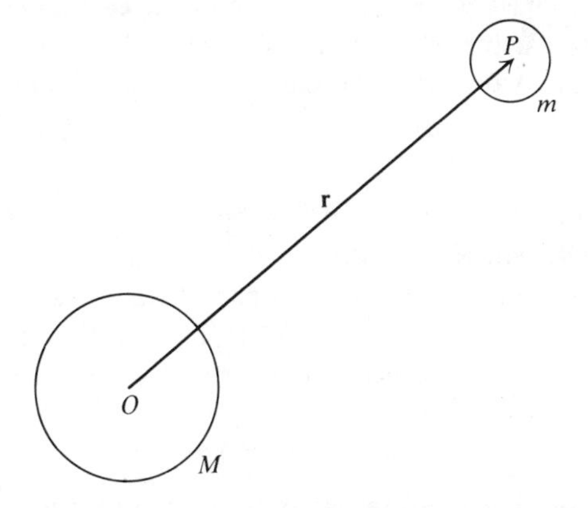

Fig. 5.3. Two bodies exerting gravitational forces on each other.

symmetry. Figure 5.3 shows two spheres with centres at O and P. The center P is considered to be located with respect to O as origin by the position vector $\mathbf{r}$. Then, the force which sphere O exerts on sphere P is given mathematically as

$$\mathbf{F}_{OP} = -G\frac{Mm}{r^3}\mathbf{r}, \tag{5.9}$$

where M and m are the respective masses. The force which sphere P exerts on sphere O is $\mathbf{F}_{PO} = -\mathbf{F}_{OP}$, by the third law. The negative sign in Eq. (5.9) is necessary, since the force is attractive, i.e. from P to O in the direction opposite to $\mathbf{r}$. The constant of proportionality in Eq. (5.9) is called the *gravitational constant* and has the value $G = 6.67 \times 10^{-11}$ N m^2 kg^{-2} (or m^3 kg^{-1} s^{-2}). It is measured with a torsion balance entirely analogous to that used to verify Coulomb's law in electrostatics (cf. Section 22.4), but the experiment is difficult to perform and G is the least accurately known of all the constants of nature.

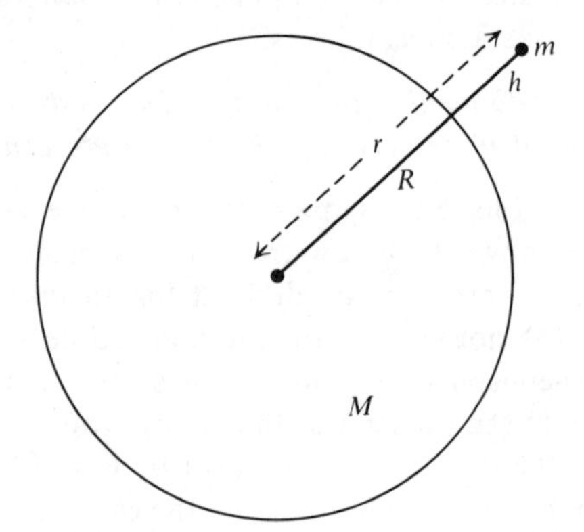

Fig. 5.4. A small body and the earth exerting gravitational forces on each other.

Consider now a mass m near the surface of the earth. The gravitational force exerted on it by the earth has a magnitude GMm/r^2, where M is here the mass of the earth. The correct value of r to be substituted into this expression for the force is

$$r = R + h,$$

where R is the radius of the earth and h is the height of the object above the earth, as shown in Fig. 5.4. Since the radius of the earth is 6.37×10^6 m, the difference between r and R is negligible for all ordinary heights h. So the force which the earth exerts on the object is essentially GMm/R^2, and this, by the second law, will cause the object to accelerate towards the surface of the earth with an acceleration g. Thus

$$F = G(Mm/R^2) = mg, \tag{5.10}$$

where, for simplicity, the equation has been written in terms of the magnitudes of the vectors only. This simplification will rarely cause any ambiguity. The mass of the object is seen to cancel out from Eq. (5.10), giving

$$g = GM/R^2. \tag{5.11}$$

This is the gravitational acceleration of any object near the earth, and is seen to be a constant. Its value is 9.8 m s^{-2}. In fact, g varies slightly over the surface of the earth for two main reasons: one, the radius of the earth at the equator is greater than the radius at the poles; two, the earth is spinning about its axis. The maximum difference in g between any two points on the earth only amounts to about 0.05 m s^{-2}, and this is negligible for most purposes. At any one place on earth the gravitational acceleration is rigorously independent of the mass of the object, and this is the justification for the use of the beam balance for comparisons of mass.

The force exerted on the object by the earth is called the *weight* of the object, and is denoted by **W**, or more simply by W when the vector nature of this force need not be taken into account. By Eq. (5.10)

$$W = mg. \tag{5.12}$$

The reaction to this force, in the sense of the third law, is an equal and opposite force exerted by the object on the earth, and acting at the center of the earth. Thus every time an apple falls to the ground, the earth accelerates upwards toward the apple. Fortunately, the enormous mass of the earth renders this real mechanical effect of no practical importance. It must be stressed again that the question of how the earth exerts its force on an object over the intervening distance (and vice versa) is left open. The physicist is satisfied with the quantitative law of gravitation expressed in Eq. (5.9) without worrying about the philosophical implications of action at a distance.

Example 1. What is the acceleration of gravity at a height of 1000 km above the surface of the earth?

Solution. By Newton's second law, the acceleration of gravity at any height is the force exerted on any object at that height divided by the mass of the object. Let us denote it by $\mathbf{g}'$ to distinguish it from $\mathbf{g}$, the acceleration of gravity at the surface of the earth. Then the magnitude of $\mathbf{g}'$ is given by

$$g' = \frac{GM}{r^2} = \frac{GM}{(R+h)^2} = \frac{GM}{R^2}\left(\frac{R}{R+h}\right)^2 = g\left(\frac{R}{R+h}\right)^2.$$

This equation tells us the manner in which g' varies with height. Using the value of R in the text and $h = 10^6$ m, we get

$$g' = 9.8 \text{ m s}^{-2}\left(\frac{6.37}{7.37}\right)^2 = 9.8 \text{ m s}^{-2} \times 0.747 = 7.3 \text{ m s}^{-2}.$$

Even at this great height the gravitational acceleration is still very large.

Example 2. What is the weight of a man of mass 100 kg at the surface of the earth?

Solution. From Eq. (5.12), $W = mg = 100 \text{ kg} \times 9.8 \text{ m s}^{-2}$

$$= 980 \text{ N}.$$

Correct scientific usage is to express weights in newtons since they are forces. The conversion factor g is the gravitational force acting per unit mass, and therefore could equally well be written as 9.8 N kg^{-1}. The two units, m s^{-2} and N kg^{-1}, are equivalent. In everday life we usually speak of a man "weighing" 100 kg. Although, strictly speaking, this is incorrect, it is so deeply rooted in our language that we shall probably never be able to eradicate it.

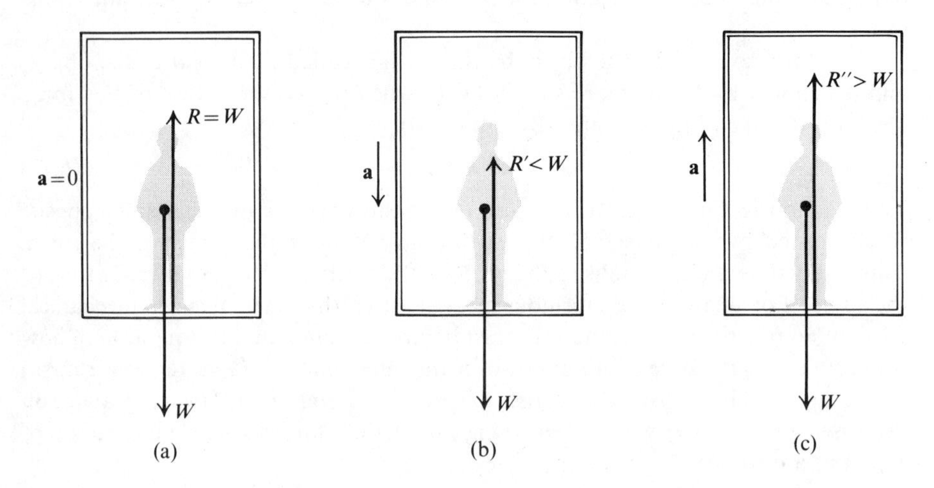

Fig. 5.5. (a), (b), and (c) Illustrating Example 3.

Example 3. A 100-kg man is standing in an elevator. What force does the floor exert on his feet (a) when the elevator has a constant velocity, (b) when it has an acceleration of 1.0 m s^{-2} downwards, and (c) when it has an acceleration of 1.0 m s^{-2} upwards?

Solution. a) When the velocity is constant, the elevator and all its contents are in equilibrium. If we denote by R the magnitude of the force exerted on the man's feet by the floor, then, by Newton's first law,

$$R = W.$$

Since $W = 980$ N, we see that $R = 980$ N also, and this force obviously acts upwards as shown in Fig. 5.5(a).

b) Consider now the situation when the elevator and its contents are being accelerated downwards. Let us focus our attention on the man, and ask what force is causing his acceleration. Since the only forces acting on him are his weight of

980 N and the force R' exerted by the floor on his feet, it will be clear that R' must now be less than 980 N, and the second law gives

$$W - R' = ma$$

or

$$R' = W - ma. \tag{5.13}$$

This situation is illustrated in Fig. 5.5(b). Since $a = 1.0 \text{ m s}^{-2}$,

$$R' = 980 \text{ N} - (100 \text{ kg} \times 1.0 \text{ m s}^{-2})$$

$$= 980 \text{ N} - 100 \text{ N} = 880 \text{ N}.$$

c) The situation when the elevator is accelerating upwards is shown in Fig. 5.5(c). Since the man has an acceleration upwards, the unbalanced force on him must also be upwards, and this unbalanced force is

$$R'' - W = ma$$

or

$$R'' = W + ma$$

$$= 1080 \text{ N}. \tag{5.14}$$

In all three situations the weight W of the man is the same, but his *sensation* of weight is very different. In (a) the floor exerts a force of 980 N upwards on the soles of his feet. In such a case the man feels quite normal and at ease. The earth is exerting a force on every part of his body, on every atom of every cell, and the floor is exerting a force equal to his weight distributed over the soles of his feet. He accepts this as a normal state of affairs because he has always lived with it, and all his bodily functions operate normally. The evolutionary process has produced on earth life-forms conditioned to the all-pervasive pull of gravity.

In (b), the man has a sensation of lightness, as if he were floating, while in (c) he feels himself pressed into the floor. He describes these two sensations respectively as a decrease and an increase in his weight. Of course his terminology is scientifically wrong: there has been no change in his *true* weight—only in his experience, both physiological and psychological, of that weight. Once again we find the word "weight" being used in a sense different from its precise, scientific meaning. As with the phrase "a man weighing 100 kg" mentioned above, there is nothing we can do but be constantly on our guard. From Eqs. (5.13) and (5.14) it will be clear that the apparent change of weight experienced is of magnitude ma. This apparent change is an increase if $\mathbf{a}$ and $\mathbf{g}$ are in opposite directions as in (c), and is a decrease if $\mathbf{a}$ and $\mathbf{g}$ are in the same direction as in (b).

The apparent change of weight is most dramatically shown during the take-off of a manned rocket. If the rocket has, for example, an acceleration during take-off of 5 g vertically *upwards*, then the astronaut will experience this as a change of weight of amount 5 mg downwards, where m is his mass. He will therefore experience a total weight of 6 mg. The muscles of a man reared on earth are not able to operate under these conditions, and the astronaut lies down on a specially designed couch during the acceleration period. Even when lying down his arms and legs feel leaden and he has difficulty raising them; his internal organs are compressed; breathing requires great effort; and the loose skin of his face is drawn tight against his skull and hangs down in folds at the side. The statement that his weight increases markedly during take-off is one that explains the facts perfectly.

Example 4. A space capsule is in a circular orbit at a height of 1000 km above the surface of the earth. What is the apparent change of weight of an astronaut of mass 100 kg?

Solution. As discussed in Section 4.8, a particle moving around a circle at constant speed has a centripetal acceleration which, in the case of an orbiting space capsule and its contents, is caused by the gravitational pull of the earth. At a height of 1000 km this pull has been reduced to 7.3 N per unit mass, as shown in Example 1 above. Since there is no atmosphere at such a great height, the gravitational pull is the only force acting, and therefore the centripetal acceleration is $g' = 7.3 \text{ m s}^{-2}$. The capsule and its contents are in what is known as *free fall* towards the center of the earth, and the situation is analogous to the elevator accelerating downwards. Applying Eq. (5.13), with $W = mg'$ and $a = g'$, we find $R' = 0$. The apparent loss of weight of the astronaut is equal to his actual weight at that height. The result is that he experiences no weight at all, and his state is popularly referred to as one of "weightlessness". Notice, however, that a 100-kg astronaut has a *true* weight of 730 N at that height.

5.5 THE MOTION OF CHARGED PARTICLES IN AN ELECTRIC FIELD

A very direct, simple, and important application of Newton's laws of motion can be made in the case of charged particles traveling between the plates of a parallel-plate capacitor in a vacuum. Although the electrical concepts involved will not be fully discussed until Chapters 22 and 23, the existence of such particles and the general effect of their motion in an electric field are today as commonplace as tractors pulling logs. The TV set in the home, and the cathode-ray oscilloscope in the laboratory, have familiarized everyone with the fact that beams of electrons are deflected into curved paths when the proper electrical connections are made. For the present purpose the essential device is a pair of parallel plates maintained at a potential difference V, as shown in Fig. 5.6, by connecting the upper plate to the positive terminal of a high-voltage supply whose negative terminal is grounded, and connecting the lower plate to ground. The separation of the plates is d and the length of the plates is l. The whole apparatus is enclosed in a glass envelope which is exhausted to an extremely low pressure. It is found experimentally that small charged particles between such a pair of plates behave in a way that cannot be explained without the assumption that a force of electrical origin is acting on them. In particular, if a beam of fast-moving electrons enters the region between the plates with velocity v_0 along the x-axis, as shown, it is found that the beam is deflected upwards. This can be detected by allowing the beam finally to strike a TV-type screen at the right-hand end of the glass envelope. Hence, the force acting cannot be gravity. Indeed, if the voltage across the plates is disconnected, so that only gravity acts, there is no observable deflection of the beam at all. The time of transit is so short that the distance fallen under gravity is quite negligible. Since a deflection necessarily involves an acceleration, it follows that some electrical force must be exerted when the voltage is on. As will be discussed in Chapter 22, the electrical state of the space between the plates is specified by the ratio V/d, which is defined as the electric field intensity, E. The force experienced by an electron between the plates is vertically upwards along the y-axis and of magnitude $F = eE$, where $e = 1.602 \times 10^{-19}$ C is the magnitude of the charge on the electron. The electrical details are not at this stage important. What *is* important is that the force is constant so long as V and d are constant, and therefore Newton's laws can be used to predict the motion.

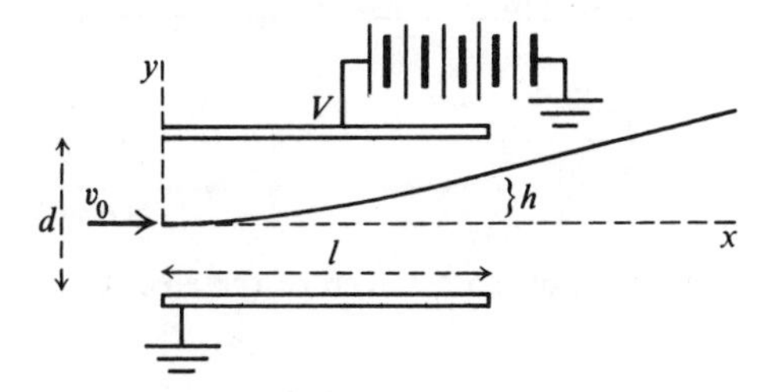

Fig. 5.6. The deflection of a beam of electrons in passing between charged parallel plates.

Since the force acting on the electron is constant, it follows that the acceleration is constant and in the same direction as the force, in this case along the y-axis toward the upper plate. The magnitude of this constant acceleration is just eE/m, where m is the mass of an electron, and this constant value may be denoted by α. Then the equations of motion for the x- and y-directions may be written down as follows:

x-direction	y-direction
$\ddot{x} = 0$	$\ddot{y} = \text{constant} = \alpha$
$\dot{x} = \text{constant} = v_0$	$\dot{y} = \alpha t$
$x = v_0 t$	$y = \frac{1}{2}\alpha t^2$

This form of the equations fits the initial conditions that at $t = 0$ the electron is at the origin and has no velocity in the y-direction. These results should be compared with the similar results in Section 4.6 derived for a projectile near the earth.

The two problems are essentially the same in structure, although the nature of the force is very different. If h denotes the vertical displacement of the electron when it leaves the plates, this quantity may be found by first finding the time it takes the electron to travel a horizontal distance l, which is just l/v_0, and then substituting this time into the equation for y. This gives

$$h = \tfrac{1}{2}\alpha(l/v_0)^2. \tag{5.15}$$

When the particle leaves the space between the plates, it is no longer acted on by the electrical force, but only by gravity, and the effect of this was demonstrated to be negligible. Hence, it travels in a straight line which is a tangent line to the parabola at the point of exit, as shown in Fig. 5.7. It is a property of the parabola

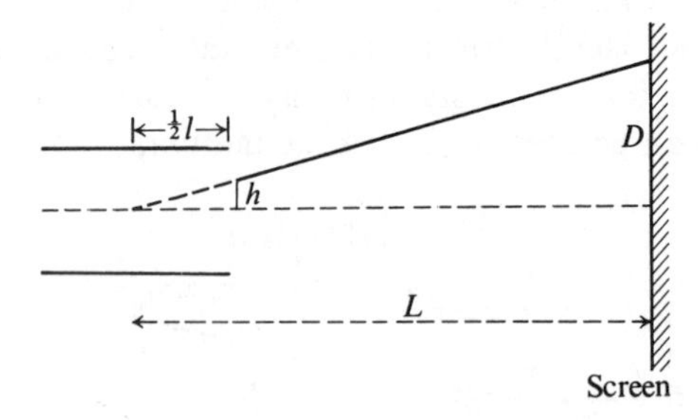

Fig. 5.7. The deflection on a screen of the electrons of Fig. 5.6.

that this tangent line when produced backwards cuts the x-axis exactly at $x = \frac{1}{2}l$, the mid-point of the plates. Simple proportion then yields the relation $D/L = h/\frac{1}{2}l$, where D is the final deflection observed on the screen, and L is the distance between the screen and the mid-point of the plates. If we use the calculated value of h from Eq. (5.15), the value of the final deflection is

$$D = 2Lh/l = Ll\alpha/v_0^2$$

$$= Ll\left(\frac{e}{m}\right)\left(\frac{1}{v_0^2}\right)E. \tag{5.16}$$

The deflection is therefore proportional to the electric intensity E, which in turn is proportional to the voltage across the plates. This is very important indeed, since the very small mass of the electrons allows the device to be adapted for use in following the changes in a rapidly varying voltage.

5.6 THE MOTION OF CHARGED PARTICLES IN A MAGNETIC FIELD

Charged particles are found to follow curved paths even when no electric field is present, and so there must be some other force acting. The nature of this force will be discussed in Chapter 25. For the present it suffices to assume the results (a) that the force acts perpendicularly to the direction of motion of the particles, and

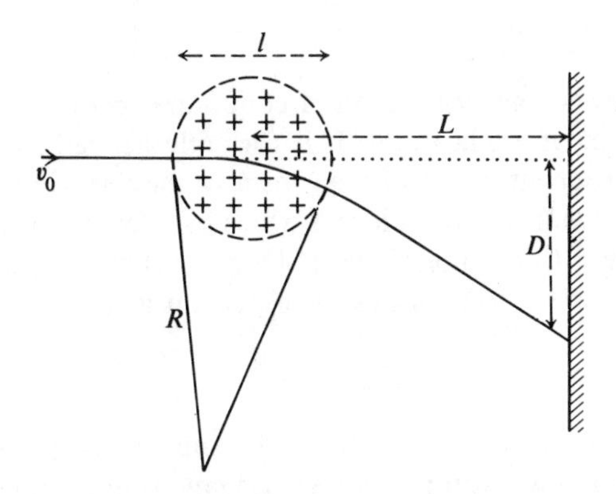

Fig. 5.8. The deflection of a beam of electrons in passing through a magnetic field.

(b) that the magnitude of the force is given by Bev_0 for electrons. The quantity B is the magnitude of the magnetic induction which is the magnetic quantity analogous to the electric field intensity. Figure 5.8 shows a beam of electrons, of initial speed v_0, being deflected by a magnetic field which is confined to a circular region and which acts perpendicularly into the page, as indicated conventionally by the crosses. As stated in Section 25.3, the path of the electrons is circular in the region of the magnetic field, and outside this region the electrons travel in straight lines as in the electrical case. The radius of the circular path is R, and the other dimensions of the apparatus are as they were in the electrical case. By simple proportion

$$D'/L = l/R \tag{5.17}$$

if the deflection D' is not too large. To determine the radius R the result of Section 4.8 may be used, that the acceleration of a particle traveling with speed v_0 in a circular path of radius R is toward the center of the circle and has magnitude v_0^2/R. Newton's second law may now be applied to give

$$Bev_0 = mv_0^2/R,$$

which yields

$$R = mv_0/Be. \tag{5.18}$$

Combining Eqs. (5.17) and (5.18) gives the desired expression for the magnetic deflection,

$$D' = Ll\left(\frac{e}{m}\right)\left(\frac{1}{v_0}\right)B. \tag{5.19}$$

Comparison with Eq. (5.16) shows that both the electric and magnetic deflections are proportional to the ratio e/m and to the respective fields E and B. But their dependence on the velocity v_0 of the electrons is different, and it was this difference that proved so useful in the early days of atomic physics. The applications of these results are discussed further in Chapter 28.

5.7 WORK

The concept of *work* is a generalization of common experience and is used when a force moves its point of application. It is formally defined as the product of the force times the component of the displacement in the direction of the force. Consider an object being pulled along a level surface by a force F whose line of action is inclined at an angle θ to the horizontal (Fig. 5.9). If the object moves through a distance x, then the work done in this displacement is defined to be

$$\mathscr{W} \equiv Fx \cos \theta. \tag{5.20}$$

The units of work are those of force times distance, or N m, and the name joule (J) is given to the unit of work in the MKSA system. It must be noted that work is a scalar quantity and has no directional properties at all. It is formed from two vector quantities, the force and the displacement, but involves only their magnitudes and the angle between their directions. Equation (5.20) can be rewritten in two alternative ways, as

$$\mathscr{W} \equiv (F \cos \theta)x \qquad \text{or} \qquad F(x \cos \theta).$$

The first shows the work as the resolved part of the force along the line of the displacement times that displacement; the second shows the work as the force times the resolved part of the displacement along the line of action of the force. The two ways of looking at it are entirely equivalent.

If the force is constant both in magnitude and direction, and if the distance x is traversed in time t, then the rate at which the force is doing work is

$$P = Fx \cos \theta / t = Fv \cos \theta, \tag{5.21}$$

where v is the velocity of the point of application. To this rate of doing work, P, is given the name *power*. The unit of power is the joule per second (J s^{-1}) to which is given the name watt, abbreviated W.

If the force is not constant, because either its magnitude or its direction is changing, or both, then the power is still given by Eq. (5.21), where F, v and $\cos \theta$ all apply to one instant of time. But to calculate the work done we must divide the displacement up into small parts over which the force can be considered as constant, evaluate the work for each small subdivision of the displacement, and then

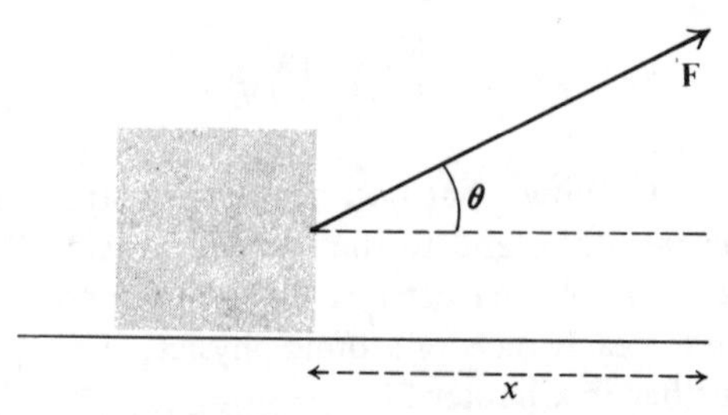

Fig. 5.9. A force moving its point of application does work.

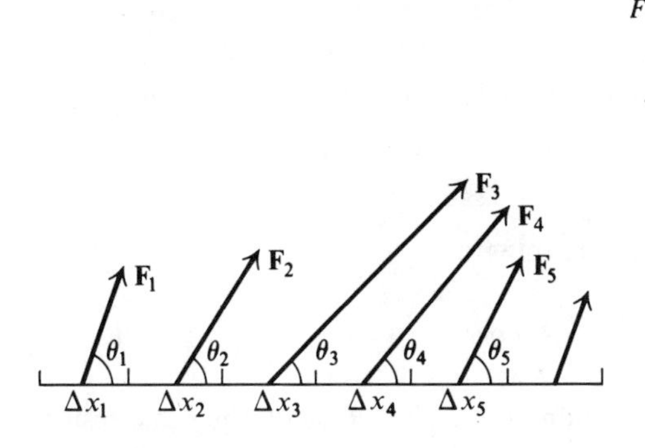

Fig. 5.10. Work being done by a variable force.

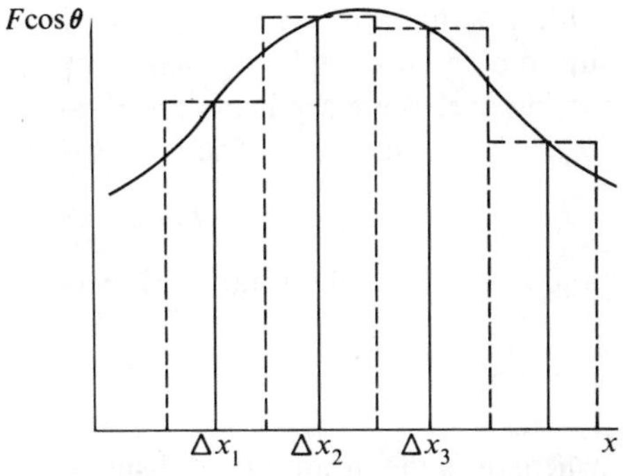

Fig. 5.11. The component of the force of Fig. 5.10 in the direction of the displacement as a function of the displacement.

sum all the individual small amounts of work. Thus, consider an object being moved along the x-axis by a variable force. Figure 5.10 shows how the force changes as the object moves, both in magnitude and direction. For the elementary displacements Δx_1, Δx_2, Δx_3, . . . , etc., shown, the amounts of work done are $F_1 \cos \theta_1 \Delta x_1$, $F_2 \cos \theta_2 \Delta x_2$, $F_3 \cos \theta_3 \Delta x_3$, . . . , and so on, the approximation being quite good if the Δx are small enough. The total work done in a finite displacement is then given by

$$\mathscr{W} = F_1 \cos \theta_1 \Delta x_1 + F_2 \cos \theta_2 \Delta x_2 + F_3 \cos \theta_3 \Delta x_3 + \ldots$$

If a graph of $F \cos \theta$ is plotted against x as shown in Fig. 5.11, this work is seen to be approximately equal to the area under the curve. As the intervals Δx are made smaller and smaller, the approximation gets better and better until, in the limit, the total work is obtained exactly as the area under the curve. Using the symbolism of integral calculus as explained in Section 3.10, we may write for the work done by a variable force in moving its point of application from x_1 to x_2

$$\mathscr{W} = \int_1^2 F \cos \theta \, dx. \tag{5.22}$$

For motion in two or three dimensions the displacement will no longer be Δx, but will be along some general direction. Figure 5.12 shows the curved path of an

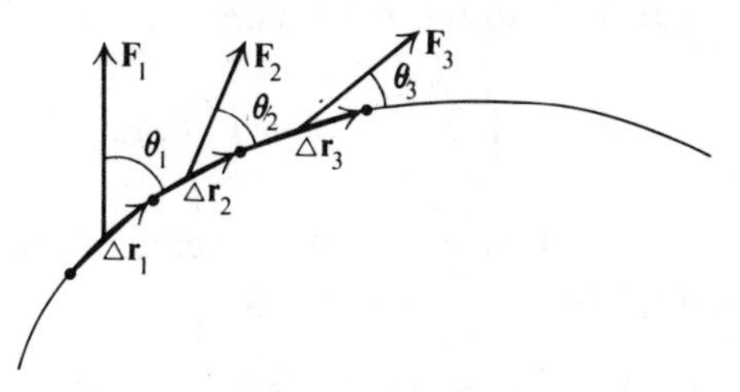

Fig. 5.12. Work being done by a variable force on an object moving in two dimensions.

object being moved in the plane of the page. This curved path has been broken up into small straight segments $\Delta \mathbf{r}_1$, $\Delta \mathbf{r}_2$, $\Delta \mathbf{r}_3$, . . . , etc., and the average forces acting over these segments are $\mathbf{F}_1$ at an angle θ_1, $\mathbf{F}_2$ at an angle θ_2, $\mathbf{F}_3$ at an angle θ_3, . . . , and so on. The total work done is

$$\mathscr{W} = F_1 \cos \theta_1 \Delta r_1 + F_2 \cos \theta_2 \Delta r_2 + \dots$$

and in the limit the total work becomes the integral

$$\mathscr{W} = \int_1^2 F \cos \theta \, dr, \tag{5.23}$$

where dr is the infinitesimal displacement along the curve. Exactly the same result holds for three dimensions.

5.8 WORK DONE BY THE UNBALANCED FORCE

The procedure described above enables us to calculate, at least in principle, the work done by any force in any displacement. It is of particular interest to calculate the work done by the resultant, unbalanced force $\mathbf{F}$, which, by Newton's second law, is just $m\mathbf{a}$. Consider for simplicity an object being accelerated, not necessarily uniformly, along the x-axis by a force that also acts along the x-axis, so that there is no $\cos \theta$ factor to worry about. Then, the work done between any two points 1 and 2 by the unbalanced force is

$$\int_1^2 F \, dx = \int_1^2 ma \, dx. \tag{5.24}$$

Consider a very small part of the whole displacement, Δx, so small, in fact, that the acceleration may be considered constant over it. Then, the following approximate relations hold:

$$\Delta x = v \Delta t, \qquad \text{and} \qquad \Delta v = a \Delta t. \tag{5.25}$$

Dividing these gives

$$a = v \frac{\Delta v}{\Delta x}$$

and therefore the small amount of work done by the unbalanced force over Δx is $ma\Delta x = mv\Delta v$. Adding up all the small amounts of work and proceeding to the limit in the usual way leads to the expression

$$\mathscr{W} = \int_1^2 mv \, dv = m \int_1^2 v \, dv.$$

The integral may be evaluated by use of the general rule stated in Section 3.10 and leads to the fundamentally important result

$$\mathscr{W} = \tfrac{1}{2}mv_2^2 - \tfrac{1}{2}mv_1^2. \tag{5.26}$$

Although this has been proved only for the special case of one-dimensional motion, it holds for any motion at all. In fact, it follows logically from the definition of work and from Newton's second law. The quantity $\frac{1}{2}mv^2$ which appears here is given the name *kinetic energy*, and will be represented by the symbol K. It is a scalar quantity and its units are those of work, namely joules. The result may be formulated in words as follows:

> *the work done by the unbalanced force in any displacement is equal to the increase in kinetic energy in that displacement,* $K_2 - K_1$.

This result is known as the *work-energy principle.*

5.9 POTENTIAL ENERGY

Although the work done by the unbalanced force in any displacement is given by the increase in kinetic energy, the unbalanced force is the resultant of all the actual forces acting, and it is of great interest to evaluate the separate amounts of work done by each of the forces. In particular, the work done by gravity merits special attention. Consider an object A that falls freely from rest through a height h. The earth exerts a force on the body, $\mathbf{W}$, which acts downwards, and may be considered constant so long as h is not too great. In free fall this weight is the only force acting if we neglect the frictional resistance of the air for the moment, and so the work done by gravity is equal to the increase in kinetic energy of the object A. Since the weight is constant, the work done is simply Wh, and we get

$$Wh = mgh = \tfrac{1}{2}mv^2, \tag{5.27}$$

where v is the speed with which the object strikes the ground. This equation may be rearranged to read $v^2 = 2gh$, a relation we have already derived in Section 3.11. Now consider a second object B that is projected horizontally with velocity $\mathbf{u}$ from the same height h, as shown in Fig. 5.13. It will follow a parabolic path so

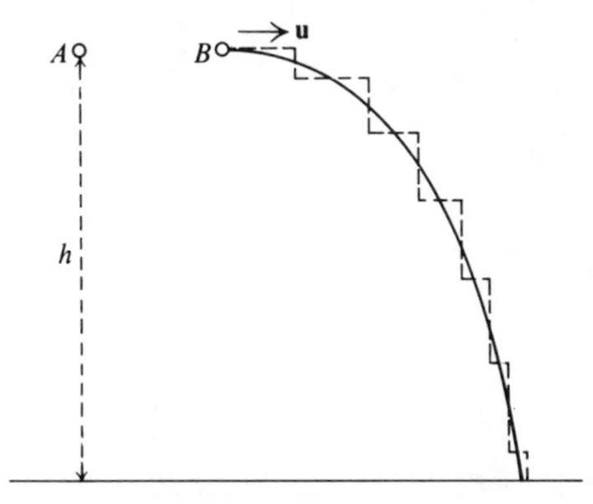

Fig. 5.13. The trajectory of a projectile and the approximation to the trajectory by which the work done by gravity may be calculated.

that its weight is no longer parallel to the displacement at any time. However, if the actual smooth curve is approximated to by a series of steps as shown by the dotted line, then gravity does no work in moving the object horizontally along any step, since in such a movement the force and displacement are at right angles, and the $\cos\theta$ term in the expression for the work is zero. So the total amount of work done by gravity is just W times the sum of the heights of all the vertical steps, and this is just Wh as before. The steps can be made smaller and smaller until in the limit they coincide with the original smooth curve. The argument remains unchanged, that the work done by gravity depends only on the vertical height fallen, not at all on the horizontal displacement. If an object is lifted through a height h, starting from rest and finishing at rest so that there is no increase in kinetic energy, then some external force must have acted against gravity during the lifting. If this force is of magnitude F, then the unbalanced upward force is $F - W$, and the result of the previous section tells us that

$$\int_1^2 (F - W)dy = 0.$$

This means that

$$\int_1^2 F\,dy = \int_1^2 W\,dy = W\int_1^2 dy = Wh,$$

and we see that, no matter how the external force F happens to vary during the lifting, it performs an amount of work Wh. The only condition is that the object starts from rest and finishes at rest. This amount of work that must be performed *against* gravity in lifting a body from the earth to a height h is called the *gravitational potential energy* of the body relative to the earth. It is a joint property of the object and the earth, and is quite independent of the path actually taken during the lifting. Thus Fig. 5.14 shows a possible, if very involved, path between point 1 on the earth and point 2 at a height h. The object has been lifted much higher than h

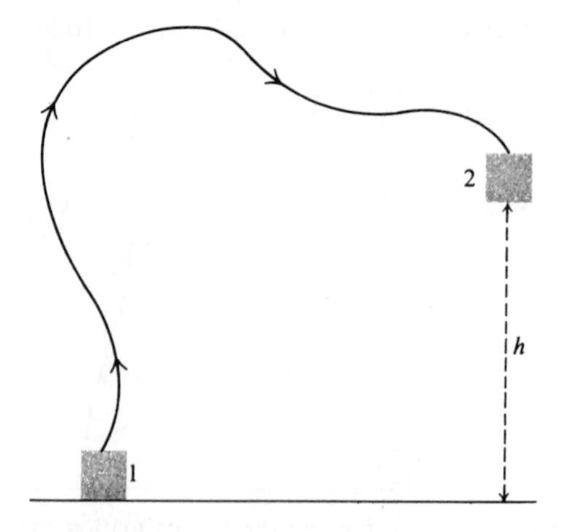

Fig. 5.14. A general path which may be taken by a body in rising a vertical height h.

during transit, and has moved laterally as well. It makes no difference: so long as it starts from rest and finishes up at rest, the total work performed by the outside agency that does the lifting is just Wh, the potential energy of the object at that height. Gravitational potential energy has here been defined relative to the earth, so that an object on the earth is considered to have zero potential energy. This is the natural thing to do in all circumstances connected with processes near the surface of the earth, but it is not really essential, and any other level may be chosen as the zero if the particular problem seems to warrant it. The work done against gravity only depends on the difference in level between the starting point and the finishing point, not on the absolute heights of these two levels. An object nearer the center of the earth than the arbitrarily chosen zero level will have a negative potential energy, since no external force is needed to overcome gravity in moving it there, gravity doing the job itself. Potential energy will be denoted by the symbol U.

5.10 CONSERVATION OF MECHANICAL ENERGY

Consider an object that falls from rest from a height h. At any instant in its flight its speed v is given by $v^2 = 2g \times$ (distance fallen) by the equations of uniformly accelerated motion. Consequently, when it has reached a height y above the ground, its kinetic energy is

$$\begin{aligned} K &= \tfrac{1}{2}mv^2 = \tfrac{1}{2}m\{2g(h - y)\} \\ &= mgh - mgy. \end{aligned}$$

Its potential energy U at that height is mgy, and therefore at any point

$$K + U = mgh = \text{constant}. \tag{5.28}$$

The constant value of $K + U$ is called the *total mechanical energy*, and is denoted by the symbol E. Thus $K + U = E$. As the object falls, its initial mechanical energy is entirely potential and this is gradually converted into kinetic energy until on reaching the earth the mechanical energy is entirely kinetic. At any stage the sum of the two kinds of energy is a constant, and this constant is determined by the initial conditions of the problem alone. Once the motion starts there is no way of increasing E. Equation (5.28) expresses the *principle of conservation of mechanical energy*. It arises because of the nature of the gravitational force, the work done by or against this force being independent of the actual path taken between any two points and depending only on the end-points of the path. Such a force is called a *conservative* force, and, in addition to gravity, elastic forces (see Chapter 10) and electrical forces (see Chapter 22) are of this kind. The principle of conservation of mechanical energy is therefore a restricted principle, but it is, nevertheless, of great utility, since intermolecular and interatomic forces are also conservative, and the general features of the motion of quite complicated systems can be discussed without actually having to solve Newton's laws of motion. The basic idea is as follows. If the motion is one-dimensional, or can be reduced to a problem in one dimension, then the principle tells us that

$$\tfrac{1}{2}mv^2 + U(x) = E, \text{ a constant,} \tag{5.29}$$

which may be rewritten as

$$v = \sqrt{\frac{2}{m}\{E - U(x)\}}. \tag{5.30}$$

If the functional form of the potential energy is known, the value of the speed v at any point can be found, and this, in principle, is the solution to the problem. How the functional form of $U(x)$ is known is not our concern here. Mechanics limits itself to studying the motion under a *given* potential energy function. As an instructive example, let us consider the relative motion of the two constituent atoms of a diatomic molecule. This is a complicated three-dimensional problem, but if we are interested only in the relative motion along the line of centers of the atoms, it reduces to a problem in one dimension which we may solve by using Eq. (5.30). The problem of determining the interatomic potential function of a diatomic molecule is a matter for the chemists, and we shall assume a curve of the general shape shown in Fig. 5.15(a). The interatomic potential energy is plotted against the separation x of the atoms. The zero of potential energy is chosen to be when the molecule is completely dissociated, that is when its constituent atoms are infinitely far apart and at rest. This results in the energy being negative for nearly all the range of x. Figure 5.15(b) shows the interatomic force $F(x)$ at any value of the separation. Repulsive forces are positive, since they tend to increase the separation; attractive forces are negative, since they tend to decrease the separation. The force is zero at the equilibrium separation $x = x_0$. To understand how the atoms move relative to one another along their line of centers it is necessary to know the value of the total mechanical energy E. Since we cannot take the square root of a negative quantity, Eq. (5.30) tells us that E must be greater than $U(x)$ everywhere. A situation where E was less than U would imply a negative kinetic energy, which is physically impossible. Thus the minimum value of E is E_0, corresponding to the lowest point of the curve, and the physical interpretation of this situation is that the atoms are at rest relative to one another. In other words, they are moving as a rigid body. Note that the molecule might well be rotating and translating; we are interested here only in the internal relative motion of the two atoms. If the mechanical energy has a slightly greater value, corresponding to the line $E = E_1$, then relative motion is possible between the limits of separation $x = x_1$ and $x = x_1'$, since only between these limits does the potential energy curve lie entirely below the line $E = E_1$. Outside these limits E_1 is less than U, and the kinetic energy would be negative. With this value of the total mechanical energy, the separation of the atoms oscillates between the values x_1 and x_1'. At $x = x_1$, $E_1 = U$, and so the particles are instantaneously at rest. But the force at $x = x_1$ is repulsive, so the separation immediately begins to increase, and the relative velocity of the atoms increases until it reaches a maximum value at $x = x_0$, since at that point $E_1 - U$ has its maximum value. Thereafter the relative velocity decreases again until the atoms are at rest at $x = x_1'$. At any value of x between x_1 and x_1' the relative velocity is proportional to the square root of the depth of the U-curve below the line $E = E_1$. It will be noticed that the equilibrium value $x = x_0$ lies very nearly half-way between x_1 and x_1'. This means that for this value of the total energy, x_0 is the mean separation of the atoms.

For a greater value of the total mechanical energy, $E = E_2$, the general features of the motion may be deduced in the same way. The limits of the atomic separation are now x_2 and x_2', and the separation oscillates between these values. Now, however, the mean position is no longer x_0, but has been shifted to a higher value, as shown by the heavy dot in Fig. 5.15(a). This is because the potential energy curve is not symmetrical. This shift in the mean separation of the atoms is what gives rise to the expansion of solids and liquids when heated. Such expansion is not due to the increased amplitude of oscillation but to the asymmetric nature of such oscillation.

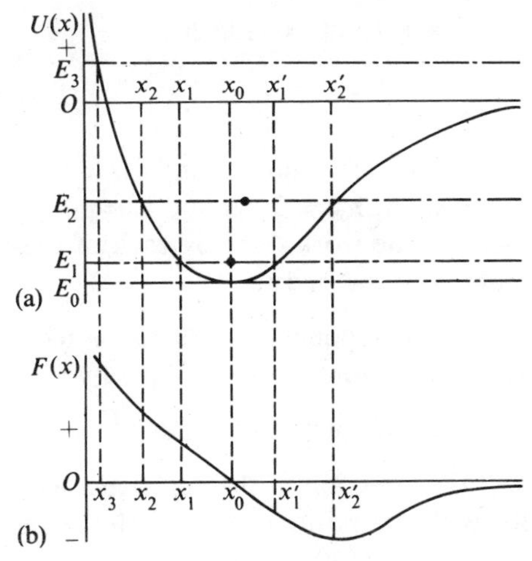

Fig. 5.15. Interatomic potential energy and force curves for a diatomic molecule.

When the total mechanical energy is positive, as shown by the line $E = E_3$, the situation is very different indeed. The line $E = E_3$ now intersects the potential energy curve once only, at $x = x_3$. The interpretation of this is as follows. If the particles are initially approaching each other, they will continue to do so with increasing relative velocity until the point $x = x_0$, where the relative velocity begins to decrease. The atoms come to a closest separation of x_3, and then begin to recede from each other again with increasing relative velocity up to $x = x_0$ and then with decreasing relative velocity out to infinite separation. No molecule exists at all. The same general interpretation holds for the particular value $E = 0$, but here the atoms reach a state of rest when infinitely far apart. The depth E_0 of the potential energy curve below zero is called the *dissociation energy* of the molecule: it represents the minimum amount of energy needed to separate the molecule into its constituent atoms.

This type of analysis of one-dimensional problems is of great importance. In many situations it is enough to be able to describe the qualitative features of the motion, without actually solving the equations of motion, and for this purpose the consideration of the potential energy curve is of enormous assistance. Thus where the potential energy curve shows a trough or well, as in the above example, oscillatory motion will be possible so long as the total energy line cuts both walls of the well. This matter is discussed again in Chapter 10, where the special properties of the harmonic potential curve are investigated.

Example 1. What constant force is needed to stop a bullet of mass 25 g in a distance of 10 cm if the bullet has a speed of 300 m s^{-1}?

Solution. The initial kinetic energy of the bullet is $\frac{1}{2} \times 0.025 \text{ kg} \times (300 \text{ m s}^{-1})^2 = 1125 \text{ J}$, and this has to be reduced to zero by a force F acting through a distance of 0.10 m. Therefore, by the work-energy principle,

$$F \times 0.10 \text{ m} = 1125 \text{ J},$$

and

$$F = 1125 \text{ J}/0.10 \text{ m} = 11{,}250 \text{ N}.$$

Example 2. A body is constrained to move along the x-axis through a distance of 20 m by three forces $\mathbf{F}_1 = 80$ N, $\mathbf{F}_2 = 30$ N and $\mathbf{F}_3 = 40$ N whose directions are shown in Fig. 5.16. What is the work done by each of these forces, and what is the change in kinetic energy of the body?

Solution. The force $\mathbf{F}_1$ has a component $F_1 \cos 30° = 69.3$ N along the positive direction of the x-axis, and therefore the work done by $\mathbf{F}_1$ is

$$\mathscr{W}_1 = (F_1 \cos 30°)\, x = 69.3 \text{ N} \times 20 \text{ m} = 1386 \text{ J}.$$

The second force $\mathbf{F}_2$ is in the direction of the negative x-axis, which means that it makes an angle of 180° with the positive x-axis. Hence the work done by $\mathbf{F}_2$ is

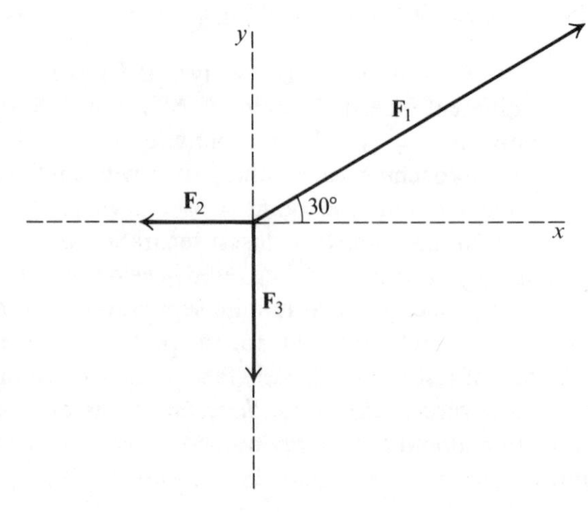

Fig. 5.16

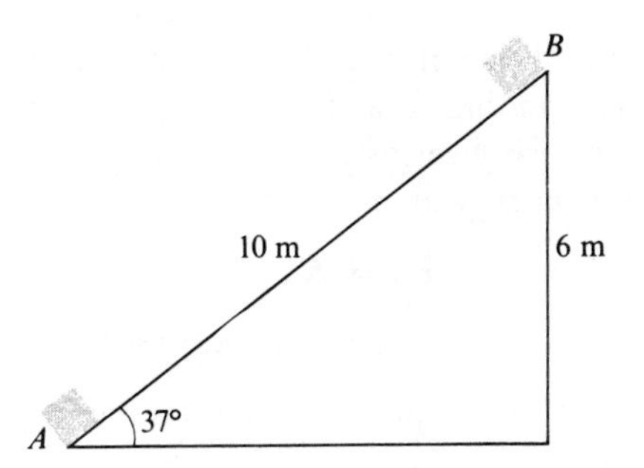

Fig. 5.17

$\mathscr{W}_2 = (F_2 \cos 180°)\, x = -F_2 x = -30 \text{ N} \times 20 \text{ m} = -600 \text{ J}$. The negative sign means that work has to be done *against* the force $\mathbf{F}_2$. The third force $\mathbf{F}_3$ acts in a direction perpendicular to the x-axis, and so

$$\mathscr{W}_3 = (F_3 \cos 90°)\, x = 0.$$

The total work done by all three forces is

$$\mathscr{W}_1 + \mathscr{W}_2 + \mathscr{W}_3 = +\ 1386 \text{ J} - 600 \text{ J} = +\ 786 \text{ J}.$$

Therefore the increase in kinetic energy is also 786 J.

Example 3. A coconut falls to earth from a height of 10 m. What is its kinetic energy when it is 4 m from the ground if its mass is 1.0 kg? What is its speed at that height?

Solution. At a height of 10 m the kinetic energy is zero and so the total mechanical energy is just equal to the potential energy. Therefore

$$E = 1.0 \text{ kg} \times 9.8 \text{ m s}^{-2} \times 10 \text{ m} = 98 \text{ J}.$$

At a height of 4 m above the ground the coconut has both potential energy and kinetic energy, hence

$$E = \tfrac{1}{2}mv^2 + mgh = \tfrac{1}{2} \times 1.0 \text{ kg} \times v^2 + 1.0 \text{ kg} \times 9.8 \text{ m s}^{-2} \times 4 \text{ m}$$

$$= 0.5 \text{ kg} \times v^2 + 39.2 \text{ J} = 98 \text{ J}$$

since we assume the total energy has not changed. Hence

$$K = 0.5 \text{ kg} \times v^2 = 58.8 \text{ J}.$$

The speed v may be immediately be calculated as

$$v = \sqrt{58.8 \text{ J}/0.5 \text{ kg}} = 10.8 \text{ m s}^{-1}.$$

Example 4. A block of mass 5.0 kg is projected up a frictionless plane inclined at 37° to the horizontal. It slides a distance of 10 m along the surface, stops, and then slides down to its original starting point. With what speed did it start up the plane, and with what speed does it regain its starting point?

Solution. The situation is shown in Fig. 5.17. Since the angle of the plane is 37°, the vertical height corresponding to a slant height of 10 m is just 6 m. We may conveniently choose our zero level of potential energy at the lowest point A, so that the total mechanical energy at A is

$$E_A = K_A + U_A$$
$$= \tfrac{1}{2}(5.0 \text{ kg})\, v^2 + 0.$$

Since there are no frictional forces present, the total mechanical energy at B is also E_A. Therefore, since $K_B = 0$,

$$E_B = 0 + (5.0 \text{ kg} \times 9.8 \text{ m s}^{-2} \times 6 \text{ m}) = E_A,$$

or

$$v^2 = (5.0 \text{ kg} \times 9.8 \text{ m s}^{-2} \times 6 \text{ m})/2.5 \text{ kg} = 117.6 \text{ m}^2 \text{ s}^{-2}.$$

Therefore

$$v = 10.8 \text{ m s}^{-1}.$$

Since energy is conserved throughout the motion, this result also gives us the speed with which the block regains its starting point.

5.11 FRICTIONAL FORCES

In order to gain some insight into the nature of frictional forces, consider the following situation. A block of mass m is on a rough, level surface, and a force parallel to the surface, of magnitude F, is applied to the block. If we start with a very small value of F and gradually increase it, we find that the block does not begin to move until the applied force has reached a critical value. If F is less than this critical value, the block remains in equilibrium, and therefore we know that there is no unbalanced force acting on it. Hence there must be some other force acting in the opposite direction which just cancels the effect of F. This force, denoted by f, acts between the surfaces in contact and is known as the *force of friction.* The frictional force obeys two basic laws which are illustrated in Fig. 5.18. The first law states that the frictional force is independent of the area of contact of the solids: the second law states that the frictional force increases steadily until it reaches a maximum value $f_{\max}$ when slipping is just about to occur. If N is the normal force between the bodies, then the maximum frictional force is given by

$$f_{\max} = \mu N, \tag{5.31}$$

where μ is known as the *coefficient of friction.* These laws were discovered in the 15th century by Leonardo da Vinci. It must be stressed that they are *empirical laws* and do not have the universal validity of laws such as Newton's laws. When slipping has commenced, the frictional force is very nearly independent of the speed of sliding. In some cases the value of μ for sliding friction is slightly less than that given by Eq. (5.31), but the difference is not important for our purposes.

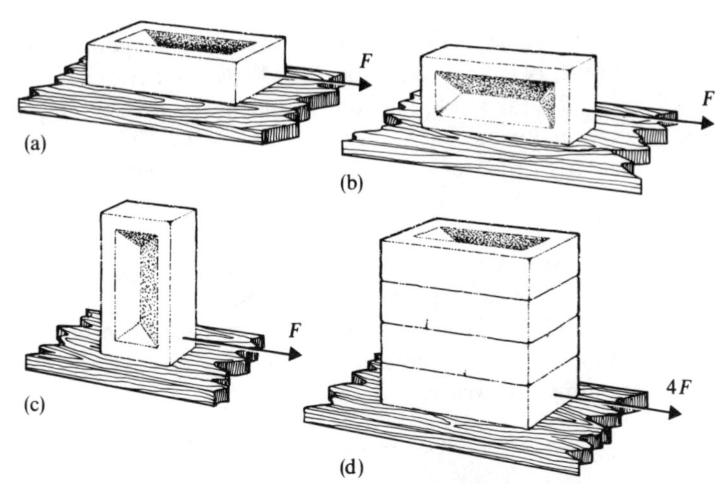

Fig. 5.18. Illustration of the two basic laws of friction. The friction is independent of the area of contact between the solids, and is directly proportional to the load. For a single brick, the friction is the same whether the brick is lying flat (a), or on its side (b), or standing on its end (c). Increasing the load, however, increases the friction proportionately (d). (After F. P. Bowden and D. Tabor, *Friction and Lubrication*, Methuen, London, and John Wiley, New York.)

Frictional forces of this type are of great importance in everyday life. Without the frictional force between the ground and our feet we should be incapable of walking, since it is the frictional force that propels us forward. In the action of walking, we attempt to slide our feet backwards along the ground, and the frictional force prevents this sliding motion from taking place and so acts forward. In the same way, it is the frictional forces between the road and the driving wheels that propel an automobile. The motor acts to rotate the wheels so that the surface of the tire tries to slide backwards. In this case also the frictional force acts forward to prevent the tire from slipping, and so provides the propelling force.

Indeed the tires *will* slide backwards on surfaces which produce too small a frictional force, such as ice. In such a case the wheels rotate but the automobile does not move. The front wheels are quite different, as can be seen by considering what happens to them when the driving wheels are slipping on the ice: they do not rotate at all. When the automobile is in motion the front wheels are rotated by a frictional force that must act backwards, but this frictional force is quite small—just large enough, in fact, to cause the rotation. From Eq. (5.31) we see that the driving force exerted by the road is proportional to the part of the weight of the vehicle that is supported by the rear wheels.

The type of friction discussed above occurs when solid surfaces are in more or less dry contact. If the surfaces are lubricated the friction is very substantially reduced and the coefficient of friction may drop as low as 0.001. When this is so, we may take the frictional force to be zero without making an appreciable error, and we say that the surface is *smooth*. A smooth surface can exert only a normal force on an object in contact with it.

Consider now a block being dragged up a rough inclined plane from a position 1, where its kinetic energy is K_1, to a position 2, where its kinetic energy is K_2. From the result of Section 5.8, the work done on the block by the resultant, unbalanced force is the change in kinetic energy, or $\mathscr{W} = K_2 - K_1$. The work $\mathscr{W}$ can be split up into three parts: the work done by the external agent doing the dragging, which we may denote by $\mathscr{W}_{\text{ext}}$; the work done by gravity, which by definition is equal to the decrease in gravitational potential energy, $-(U_2 - U_1)$; and the work done by the frictional forces exerted by the rough plane and by the air also. This last amount of work is always negative, since friction acts in a direction opposite to the motion of the block. It is consequently more natural to talk of the work done *against* friction, and we denote this essentially positive quantity by $\mathscr{W}_f$. Hence, the equation may be written

$$\mathscr{W} = \mathscr{W}_{\text{ext}} - (U_2 - U_1) - \mathscr{W}_f = K_2 - K_1,$$

or, by a rearrangement of terms,

$$\mathscr{W}_{\text{ext}} = (K_2 - K_1) + (U_2 - U_1) + \mathscr{W}_f. \tag{5.32}$$

In words, the work done in any displacement by an external force is equal to the change in kinetic energy plus the change in potential energy plus the work needed to overcome the frictional forces. If the sum of the kinetic and potential energies is written as the total mechanical energy, $K + U = E$, then Eq. (5.32) becomes

$$\mathscr{W}_{\text{ext}} = E_2 - E_1 + \mathscr{W}_f. \tag{5.33}$$

It is seen that the presence of frictional forces will always cause a decrease in the mechanical energy. For this reason they are referred to as *dissipative forces.* In such a situation as the above, the mechanical energy E is not conserved. Only if both $\mathscr{W}_{\text{ext}}$ and $\mathscr{W}_f$ are zero do we have $E_1 = E_2 =$ constant.

Example 1. The coefficient of friction between the tires and the road for a particular automobile is 0.8 and its rear wheels support 40% of its weight. What is the maximum possible acceleration that can be achieved on starting?

Solution. If m is the mass of the automobile, then 0.4 mg is the force that the rear wheels exert downwards on the road surface, and so, by Newton's third law, this is also the force that the road exerts normally on the rear wheels. Hence the maximum frictional force that the road can exert on the tires, which occurs when the tires are on the point of slipping, is given by Eq. (5.31) as

$$f_{\text{max}} = \mu \times 0.4\, mg = ma_{\text{max}}.$$

The mass of the automobile cancels out, and we get

$$a_{\text{max}} = 0.4\, \mu g = 0.32\, g = 3 \text{ m s}^{-2}.$$

Example 2. What constant horizontal force is needed to drag a block of mass 8.0 kg along a rough, horizontal table ($\mu = 0.5$) with an acceleration of 1.8 m s^{-2}?

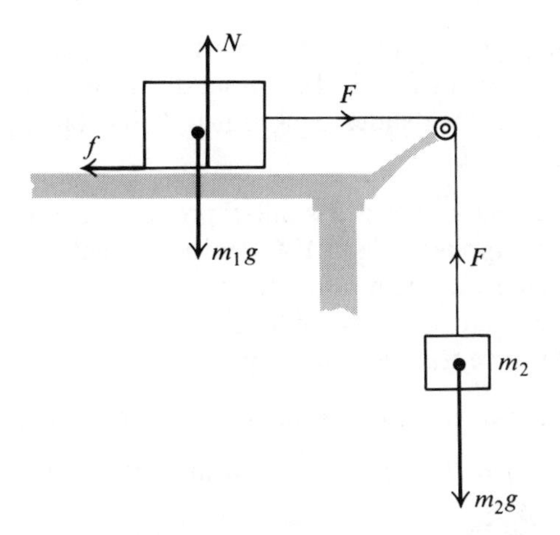

Fig. 5.19

What mass, hanging from a light cord attached to the 8.0-kg block and passing over a small frictionless pulley, will produce this acceleration?

Solution. The situation is shown in Fig. 5.19. Since the block stays on the table it has no acceleration perpendicular to the surface, therefore

$$N = m_1 g,$$

and so the frictional force is

$$f = \mu N = \mu m_1 g.$$

The unbalanced force parallel to the surface is $F - f = m_1 a$, by Newton's second law, therefore

$$F = m_1 a + \mu m_1 g = (8.0 \text{ kg} \times 1.8 \text{ m s}^{-2}) + (0.5 \times 8.0 \text{ kg} \times 9.8 \text{ m s}^{-2})$$

$$= 14.4 \text{ N} + 39.2 \text{ N} = 53.6 \text{ N}.$$

If the force F is supplied by the weight of a mass m_2 hanging as shown, then the force F is known as the *tension* in the cord. If the mass of the cord can be ignored (this is indicated by the word "light" in the question), then the cord serves merely to transmit the force F without change. Since m_2 must have the same acceleration downwards as the mass m_1 has horizontally, Newton's second law applied to m_2 gives

$$m_2 g - F = m_2 a.$$

Adding this to the previous equation, the tension F cancels out and we find

$$m_2 g = (m_1 a + \mu m_1 g) + m_2 a,$$

or

$$m_2 = m_1 \frac{a + \mu g}{g - a}$$

$$= 8.0 \text{ kg} \times \frac{6.7 \text{ m s}^{-2}}{8 \text{ m s}^{-2}} = 6.7 \text{ kg}.$$

Example 3. A 5.0-kg block is projected up a rough plane inclined at 37° to the horizontal with a speed of 14 m s^{-1}. It slides a distance of 10 m along the surface of the plane, stops, and then slides back down to its starting point. With what speed does it reach there?

Solution. We have already solved a similar problem in the absence of friction, and the same diagram applies (Fig. 5.17). In the present situation, we must treat the motion up the plane and the motion down the plane separately. This is necessary because the frictional forces are not given: they must be calculated from a knowledge of the distance traveled up the plane.

Motion up. The values of the various quantities of interest are as follows:

$$K_A = \tfrac{1}{2} \times 5.0 \text{ kg} \times (14 \text{ m s}^{-1})^2 = 490 \text{ J};$$

$$K_B = 0;$$

$$U_A = 0$$

$$U_B = 5.0 \text{ kg} \times 9.8 \text{ m s}^{-2} \times 6 \text{ m} = 294 \text{ J};$$

$$\mathscr{W}_{\text{ext}} = 0.$$

Applying Eq. (5.32), we get

$$0 = (0 - 490 \text{ J}) + (294 \text{ J} - 0) + \mathscr{W}_f;$$

therefore

$$\mathscr{W}_f = 490 \text{ J} - 294 \text{ J} = 196 \text{ J}.$$

Motion down. All the quantities of interest retain their previous values except for K_A which is reduced to

$$K_A = 2.5 \text{ kg} \times v^2,$$

where v is the speed of the block on regaining its starting point. Again substituting into Eq. (5.32) gives

$$0 = (2.5 \text{ kg} \times v^2 - 0) + (0 - 294 \text{ J}) + 196 \text{ J},$$

whence $(2.5 \text{ kg})v^2 = 98$ J, giving $v = 6.3$ m s^{-1}.

PROBLEMS

5.1 A ball of mass 1.5 kg is moving northwards on a smooth table with a velocity of 2 m s^{-1}. It collides with a stationary ball and after the collision moves eastwards with a velocity of 1.5 m s^{-1}. What are the magnitude and direction of the change in momentum of the ball?

5.2 If the collision in Problem 5.1 lasts for 0.002 s, what are the magnitude and direction of the average force acting on the ball during the impact?

5.3 What are the force acting on, and the change of momentum of, the initially stationary ball in Problem 5.1? If the stationary ball has a mass of 6 kg, with what velocity does it move off?

5.4 A skier, starting from rest, slides down a uniform slope of 1 in 100. Assuming there to be no frictional force between the skis and the snow, compute the time required to travel a distance of 49 m.

5.5 Repeat Problem 5.4 for the case where there is a frictional resisting force equal to 0.5% of the weight of the skier.

5.6 A body of mass 5 kg is subjected simultaneously to two forces: a force of 3 N to the east, and a force of 4 N to the north. What is the acceleration of the body?

5.7 Given that the moon takes 27.3 days to circle the earth, calculate the distance between the earth and the moon on the assumption that the orbit is a circle.

5.8 Prove that for any earth satellite in a circular orbit the square of its period is directly proportional to the cube of its distance from the center of the earth.

5.9 Assuming that the earth travels round the sun in a circular orbit of radius 1.5×10^{11} m in a time of 365 days, estimate the mass of the sun.

5.10 Prove that the acceleration of gravity on any planet is proportional to the product of the radius and the mean density of the planet.

5.11 The engines of a rocket of initial mass 2×10^5 kg provide a thrust of 2.5×10^6 N. What is the initial upward acceleration of the rocket?

5.12 An elevator of mass 2.5×10^3 kg is accelerating upwards at 2 m s^{-2}. What is the tension in the cable? What will the tension be when the elevator is accelerating downwards at 2 m s^{-2}?

5.13 Consider a space-ship during the whole period from lift-off to landing on the moon and assume that its rockets are only fired during actual lift-off and landing. Discuss the sensation of weight experienced by the astronauts during the various stages of the flight. How would this be affected if the destination were Mars rather than the moon?

5.14 The acceleration of gravity is measured in an aircraft flying low over the equator at a speed of 300 m s^{-1} relative to the ground, first in an east-west direction, and then in a west-east direction. What difference will there be in the two values so obtained? Which will be greater?

5.15 How much work is done when a bucket weighing 1.5 kg is pulled up from the bottom of a well 8 m deep with 10 kg of water in it? If there is a hole in the bucket so that water leaks out at a steady rate as it rises, leaving only 8 kg when it reaches the top, how much work is needed?

5.16 A man pushes a cart along a road for 1000 m and does 0.25 MJ of work. What force (assumed constant) does he exert?

5.17 If the cart in the previous problem travels at a constant speed, how much work is done by the frictional forces over the 1000 m?

5.18 A block weighing 200 N is pushed 2 m up an incline of 30° by the application of a constant horizontal force of 150 N. How much work is done by this force? How much work is done by the weight of the block? If the block moves with constant velocity, what is the frictional force acting, and how much work is done by it?

5.19 An object is moved along the x-axis from the origin to the point $x = 5$ m by a variable force whose component along the positive direction of the x-axis is $(6 \text{ N m}^{-1})x + 8$ N. How much work is done by this force?

5.20 A body is attracted toward the origin by a force whose component along the x-axis is $(-4 \text{ N m}^{-1})x$, the negative sign being required since the force is opposite in sense to the displacement, x. How much work does this force do in moving the body from $x = 5$ m to $x = 2$ m?

5.21 An object falls to earth from a very great height so that the gravitational attraction of the earth cannot be taken as constant. Prove that, if the distance of the object from the center of the earth is r at the moment it is dropped, the work done by gravity during the fall is $GMm[(1/R) - (1/r)]$, where m is the mass of the object, M is the mass of the earth, R is the radius of the earth, and G is the gravitational constant.

Is the answer to this problem changed if air-resistance is taken into account?

5.22 An object falls to earth from a height h above the surface which is large enough for the ratio (h/R) to be appreciable but for the square of this ratio to be negligible. Show in this case that the work done by gravity is $mg_0h[1 - (h/R)]$, where $g_0 = GM/R^2$ is the standard gravitational acceleration at the surface of the earth.

5.23 How much work is done by gravity on an earth satellite during one revolution in a circular orbit? How much is done during one revolution in an elliptical orbit?

5.24 For an earth satellite in a circular orbit or radius r, prove that the kinetic energy is $K = GMm/2r$.

5.25 A ball is suspended from a fixed support by a string 1.5 m long. The ball is moved to the side until the string is taut and makes an angle of 60° with the vertical. If the ball is now released, find the speed with which it passes through its lowest point.

5.26 A light string is hung over a frictionless pulley and carries masses of 1.00 kg and 1.02 kg at its ends. Initially, the masses are at rest on the same level. If they are now released, calculate the velocity of either of them after the heavier has fallen through 0.5 m.

5.27 About 7000 m^3 of water per second flow over the Horseshoe at Niagara Falls and fall through a distance of 50 m approximately. How much work is done by gravity on this water in one day?

5.28 A body is projected with a velocity of 4 m s^{-1} along a rough horizontal surface. If the mass of the body is 0.5 kg and it is brought to rest in a distance of 2.0 m, what is the frictional force acting on it and what is the coefficient of sliding friction?

5.29 Discuss qualitatively the motion of a body that moves along the axis of x with a potential energy that varies with x, as shown in Fig. 5.20, for the following values of its total mechanical energy: (a) -4 J, (b) 0 J, and (c) $+1$ J.

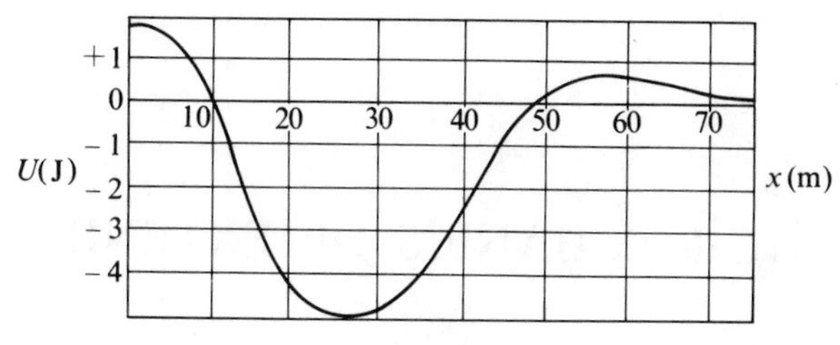

Figure 5.20

5.30 Find the velocity of the body in the previous problem when its total energy is +1 J at the points $x = 10$ m and $x = 20$ m, assuming the mass of the body to be 1 kg.

5.31 A block of mass 10 kg is pushed 25 m up the sloping surface of a plane inclined at 37° to the horizontal by a constant force of 100 N acting parallel to the plane. The coefficient of friction between the block and the surface is 0.3. How much work is done by the agency exerting the force? What is the increase in kinetic energy of the block? How much work is done against friction? What is the increase in potential energy of the block?

CHAPTER 6

THE MECHANICS OF EXTENDED BODIES

6.1 INTRODUCTION

In the previous chapter attention was restricted to those features of the motion of objects that did not involve their size. We now wish to relax this restriction, and shall initially consider the mechanics of rigid bodies. A rigid body is defined as one for which the distance between any two points of the body is constant. This excludes from consideration the elastic nature of real bodies, a subject dealt with later in the chapter. So long as size-dependent effects were being ignored, it was a matter of no importance where we drew the various forces acting on an object. This is now a relevant question, and it is necessary first of all to define the *center of mass* of a rigid body, or *center of gravity*, as it is often referred to. An extended body may be thought of as consisting of an enormous number of particles, and the gravitational forces acting on all of these particles separately can be replaced by a single force equal to the total weight of the body. This acts through a point defined as the center of mass. For a symmetrical body the position of this point is usually obvious by inspection of the symmetry, and calculations of mass centers for more complicated shapes will not be dealt with in this book. In all questions of translational motion, the body may be considered to be replaced by a particle of the same mass as the body located at the center of mass.

6.2 TORQUE

The most obvious and important effect of the actual size of real objects is that when forces are applied to them they may cause not only translational motion but also rotational motion. Consider a large crate resting on the ground and being pushed by a horizontal force F exerted at a distance h from the ground. The crate is prevented from sliding by a projecting rib at O. This obstruction exerts a force on the crate, and the components of this force in the horizontal and vertical directions have magnitudes f and N, as shown in Fig. 6.1. The weight W of the crate acts through the center of mass G. As far as the translational motion of the crate is concerned, all the forces may be considered to act at G, and the conditions for translational equilibrium are simply

$$F = f \qquad \text{and} \qquad W = N. \tag{6.1}$$

However, common experience tells us that under certain conditions the force F will cause the crate to tip up, and the effectiveness of the force in this tipping action

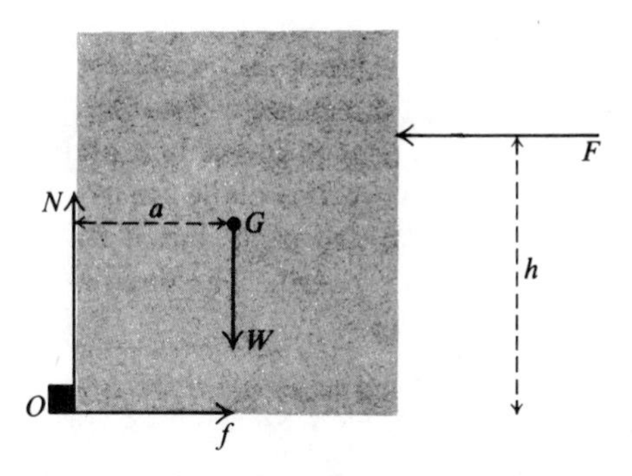

Fig. 6.1. The forces on a crate which is being acted upon by a horizontal force.

depends on the distance h. The measure of the ability of a force to cause rotation is called the *torque*, and in this case, for rotation about O, is just the product of F and h. As the right-hand end of the crate lifts off the ground, the weight W also has a torque, of magnitude Wa, where a is the distance between G and the left-hand edge of the crate. Whether or not the crate will tip up will depend on the relative magnitudes of these two torques. When the crate is just on the point of moving, the torques are equal,

$$Fh = Wa. \tag{6.2}$$

The two equations (6.1) and (6.2) are the conditions for equilibrium of the crate. Equation (6.1) expresses the fact that the crate is at rest or moving with uniform translational velocity; Eq. (6.2) expresses the fact that the crate is not rotating. The two forces N and f have no tendency to cause rotation about O, since they act through that point. The general specification of torque in a plane can be seen from Fig. 6.2. To specify the point P at which a force $\mathbf{F}$ acts, an origin O is needed. The point P is located by the position vector $\mathbf{r}$ from O to P. The torque about O is a physical quantity that depends on the two vectors $\mathbf{F}$ and $\mathbf{r}$ jointly, and, so long as motion in a plane only is being considered, may be represented by the product of the scalar magnitude of the force, F, and the perpendicular distance p from O to the line of action of the force. Thus the torque is just Fp in this case. A torque is taken as positive when it tends to cause a counterclockwise rotation of the body and negative for a clockwise rotation. Although all problems where the forces act in the same plane may be treated by considering the torques to be scalar quantities, the situation is very different in three dimensions. Then it is necessary to generalize the vector algebra and write torque as a vector quantity depending on both $\mathbf{F}$ and $\mathbf{r}$. Such refinements will not be needed in this book. One other result of restricting the treatment to forces in a plane is that the dimensions of the physical objects perpendicular to this plane are irrelevant. Thus the crate in Fig. 6.1 actually has extension into the page, but it may be treated as being a lamina—i.e. a very thin rectangular board. The conditions for equilibrium of a rigid body may now be stated as follows.

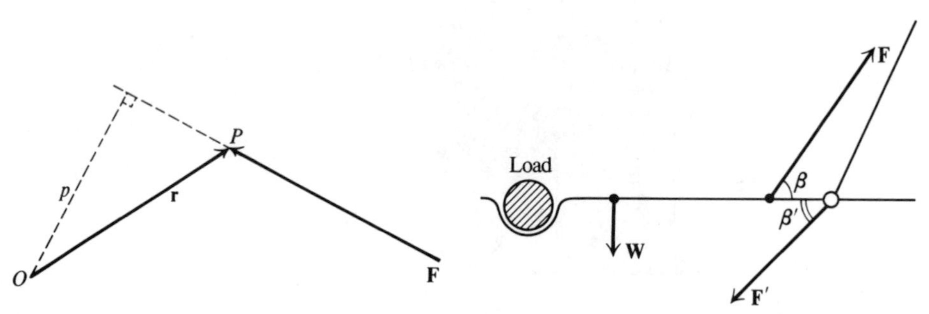

Fig. 6.2. A force **F** produces a torque about the point O.

Fig. 6.3. The forces acting on a horizontal forearm when the hand supports a load

A rigid body will be in equilibrium when (a) there is no unbalanced force acting on it, and (b) there is no unbalanced torque acting on it.

This may be considered to be a generalization of Newton's first law of motion. As an example of the application of these conditions of equilibrium, consider a heavy object being held in the palm of the hand with the forearm being held horizontal. The whole forearm, including the load in the hand, can be considered as a rigid body in equilibrium under the action of various forces. A schematic diagram is shown in Fig. 6.3. The principal muscular force exerted is represented by **F**, which acts at a distance a from the elbow joint and at an angle β with the line of the forearm. The weight of forearm plus load is **W**, and this acts at a distance of b from the elbow. A third force **F′** must be exerted on the forearm for equilibrium to be possible. Equating the torques of **F** and **W** about the elbow gives

$$aF \sin \beta = bW.$$

Since the distance a is very much less than the distance b, it is seen that the muscle exerts a force very much larger than the weight to be supported. As the angle between the upper arm and the vertical is increased, the angle β decreases, and a correspondingly larger force **F** must be exerted by the muscle to compensate for the decrease in $\sin \beta$, a result in accord with common experience. The equilibrium conditions for translational equilibrium are

$$W + F' \sin \beta' = F \sin \beta$$

and

$$F' \cos \beta' = F \cos \beta.$$

If the values of a, b, and β are known, these may be solved for the two components of **F′**, but the determination of this force is rarely of interest.

Example 1. A uniform plank of length 6 m and weight 320 N is supported on two trestles each 1 m from an end of the plank, as shown in Fig. 6.4. How close to an end of the plank can a man weighing 800 N walk before it tips?

Solution. The plank will tip up when the force R_1 exerted on the trestle farthest

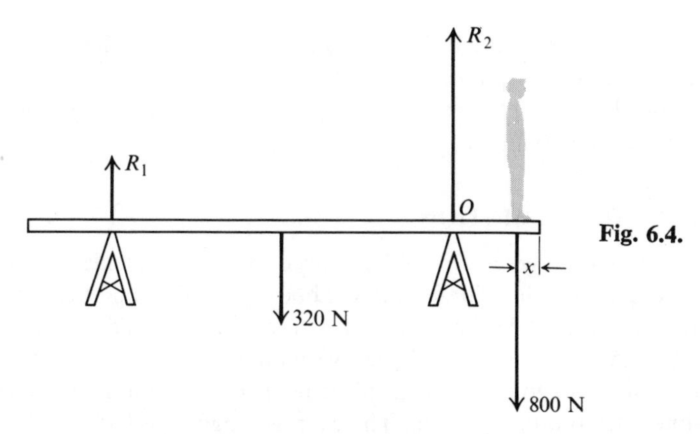

Fig. 6.4.

from the man is zero. If the plank tips when the man is at a distance x from one end, then in that position the counter-clockwise torque of the weight of the plank about O just equals the clockwise torque of the man's weight. Therefore

$$320\text{ N} \times 2\text{ m} = 800\text{ N} \times (1\text{ m} - x),$$

or

$$\begin{aligned} x &= (800\text{ N m} - 640\text{ N m})/800\text{ N} \\ &= 0.2\text{ m}. \end{aligned}$$

Example 2. A mirror of weight 75 N is hung by a cord from a hook on the wall, the cord making an angle of 15° with the horizontal as shown in Fig. 6.5. Calculate the tension in the cord.

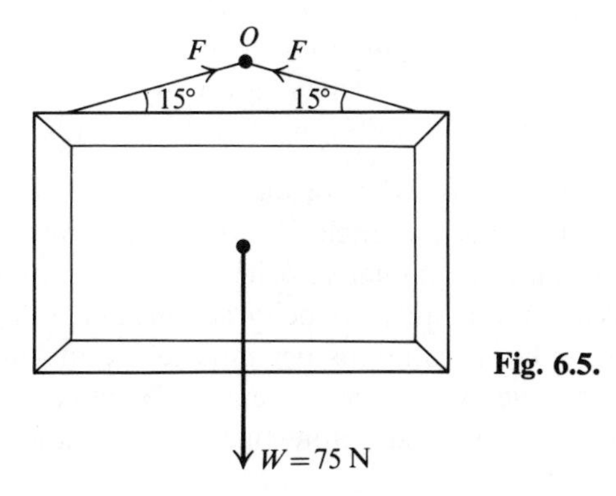

Fig. 6.5.

Solution. Let us consider the equilibrium of the mirror. The only forces acting on the mirror are its weight and the two tensions in both parts of the cord. Resolving these tensions horizontally, we get $F \cos 15°$ for the magnitude of each, and the resultant effect is zero since these components act in opposite directions. Resolving vertically we get

$$F \sin 15° + F \sin 15° = W,$$

or

$$F = 75\text{ N}/2 \sin 15° = 75\text{ N}/2 \times 0.176 = 216\text{ N}.$$

It is obvious in this case that there is no tendency for the mirror to rotate. This may be seen by taking torques about the fixed point O. So long as the mirror hangs symmetrically, all the forces pass through this point, and therefore none can produce any torque about it.

6.3 ROTATIONAL MOTION

Consider a rigid body rotating about a fixed axis perpendicular to the page through O, and consider any line OP fixed in the body, as in Fig. 6.6. Let θ be the angle that the line OP fixed in the body makes with a line Ox fixed in the page. Then, as the body rotates, the angle θ will vary with time, and an angle-time diagram may be drawn quite analogous to the displacement-time diagrams discussed in Chapter 3 for linear translation motion. The average angular velocity of the rigid body is defined as the change in the angle θ during an interval of time divided by the interval, and the instantaneous angular velocity $\omega \equiv \dot{\theta}$ is defined as the limiting value of this average velocity as the interval is shrunk to zero. The whole procedure is entirely analogous to that adopted for linear translational motion. Thus angular acceleration α is defined as the rate of change of angular velocity: $\alpha \equiv \dot{\omega} \equiv \ddot{\theta}$. If another point P' in the body is chosen, the angular velocity and the angular acceleration of the line OP' will be identical with those of the line OP. Only the angular coordinate θ will be different. If the angular acceleration is constant, then equations similar to Eqs. (3.14), (3.15), and (3.16) may be obtained. If at time $t = 0$ a line OP in the body makes an angle θ_0 with Ox and has an angular velocity ω_0, then at any later time t the angular velocity and position of the line will be given by

$$\left.\begin{aligned} \omega &= \omega_0 + \alpha t, \\ \theta &= \theta_0 + \omega_0 t + \tfrac{1}{2}\alpha t^2, \\ \omega^2 &= \omega_0^2 + 2\alpha(\theta - \theta_0). \end{aligned}\right\} \tag{6.3}$$

with

These equations follow from the constancy of the acceleration α.

In all such equations the angle θ must be in units of radians, as defined in Appendix A. The units of angular velocity and acceleration are rad s^{-1} and rad s^{-2}, respectively. When the angular velocity is constant, other units than rad s^{-1} are commonly used, either revolutions per minute (r.p.m.) or cycles per second (c/s). One revolution or one cycle is equivalent to 2π radians, so these are easily converted into radians per second. Note that although the radian is a dimensionless

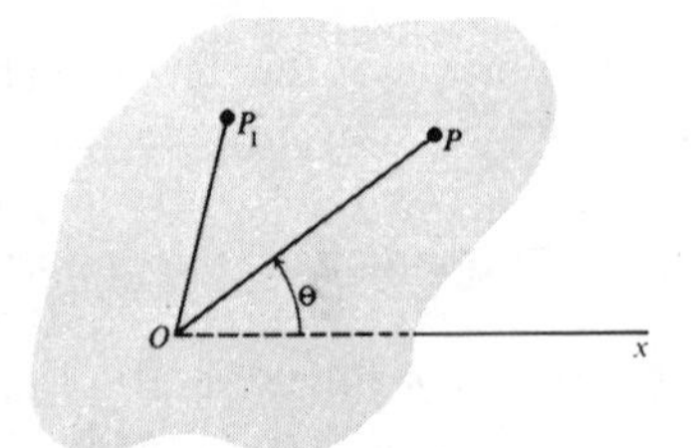

Fig. 6.6. A rigid body rotating about a fixed axis.

quantity it is convenient to retain it in the units of angular velocity and acceleration.

Example 1. An LP disk takes 1 s to reach its constant angular velocity of $33\frac{1}{3}$ r.p.m. What is its angular acceleration, assumed constant, while it is speeding up?

Solution. From the first of Eqs. (6.3) above, the constant angular acceleration is $\alpha = (\omega - \omega_0)/t$. In this example

$$\omega_0 = 0,\ t = 1 \text{ s, and } \omega = 33.3 \text{ rev min}^{-1} \times \left(\frac{2\pi \text{ rad}}{1 \text{ rev}}\right) \times \left(\frac{1 \text{ min}}{60 \text{ s}}\right)$$
$$= 3.49 \text{ rad s}^{-1},$$

whence $\alpha = 3.49 \text{ rad s}^{-2}$.

Example 2. A body is rotating about a horizontal axis with a constant angular acceleration of 0.2 rad s^{-2}. At $t = 0$ its angular velocity is 0.05 rad s^{-1}, and a marked line in the body is horizontal. What will the angular velocity be after (a) 4 s, and (b) 12 s, and what angles will the marked line make with the horizontal after these times?

Solution. a) From Eq. (6.3), with $\alpha = 0.2 \text{ rad s}^{-2}$ and $\omega_0 = 0.05 \text{ rad s}^{-1}$, we get for $t = 4$ s:

$$\omega = 0.05 \text{ rad s}^{-1} + (0.2 \text{ rad s}^{-2} \times 4 \text{ s})$$
$$= 0.85 \text{ rad s}^{-1}$$
$$\theta - \theta_0 = (0.05 \text{ rad s}^{-1} \times 4 \text{ s}) + (\tfrac{1}{2} \times 0.2 \text{ rad s}^{-2} \times 16 \text{ s}^2)$$
$$= 1.8 \text{ rad}.$$

If we want the angle in degrees, the conversion $2\pi \text{ rad} \equiv 360°$ may be used, giving

$$\theta - \theta_0 = 1.8 \text{ rad} \times \frac{360°}{2\pi \text{ rad}} = 103°.$$

This gives the angular displacement from starting. Since the marked line was initially horizontal, it will also give us the angle that the line makes with the horizontal at $t = 4$ s.

b) The calculation for $t = 12$ s proceeds exactly as above, and the details are left as an exercise. But some comment is needed about the final value of the angular displacement, which turns out to be

$$\theta - \theta_0 = 15 \text{ rad} = 859°.$$

This is certainly the total angular displacement, but it cannot be the angle made with the horizontal by the marked line of the body. Angles greater than 360° have no meaning in that sense. What the solution means is that the body has rotated twice through a complete rotation (2 × 360°) and then rotated a further angle of 859° − 720° = 139°.

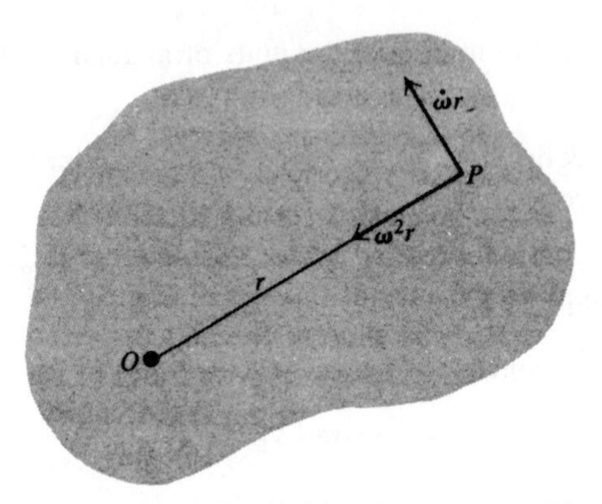

Fig. 6.7. A rotating rigid body considered as an assemblage of particles.

6.4 MOMENT OF INERTIA

Any rigid body can be thought of as an assemblage of particles, the separation of any two particles being constant. Thus in Fig. 6.7 the point P may be considered to be the site of a material particle of mass m at a distance r from the axis of rotation through O. As the body rotates, the particle at P will travel in a circle with center O and radius r, and at any instant its linear velocity will be along the tangent to this circle. The magnitude of this tangential velocity is $v = \omega r$, as discussed in Section 4.8. If the angular velocity is constant, the only component of the acceleration is the centripetal one, of magnitude $v^2/r = \omega^2 r$, but when, in addition, the tangential speed of the particle is changing, there is a tangential component of acceleration of magnitude $\dot{\omega} r = \alpha r$. Similar expressions hold for every constituent particle of the rigid body, different particles having different values of r in general, but all having common values of ω and α. The two components of acceleration for the particle at P are shown in Fig. 6.7. The momentum of the particle at P is $mv = m\omega r$, and this is in a direction perpendicular to the line OP. The *angular momentum* of this particle is defined as the product of this linear momentum and the distance r from the axis, or $m\omega r^2$. The total angular momentum of the whole rigid body is the sum of the angular momenta of all the constituent particles of the body. In symbols

$$L = \sum m\omega r^2 = \omega \sum mr^2,$$

the factor ω coming outside the summation sign, since it is the same for all the particles of the body. The sum of all mr^2-terms is defined as the *moment of inertia* of the rigid body and is denoted by the symbol I. Thus

$$I \equiv \sum mr^2. \tag{6.4}$$

Moment of inertia is a scalar quantity, and it depends on the axis of rotation chosen. The moments of inertia of a few simple bodies are listed for convenience in Fig.

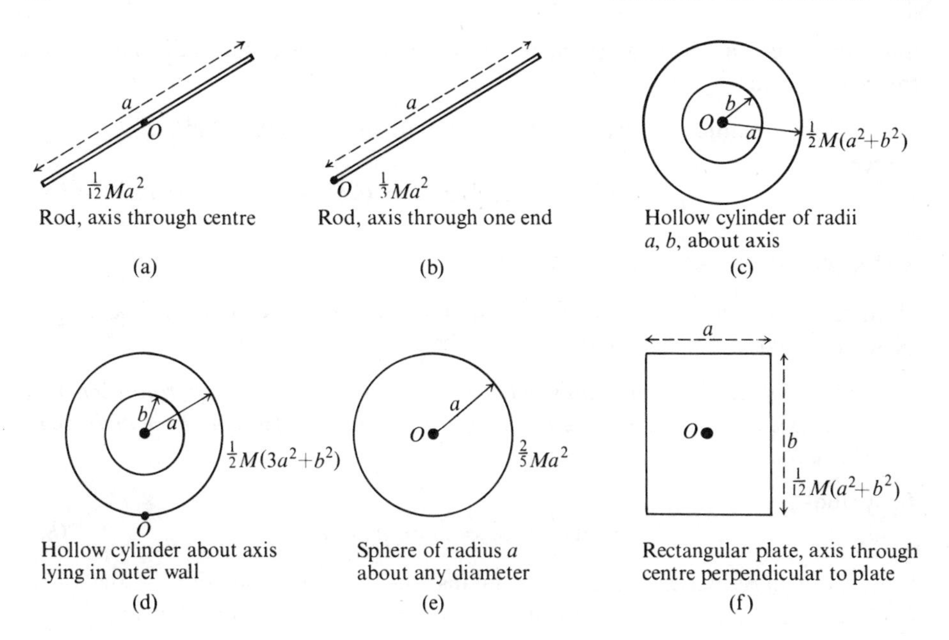

Fig. 6.8. The moments of inertia of a few simple bodies.

6.8, the axis of rotation in each case being perpendicular to the page through the point O. In all of these diagrams M is the total mass of the object. Moment of inertia plays exactly the same role in rotational motion as mass plays in translational motion. Whereas an object of mass M having a translational velocity of magnitude v has a linear momentum of magnitude $p = Mv$, the same object rotating with angular velocity ω about some axis has an angular momentum $L = I\omega$, where I is the moment of inertia of the body for that axis. The units of moment of inertia are kg m^2. Moment of inertia always appears as the total mass M of the body times the square of some characteristic linear dimension of the body. This linear dimension is given the name *radius of gyration* and is denoted by the symbol k. Then, $I = Mk^2$. Thus for a cylinder of internal and external radii b and a rotating about its central axis, the radius of gyration is given by

$$k^2 = \tfrac{1}{2}(a^2 + b^2).$$

Example. A small object of mass 0.5 kg is fastened to the edge of a circular disk of mass 2 kg and radius 10 cm. What is the total moment of inertia of the system about the center of the disk?

Solution. The moment of inertia of the disk may be found from Fig. 6.8(c) by setting $b = 0$, since a disk is simply a solid cylinder of small height, and the height is not involved in the moment of inertia. Therefore

$$I(\text{disk}) = \tfrac{1}{2}Ma^2 = \tfrac{1}{2} \times 2 \text{ kg} \times (0.1 \text{ m})^2 = 0.01 \text{ kg m}^2.$$

Since the object at the edge is small, the whole of its mass is at a distance a from the center, hence

$$I(\text{object}) = ma^2 = 0.5 \text{ kg} \times 0.01 \text{ m}^2 = 0.005 \text{ kg m}^2.$$

Hence

$$I = I(\text{disk}) + I(\text{object}) = (\tfrac{1}{2}M + m)a^2 = 0.015 \text{ kg m}^2.$$

6.5 NEWTON'S SECOND LAW

With the moment of inertia of a rigid body defined, the rotational analog of Newton's second law of motion can now be stated.

> *The unbalanced torque acting on a rigid body that is free to rotate about a fixed axis is equal to the rate of change of angular momemum of the body about that axis.*

In symbols,

$$\text{Torque} = N = \dot{L} = d(I\omega)/dt. \tag{6.5}$$

If the moment of inertia of the body remains constant, the expression for the torque reduces to the much simpler form $N = I\dot{\omega} = I\ddot{\theta} = I\alpha$. If, in addition, the unbalanced torque acting on the body is constant, then the angular acceleration is constant, and Eqs. (6.3) may be used. If there is no unbalanced torque present, then the angular momentum is constant. A simple but important example is provided by the motion of a planet around the sun. The only force acting on the planet is toward the sun as center, and, hence, the torque on the planet is zero. It follows that the angular momentum of any planet is a constant. Another example is provided by a skater going into a fast spin (Fig. 6.9). To do this he starts to spin with his arms outstretched. In this position his moment of inertia about the

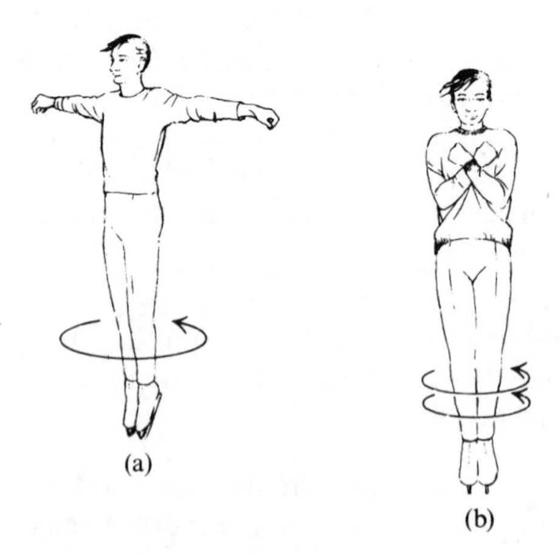

Fig. 6.9. A skater spins with arms outstretched in (a). When the arms are drawn in to his chest in (b), the skater's angular velocity increases.

vertical axis through his center of mass is large. When he brings his arms in to his chest, the moment of inertia is considerably reduced, since the mass of his arms is now nearer the vertical axis. But the torque exerted by the ice on him through contact with his skates is very small indeed, and so to a very good approximation his angular momentum is constant. It follows that his angular velocity must increase to compensate for the decrease in his moment of inertia. In the same way as it requires a large force to change the state of linear motion of a large mass, so it requires a large torque to change the state of rotational motion of a body with a large moment of inertia. This property is of vital importance in all forms of vehicle, for the basic drive from the engine is always applied in a more or less irregular manner, coming as it does from a small number of cylinders each of which supplies a fluctuating drive. A flywheel with a very large moment of inertia is used to smooth out the fluctuations and provide a steady torque to the driving wheels.

Example. A light cord is wrapped around the rim of a wheel of radius 0.5 m and moment of inertia 3.5 kg m², and a mass of 0.5 kg is hung from the cord. What is the angular acceleration of the wheel? What is the angular velocity of the wheel when the mass has fallen through a distance of 6 m? Frictional effects may be neglected.

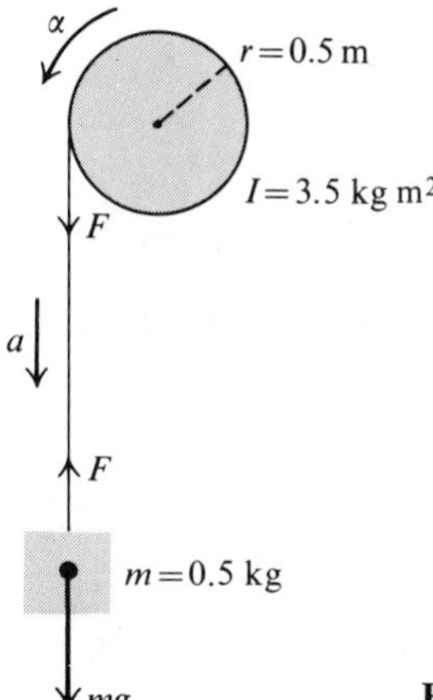

Fig. 6.10.

Solution. The situation is shown in Fig. 6.10. Since the cord does not slip on the wheel, the linear acceleration of the mass and the angular acceleration α of the wheel are related by

$$a = \alpha r.$$

The equation of motion of the mass is

$$mg - F = ma,$$

where F is the tension in the cord. Multiplying this equation by the radius r of the wheel gives

$$mgr - Fr = mar = m\alpha r^2.$$

The equation of motion of the wheel is

$$\text{torque} = Fr = I\alpha.$$

Adding these last two equations eliminates the tension F, and we get

$$mgr = m\alpha r^2 + I\alpha = (mr^2 + I)\alpha.$$

Therefore

$$\alpha = mgr/(mr^2 + I) = \frac{0.5 \text{ kg} \times 9.8 \text{ m s}^{-2} \times 0.5 \text{ m}}{3.5 \text{ kg m}^2 + 0.5 \text{ kg} \times (0.5 \text{ m})^2}$$

$$= \frac{2.45}{3.625} \text{ s}^{-2} = 0.676 \text{ rad s}^{-2}.$$

A 6-m displacement of the mass corresponds to an angular displacement of the wheel of

$$\theta - \theta_0 = 6 \text{ m}/0.5 \text{ m} = 12 \text{ rad}.$$

Hence, from Eq. (6.3), since $\omega_0 = 0$,

$$\omega^2 = 2\alpha(\theta - \theta_0) = 2 \times 0.676 \text{ rad s}^{-2} \times 12 \text{ rad}$$

$$= 16.2 \text{ rad}^2 \text{ s}^{-2}$$

$$\omega = 4.0 \text{ rad s}^{-1}.$$

6.6 THE KINETIC ENERGY OF ROTATION

Since the kinetic energy of a mass m moving with speed v is $\frac{1}{2}mv^2$, the kinetic energy of a rotating body can easily be inferred by simple substitution of the rotational analogs of the mass and the linear speed. Thus we expect the expression to be

$$K = \tfrac{1}{2}I\omega^2. \tag{6.6}$$

That this expression is indeed correct can be seen by considering the kinetic energy of any particle of the rigid body, and then summing over all particles. For any particle has a kinetic energy $\frac{1}{2}mv^2 = \frac{1}{2}m(\omega r)^2 = \frac{1}{2}\omega^2(mr^2)$. The expression mr^2 is just the moment of inertia of that particular particle about the axis of rotation, and so, on summing over all particles, Eq. (6.6) is obtained. The moment of inertia of a collection of objects about a common axis is just the sum of their separate moments of inertia, since moment of inertia is a scalar quantity. This rotational kinetic energy is equal to the work done by the unbalanced torque in rotating the body through an angle θ, just as the translational kinetic energy is the work done by the unbalanced force in the corresponding displacement.

Example. If the rotation of the wheel in Fig. 6.10 is opposed by a frictional torque of 0.5 N m, what will be the angular velocity after the mass has fallen a distance of 6 m?

Solution. When the mass m falls through a distance h, gravity does an amount mgh of work, and this work goes to increase the kinetic energy of both the mass itself and the wheel, and to overcome friction. If $\mathscr{W}_f$ denotes the work that has to be done to overcome friction, then

$$mgh = \tfrac{1}{2}mv^2 + \tfrac{1}{2}I\omega^2 + \mathscr{W}_f.$$

Since $v = \omega r$, this may be rewritten as

$$mgh = \tfrac{1}{2}(mr^2 + I)\omega^2 + \mathscr{W}_f.$$

Therefore

$$\omega^2 = 2(mgh - \mathscr{W}_f)/(mr^2 + I).$$

The angular displacement of the wheel is (by the result of the Example of the last Section) 12 rad, and consequently the work done against the frictional torque during this displacement is $\mathscr{W}_f = 0.5\ \text{N m} \times 12\ \text{rad} = 6.0\ \text{J}$. The work done by gravity is $mgh = 0.5\ \text{kg} \times 9.8\ \text{m s}^{-2} \times 6\ \text{m} = 29.4\ \text{J}$, and the value of $mr^2 + I$ is $3.625\ \text{kg m}^2$. Hence

$$\omega^2 = 2(29.4\ \text{J} - 6.0\ \text{J})/3.625\ \text{kg m}^2$$

$$= \frac{46.8\ \text{J}}{3.625\ \text{kg m}^2} = 12.9\ \text{rad}^2\ \text{s}^{-2}.$$

Therefore

$$\omega = 3.6\ \text{rad s}^{-1}.$$

6.7 THE MECHANICS OF ELASTIC SOLIDS

Consider a rectangular rod of some solid material to the ends of which equal and opposite forces **F** and −**F** are applied (Fig. 6.11). The weight of the rod will be ignored. If the rod is considered as a rigid body, then its motion is determined by considering all the external forces to act at its center of mass, which in this case merely gives the uninteresting result that the two applied forces cancel. For a real solid this result, although of course still true as far as the motion of the rod as a whole is concerned, misses the essential point that the state of the rod under the action of these two equal and opposite forces is not at all the same as its state when no forces act. The separation of any two points in a rigid body never changes; but, in a real body, the two forces set up stresses that result in a deformation of the body. It is not sufficient merely to add the two vector forces acting

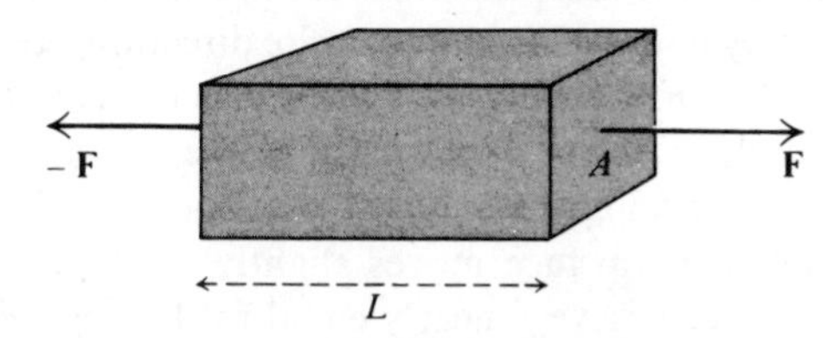

Fig. 6.11. Tensile stress being applied to a rod.

and conclude that there is no resultant. In fact, the mathematical language of vectors is insufficiently powerful to describe what we mean by the state of stress inside such a rod, and for this reason vector notation will not be used in the remainder of this chapter. It turns out that the deformation of a body depends not on a force alone but on the force per unit area acting anywhere in the body. A force per unit area is a more complicated quantity than a vector, since the unit area can be chosen in any orientation. *Stress* is the technical name given to a force per unit area. The stress at any point in a body depends ultimately on the mutual interactions between the basic units of the solid, atoms or ions or molecules or even groups of molecules. The units of stress are $N\,m^{-2}$. The effect of stressing a body is to change its size and shape, and the deformation is measured by the *strain* in the body. In physics the word "strain" carries no connotation of force; it is a purely geometrical quantity. There are three different kinds of stress, each with its own type of strain.

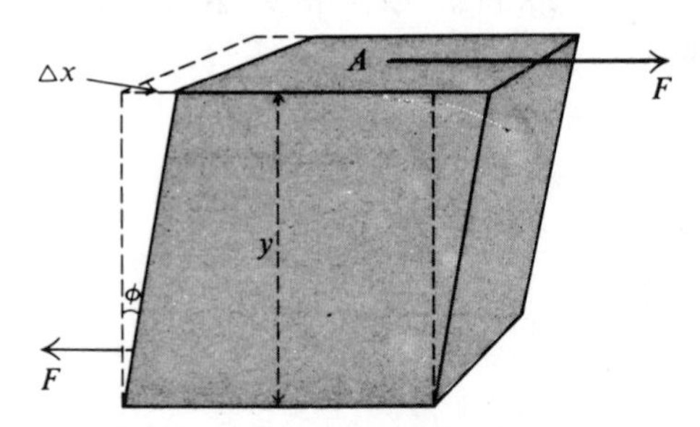

Fig. 6.12. Illustration of shear stress and strain.

For the rod shown in Fig. 6.11 the stress is F/A, where A is the cross-sectional area of the rod. This is known as the *linear tensile stress*. The main effect of applying such a stress is to increase the length of the rod, and the fractional increase of length, $\Delta L/L$, is known as the *linear tensile strain*. It is a dimensionless quantity. If the directions of both forces are reversed, the rod will decrease in length and the quantities are then called the *linear compressive stress and strain*.

Shear stress and shear strain arise when a torque is applied to a solid body. In Fig. 6.12 a rectangular block is shown, rigidly fixed to a plane support. A force F is applied tangentially to the upper surface and the shear stress is defined as the ratio F/A, where A is the area of the top surface. This differs from linear tensile stress, where the applied force was perpendicular to the surface. The lower surface of the block is acted on by a force F in the opposite direction exerted by the support. The torque on the block is Fy, where y is the distance between the top and bottom faces. The block would, of course, rotate under the action of this torque if the support did not exert on it an equal and opposite torque. The result of all these complex actions is that the top face moves slightly to the right through a small distance Δx. The ratio $\Delta x/y$ is very nearly equal to the angle ϕ, and this angle is defined as the shear strain. It is, of course, dimensionless.

Finally, a body may experience a bulk stress, which is a uniform compression in all directions. Such a bulk stress may, for example, be produced by immersing the body in a liquid where (as discussed in Section 9.2) it will be subjected to a pressure p which depends on the depth of the liquid and which is approximately constant over the surface of the body so long as the body is small. Before immersion the body was subjected to a pressure p_0 due to the atmosphere, and the bulk stress is defined as the excess pressure $\Delta p = p - p_0$. The effect of this excess pressure is to reduce the volume of the body without changing its shape, and the bulk strain is defined by the fractional decrease in volume, $-\Delta V/V$. The negative sign is incorporated in the definition because an increase of pressure always tends to decrease the volume.

The concepts of linear stress and strain and shear stress and strain are applicable only to solids, but bulk stress and strain are possible in liquids and in gases also.

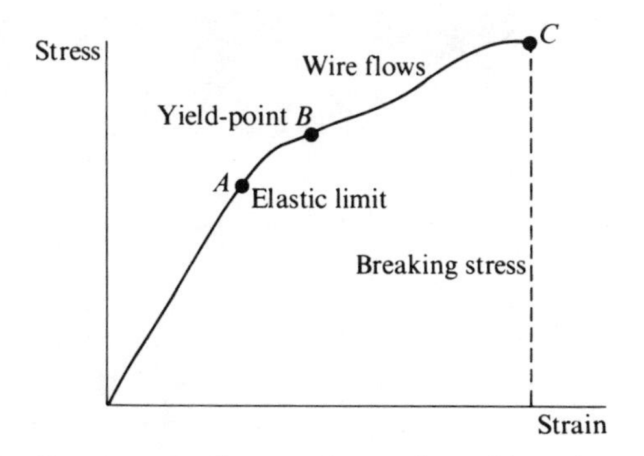

Fig. 6.13. Stress-strain diagram for a wire subjected to tensile stress.

It must not be thought that a stress in a solid is always of one of the types described. Usually the different kinds of stress and strain are intermingled, but the above classification is convenient.

If the stress is not too large, it is found to bear a linear relation to the resulting strain, and this is known as *Hooke's law*. Figure 6.13 shows a general stress-strain diagram for a metal wire loaded at one end and suspended from a fixed support. It follows a straight line for part of the way, and this is the region where Hooke's law is obeyed. At A, the *elastic limit*, the straight line becomes a curve. If the stress on the wire is increased to a point beyond A, the wire will not, on removal of the stress, return to its original length. It has been permanently strained. For a stress greater than that corresponding to point B, the wire begins to exhibit plastic flow until it finally breaks at C. Such curves as this need care in their interpretation, because the cross-sectional area of the wire decreases as the strain increases and therefore the relation between load and stress continually changes. The processes taking place are enormously complex, and in any event different materials show widely differing behavior. The important point is that for small stresses the stress-strain diagram is a straight line. The factor of proportionality between stress and strain is called a *modulus of elasticity*, and there is a modulus for each of the three types of stress described above.

a) Young's modulus, Y, is defined as: $\dfrac{\text{linear stress}}{\text{linear strain}} = \dfrac{F/A}{\Delta L/L}$.

b) The rigidity modulus, n, is defined as: $\dfrac{\text{shear stress}}{\text{shear strain}} = \dfrac{F/A}{\phi}$.

c) The bulk modulus, B, is defined as: $\dfrac{\text{pressure change}}{\text{bulk strain}} = \dfrac{\Delta p}{-\Delta V/V}$.

The inverse of the bulk modulus is known as the compressibility, $1/B = k$. Since strain is a dimensionless quantity, the units of all the elastic moduli are simply those of stress, N m^{-2}. Table 6.1 gives the three elastic moduli for some typical solids.

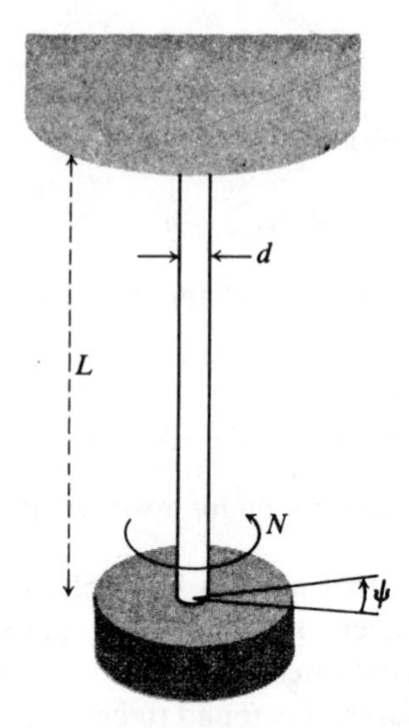

Fig. 6.14. A wire being subjected to shearing stress.

A very important example of a situation in which shearing stress is involved is in the twisting of a thin rod. Consider such a rod, of length L and diameter d, with its upper end rigidly fixed, as shown in Fig. 6.14. Suppose that the lower end is subjected to a torque N by some means. Then the lower end of the rod will rotate through an angle ψ, which depends on the length, on the diameter, and on the rigidity modulus of the material of the rod. The full mathematical analysis of this problem is beyond the scope of this book, but a dimensional analysis is done in Appendix B, and this leads easily enough to the correct functional dependence. The full result is

$$N = \left\{\frac{\pi}{32}\frac{nd^4}{L}\right\}\psi = \tau\psi, \tag{6.7}$$

Table 6.1. Elastic moduli of solids

Substance	Y	B (10^{10} N m^{-2})	n
Aluminum	7.1	7.7	2.5
Brass	10	10.5	3.5
Copper	12	14	4.5
Glass	~9	~5	~3
Lead	1.8	4.3	0.8
Nickel	21	18	7.6
Steel	21	18	8.3

where τ is called the *torsional constant* of the rod and represents the torque needed to produce a twist of 1 rad. Because the diameter of the wire occurs to the fourth power, the torsional constant can be made very small by choosing a fine fiber instead of an actual rod or wire. This arrangement is widely used in measuring instruments of all kinds where very small twists can be detected optically if necessary.

Example 1. A sample of bone in the form of a cylinder of cross-sectional area 1.5 cm² is loaded on its upper end by a mass of 10 kg. By careful measurement with a traveling microscope, the length of the cylindrical sample is observed to decrease by 0.0065%. Calculate the value of Young's modulus for the specimen

Solution. A decrease in length of 0.0065% is a linear compressive strain of 6.5×10^{-5}. The compressive stress exerted on the sample is

$$\frac{10 \text{ kg} \times 9.8 \text{ m s}^{-2}}{1.5 \times 10^{-4} \text{ m}^2} = 6.53 \times 10^5 \text{ N m}^{-2}.$$

Hence Young's modulus is

$$Y = \frac{\text{stress}}{\text{strain}} = \frac{6.53 \times 10^5 \text{ N m}^{-2}}{6.5 \times 10^{-5}} = 1.0 \times 10^{10} \text{ N m}^{-2}.$$

Example 2. A brass wire of diameter 0.82 mm and a steel wire of the same length have identical torsional constants. What is the diameter of the steel wire?

Solution. From Eq. (6.7) we see that, for two wires of the same length,

$$\tau_1/\tau_2 = n_1 d_1^4 / n_2 d_2^4.$$

Hence, if $\tau_1 = \tau_2$ as required, then $n_1 d_1^4 = n_2 d_2^4$, which leads to

$$\begin{aligned} d_{\text{steel}}^4 &= n_{\text{brass}} d_{\text{brass}}^4 / n_{\text{steel}} \\ &= \frac{3.5 \times 10^{10} \text{ N m}^{-2}}{8.3 \times 10^{10} \text{ N m}^{-2}} \times (0.82 \text{ mm})^4 \\ &= 0.422 \times (0.82 \text{ mm})^4. \end{aligned}$$

Therefore $$d_{\text{steel}} = (0.422)^{1/4} \times 0.82 \text{ mm}$$
$$= 0.806 \times 0.82 \text{ mm} = 0.66 \text{ mm}.$$

Example 3. A block of lead of volume 0.493 m^3 is dropped into the sea. What is its new volume when it is on the sea bed where the pressure is 8.6×10^7 N m^{-2} greater than at the surface?

Solution. From the equation defining the bulk modulus we get

$$-\Delta V/V = \Delta p/B,$$

therefore

$$-\Delta V = V\Delta p/B$$

$$= 0.493 \text{ m}^3 \times \frac{8.6 \times 10^7 \text{ N m}^{-2}}{4.3 \times 10^{10} \text{ N m}^{-2}}$$

$$= 0.493 \text{ m}^3 \times 2 \times 10^{-3}$$

$$= 0.986 \times 10^{-3} \text{ m}^3 = 0.001 \text{ m}^3.$$

Hence the new volume of the lead block is 0.492 m^3.

PROBLEMS

6.1 A uniform rectangular gate $ABCD$ weighs 100 N and is hinged at A and D, where A is vertically above D. The dimensions of the gate are $AB = 4$ m, $BC = 1.5$ m. If the force exerted on the gate at A is horizontal, what is its magnitude? What are the magnitude and direction of the force exerted on the gate at hinge D?

6.2 A mass M is being weighed in an analytical balance the arms of which are not exactly equal. When M is in the left-hand pan, the standard mass in the right-hand pan is M'; when M is in the right-hand pan, it is M''. Show that the correct value of the mass is $(M'M'')^{\frac{1}{2}}$.

6.3 Two forces with the same magnitude F but acting in opposite directions act along the sides AB and CD of a square, the side of the square being a. Calculate the resultant torque of this pair of forces about any point in the plane of the square, and show that it is dependent only on F and a. Such a pair of equal and opposite forces is known as a *couple*.

6.4 A uniform ladder of mass 40 kg and length 5 m leans against a wall at a height of 4 m. If the force exerted by the wall on the ladder is perpendicular to the ladder, find the magnitude of this force. Find also the magnitude and direction of the force exerted on the ladder by the ground. If a 60-kg man begins to climb the ladder, how far up can he go before it slips, if the coefficient of friction between ladder and ground is 0.45?

6.5 A uniform meter stick, of mass 100 g, has a 250-g mass suspended from the 25-cm mark and a 500-g mass suspended from the 80-cm mark. The whole is suspended so as to be horizontal by laboratory spring-scales attached to the ends of the meter stick. What are the readings on these scales?

6.6 A roller of diameter 0.8 m weighs 500 N. What horizontal force is needed to pull the roller over a brick of height 5 cm?

6.7 A cylinder of radius 20 cm is rotating about its axis at 800 r.p.m. What is the tangential velocity of a point on the surface of the cylinder?

6.8 A wheel requires 4 s to rotate through 48 rad, its angular velocity at the end of this time being 20 rad s^{-1}. Calculate its angular velocity at the start of this interval, and also its constant angular acceleration.

6.9 A body starts from rest and rotates about a fixed axis with constant angular acceleration. Prove that, when it has rotated through 0.5 rad, the tangential acceleration of any point of the body is equal in magnitude to the centripetal acceleration.

6.10 What is the angular momentum of a body that is rotating about an axis with a constant angular velocity of 6 rad s^{-1} if its moment of inertia about that axis is 250 kg m^2? If the mass of the body is 10 kg, what is its radius of gyration about the same axis?

6.11 What is the rotational kinetic energy of the body in the previous question?

6.12 Calculate the moment of inertia of the wheel and axle shown in Fig. 6.15 on the assumption that it is constructed throughout of a material of density $(2/\pi) \times 10^4$ kg m^{-3}. Calculate also the radius of gyration of the wheel and axle.

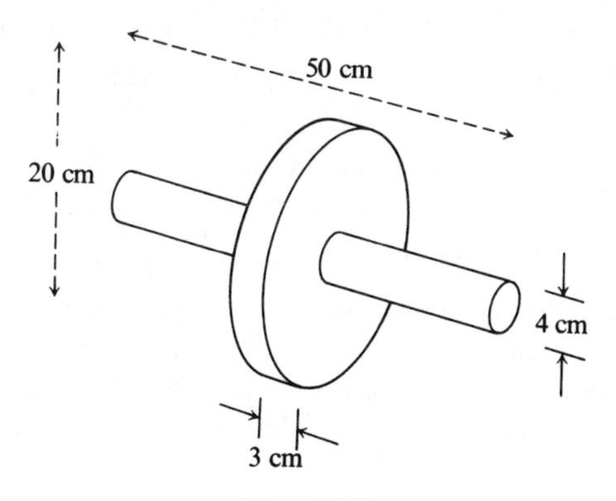

Fig. 6.15

6.13 A wheel and axle as shown in Fig. 6.15 has a light string wrapped round its axle supporting a mass of 1 kg. This mass is initially held so that the string is just taut, and then it is released. It accelerates downwards under gravity, falling a distance of 6.20 m in 10 s from rest. Calculate the moment of inertia of the wheel and axle and compare the result with the answer to the previous question. Give a physical explanation of the slight disparity.

6.14 Using the same data as in Problems 6.12 and 6.13, calculate the angular acceleration of the wheel and axle, and the tension in the string. Calculate also the kinetic energy of the system after the 1-kg mass has fallen through 6.20 m.

6.15 A large circular disk, of radius 4 m and mass 100 kg, is pivoted so that it can rotate freely without friction in a horizontal plane about a vertical axis through its center. A boy of mass 50 kg walks once completely round the perimeter of the disk, returning to the same point of the disk as he started from. He then finds that he is not in the same position with respect to the ground. Explain why this is so, and calculate the angle through which the disk has rotated due to the motion of the boy.

6.16 A common piece of demonstration equipment in physics classrooms is a chair that can rotate with negligible friction about a vertical axis. The professor seats himself in such a chair and holds out his arms with a weight in each hand. His moment of inertia about the axis of rotation is 2.0 kg m^2. His assistant sets him rotating with an angular velocity of 1.5 rad s^{-1}. While he is spinning, he suddenly brings his arms in to his chest, thereby reducing his moment of inertia to 0.3 kg m^2. What is the new value of the angular velocity?

6.17 A sphere and a solid cylinder roll without slipping down two separate inclined planes, and are found to travel the same distance in the same time. If the angle of the plane down which the sphere rolls is 30°, what is the angle of the other plane?

6.18 A mass of 20 kg is supported by a nickel wire of length 5.0 m and of cross-sectional area 2.5 mm^2. What is the resulting elongation of the wire?

6.19 If the length of a steel rod 2.5 m long is reduced by 0.5 mm, what is the compressional stress in the rod?

6.20 What increase in pressure would be needed to decrease the volume of 1 m^3 of water by 10^{-4} m^3? The bulk modulus of water is 2.1×10^9 N m^{-2}.

6.21 What pressure would be needed to increase the density of copper by 0.1%?

6.22 When a hole is punched in a metal plate, the metal yields under the shear stress imposed. If a shear stress of 3.5×10^8 N m^{-2} is needed to cause rupture in mild steel, determine the force required to punch a hole of diameter 1.5 cm in a steel plate of thickness 0.5 cm. [*Hint.* The stress acts over the cylindrical wall of the hole.]

6.23 Calculate the torsional constant of a brass rod of length 31.42 cm and diameter 2.0 cm. If one end of this rod is held rigidly, what torque must be applied to the other end in order to twist it through 5°?

6.24 If the diameter of the rod in the previous problem is increased by 0.1 mm, what is the new torque that must be applied to produce the same twist?

CHAPTER 7

TEMPERATURE AND HEAT

7.1 INTRODUCTION

As discussed in Section 5.11, mechanical energy is not conserved when frictional forces are present. It is a matter of common experience that frictional forces cause a rise of temperature, and we talk of heat being produced. The notions of heat and temperature are primary human notions and, like nearly all such notions, are lacking in precision. For a long time, indeed, the two separate concepts of heat and temperature were not even distinguished, and the realization that there was an exact numerical connection between the mechanical work put into a system and the heat required to produce the same final result took even longer. Count Rumford* and Sir Humphry Davy at the close of the eighteenth century sounded the death-knell of the old caloric theory, which regarded heat as a massless, indestructible substance, by showing experimentally that limitless quantities of heat could be produced by friction. Speculation by various philosophers and scientists, and preliminary experimental work by Mayer, was followed by forty years of careful, painstaking experimentation by Joule, who, finally, by 1878 had proved beyond all question the equivalence in their effects of mechanical work and heat. Subsequent work by Rowland in Baltimore, and by Laby and Hercus in Melbourne, fully confirmed Joule's results, although improving on his accuracy, and other work using electrical methods of heating led to the belief that the quantity of heat produced in any process is equivalent to the work, mechanical or electrical, put into the system. This belief is so firmly held today that details of the experimental work that led to its acceptance are of historical interest only. By our present-day standards, Joule's work was really nothing more than the determination of the numerical relation between two different units of energy, the calorie and the joule, and in this respect was akin to determining the length of a foot-rule in terms of the meter. However, it is easy to be wise in retrospect, and it should always be remembered that the acceptance of heat as a mode of transference of energy equivalent to mechanical work was not won without opposition and effort. Joule's career is a salutary reminder to young scientists of the importance of single-minded persistence, patience, and skill.

---

* Benjamin Thompson, an American who took Britain's side in the War of Independence; he left the United States and eventually settled in Bavaria, rising rapidly in the service of the Elector until he was eventually made Counsellor, with the title of Count Rumford.

7.2 TEMPERATURE

In dealing with thermal effects it is important to realize that we are faced with a new class of problem. Mechanics is mainly concerned with objects possessing simple individuality and with the external relations of such objects. In thermodynamics (as the study of the processes involving heat is called) we are concerned with systems regarded as a complex whole, and with the internal changes of such systems. When the notion of simple individuality is given up in mechanics (as, for example, when the motion of a liquid or a gas is considered), the results obtained become more like the results of thermodynamics, and indeed it becomes impossible not to introduce the concept of temperature if realistic agreement with experiment is to be achieved. A system is any portion of the real world that we choose to isolate conceptually, and it may or may not be physically isolated from its surroundings. A gas inside a vacuum bottle is essentially isolated from its environment, whereas an animal is a system that interchanges energy, matter, and work with its surroundings. In thermodynamics we are concerned with systems in equilibrium, which means that the vastly complex nature of any real system can be described fully by a small number of variables. Thus the state of a gas in an enclosure is completely determined by its pressure and its volume. However, if rapid changes are occurring in a system, the simple description breaks down, since concepts such as the pressure can no longer be unambiguously defined. This is a very important point for biomedical scientists to bear in mind. Of course, if the changes take place slowly enough it may be possible in practice to use the simple description in terms of pressure and volume, but a stage is eventually reached at which random fluctuations in the measurements obtained indicates that the system is no longer describable in such terms.

The criterion for equilibrium is that all parts of the system be at the same temperature. If two different systems are allowed to exchange energy, then they will reach a state of common equilibrium, and we say they are at the same temperature. A thermometer is a system that can be easily transported, that very quickly comes into equilibrium with any other system with which it is put into contact, and that carries a scale of some sort on which an indicator moves as the thermal state of the thermometer is altered. Suppose a thermometer T is put into thermal contact with a system A, and then into contact with a second system B isolated from A. If the scale of the thermometer indicates the same reading in both cases, then we say that the temperatures of A and B are the same. This is the basic law of thermometry and has been dignified with the name *zeroth law of thermodynamics*. Any property of matter that changes with the thermal state of a system may be chosen as the indicator in a thermometer, the so-called *thermometric property*. For medical and biological purposes the three important types of thermometer are:

a) liquid-in-glass thermometers, in which the thermometric property is the volume of a liquid relative to its container;
b) resistance thermometers, in which the thermometric property is the electrical resistance of a small coil of wire; and

c) thermocouples, in which the thermometric property is the thermoelectric voltage produced at the junction of two dissimilar wires.

The principles of operation of types (b) and (c) are fully discussed in Sections 23.8 and 24.6 and the method of construction and use of type (a) are too familiar to require description. The ultimate standard of temperature measurement uses none of these thermometric properties. Instead it uses the pressure of a fixed volume of hydrogen or helium gas. The necessary equipment is bulky, and elaborate corrections have to be computed in obtaining a temperature reading, so that these instruments are only really used in the national standardization laboratories throughout the world. It is sufficient to know that the thermometers commercially available give reliable measurements of temperature, without worrying unduly about how the ultimate standardization is carried out.

Thermometers as so far discussed only give a qualitative measure of temperature. That is, they indicate when one temperature is greater or less than another. To make them quantitative instruments it is necessary to introduce a temperature scale, and for this it is essential to have agreement on some one standard temperature. An internationally agreed standard temperature must be reproducible with great accuracy and this requires the specification of a system. The choice is a system in which pure water exists simultaneously as a solid (ice), a liquid, and a gas (steam). This system is called a *triple-point cell*, and the temperature of the system the triple-point temperature. This temperature is assigned the value 273.160K (read kelvin). Any other temperature can then be uniquely determined as follows. A constant-volume hydrogen thermometer is allowed to come into equilibrium with the triple-point cell and the pressure of the hydrogen is read as p_{tp}. The thermometer is then allowed to come into equilibrium with the system whose temperature it is desired to measure, and gives a reading of p. Then the unknown temperature is defined to be

$$T(\mathrm{K}) \equiv 273.160\,(p/p_{tp}). \tag{7.1}$$

This procedure defines in an unambiguous fashion the absolute gas scale of temperature. The more familiar scale of temperature in common use is defined in terms of the absolute gas scale by the relation

$$\theta(^\circ\mathrm{C}) \equiv T - 273.15, \tag{7.2}$$

where the temperature is now in degrees Celsius. It is convenient to have easily reproducible systems of accurately known temperatures, and a short list of these is given in **Table 7.1.** In this table, NBP refers to the normal boiling point of a liquid, and NMP to the normal melting point of a solid, the word *normal* meaning that the values refer to a standard pressure of 101,325 N m^{-2}, which is 1 standard atmosphere. That the NMP of ice and the NBP of water come out so conveniently as 0°C and 100°C, respectively, is, of course, no accident. Previously these two fixed points were used as the absolute standards, and the new standard was carefully chosen to give the same result as before.

Table 7.1. Standard fixed points of temperature relative to a value of 273.160K for the triple-point of water

Fixed point	T (K)	θ (°C)
NBP of oxygen	90.18	−182.97
NMP of ice	273.15	0.00
NBP of water	373.15	100.00
NBP of sulfur	717.75	444.60
NMP of antimony	903.65	630.50
NMP of silver	1233.95	960.80
NMP of gold	1336.15	1063.00

7.3 HEAT AND THE FIRST LAW OF THERMODYNAMICS

When a bunsen flame is applied to a can of water, the temperature of the water rises; when the water in the can is vigorously stirred, its temperature rises. In both cases there has been an increase in the thermal energy of the water, and the thermometer reading gives the measure of that thermal energy. The name *heat* is given to energy as it is transferred by one of the processes of conduction, convection, and radiation discussed in Sections 7.6, 7.7, and 7.8. Work is the name given to energy as it is transferred by a mechanical process. Properly speaking, a system cannot contain heat any more than it can contain work. It can only contain energy of various kinds, including thermal energy. It is worth making this precise distinction between heat as a mode of transference of energy and thermal energy as a property of the system, since not all the heat transferred to a system necessarily goes into thermal energy. Some of it may be locked up in the form of elastic potential energy, or chemical energy. If we write the symbol Q for the heat supplied to the system and $\mathscr{W}$ for the mechanical work done on the system, then

$$Q + \mathscr{W} = E_2 - E_1 \tag{7.3}$$

is the generalization of the conservation of energy principle applicable to the kind of system considered in thermodynamics. It is called the *first law of thermodynamics*. The quantity E, which was previously called the total energy, is now better labeled the *internal energy* of the system. On the sub-microscopic scale it is indeed still the sum of the kinetic and potential energies of all the constituent particles of the system, and so no contradiction is involved. The internal energy E is a perfectly definite function of the state of the system, like its pressure, volume, or temperature, and so the difference $E_2 - E_1$ does not depend on the details of the process used to effect the change. But the same change can be effected by various combinations of Q and $\mathscr{W}$, since both of these depend on how the change is brought about. Thus one could heat vigorously and hardly stir at all, or stir vigorously and not heat so strongly. It is the sum of the two terms, $Q + \mathscr{W}$, that depends only on the initial and final states, not the individual quantities themselves. In Eq. (7.3) $\mathscr{W}$ must be taken as the net work done on the system. Thus when water in a can is heated and stirred, it expands, and in so doing a certain amount of work is done by the water against the pressure of the atmosphere. This must be subtracted from the work of stirring to give the net amount.

7.4 HEAT CAPACITY

Quantity of heat is measured by the observable effects it produces, the most obvious of which is rise in temperature. If a system receives an amount of heat Q so that its temperature rises a small amount $\Delta\theta$, then the ratio $Q/\Delta\theta$ is called the *heat capacity* of the system. The usual quantity tabulated is the heat capacity of unit mass of a material. This is called the *specific heat* when the unit of mass is chosen as the gram, and is denoted by the symbol c. It is conventional in thermodynamics to refer specific heats to the gram and not to the kilogram, although this constitutes a mild departure from MKSA orthodoxy. If it is necessary to specify how the measurement was carried out, whether at constant volume or constant pressure, the corresponding symbol is written as a subscript, thus: c_p or c_v. It is often convenient to choose the gram-molecular mass, or gram-atomic mass, as the unit of mass. One *mole* of an element of atomic mass M is M grams of the element; one mole of a compound substance having molecular mass M is M grams of the substance. The word "mole" can be used to indicate either the gram-atomic mass or the gram-molecular mass. One mole always contains the same number of particles and this number is known as Avogadro's constant

$$N_0 = 6.0249 \times 10^{23}\ \text{mol}^{-1}.$$

The heat capacity of one mole of a substance is referred to as the *molar heat*, and is denoted by the symbol C. The more specific symbols C_p and C_v can be used if necessary. If these symbols are used, the relation between the amount of heat supplied and the consequent rise in temperature can be written

$$Q = mc\Delta\theta = nC\Delta\theta, \tag{7.4}$$

where $n = m/M$ is the number of moles present. Notice that the change in temperature in this equation could equally well be written as ΔT, since the only difference between the Celsius scale and the Kelvin scale is an additive constant. Changes of temperature will be given the unit deg,* and particular temperatures will have units °C or K. The units of specific and molar heat are J g^{-1} deg^{-1} and J mol^{-1} deg^{-1}, respectively. Table 7.2 shows the specific heats of some typical substances. For the gases listed it is the specific heat at constant pressure that is given, and the values are for a temperature of 0°C and a pressure of one standard atmosphere.

---

* The unit "deg" is not in accordance with the S.I. system, but is used to avoid confusion between temperature and temperature difference.

Table 7.2, Typical values of specific heat

Substance	Specific heat ($J\ g^{-1}\ deg^{-1}$)	Substance	Specific heat ($J\ g^{-1}\ deg^{-1}$)
Aluminum	0.91	Alcohol (ethyl)	2.52
Concrete	~3	Chloroform	0.97
Copper	0.39	Glycerine	2.44
Glass (crown)	0.67	Mercury	0.14
Ice	2.11	Water	4.186
Lead	0.13	Air	0.95
Paraffin wax	2.90	Ammonia	2.20
Silver	0.23	Steam (at 100°C)	2.02

It was previously the custom to express specific heats in terms of the specific heat of water, which was taken as the standard substance. The unit of heat, the calorie, was defined as the quantity of heat necessary to raise the temperature of one gram of water from 14.5°C to 15.5°C. The kilogram calorie (kcal) was similarly defined in terms of one kilogram of water. The experiments of Joule and others may be regarded as determinations of the specific heat of water in terms of the mechanical units of energy, and they led to the conversion relation

$$\begin{aligned} 1 \text{ calorie} &\equiv 4.186 \text{ J}, \\ 1 \text{ kcal} &\equiv 4186 \text{ J}. \end{aligned} \tag{7.5}$$

Although the kcal is a common energy unit in biology, we shall not use it in this book.

The molar heats of many solid elements are very close to 25 J mol^{-1} deg^{-1}, a fact known as the law of Dulong and Petit. Thus silver has an atomic weight of nearly 108, and its specific heat from Table 7.2 is 0.23 J g^{-1} deg^{-1}. Its molar heat is the product of these, or 24.8 J mol^{-1} deg^{-1}, in very good agreement with the law. However, the specific heats of solids fall off with decrease of temperature and they tend to zero as the temperature approaches absolute zero. Figure 7.1 shows the variation in molar heat with temperature of lead, silver, aluminum, and diamond. It will be seen that lead has already reached the Dulong-Petit value at just over 100K, whereas diamond has scarcely any molar heat at that temperature. Nevertheless, all four curves shown have precisely the same form and can be made to coincide if a separate horizontal scale of temperature is used for each. The form of such curves as these presented a great challenge to theoretical physicists, and the complete elucidation of their shape proved to be impossible on the basis of classical physics. New ideas involving the quantum of energy had to be used before the behaviour of the metals could be understood. For our purpose it is enough to comment that the fact that all specific or molar heats tend to zero as

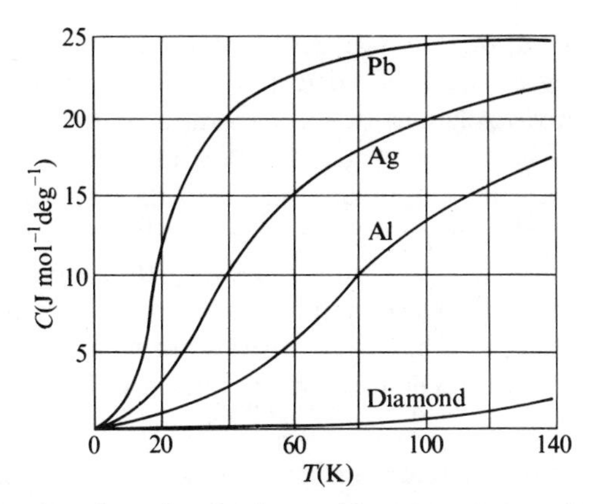

Fig. 7.1. The variation of molar heat with temperature at low temperatures.

the absolute temperature tends to zero is a consequence of the so-called *third law of thermodynamics*, which states essentially that the absolute zero of temperature is unattainable. The difficulty of reaching ever lower temperatures increases exponentially the lower one gets. With special equipment, temperatures only a very small fraction of a degree away from absolute zero have been attained.

7.5 LATENT HEAT

All substances can exist in one of three *phases*, as they are called, the solid phase, the liquid phase, and the gaseous phase; and the change from one phase to another is marked by a sudden increase in the internal energy. Thus one gram of steam at 100°C has a much higher internal energy than one gram of water at 100°C. It also has a much larger volume, 1673 cm^3 as opposed to 1 cm^3. The heat needed to change one gram of water at 100°C to steam is found to be 2256 J, and the work done *on* the substance during this change is the constant pressure of the atmosphere, which we may take roughly as 10^5 N m^{-2}, times the change in volume which is -1672 cm^3. The negative sign is because in becoming steam the water does work *on* the atmosphere, not vice versa. Hence, substituting into Eq. (7.3),

$$+2256 \text{ J} - (10^5 \text{ N m}^{-2} \times 1672 \times 10^{-6} \text{ m}^3) = E_2 - E_1,$$

which gives

$$E_2 - E_1 = 2089 \text{ J}.$$

The amount of heat required to change unit mass of a substance from the solid phase to the liquid phase at constant temperature and pressure is called the *latent heat of fusion* and is denoted by the symbol l_f if the unit of mass is the gram, and L_f if it is the mole. The *latent heat of vaporization* l_v (or L_v) for the phase change liquid $\leftrightarrow$ vapor, and the *latent heat of sublimation* l_s (or L_s) for the phase change solid $\leftrightarrow$ vapor are similarly defined. Latent heats for a few common substances are shown in Table 7.3.

The latent heat of vaporization of body sweat is the basis for the temperature regulation of the body of warm-blooded animals. One region of the brain, the *hypothalamus*, is sensitive to changes in the temperature of the blood passing through it, and this organ controls the sweat glands. When it detects an increase in blood temperature, it increases the output of sweat, and the vaporization of this sweat from the skin requires the supply of the necessary latent heat. This latent heat comes from the body, which is therefore cooled. As the surface of the body cools, the hypothalamus detects a decrease in the temperature of the blood and reduces the output of sweat accordingly. The system operates to maintain human body temperature near 37°C and is an excellent example of negative feedback **(Section 33.11).**

Table 7.3 Latent heats of fusion for some common solids, and latent heats of vaporization for some common liquids. The normal melting point (NMP) and normal boiling point (NBP) are also shown.

Substance	NMP (°C)	l_f (J g^{-1})	Substance	NBP (°C)	l_v (J g^{-1})
Copper	1083	212.5	Alcohol (ethyl)	78.3	857
Ice	0	334.8	Chloroform	61.5	246
Lead	327	26.5	Ether	34.6	352
Naphthalene	80	149.5	Turpentine	156.0	294
Silver	961	102.1	Mercury	356.6	294
Sulfur	113	39.1	Water	100.0	2256

Whenever there is a change in the internal potential energy of a substance, a latent heat exists. When substances are mixed, go into solution, crystallize, or take part in a chemical reaction, heat is in general evolved or absorbed, and a heat of mixture, or a heat of solution, or a heat of crystallization, or a heat of formation is involved. These various heats of reaction are defined in the same way as latent heats of phase change. Even when a solid undergoes a small change in its crystalline arrangement, there will be a corresponding latent heat.

Example. How much heat is required to change 5 g of ice at −5°C into water at 50°C?

Solution. The total heat required Q may be calculated as the sum of three separate amounts of heat, $Q_1 + Q_2 + Q_3$, so long as the specific heats involved are assumed to be constant.

i) Q_1 is the heat needed to raise the temperature of ice to 0°C, namely $Q_1 = mc(\theta_2 - \theta_1) = 5 \text{ g} \times 2.11 \text{ J g}^{-1} \text{ deg}^{-1} \times 5 \text{ deg} = 53 \text{ J}$;
ii) Q_2 is the heat required to melt the ice at 0°C, namely $Q_2 = ml_f = 5 \text{ g} \times 334.8 \text{ J g}^{-1} = 1674 \text{ J}$;

iii) Q_3 is the heat required to warm the water up to 50°C, namely $Q_3 = 5$ g $\times$ 4.186 J g^{-1} deg^{-1} $\times$ 50 deg = 1046 J.

Hence, the total heat needed is $Q = 2773$ J.

7.6 TRANSFER OF HEAT BY CONDUCTION

The mode of transfer of energy known as *heat conduction* does not involve any gross motion of the body through which the energy is transferred. For example, if a spoon is put into a cup of very hot coffee, the end of the spoon becomes warm, although there has been no motion of the spoon. The part of the spoon in contact with the coffee has its thermal energy increased, and the thermal energy is shared progressively with neighboring parts of the spoon until it reaches the free end.

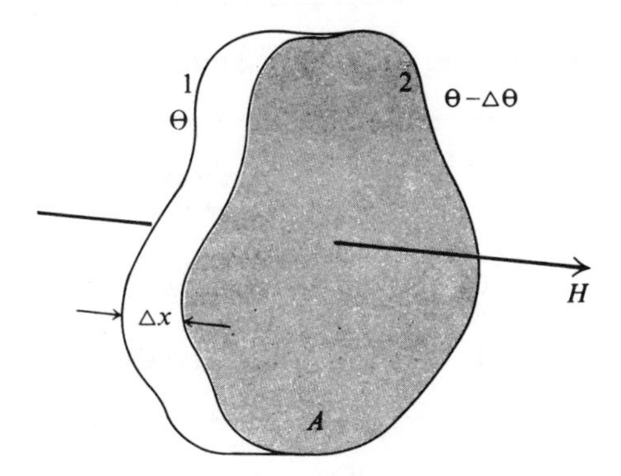

Fig. 7.2. Illustrating Fourier's law.

Conductivity is a mechanism of transfer in solids, liquids, and gases that depends on the existence of a temperature difference, and is essentially a non-equilibrium process. As such, it falls outside the strict territory of thermodynamics. The details of the transfer process on the submicroscopic level vary from material to material. Thus in metals, which are very good conductors of heat, the energy transport is effected by the free electrons in the same way as electricity is conducted, and the two are in fact proportional. But in insulating materials the transfer is more complicated and depends on the molecular vibrations themselves. On the macroscopic (or large-scale) level, it is possible to ignore the mechanism of transfer completely and to specify the conductive property of any material as follows. Consider a thin, parallel-sided slab of material as shown in Fig. 7.2, of thickness Δx and area A, with face 1 maintained at a steady temperature θ and face 2 at a steady temperature of $\theta - \Delta\theta$. In this steady state the heat energy flows at a steady rate from face 1 to face 2, and Fourier's law of thermal conduction gives the rate of flow of heat, or heat current, H, as proportional to A, proportional to the drop in temperature $-\Delta\theta$, and inversely proportional to Δx. In symbols,

$$H = -KA(\Delta\theta/\Delta x). \tag{7.6}$$

The constant of proportionality, K, is known as the *thermal conductivity* of the material, and some typical values are given in Table 7.4. The units of the heat current H are watts (W), and so the units of K are W m^{-1} deg^{-1}. The enormous variability of the thermal conductivity is apparent from Table 7.4. Silver is the best conductor of heat known, as it is the best conductor of electricity. As the crystalline perfection of the substance decreases, so its thermal conductivity decreases. The value of 72 W m^{-1} deg^{-1} for cast-iron is a reflection of the fact that its crystal structure is very much less perfect than that of silver; and mercury, though a metal, shows the small value to be expected on the basis of its liquid structure.

Table 7.4. Thermal conductivities of some common substances

Substance	K (W m^{-1} deg^{-1})	Substance	K (W m^{-1} deg^{-1})
Brick	0.6	Iron (cast)	72
Concrete	0.1	Silver	418
Copper	365	Wood	0.15
Cork	0.05	Mercury	8.0
Glass (crown)	1.0	Water	0.59
Ice	2.1	Air (at 0°C)	0.024

Consider a long cylindrical bar of length L and uniform cross-sectional area A, the ends of which are maintained steadily at temperatures θ_1 and θ_2, θ_1 being the greater. Then, if there are no losses of heat from the sides of the bar, the temperature drop in uniform, and the graph of θ against x, the distance from the hot end, is a straight line. Under these circumstances the heat current is just

$$H = KA\,\frac{\theta_1 - \theta_2}{L} \tag{7.7}$$

This situation is pictured in Fig. 7.3(a), where the heat current has been indicated by streamlines as if it were a fluid flowing. But if there are losses from the sides of the bar, then the graph of temperature against distance is no longer a straight line, and the differential form of Fourier's law would have to be used if the heat current were to be calculated. In practice the heat losses from the sides can be made very small by suitable lagging of the bar, and the measurement of the heat current along such a bar is the most common method of measuring the thermal conductivities of solids.

Example. The air in a room is at 25°C and the outside air temperature is 0°C. The window of the room has an area of 2 m^2 and a thickness of 2 mm. Calculate the rate of loss of heat by conduction through the window.

Solution. Equation 7.7 can be applied in a straightforward way by use of the values $K = 1.0 \text{ W m}^{-1} \text{ deg}^{-1}$, $A = 2 \text{ m}^2$, $(\theta_1 - \theta_2) = 25$ deg, and $L = 0.002$ m. This gives

$$H = 1.0 \text{ W m}^{-1} \text{ deg}^{-1} \times 2 \text{ m}^2 \times \frac{25 \text{ deg}}{0.002 \text{ m}}$$
$$= 25 \times 10^3 \text{ W} = 25 \text{ kW}.$$

This is a colossal rate of loss of heat, and the cost of keeping such a room warm would be prohibitive. Clearly, there is something wrong with the calculation, and exactly what is wrong will become clear in the next section.

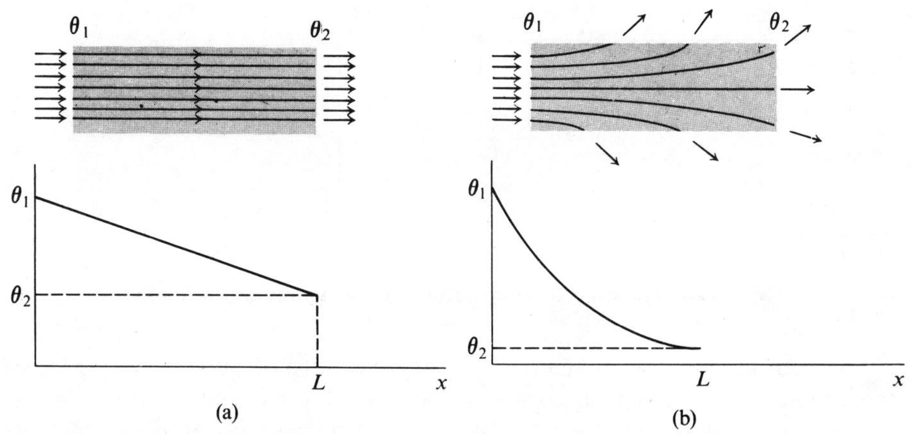

Fig. 7.3. Steady heat flow and temperature distribution for (a) a lagged, and (b) an unlagged bar.

7.7 TRANSFER OF HEAT BY CONVECTION

Transfer of heat by convection in a fluid is said to take place when the fluid itself moves. If we consider the heat loss from a solid immersed in a fluid, there are two distinct cases. In *natural convection* the motion of the fluid, is due entirely to the presence of the hot body. As the temperature of the fluid near the body increases, its density decreases and the fluid moves upwards, to be replaced by other fluid moving downwards. Gravity is the operative factor in natural convection. In *forced convection* the relative motion of the fluid and the hot body is maintained by some external agency—for example, by a fan. Convection is a very complex physical process and is not amenable, except in very special circumstances, to exact mathematical analysis. For situations of practical importance to engineers and builders semiempirical results have been established by a combination of experiment and dimensional analysis. The basic assumption made is that near any hot surface there will be a layer of stagnant fluid and that over this layer there is a nonlinear drop of temperature. The rate of loss of heat by convection per unit area of surface is proportional to $(\theta - \theta_0)^{1.25}$, where θ is the temperature of the surface and θ_0 the temperature of the main body of the fluid well clear of the stagnant layer. For the example of the window discussed in the last section the drop in temperature down the outside stagnant layer, the window, and the inside stagnant layer is as shown in Fig. 7.4. The drop over the window itself is quite negligible compared to the drop over the stagnant layers, and calculation gives a rate of loss of heat of about 80 W, a much more realistic figure. This state of affairs holds, of course, for natural convection, and it is essential that the air near the window is not moved violently by a fan or some such device.

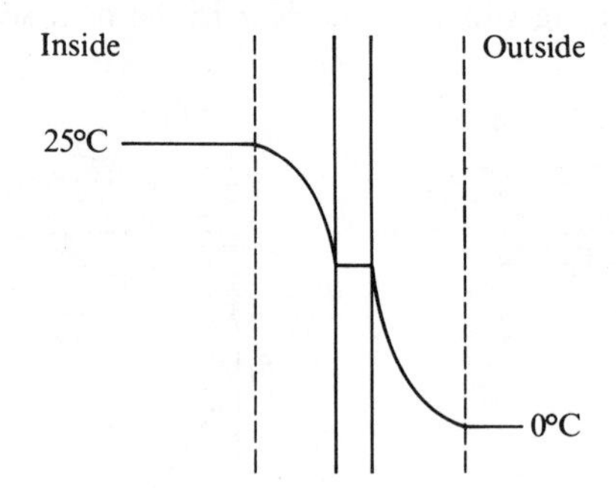

Fig. 7.4. The drop of temperature in the vicinity of a window.

A much simpler state of affairs holds for the cooling of a body in air by forced convection. No build-up of a stagnant layer is possible and Newton's law of cooling holds, namely that the rate of loss of heat is proportional to the difference of temperature between the body and its surroundings.

$$H = hA(\theta - \theta_0). \tag{7.8}$$

So long as the formation of a stagnant layer is prevented, by blowing air vigorously round the object by means of a fan, Newton's law holds over a wide range.

The two laws of convection may be summarized by saying that the rate of loss of heat by convection per unit area of the surface is proportional to $(\theta - \theta_0)^n$, where $n = 1.25$ for natural convection and $n = 1$ for forced convection.

7.8 TRANSFER OF HEAT BY RADIATION

Every body emits energy in the form of electromagnetic radiation, and thermal radiation, or infrared radiation, lies in the wavelength range from about 1 μm to about 100 μm, where 1 $\mu\text{m} = 10^{-6}$ m is often written simply as 1 μ and given the name *micron*. Any two bodies A and B will exchange energy in the form of radiant heat until their temperatures equalize, even when there is no possibility of conduction or convection being involved. Suppose that A emits more radiation than B; then, if equalization of temperature is to be possible, it must also absorb more radiation than B, and the general result holds that good radiators are good absorbers. The most efficient absorber possible is one that absorbs all the electromagnetic radiation that falls on it, and to this is given the name *black body.* A black body may be experimentally realized by making a very small hole in a hollow container. Then all radiation falling on this hole will pass into the container and none will get out, so that the hole is as good a black body as it is possible to get. Strange as it may seem, the human skin, quite irrespective of pigmentation, is an excellent approximation to a black body, although it shows anomalously high reflectivity for certain narrow ranges of wavelength. Since a black body is a perfect absorber, it is also the best possible emitter of radiation, and all other bodies are classified with respect to their ability to emit by comparing them to a black body. The *total emissive power* of a body is defined as the total radiant energy of all wavelengths emitted by the body per square meter of its surface per second. For a black body the total emissive power, written $\mathscr{E}_0$, is proportional to the fourth power of the absolute temperature,

$$\mathscr{E}_0 = \sigma T^4. \tag{7.9}$$

This is the Stefan-Boltzmann law, and the universal constant σ is called Stefan's constant. It has the value

$$\sigma = 5.685 \times 10^{-8}\ \mathrm{W\ m^{-2}\ K^{-4}}.$$

The total emissive power of any other body is a fraction of this

$$\mathscr{E} = e\mathscr{E}_0, \tag{7.10}$$

where e is called the *emissivity* of the body.

The radiation from a body occurs whether or not there is a difference of temperature between the body and its surroundings. If there is no such temperature difference, then the body is absorbing exactly as much radiation as it emits and there is no change in temperature observable. If, however, a body at temperature T is in an enclosure of temperature T_0 less than T, then there is a net outflow of radiation from the body of amount

$$e\sigma T^4 - e\sigma T_0^4 = e\sigma(T^4 - T_0^4) \tag{7.11}$$

per unit area per second.

The strong dependence of the rate of emission of radiant heat on the temperature makes thermal radiation from the skin a most useful diagnostic instrument. The temperature of the human skin varies from place to place, and changes of skin temperature occur when there is any circulatory or cellular abnormality. Using thermoelectric devices pressed against the skin is not very satisfactory, since they change the pattern of circulation in the area of contact and so affect the readings. The first use of thermal radiation as a means of observing local variations in skin temperature was by R. N. Lawson in Montreal, and since then more and more workers have done research on the potentialities of the method. It is interesting to

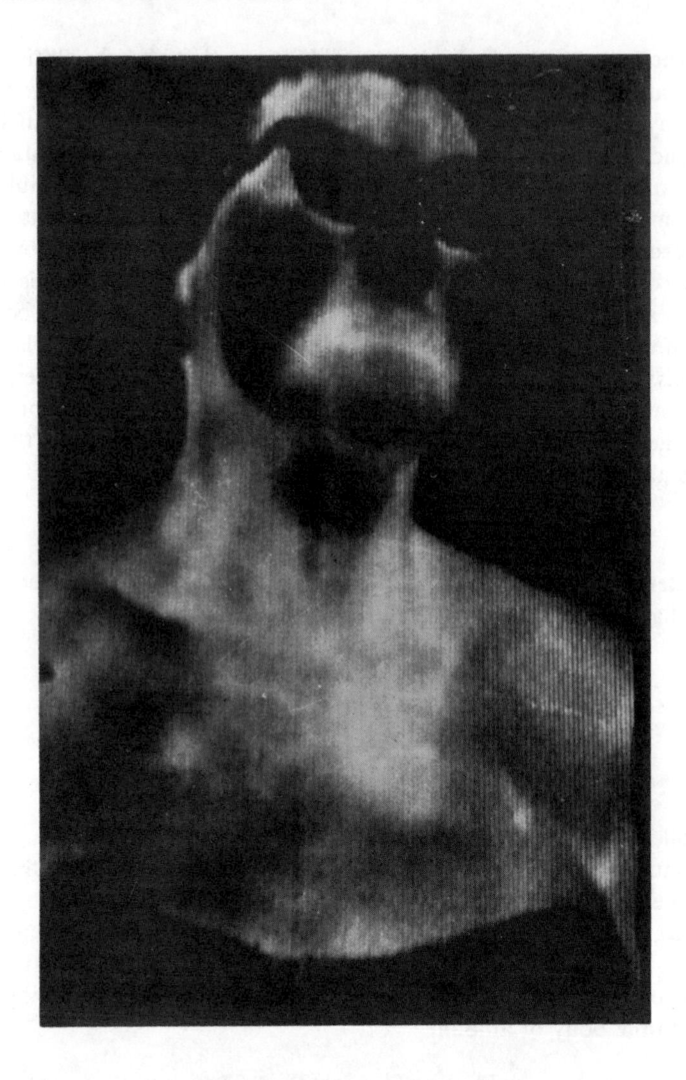

Fig. 7.5. The thermogram of a man suffering from a thyroid tumor. (Reprinted by permission from "Medical Thermography" by J. Gershon-Cohen, *Scientific American*, Feb., 1967.)

note that this development, like many others, is a spill-off from military expenditure, since the infrared detectors used were originally intended for the use of the armed services. Several different kinds of detector are in use. One group of instruments uses temperature-sensitive devices such as thermistors and thermocouples; the other uses radiation-sensitive devices which may be photoconductive, photovoltaic, or photoemissive. The details are not important for our purpose. Study of Eq. (7.9) shows that a temperature change of 0.7 deg will result in a change of 1% in the total emissive power of a body at 300K. Since the temperature changes associated with abnormalities are usually several degrees, it should be clear that the potentialities of detecting these are great. Figure 7.5 shows the *thermogram*, as it is called, of a man suffering from a thyroid tumor. Such a picture is obviously of enormous diagnostic significance, and it can be made with minimal inconvenience to the patient. The use of thermography is bound to increase over the next few years.

Example 1. In the English-speaking world the Fahrenheit scale of temperature is still used. On this scale the NMP of ice is given the value 32°F and the NBP of water the value 212°F. Find the conversion equation between temperatures in degrees Celsius and temperatures in degrees Fahrenheit. At what unique temperature would a Celsius and a Fahrenheit thermometer record the same numerical value?

Solution. Let us denote the readings on the two types of thermometer by the symbols C and F respectively. Then we may write the connecting equation in the form

$$C = aF + b,$$

where a and b are constants to be determined. Using the known values of C and F at the two fixed points we get:

NMP of ice: $$0 = a.32 + b,$$

NBP of water: $$100 = a.212 + b.$$

Subtracting gives

$$100 = a.180,$$

therefore $a = 5/9$ and the value of b is therefore $b = -32(5/9)$.
Hence

$$C = \frac{5}{9}(F - 32)$$

which is the desired conversion equation. If $C = F$, then

$$F = \frac{5}{9}(F - 32),$$

$$\therefore \quad 9F = 5F - 160, \text{ leading to } F = C = -40.$$

So temperatures of −40°C and −40°F are equivalent.

Example 2. An 800-g block of brass is heated in a steam chamber, and is then quickly transferred to a brass container of mass 200 g which contains 320 g of water at 20°C. The temperature of the water rises to 34.5°C as a result of adding the hot brass block. Determine the specific heat of brass from these data.

Solution. Denoting by c the unknown specific heat of brass, the heat given out by the block in cooling from its initial temperature of 100°C to the final temperature of 34.5°C is, from Eq. (7.4),

$$Q_1 = 800 \text{ g} \times c \times (100 - 34.5)\text{deg}.$$

This heat goes to raising the temperature of both the water and its brass container. The heat needed to raise the temperature of the water from 20°C to 34.5°C is

$$Q_2 = 320 \text{ g} \times 4.186 \text{ J g}^{-1} \text{ deg}^{-1} \times 14.5 \text{ deg}$$
$$= 1.94 \times 10^4 \text{ J}.$$

The heat needed to raise the temperature of the brass container through 14.5 deg is

$$Q_3 = 200 \text{ g} \times c \times 14.5 \text{ deg}.$$

Assuming that there are no losses of heat, then

$$Q_1 = Q_2 + Q_3.$$

This is the basic *calorimetric equation*, which may be expressed in words:

The heat given out by the hot bodies in cooling = *the heat absorbed by the cold bodies in warming up.*

Therefore $800 \text{ g} \times c \times 65.5 \text{ deg} = 1.94 \times 10^4 \text{ J} + 200 \text{ g} \times c \times 14.5 \text{ deg}$ or

$$c = \frac{1.94 \times 10^4}{800 \times 65.5 - 200 \times 14.5} \text{ J g}^{-1} \text{ deg}^{-1}$$
$$= 0.39 \text{ J g}^{-1} \text{ deg}^{-1}.$$

In practice, some of the heat will be lost, despite careful attempts at insulation. Calorimetric experiments like this are very difficult to perform with precision.

Example 3. A rod of copper of length 10 cm and cross-sectional area 8 cm² has one end in a steam chamber and the other in a mixture of ice and water. The sides of the bar are well lagged so that heat losses through them may be ignored. How much ice will melt in 15 min?

Solution. The heat current through the bar is

$$H = KA\frac{\theta_1 - \theta_2}{L} = 365 \text{ W m}^{-1} \text{ deg}^{-1} \times 8 \times 10^{-4} \text{ m}^2 \times \frac{100 \text{ deg}}{0.1 \text{ m}}$$
$$= 292 \text{ W}.$$

In 15 min, an amount of heat $292 \text{ W} \times 900 \text{ s} = 2.70 \times 10^5 \text{ J}$ is transferred. Since 335 J are needed to melt one gram of ice, the amount melted is

$$\frac{2.70 \times 10^5 \text{ J}}{335 \text{ J g}^{-1}} = 80.6 \text{ g}.$$

Example 4. If a physician is interested in fluctuations in skin temperature of 0.1 deg about a normal value of 300 K, what must be the sensitivity of his equipment for producing thermograms?

Solution. Since $\mathscr{E} = e\sigma T^4$, the rate of change of $\mathscr{E}$ with T is given by (see Table 3.2)

$$\frac{d\mathscr{E}}{dT} = 4e\sigma T^3 = (e\sigma T^4)\frac{4}{T} = \frac{4\mathscr{E}}{T}.$$

Therefore, for small changes $\Delta\mathscr{E}$ and ΔT we have approximately

$$\frac{\Delta\mathscr{E}}{\Delta T} \approx \frac{4\mathscr{E}}{T}$$

or

$$\frac{\Delta\mathscr{E}}{\mathscr{E}} = 4\frac{\Delta T}{T} = 4\frac{0.1}{300}$$

$$= \frac{0.4}{3}\% = 0.13\%.$$

So his detectors must be able to reliably detect fluctuations of 0.13% in the radiation emitted by the skin.

PROBLEMS

7.1 In central Europe the Reaumur scale is sometimes found, with the ice- and steam-points having the values 0°R and 80°R, respectively. What is the Kelvin temperature corresponding to 30°R?

7.2 A platinum resistance thermometer has a resistance of 10 Ω at 0°C and a resistance of 13.95 Ω at 100°C. Assuming that the resistance changes uniformly with temperature, what temperature corresponds to a resistance of 10.79 Ω?

7.3 The resistance of platinum does not actually increase uniformly with temperature, and at 50°C the resistance thermometer of the previous problem has a resistance of 12.018 Ω. Find the values of the constants a and b in the formula $R = R_0(1 + a\theta + b\theta^2)$ that adequately describes the variation of the resistance of platinum over the range 0°C–660°C, where R_0 is the resistance at the ice-point.

7.4 In a certain process 2000 J of heat are supplied to a system and 400 J of work are done on the system. What is the change in the internal energy of the system?

7.5 In a cyclic process a substance begins and ends in the same state, with no net change in its internal energy. If during such a process 5000 J of heat are absorbed by the substance and 3500 J of heat are rejected, what is the net amount of work done by the substance during the cycle? If the cyclic process repeats itself 10 times per second, what power is being generated?

7.6 A copper can of mass 200 g contains 80 g of water at 20°C. Lead shot of mass 100 g at a temperature of 90°C is rapidly transferred into the water. Assuming no heat losses to the surroundings, calculate the final temperature.

7.7 A copper vessel of mass 50 g contains 100 g of water at 15°C. A piece of metal of mass 80 g at a temperature of 95°C is dropped in the water, and the final temperature is found to be 17.5°C. Calculate the specific heat of the metal.

7.8 A lead bullet moving at a speed of 400 m s^{-1} strikes a target and is brought to rest. If half of the heat produced goes to raising the temperature of the bullet, by how many degrees will its temperature rise?

7.9 An object of specific heat 0.98 J g^{-1} deg^{-1} falls freely under gravity through a height of 100 m and is then brought to rest by colliding with the earth. Assuming that all the heat produced goes to increasing the temperature of the object, calculate what the rise in temperature will be.

7.10 Prove that the difference in temperature between the top and the bottom of a waterfall is given by the expression $(2.34 \times 10^{-3} \text{ deg m}^{-1})H$, where H is the height of the waterfall in meters and it is assumed that all the kinetic energy of the water on impact at the bottom is used up in heating it.

7.11 An aluminum container of mass 400 g contains 600 g of water and 200 g of ice. If 5000 J of heat are supplied to the system, how much ice will melt? How much more heat must be supplied to melt the rest of the ice? If 100,000 J had been supplied initially, what would the final temperature have been?

7.12 A pond of area 100 m^2 is covered with ice at 0°C. If heat is absorbed from the sun at a rate of 0.25 cal cm^{-2} min^{-1}, how much ice will melt in 1 h?

7.13 How much steam must be passed into a mixture of ice and water in order to melt 20 g of ice?

7.14 Near absolute zero the specific heat of some solids obeys the equation $c = AT^3$, where T is the Kelvin temperature and A is a constant that depends on the nature of the solid. Calculate the heat needed to raise the temperature of 1 g of solid from 1K to 10K.

7.15 A hollow glass cube, with walls 1.5 mm thick and edge 10 cm, is filled with ice and placed in a container maintained at 100°C. At what rate will the ice melt when the flow of heat across the walls has become steady?

7.16 A compound wall consists of parallel layers of two different materials, 5 cm of brick and 1 cm of wood. If the difference of temperature across the compound wall is 20 deg, calculate the heat current per square meter of wall.

7.17 A well-lagged cast-iron rod of length 0.5 m conducts heat at a rate of 50 J s^{-1} from a reservoir at 400°C to a reservoir at 100°C. What is the diameter of the rod?

7.18 Repeat Problem 7.17 for a silver rod.

7.19 A 25-W lamp bulb is placed inside a closed metal box, and the temperature of the box rises until it is 45°C, the ambient temperature of the surroundings being 15°C. What will the temperature of the box be if the bulb is replaced by a 50-W bulb (a) when conditions of forced convection hold, and (b) when conditions of natural convection hold? It may be assumed that the entire power output of the bulbs goes toward heating the surface of the box by radiation.

7.20 The filament of a 100-W light bulb is made of tungsten of emissivity 0.3 and has a total length of 0.2 m. What is the diameter of the filament if its operating temperature is 3000K?

7.21 The filament of a light bulb operates at 2500K. It has a diameter of 0.1 mm and is made of metal of emissivity 0.35. What length of filament is needed if the power of the bulb is to be 40 W?

7.22 From measurements of the solar radiation received on earth it may be computed that the sun radiates energy at a rate of 62.5 MW m^{-2} of its surface. Assume that the sun radiates as a perfect black body and calculate the surface temperature.

CHAPTER 8

THE KINETIC THEORY OF GASES

8.1 INTRODUCTION

In this chapter the simple rules of elementary mechanics, coupled with a few simple statistical ideas, will be applied to the problem of obtaining a molecular explanation of the known gas laws. These laws are (a) Boyle's law, which states that for a fixed mass of gas at constant temperature the product of the pressure and the volume is a constant; and (b) Charles' law, which states that for a fixed mass of gas heated at constant pressure the volume increases by a constant fraction of the volume at 0°C for each degree rise in temperature. Boyle's law is obeyed by most gases if the pressure is sufficiently low, and is an acceptable basis for rough calculations for all gases. For many purposes it is convenient to imagine an *ideal gas* which would obey Boyle's law under all circumstances. In symbols,

$$pV = \text{constant}, \tag{8.1}$$

where p is the pressure of the gas, V is the volume, and the constant depends on the total mass of gas present. Since the total mass of gas is the product of its density and its volume, ρV, Boyle's law may be rewritten as

$$p/\rho = \text{constant}, \tag{8.2}$$

where now the constant is independent of the mass of gas. Charles' law is concerned with the increase in volume of a gas at constant pressure, and may be written

$$\frac{V - V_0}{V_0 \theta} = \text{constant} = \beta, \tag{8.3}$$

where V and V_0 are the volumes of the gas at θ and at 0°C, respectively, and β, called the volume expansivity of the gas, has the value 1/273 for most gases. If Boyle's law and Charles' law are both obeyed by a gas, then the pressure of a constant volume of a gas will increase by 1/273 of its pressure at 0°C for each degree rise in temperature. All of these results are jointly referred to as the gas laws, and an ideal gas is one that obeys them all. The student may be tempted to think that the definition of the absolute scale of temperature given in Section 7.2 depends on the existence of such a thing as the ideal gas. This is not so.

The operational definition of the absolute temperature scale is made in terms of the real gases hydrogen and helium, and does not depend at all on the gas laws.

The gas laws, which are approximately obeyed by real gases under certain circumstances, may be summarized in the single equation

$$\frac{pV}{T} = \text{constant}, \tag{8.4}$$

where T is the temperature in kelvins.

8.2 CALCULATION OF THE PRESSURE OF AN IDEAL GAS

It is first of all necessary to decide on the molecular model to be used in deriving the results of kinetic theory. Initially we shall assume the simplest possible model, and this may be progressively refined if the need should arise. Let us assume, then, that a gas consists of a very large number of vanishingly small particles that exchange no energy with the walls of the containing vessel when they collide with them. Collisions among the particles themselves are not assumed to occur, since they are essentially regarded as points. Consider the motion of one particle in a cubical container of side L. Because the collisions with the walls are perfectly elastic, the particle rebounds from the wall at the same angle at which it approached the wall, as shown in Fig. 8.1, and the speed of the particle is unchanged by the collision. Let the speed of the particle be denoted by c and the components of its velocity by u, v, and w along the axes of x, y, and z, respectively. The diagram shows a particle traveling in a plane perpendicular to the z-axis for convenience of representation. At the collision with the right-hand wall shown, the component u of the velocity is reversed in direction, and consequently the particle suffers a change of momentum in the x-direction of $2mu$, where m is the mass of the particle. The time taken for the particle to return to the right-hand wall after its collision is

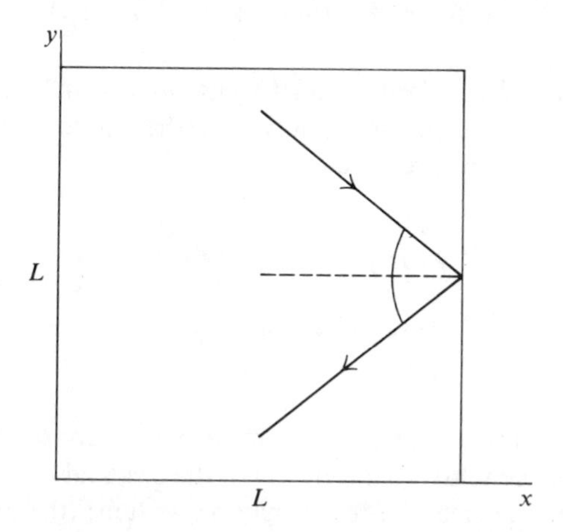

Fig. 8.1. The collision of an ideal gas particle with a wall of the containing vessel.

just $2L/u$, and it will continue to strike this wall at intervals of $2L/u$, quite independently of its motions in the y- and z-directions. These motions will only affect the place on the right-hand wall where the successive collisions occur. Hence, the rate of change of momentum of the particle due to collisions with the right-hand wall is $(2mu)(u/2L) = mu^2/L$. By Newton's second law this must be the force exerted on the particle by the right-hand wall, and by his third law it must also be the force exerted on the right-hand wall by the particle. The analysis for the left-hand wall is precisely similar, and the force exerted by the particle on both these walls is $2mu^2/L$. Note that this is the total force exerted by the particle on the *pair* of walls perpendicular to the x-axis. In exactly the same way, the force on the pair of walls perpendicular to the y-axis is $2mv^2/L$ and that on the pair of walls perpendicular to the z-axis is $2mw^2/L$. Hence, the total force exerted by this one particle on all 6 walls of the cube is

$$F = \frac{2m}{L}(u^2 + v^2 + w^2).$$

In Section 4.5 we saw that, in two dimensions, the square of the magnitude of a vector was equal to the sum of the squares of its two components. In three dimensions a similar result holds for the sum of the squares of the three components, and we have $u^2 + v^2 + w^2 = c^2$. Hence

$$F = \frac{2m}{L}c^2. \tag{8.5}$$

Let us now assume that we have a very large number of particles, each moving quite independently of all the others, and of masses $m_1, m_2, \ldots$ and speeds $c_1, c_2, \ldots$ Then, each particle will exert on the 6 walls a force given by Eq. (8.5), and consequently the total force exerted by all the particles will be

$$\text{total force} = \frac{2}{L}(m_1c_1^2 + m_2c_2^2 + \ldots). \tag{8.6}$$

If the number of particles is large enough, this force will be experienced by the walls as a uniform pressure spread equally over the whole area $6L^2$ of the walls, and, hence, the value of this pressure is

$$\begin{aligned} p &= \frac{1}{6L^2}\cdot\frac{2}{L}(m_1c_1^2 + m_2c_2^2 + \ldots) \\ &= \frac{2}{3}\,\frac{(\frac{1}{2}m_1c_1^2 + \frac{1}{2}m_2c_2^2 + \ldots)}{L^3}. \end{aligned} \tag{8.7}$$

The expression in parentheses in the numerator is the sum of the kinetic energies of all the particles, or the total kinetic energy of the gas, and L^3 is just the volume of the cubical container. Hence, the result may be written: the pressure is equal to two-thirds of the total kinetic energy per unit volume.

Since kinetic energy is a scalar quantity, the total pressure exerted by a mixture of different gases is just the sum of the pressures each would exert if it alone occupied the container. This is known as Dalton's law of partial pressures. It is implicit in our derivation, since we have not assumed that the particles all have the same mass.

It is convenient to write

$$m_1c_1^2 + m_2c_2^2 + m_3c_3^2 + \ldots = (m_1 + m_2 + \ldots)C^2. \tag{8.8}$$

This equation serves to define the mass-weighted mean-square speed of the particles, C^2. If the particles all have the same mass, C^2 is just the mean of the squares of the various speeds. The parameter C is called the root-mean-square speed, or r.m.s. speed, and is the most suitable mean to use in kinetic theory. The sum $m_1 + m_2 + m_3 + \ldots$ is the total mass of gas present and is therefore given by the product ρV. Substituting the definition of the mean-square speed into Eq. (8.7) gives

$$p = \frac{2}{3}\frac{\frac{1}{2}\rho VC^2}{V} = \frac{1}{3}\rho C^2. \tag{8.9}$$

This equation is convenient for calculating the root-mean-square speed for any gas at a definite pressure. Thus air at room temperature and 1 atm pressure has a density of 1.23 kg m^{-3}. Hence,

$$C^2 = \frac{3 \times 10^5 \text{ N m}^{-2}}{1.23 \text{ kg m}^{-3}} = 25 \times 10^4 \text{ m}^2 \text{ s}^{-2},$$

$$C = 500 \text{ m s}^{-1}.$$

This is about the speed of a rifle bullet.

8.3 CORRELATION WITH EXPERIMENT

We have now established certain results on the basis of our molecular model, and we must at this stage make a correlation with the known experimental results for gases. In particular, we have shown that the product of the pressure and the volume is proportional to the total kinetic energy of the molecules. Since Boyle's law states that this product is constant if the temperature is kept constant, agreement with experiment can be obtained if we simply correlate the total kinetic energy with temperature in some simple way. Now, temperature is not an extensive property of a gas. It does not depend, that is to say, on the amount of gas present. So we correlate the *average* kinetic energy of the molecules with the temperature through the equation

$$\text{average kinetic energy} = \frac{3}{2}kT, \tag{8.10}$$

where k is a new universal constant called Boltzmann's constant. Its value is

$$k = 1.3804 \times 10^{-23} \text{ J K}^{-1}.$$

If N is the total number of particles in the gas, the total kinetic energy is $\frac{3}{2}NkT$, and consequently

$$pV = NkT. \tag{8.11}$$

This equation contains all the gas laws, and we see that our simple model is sufficient to achieve agreement with the behaviour of an ideal gas. From Eq. (8.11) the number of particles in a gas is seen to depend only on the values of p, V, and T. This is known as Avogadro's law.

If one mole of the gas is considered (see Section 7.4), then the number of particles is $N_0 = 6.0249 \times 10^{23}$, Avogadro's constant, and the equation becomes

$$pV = N_0kT = RT, \tag{8.12}$$

where R is the universal gas constant and has the value

$$R = 8.3170 \text{ J mol}^{-1} \text{ K}^{-1}$$

If there are n moles of gas present, the gas-law equation may be written

$$pV = nRT. \tag{8.13}$$

The enormous size of N_0 is impossible to imagine, but if you reckon that we breathe out about 400 cm^3 of air at each breath, then each breath contains about 10^{22} molecules. It has been estimated that the whole atmosphere of the earth contains about 10^{44} molecules, so one molecule bears to one breath the same relation as one breath to the whole atmosphere. If we assume that the last breath of George Washington, say, has by now become completely scattered throughout the atmosphere, then the chances are that we each inhale one molecule of this with each breath we take. Since our lungs hold about 2000 cm^3, we probably each have about five molecules of George Washington's dying breath already in our lungs.*

Our derivation of the pressure in the gas was made on the assumption that the particles of the gas do not collide among themselves. This is clearly a false assumption, since it is only through the mechanism of collisions that equilibrium can be established at all. However, it does not affect the result if we take intermolecular collisions into account so long as we regard them as elastic, for they then cannot affect the kinetic energy of the gas, nor can a collision in the interior produce of itself any pressure on the boundary. The assumption that there is no change of kinetic energy on collision with the walls has to be modified when heat exchange between the container and the gas is considered, but such heat exchanges take place so slowly that very little energy is transferred at a single collision.

8.4 INTERNAL ENERGY AND THE SPECIFIC HEATS OF A GAS

For our very simple model of a gas the internal energy is simply the total kinetic energy, since there are no interactions at all between the molecules. Hence, by

---

* This striking calculation is due to Sir James Jeans.

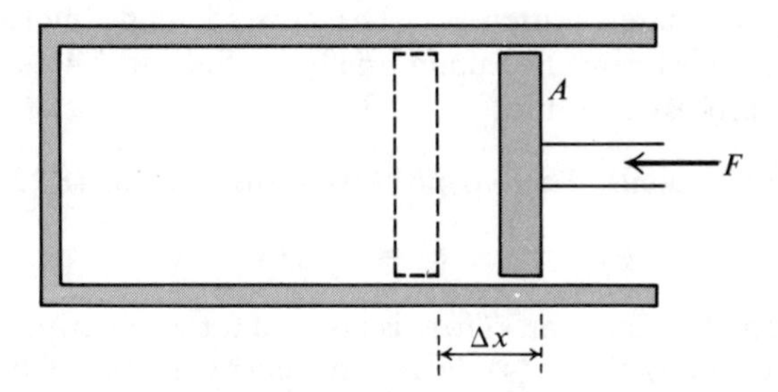

Fig. 8.2. Energy being given to a gas by compression.

the results of the previous section we have for the internal energy the expression

$$E = \frac{3}{2}NkT = \frac{3}{2}nRT. \tag{8.14}$$

Let us now consider the rise in temperature that occurs when a container of gas is heated. To make the situation as general as possible, let the container be fitted with a movable piston that can slide without friction, as shown in Fig. 8.2. Energy can be communicated to the gas in two distinct ways: mechanical work can be done on the gas by moving the piston in, or the gas can be heated. The equation governing both of these processes is the first law of thermodynamics,

$$Q + \mathscr{W} = E_2 - E_1,$$

as discussed in Section 7.3 If the pressure outside the piston is taken to be zero, the work done on the gas when the piston moves in a small distance Δx is $F \cdot \Delta x = pA\Delta x$, where A is the cross-sectional area of the piston. The product $A\Delta x$ is the magnitude of the change in volume, ΔV, of the gas. Since the change in volume is negative, corresponding to a compression,the work done on the gas is $\mathscr{W} = -p\Delta V$. This is a positive quantity. The work done on the gas would be negative if the change in volume were positive,corresponding to an expansion. The first law may be recast in the form

$$Q = E_2 - E_1 + p\Delta V, \tag{8.15}$$

where the work term has been transferred to the right-hand side. Many textbooks, in fact, introduce the first law in this form.

The amount of heat required to raise the temperature of the gas will depend on the amount of work done. There are an infinite number of ways of intermingling the heat term and the work term in the first law to produce the same change in internal energy, and so there are an infinite number of specific heats. Two of these of special simplicity are chosen. The first, where no work is done at all and so the volume of the gas does not change, is the specific heat at constant volume and is given the symbol c_v; the second, where the pressure of the gas is kept constant by allowing it to expand during the transfer of heat, is the specific heat at constant pressure, written c_p. The corresponding molar heats are C_v and C_p. This matter was touched upon briefly in Section 7.4 Let us apply the first law to these processes in turn.

Heating at constant volume. There is no work term and Eq. (8.15) is simply

$$Q_v = E_2 - E_1 = nC_v(T_2 - T_1) \tag{8.16}$$

from the definition of molar heat. There is no need for a subscript v on the internal energy terms, because by Eq. (8.14) the internal energy in our model depends only on the temperature.

Heating at constant pressure. We assume the same rise of temperature as in the constant volume case, and so $Q_p = nC_p(T_2 - T_1)$ by the definition of the molar heat at constant pressure. In this case, since the pressure would rise if the volume did not increase by a compensating amount, the piston must be allowed to move outwards. The work term $p\Delta V$ is now simply $p(V_2 - V_1)$ since the pressure is constant. So the first law gives

$$Q_p = E_2 - E_1 + p(V_2 - V_1) = nC_p(T_2 - T_1). \tag{8.17}$$

The change of internal energy is the same as it was in the constant volume process, since the initial and final temperatures are the same in the two cases. Substituting the value of $E_2 - E_1$ from Eq. (8.16) into Eq. (8.17) gives

$$nC_v(T_2 - T_1) + p(V_2 - V_1) = nC_p(T_2 - T_1). \tag{8.18}$$

But

$$p(V_2 - V_1) = nR(T_2 - T_1)$$

by Eq. (8.13), so Eq. (8.18) reduces to

$$nC_v(T_2 - T_1) + nR(T_2 - T_1) = nC_p(T_2 - T_1),$$

or

$$C_p - C_v = R. \tag{8.19}$$

Now return to Eqs. (8.16) and (8.14). Since $E = 3nRT/2$, we get

$$E_2 - E_1 = \frac{3}{2}nR(T_2 - T_1) = nC_v(T_2 - T_1),$$

and consequently

$$C_v = \frac{3}{2}R. \tag{8.20}$$

From Eqs. (8.19) and (8.20), the ratio C_p/C_v is seen to be 5/3. This ratio is usually denoted by the symbol γ.

All these relations involving the molar heats of a gas can be checked by experiment. It is found that the noble gases such as argon and helium fit the simple theory very well indeed at all temperatures but that other gases give very different results for the molar heats at high temperatures. Thus for oxygen at room temperature, $C_v = 2.54R$ instead of the theoretical value $1.5R$, and $\gamma = 1.39$ rather than

1.67 as the above theory predicts. In other words, our simple model has broken down. To explain the high values of the molar heats of real gases, it is necessary to introduce structure into the simple particles. For our molecules the internal energy per particle was $3kT/2$, corresponding to the fact that only kinetic energy of translation is involved, namely $\frac{1}{2}m(u^2 + v^2 + w^2)$. The fact that there are three squared terms is described by saying that our molecules have three *degrees of freedom* and there is an energy of $\frac{1}{2}kT$ associated with each, making up the total of $3kT/2$. If we consider a diatomic molecule such as oxygen, it has a dumb-bell configuration as shown in Fig. 8.3, and can have other motions besides simple translation. Thus it can rotate about any one of three perpendicular axes through the center of the molecule, one axis being aligned with, and the other two being perpendicular to, the line of centers OO'. The moment of inertia about this line of centers is so small that this particular rotation is ineffective, so we are left with two possible rotations, or two extra degrees of freedom. Assigning an amount of energy $\frac{1}{2}kT$ to each increases the total internal energy per molecule to $5kT/2$, which immediately leads to the value $2.5R$ for the molar heat at constant volume, in good agreement with the experimental result. But a diatomic molecule such as oxygen can also vibrate to and fro along the line OO', the two atoms alternately approaching each other and receding from each other. For such a motion the energy is in two parts, one kinetic and one potential, as discussed in Section 10.3, and again an energy $\frac{1}{2}kT$ is assigned to each, bringing the total energy for a molecule that is translating, rotating, and vibrating up to $7kT/2$. This gives a value of $7R/2$ for the molar heat at constant volume, and this has been verified at high temperatures.

The principle of assigning an amount $\frac{1}{2}kT$ of energy to every degree of freedom of a system is known as the classical theorem of equipartition of energy. It at once provides an explanation of the Dulong-Petit law for the specific heats of solids (see Section 7.4), for the atoms in a crystal can only vibrate, although their vibrations can be along three directions in space. Hence, there are 6 degrees of freedom for each atom, and we should expect the specific heat at constant volume to be given by $3R$, which is about 25 J mol^{-1} deg^{-1}, the Dulong-Petit value.

The really interesting feature of the results for a gas lies in the fact that at low temperatures, at which the average energy per molecule is low, the energy is entirely translational. It is not shared between translation, rotation, and vibration. There is a definite threshold of energy below which rotation and vibration cannot be

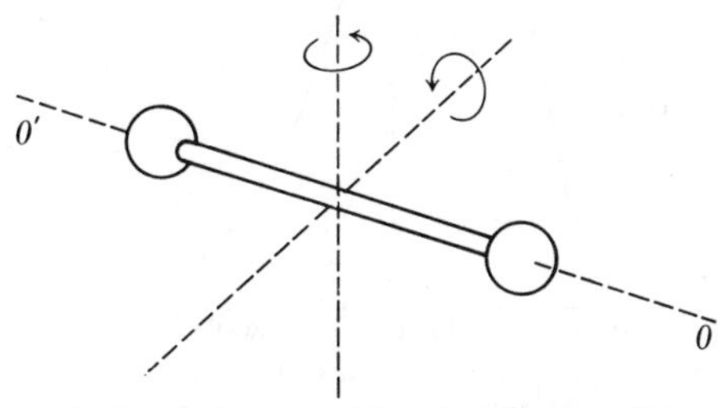

Fig. 8.3. A diatomic molecule, showing possible rotations.

activated. When this threshold has been passed, both translation and rotation occur, but there is another higher threshold which has to be passed before vibration can occur. It is rather like slowly increasing the income of a poor family. When the general income is low, it is spent on the necessities of life only. Before an automobile can be bought, a definite threshold of income must be passed, and then there is another threshold before a house is bought, and so on. Classical physics could not explain these thresholds of energy, and a new theory, the quantum theory, had to be introduced.

8.5 MAXWELL'S LAW

So far in our discussion of the kinetic theory of a gas we have been concerned only with the average values of kinetic energy and the root-mean-square value of the molecular speed. However, it is obvious that not all the molecules can be traveling with the same speed, since collisions among them would soon destroy that equality. The problem of determining how the speeds of the molecules are distributed is much too difficult for us to discuss here, but it is worth looking at the result in graphical form. Figure 8.4 shows what is known as the Maxwell distribution curve for the molecular speeds. The form of this curve is given by the equation

$$y = f(c) = Ac^2 \exp(-Bc^2), \tag{8.21}$$

where A and B depend on the temperature and on the nature of the gas. Figure 8.4 has been drawn for nitrogen at 0°C. The meaning of the curve is as follows: the probability that a molecule of the gas will have a speed that lies in the interval c to $c + \Delta c$ is given by $f(c)\Delta c$. Since this is the area of a very narrow vertical strip at c, it follows that the whole area under the curve is the probability that a molecule of the gas has some speed, and this is of course a certainty. Hence, the area under the curve is unity. It will be seen that the Maxwell distribution curve is not symmetrical, and as a consequence of this we have a choice in what we take as the average speed. The peak of the curve corresponds to what is known as the

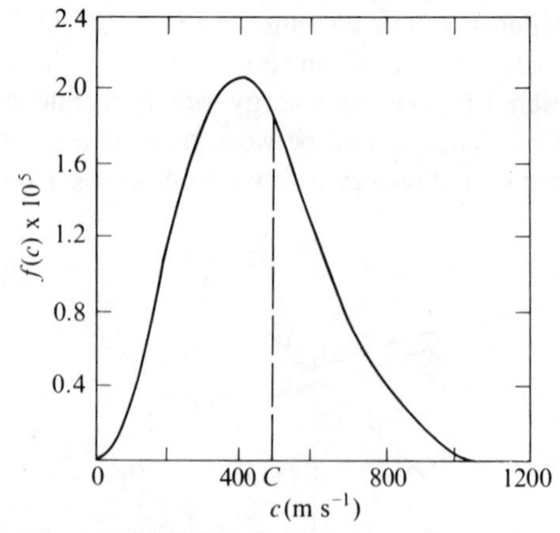

Fig. 8.4. The Maxwell distribution curve for molecular speeds.

most probable speed, and the root-mean-square speed which we have used is somewhat higher than this. The root-mean-square speed, C, is shown by the dotted line and is in fact $22\frac{1}{2}\%$ higher than the most probable speed.

As the temperature rises, the hump of the Maxwell curve moves to the right, corresponding to a higher root-mean-square speed. The general form of the curve has been amply verified by experiment.

8.6 DISORDER AND THE SECOND LAW OF THERMODYNAMICS

The first law of thermodynamics was introduced in the last chapter and the third law was mentioned. The second law of thermodynamics cannot be introduced in an elementary way without a rather detailed discussion of heat engines, which are of little relevance to biomedical scientists. But the meaning and content of the second law can be discussed in terms of our model of a gas. The distinguishing feature of the molecular motions in the kinetic theory of gases is their randomness. Any molecule is traveling in an entirely haphazard fashion, making collisions with other molecules and with the walls of the container, and its velocity is constantly changing discontinuously as a result of all these collisions. The molecules are evenly distributed throughout the container and their speeds are distributed according to the Maxwell distribution. This is what we mean by an equilibrium state of a gas. It is a statistical equilibrium.

The elementary ideas needed to appreciate the content of the second law are most easily grasped from a consideration of two simple games: dice and bridge. Consider throwing two dice and turning up a total of 9. This total can be arrived at in four different ways, as shown below. Each of these four arrangements is called

First die	Second die
6	3
3	6
5	4
4	5

a *microscopic complexion* and the total of 9 is called the *macroscopic complexion*. The number of microscopic complexions that correspond to any macroscopic complexion is given the symbol W. Here we have $W_9 = 4$. If we throw a 12 with the two dice, then clearly this can happen in only one way, with a 6 on each die, so we conclude that $W_{12} = 1$. We would say that throwing a 9 is four times as likely as throwing a 12, but the important point is that any specified microscopic complexion is just as likely as any other. Thus a 6 on the first die and a 3 on the second is no more probable than a 6 on each. The greater probability of throwing a 9 arises from the fact that the macroscopic complexion 9 has four microscopic complexions corresponding to it, whereas 12 has only one.

Consider now the deal of a hand at bridge or whist. Each player receives 13 cards, and he puts a value on his hand depending on what particular cards he receives. Suppose that a player is dealt all 13 spades; he properly regards this as an exceptional circumstance and puts an enormous value on his hand. More usually he will find all four suits, and only a sprinkling of court cards, as shown below.

	Exceptional hand	Typical hand		
S.	A K Q J 10 9 8 7 6 5 4 3 2	K 9 4		K × ×
H.	—	J 7 5 3	or	J × × ×
D.	—	A 8 6		A × ×
C.	—	Q 3 2		Q × ×

The typical hand is more often written with crosses indicating the cards lower than 10, since the actual values of these low cards is rarely important. Here the actual 13 cards dealt is the microscopic complexion, and the corresponding macroscopic complexion is the value of the hand. A little thought will soon convince the reader that the exceptional hand shown is no less probable than the typical hand *so long as the precise values of the 13 cards in the typical hand are named.* However, if the 8 of spades is substituted for the 9 of spades in the typical hand, it still remains a typical hand. The microscopic complexion has been changed without altering the macroscopic complexion. Obviously there are an enormous number of typical hands of the type shown, and the bridge player does not distinguish between them. To him they are all the same value, or macroscopic complexion. The fact that a typical hand is more probable than an exceptional hand depends on the fact that there are many more microscopic complexions underlying it.

We may now apply these ideas to our model of a gas. Suppose that at a given instant of time all the molecules of the gas occupy only one corner of the container; we should regard this as a very exceptional circumstance indeed, a very improbable macroscopic complexion for the gas. The normal state of the gas, with the particles equally distributed throughout the volume, we regard as a very much more probable macroscopic complexion. But, exactly as with the dice and the cards, all possible microscopic complexions are equally probable, and the overwhelming probability of the equilibrium configuration arises because to it there corresponds an overwhelmingly large number of microscopic complexions. The number of molecules in the container is so enormous that the chance of any deviation from equilibrium is vanishingly small. But it is not impossible; only very improbable. In practice the word "impossible" is not really inappropriate. To specify any microscopic complexion would involve knowing the positions and velocities of all the molecules at any instant, and all such complexions are equally probable. The second law of thermodynamics is concerned with the statistical fact that, starting from any such microscopic complexion, the molecules will very quickly (almost instantaneously, in fact) redistribute themselves so that they are in a set of microscopic complexions all of which we indiscriminately recognize as the macroscopic complexion of equilibrium. It is the enormous size of the numbers involved that makes the statistical probability of equilibrium a practical certainty. The equilibrium state of a gas is a state of maximum disorder of the molecules. On the molecular level, all the processes are entirely reversible, and in any collision between two particles of the gas the relative velocity after the collision is uniquely determined by the relative velocity before the collision. This can result in the relative speed increasing or decreasing. But there is an overwhelming statistical tendency for collisions to equalize the speeds, and this leads to the complete randomness we call equilibrium. The second law may be stated as follows.

An isolated system, free of external influence, will always pass from states of relative order to states of relative disorder until eventually it reaches the state of maximum disorder.

This law is of crucial importance, because it indicates a unique time direction. Newton's laws of mechanics do not depend on the direction of time, and any process that is dynamically possible could equally well be run backwards in time. But when there are large numbers of molecules present, there is an overwhelming probability that the molecular collisions will tend to equalize the energies of the particles, and statistically the tendency is always toward maximum disorder. If we witness any natural process in which randomness decreases and there is no obvious outside agency causing this, then we are witnessing time running backwards. Since time never runs backwards, we are the subject of a trick.

An excellent example is afforded by the constantly recurring scene in the old movies of the 1920's. The heroine is bound to the railroad track in the path of an oncoming train. Always the engineer is warned in time, and manages to pull up with the cowcatcher actually touching the damsel's dress. The well-trained physicist spots the trick at once, for careful scrutiny shows the smoke going backwards into the smokestack. Here is a marked decrease in randomness, and he at once realizes that the scene is being run backwards in time. The train starts from the heroine and reverses down the line with the smoke pouring naturally from its stack. When the scene is reversed in time, everything looks quite natural except the smoke.

The physicist has a special measure of randomness called *entropy*, denoted by the symbol S. This is defined by the equation

$$S = k \ln W, \tag{8.22}$$

where W is the number of microscopic complexions of the system in question. Clearly, the more random the system the greater the entropy, and we may restate the second law as follows.

The entropy of an isolated system never decreases.

It is important that the system be isolated, because if heat energy can enter from outside, then the entropy can decrease. Living creatures are examples of systems of high organization which maintain their lack of randomness by making energy contacts with their environment; that is they breathe and eat. When such contact is broken completely at death, the entropy begins to increase and the high degree of organization associated with life immediately begins to break down.

8.7 THE PROPERTIES OF REAL GASES

Gases at ordinary and high pressures depart seriously from the simple gas laws given at the beginning of this chapter. According to Boyle's law, a graph of pressure against volume should be a rectangular hyperbola, there being one curve

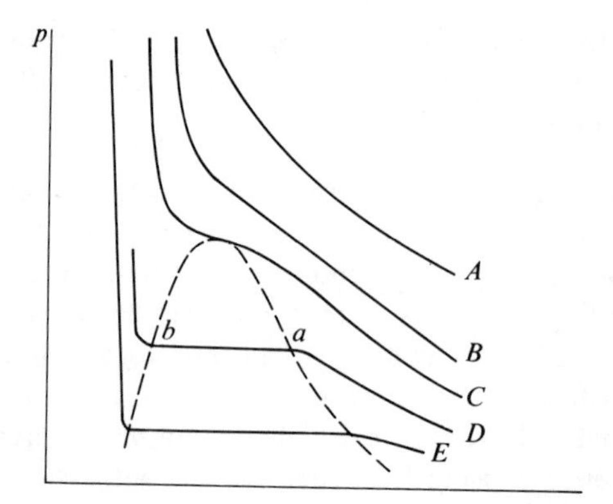

Fig. 8.5. The isotherms of a real gas.

for each value of the temperature. Such curves are called *isotherms*. The isotherms of a real gas are as shown in Fig. 8.5. The upper curve corresponds to ideal-gas behavior according to Boyle's law. The substance remains a gas at all pressures. The second isotherm, *B*, shows marked departures from the ideal hyperbolic shape, but again the substance remains a gas at all pressures. For curve *D* the situation is very different. If we start at very low pressures and gradually increase the pressure at constant temperature, it will be seen that the volume decreases until the point *a* is reached, at which point the volume decreases without any further increase in pressure, some of the gas liquefying, until at point *b* the liquefaction is complete. Thereafter the isotherm rises almost vertically, since liquids are highly incompressible. Isotherm *C* is called the *critical isotherm*, because it marks the boundary between isotherms where liquefaction has occurred, such as *D* and *E*, and isotherms where no liquefaction has occurred, such as *A* and *B*. At temperatures higher than this critical temperature no liquefaction of a gas is possible. The dashed curve in Fig. 8.5 encloses the region where the gas may exist in equilibrium with its liquid. It is conventional to refer to a gas at a temperature below its critical temperature as a *vapor*, and to the pressure corresponding to such a line as *ab* as the *saturated vapor pressure* at that temperature.

It is possible to derive equations that approximately reproduce the shape of the p-V curves for real gases, but this is scarcely worth the effort, since the approximations involved are always more or less dubious, and the actual behaviour of any gas is better described by its experimentally determined isotherms. It is very important to know where the critical isotherm is; because if it is desired to liquefy a gas, then the gas must first be cooled below its critical temperature.

The saturated vapor pressure is the maximum pressure that can be exerted by the vapor at that temperature and is dependent only on the temperature. Water vapor plays a very important role in biology, and its presence in the atmosphere is essential to life as we know it. The partial pressure of the water vapor in air at any temperature is usually less than the saturated vapor pressure for the same temperature, and the ratio of the two, expressed as a percentage, is known as the relative humidity.

$$\text{Relative humidity } (\%) = 100 \times \frac{\text{partial pressure of vapor}}{\text{saturated vapor pressure at same temperature}}$$

It is important to remember that relative humidity values are meaningless by themselves and that some other information, such as air temperature, or one of the actual vapor pressures, is needed. Saturation can be achieved in two ways: either the total water content can be increased until the pressure of the vapor is the saturated vapor pressure, or the temperature can be lowered until the actual amount

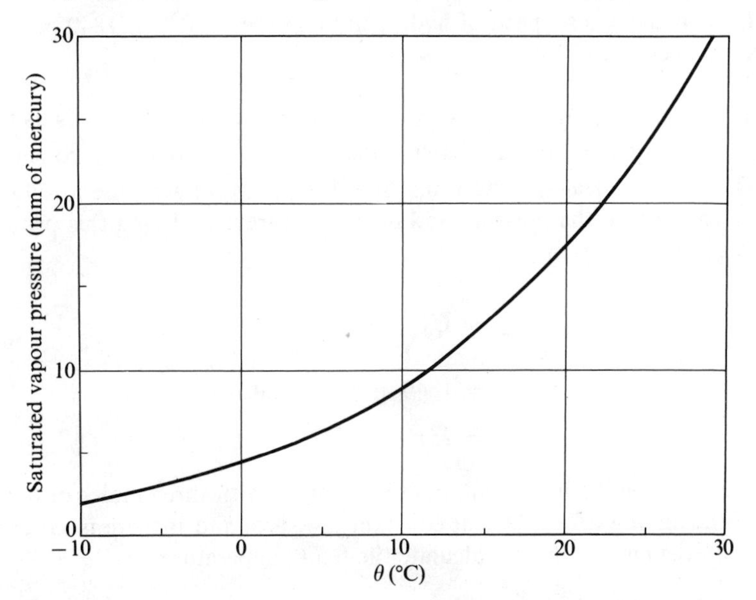

Fig. 8.6. The variation of the saturated vapor pressure of water with temperature near ordinary room temperature.

of water vapor in the air is enough to cause saturation at the new temperature. It is this process that causes the formation of clouds, fog, and rain. The temperature at which moist air would be saturated is called the *dew-point*, and constitutes an absolute, important, and easily measured parameter. All that is necessary is to cool a brightly polished metal surface and observe the temperature at which it becomes clouded with condensed moisture. The relative humidity is just the saturated vapor pressure at the dew-point divided by the saturated vapor pressure at the actual air temperature, times 100. Figure 8.6 shows the variation of the saturated vapor pressure of water with temperature.

Example 1. A fixed mass of gas occupies a container of volume 4 liters. Its temperature is 27°C, and its pressure is 1.5 atm. If it is heated to a new temperature of 87°C and allowed to expand to a new volume of 6 liters, what is the new pressure of the gas?

Solution. Indicating the initial and final states of the gas by subscripts 1 and 2, the gas laws give

$$p_1V_1/T_1 = p_2V_2/T_2,$$

therefore

$$p_2 = p_1(V_1/V_2)(T_2/T_1) = 1.5 \text{ atm} \times (4/6)(360/300),$$

since $T = \theta + 273$. Therefore

$$p_2 = 1.2 \text{ atm}.$$

Example 2. If the r.m.s. speed of hydrogen molecules at 0°C is 1840 m s^{-1}, what is the r.m.s. speed at 100°C?

Solution. Equation (8.9) gives $C = \sqrt{3p/\rho}$, and Eq. (8.2) further tells us that the ratio p/ρ is proportional to the absolute temperature T. Hence we conclude that $C \propto \sqrt{T}$. For an ideal gas the r.m.s. speed depends only on the absolute temperature, and not on the pressure and density separately. Using this proportionality, we get for hydrogen

$$\begin{aligned} C_{100} &= C_0\sqrt{\frac{373\text{K}}{273\text{K}}} \\ &= 1840 \text{ m s}^{-1} \times 1.367 \\ &= 2510 \text{ m s}^{-1}. \end{aligned}$$

Example 3. One mole of helium at 300K is mixed with three moles of oxygen at 400K. If the mixing takes place at constant pressure, and if there is no exchange of heat with the environment, calculate the final temperature.

Solution. The basic calorimetric equation holds, so that the heat given out by the oxygen equals the heat absorbed by the helium. Since the molar heats of monatomic and diatomic gases are $5R/2$ and $7R/2$ respectively, we have

$$1\ \text{mol} \times \frac{5}{2} R \times (T - 300\text{K}) = 3\ \text{mol} \times \frac{7}{2} R \times (400\text{K} - T).$$

Canceling $\frac{1}{2}R$ from both sides of this equation, we get

$$5(T - 300\text{K}) = 21(400\text{K} - T),$$

or

$$26T = 9900\text{K}.$$

Therefore

$$T = 381\text{K}.$$

Example 4. A room is at a temperature of 20°C and a relative humidity of 60%. What is the dew point? If the temperature drops to 16°C, what is the new value of the relative humidity?

Solution. From Fig. 8.6, the saturated vapor pressure at 20°C is 17.6 mm of mercury, and the partial pressure of the vapor is 60% of this. Hence the partial pressure of the vapor is $0.6 \times 17.6 = 10.6$ mm of mercury. The dew point is that temperature for which 10.6 mm of mercury is the saturated vapor pressure, and Fig. 8.6 shows this temperature to be 12°C.

For a lower temperature of 16°C we may read off a value of 13.6 mm of mercury for the saturated vapor pressure, and therefore the new relative humidity is

$$100 \times \frac{10.6}{13.6} = 78\%.$$

PROBLEMS

In the following problems the approximation 1 standard atmosphere (1 atm) $= 10^5\ \text{N m}^{-2}$ may be used. This pressure is also sometimes expressed as 760 mm of mercury, referring to the barometric height. In dealing with gases a volume unit of 1 liter $= 10^{-3}\ \text{m}^3$ is very convenient.

8.1 A capillary tube of uniform bore contains a thread of mercury of length 14 cm which encloses dry air between it and the closed end of the tube, the other end being open to the atmosphere. When the tube is placed vertically with the open end downwards, the air column is 22 cm long; when placed vertically with the open end upwards, the air column is only 15 cm long. Calculate the atmospheric pressure in cm of mercury.

8.2 A sample of gas at temperature 27°C, pressure 1 atm, and volume 0.4 liter is heated until its temperature is 177°C and its volume has increased to 0.5 liter. What is its new pressure?

8.3 A sample of gas has a volume of 5 liters at −73°C. What will be its new volume at 27°C if the pressure is unchanged?

8.4 At the bottom of a lake where the temperature is 7°C the pressure is 2.8 atm. An air bubble of diameter 4 cm at the bottom rises to the surface, where the temperature is 27°C. What is its diameter at the surface?

8.5 A container of volume 10 liters holds a gas at a pressure of 2 atm. Compute the total kinetic energy of all the particles of the gas.

8.6 If the r.m.s. speed of the molecules of a gas is 400 m s^{-1} when the pressure is 700 mm of mercury, what is the density of the gas?

8.7 The unit of energy or work corresponding to units of atmospheres and liters for pressure and for volume is the liter-atm. Compute the value of the universal gas constant in liter-atm mol^{-1} deg^{-1}.

8.8 Benzene vapor (C_6H_6) has a molecular weight of 78. Calculate the average translational kinetic energy of a molecule of benzene vapor at 100°C and the r.m.s. speed of the molecules at the same temperature. Mass of one atom of hydrogen is 1.66×10^{-27} kg.

8.9 A vessel contains 12 g of methane (CH_4) and 55 g of carbon dioxide (CO_2) under a total pressure of 2 atm and at a temperature of 27°C. What are the partial pressures of the two gases and what is the volume of the vessel? Take the atomic masses of H, C, and O as 1, 12, and 16.

8.10 Calculate the total translational kinetic energy of 3 mol of an ideal gas at 227°C.

8.11 Gas in a container is in the gravitational field of the earth, and so the molecules are moving slower near the top of the container than near the bottom. Show, using any gas you choose at about room temperature, that the change in velocity between roof and floor in a container of height 10 m is completely negligible in comparison with the r.m.s. speed of the molecules.

8.12 Two moles of a monatomic gas at 300°C are mixed with 1 mol of a diatomic gas at 100°C. Calculate the final temperature of the mixture, assuming that the mixing takes place at constant volume and that there is no exchange of heat with the environment.

8.13 A monatomic gas is heated so that it expands at constant pressure. What percentage of the heat supplied goes into increasing the thermal energy of the gas? What percentage is used up in doing external work?

CHAPTER 9

PROPERTIES OF LIQUIDS

9.1 INTRODUCTION

The properties of liquids are of paramount importance to the biologist and the medical scientist. The most common of all liquids—water—is absolutely indispensable for life as we know it. All living organisms originated in an aqueous environment and they have through the evolutionary process become dependent on water in many ways. Water accounts for between 60 % and 95 % of the material of nearly all organisms, including the human, and it provides the means of transporting nutrients dissolved or suspended in it to all parts of plants and other biological structures. The distinction between a solution and a suspension is worth making at the very beginning. In a solution, a chemical species (the solute) breaks up into its molecules or into portions of its molecules when it is added to a pure liquid (the solvent) and the result is a homogeneous solution in which it is impossible to detect the physical presence of the solute. In a suspension, one substance in the form of very small particles is present in a pure liquid, but the particles are of a size that makes them physically distinguishable from the liquid. As the size of particles in a suspension is reduced we reach a state referred to as the *colloidal* state which is half way between a true solution and a true suspension. Water is a very unusual liquid with unique physical properties that depend on the fact that its molecules form strong bonds with one another, so that they exist in a partially ordered state that has certain crystalline properties. As a chemical it is unique. It is a compound of great stability, a remarkable solvent and a powerful source of chemical energy. The physical properties of water, and of solutions and suspensions in water, are crucial for the whole of biology, and the purpose of this chapter is to explain these physical properties. The concepts introduced, of course, apply to other liquids as well, and some of them apply also to gases, which were discussed in Chapter 8, so that we shall use the word "fluid" on occasion when either a liquid or a gas is being discussed.

9.2 PRESSURE

A liquid is characterized by the fact that it has a definite volume but no definite shape. A given quantity of liquid will always at the same temperature fill a specific container to the same level. When applying the laws of mechanics to liquids, we do not usually consider the liquid as a whole, but rather concentrate on some arbitrary small portion of it, and we customarily refer all our physical quantities to unit volume. Thus, instead of talking about the mass m of the whole liquid whose volume is V, we work with the density ρ, which is defined as the mass per unit volume:

$$\rho = m/V. \tag{9.1}$$

The units of density are kg m^{-3}, and the densities of some common liquids are given in Table 9.1. It is often more convenient to work with the *relative density*

Table 9.1. Properties of liquids near room temperature

Substance	Density, ρ (kg m^{-3})	Viscosity, η (10^{-5} kg m^{-1} s^{-1})	Surface tension, S (10^{-3} N m^{-1})
Benzene	879	65	29
Chloroform	1,527	58	27
Ether	736	234	17
Mercury	13,600	156	465
Olive Oil	920	8,400	32
Turpentine	870	149	27
Water (at 4°C)	1,000	100	73

(sometimes called *specific gravity*) which is defined as the density of the liquid divided by the density of pure water at 4°C. For example, the density of benzene is 879 kg m^{-3}, and so its relative density is

$$\frac{879 \text{ kg m}^{-3}}{1000 \text{ kg m}^{-3}} = 0.879.$$

Relative density, being a ratio, has no dimensions.

Similarly any other mechanical property of a liquid may be referred to unit volume by using the density ρ instead of the mass. Thus the kinetic energy per unit volume, or kinetic energy density, is $\frac{1}{2}\rho v^2$ and the gravitational potential energy density at a height y above the reference level is $\rho g y$.

A fundamental property of any fluid is that it exerts a pressure. The pressure exerted by a gas on the walls of its container has been analyzed in the previous chapter in terms of the bombardment of the walls of the container by the gas molecules. We on earth live permanently at the bottom of the atmosphere which exerts a pressure on everything. Since pressure in a liquid exhibits some features not usually taken into account in gases, it is convenient to illustrate here the precise meaning of pressure in a liquid. Consider a small disk of area A placed in a liquid. The liquid exerts a force F on both sides of the disk as shown in Fig. 9.1, and it is found that the force is quite independent of the orientation of the disk, and always acts perpendicularly to the surface of the disk. The pressure at the point occupied by the disk is defined as

$$p = F/A. \tag{9.2}$$

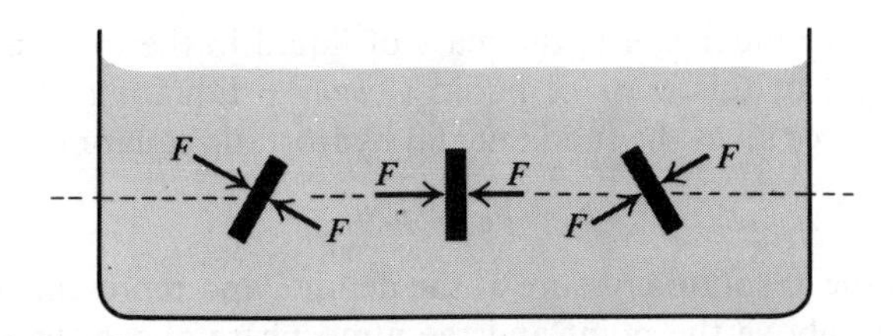

Fig. 9.1. The pressure in a liquid at a constant depth is the same in all directions.

The disk must be small enough so that the pressure does not vary appreciably over its surface. The dimensions of pressure are

$$\text{pressure} = \frac{\text{force}}{\text{area}} = \frac{\text{force} \times \text{distance}}{\text{area} \times \text{distance}} = \frac{\text{energy}}{\text{volume}},$$

and we see that we may alternatively regard pressure as an energy density. This is a most fruitful way of regarding it.

The pressure in a liquid depends only on the depth at which we measure it. Consider a tank of liquid where H is the height of the surface above the base. Consider a point in the liquid at a depth h below the surface, and imagine a cylindrical column of liquid above this point reaching to the surface, as shown in Fig. 9.2. All the rest of the liquid in the tank exerts pressure normally over the surface of the cylindrical column, and the atmosphere is pressing down on the top of the column with a pressure p_0 referred to as *atmospheric pressure*. All the forces resulting from the horizontal pressures cancel out, as is obvious by symmetry, and we are concerned only with the vertical forces. If the cross-sectional area of the cylinder is A, then the pressure p in the liquid at the depth h exerts on the cylinder an upward force pA, while the atmosphere exerts a downward force on the top of the cylinder of amount p_0A. The resultant upward force is $(p - p_0)A$, and this force must exactly balance the weight of liquid in the cylinder since the whole liquid is at rest.

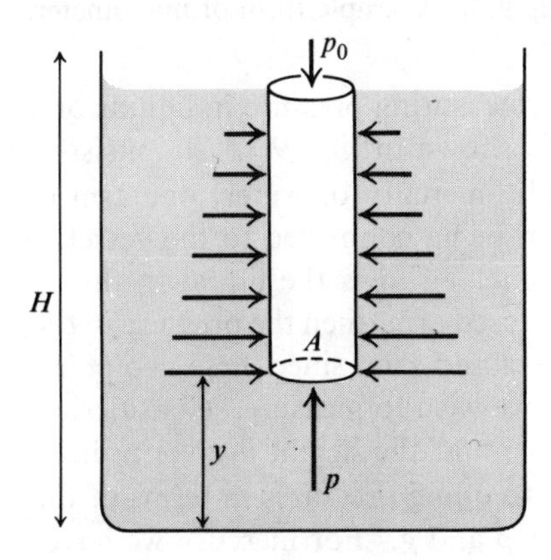

Fig. 9.2. The pressure at a depth h in a liquid.

The volume of the cylinder is hA; the mass of liquid in the cylinder is ρhA; and therefore the weight of this mass of liquid is ρghA. Equating the weight to the resultant upward force gives the fundamental hydrostatic equation

$$p - p_0 = \rho gh. \tag{9.3}$$

The pressure p is the absolute pressure at the depth h and represents the combined effect of the liquid above the point and the atmosphere above the surface of the liquid. Often we are more interested in the effect of the liquid alone, without bothering about p_0 which is a constant in any given situation. The difference $p - p_0$, which is the pressure at depth h expressed as an excess over atmospheric pressure, is called the *gage pressure* at that point. The gage pressure depends *only* on the depth, and all points at the same level in a liquid, no matter the shape of the container, are at the same pressure.

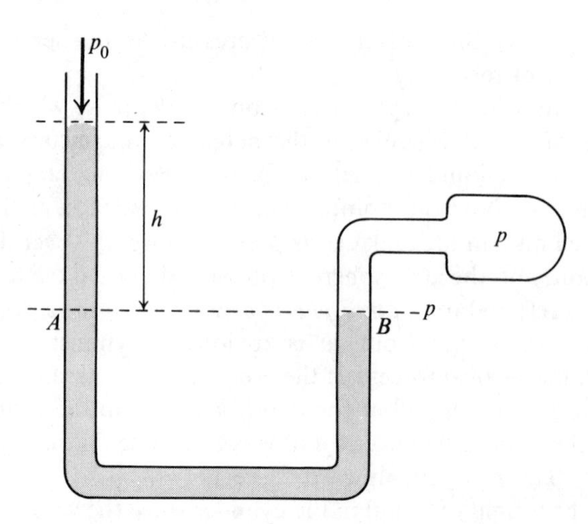

Fig. 9.3. A simple form of manometer.

A common device for measuring pressure in liquids or gases is the *manometer*, a simple form of which is shown in Fig. 9.3. It consists essentially of a U-tube containing a liquid (usually mercury or water) one arm of which is open to the atmosphere, the other arm being connected to the vessel whose pressure we wish to measure. If the manometer shows the liquid in the open arm standing at a height h above that in the closed arm, then the pressure in the open arm at A is equal to the pressure at B in the closed arm, since these two points are at the same level, and this is p, the pressure we wish to measure. Considering the open arm, we get $p - p_0 = \rho gh$ by Eq. (9.3), and the height h gives a direct measure of the gage pressure. It is customary to quote pressures in terms of the manometric height h, using the known values of ρ and g. For mercury we have

$$\rho gh = 13{,}600 \text{ kg m}^{-3} \times 9.8 \text{ m s}^{-2} \times h = (1.33 \times 10^5 \text{ N m}^{-3})h$$

The pressure exerted by a column of mercury of height 1 mm is called 1 torr, and is seen to have the value:

$$1 \text{ torr} = (1.33 \times 10^5 \text{ N m}^{-3}) \times 10^{-3} \text{ m} = 133 \text{ N m}^{-2}.$$

Standard atmospheric pressure is 760 torr, therefore

$$1 \text{ atm} = 760 \text{ torr} = 1.01 \times 10^5 \text{ N m}^{-2}.$$

It will be noticed that 1 atm is very nearly equal to 10^5 N m^{-2}, a useful result to remember. When small pressures are to be measured (for example pressures in the venous system) water may be used in the manometer instead of mercury. This gives much more sensitivity, since 1 mm of water corresponds to a pressure of only 9.8 N m^{-2} instead of the 133 N m^{-2} for mercury.

Modern physiologists use *electromanometers* in preference to the cumbersome U-tube types. These instruments utilize a *pressure transducer*. A transducer is a device that converts a signal of one type into a signal of another type (usually electrical). In electromanometers, the transducer element deforms slightly under the action of the pressure to be measured, and the deformation is transformed into an electrical signal that is amplified and recorded. These are much more convenient, but it must always be remembered that they are ultimately calibrated in terms of the mercury manometer.

A simple example of the importance of appreciating how hydrostatic pressure depends on height is provided by considering the correct positioning of the control clamps on an intravenous transfusion apparatus. Blood moves under the influence of gravity from the supply bottle through a fine nozzle into the drip chamber, and from there through a long tube to the exit point O, as shown in Fig. 9.4. Consider first the situation where the control clamp is placed at A, and is closed. The pressure at the outlet O is essentially atmospheric, and therefore the pressure at any point P below the drip chamber is below atmospheric pressure since P is higher than O. If there should be a small hole in the tube at P, then air will enter the tube, and this air will be carried down into the vein when the control clamp is opened. Cases of fatal aeroembolisms from this cause have been reported. Consider now the control clamp placed at B, very close to the outlet. Since the pressure at C at the blood surface is atmospheric, the pressure at any point in the tube above B is equal to atmospheric pressure plus the pressure of the column of liquid above that point. In this case, if there is a small hole in the tube, blood will be forced out rather than air being sucked in, and no aeroembolisms are possible.

Equation (9.3) may be expressed in a more general form as follows. Let the point at depth h be at a vertical height y above the bottom of the tank (Fig. 9.2), where H is the total height of liquid in the tank. Then

$$p - p_0 = \rho g h = \rho g(H - y),$$

and re-arranging this equation leads to

$$p + \rho g y = p_0 + \rho g H = \text{constant}. \tag{9.4}$$

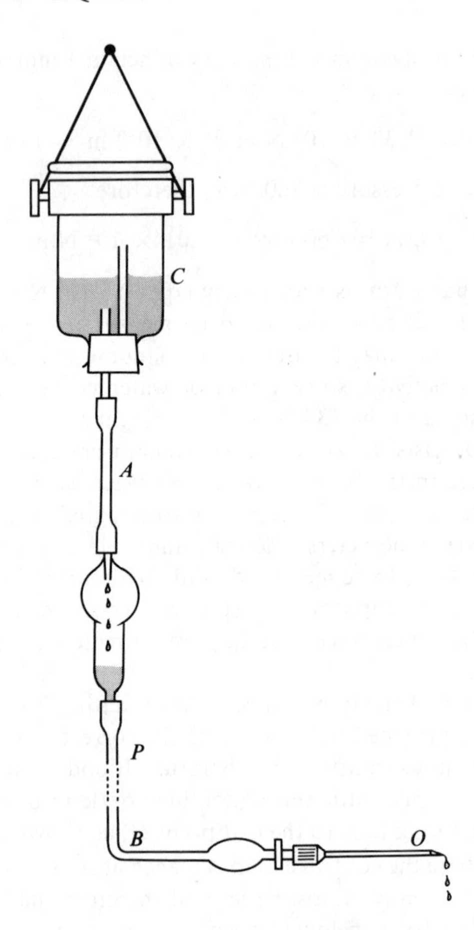

Fig. 9.4. An intravenous transfusion apparatus.

The sum of the two terms is seen to be a constant, since all the terms on the right hand side of the equation are constant. In physics we are always very interested in quantities that remain constant, and Eq. (9.4) gives us just such a quantity. How can we interpret it? Well, the second term, ρgy, is the gravitational potential energy density of the liquid at a height y above the bottom of the tank, and we have already seen that we may interpret p as a pressure energy density. So Eq. (9.4) is seen to be a special case of the principle of conservation of energy, and we may properly label the constant in it as E, the total energy density:

$$p + \rho gy = E, \text{ a constant.} \tag{9.5}$$

This is a special case of Bernoulli's equation which is discussed in Section 9.4 below.

Eq. (9.3) shows that the pressure at any point in a liquid at rest is the sum of the pressure due to the liquid itself, together with the pressure exerted on the surface by the atmosphere. The atmospheric pressure is transmitted to every point of the liquid, and this is a special case of *Pascal's law:*

> *Whenever an external pressure is applied to any confined fluid at rest, the pressure is increased at every point in the fluid by the amount of the external pressure.*

An example of the application of Pascal's law is the water mattress used to cut down bed-sores by distributing the weight of the body uniformly. Another example is the protection afforded to anything entirely surrounded by a liquid, for example the fetus enclosed in its liquid sac. Any sudden jar on the outside of the sac becomes distributed as a uniform pressure over the fetus with very much less risk of damage.

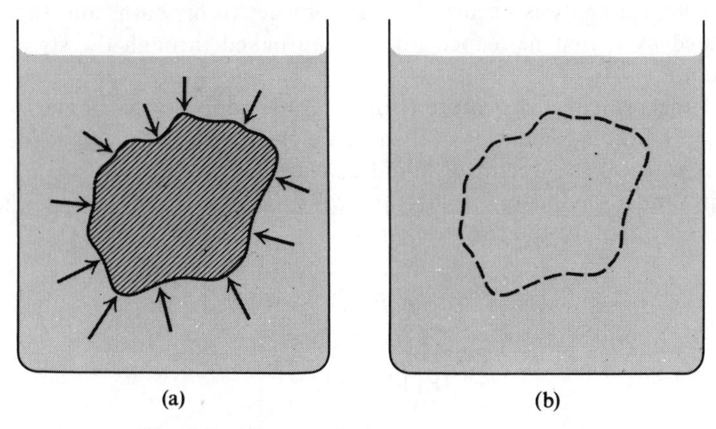

Fig. 9.5. The proof of Archimedes' principle.

9.3 BUOYANCY AND THE PRINCIPLE OF ARCHIMEDES

This most important principle may be stated quite generally as follows:

> *If a body is wholly or partially immersed in a fluid, it experiences an upthrust equal to the weight of fluid displaced, and this upthrust acts through the center of gravity of the displaced fluid.*

The principle has been stated for a fluid, since it holds equally well for gases as for liquids. The upthrust is known as the *buoyancy force*. To prove Archimedes' principle, consider an object of arbitrary shape immersed in a liquid (Fig. 9.5(a)). The surrounding liquid exerts pressure at all points on the surface of the object, and we wish to calculate the resulting force that acts upwards on it. That the resultant force must act upwards is clear, since pressure increases with depth, so that the force must be greater on the underside of the object than on the upper side.

Now imagine that the irregular object is replaced by liquid. This liquid will occupy precisely the same space as was previously occupied by the solid, as indicated by the dashed outline in Fig. 9.5(b). Since the replacement liquid has the same boundary as the solid object had, the remainder of the liquid must exert upon it precisely the same forces as it exerted on the solid object. But the lump of liquid that has replaced the solid is at rest, and therefore there is no unbalanced force acting on it, and no unbalanced torque acting on it. The weight of the lump of liquid acts downwards through its center of gravity, and consequently we deduce that the rest of the liquid exerts an upwards force on it equal to its weight and that this upward force acts through its center of gravity. Hence, when the solid was there, the liquid exerted upon it an upward force equal to the weight of liquid displaced, this force acting through the center of gravity of the displaced liquid. It will be noticed that no mathematics at all is needed for this proof, even although we have considered a body of quite arbitrary shape. Legend has it that this beautiful piece of logical analysis occurred to Archimedes in his bath, and that he was so exhilarated by it that he leaped up and ran naked through the streets to tell people about it.

To put the principle into more formal terms, suppose the immersed object has a volume V and is of density ρ; then its weight is $\rho g V = W$. If the density of the liquid is $\rho' < \rho$, the weight of liquid displaced is given by $W' = \rho' g V$, since a solid object must obviously displace its own *volume* of liquid. Hence the resultant downward force, or *effective weight* is

$$W'' = W - W' = (\rho - \rho')gV = \rho g V\left(1 - \frac{\rho'}{\rho}\right)$$

$$= W\left(1 - \frac{\rho'}{\rho}\right). \tag{9.6}$$

This correction to the weight of an object must be taken into account in high precision work, even when the fluid is air.

If $\rho' > \rho$, then the body will rise to the surface of the liquid under the action of the resultant upward force, and it will settle in equilibrium when just so much of it is immersed as to give a buoyancy upthrust equal to its own weight. When $\rho' = \rho$, there will be no resultant force acting on the body, and it will remain at rest at any height in the liquid. This is a most convenient method of measuring the density of very small objects. The procedure is to mix two liquids of different density until the density of the mixture is just equal to that of the unknown object. The point of equality is observed when the object shows no tendency to either rise or sink.

Example 1. Ice has a density relative to sea-water of 0.90. What proportion of an iceberg is submerged?

Solution. Let V be the volume of the iceberg, and k the proportion submerged, so that the volume submerged is kV. The weight of the iceberg is $\rho g V$, and the upthrust due to the weight of sea-water displaced is $\rho' g(kV)$. Since the iceberg floats, these must be equal, therefore

$$\rho g V = \rho' g k V.$$

Hence

$$k = \rho/\rho' = 0.90.$$

So 90% of the iceberg is submerged.

Example 2. In order to determine the density of a small marine animal, benzene and chloroform were mixed in a proportion of 78% and 22% by volume respectively. The animal was observed to remain suspended in the resulting mixture. Calculate the average density of the animal.

Solution. Let V_1, V_2 be the actual volumes of benzene and chloroform used, and let ρ_1, ρ_2 be their densities. Then the total volume of liquid is $V = V_1 + V_2$, and the total mass of liquid is $M = \rho_1 V_1 + \rho_2 V_2$. Hence the density of the mixture is

$$\rho = M/V = (\rho_1 V_1 + \rho_2 V_2)/(V_1 + V_2)$$

$$= \rho_1 \left(\frac{V_1}{V_1 + V_2} \right) + \rho_2 \left(\frac{V_2}{V_1 + V_2} \right).$$

The fractions in brackets are the proportions by volume of benzene and chloroform, which are given, and the densities are listed in Table 8.1. Hence the density of the marine animal is,

$$\rho = 0.78 \times 879 \text{ kg m}^{-3} + 0.22 \times 1527 \text{ kg m}^{-3} = 1022 \text{ kg m}^{-3}.$$

The determination of the densities of reasonably large animals is also of zoological importance, and for this purpose Archimedes' principle is again utilized. To take a concrete example of the technique, consider the problem of determining the density of a cuttlefish, which is known to have a density just greater than pure water. The cuttlefish is first weighed in air giving a result W (there is no need to correct for the buoyancy of the air in this case), and then in water giving a very much lower apparent weight W''. From Eq. (9.6) by a slight re-arrangement the relative density of the cuttlefish is found to be

$$\rho/\rho' = W/(W - W''). \tag{9.7}$$

Example. A fish weighs 348 g in air, and 23 g in pure water. Calculate the relative density of the fish.

Solution. From Eq. (9.7) the required density is

$$\rho/\rho' = 348 \text{ g}/(348 - 23) \text{ g} = 348/325 = 1.07.$$

9.4 THE FLOW OF IDEAL FLUIDS

In beginning our study of the motion of fluids, it is convenient to introduce the concept of an *ideal* fluid. Such a fluid is characterized by being incompressible, and by there being no internal frictional forces acting in it. Real liquids are very slightly compressible. When real fluids move, there are frictional forces (known as *viscous* forces) that act between adjacent portions of the fluid, and such forces lead to a loss of mechanical energy. Let us consider to begin with the steady flow of an ideal liquid through a pipe, as shown in Fig. 9.6. For steady flow, any small volume of liquid that passes a point such as P follows exactly the same path as the preceding particles that passed the same point. The lines of motion of the various small elements of volume can be joined up into streamlines as shown. If the flow of an ideal fluid is too fast, or if there are obstructions or sharp bends to be negotiated, then the flow becomes *turbulent*, and is characterized by the presence of whirls and eddies, such as those seen in a cloud of cigarette smoke.

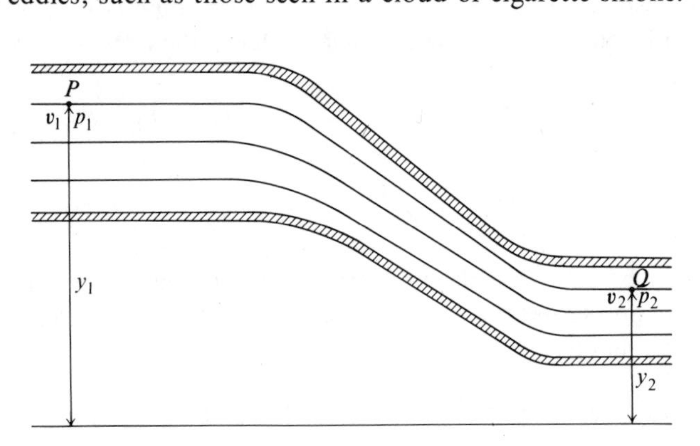

Fig. 9.6. The steady flow of an ideal liquid through a pipe.

For steady, streamline flow, since there are assumed to be no viscous forces present, mechanical energy is conserved, and for any element on a streamline we may write

$$p + \rho g y + \tfrac{1}{2}\rho v^2 = E, \text{ a constant.} \tag{9.8}$$

This will be seen to be Eq. (9.5) with the addition now of a term $\frac{1}{2}\rho v^2$ which is the kinetic energy density. This is Bernoulli's equation, and is of fundamental importance in all discussions of the flow of fluids. We may note that, although gases are highly compressible, Bernoulli's equation may safely be applied to them so long as the velocity of the gas is small compared to the velocity of sound.

The volume of liquid flowing through the tube per unit time is easily found. At point P of the tube the velocity of the liquid is v_1, which means that a length v_1 of liquid passes P every second. If the cross section of the tube at P has an area A_1, this means that a volume $v_1 A_1$ per second passes P. Similarly a volume $v_2 A_2$ passes point Q every second, where A_2 is the cross-sectional area at Q. Since no

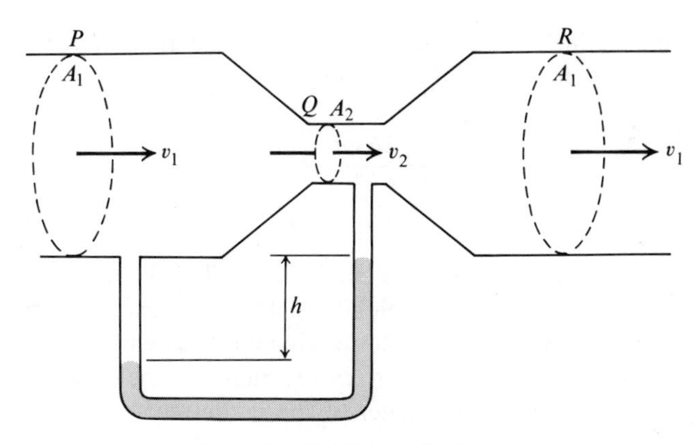

Fig. 9.7. The Venturi tube.

liquid is gained or lost between these points, and since we are assuming that the liquid is incompressible, it follows that

$$v_1 A_1 = v_2 A_2. \tag{9.9}$$

So the narrower the tube, the faster the liquid flows. For ideal liquids the velocity is constant over the cross section of the tube at any point. As we shall see in Section 9.5 this is not true when viscous forces are taken into account, but even then Eq. (9.9) provides a very useful means of defining the *average* velocity of flow of the liquid over any cross section:

$$\text{average velocity of flow, } \bar{v} = \frac{\text{volume rate of flow}}{\text{cross-sectional area}}. \tag{9.10}$$

An important application of Eqs. (9.8) and (9.9) is the Venturi tube, as shown in Fig. 9.7. It consists essentially of a conical constriction in a pipe from the full bore of cross-sectional area A_1 at P to a much narrower bore A_2 at Q (called the *throat* of the Venturi tube), and a gradual widening out to the full bore again at R. Very careful engineering design is necessary to avoid the formation of eddies in the liquid. Let us apply our equation of flow to the points P and Q. Since there is no change of level the $\rho g y$-term is not involved, and we get

$$p_1 + \tfrac{1}{2}\rho v_1^2 = p_2 + \tfrac{1}{2}\rho v_2^2. \tag{9.11}$$

The difference of pressure $p_1 - p_2$ is measured with a manometer, and we find, using Eq. (9.9) to express v_2 in terms of v_1,

$$\begin{aligned} p_1 - p_2 &= \tfrac{1}{2}\rho v_2^2 - \tfrac{1}{2}\rho v_1^2 \\ &= \tfrac{1}{2}\rho v_1^2\left\{\left(\frac{A_1}{A_2}\right)^2 - 1\right\}, \end{aligned}$$

giving finally

$$v_1^2 = \frac{2(p_1 - p_2)}{\rho\left\{\left(\dfrac{A_1}{A_2}\right)^2 - 1\right\}}. \tag{9.12}$$

If h is the difference in levels of the two arms of the manometer, then $p_1 - p_2 = \rho' gh$, where ρ' is the density of the mercury. Combining this with Eq. (9.12) we obtain

$$v_1^2 = \frac{2\rho' gh}{\rho\left\{\left(\frac{A_1}{A_2}\right)^2 - 1\right\}} = Kh, \tag{9.13}$$

where K is a constant for any given Venturi tube. The Venturi tube method has been adapted to measure the flow velocity of blood in arteries. This is done by introducing a tube (called a *cannula*) containing a narrowed orifice into the artery, or alternatively by constricting the artery externally. The pressure difference is measured with a differential manometer of special design.

Example. A Venturi tube has a main diameter of 4.0 cm and a throat diameter of 2.0 cm, and is measuring the rate of flow of water. The pressure difference indicated by the manometer is 22 torr. Calculate (a) the velocity of the liquid in the main tube, and (b) the volume rate of flow of liquid.

Solution. The area of cross section of the tube is proportional to the square of the diameter, hence

$$A_1/A_2 = (4.0 \text{ cm}/2.0 \text{ cm})^2 = 4, \qquad \text{and} \qquad (A_1/A_2)^2 - 1 = 15.$$

From Eq. (9.13) therefore

$$v_1^2 = \frac{2 \times 1.36 \times 10^4 \text{ kg m}^{-3} \times 9.8 \text{ m s}^{-2} \times 22 \times 10^{-3} \text{ m}}{10^3 \text{ kg m}^{-3} \times 15}$$

$$= 0.39 \text{ m}^2 \text{ s}^{-2}.$$

Therefore (a) $v_1 = 0.62 \text{ m s}^{-1}$

and (b) the volume rate of flow is

$$A_1 v_1 = \pi \times \left(\frac{0.04 \text{ m}}{2}\right)^2 \times 0.62 \text{ m s}^{-1} = 7.8 \times 10^{-4} \text{ m}^3 \text{ s}^{-1}.$$

The Venturi tube operates on the fundamental principle that the pressure in any constricted region is less than the pressure in the main pipe. If the rate of flow through the pipe is made large enough, the pressure in the restricted throat may be reduced below atmospheric pressure, and this fact is utilized in the design of injectors. A simple form of injector is shown in Fig. 9.8. Fluid A flows into a constriction in a tube through a nozzle, and the pressure at P is lower than atmospheric. A side tube opening into a region at atmospheric pressure permits a second fluid B to be pulled into the tube, and the mixture of the two fluids proceeds into the main tube. We have talked of fluids here, rather than simply liquids, because injectors of this sort are widely used with gases as well as liquids—for example to control the air–oxygen mixture to be administered to a patient.

The Venturi effect is of medical importance whenever there is a narrowing of

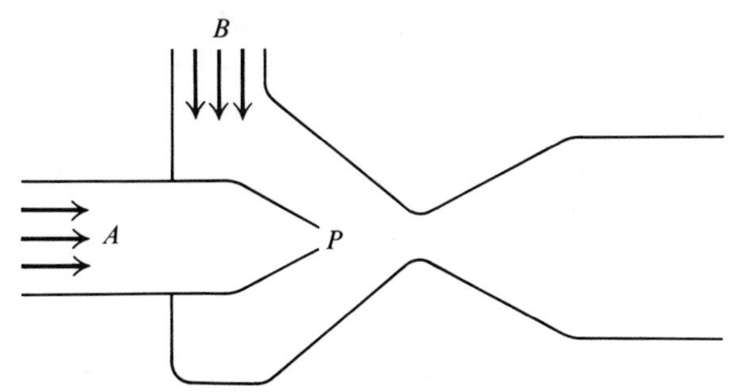

Fig. 9.8. The principle of the injector.

any essential liquid-carrying body tube. For example, if an artery is narrowed by internal lesions (*atherosclerotic plaques*) or by external pressure resulting from a tumor, then the blood pressure in the constricted region of the artery will fall very dramatically. Normally the blood pressure is balanced by an external force, but now this force is unbalanced, and it may even cause the artery to close completely. Once complete closure has taken place, of course, the pressure in the artery will immediately increase again, and there will result a rhythmic opening and closing known as *flutter*. It is possible that such flutter in the bronchial passages when narrowed is responsible for the *rales*, the characteristic sound of respiratory disorder.

One further application of the Venturi effect is in the common bunsen burner, where the amount of air being mixed with the gas is controlled by varying the bore of the side inlet tube.

Another very important flow-measuring device widely used in physiology is the Pitot tube (Fig. 9.9). A central tube opens through a narrow orifice in the upstream direction and is connected to the left arm of a manometer as shown. The surrounding tube has openings at C perpendicular to the flow, and this outside tube connects with the right arm of the manometer. Consider the central streamline at A. This enters the orifice in the Pitot tube, and the fluid is brought to rest at some point B. Applying Eq. (9.11) to points A and B, we have

$$p_A + \tfrac{1}{2}\rho v^2 = p_B \tag{9.14}$$

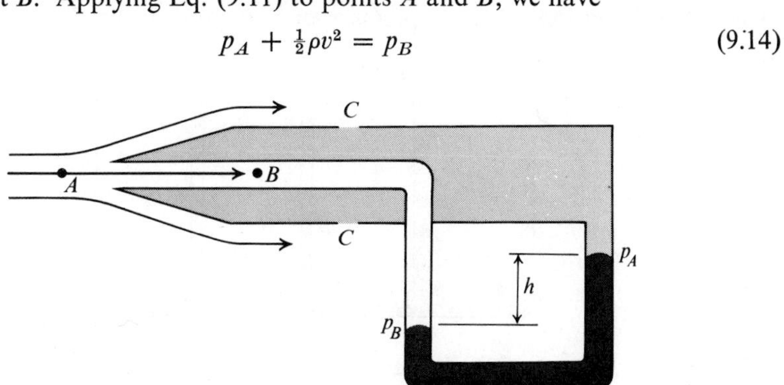

Fig. 9.9. The Pitot tube.

and this is the pressure exerted on the mercury in the left arm as shown. The pressure just inside the openings at C is just p_A, and the manometer therefore records directly the pressure difference

$$p_B - p_A = \tfrac{1}{2}\rho v^2.$$

If the difference in level of the two arms is h, the manometer liquid having a density ρ', then

$$\tfrac{1}{2}\rho v^2 = \rho' g h$$

$$v = \sqrt{2\rho' g h/\rho}. \tag{9.15}$$

Like Venturi tubes, Pitot tubes must be engineered with great precision if serious errors due to eddies are to be avoided. They have been used with success to measure the velocity of blood flow in arteries. One difficulty here is that blood is by no means an ideal fluid, and the velocity measured by the Pitot tube depends on precisely where in the cross-section of the artery the entrance orifice is situated. Indeed, very small Pitot tubes have actually been used to measure the variation in velocity of the blood across an artery.

9.5 VISCOSITY AND THE FLOW OF REAL FLUIDS

If an ideal fluid is flowing in a horizontal tube, it requires no external pressure to maintain the flow. The situation is similar to that of an object moving with constant velocity over a perfectly smooth surface; by Newton's first law there is no unbalanced external force acting upon it. But if the horizontal surface is rough so that there exists a frictional retarding force, then a condition of constant velocity can be maintained only if there is a driving force equal to the frictional force. So it is with liquids flowing through tubes. Figure 9.10 shows a horizontal tube connected to the base of a large reservoir in which the level is kept constant by adjusting the rate of inflow at A to be slightly greater than the rate of outflow at B: under these conditions the level in the reservoir cannot fall below that of the outflow tube C. Three vertical side tubes serve as simple manometers, and the height of liquid in them indicates the pressure at that point in the horizontal tube. It

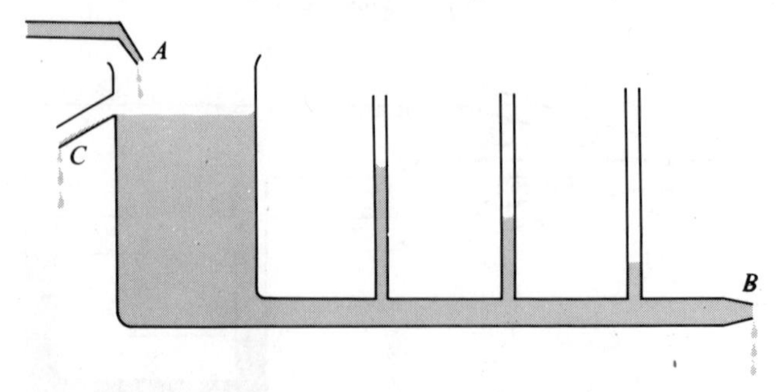

Fig. 9.10. The fall in pressure along a horizontal tube in which a real liquid flows. This fall shows that energy is being dissipated.

will be seen that there is a steady drop in pressure as we move from left to right. If the pressure drops by an amount p over a length l of the tube, the ratio p/l is known as the pressure gradient down the tube. Its units are N m^{-2} per meter, or N m^{-3}. Since all points of the horizontal tube are at the same level, and since the cross section is of constant area, then the decrease in pressure means a loss of mechanical energy. Such a loss of mechanical energy is typical of frictional, or dissipative, forces. The property of a liquid that determines the magnitude of these dissipative forces is known as the *viscosity*, and is given the symbol η. One of the effects of viscosity is that the velocity of a particle of liquid depends on how far it is from the axis of the tube. If some dye is injected into the flowing liquid, it will be seen that the dye particles (which we may assume to travel with the local liquid velocity) on the axis move faster than dye particles off the axis, and that in particular the dye particles in contact with the walls of the tube are at rest. The distribution of velocity as a function of the radial distance from the axis is shown in Fig. 9.11 for a cross section of the tube not too close to the reservoir. This is a necessary qualification, since the velocity profile is flat at the entry point, and only assumes its typical parabolic shape at a distance from it.

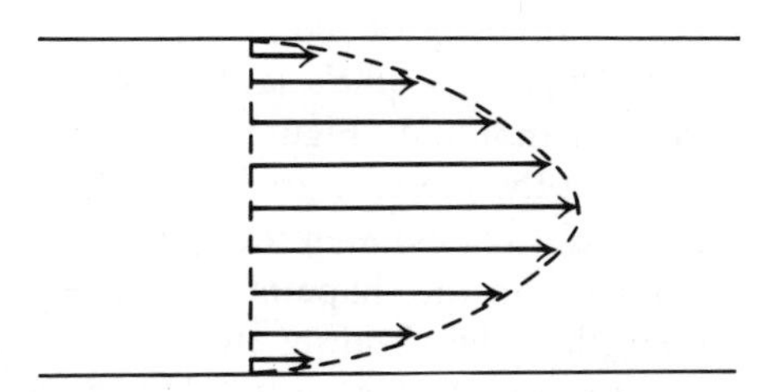

Fig. 9.11. The velocity profile across a tube. The liquid in contact with the wall of the tube does not move, and the velocity increases as we move to the center of the tube as shown schematically.

If the radius of the tube is a, then Poiseuille's law states that the volume rate of flow Q through the tube of a liquid of viscosity η is given by

$$Q = \frac{\pi a^4 p}{8\eta l}. \tag{9.16}$$

The pressure gradient p/l down the tube may be measured using two of the manometers as shown. If l is their distance apart, and if they show a difference of height h, then $p/l = \rho g h/l$, where ρ is the density of the liquid. From Eq. (9.16) it may be deduced that the units of viscosity are kg m^{-1} s^{-1}, and some typical values are listed in Table 9.1. The old c.g.s. unit was the poise (P), and 1 kg m^{-1} s^{-1} = 10 P.

The fact that the radius of the tube occurs to the fourth power in Poiseuille's formula is of importance if it is necessary to transfer large quantities of liquid quickly through a tube. For example, in blood transfusions, it is much preferable to increase the bore of the needle rather than to increase the height of the bottle. A two-fold increase in bore will increase the flow rate sixteen-fold.

From the Poiseuille formula and from the definition of average fluid velocity given in Eq. (9.10) we may calculate the average velocity of flow through the tube as follows:

$$\bar{v} = \frac{\text{volume rate of flow}}{\text{cross-sectional area}} = \frac{\pi a^4 p}{8\eta l} \div \pi a^2 = \frac{a^2 p}{8\eta l}. \tag{9.17}$$

The type of viscous flow we have been discussing is a steady streamline flow, and is called *laminar* flow, since we may think of the liquid in the tube as consisting of concentric, thin, hollow tubes of gradually decreasing radius. The motion is similar to what would occur if these tubes were moved relatively to one another, by pulling on the narrowest one on the axis. It would drag along the next widest tube, which in turn would drag along the next widest, and so on. The slipping of the tubes relative to one another would result in the velocity profile shown in Fig. 9.11.

Although Poiseuille's formula holds well for pure liquids, it does not hold for suspensions or dispersions, mixtures of different kinds of material. Blood is an anomalous liquid of this type. For blood, doubling the pressure difference between the ends of the tube does not double the flow; it may triple or even quadruple it. In other words, the viscosity of blood decreases as the shear stress in it increases. A liquid that obeys Poiseuille's law of flow is called a Newtonian liquid: blood is a non-Newtonian liquid. Figure 9.12 shows a plot of the rates of flow of water and blood against the pressure gradient. The characteristic straight-line plot for water becomes a distinct curve in the case of blood. The explanation for this anomalous behavior lies in the shape of the molecules in suspension. If these are long and rod-like, then the shearing stresses in the liquid will tend to align them parallel to the direction of flow, thus reducing the measured viscosity.

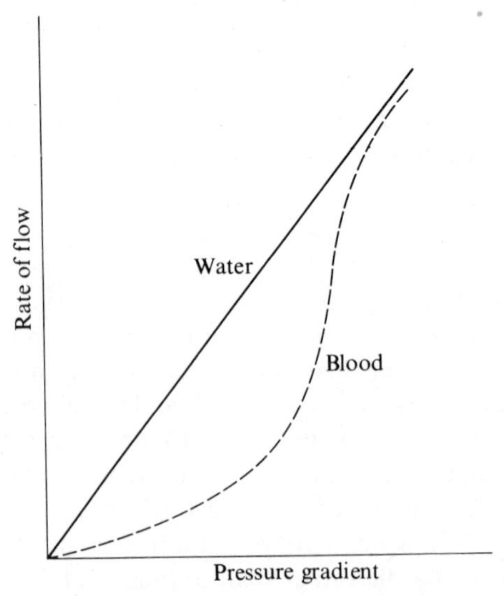

Fig. 9.12. The rates of flow of water and blood as functions of the pressure gradient.

Synovial fluid, which helps in the lubrication of the human joints, also behaves in this anomalous fashion.

The viscosity of a fluid exerts a retarding force on an object that moves through it, and in Appendix B Stokes' law for the force acting on a small sphere of radius a moving with velocity v is derived by a dimensional argument. The complete expression is

$$F = 6\pi a \eta v. \tag{9.18}$$

If such a small sphere falls under gravity in a fluid, it will rapidly reach a terminal velocity when the retarding forces are equal to the weight. If ρ is the density of the material of the sphere and ρ_0 is the density of the fluid, then the weight of the sphere is $4\pi a^3 \rho g/3$ and the upward buoyancy force, which is equal to the weight of fluid displaced, is $4\pi a^3 \rho_0 g/3$. Hence, the acceleration of the drop will vanish and it will reach its terminal velocity when

$$\frac{4}{3}\pi a^3 \rho g = \frac{4}{3}\pi a^3 \rho_0 g + 6\pi a \eta v$$

or

$$v = 2a^2(\rho - \rho_0)g/9\eta. \tag{9.19}$$

The terminal velocity is called the *sedimentation velocity*, and experiments on sedimentation can give valuable information about the size and weight of very small particles. The sedimentation velocity is proportional to g; and if it is desired to increase the sedimentation rate a *centrifuge* can be used. If the container is set spinning about an axis with a very high angular velocity, the particles in suspension are subjected to a very large centripetal acceleration. This acceleration is $\omega^2 r$, where ω is the angular velocity and r is the distance of a particle from the axis of rotation. In Examples 3 and 4 of Section 5.4, it was seen that, in an accelerating environment, the acceleration is experienced as an apparent change of weight acting in a direction opposite to that of the acceleration. Consequently, in the centrifuge, the apparent change of weight is $m\omega^2 r$ in a direction radially outwards, and this is known as the *centrifugal force*. Sedimentation takes place radially outwards under this force, and Eq. (9.19) holds with $\omega^2 r$ in place of the true gravitational term, g. The angular velocity can be made very large so that a much higher sedimentation rate is achieved.

Example. An air bubble of radius 5 mm rises through a vat of syrup at a steady speed of 2 mm s^{-1}. If the syrup has a density of 1.4×10^3 kg m^{-3}, what is its viscosity?

Solution. In this case the sedimentation velocity is upwards, since $\rho_0 > \rho$. Indeed, the density of the air inside the bubble may be neglected completely. Eq. (9.19) applies, with $a = 0.005$ m, $(\rho - \rho_0) = 1.4 \times 10^3$ kg m^{-3}, and $v = 0.002$ m s^{-1}. Re-arranging the equation we get

$$\eta = 2a^2(\rho - \rho_0)g/9v = \frac{2 \times 25 \times 10^{-6}\text{ m}^2 \times 1.4 \times 10^3\text{ kg m}^{-3} \times 9.8\text{ m s}^{-2}}{9 \times 0.002\text{ m s}^{-1}}$$

$$= 38\text{ kg m}^{-1}\text{ s}^{-1}.$$

9.6 DIFFUSION AND OSMOSIS

If water is poured carefully on to a solution of copper sulfate in water so that the separation between the water and the solution is visible, it will be found that the characteristic blue color of the solution gradually spreads until the whole liquid is uniformly blue. On this gross scale it will take many days for the mixing to be complete, but similar processes in biological cells take only milliseconds. The *solute*, in this case copper sulfate, is said to *diffuse* through the liquid, and the water also diffuses downwards into the original solution. One usually concentrates on the movement of the solute. The diffusion of a solute can be regarded as analogous to the flow of heat, and *Fick's law* states that the rate of diffusion per unit area in a direction perpendicular to the area is proportional to the gradient of concentration of solute in that direction. The concentration is the mass of solute per unit volume. Consider as a one-dimensional example a pipe of cross-sectional area A down which a solute is diffusing (Fig. 9.13), the concentration of solute being assumed constant over any cross section of the pipe. If the change of concentration over a short length Δx of the pipe is Δc, then the mass Δm of solute diffusing down the pipe in time Δt is given by

$$\frac{\Delta m}{\Delta t} = -DA\frac{\Delta c}{\Delta x}, \tag{9.20}$$

the negative sign indicating that the direction of motion of solute is opposite to the direction of increase of concentration. D is the diffusion constant, and typical values of D for biologically important molecules diffusing through water at room temperature range from 1 to $100 \times 10^{-11}\ \text{m}^2\,\text{s}^{-1}$, the corresponding range of molecular weights being about 10^4. It may be shown that the diffusion constant is related to the temperature and the viscosity of the liquid by the equation

$$D = kT/6\pi a\eta,$$

where a is the radius of the particle of solute assumed spherical and k is Boltzmann's constant. Since the radius of a sphere is proportional to the cube root of its mass, we have the important result that D is also inversely proportional to the cube root of the mass. This explains the fact that such a wide range of molecular weights gives a comparatively small range of values of D. For diffusion in gases the above

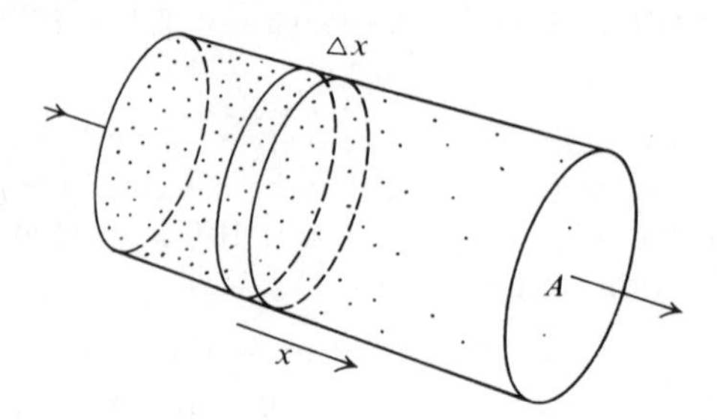

Fig. 9.13. Diffusion due to a concentration gradient.

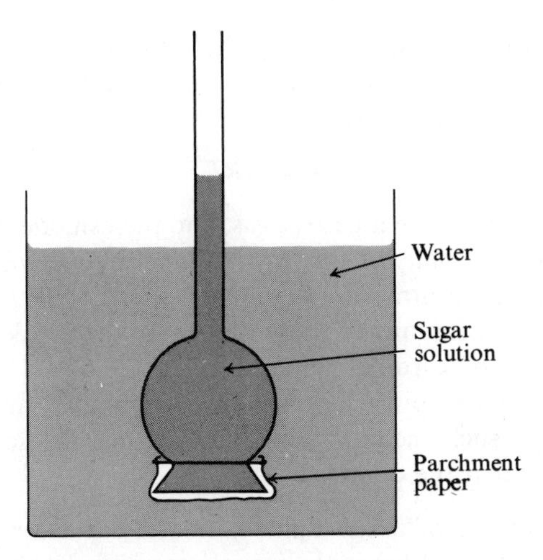

Fig. 9.14. Osmosis through a semipermeable membrane.

result does not hold, and D in gases is inversely proportional to the square root of the mass of the solute particles, not the cube root.

If a strong cane-sugar solution is poured into a vessel of water, the mixture gradually becomes homogeneous through the joint diffusion of the solute molecules into the region of pure water and the diffusion of water molecules in the opposite direction. If instead of placing the sugar solution directly into the water it is placed in an inverted thistle funnel, the lower end of which is closed with parchment paper, as shown in Fig. 9.14, the outward diffusion of the solute is prevented. The paper is said to be impermeable to the sugar solute. However, the water molecules can freely diffuse in the opposite direction, and because of this the level of the solution rises in the stem of the thistle funnel. The parchment paper is said to be a *semipermeable membrane* and the process of selective diffusion is called *osmosis*. The final increase in pressure of the solution in the thistle funnel is the *osmotic pressure* P of the solution. It appears strange at first sight that the water should pass from a region of low pressure to a region of high pressure. But it must be borne in mind that before the process starts the pressure of water in the thistle funnel is less than the pressure of water outside, since the total pressure of the solution is the same as that of the water, and the solute makes a contribution to this total pressure. The osmotic process equalizes the water pressures inside and outside, and consequently the final osmotic pressure is the pressure due to the molecules of solute alone.

The osmotic pressure P is found by experiment on weak solutions to be proportional to the concentration of solute, which means inversely proportional to the volume of the solution. It is also proportional to the absolute temperature. Hence, the laws of osmosis may be written

$$PV = R'T,$$

where R' is a constant depending only on the mass of solute present. The value of R' turns out to be just nR, where R is the universal gas constant and n is the number of moles of solute present; so

$$PV = nRT. \tag{9.21}$$

The solute behaves as if it were a perfect gas, and the osmotic pressure arises from the bombardment of the walls of the thistle funnel by the molecules of sugar in the solution. At high concentration the simple law of Eq. (9.21) breaks down for reasons that are quite analogous to those that cause the breakdown of the simple gas laws. Osmosis is of cardinal importance in biological processes, since all living cells are surrounded by semipermeable membranes, and are further subdivided internally by such membranes. These permit the selective diffusion of only those molecules that the cell needs.

Example. A mass of 5 g of a substance of molecular mass 250 is dissolved in 600 cm³ of water at 27°C. What is the osmotic pressure of the solution?

Solution. Directly from Eq. (9.21) we get

$$P = \left(\frac{5}{250}\text{ mol} \times 8.3\text{ J mol}^{-1}\text{ K}^{-1} \times 300\text{K}\right) \Big/ 6 \times 10^{-4}\text{ m}^3$$
$$= 8.3 \times 10^4\text{ N m}^{-2}.$$

9.7 SURFACE TENSION

One of the most important properties of a liquid is the tendency for its surface to contract. The surface behaves, in fact, as if it were an elastic skin that constantly tried to decrease its area. One result of this tendency is that small globules of liquid are as nearly spherical as gravity will allow. The symmetry of the rainbow, for example, proves that raindrops in the atmosphere are spheres. Under conditions of weightlessness in space ships even large masses of liquids will adopt a spherical configuration. The analogy with an elastic skin is not close, because, although the tension of such a skin increases with stretching, the tension in the surface of a liquid is independent of the area. It is called the *surface tension*, and is defined as the force per unit length acting across any line drawn in the surface and tending to pull the surface apart across the line. Surface tension arises because the molecules near the surface are closer together than those deep inside the liquid, and this implies a certain surface energy. The surface tension S of a liquid can be regarded as the potential energy per unit area of the surface. The units of S are either N m^{-1} or J m^{-2}, which are of course equivalent, and some typical values are listed in Table 9.1. The tendency for a liquid to assume a configuration of minimum surface area is a consequence of the general principle of conservation of energy. Either work or heat or both must be supplied to cause an increase in the surface, and the latent heat of vaporization of a liquid can be thought of in this way. When a liquid evaporates, there is an enormous increase in its surface area,

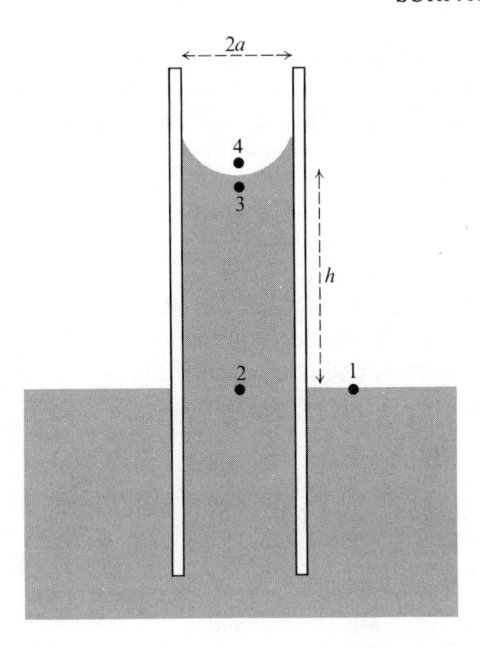

Fig. 9.15. The rise of liquid in a capillary tube.

and heat must be supplied to accomplish this. Surface energy, or surface tension, is a mutual property of the two materials that share the common surface. When no indication is given to the contrary, surface tensions may be assumed to be relative to air.

A very important consequence of the existence of surface tension is that there exists a difference of pressure across any curved surface separating two fluids. The pressure on the concave side exceeds the pressure on the convex side by an amount

$$p_1 - p_2 = 2S/r, \tag{9.22}$$

where r is the radius of the surface. This relation may be shown to be a consequence of energy conservation. It may be used to determine the rise of a liquid in a capillary tube. For if a very narrow tube is placed vertically into a liquid, as shown in Fig. 9.15, then the system will come to equilibrium when the pressures at the points 1 and 2 are equal, since these two points are on the same level. But the pressure p_2 at the point 2 is greater than the pressure p_3 at the point 3 just inside the upper curved surface by an amount ρgh, where ρ is the density of the liquid (see Section 9.2), and the pressure p_3 is less than the pressure p_4 at the point 4 on the concave side of the upper surface by an amount $2S/r$, by Eq. (9.22). But $p_1 = p_4$ = atmospheric pressure; therefore we have

$$p_1 = p_2 = p_3 + \rho gh = \left(p_4 - \frac{2S}{r}\right) + \rho gh,$$

which simplifies to

$$h = \frac{2S}{\rho gr}.$$

The radius of curvature of the upper surface of the liquid is proportional to the radius of the capillary tube, the constant of proportionality depending on the nature of the contact interaction between the liquid and the material of the tube. If we write $a = Cr$, where C is this constant, then the height of liquid in the capillary tube is given by

$$h = \frac{2CS}{\rho g a} . \tag{9.23}$$

For water and most other liquids that wet glass, $C = 1$. But for mercury $C = -0.77$, the negative sign indicating that the mercury-air interface in the tube is concave toward the mercury, and consequently the mercury level falls because of capillary action. Since mercury is so widely used, it is of interest to see how big the capillary effect is with this liquid. For mercury $S = 0.465\ \mathrm{N\ m^{-1}}$ and $\rho = 1.36 \times 10^4\ \mathrm{kg\ m^{-3}}$, giving

$$h = \frac{5.34 \times 10^{-6}\ \mathrm{m^2}}{a} .$$

Clearly, the radius of the tube must be very small before any appreciable depression of the mercury surface will be observed.

Example. A capillary in the trunk of a tree has a diameter of 0.01 mm. How high will surface tension pull up a column of water?

Solution. Using the value $S = 0.073\ \mathrm{N\ m^{-1}}$ from Table 9.1, Eq. (9.23) gives

$$\begin{aligned} h &= \frac{2 \times 0.073\ \mathrm{N\ m^{-1}}}{1000\ \mathrm{kg\ m^{-3}} \times 9.8\ \mathrm{m\ s^{-2}} \times 0.5 \times 10^{-5}\ \mathrm{m}} \\ &= 3.0\ \mathrm{m}. \end{aligned}$$

Since trees grow to heights much greater than this, clearly surface tension is not responsible for the transport of water from the roots to the leaves.

9.8 SOME BIOLOGICAL APPLICATIONS

Since solutions and suspensions of solids in liquids play a major role in biology, all of the properties of liquids discussed have to be considered in any given situation. For example, when a centrifuge containing a solution of biological interest is spun very rapidly, the phenomenon of diffusion interacts with the sedimenting process described in Section 9.5 above. Consideration of Eq. (9.19) would indicate that after a sufficient time all the particles would end up at the outside of the centrifuge in a hard pellet. And indeed, if the particles in solution are large enough, this is precisely what happens, and one says that the particles have been *pelleted* or *sedimented.* This is a routine method of separating large particles from solution or suspension. But with smaller particles the process of diffusion will tend to spread the particles back toward the axis of the centrifuge, and the two

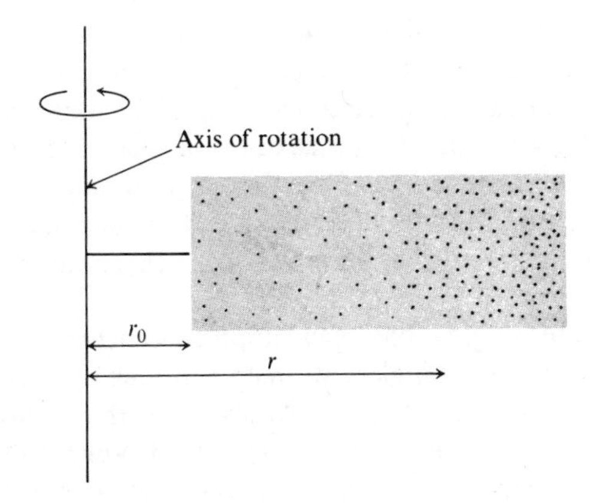

Fig. 9.16. The concentration of particles in suspension increases with distance from the axis of rotation of a centrifuge. This is the basis of the sedimentation equilibrium method.

conflicting tendencies will result in an equilibrium gradient of concentration, as shown in Fig. 9.16. If r_0 is the radial distance from the axis of rotation to the inner edge of the tube and if c_0 is the concentration there, then the concentration at a radial distance r may be shown to be

$$c = c_0 \exp \{A(r^2 - r_0^2)\}, \tag{9.24}$$

where $A = (m\omega^2/2kT)(1 - \rho_0/\rho)$ is seen to depend on the angular velocity of the centrifuge, on the temperature, on the densities of particles and of liquid, and on the mass of the particles; k is Boltzmann's constant. The concentration gradient given by Eq. (9.24) can be measured optically, and the value of the parameter A can be determined. In this way the mass m can be found. In practice the particles will be large molecules of biological interest and m will give the molecular weight. It is an important advantage of this method that the determination of m does not depend in any way on the *shape* of the particle. This method is known as the *sedimentation equilibrium method.*

An alternative method is the *density gradient sedimentation method.* The physical basis of this method is that if a particle of density ρ is in a liquid in which there is a density gradient from ρ' to ρ'', where $\rho' < \rho < \rho''$, then the particle will move until finally it comes to rest at a point where the liquid density matches its own density. In this position the particle has minimum energy. The appropriate density gradient is prepared by spinning a solution of cesium chloride in a centrifuge, and the particles whose density is required are added. If the concentration gradient of the CsCl solution is first accurately measured, then the density of the particles is read off. The particles will not actually come to equilibrium in an infinitely thin band but will occupy a band of finite thickness, owing to the random thermal motion of the particles. Indeed, the thickness of the band is inversely related to the mass of the particles, and so from a single photograph both the density and the molecular weight of the particles can be estimated.

If a mixture of particles is inserted into the CsCl solution in the centrifuge, the different particles will separate out at different places. This is the basis of the *density gradient separation method.* For this purpose sugar solution is actually preferred to the CsCl solution because of its higher viscosity. This prevents any casual agitation of the tube after removal from the centrifuge causing serious remixing.

As another example of great biological importance where the various properties exhibited by liquids are intermingled, we may consider the passage of substances through the plasma membrane of a cell. At first sight this seems to be a comparatively simple problem. The membrane is permeable to some substances and not to others: electron microscope photographs (see Section 20.6) show that the membrane is perforated by pores whose diameters are of molecular dimensions; hence, we may expect the membrane to be permeable to molecules smaller than the pores and impermeable to molecules larger than the pores. Indeed molecules of water, which are small, pass freely through the membrane, whereas protein molecules, which are very large, do not. But further experiments soon bring to light cases where molecules larger than the pores are being transported across, and other cases where smaller molecules are being barred. It appears that the membrane is exercising a choice, and it is not clear how such a process can be explained on a purely mechanistic basis. Support for this hypothesis of *active transport*, as it is called, comes from the observation that a significant difference in concentration of molecules can be maintained by the cell across the membrane that is permeable to the molecules. On a purely mechanistic theory the concentrations should equalize by diffusion through the membrane, and such equalization does in fact take place across the membrane of a *dead* cell.

The passage of molecules through the pores of the plasma membrane is by no means an easily understood process. If the diameter of the pores is only slightly larger than that of the molecules, then the process is one of diffusion. But if the molecules are very much smaller than the pores, then one might suspect that bulk flow according to the Poiseuille law would occur. In intermediate cases one might expect a mixture of the two phenomena. In osmosis, where it is water that passes through the membrane, a rate of flow that is higher than would be expected on the basis of diffusion alone is found, and it is concluded that osmotic flow is, partially at least, caused by pressure differences and not entirely by diffusion. It should be obvious that a clear understanding of the various possible mechanisms is essential for any biologist working in such a field.

PROBLEMS

9.1 Calculate the gage pressure at the bottom of a tank filled with benzene to a depth of 2.5 m. If the bottom of the tank is rectangular of dimensions 3.0 m by 1.5 m, what force does the benzene exert on it?

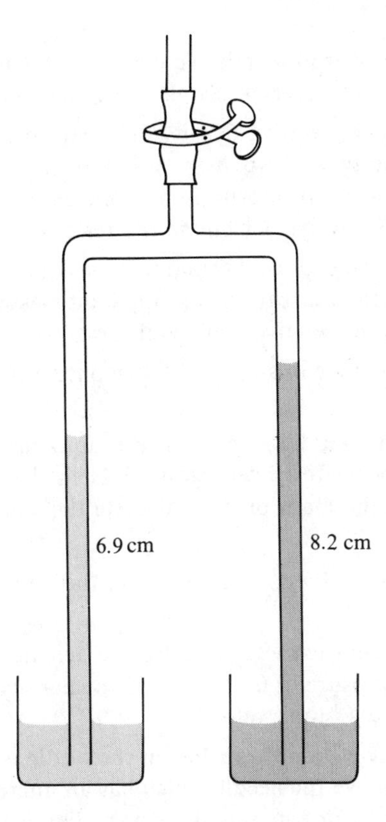

Fig. 9.17. Hare's apparatus for the determination of relative density.

9.2 Calculate the gage pressure at a depth of 1 km under the sea. The relative density of sea water may be assumed constant and equal to 1.05.

9.3 Hare's apparatus (Fig. 9.17) is a convenient method of measuring the relative density of a liquid. Pure water and the liquid under test are sucked up into the branched tube, and the levels of the surfaces recorded when the clip is closed. If the water stands at a height of 8.2 cm while the liquid level is 6.9 cm, what is the relative density of the liquid?

9.4 A vertical U-tube is partly filled with mercury, and a liquid of unknown density is poured into one arm. The surface of the mercury is at 7.7 cm, and the free surface of the liquid is at 23.6 cm. Calculate the density and relative density of the liquid, if the liquid–mercury surface is at 6.2 cm.

9.5 A polar bear of mass 160 kg jumps on to an ice floe, and its weight is just enough to cause the floe to sink. If the densities of the sea and the ice are 1025 kg m^{-3} and 922 kg m^{-3} respectively, calculate the mass of the ice floe.

9.6 What minimum volume of timber of relative density 0.75 must be used to construct a raft if 6 kg of iron of relative density 7.9 is used in the construction, and if the raft is designed to carry 8 people of average mass 85 kg in fresh water?

9.7 A mixture of xylene (relative density 0.87) and brombenzene (relative density 1.50) is used to determine the density of blood. Drops of blood are found to remain suspended when the proportions by volume of xylene and brombenzene are 72% and 28% respectively. Calculate the relative density of blood.

9.8 A piece of glass is weighed and found to have a mass of 0.84 kg in air. When weighed in water and then in turpentine the apparent masses are found to be 0.45 kg and 0.56 kg. Calculate the relative density of the glass and of the turpentine.

9.9 An animal weighs 467 g in air, and 34 g in pure water. Calculate the relative density of the animal.

9.10 A liquid of density 950 kg m^{-3} flows in a horizontal pipe of radius 4.5 cm. In a section of the tube of restricted radius 3.2 cm the liquid pressure is 1.5×10^3 N m^{-2} less than in the main pipe. Calculate the velocity of the liquid in the pipe.

9.11 Using the data of Problem 9.10, calculate the volume rate of flow of liquid through the pipe.

9.12 A Pitot tube is being used to measure the velocity of blood flow, and the manometer records a pressure of 0.25 torr. If the density of blood is taken to be 1020 kg m^{-3}, calculate the blood velocity.

9.13 In giving a patient a blood transfusion the bottle is set up so that the level of blood in it is 1.3 m above the needle which has an internal diameter of 0.36 mm and is 3 cm in length. In one minute 4.5 cm^3 of liquid pass through the needle. Use these data to calculate the viscosity of blood.

9.14 What excess pressure is required to send water through a hypodermic needle of length 2 cm and diameter 0.3 mm at a rate of 1 cm^3 s^{-1}?

9.15 The radius of the aorta in humans is about 1 cm and the cardiac output is about 5×10^{-3} m^3 per minute. What is the average velocity of flow in the aorta?

9.16 The diffusion coefficient of sucrose in water is 5.2×10^{-10} m^2 s^{-1}. Calculate how much sucrose will diffuse down a horizontal pipe of cross-sectional area 5 cm^2 in 10 s under a concentration gradient of 0.25 kg m^{-3} per meter.

9.17 Powdered chalk is sprinkled on the surface of water in a beaker. Assuming the particles to be spherical, calculate the radius of the largest particles remaining in suspension after 24 hours if the depth of the water is 10 cm. The density of chalk may be taken as 4000 kg m^{-3}.

9.18 A mass of 4.5 g of a substance of molecular mass 382 is dissolved in 750 cm^3 of water at a temperature of 17°C. Calculate the resulting osmotic pressure of the solution.

9.19 Calculate the osmotic pressure at 15°C of a solution of 10 g of sugar dissolved in 1000 cm^3 of water, the molecular mass of the sugar being 360.

9.20 A clean glass tube of internal diameter 0.5 mm is standing vertically in a dish of water. How far does the water rise in the tube. ?

9.21 What would the diameter of the capillaries in the xylem of trees have to be if surface tension were to be a satisfactory explanation of how sap gets to the top of a 100-m redwood? The surface tension of sap may be assumed to be the same as that of water.

CHAPTER 10

VIBRATIONS

10.1 INTRODUCTION

Vibratory to-and-fro motion is very common in nature, and in many cases this motion is of a particularly simple kind. These cases occur when there exists a restoring force on the system whose magnitude is proportional to the displacement of the system from its equilibrium position. We shall begin by looking at a few such cases by way of example.

a) A spring of force constant k carries a mass m at one end: it is pulled down an extra distance x and released. Find the equation of motion. Let L_0 be the unstretched length of the spring and let L be the length when the mass m is hanging from it in equilibrium, as shown in Fig. 10.1. Then, since the force exerted by a stretched spring is proportional to the amount of stretching, we have

$$mg = k(L - L_0).$$

When the mass is pulled down an extra distance x, the tension increases to $k(x + L - L_0)$. In this situation there is an unbalanced force on the mass equal to the tension in the spring less the weight of the mass, and this resultant force acts upwards, i.e. it is a restoring force tending to pull the mass back to its equilibrium position. Writing Newton's second law for this case,

$$\text{unbalanced force} = k(x + L - L_0) - mg = m\,(\text{upward acceleration}).$$

If x is reckoned as positive when the mass is below the equilibrium position as shown in Fig. 10.1, then the upward acceleration must be reckoned negative. We use the convenient dot notation and write it as $-\ddot{x}$. Newton's second law then reads, subtracting out the term mg,

$$kx = -m\ddot{x}, \qquad \text{or} \qquad \ddot{x} + \left[\frac{k}{m}\right] x = 0. \tag{10.1}$$

The dimensions of the ratio k/m are important. Since k is a force per unit length, they come out to be dimensionally $(\text{time})^{-2}$. The units of k/m therefore are s^{-2}.

b) The mercury in a narrow U-tube is pushed down and released. The U-tube is shown in Fig. 10.2, the equilibrium situation with the mercury at the same level in each arm being shown by the dashed line. Let the total length of the mercury column be L, the density of mercury be ρ, the cross-section of the U-tube be A, and the depth below equilibrium that one arm is depressed be x. Then, the total mass of mercury is $m = LA\rho$.

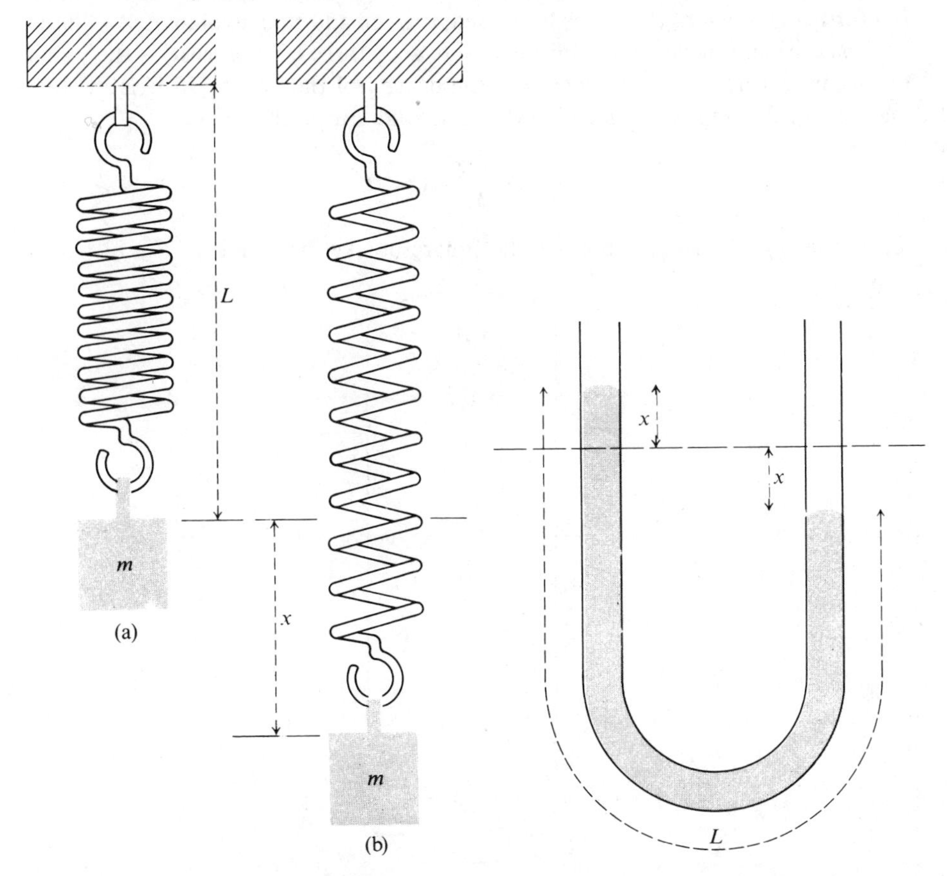

Fig. 10.1. A spring (a) in equilibrium when a mass is suspended from it, and (b) stretched beyond this position.

Fig. 10.2. An oscillating column of mercury in a U-tube.

When the right-hand arm is depressed as shown, the restoring force is clearly just the weight of a length $2x$ of mercury, since the left-hand arm mercury is at a height $2x$ above the other. This weight is $2xA\rho g$, and so Newton's second law applied to the whole mass of mercury gives $(LA\rho)\ddot{x} = -2A\rho g x$, or

$$\ddot{x} + \left[\frac{2g}{L}\right] x = 0. \tag{10.2}$$

Again the negative sign arises because the force tends to reduce x and again the square bracket has units s^{-2}.

c) A cylindrical rod weighted at the bottom floats vertically in a liquid, and the rod is pushed down a distance x. When the rod is floating normally, it is in equilibrium under the joint action of its weight and the upthrust of the liquid. If a further length x of the rod is immersed, then an additional volume xA of liquid is displaced; and, hence, the additional buoyancy force (or upthrust) is the weight of this extra volume, $xA\rho g$, where ρ is the density of the liquid. If M is the total mass of the floating body, then Newton's second law gives

$$\ddot{x} + \left[\frac{A\rho g}{M}\right] x = 0. \tag{10.3}$$

Again the negative sign, and again the square bracket has units s^{-2}.

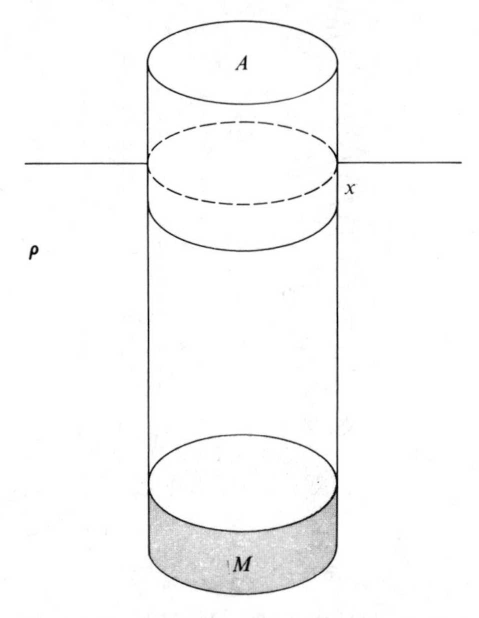

Fig. 10.3. A vertical rod oscillating in a liquid in which it floats.

If the cylinder can float upright without the help of the extra weights at the bottom, then it is left as an exercise to show that the equation of motion is

$$\ddot{x} + \left[\frac{g\rho}{L\rho'}\right] x = 0,$$

where L and ρ' are the length and density of the rod.

In all the above examples the final form of the equation of motion has been the same, viz.

$$\ddot{x} + [\quad]x = 0,$$

where the bracket has held a combination of the physical constants that define the problem and has in every case the dimensions (time)$^{-2}$. Thus the quantity in the bracket is a (frequency)^2, and we denote it always by the symbol

$$[\quad] \equiv \omega^2,$$

where ω is the angular frequency of the problem. Other constants can be defined in relation to ω as follows:

$$\text{the frequency } f = \omega/2\pi,$$
$$\text{the period } T = 1/f = 2\pi/\omega.$$

Clearly, which constant is used is immaterial and only usage dictates the form. Thus in ordinary a.c. electricity the frequency f is always used and discussions of pendulums are usually made in terms of the period T. The angular frequency ω is most often used theoretically and the general equation of all the above motions may be written in the form

$$\ddot{x} + \omega^2 x = 0. \tag{10.4}$$

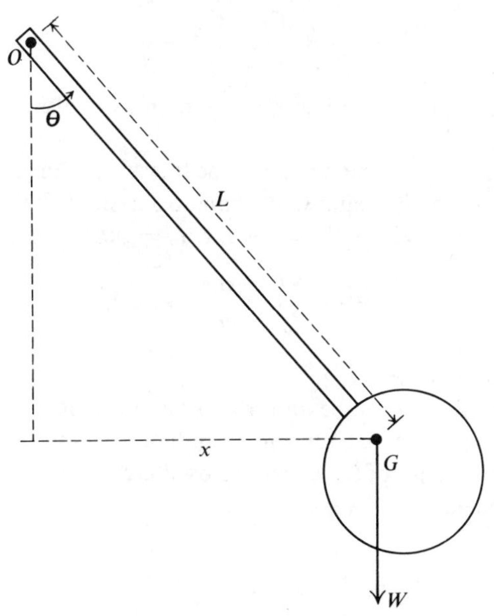

Fig. 10.4. The pendulum.

Any system that has this equation of motion is said to be executing simple harmonic vibrations along the x-axis. Other variables are possible. Thus a system moving according to the equation $\ddot{y} + \omega^2 y = 0$ is executing simple harmonic motion along the y-axis. Equally, a system in rotational motion according to the equation $\ddot{\theta} + \omega^2 \theta = 0$ is executing rotational simple harmonic motion.

For example, consider the case of a pendulum. A pendulum consists of a rigid body, often in the form of a rod to which a heavy disc is attached, as shown in Fig. 10.4. The pendulum is free to swing in a plane about a fixed point O. If the center of mass of such a pendulum is at G, the only torque on it about O is Wx, where W is the total weight of the pendulum. If L is the distance between O and G, then $x = L \sin \theta$, and the torque is $WL \sin \theta$. Because of this torque the angle θ will decrease and the angular acceleration will be $-\ddot{\theta}$. The negative sign again is necessary because the acceleration is opposite in sense to the angle. Newton's second law must here be applied in its rotational form "torque = moment of inertia times angular acceleration", and this leads to

$$I_0\ddot{\theta} + WL \sin \theta = 0,$$

where I_0 is the moment of inertia of the pendulum about O. As it stands, this is *not* in the required simple form. However, as discussed in Appendix A, the values of $\sin \theta$, $\tan \theta$ and θ do not differ very much if the angle θ is small (less than about 0.15 rad), and to this degree of approximation the equation may be rewritten

$$\ddot{\theta} + \left[\frac{WL}{I_0}\right] \theta = 0. \tag{10.5}$$

Hence, a pendulum oscillates with simple harmonic angular motion so long as the amplitude is small.

A simple pendulum is one with all the mass concentrated at the point G. This can be approximately realized by hanging a small bob of mass m from a string of length L. In this case $W = mg$ and $I_o = mL^2$, which gives

$$\omega^2 = \frac{WL}{I_o} = \frac{mgL}{mL^2} = g/L, \tag{10.6}$$

and so the period is $T = 2\pi\sqrt{L/g}$.

The variable need not even be an ordinary coordinate: it might be the electric charge, as in the case of a capacitor and an inductor connected in series, where the electric charge on the plates of the capacitor oscillates between positive and negative with a certain frequency.

10.2 THE SOLUTION OF THE EQUATION OF MOTION

The equation of motion requires us to find a function whose second rate of change with respect to time is simply $-\omega^2$ multiplied by itself. We look back at Table 3.2, where first and second derivatives for all the common functions are listed, and we see that both $x = A \sin \omega t$ and $x = A \cos \omega t$ fit our requirements. The corresponding velocities, listed in column II of the table, are $\dot{x} = +\omega A \cos \omega t$ and $\dot{x} = -\omega A \sin \omega t$. But it is not difficult to see that neither of these expressions is satisfactory in itself. For consider the situation where we are observing a pendulum, and suppose firstly that we start our stop-watch as the pendulum swings through the vertical position. Then the solution $x = A \sin \omega t$ is suitable, because at $t = 0$ (when we start the watch) the sine function is zero which fits the situation. But the other solution $x = A \cos \omega t$ is not zero at $t = 0$, and consequently this is not a suitable solution. Now suppose that we choose to start our stop-watch when the pendulum is at its maximum displacement. Clearly the cosine solution is now satisfactory, but the sine solution is not. And neither suits the general situation where we start the watch at some intermediate point. However, there is an easy way out. For the rate of change of the sum of two functions is just the sum of the rates of change of the functions separately. Hence, since both the sine and the cosine forms of the solution satisfy the original equation, so will their sum, and we conclude that the most general solution of the equation of motion is

$$x = A \sin \omega t + B \cos \omega t. \tag{10.7}$$

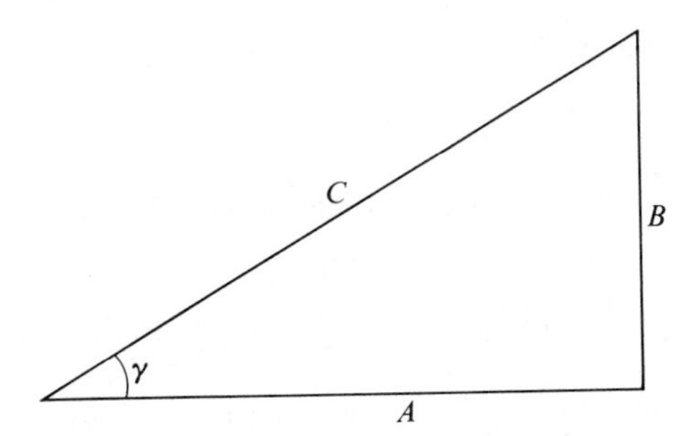

Fig. 10.5. The relation between A, B and C, γ in the two forms of the solution of the simple harmonic equation.

A second arbitrary constant, B, is now used with the cosine term to make it as general as possible. There are now two arbitrary constants which can be adjusted to suit the two initial conditions as we know to be necessary. An alterna-tive (but equivalent) form of solution can be obtained in terms of two other con-stants C and γ. These are defined in terms of A and B as shown in Fig. 10.5,

$$A = C \cos \gamma \qquad \text{and} \qquad B = C \sin \gamma.$$

This gives

$$\begin{aligned} x &= (C \cos \gamma) \sin \omega t + (C \sin \gamma) \cos \omega t \\ &= C \sin (\omega t + \gamma). \text{ (See Appendix A.)} \end{aligned} \tag{10.8}$$

This is often the most useful form.

In Eq. (10.8) the whole term $(\omega t + \gamma)$ is called the *phase* of the motion and γ is called the *phase constant*. To see how to handle this form of the equation, let us look once again at the different initial conditions possible with the pendulum.

Case (a). Stop-watch started as bob passes through equilibrium. Hence, we impose the initial conditions $x = 0$ and $v = +V$ at $t = 0$. Our solution Eq. (10.8) gives $0 = C \sin \gamma$; hence, $\gamma = 0, \pi, 2\pi, 3\pi, \ldots$ The velocity in general is

$$v = \dot{x} = \omega C \cos(\omega t + \gamma).$$

So at $t = 0$,

$$\begin{aligned} +V &= \omega C \cos(\omega 0 + 0) \quad &\text{or} \quad &\omega C \cos(\omega 0 + \pi) \\ &= +\omega C \quad &\text{or} \quad &-\omega C. \end{aligned}$$

The easiest way to write the solution is therefore to choose $\gamma = 0$ and then

$$x = \frac{V}{\omega} \sin \omega t.$$

Case (*b*). Stop-watch started as bob is at $x = +X$ and stationary. Hence, we impose the initial conditions $x = +X$ and $v = 0$ at $t = 0$. These require that $\gamma = \pi/2$ and $C = X$, giving $x = X \cos \omega t$, the details being left as an exercise.

These two cases should be worked out also using the solution in the form of Eq. (10.7).

10.3 SIMPLE HARMONIC MOTION FROM THE STANDPOINT OF ENERGY

Consider a block of mass m free to slide on a frictionless table and attached to a horizontally mounted spring of force constant k. If the spring is extended by an amount x, then it exerts a restoring force $F = kx$ on the block and this, by Newton's second law, must be equal to $-m\ddot{x}$. So the equation of motion becomes

$$\ddot{x} + \frac{k}{m} x = 0 \qquad \text{or} \qquad \ddot{x} + \omega^2 x = 0 \qquad \text{with} \qquad \omega^2 = \frac{k}{m}.$$

We introduce the concept of the potential energy of the stretched spring. This is defined as the work that has to be done by an external force against the tension in extending the spring by an amount x from its unstretched position. If F' denotes the required external force, then in the absence of friction the unbalanced force acting on the block during the extension is $F' - F$. By the work-energy principle, the work done by the unbalanced force is equal to the increase in the kinetic energy of the block. Hence, if the block starts from rest and finishes at rest, the work done is zero, and we conclude that the work done by the external force F' (which is defined as the potential energy of the stretched spring) is equal to the work done against the elastic force $F = kx$. This latter amount of work may be found by calculating the area under the graph of F against distance (Fig. 10.6). This graph is a straight line through the origin of slope k. The area under the line is just the area of a triangle, half the base times the height, and this gives the work performed, in the extension of amount x, as $\frac{1}{2}x(kx) = \frac{1}{2}kx^2$. This is defined as the potential energy of the spring for that extension, $U(x)$; hence,

$$U(x) = \tfrac{1}{2}kx^2. \tag{10.9}$$

The whole process of vibration of the block is conservative, since there are no

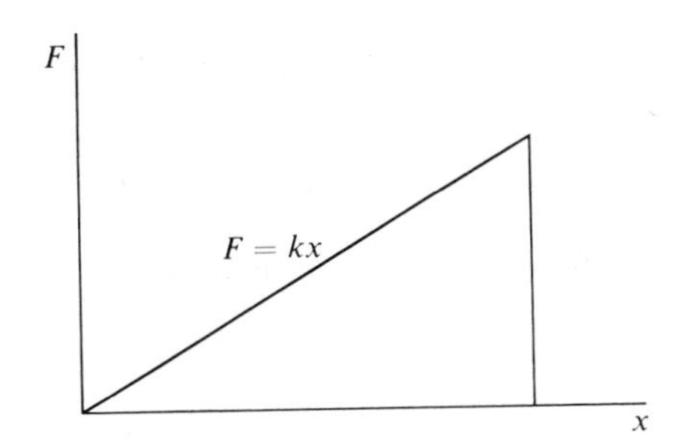

Fig. 10.6. The elastic force exerted by a spring increases linearly with the extension x.

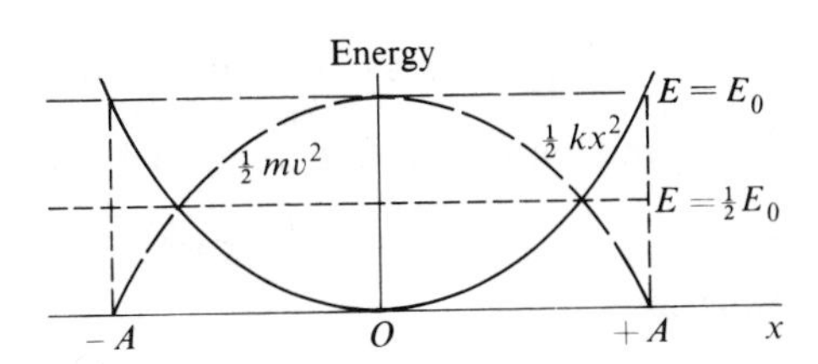

Fig. 10.7. Plots of the kinetic energy, the potential energy, and the total energy in simple harmonic motion.

frictional losses (we ignore losses due to air-friction), and so the mechanical energy of the block is a constant $= E$. This means that

$$\tfrac{1}{2}mv^2 + \tfrac{1}{2}kx^2 = E \text{ (constant)}. \tag{10.10}$$

The first term is the usual kinetic energy, and the nature of the solution can be found if we plot the potential energy as a function of x and draw in horizontal lines for various values of the total mechanical energy E, as discussed in Section 5.10 and shown in Fig. 10.7. The potential energy curve is an inverted parabola forming an infinite potential well. In other words, the curve can be extended indefinitely upwards. Suppose that the value of the total mechanical energy of the system is $E = E_0$. Then, since $\frac{1}{2}mv^2$ cannot be negative, the motion must be restricted to that region of the x-axis where $E_0 - \frac{1}{2}kx^2$ is positive, which means to that region where the $V(x)$ curve lies entirely below the line $E = E_0$. Where this line cuts the $V(x)$ curve, $E_0 = \frac{1}{2}kx^2$ and the block must be at rest. These two points represent the maximum possible displacement of the block, $x = \pm A$, where A is called the *amplitude*. Since $\frac{1}{2}mv^2 + \frac{1}{2}kx^2 = E_0$, the curve of the kinetic energy as a function of position is the same as the $V(x)$ curve, but reflected in the line $E = \frac{1}{2}E_0$. So where the potential energy is large, the kinetic energy is small, and vice versa. At the extreme points of the motion, where the kinetic energy is zero, all the energy is potential and $E_0 = \frac{1}{2}kA^2$. Hence, at a general point, $\frac{1}{2}mv^2 + \frac{1}{2}kx^2 = \frac{1}{2}kA^2$, from which we easily derive the important relation

$$v^2 = \frac{k}{m}(A^2 - x^2) = \omega^2(A^2 - x^2). \tag{10.11}$$

The maximum speed of the block occurs at $x = 0$, where the potential energy is zero, and is given from the above equation by

$$v_{\max} = V = \pm\omega A. \tag{10.12}$$

The two possible signs arise when we take the square root and correspond to the two possible directions of motion of the block.

These relations may also be derived from the solution Eq. (10.8) to the equation of motion. For then we have

$$x = C \sin(\omega t + \gamma) \qquad \text{and} \qquad v = \dot{x} = \omega C \cos(\omega t + \gamma).$$

Using the Pythagoras theorem, $\sin^2 + \cos^2 = 1$ for any angle, we easily get

$$\sin^2(\omega t + \gamma) + \cos^2(\omega t + \gamma) = (x/C)^2 + (v/\omega C)^2 = 1,$$

whence

$$v^2 = \omega^2(C^2 - x^2)$$

as before. The result is identical, although we use a different symbol for the amplitude. The value of the maximum velocity is immediately apparent from the form of the expression for v as a function of time, since the maximum value of the cosine term is unity.

10.4 THE ROTOR DIAGRAM REPRESENTATION OF SIMPLE HARMONIC MOTION

Simple harmonic motion, which we have been discussing at length, may be defined in quite another way, as follows: it is the projection on a diameter of uniform motion in a circle. This definition is very useful indeed, but carries the risk that the unwary reader will think of simple harmonic motion as a circular motion, whereas it is (usually) a linear motion. So long as this risk is clearly recognized, we may learn a great deal from this new definition. Consider, then, a circle of radius A and consider a point moving round the circle in a counterclockwise direction with uniform speed V and angular velocity ω, where $V = \omega A$, as shown in Fig. 10.8. Measure all angles from the positive direction of the x-axis as usual, and suppose that the particle is at an angle γ at $t = 0$ when the stop-watch is started. In a subsequent time t it will move through a further angle ωt and be at an angular position $\omega t + \gamma$. The velocity of the particle is along the tangent at Q and the acceleration ($= V^2/A = \omega^2 A$) is along the radius toward the center, as we discussed previously. Consider now the projection of the motion of this particle on the y-axis. Physically, this would correspond to putting a dab of Plasticine on the edge of a record and looking at it from a distance with the eye at the level of the Plasticine; it would be seen to vibrate to and fro with simple harmonic motion. To find the projection of the motion, we resolve along the y-direction. The projection of the point Q is at P, whose y-coordinate is

$$y = A \sin(\omega t + \gamma), \tag{10.13}$$

which we recognize immediately as the proper form of simple harmonic motion. The tangential velocity V of the particle at Q has a vertical component

$$\begin{aligned} \dot{y} &= V \cos(\omega t + \gamma) \\ &= \omega A \cos(\omega t + \gamma), \end{aligned} \tag{10.14}$$

and the acceleration $\omega^2 A$ along QO has a vertical component

$$\begin{aligned}\ddot{y} &= -\omega^2 A \sin(\omega t + \gamma) \\ &= -\omega^2 y. \end{aligned} \tag{10.15}$$

Everything tallies exactly, and we have no doubt that this description of simple harmonic motion is entirely equivalent to the others.

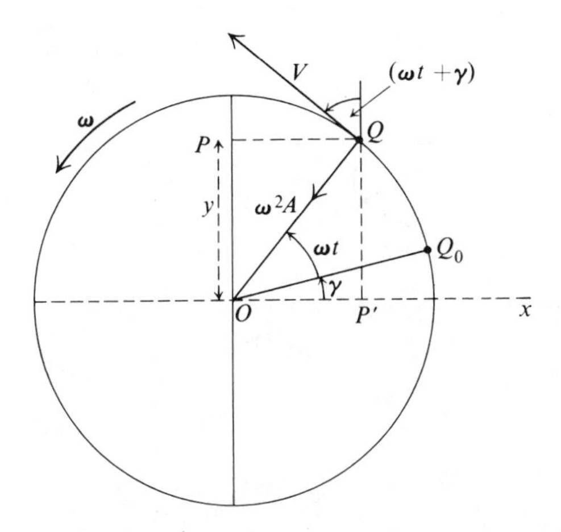

Fig. 10.8. The rotor diagram.

If we look at the projection of the circular motion of Q on the x-axis rather than on the y-axis, the following results may be deduced:

$$x = A \cos(\omega t + \gamma), \tag{10.16}$$

$$\dot{x} = -\omega A \sin(\omega t + \gamma), \tag{10.17}$$

and

$$\ddot{x} = -\omega^2 x. \tag{10.18}$$

This is also simple harmonic motion, but it differs in phase by $\pi/2$ from the motion along the y-axis. This follows from the fact that

$$\cos(\omega t + \gamma) = \sin\left(\omega t + \gamma + \frac{\pi}{2}\right).$$

Hence the very important result is obtained that uniform circular motion is equivalent to two mutually perpendicular simple harmonic motions of the same frequency whose phases differ by $\pi/2$. A particle which possesses simultaneously these two motions along the y- and x-axes lies on a circle. The equation of this circle may be obtained by squaring and adding Eqs. (10.13) and (10.16), which gives

$$x^2 + y^2 = A^2[\sin^2(\omega t + \gamma) + \cos^2(\omega t + \gamma)] = A^2.$$

More generally, if a particle has simultaneously two simple harmonic motions, $y = B \sin(\omega t + \gamma)$ along the y-axis and $x = A \cos(\omega t + \gamma)$ along the x-axis, then it will move in two dimensions along the curve obtained by eliminating the time between these two equations. This gives

$$\cos^2(\omega t + \gamma) + \sin^2(\omega t + \gamma) = 1 = x^2/A^2 + y^2/B^2. \tag{10.19}$$

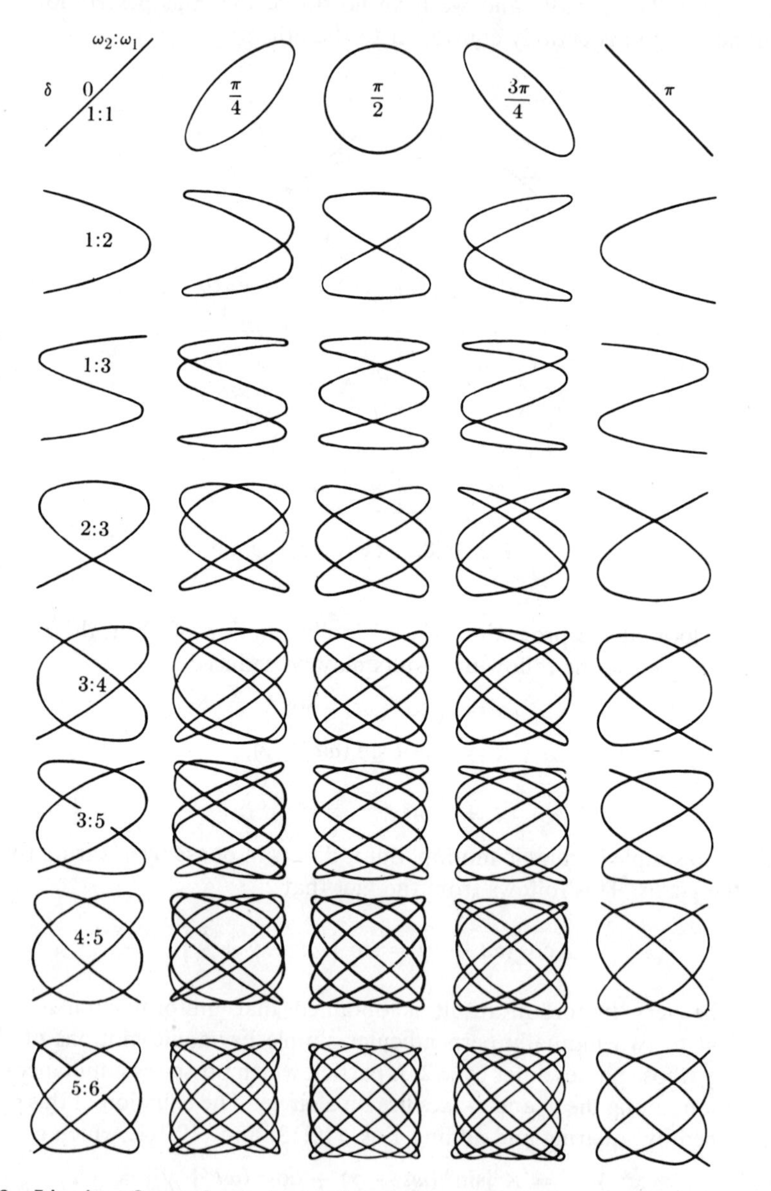

Fig. 10.9. Lissajous figures for various ratios of the two frequencies involved. (Reprinted by permission from *Fundamental University Physics*, Vol. I by M. Alonso and E. J. Finn (Addison-Wesley).)

This is the equation of an ellipse. In this case also the motions have the same frequency and differ in phase by $\pi/2$. For other phase differences the path becomes a tilted ellipse that degenerates into a straight line for certain phase differences, even when the two amplitudes are equal. If the frequencies are not the same, the path is even more complex and is called a *Lissajous figure*. Use is made of Lissajous figures in calibrating frequencies in a.c. circuits. The result of combining two simple harmonic motions of the same frequency at right angles has an important application in optics, for light is an electromagnetic radiation, and the electric field at any point in space is what gives rise to the optical effects. In plane-polarized monochromatic light the electric field at any point in a beam of light is varying sinusoidally with simple harmonic motion, $E = E_0 \sin(\omega t + \gamma)$. In circularly and elliptically polarized light each component of the electric field perpendicular to the direction of propagation of the light vibrates with simple harmonic motion of the same frequency, and the tip of the electric vector will traverse an ellipse in general, and a circle in the special case where the amplitudes in the two directions are equal and the phase difference is $\pi/2$.

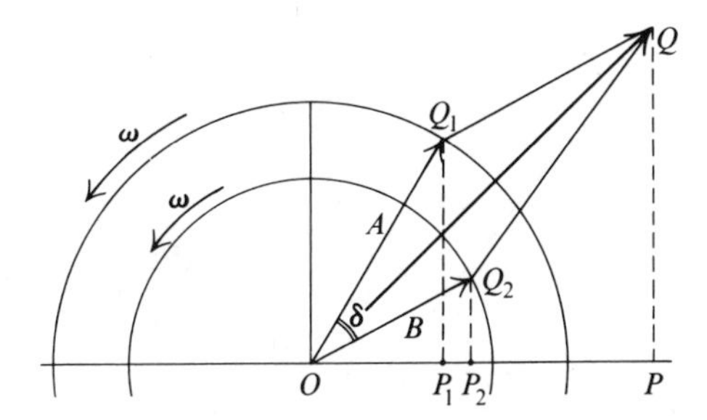

Fig. 10.10. The combined effect of two coherent simple harmonic vibrations given by use of the rotor diagram.

10.5 THE COMBINATION OF SIMPLE HARMONIC MOTIONS ALONG A LINE

Another optical effect, interference, can be explained by considering the combination of two simple harmonic motions along the same straight line. Let the first motion be represented by the rotation of point Q_1 in the rotor diagram of Fig. 10.10 and the second motion by the point Q_2. The two motions have the same frequency and, although they do not have the same phase, they preserve the same phase relation throughout the motion. In other words, there is a constant difference of phase between them. The motions in such a case are said to be *coherent*. On the rotor diagram the difference of phase is simply the angle between OQ_1, and OQ_2. This difference of phase is $(\omega t + \gamma_1) - (\omega t + \gamma_2) = \gamma_1 - \gamma_2 = \delta$, say. We consider the projection of these motions on the x-axis. The projection of Q_2 is P_2 at a displacement of x_2 along the axis. The projection of Q_1 is P_1 at a displacement of x_1. Since OQ_1 and Q_2Q are opposite sides of a parallelogram, it is clear that $OP_1 = P_2P$. Hence, the total displacement of the particle is $OP_2 + OP_1 = OP_2 + P_2P = OP$, and the resultant motion is the projection on the x-axis of the uniform circular motion of Q, the opposite corner of the parallelogram formed with OQ_1 and OQ_2 as contiguous sides. This is merely the parallelogram rule for combining two vectors, and OQ_1 and OQ_2 may indeed be thought of as vectors. Thus the resultant motion is also a simple harmonic motion and, since the point Q rotates with the same angular velocity ω as Q_1 and Q_2, the frequency is unchanged. If the difference of phase δ is zero, or 2π, or 4π, . . . , then OQ_1, OQ_2 and OQ all lie along the same direction and the resultant motion has an amplitude which is the sum of the component amplitudes. Thus the resulting motion is $x = (A + B) \cos (\omega t + \gamma)$. If, on the other hand, the phase difference is π, 3π, 5π, . . . , then the lines OQ_1 and OQ_2 are oppositely oriented and the resultant amplitude is $A - B$. The first case is called *constructive interference*; the second case, *destructive interference.* If the amplitudes of the two constituent motions are the same, then the resultant motion in the second case is zero motion, corresponding to complete darkness in the optical case.

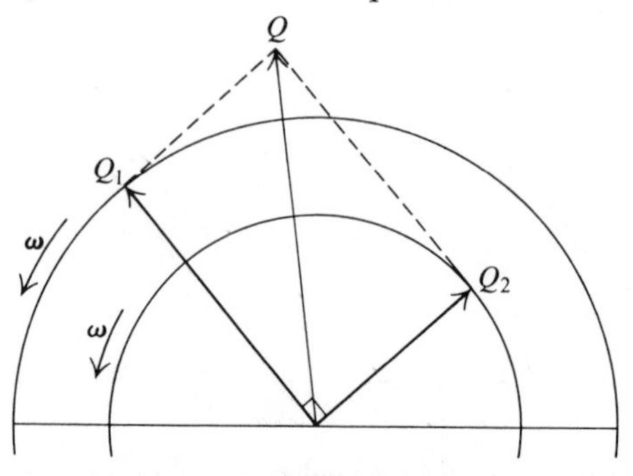

Fig. 10.11. The combined effect of two coherent simple harmonic vibrations $\pi/2$ out of phase.

Note the importance of the concept of coherence in all this. If the two motions are not coherent, the phases of the two components will jump about in a random way and it will be impossible to have a steady resultant motion. Similarly, it should be abundantly clear from the above discussion why it is impossible to produce interference effects with two beams of light of different frequencies; for if Q_1 and Q_2 are rotating round the circle with different angular velocities, then no steady resultant state is possible.

If the difference of phase is $\pi/2$, $3\pi/2$, $5\pi/2$, . . . , etc., then the resultant amplitude is easily found by Pythagoras, as shown in Fig. 10.11:

$$OQ^2 = OQ_1^2 + OQ_2^2.$$

This is the basic equation of a.c. theory, where the sinusoidal voltage across an inductor is $\frac{1}{2}\pi$ ahead of the sinusoidal voltage across a resistor. The addition of these two voltages is performed with a rotor diagram, as explained in Chapter 27. Notice that in order to get the resultant amplitude it is not necessary to draw the whole rotor diagram since the rotation is common to everything in it. Thus only the vectors OQ_1 and OQ_2 need be drawn. One simply remembers in the back of one's mind that the whole thing is really rotating.

10.6 DAMPING

All processes, including the vibratory processes described in this chapter, are subject to energy losses due to friction in one form or another. If the displacement from equilibrium of a vibrating system is plotted as a function of the time, one or other of the two curves shown in Fig. 10.12 is obtained.

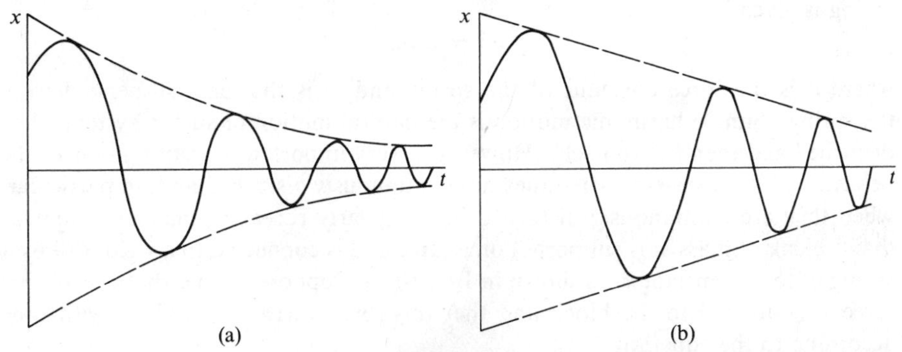

Fig. 10.12. Simple harmonic vibrations (a) exponentially damped and (b) linearly damped.

The amplitude gradually decreases, which corresponds to the energy being dissipated, since the energy is always proportional to the square of the amplitude. In Fig. 10.12(a) the dotted lines show the exponential decrease in amplitude (see Appendix A, Section 1) typical of systems subject to fluid friction. This is the type of behavior shown by pendulums swinging in air, for example. In fluids the frictional forces are dependent on the velocity. If the frictional forces are constant in magnitude, as they are when two solid surfaces rub together, the dotted lines joining the peaks in the curve are straight lines, as shown in Fig. 10.12(b). The frictional forces change the frequency of the vibration. In normal circumstances the change is very slight; but if the damping is large enough, the system will cease to vibrate all together. Figure 10.13 shows three displacement-time curves for a system that has been displaced initially and then released. In curve A the system shows a small amount of vibratory motion, in that it overshoots the equilibrium position and then returns. In curve C the damping is so heavy that the system takes a very long time to reach equilibrium. This would correspond, perhaps, to a pendulum trying to swing in molasses. Curve B is the situation of *critical damping*, which divides the two cases. Many vibrating systems in measuring instruments are used in a critically damped state or in a very slightly underdamped state. Thus a galvanometer for measuring very small currents is a vibrating system, although here the damping is electromagnetic in kind (see Section 26.6). If the damping is too light, the needle will vibrate to and fro about the desired reading, making the instrument virtually useless. If the damping is too heavy, the needle will creep up to its final reading so slowly that again the usefulness of the instrument is impaired. When the damping is about critical, the needle comes up to its final reading quickly and without any oscillation.

10.7 RESONANCE

In this chapter various systems have been discussed that exhibit simple harmonic motion. For each system there is a characteristic frequency that depends on the physical quantities involved. Thus the characteristic frequency of vibration of a spring is given by

$$\omega^2 = k/m,$$

where k is the force constant of the spring and m is the mass suspended from the spring. Simple harmonic motion is the natural motion of such a system when disturbed and then left to itself. However, a very important question involves the behavior of such systems when they are continuously disturbed and, in particular, when they are continuously disturbed in a regularly repeating manner. Suppose that a block of mass m is supported on a table and is connected to a rigid wall by a spring of force constant k, as shown in Fig. 10.14. Suppose further that an external force F is applied to the block and that this force varies sinusoidally with time according to the equation

$$F = F_0 \sin \omega' t,$$

where ω' is the angular frequency of the force. The force changes its direction regularly with time, and tends to compress and extend the spring alternately.

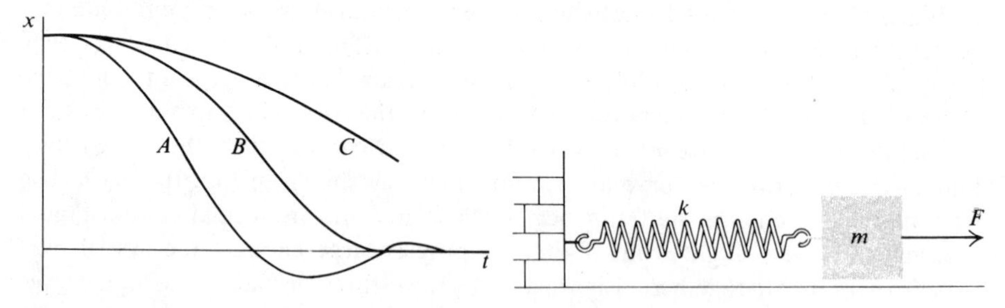

Fig. 10.13. Heavily damped systems returning to equilibrium, B showing the case of critical damping.

Fig. 10.14. A system for producing forced vibrations.

The result of applying such a force is to compel the mass to vibrate with the frequency ω' of the force, not with its own natural frequency ω. The amplitude and therefore the energy of the vibrating mass depend in a complicated way on the difference between ω' and ω. As the forcing frequency ω' becomes more equal to

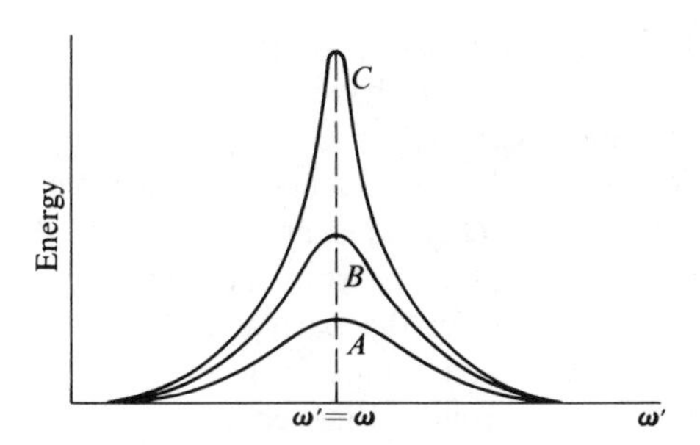

Fig. 10.15. Resonance curves with different amounts of damping in the system.

Fig. 10.16. The Tacoma bridge in process of collapsing due to resonance. (Photograph by courtesy of U.P.I.)

ω, more and more energy is transferred to the mass, until, when $\omega = \omega'$, the maximum amount of energy is transferred. This maximum transfer of energy is known as *resonance*. Figure 10.15 shows how the energy E of the block varies with the frequency ω' of the applied force. The three resonance curves A, B, C correspond to heavy, medium, and light damping, respectively. It will be seen that when the damping is light, the response of the system is very sharp indeed. It hardly vibrates at all until the forcing frequency ω' is quite close to the natural frequency. Such a system is said to have a high Q-factor, where the Q-factor is a parameter that measures essentially the sharpness of the resonance peak. The Q-factor of an electrical system is discussed in Section 27.7. Sharp resonance curves, corresponding to high-Q systems, are very useful. For example, the tuning of a radio involves absorbing energy from the incoming radio waves, and one is interested only in radio waves of the particular frequency corresponding to the station one wants to hear. The tuning is accomplished by varying the natural frequency ω of an electrical circuit until it equals ω'. This is done by rotating the tuning knob on the case. If the circuit does not have a high Q-factor, then other radio stations whose frequencies are not too far from ω' will also be picked up and cause interference. To be selective, a high Q-factor is essential. But some care is needed in dealing with mechanical systems with high Q-factors, for the energy absorbed at resonance from the source can be destructively high. Figure 10.16 shows the Tacoma bridge in process of destroying itself on 7 November 1940. Such a structure, of course, is not a simple vibrator, but essentially the situation is as described above. The bridge is absorbing excessive energy because of resonance, and the amplitude of vibration becomes so large that the structure collapses.

Example 1. A thin rod of length 1.5 m is pivoted at one end so that it can oscillate in a vertical plane as a pendulum. What is its period, and what would be the length of a simple pendulum with the same period?

Solution. From Fig. 6.8(b) the moment of inertia of a thin rod of length a is $I_0 = \frac{1}{3}Ma^2$ about one end. The center of mass of the rod is at $L = \frac{1}{2}a$, and substitution into Eq. (10.5) gives

$$\omega^2 = \frac{WL}{I_0} = Mg(\tfrac{1}{2}a)/\tfrac{1}{3}Ma^2 = 3g/2a.$$

For a simple pendulum the corresponding expression is g/L, and therefore the length of the equivalent simple pendulum is

$$L = 2a/3 = 1 \text{ m}.$$

The period of the rod is

$$T = 2\pi/\omega = 2\pi\sqrt{2a/3g} = 2\pi\sqrt{\frac{2 \times 1.5 \text{ m}}{3 \times 9.8 \text{ m s}^{-2}}}$$

$$= 2.0 \text{ s}.$$

Example 2. A 5-kg mass oscillates with an amplitude of 25 cm when attached to a spring of force constant 125 N m^{-1}. Find (a) the period, (b) the maximum speed, (c) the speed when the displacement from equilibrium is 20 cm, and (d) the kinetic and potential energies at that point.

Solution. From Eq. (10.1), $\omega^2 = 125\ \text{N m}^{-1}/5\ \text{kg} = 25\ \text{rad}^2\ \text{s}^{-2}$

therefore $\omega = 5\ \text{rad s}^{-1}$.

a) The period $T = 2\pi/\omega = (2\pi/5)\ \text{s} = 1.26\ \text{s}$.

b) The maximum speed is $\omega A = 5\ \text{rad s}^{-1} \times 25\ \text{cm} = 125\ \text{cm s}^{-1} = 1.25\ \text{m s}^{-1}$.

c) When $x = 20$ cm, we use Eq. (10.11) to give

$$v^2 = \omega^2(A^2 - x^2) = 25\ \text{rad}^2\ \text{s}^{-2}\ (25^2 - 20^2)\ \text{cm}^2$$

$$= 25\ \text{rad}^2\ \text{s}^{-2} \times 15^2\ \text{cm}^2.$$

Therefore

$$v = 5\ \text{rad s}^{-1} \times 15\ \text{cm} = 75\ \text{cm s}^{-1} = 0.75\ \text{m s}^{-1}.$$

d) The kinetic energy is $\frac{1}{2}mv^2 = \frac{1}{2} \times 5\ \text{kg} \times (0.75\ \text{m s}^{-1})^2 = 1.4\ \text{J}$ and the potential energy is $\frac{1}{2}kx^2 = \frac{1}{2} \times 125\ \text{N m}^{-1} \times (0.2\ \text{m})^2 = 2.5\ \text{J}$. The total mechanical energy is therefore 3.9 J which agrees with the value calculated directly from $E = \frac{1}{2}kA^2 = \frac{1}{2} \times 125\ \text{N m}^{-1} \times (0.25\ \text{m})^2$.

Example 3. A horizontal board is made to perform vertical oscillations of angular frequency 7.0 rad s^{-1}. If an object placed on the board is to remain in contact with it at all times, what is the maximum possible amplitude of the vibrations?

Solution. If the object remains in contact at all times, then it too is performing the simple harmonic vibrations, and consequently its maximum acceleration, which occurs at the extremities of the motion, is $\omega^2 A$. At the higher extreme point, the object will be on the point of leaving the board when the normal force between them is zero. When this is so, the only force acting will be gravity, and therefore the maximum possible amplitude must satisfy the equation:

$$\omega^2 A_{\text{max}} = g,$$

which gives

$$A_{\text{max}} = g/\omega^2 = 9.8\ \text{m s}^{-2}/49\ \text{rad}^2\ \text{s}^{-2} = 0.20\ \text{m}.$$

Example 4. Calculate the sum of $4 \sin \omega t + 18 \sin(\omega t + \frac{1}{2}\pi) + 15 \sin(\omega t - \frac{1}{2}\pi)$.

Solution. The rotor diagram for this sum is shown in Fig. 10.17. The first term is represented by a line of length 4 units along the positive x-axis; the second term by a line of length 18 units along the positive y-axis, since $\gamma = +\frac{1}{2}\pi$; and the third term by a line of length 15 units along the negative y-axis, since $\gamma = -\frac{1}{2}\pi$. The second and third terms are along the same line, and therefore their sum is represented by a line of length 3 units along the positive y-axis. The sum of all three terms is represented by a line of length $\sqrt{3^2 + 4^2} = 5$ units at 37° to the x-axis as shown.

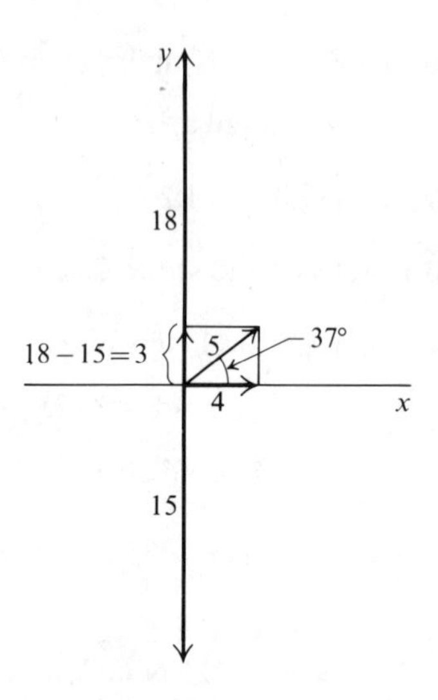

Fig. 10.17.

PROBLEMS

(In many numerical problems on vibrations the arithmetic can be much simplified by noting that the approximation $g = \pi^2$ m s^{-2} is very good, better than 1%.)

10.1 A steel spiral spring has an unstretched length of 8 cm. When a weight is hung on it, its length becomes 14 cm. If this weight is now pulled down a further distance and released, find the period of the ensuing motion.

10.2 The scale of a spring balance reading from 0 to 10 kg is 25 cm long and a body hung from the balance has a period of vibration of 0.8 s. What is the mass of the body?

10.3 When a metal cylinder of height 0.15 m, floating upright in mercury of density 1.36×10^4 kg m^{-3}, is set into vertical oscillation, the period is 0.62 s. Calculate the density of the metal.

10.4 A test-tube of mass 6 g and of external diameter 2 cm is floated vertically in water by putting 10 g of mercury at the bottom of the tube. Calculate the period of oscillation of the system.

10.5 If the length of a simple pendulum is increased by 2%, calculate the percentage change in the period of the pendulum.

10.6 A circular hoop oscillating about a point on its circumference has the same period as a simple pendulum of length equal to the diameter of the hoop. What is the moment of inertia of the hoop in this arrangement?

10.7 A steel seconds pendulum (i.e. a pendulum with $T = 2$ s) is correct at 20°C. How many beats a day will it lose if the temperature is 25°C and the pendulum has increased in length by 0.006%?

10.8 A flat circular disk of diameter 0.50 m is pivoted so that it is free to oscillate about an axis passing through a point on its edge and perpendicular to its face. The radius of gyration of the disk in this configuration is 0.306 m. What is the length of a simple pendulum of the same period?

10.9 A long thin wire of torsional constant 0.16 N m rad^{-1} supports at its lower end a circular metal disk of mass 2 kg and radius 0.2 m; its upper end is securely clamped. The arrangement is shown in Figure 6.14 (p. 140). Show that twisting the wire and then releasing it causes the disk to execute simple harmonic vibrations, and calculate the period of these oscillations.

10.10 A point moving with simple harmonic motion along the axis of x has velocities of 20 and 25 m s^{-1} at distances of 10 m and 8 m, respectively, from its center of attraction. Calculate the period of the oscillations and the acceleration of the point at a distance of 1 m from the center.

10.11 A body is executing simple harmonic motion. When the displacement is 6 cm, the speed is 16 cm s^{-1}; when the displacement is 8 cm, the speed is 12 cm s^{-1}. Find the amplitude and period of the motion.

10.12 A point moves with simple harmonic motion along the axis of x. Its amplitude is 20 cm and its frequency is 5 c/s. At time $t = 0$ the point is at $x = 0$ and is traveling from right to left. Write down mathematical expressions for the displacement, velocity, and acceleration of the point as a function of time.

10.13 A body is vibrating with simple harmonic motion of amplitude 20 cm and period 0.5 s. Calculate the maximum values of the acceleration and velocity, and the values of acceleration and velocity when the body is 10 cm away from its force center. How long does the body take to move from the force center to a point 15 cm away?

10.14 On a given day the depth of high water over a harbor bar is 10 m, and at low water $6\frac{1}{4}$ h earlier it is 7 m. If high water is due at 3.20 p.m., what is the earliest time in the afternoon that a ship drawing 9 m can cross the bar, assuming the rise and fall of the tide to be simple harmonic.

10.15 A horizontal board is made to perform simple harmonic oscillations horizontally, with an amplitude of 0.3 m and a period of 4 s. Calculate the minimum value of the coefficient of friction between a heavy body placed on the board and the board if the body is not to slip.

10.16 A mass of 0.5 kg is attached to the lower end of a spring of force constant 50 N m^{-1} and supported in the hand so that the spring is not extended by the weight of the mass. If the mass is now let fall, what will be the maximum extension of the spring? What will be the period of the ensuing simple harmonic vibrations?

10.17 A particle of mass 0.10 kg has a velocity given by the expression

$$v = (4 \text{ m s}^{-1}) \sin\left(\frac{\pi t}{2 \text{ s}} + \frac{\pi}{3}\right).$$

Write down the expression for the displacement of the particle as a function of time. What is the kinetic energy of the particle at $t = 0.5$ s?

10.18 Prove that in simple harmonic motion the time average of the kinetic energy and the time average of the potential energy are both equal to $\frac{1}{4}kA^2$. [*Hint*. Use the relation $\cos 2\theta = \cos^2\theta - \sin^2\theta$.]

10.19 Two springs of force constants k_1 and k_2 are joined end to end. Show that they behave like a single spring of force constant k given by

$$\frac{1}{k} = \frac{1}{k_1} + \frac{1}{k_2}.$$

10.20 A mass of 1 kg is attached to the end of a spring of force constant 4000 N m^{-1}, and the other end of this spring is connected in turn to the lower end of a second spring of force constant 2000 N m^{-1} whose upper end is securely fastened to a rigid beam. Calculate the period of the simple harmonic vibrations of the mass when it is displaced. Calculate also the total mechanical energy of the system if the energy is taken as zero when the mass hangs in equilibrium, and the amplitude of vibration is 10 cm.

10.21 A mass of 5 kg hangs from a long elastic spring of force constant 5000 N m^{-1}, and both spring and mass are traveling vertically upwards with uniform velocity of 1 m s^{-1}. The upper end of the spring is suddenly brought to rest. Find the amplitude of the ensuing simple harmonic motion of the mass.

10.22 A point moves so that its displacement from the origin at any time is given by

$$x = (3 \text{ cm}) \sin 2\pi ft + (4 \text{ cm}) \cos 2\pi ft.$$

What is the amplitude of the motion of the point?

10.23 A particle is subjected simultaneously to two simple harmonic motions of the same frequency and direction. Their equations are

$$x_1 = (8 \text{ cm}) \sin\left(\omega t + \frac{\pi}{3}\right) \quad \text{and} \quad x_2 = (12 \text{ cm}) \sin\left(\omega t + \frac{2\pi}{3}\right).$$

Find the amplitude of the resulting motion graphically.

10.24 A particle is simultaneously subjected to simple harmonic motions $x_1 = A \sin 2\pi ft$ and $x_2 = A \sin(2\pi ft + 2\theta)$, where θ is a constant. Determine graphically the resultant amplitude for several values of the phase difference 2θ and verify that the expression $A' = 2A\cos\theta$ correctly gives the resultant amplitude.

CHAPTER 11

WAVE MOTION

11.1 INTRODUCTION

Consider a familiar situation, a pendulum hanging from a stand on the laboratory bench. If the pendulum is set into vibration, then the amplitude of the vibration gradually decreases until eventually the pendulum is brought to rest. Our discussion of the energy of the pendulum showed that the total mechanical energy was proportional to the square of the amplitude, and consequently we see that the energy of the vibrating pendulum is gradually dissipated. Where does it go? Well, clearly, the friction of the air has a lot to do with it, so let us imagine the experiment conducted in a vacuum. Would the vibration now persist for ever? The answer is no. There is always an energy loss through the support and the laboratory bench itself and thence down into the floor, and there is nothing we can do to reduce this loss to zero. To show the leakage of energy through the supporting system, we consider two pendulums hanging from the same wooden support. If one of them is set swinging while the other is not, we can see how the initially stationary pendulum gradually begins to swing, admittedly with a small amplitude, but with some energy that could only come from the other pendulum. If the common support is held in the hand, the transfer of energy is much more apparent, since the human frame is more yielding than wood. If a much more direct connection is made between the pendulums, by joining them with a weak spring, for instance, then the interchange of energy is very striking indeed.

The general point is this: any vibrating body that is connected to its environment will communicate energy to that environment; and if the environment consists of a set of similar bodies, then they will gradually be set into vibration by the original body. In this way energy flows through matter. One very important point emerges from a study of the simple two-pendulum system: that although the frequencies of the two pendulums are the same, the phases are not. Put another way, it takes time for the energy to be transferred, and this time manifests itself as a phase difference between the various parts of the compound system. Consider a very large number of identical pendulums all hanging from the same support and linked elastically with one another. If the first pendulum is set swinging, this motion will gradually travel down the line of pendulums, and this constitutes what is called a wave. Each pendulum will vibrate with simple harmonic motion, but there will be a progressive change of phase as we proceed along the line. If we plot a graph of the pendulums' displacements at any instant of time as a function

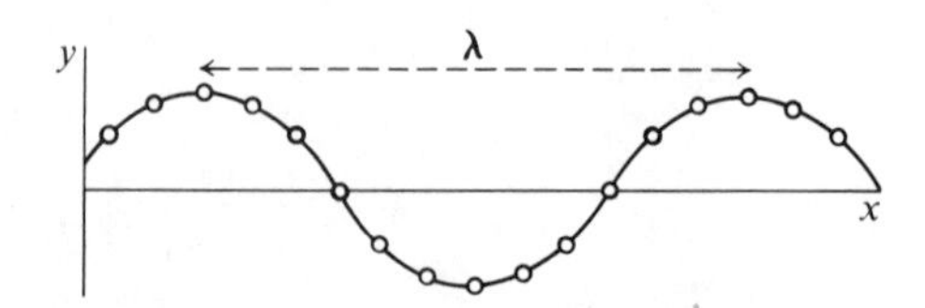

Fig. 11.1. The displacements of linked pendulums at any instant as a function of distance.

of their distance from the starting point, then this will be the familiar sine or cosine curve, as shown in Fig. 11.1. The x-axis has been chosen as the line through the equilibrium positions of all the pendulums, and the y-axis is taken as the direction of vibration of the pendulums. As time goes on, successive pendulums reach their maximum displacement, and the crests and troughs of the curve in Fig. 11.1 move from left to right. The individual pendulums do not change their positions along the x-axis; they merely vibrate along the y-axis. It is the *shape* of the sine curve that moves. To maintain all the pendulums swinging, of course, it will be necessary to feed energy continuously into the system by maintaining the first pendulum at a constant amplitude. The shortest distance between any two of the pendulums that are vibrating in phase is called the *wavelength*, and the velocity of propagation of the wave is just this distance divided by the period of the simple harmonic motion of any pendulum:

$$V = \lambda/T = f\lambda. \tag{11.1}$$

If we imagine the number of pendulums to be indefinitely increased and their separation reduced, then we approach a continuous medium, and this is a suitable model for the propagation of waves through gases, liquids, and solids.*

11.2 THE EQUATION OF A SINUSOIDAL WAVE

Let us consider an elastic string of infinite length, and consider what happens when one end of the string is vibrated with simple harmonic motion in a direction perpendicular to the length. The energy of the vibrating end is gradually transmitted down the string, and progressively different portions of the string take up the simple harmonic motion of the vibrating end. As seen above, the phase varies continuously along the string and the appearance of the string at any instant is a sine wave. This wave travels along the string with a velocity V and

* The alert reader might wonder why, in the light of the atomistic structure of all matter, the discrete pendulum model is not preferred to the continuous model, since real continuity is obviously incompatible with an atomistic hypothesis. The answer is that it is mathematically much simpler to deal with an infinite set of vibrating bodies at zero separation than it is to deal with a discrete set. In any event, the interatomic distances are so small that both methods always give the same answer.

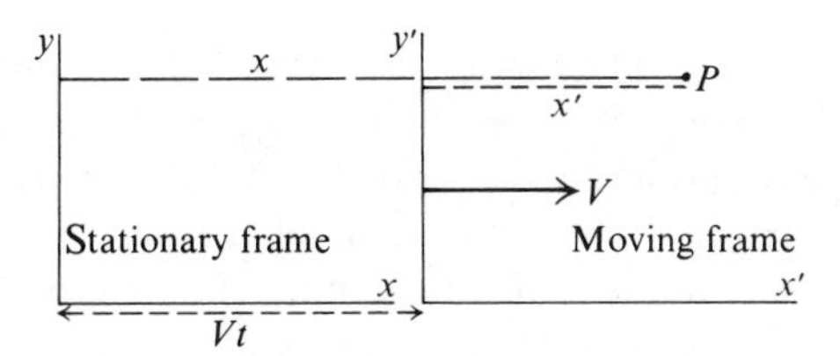

Fig. 11.2. The relation between the frames of reference of a stationary observer and of an observer moving relative to the first with constant velocity.

the points of the string separated by a distance λ are at the same phase. Notice the twin characteristics of such a wave:

i) at any instant of time the profile of the string is a sine curve;
ii) at any point along the string the motion is simple harmonic with equation

$$y = A \sin (\omega t + \gamma).$$

To find the equation of the wave, let us consider an observer traveling with the same velocity as the wave. Clearly, he will see, not a traveling wave, but a stationary sine curve, since his velocity relative to the wave is zero. In other words, in his frame of reference the equation of the wave is just $y = A \sin kx'$, where k is a constant and x' is referred to the moving observer. It is easy to find k in terms of λ, since the sine curve repeats at intervals of λ in x' and the angle kx' repeats at intervals of 2π. So, clearly, $k(x' + \lambda) = kx' + 2\pi$; hence, $k = 2\pi/\lambda$. The relation between the observer's frame of reference and a stationary frame is shown in Fig. 11.2. We assume the wave to be traveling from left to right. Then, by considering any point P we see that its coordinate in the stationary frame is x and in the moving frame it is $x' = x - Vt$, where t is the time that has elapsed since the vertical axes of the two frames were coincident. Thus any curve that has an equation $y = F(x')$ in the moving frame will have the equation $y = F(x - Vt)$ in the stationary frame, where F can be any function at all. In particular, where it is the sine function, we conclude that the equation of a sinusoidal wave traveling from left to right with velocity V as seen by a stationary observer is

$$y = A \sin k(x - Vt). \tag{11.2}$$

Using the relations $k = 2\pi/\lambda$, $V = f\lambda$, $f = 1/T = \omega/2\pi$, we may rewrite Eq. (11.2) in several equivalent forms, as follows:

$$y = A \sin 2\pi \left\{\frac{x}{\lambda} - \frac{t}{T}\right\}, \tag{11.3}$$

$$y = A \sin 2\pi \left\{\frac{x}{\lambda} - ft\right\}, \tag{11.4}$$

$$y = A \sin (kx - \omega t). \tag{11.5}$$

The equation of a sinusoidal wave traveling from right to left is simply

$y = A \sin k(x + Vt)$, obtained from Eq. (11.2) by changing the sign of V; the other forms are obtained similarly by changing the sign of the term involving the time. Looking at the above equations, we see that they indeed satisfy the twofold requirement of wave motion: for if we fix the value of the time, they represent a spatial sine curve, and if we fix the value of x they represent a simple harmonic vibration.

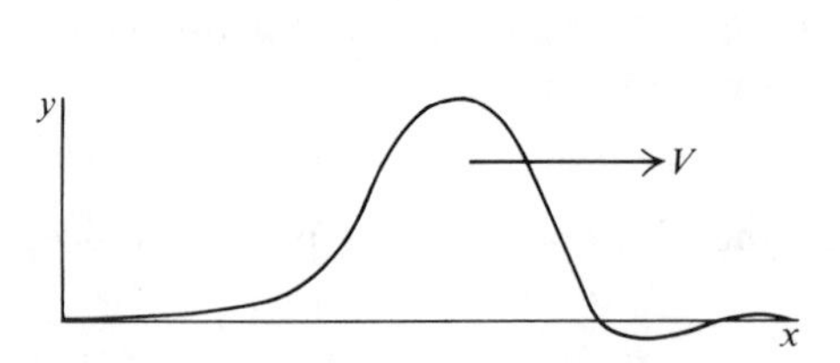

Fig. 11.3. The displacement curve for a string as a pulse moves along it.

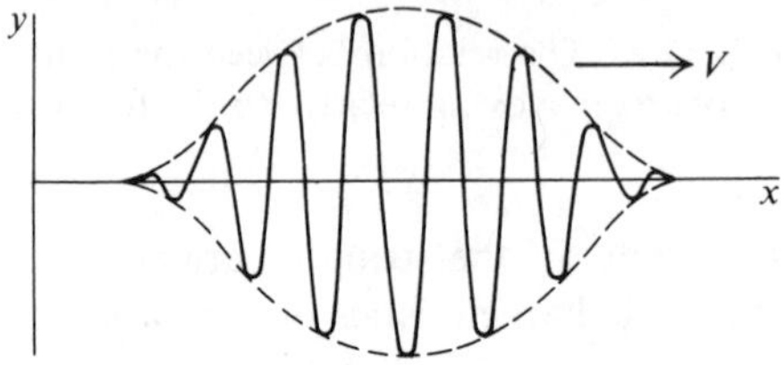

Fig. 11.4. The displacement curve for a string as a wave packet moves along it.

The sine wave discussed above is rather rare in nature, and other forms of disturbance are much more common. Thus suppose that the end of our infinite string is given a sudden flick; then energy will certainly travel along the string, but not in the form of a continuous sine wave. Rather will there be a single hump in the string that travels outwards, as shown in Fig. 11.3. This is usually referred to as a *pulse*. It is characterized by a value of the velocity V which will be the same as the velocity with which a sine wave would travel along the same string, but there is now no frequency or wavelength associated with it. Between this case and the pure sine wave there are many intermediate forms of traveling disturbance. The type shown in Fig. 11.4 is common and is called a wave group or wave packet. Here again we cannot specify one unique frequency and wavelength, although it might appear that the disturbed region enclosed by the dotted lines in the diagram does have these properties. In fact such a wave group is characterized by a range of frequencies and wavelengths: the smaller the range the more extended the group, until, for an exactly defined frequency, the group becomes an infinite sinusoidal wave. In real life a perfect sinusoidal wave is impossible, and, for example, light from a sodium lamp will have the characteristics of an extended group, since the frequency range for the yellow line is very narrow. The importance of the perfect sinusoidal wave lies in the fact that it is not a bad approximation to reasonably monochromatic light, and it can easily be dealt with theoretically.

11.3 THE VELOCITY OF WAVES IN ELASTIC MEDIA

The velocity of any kind of disturbance traveling through an elastic medium depends only on the properties of the medium. When the direction of vibration of the particles of the medium is parallel to the direction of propagation of the wave, it is said to be a *longitudinal* wave, and all such waves are generally referred to as *sound waves*. Such waves are the only kind of wave possible in a liquid or a gas. Elastic solids, however, can also transmit *transverse* waves, since solids can support a shear stress. Transverse waves, such as the waves on the string discussed above and electromagnetic waves, exhibit the phenomenon of *polarization*, which arises because there is no one unique direction perpendicular to the direction of propagation. This matter is discussed in more detail in Section 17.6. The formulae for the velocity of propagation in the three cases of importance are as follows:

$$\text{sound waves in a liquid or gas:} \quad V = \sqrt{B/\rho}, \tag{11.6}$$

$$\text{sound waves in a solid:} \quad V = \sqrt{Y/\rho}, \tag{11.7}$$

$$\text{transverse waves on strings:} \quad V = \sqrt{F/\mu}. \tag{11.8}$$

In these expressions B is the bulk modulus of the fluid, Y is Young's modulus for the solid, ρ is the density, F is the tension in the string, and μ is the mass per unit length of the string. The transverse waves in strings are usually dealt with under sound waves because of their intimate connection with stringed musical instruments.

The velocity of sound in a gas is very dependent on the temperature, and it is of interest to investigate this dependence. The bulk modulus for any fluid is defined as the increment of pressure applied to it divided by the fractional decrease in volume that results. In symbols,

$$B \equiv \frac{\Delta p}{-\Delta v/v} \rightarrow -v\frac{dp}{dv}. \tag{11.9}$$

The fluctuations in pressure as a sound wave passes through a gas are so rapid that virtually no heat flows, and the process obeys the law pv^{γ} = constant. Thus

$$p = \text{constant} \times v^{-\gamma},$$

and the rule shown in Table 3.2 allows us to find the derivative,

$$\frac{dp}{dv} = \text{constant} \times -\gamma v^{-\gamma-1}.$$

Hence, the bulk modulus for a gas is given by

$$B = \text{constant} \times \gamma v^{-\gamma} = \gamma p. \tag{11.10}$$

Hence, finally, the velocity of sound in a gas is

$$V = \sqrt{\frac{\gamma p}{\rho}} = \sqrt{\frac{\gamma RT}{M}}, \tag{11.11}$$

where M is the molar mass and R is the universal gas constant. It is therefore seen that the velocity of sound in a gas depends only on the absolute temperature and not on the pressure or the density. This expression is closely connected with the value of the root-mean-square velocity of the molecules of tne gas. In Section 8.2 it was shown that this velocity is given by

$$C = \sqrt{\frac{3p}{\rho}} = \sqrt{\frac{3RT}{M}}.$$

Fairly obviously, the velocity of sound cannot be greater than C, and this is true, since γ is less than 3 for gases that obey even approximately the perfect gas laws on which Eq. (11.11) is based.

At 300K, the velocity of sound in air ($M = 28.8 \text{ g mol}^{-1}$ and $\gamma = 1.4$) is easily calculated to be

$$V = \sqrt{\frac{1.4 \times 8.32 \text{ J mol}^{-1}\text{ K}^{-1} \times 300 \text{ K}}{0.0288 \text{ kg mol}^{-1}}} = 348 \text{ m sec}^{-1}.$$

By comparison, the velocity of sound in water is 1460 m s^{-1}.

The relatively low speed of sound makes possible the pulse-echo technique of locating objects and estimating their distance away without the need for very elaborate equipment. The principle is very simple: a pulse of sound is sent out and the time taken for the echo to come back after reflection from the object of interest is measured. A simple calculation gives the distance of the object. In recent years this method has been applied as a diagnostic tool in medicine using, not audible sound waves, but *ultrasonic waves* whose frequencies are higher than the ear can respond to. The upper limit of audibility varies from person to person, but a value of 20,000 c/s is generally accepted. The reason for choosing to work with ultrasonic waves is that the shorter the wavelength the better can small objects be detected. This is a very important physical principle which is discussed in detail in later chapters. An ultrasonic pulse of frequency about 2 Mc/s is produced by a crystal and fed into the brain, perhaps, by means of a suitable transducer. The technical details are not relevant for our purpose here. The velocity of sound in soft biological tissue varies between 1490 and 1610 m s^{-1}, and the typical to-and-fro distances involved are of the order of 0.10–0.20 m, so the electronic receiving equipment is easily able to cope with the echo times of about 100 μs. Typically the crystal will emit 1000 pulses per second, each of about 1 μs duration, at a power of about 1 W m^{-2}. The wave will suffer reflection at any boundary between materials of different acoustic properties. The determining property is the *acoustic impedance*, which is defined as ρV, where ρ is the density and V the velocity of sound in the material. The reflected pulse is detected and shows up as a blip on the screen of a cathode-ray oscilloscope, the distance of this blip from the start of the trace being calibrated directly in terms of the penetration distance. Experienced workers can interpret these traces in terms of pathological conditions of various kinds. The technique has many advantages for diagnostic work on the brain, the heart, the eye, and the abdominal region in general. It is convenient and painless in operation, and relatively inexpensive. As far as is known, it presents no hazard to health, although this must remain an open question.

Another use of ultrasonics in diagnosis also involves the reflection of a suitable pulse, but this time from a moving object. When a wave is reflected from a moving object, it returns with a slightly altered frequency, the so-called Doppler effect. The change in frequency can be made to produce audible signals, and the clinician can literally hear the movement of, for example, fetal blood.

11.4 STANDING WAVES

So far we have been speaking about the propagation of waves in infinite strings and fluids. When the string or fluid has a finite extension, our treatment requires modification. For whenever a traveling wave meets any discontinuity in the medium, a reflected wave will immediately be established, and the resulting state of the medium behind the discontinuity will be described by the superposition of the incident and reflected waves. As a simple and highly important example, consider the vibrations of a stretched string clamped firmly at both ends. The state of the string cannot be described by the equation of a traveling wave, for such a wave is reflected successively from both ends, and the state of the string is the resultant of two traveling waves traveling in both directions. Let the two waves be

$$y_1 = A \sin (kx + \omega t) \qquad \text{and} \qquad y_2 = B \sin (kx - \omega t),$$

the first traveling from right to left and the second from left to right. Let the ends of the string be rigidly clamped at $x = 0$ and $x = L$. Then, the resultant displacement of any point of the string is given by

$$y = y_1 + y_2 = A \sin (kx + \omega t) + B \sin (kx - \omega t). \tag{11.12}$$

We must now impose on this solution the boundary conditions that $y = 0$ *at all times* when $x = 0$ and when $x = L$.

First condition.

$$y = A \sin (k0 + \omega t) + B \sin (k0 - \omega t) = 0.$$

Hence,

$$A \sin (+\omega t) + B \sin (-\omega t) = 0,$$

which gives

$$A = B.$$

So the solution is reduced to

$$y = A[\sin (kx + \omega t) + \sin (kx - \omega t)].$$

At this point we invoke the trigonometrical identity

$$\sin X + \sin Y = 2 \sin \tfrac{1}{2}(X + Y) \cos \tfrac{1}{2}(X - Y),$$

where X and Y are any two angles (see Appendix A). Applying this to the above solution gives

$$y = 2A \sin kx \cos \omega t, \tag{11.13}$$

and we see that the traveling wave has vanished, since this equation no longer has the appropriate form for such a wave. This equation represents instead a stationary wave At any time ($t =$ constant) we see that it represents a sine function shape, and at any point along the string ($x =$ constant) it represents a simple harmonic vibration.

Second condition. At $x = L$ the displacement is zero at all times; hence,

$$\sin kL = 0,$$

which means that the angle $kL = 0$, or π, or 2π, or 3π, . . . , or $N\pi$, where N is any integer including zero. In other words,

$$k = 0, \frac{\pi}{L}, \frac{2\pi}{L}, \ldots, \frac{N\pi}{L}.$$

In terms of the wavelength $\lambda = 2\pi/k$, this means that

$$\lambda = \infty, 2L, L, \frac{2L}{3}, \ldots, \frac{2L}{N}. \tag{11.14}$$

This is a most striking result. The imposition of the first boundary condition reduced the two traveling waves to a stationary wave, and the second boundary condition has restricted the possible wavelengths of the vibrations to certain specific values. This process of restricting the possible values of a physical quantity to a set of discrete values is called *quantization*. It arises from the necessity of imposing boundary conditions on any real problem. Although the connection between a stretched string and an atom may seem rather remote, the quantization of energy in an atom comes from the same kind of consideration. Let us quote Erwin Schrödinger, from his epoch-making paper of 1926:

> "In this paper I wish to consider first the simple case of the hydrogen atom and show that the usual quantum conditions can be replaced by another postulate in which the notion of 'whole numbers' merely as such is not introduced. Rather, when integralness does appear, it arises in the same natural way as it does in the case of the node-numbers of a vibrating string. The new conception is capable of generalization, and strikes—I believe—very deeply at the true nature of the quantum rules."

This matter is considered further in Chapter 30.

Figure 11.5 shows the first few modes of vibration of a vibrating string. The case of infinite wavelength merely corresponds to no vibration at all and is physically uninteresting.

The points where the string is permanently at rest are called the *nodes* and the points halfway between adjacent nodes where the amplitude of vibration is a maximum are called *antinodes*. It will be clear that the possible wavelengths are easily written down once we know whether the end-points are nodes or antinodes: it is merely a question of fitting a sequence of nodes and antinodes to the available length.

Since the velocity of propagation of waves along a string is $V = \sqrt{F/\mu}$, we see that fixing the possible values of the wavelength also fixes the possible values of the frequency. For a string fixed at both ends the possible frequencies are

$$\left.\begin{aligned} f_0 &= \frac{1}{2L}\sqrt{\frac{F}{\mu}}: \text{ fundamental or 1st harmonic} \\ f_1 = 2f_0 &= \frac{2}{2L}\sqrt{\frac{F}{\mu}}: \text{ 1st overtone or 2nd harmonic} \\ f_2 = 3f_0 &= \frac{3}{2L}\sqrt{\frac{F}{\mu}}: \text{ 2nd overtone or 3rd harmonic} \\ &\;\vdots \\ f_{n-1} = nf_0 &= \frac{n}{2L}\sqrt{\frac{F}{\mu}}: \text{ } (n-1)\text{st overtone or } n\text{th harmonic.} \end{aligned}\right\} \quad (11.15)$$

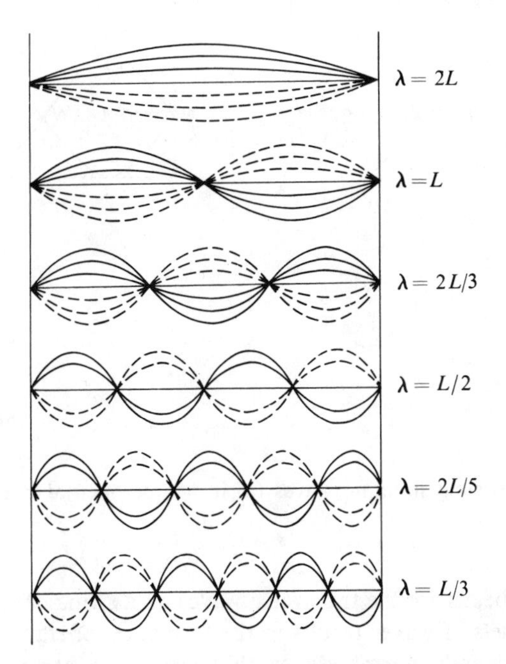

Fig. 11.5. Some possible standing waves on a string clamped at both ends.

The terminology is a little confusing but is sanctioned by usage. The set of overtones are only referred to as harmonics when they are integral multiples of the fundamental frequency.

Longitudinal waves traveling along a tube of finite length are reflected from

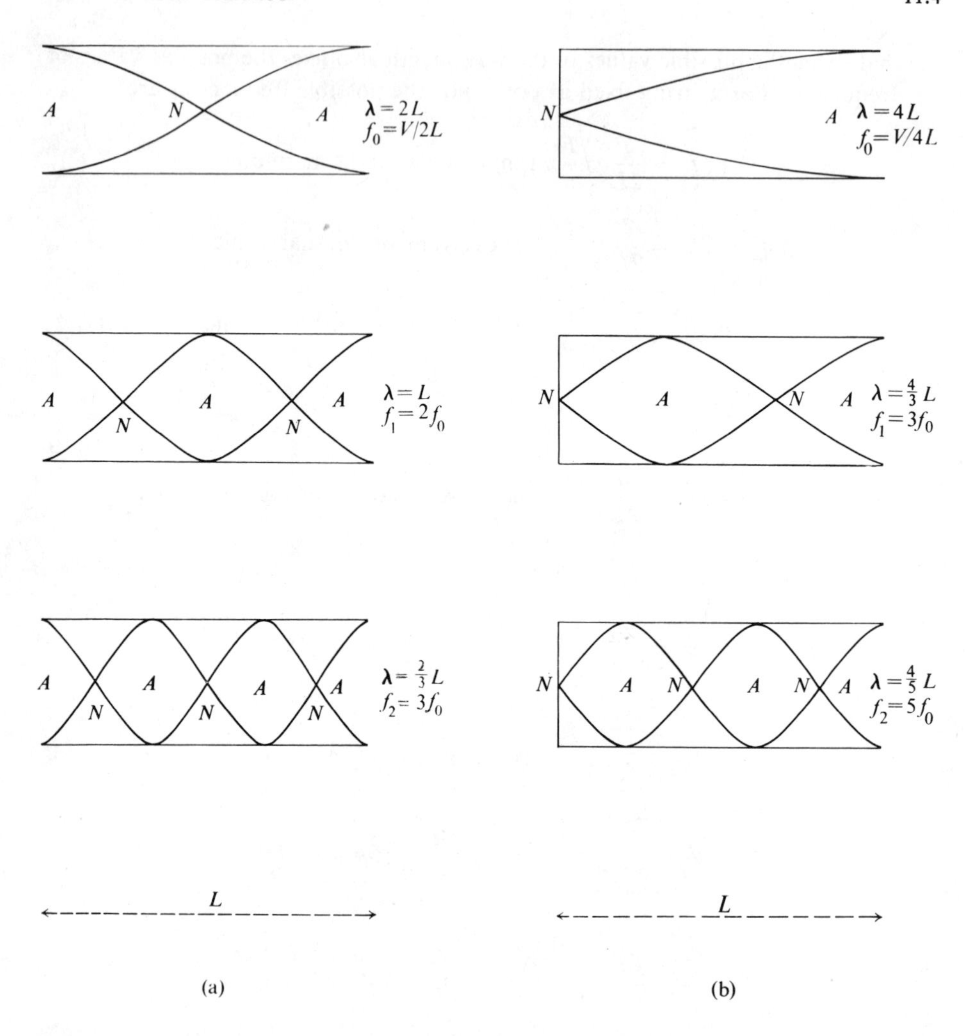

Fig. 11.6. Some possible standing waves (a) in a tube open at both ends and (b) in a tube open at one end.

the ends of the tube, as are the transverse waves on a string, and again interference between the two sets of waves results in the formation of stationary waves. For a tube closed at both ends the analysis would be exactly as above, since the displacements of the gas particles at a closed end must be zero. However, tubes closed at both ends are rather useless and we are more interested in tubes open at one or at both ends. When a longitudinal wave meets the open end of a tube, it is reflected and the gas particles have their maximum amplitude there (or, better, slightly outside the open end). So there is always an antinode of displacement near the open end of a pipe. At a closed end there is, of course, a node. Knowing this it is

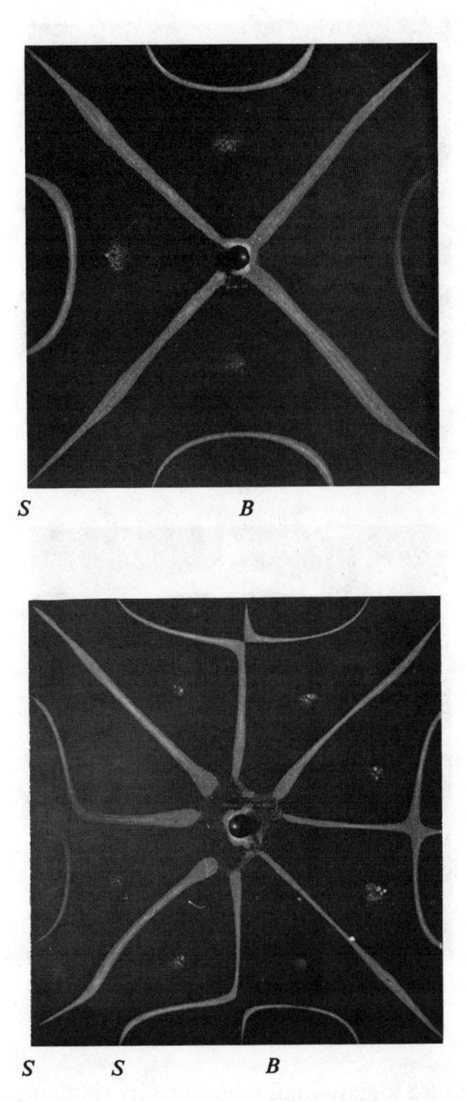

Fig. 11.7. Chladni figures for a square plate bolted at the centre. The plate was bowed at the points B and held lightly at points S. The node lines were rendered visible with salt.

easy to sketch in the possible modes of vibration, and the first few are shown in Fig. 11.6. We see that, for the pipe open at both ends (open pipe), the fundamental frequency is $V/2L$ and all harmonics are present. For the pipe open at one end and closed at the other (closed pipe), the fundamental frequency is $V/4L$ and only the odd harmonics are present. In musical terminology, the pitch of a closed pipe is an octave lower than the pitch of an open pipe of equal length, and the quality of the musical note emitted is different.

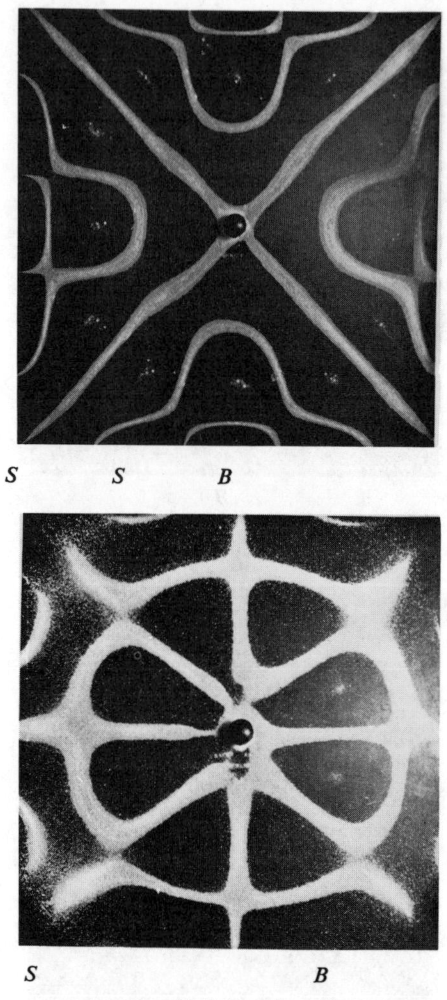

Fig. 11.7 (continued)

A rod may be set into longitudinal vibrations by clamping it at some point and stroking it vigorously with a chamois leather sprinkled with resin. The point of clamping must obviously be a node and the free ends must be antinodes. Hence, the possible wavelengths and frequencies are easily inferred. The case of a plate or a membrane (for example, a drum) is much more complex, since it is two-dimensional; but patterns of node lines and antinode lines exist in any vibrating plate, the exact pattern depending on the shape of the plate and on how it is set vibrating. Figure 11.7 shows four of the possible patterns of node lines on a square plate bolted at its center. The node lines have been rendered visible by sprinkling salt on the plate. The vibrations are induced by bowing the edge of the plate at the position marked B in Fig. 11.7 with a violoncello bow and lightly holding the edge at the positions marked S. Such patterns of node lines on a plate are known as Chladni figures. The crucial difference between the resonant frequencies of a plate and those of a wire or rod is that, in the case of the plate, the overtone frequencies are not integral multiples of the fundamental frequency, and consequently the word "harmonic" is inappropriate. This inharmonicity shows up in the discordant nature of the sound produced by a vibrating plate.

The various kinds of standing waves discussed represent possible ways in which the systems can vibrate. If energy is supplied to a system in the form of vibration, the system will only respond if the applied frequency is close to one of its own natural frequencies. This is the phenomenon of resonance discussed in the last chapter. This may easily be demonstrated by sounding a tuning-fork inside a piano with the dampers off. Each string of the piano has its own natural frequencies given by Eqs. (11.15), and those strings which have a natural frequency close to that of the fork vibrate in sympathy. Resonance is, in fact, basic to the design of all musical instruments.

11.5 BEATS

In the discussion of the rotor diagram interpretation of interference it was stated that interference between vibrations of different frequencies was impossible. For light this is true but for sound waves the statement needs modification. Let us consider two sound waves being propagated through the air, and let them have very nearly equal frequencies f and $f + \varepsilon$, where ε is a small quantity of the order of 1 or 2 c/s. (The smallness is relative to a typical sound frequency of hundreds of cycles per second.) Consider the vibration of the air at any point in space. Then, if the amplitudes of the two waves are equal, we can represent the situation by the rotor diagram of Fig. 11.8. Let OA represent the vibration of frequency f (angular frequency $\omega = 2\pi f$), OB the vibration of frequency $f + \varepsilon$, and OC the resultant of these two, so that the instantaneous amplitude of the resultant is OC'. Since the point B is rotating round the reference circle with angular velocity $2\pi(f + \varepsilon)$ while the point A is rotating with angular velocity $2\pi f$, it follows that B is rotating *relative to the line* OA with angular velocity $2\pi\varepsilon$. When OA and OB coincide, the resultant amplitude is just $2OA$; and when OA and OB point in opposite directions, the resultant amplitude is zero. So the resultant amplitude is a maximum every time OB catches up and passes OA, which it does once every $1/\varepsilon$ seconds. At that point of space, therefore, the ear will hear a succession of maximum sounds of frequency ε, and these are called *beats*. From the above analysis we see the general result

$$\text{beat frequency} = \text{difference in frequency of the two waves.}$$

This is a very good example of the use of the rotor diagram. If the difference ε between the constituent frequencies is too great (more than about 6 or 7 c/s), beats are not heard. Instead the ear distinguishes the separate frequencies as a discord. In other words, the ear *resolves* the two components.

The displacement with time of the air particles at a point where beats are being heard has the form shown in Fig. 11.9, obtained by summing two sine waves of slightly different frequency.

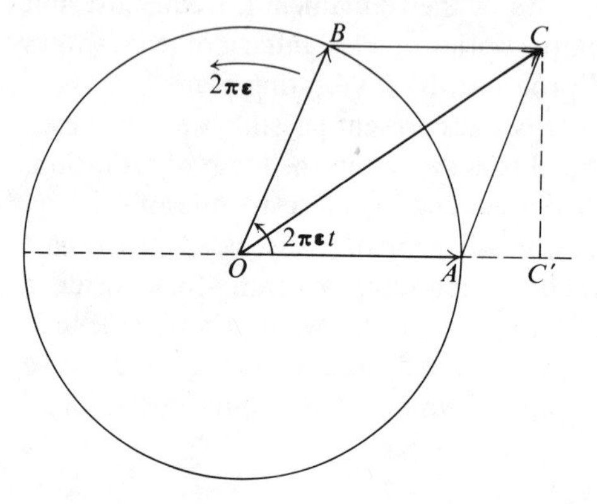

Fig. 11.8. A rotor diagram for two waves of equal amplitude and almost equal frequency.

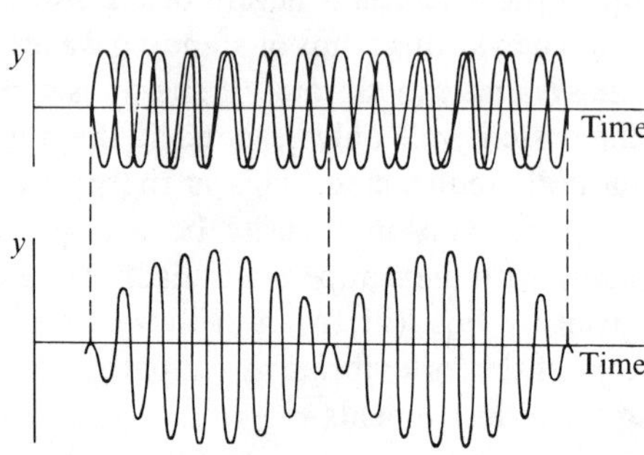

Fig. 11.9. The displacements of air particles at a point as a function of time for two separate disturbances of nearly equal frequency, and the combined effect of the two.

11.6 INTENSITY AND INTENSITY LEVEL

From a geometrical point of view it is the wave-form that is propagated by a traveling wave, but physically what is propagated is energy. The *intensity* of a traveling wave, I, is defined as the time average of the power transmitted across unit area normal to the direction of flow. The dimensions of intensity are energy area^{-1} time^{-1} and the units are watts per square meter, or W m^{-2}. The intensity of a sound wave may be expressed in terms of either the maximum amplitude A of the vibratory motion of the individual particles of the fluid or the maximum excess of pressure in the fluid, P. The fluid through which the wave is traveling will be at some average pressure p, but the sound wave will manifest itself as regions of alternately high and low pressure ranging between $p + P$ and $p - P$. The two expressions for the intensity are

$$I = \tfrac{1}{2}\omega^2 A^2(\rho V) = \frac{P^2}{2\rho V}. \tag{11.16}$$

The combination (ρV) is known as the acoustic impedance, as was mentioned above. It will be noticed particularly that in both forms of the expression the intensity is proportional to the square of the amplitude, and this is a general result for all kinds of wave motion. An intensity of 1 W m^{-2} corresponds to about the loudest tolerable sound, and the faintest sound detectable by the ear is about 10^{-12} W m^{-2}. This range of intensities is enormous and presents problems that are best solved by resorting to a logarithmic scale. Thus the *intensity level* β of a sound wave is defined by the equation

$$\beta \equiv 10 \log_{10} (I/I_0), \tag{11.17}$$

where I_0 is an arbitrary reference level that is taken to be 10^{-12} W m^{-2} corresponding roughly to the faintest sound that can be heard. The intensity level of a sound is a dimensionless quantity and the unit is the *decibel*, abbreviated dB. For a long time it was believed that the logarithmic scale of intensity level accurately reflected the way in which the human ear responded to sound stimuli but this is not true. Nevertheless, the decibel scale of relative power is now established in scientific and engineering practice where questions of human response are irrelevant. The important subject of human estimation of loudness is deferred until the next chapter.

Table 11.1. Approximate intensity levels for a number of familiar sounds

Type of sound	Intensity level at ear (dB)
Threshold of hearing	0
Rustle of leaves	10
Very quite room	20
Average room	40
Ordinary conversation	60
Busy street traffic	70
Loud radio	80
Train going through station	90
Riveter	100
Threshold of discomfort	120
Threshold of pain	140
Damage to eardrum	160

Example 1. A wave motion has the equation

$$y = (8 \text{ cm}) \sin 2\pi\left\{\frac{x}{2.5 \text{ m}} - \frac{t}{0.08 \text{ s}}\right\}.$$

What are the amplitude, the wavelength, the frequency and the speed of this wave motion?

Solution. Equation (11.3) is the appropriate form for comparison, and we see that

$$A = 8 \text{ cm}; \qquad \lambda = 2.5 \text{ m}; \qquad T = 0.08 \text{ s}$$

whence

$$f = 1/T = 1/0.08 \text{ s} = 12.5 \text{ c/s}.$$

The speed of propagation is $V = f\lambda = 12.5 \text{ s}^{-1} \times 2.5 \text{ m} = 31.25 \text{ m s}^{-1}$.

Example 2. A piano string is 1.21 m long and has a mass of 150 g. If it is stretched with a tension of 6000 N, what is the velocity of the wave set up when the hammer strikes it? What is the fundamental frequency of the note emitted?

Solution. The mass per unit length is $\mu = 0.15$ kg/1.21 m, and therefore the speed of propagation is

$$V = \sqrt{\frac{6000 \text{ N} \times 1.21 \text{ m}}{0.15 \text{ kg}}} = 220 \text{ m s}^{-1}.$$

For the fundamental frequency, $\lambda = 2L = 2.42$ m, and so the frequency of this fundamental is $f_0 = 220 \text{ m s}^{-1}/2.42 \text{ m} = 90.9$ c/s.

The piano string will not, of course, sound a pure note of this frequency. The sound produced will be the resultant of this fundamental together with many harmonics.

Example 3. Two identical guitar strings are tuned to the same frequency of 300 c/s. If the tension of one of the strings is increased by 2%, how many beats per second will be heard when the two strings are sounded together?

Solution. The two strings are identical apart from their tensions, and so we may simply use the fact that frequency is proportional to the square root of the tension. If F is the original tension and F' the increased tension of one string, and if f and f' are the corresponding frequencies, then

$$f'/f = \sqrt{F'/F} = \sqrt{1.02\ F/F} = \sqrt{1.02} = 1.01.$$

Hence $f' = 1.01 \times 300$ c/s $= 303$ c/s, and 3 beats per second will be heard.

Example 4. If two sound waves have pressure amplitudes of P_1 and P_2, find an expression for the difference in their intensity levels.

Solution. Let I_1 and I_2, β_1 and β_2 be the intensities and intensity levels of the two sound waves. From the general result for logarithms that $\log A - \log B = \log (A/B)$, the definitions of Eqs. (11.16) and (11.17) lead to

$$\begin{aligned} \beta_2 - \beta_1 &= 10 \log (I_2/I_0) - 10 \log (I_1/I_0) \\ &= 10 \log (I_2/I_1) \\ &= 10 \log (P_2^2/P_1^2) = 20 \log (P_2/P_1). \end{aligned}$$

Example 5. An open window has dimensions 0.5 m × 2.0 m and the intensity level at the window is 60 dB. How much acoustic power enters the room?

Solution. Since $10 \log (I/I_0) = 60$ dB, $\log (I/I_0) = 6$, which gives $I = 10^6 I_0 = 10^6 \times 10^{-12} \text{ W m}^{-2} = 10^{-6} \text{ W m}^{-2}$. The area of the open window is just 1 m² and so the acoustic power entering the room is 1 μW.

PROBLEMS

11.1 Draw a graph of

$$y = (6 \text{ cm}) \cos [(3 \text{ m}^{-1})x + (100 \text{ s}^{-1})t]$$

against x at $t = 0$ and at $t = 1/800$ s. What are the amplitude, the wavelength, the frequency, and the speed of this wave motion?

11.2 A wave obeys the equation of motion

$$y = (0.5 \text{ m}) \sin 24\pi \left(\frac{t}{1 \text{ s}} - \frac{x}{36 \text{ m}}\right).$$

What are the amplitude, the frequency, the wavelength, and the speed of this wave motion? In which direction is the wave moving?

11.3 Repeat Problem 11.2 for the wave

$$y = (11.2 \text{ m}) \sin \pi \left(\frac{t}{1 \text{ s}} + \frac{x}{8 \text{ m}}\right).$$

11.4 The speed of transverse waves in a string of length 25 m is 50 m s^{-1} when the tension in the string is 200 N. What is the mass of the string?

11.5 A string of length 10 m has a mass of 0.5 kg. If the tension is 20 N, what is the velocity of transverse waves in the string?

11.6 At a temperature of 27°C what is the velocity of sound in (a) argon and (b) hydrogen?

11.7 If the temperature of air changes by 5 degrees, by what percentage amount does the velocity of sound change (a) at 20°C and (b) at −20°C?

11.8 A metal wire is stretched by a force of 135 N between rigid supports 50 cm apart. The diameter of the wire is 0.35 mm and the density of the metal is 8.8 g cm^{-3}. What is the fundamental frequency of the note emitted by the wire when it vibrates transversely?

11.9 A stretched wire gives two beats per second with a tuning-fork when its length is 1.43 m and also when its length is 1.45 m, the tension being the same in each case. What is the frequency of the tuning-fork?

11.10 If in the previous problem the tension is increased by 2%, how many beats per second will be heard when the length of the wire is 1.43 m?

11.11 Two stretched wires vibrate with the same frequency of 400 c/s. By what percentage amount must the tension in one of the wires be changed so that 2 beats per second will be heard when both wires are sounded together?

11.12 A uniform pipe of length 32 cm has a fundamental resonant frequency of 500 c/s when open at both ends. If the displacement antinodes occur at a distance of 1 cm from the open ends, calculate the velocity of sound.

11.13 If the pipe of the previous problem is now closed at one end, what will be the new fundamental frequency?

11.14 A tuning-fork of unknown frequency gives 3 beats per second when sounded with a fork of frequency 256 c/s. What are the two possible frequencies of the unknown fork? What simple practical test will distinguish between these possibilities?

11.15 Calculate the amplitude of vibration of the molecules of air in a sound wave of intensity 10^{-12} W m^{-2} and frequency 3000 c/s, given the acoustic impedance of the air to be 430 kg m^{-2} s^{-1}.

11.16 Repeat Problem 11.15 for a sound wave of intensity 1 W m^{-2}, corresponding to a painfully loud sound, the frequency being unaltered.

11.17 One violin playing alone produces an intensity level of 60 dB at a particular seat in a concert hall. How many decibels will be produced by 25 violins playing together?

11.18 Referring to the previous problem, how many violins would be required to produce at the same seat an intensity level of 120 dB?

11.19 If both the frequency and the displacement amplitude of a sound wave are doubled, what is the increase in intensity level of the wave?

CHAPTER 12

THE EAR AND HEARING

12.1 INTRODUCTION

The human ear is in some ways the most remarkable organ in the human body, and the processes that lead up to the stimulation of the auditory nerves show in a very direct way the need for a sound grasp of basic physical principles in understanding problems in biology and medicine. It is worthy of note that of the two most important contributors to our knowledge of the mechanisms of the ear, von Helmholtz and von Békésy, the first was one of the great all-rounders of all time, being equally at home in mathematics, physics, physiology, and anatomy, and the second was a communications engineer by training who won the Nobel prize for medicine in 1961.

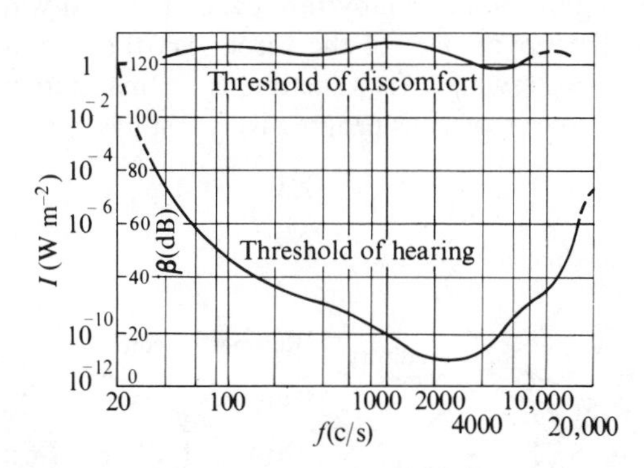

Fig. 12.1. The audibility of sound to the human ear as a function of frequency.

The primary stimulus that gives rise to the sensation of hearing is a sound wave in air. The important range of frequencies lies between 20 c/s and 20,000 c/s but the sensitivity of the ear is not uniform over the whole range. Figure 12.1 shows the correlation of audibility with frequency. Intensity level is plotted vertically and frequency horizontally, the latter on a logarithmic scale. The solid curve shows how an average human ear responds. Thus at a frequency of 1000 c/s roughly one person in two needs an intensity level of 20 dB before they can detect the note, whereas at around 3500 c/s 10 dB is sufficient. This is known as the threshold of hearing and is seen to be very dependent on frequency. Only about 10% of the population can hear a 0-dB sound, and then only in the frequency range 2000–4000 c/s. The threshold of discomfort is not noticeably frequency-dependent, as the upper curve shows, and remains pretty constant at around 120 dB. The sensitivity of the human ear in the frequency range 2000–4000 c/s is remarkable. For an intensity level of 0 dB corresponds to a maximum pressure amplitude of about 3×10^{-5} N m^{-2}, and this small fluctuation is superimposed on a general atmospheric pressure of 10^5 N m^{-2}. The amplitude of vibration of the air molecules at 0 dB is less than the diameter of an atom. A mechanism so sensitive is clearly worthy of our close study.

12.2 THE ANATOMY OF THE EAR

The principal anatomical features of the human ear are shown in Fig. 12.2. The convoluted flaps of skin and cartilage on the outside serve very little purpose and the hearing process proper begins when the pressure wave in the air enters the ear canal leading to the eardrum. The eardrum separates the ear canal from the middle ear, a small air-filled cavity that houses the ossicles, a system of three articulated bones named the hammer, the anvil, and the stirrup. The hammer is attached to the inner surface of the eardrum and is further attached to the bony surround by a muscle. The anvil articulates the hammer to the stirrup, which has a foot-plate attached to an opening in the bony labyrinth called the oval window. The middle ear is connected to the upper part of the throat via the Eustachian tube, which opens in the act of swallowing. Its function is to maintain the middle ear at atmospheric pressure, since oxygen is progressively lost by absorption at the surface.

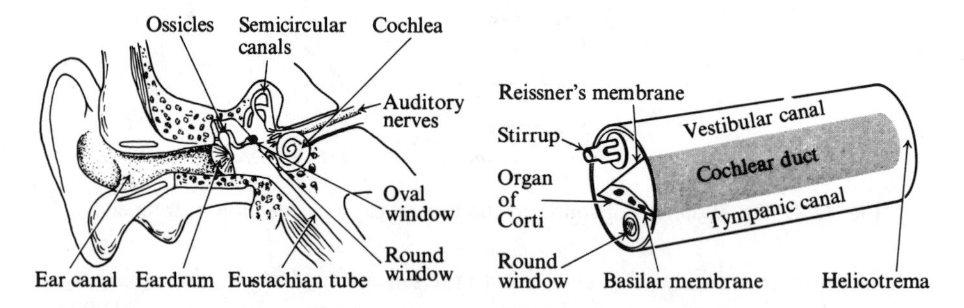

Fig. 12.2. The main features of the ear.

Fig, 12.3. An uncoiled view of the cochlea.

The oval window is the entry to the inner ear, which consists of a series of cavities heavily protected by the temporal bone, the hardest in the body. The two main parts of the inner ear are the semicircular canals which form the body's natural frame of reference and control balance, and the cochlea, a snail-shaped cavity where the main process of hearing takes place. An uncoiled view of the cochlea is shown in Fig. 12.3. It actually has about $2\frac{3}{4}$ turns over a length of about

5 mm. It is divided for most of its length into two canals, the vestibular canal leading from the oval window and the tympanic canal that ends in the round window, a membrane leading back into the middle ear. These two canals are filled with a liquid called perilymph that is virtually the same as spinal fluid, and they join up at the apex of the cochlea through a small opening called the helicotrema. The sound disturbance travels from the oval window along the vestibular canal, through the helicotrema and back down the tympanic canal, any residual energy being dissipated at the round window. Cutting a wedge-shaped slice out of the cochlea is the cochlear duct, separated from the vestibular canal by a thin membrane (Reissner's membrane) and from the tympanic canal by the basilar membrane, which is much thicker. The cochlear duct is filled with a liquid called the endolymph, which is permanently separated from the perilymph. The basilar membrane supports the organ of Corti, which contains the endings of the auditory nerves. There are some 30,000 nerve terminals distributed along the basilar membrane and occupying an area of only 30 mm long by 0.3 mm wide, a truly remarkable feat of engineering.

12.3 THE MECHANISM OF HEARING

The ear is essentially a pressure-sensing device, and specifically it senses systematic fluctuations of air pressure. The ear canal resonates slightly to frequencies in the range 3000–4000 c/s and the pressure at the eardrum is perhaps twice that outside. This resonance effect explains why the ear is most sensitive to frequencies in that range. The fluctuations of pressure at the eardrum are still very small and further amplification is necessary. The ossicles of the middle ear, which transmit the pressure fluctuations from the eardrum to the oval window, are connected to the walls of the inner ear by muscles that act as a kind of automatic volume control. As a loud sound builds up, the various muscles twist the bones slightly so that the stirrup rotates and is pulled away from the oval window. In addition, the muscular action stiffens the eardrum itself. All these actions serve to protect the ear from possible damage. Unfortunately, there is a time delay in this protective action and any sudden very loud sound can cause damage to the middle ear.

The main amplification that takes place in the middle ear depends on the fact that the oval window is between 20 and 30 times smaller than the eardrum. The situation is shown schematically in Fig. 12.4. The energy of the vibration in air outside the eardrum is communicated to the perilymph in the vestibular canal by the ossicles, which act to increase the pressure rather in the manner of a woman's high-heeled shoe. The force moving the eardrum is increased through the linkage mechanism of the ossicles into a force 2 or 3 times larger that the stirrup exerts on the oval window, and consequently the pressure behind the oval window is between 40 and 90 times larger than the pressure at the eardrum. With the twofold amplification produced by the ear canal, this means that sounds in the frequency range 3000–4000 c/s can be amplified 180 times in favorable cases. This is the pressure amplification. The intensity amplification is the square of this, or some 32,000 times. The relative areas of the eardrum and oval window also provide the essential matching of impedances (see Section 27.10) without which the high amplification would be profitless.

Inside the cochlea the original pressure wave in air has become a pressure wave in a liquid, but the cochlea is so narrow that the viscosity of the perilymph plays the dominant role, and the simple theory of wave motion discussed in the last chapter has little relevance. What actually happens is that the pressure wave induces a wave-like ripple in the basilar membrane. This membrane is light and under considerable tension near the oval window but becomes progressively thicker and less taut as one progresses round the canals and back to the round window.

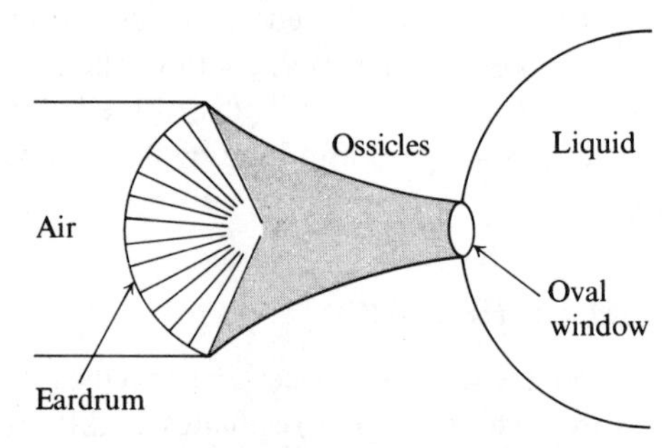

Fig. 12.4. A schematic diagram of the middle ear.

The general discussion of the propagation of waves in wires would lead us to expect that the light taut sections would respond better to high-frequency vibrations and that the heavier slacker sections would respond to the lower frequencies. This in fact is what happens, and the particular region of the basilar membrane that is stimulated depends on the frequency of the original sound wave. This is the basis of the frequency discrimination of the ear. The disturbance of the basilar membrane takes the form of a hump in a narrow region for every frequency component of a complex sound, and this produces shearing stresses in the organ of Corti, which converts the mechanical energy into electrical energy and sends the information along the auditory nerves to the brain. The organ of Corti is a gelatinous structure of some 7500 interrelated parts. It contains a mass of hairs, fixed at both ends, the complex elastic deformations of which stimulate the ends of the auditory nerves. The coded electrical messages to the auditory cortex are there translated in a way not understood into the sensation of hearing. The communication system between the organ of Corti and the auditory cortex is two-way and the brain has the capacity to suppress stimuli that are unwanted. Thus workers in particularly noisy environments soon reach a stage at which the noise around them no longer reaches the brain center at all. Such continual exposure to noise, however, causes permanent impairment of hearing that manifests itself in a sharp increase in threshold level around 4000 c/s.

12.4 PITCH AND LOUDNESS

The frequency of a pure musical note is a perfectly definite physical quantity, and this frequency produces a stimulus at a perfectly definite region of the basilar membrane. The sensation corresponding to this is called the *pitch* of the note. Certain differences of pitch give a pleasing sensation, and our musical scale is based on these intervals. Thus doubling the frequency of a sound produces a sensation of quasi-unison which we call the octave. Normal ears can easily recognize about 12 intervals in one octave, the ordinary chromatic scale, although highly trained musicians can do rather better. It is found that the sensation we call an octave is not produced by doubling the frequency of the note if the frequency is high enough. The correlation between frequency and pitch, in other words, is not linear. But what is linear is the correlation between pitch and distance along the basilar membrane at which the stimulus occurs. The pitch of a musical note is also dependent on its intensity, which would indicate that the elastic distortion of the basilar membrane is a function of the intensity also.

The loudness of a sound is roughly correlated with the intensity level of the note, and not with the intensity proper (see Section 11.6), and for a long time it was believed that the correlation was exact. But there is now no doubt that the decibel scale of intensity level no more correlates with the subjective sensation of loudness than the frequency scale correlates with the sensation of pitch. In place of the decibel logarithmic scale, a power-law relation between loudness and intensity has been proposed,

$$L = A(I/I_0)^{0.3},$$

where A is an arbitrary constant. The *sone* is defined as the unit of loudness by the equation

$$1 \text{ sone} = 40 \text{ dB at } 1000 \text{ c/s}.$$

Since at 40 dB the ratio $I/I_0 = 10^4$, this fixes the constant A as $\frac{1}{16}$, so the loudness of a note in sones is given by

$$L = \tfrac{1}{16}\,(I/I_0)^{0.3}$$

where $I_0 = 10^{-12}$ W m^{-2}.

The sone scale of loudness has been recognized internationally, and there are good indications that a similar sort of power law might hold for most of the other human sensations as well. The old Weber–Fechner law that equal increments in the logarithm of the stimulus produce equal increments in sensation can be expected to die a natural death.

CHAPTER 13

REFLECTION AND REFRACTION AT PLANE SURFACES

13.1 INTRODUCTION

We know from our everyday experiences that a hot body emits light which affects the retina of the eye and produces the sensation of seeing. This is by no means the only way of producing light. The discharge of electricity through a gas and fluorescence, for example, are both employed in common lighting systems. Light is a form of radiant energy. It must carry energy, since, if focused by a lens, it can cause paper to char and burn, and it can be turned into other forms of energy, e.g. into electrical energy in a photoelectric device. It is a particular form of what is known as electromagnetic radiation. Other forms range from radio waves through infrared and ultraviolet radiation to X- and γ-rays. All of these are wave motions which travel in a vacuum at a speed of 2.9979×10^8 m s^{-1}, the difference between one form and another being the length of the wave or the number of waves emitted per second (called the frequency of the wave motion). The whole range of electromagnetic radiation includes wavelengths with values ranging from several kilometers for long radio waves to 10^{-15} m for very penetrating γ-rays. The visible spectrum contains the wavelengths in the region roughly from 4 to 7×10^{-7} m (4000–7000 Å).

The historical development of the investigation of light is a good illustration of how models must be constantly modified to fit the advent of new experimental data. In the early days the study of light was concerned with the passage of beams of light from one medium to another and through lenses and optical instruments, the subject matter of geometrical optics. Newton believed that a source of light emitted corpuscles which traveled from the source in straight lines, being bent according to definite rules at the boundaries between media, but continuing in straight lines in the second medium. This proved a satisfactory model for all the phenomena then known. Later, however, the discovery that light traveled more slowly in material media than in air and, even more, the discovery of the phenomena of interference, diffraction, and polarization (the subject matter of physical optics), necessitated the discarding of this model. The model put in its place viewed the light-source as emitting waves in all directions just as ripples spread out when a stone is dropped into water. In this century the interpretation of the photoelectric effect and other quantum optical phenomena has only been possible on the assumption that light is emitted from a source in the form of discrete packets of waves called quanta or photons. Individually they act like particles; in large numbers they act like waves. The current model is therefore one which it is impossible to visualize in terms of normal everyday experience, but is an essentially mathematical one which allows the prediction of the result of any particular optical experiment.

13.2 THE LAWS OF REFLECTION

We shall start our discussion of optics by treating geometrical optics, which deals with phenomena which occur when the size of the apparatus is many orders of magnitude greater than the wavelength of the light and where the wave nature of the light is therefore not apparent.

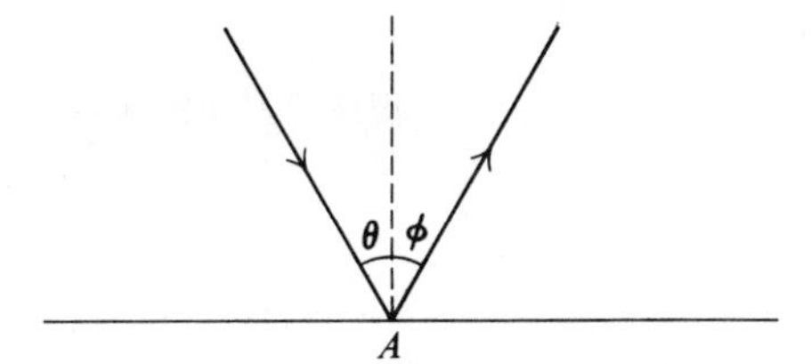

Fig. 13.1. Reflection at a plane surface.

If a screen with a pin-hole is introduced between a source of light and the point of observation, what emerges is a beam of light with negligible lateral dimensions. This is called a ray of light. Simple observation shows that this ray follows a straight path so long as the light remains in the same medium. Light, to the approximation we are dealing with, travels in straight lines.

If a beam of light is allowed to strike any rough surface, the light, as well as being partially absorbed, is scattered in all directions. If, however, the beam strikes a polished, flat, metal surface, this is no longer true; the reflection that takes place is now regular and the light after reflection all goes in a specific direction. If a ray of light strikes the surface, two very simple laws of reflection are found. If the ray of light strikes the surface at a point A, as shown in Fig. 13.1, a normal to the surface can be erected at A. The angle θ between the incident ray and the normal is called the angle of incidence and the angle ϕ between the reflected ray and the normal the angle of reflection. The laws of reflection are then the following:

a) the incident ray, reflected ray, and normal to the surface at the point of incidence all lie in one plane;

b) the angle of reflection equals the angle of incidence.

13.3 THE IMAGE IN A PLANE MIRROR

A polished flat metal surface, or any similar device, is known as a plane mirror and the rules followed by the individual rays show clearly why a plane mirror displays an image of any object held in front of it. Consider a point object O emitting rays of light in all directions. If, as is shown in Fig. 13.2, any one of these rays striking the mirror at the point A is selected, the ray after reflection follows the direction AB such that the angles of incidence and reflection are both equal to θ. If BA is produced backwards, it meets the normal to the mirror from O at the point O'.

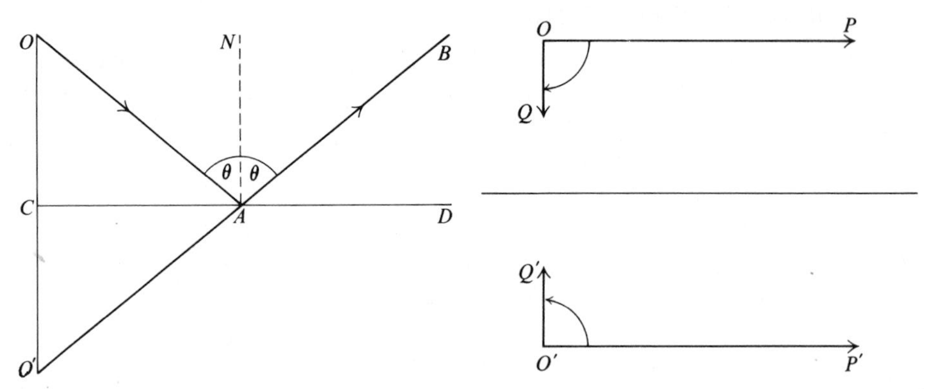

Fig. 13.2. A point object and its image in a plane mirror.

Fig. 13.3. An extended object and its image in a plane mirror.

The point at which OO' meets the mirror is labeled C. Angles OCA and $O'CA$ are both right angles, and angles BAD, OAC and $O'AC$ are all equal to $90° - \theta$. Thus triangles OCA and $O'CA$ have all corresponding angles equal and the side CA in common. They are thus congruent, whence it follows that $O'C = OC$. But OA was any ray from the object. Thus all rays after reflection appear to have come from the point O' as far behind the mirror as O is in front. To an eye viewing the rays after reflection, all rays appear to have come from O', and an image of O is therefore seen at the point O'. The real image of a point object in an optical instrument is defined as the point through which pass all possible rays from the object. The virtual image of a point object is the point from which appear to diverge all the rays from the object. In the case of the plane mirror, rays do not pass through O' but only appear to have come from there. The point O' is thus the virtual image of O in the plane mirror.

Clearly, if the object being viewed in the mirror is extended, each point of it acts like O, producing an image like O'. The pattern of images like O' produce an extended image identical with the object except that the image is perverted, as shown in Fig. 13.3. The images O', P', and Q' of O, P, and Q are labeled, and the image of every point on both arrows has been inserted. On the object it is necessary to rotate clockwise about O from P to reach Q, whereas in the image a counterclockwise rotation about O' is necessary. Note that the image must always be of the same size as the object.

13.4 THE LAWS OF REFRACTION

A ray of light passing from one medium to another is bent from its original direction at the interface. This phenomenon is known as refraction. A straight stick partially immersed in water appears bent at the surface, because the rays of light from points on the stick inside the water are bent as they enter the air. The eye looking along these rays assumes that the rays have followed a straight line path for the whole of their travel and thus places the stick at a position closer to the surface, as shown in Fig. 13.4.

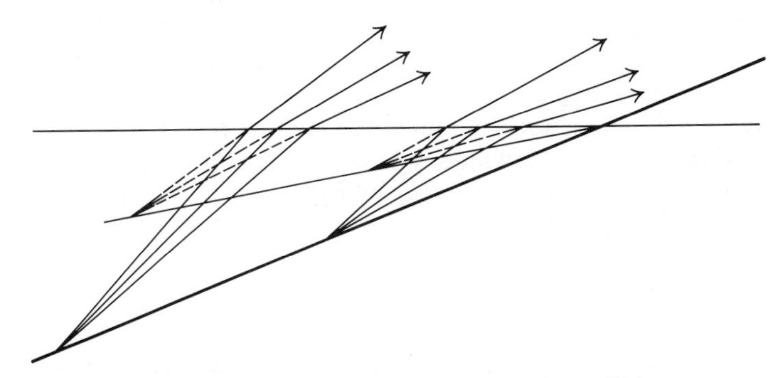

Fig. 13.4. The apparent bending of a stick due to refraction of light at a water surface.

In Fig. 13.5 is traced the path of a single ray emitted by a point object in one medium and striking the plane interface to a second medium at the point *A*. The normal to the interface at *A* is shown. The angle between the incident ray and the normal is θ_1, and the angle between the refracted ray in the second medium and the normal is θ_2. The laws governing the refraction are as follows.

a) The incident ray, the refracted ray, and the normal to the interface at the point of contact all lie in the one plane.

b) Snell's law states that

$$\frac{\sin\theta_1}{\sin\theta_2} = {}_1n_2, \tag{13.1}$$

where ${}_1n_2$ is a constant for the two media, for a particular color of light. This constant is called the relative refractive index between the two media. It is found that

$$\begin{aligned}{}_1n_2 &= \frac{\text{velocity of light in the first medium}}{\text{velocity of light in the second medium}}\\ &= \frac{\text{velocity of light in the first medium}}{\text{velocity of light } \textit{in vacuo}}\\ &\quad\times \frac{\text{velocity of light } \textit{in vacuo}}{\text{velocity of light in the second medium}}\end{aligned}$$

The ratio of the velocity of light *in vacuo* to the velocity of light in the first medium is called n_1, the absolute refractive index of medium 1. Thus

$$_1n_2 = \frac{n_2}{n_1} = \frac{1}{n_1/n_2} = \frac{1}{_2n_1}. \tag{13.2}$$

We may alternatively write Snell's law as

$$n_1 \sin \theta_1 = n_2 \sin \theta_2. \tag{13.3}$$

Note that the more optically dense a medium, the smaller is the velocity of light in that medium and the greater is its absolute refractive index. This means that a ray traveling from a less dense to a more dense medium is bent toward the normal after refraction and vice versa.

Because of the symmetry inherent in the laws of reflection and refraction, a ray of light which is reversed will always traverse its previous path. This is a general optical law known as the reversibility of light rays.

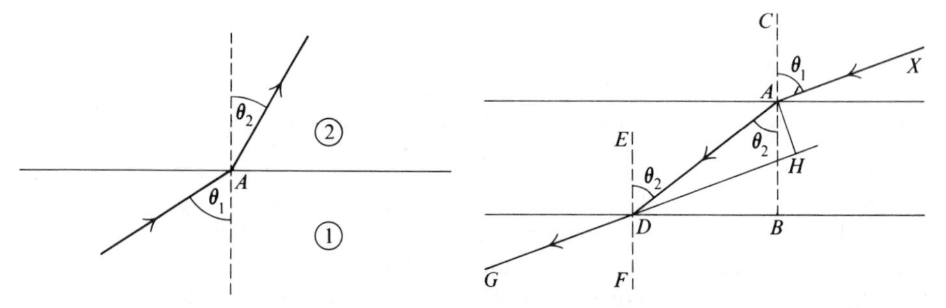

Fig. 13.5. Refraction at a plane surface.

Fig. 13.6. The displacement of a light ray in its passage through a parallel-sided glass block.

13.5 THE PASSAGE OF LIGHT THROUGH A PARALLEL-SIDED BLOCK

A ray of light traveling in a medium of refractive index n_1 strikes a parallel-sided block of material of refractive index n_2 at A, the angle of incidence being θ_1 (see Fig. 13.6). The normal to the block at A is CAB, and $AB = t$, the thickness of the block. The ray is refracted at A, the angle of refraction being θ_2, and the ray in the glass strikes the second surface at D. The normal at D is EDF. Since AB and ED are necessarily parallel, angles EDA and DAB both have the value θ_2. Since Snell's law applies at both A and D,

$$n_1 \sin \theta_1 = n_2 \sin \theta_2,$$

and

$$n_2 \sin \theta_2 = n_1 \sin GDF.$$

Therefore angle $GDF = \theta_1$. It follows that the ray is not deviated by the block but only displaced, the amount of the displacement being AH, where AH is the perpendicular from A to GD produced backwards. Since CAB is a straight line and XA is parallel to HDG, angle XAH is a right angle and angle BAH is $90° - \theta_1$. Therefore $AH = AD \cos [\theta_2 + (90° - \theta_1)]$. Also, $AB = t = AD \cos \theta_2$. Therefore

$$AH = \frac{t \cos [90° - (\theta_1 - \theta_2)]}{\cos \theta_2} = t \frac{\sin (\theta_1 - \theta_2)}{\cos \theta_2}. \tag{13.4}$$

The displacement of the ray is thus directly proportional to the thickness of the block.

13.6 TOTAL INTERNAL REFLECTION

For reasons which will be obvious, total internal reflection is a phenomenon observed only in the passage of light from a more dense to a less dense medium. A point source of light is embedded in a medium which is separated from a second, less dense medium by a plane boundary. The ray a shown in Fig. 13.7 strikes the boundary at a small angle θ_1 and is refracted away from the normal, the angle of refraction θ_2 being greater than θ_1. This follows because

$$\sin \theta_2 = \frac{n_1}{n_2} \sin \theta_1 > \sin \theta_1, \quad \text{since } n_1 > n_2.$$

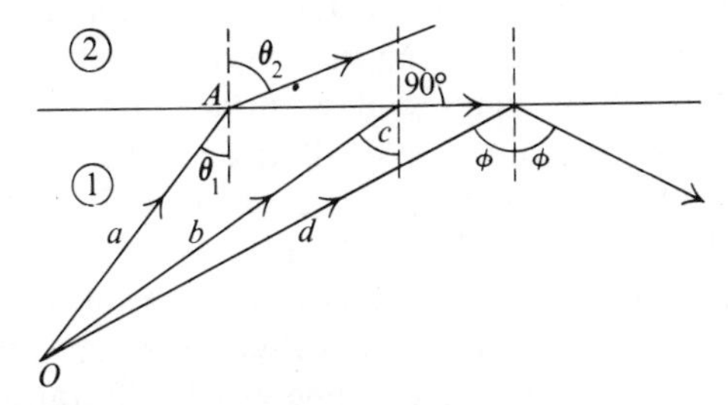

Fig. 13.7. Refraction, and total internal reflection, at a plane interface.

If rays from O striking the interface to the right of A are observed, these meet the interface at gradually increasing angles of incidence. When the situation of ray b is reached, the ray is striking the interface at such an angle that the angle of refraction is 90°. The angle at which this occurs is called the critical angle c. From Snell's law

$$\sin c = \frac{n_2}{n_1} \sin 90° = \frac{n_2}{n_1}.$$

Therefore

$$\sin c = \frac{1}{{}_2n_1}. \tag{13.5}$$

Rays such as d strike the interface at angles of incidence greater than c. If refraction took place, sin θ_2 would be greater than unity, which is clearly impossible. The ray d is thus not refracted but reflected from the interface back into the first medium according to the laws of reflection. This phenomenon is called total internal reflection.

When a beam of light strikes the interface between any two media, both reflection and refraction occur. At a metal surface a small amount of the light is actually refracted into the metal medium (and rapidly absorbed). When refraction takes place at a glass-air interface, a small amount of the light is actually reflected back to the first medium. This amount gradually increases as the angle of incidence increases. But with total internal reflection, no refraction occurs at all, which explains the name given to the phenomenon.

13.7 FIBER OPTICS

Some of the simpler applications of total internal reflection, such as its uses in prism binoculars and in periscopes, will be already familiar to the reader. Recently, interesting biomedical applications have become possible since manufacturers have found out how to extrude glass fibers of uniform thickness of the order of 2×10^{-6} m. It is clear that if light is allowed to enter one end of a solid glass rod, at least some of the light will strike the curved glass-air interface at angles greater than the critical angle and will be totally internally reflected. It will therefore continue to travel along the rod, being reflected from side to side. All other light is refracted through the glass surface into the air and is lost.

All the light entering the rod at a small angle to the axis will thus be transmitted to the other end. Even if the rod is bent out of its usual straight shape, a large amount of light will always be transmitted along the tube. If, in fact, the light is concentrated into a parallel beam sent in almost at right angles to the end face of the rod, which is polished flat, practically all the light will arrive at the other end. What has been produced is a light-guide which will transmit light from place to place.

A solid rod would break if an attempt were made to bend it. But a similar arrangement, flexible and almost unbreakable even when bent into quite tight curves, can be produced by using a tightly packed bundle of glass fibers instead of a solid rod. The fibers are normally made of glass of high refractive index coated with a thin layer of glass of low refractive index to provide a suitable boundary, which is then protected by the outer layer. If this were not done, light would pass from one fiber to another where they touched, and would eventually be absorbed in the sheathing holding the bundle together.

There are many inaccessible places which, until recently, could only be illuminated by the use of a complex optical system. An enclosure containing an explosive mixture cannot be lighted from within because of the hazard involved and is difficult to light properly from outside. The use of fibers as light-guides has made the solution of both of these problems simple. It is also easy to

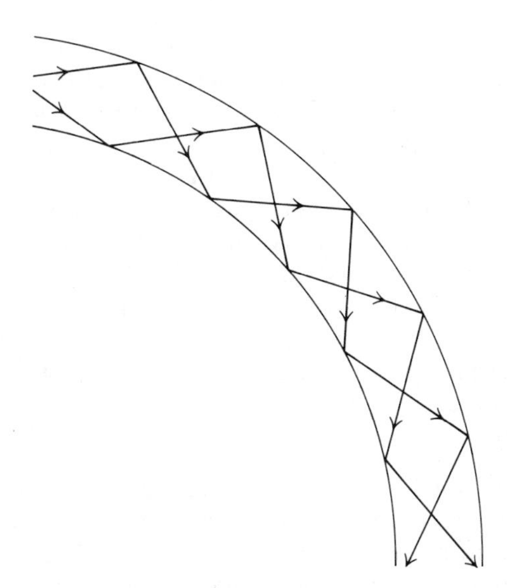

Fig. 13.8. The transmission of rays of light along a glass rod due to total internal reflection.

produce shadow-free illumination of an object by piping light to a number of illuminating positions. Research is being undertaken at the moment into the transmission (through fiber bundles) of high-intensity light from lasers (cf. Section 30.13) for use in surgery and cancer therapy. In all these applications the fibers may be oriented anyhow in the fiber bundle. However, not only light, but also pictures, may be transmitted by this device. In the latter case the fibers must be laid down coherently. In the manufacture of this type of bundle the fibers are kept parallel to one another, which ensures that both ends of an individual fiber are in the same relative position at each end of the bundle. Each fiber then transmits a small portion of a picture placed in front of it and illuminated. Light from that portion of the picture being viewed which lies straight ahead of a fiber axis comes in at the correct angle to be transmitted along it. At the other end an image is seen consisting of an assemblage of individual dots of light, each one coming from a single fiber. The image is therefore formed by a pattern of light and shade in the same way as a newspaper photograph is made up of dots of varying blackness. It is now clear why the fibers in the bundle must be of such small diameter. If the dots of light received from the individual fibers were too large, a coarse image with little detail would be received. If, however, each dot is small, a faithful reproduction of the varying pattern of light and shade over the original picture can easily be conveyed in the image.

Such arrangements are being used in bronchoscopes, cystoscopes, and similar instruments to view the internal tracts of the human body. Since the fiber bundle is flexible, it can be bent to a radius of a few inches and can therefore follow the changes of direction of the tract without difficulty. Heat-free light is passed down an outer layer of fibers to illuminate the tract ahead of the bundle, and light scattered back is transmitted through the central core of the fibers to produce an image at the other end. Since the fibers are of such small diameter, the detail in the final image is extremely good.

Difficulties of manufacture limit the length of fibers in the bundle to around 12 ft. Further, a number of the fibers are broken during the manufacturing process; this produces a certain loss of light and resolution. For this reason attempts are being made to replace the glass fibers by ones made of transparent plastic. These are much easier to handle, are less likely to break, and are more flexible. Unfortunately, it has not yet been possible to manufacture such fibers of diameters less than around 2×10^{-4} m. For transmission of light these are excellent; but this is not nearly good enough for the transmission of pictures where great detail is required.

Example 1. In the old Wild West shows, one of the tricks of the sharpshooter was to fire over his shoulder at a number of bottles which he viewed in a hand-held plane mirror. If the mirror was 15 cm wide and 50 cm from his eyes, how many bottles 20 cm apart and 6 m behind him could the sharpshooter see by reflection?

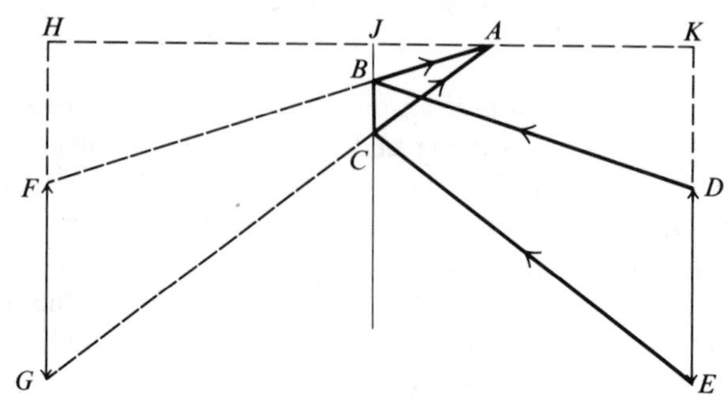

Fig. 13.9

Solution. Only the bottles in line DE in Fig. 13.9 will be seen by sharpshooter A in the mirror, since DB and EC are the rays which after reflection at the extremities of the mirror B and C go to A. The image of DE in the mirror is FG as far behind plane BC as DE is in front. Rays AB and AC produced backwards pass through F and G, the points in the image corresponding to D and E.

Triangles AFH and ABJ are similar. Thus

$$\frac{FH}{BJ} = \frac{AH}{AJ} = \frac{AJ + JH}{AJ} = 1 + \frac{JK}{AJ},$$

since $JH = JK$, therefore

$$FH = \left(1 + \frac{JK}{AJ}\right) BJ.$$

Similarly

$$GH = \left(1 + \frac{JK}{AJ}\right) CJ.$$

Therefore

$$FG = \left(1 + \frac{JK}{AJ}\right) BC$$

$$= \left(1 + \frac{6\text{ m}}{0.5\text{ m}}\right) 15\text{ cm} = 13 \times 15\text{ cm} = 195\text{ cm}.$$

The number of intervals between bottles in this length is thus

$$\frac{195\text{ cm}}{20\text{ cm}} = 9+.$$

He can therefore see 10 bottles.

Example 2. A biological solution is found to have a refractive index of 1.34. What is the velocity of light in the solution?

If a ray of light strikes the surface of the solution at an angle of incidence of 30°, what is the resulting angle of refraction?

Solution. The refractive index is given by the equation

$$n = c/v,$$

where c and v are the velocities of light in vacuum and in the biological solution respectively. Thus

$$v = \frac{c}{n} = \frac{3.00 \times 10^8\text{ m s}^{-1}}{1.34} = 2.24 \times 10^8\text{ m s}^{-1}.$$

Also, by Snell's law,

$$n_1 \sin \theta_1 = n_2 \sin \theta_2.$$

Therefore

$$\sin \theta_2 = \frac{\sin 30^\circ}{1.34} = \frac{0.50}{1.34} = 0.373.$$

Therefore

$$\theta_2 = 21^\circ 22'.$$

Example 3. A precious stone in the form of a cube of refractive index 1.60 has a tiny flaw at the center. The jeweler is unwilling to split the stone since the price drops very steeply with decreasing size. What fraction of the surface would he have to cover with the setting in order that a customer would not see the flaw?

Solution. Each face has to carry a circular setting of such a size that rays from the flaw striking the surface under the setting are blocked by the setting whereas those striking the free surface are totally internally reflected. The minimum setting size occurs when the rays striking the edge of the setting do so at the critical angle as in Fig. 13.10. It follows that

$$r = a \tan c.$$

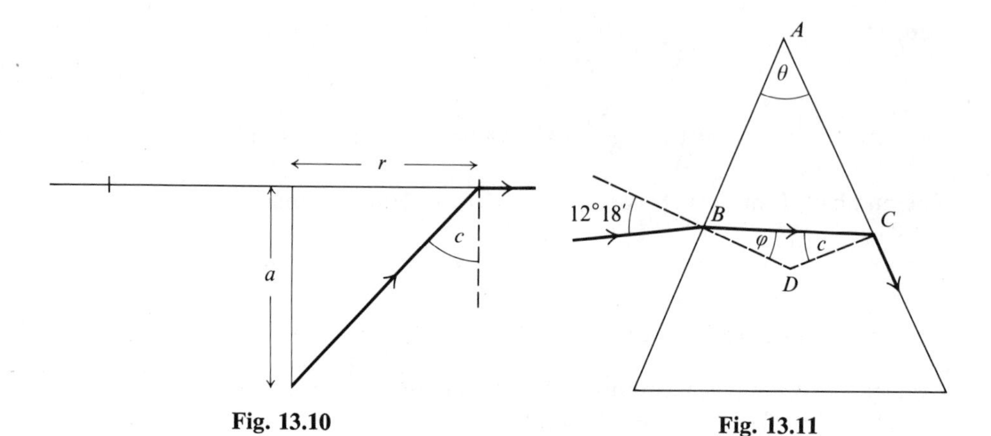

Fig. 13.10

Fig. 13.11

But

$$\sin c = 1/n, \qquad \text{or} \qquad \tan c = \frac{1}{\sqrt{n^2 - 1}},$$

therefore

$$r = \frac{a}{\sqrt{n^2 - 1}}.$$

The fraction of the area covered is thus

$$\frac{\pi r^2}{4a^2} = \frac{\pi}{4(n^2 - 1)} = \frac{\pi}{4(1.60^2 - 1)} = 0.503.$$

Example 4. A prism, a cylinder of triangular cross section, is made of glass of refractive index 1.50. A beam of light strikes one face and emerges from an adjacent face. The beam is moved in a horizontal plane to decrease the angle of incidence and, when the angle becomes 12°18′, the beam emerging from the adjacent face disappears. Calculate the angle of the prism between the two faces.

Solution. At the moment the beam disappears, the light must be falling on the second face at the critical angle c. Thereafter total internal reflection takes place at the second face and no beam emerges. But

$$\sin c = 1/n = 1/1.50 = 0.667,$$

therefore

$$c = 41°50'.$$

Also, since the angle of incidence when the beam disappears is 12°18′,

$$\sin \phi = \frac{\sin 12°18'}{1.50} = \frac{0.213}{1.50} = 0.142,$$

therefore

$$\phi = 8°10'.$$

Since the sum of the angles of a triangle is 180°, it follows that the angle BDC in Fig. 13.11 is 180° − 41°50′ − 8°10′ = 180° − 50°. But the angles of a quadrilateral must add up to 360° and the angles at B and C in $ABDC$ are right angles. Thus

$$\theta = 180° - BDC = 50°.$$

PROBLEMS

13.1 Why does the image of a light viewed by reflection from a body of water on which there are ripples appear very elongated?

13.2 A person with normal eyesight holds an object at a distance of 25 cm from his eyes when he wishes to see it as clearly as possible. Where should he hold a plane mirror if he wishes to examine his face in it?

13.3 A man whose eyes are 1.8 m from the floor stands 1.3 m from a plane mirror of height 1 m which is fixed to a vertical wall with its lower edge horizontal and 0.5 m from the floor. What length of floor can the man see by reflection in the mirror?

13.4 When a woman is standing upright, her eyes are 1.56 m from the floor and the top of her hat is 0.2 m higher. If she wishes to see herself completely in a mirror attached to the wall, what height should the mirror be and how far from the floor should its bottom edge be located? Explain why the answer is independent of how far from the mirror the woman stands.

13.5 A sight-testing chart measuring 0.6 m by 0.4 m, the longer dimension being vertical, is to be viewed by a patient by reflection in a plane mirror. The patient is seated 2.5 m from the mirror with his eyes 1.5 m from the floor, the chart being 1 m behind him with its lower edge 2.5 m from the floor. What is the smallest size of mirror which can be used and how far from the floor must its lower edge be?

13.6 If the speed of light in vacuum is 3.00×10^8 m s^{-1} and the refractive index of water is 1.33, what is the speed of light in water?

13.7 If the relative refractive indices between air and glass and between air and water are 1.50 and 1.33, respectively, show that the relative index between water and glass is 1.13.

13.8 If the speed of light in a sample of glass is 1.82×10^8 m s^{-1}, what is the refractive index of the glass?

13.9 What is the angle of incidence of a ray of light striking a water-air interface from the water side, if the angle of refraction in air is 45°, the refractive index of water being 1.33?

13.10 A rectangular tank of height 2 m is filled to the brim with water of refractive index 1.33. If a narrow beam of light strikes the water at the very edge of the tank at an angle of incidence of 60°, how far from the side of the tank does the beam of light strike the bottom?

13.11 If a plane mirror reverses left and right, why does it not also reverse up and down?

13.12 If a ray of light strikes a parallel-sided glass plate at an angle of incidence of 45°, what is the angle of refraction in the glass? If the glass plate is now covered with a layer of water, what are the angles of refraction in the water and in the glass? The refractive indices of the glass and water are 1.50 and 1.33, respectively.

13.13 Show that when a point object immersed in a liquid is viewed almost vertically from outside the liquid, the apparent depth of the object is equal to the real depth divided by the refractive index of the liquid.

13.14 A man standing symmetrically in front of a plane mirror with beveled edges can see three images of his eyes when he is 1.2 m from the mirror. The mirror is silvered on the back, is 1 m wide, and is made of glass of refractive index 1.50. Calculate the angle of bevel of the edges.

13.15 A detective in a darkened room can look through the window and watch a suspect in the sunlit street outside but the suspect cannot see him. Is the principle of the reversibility of light rays not true in this case?

13.16 What is the critical angle for carbon bisulfide which has a refractive index of 1.67?

13.17 The critical angle for diamond is 24°30′. Calculate the refractive index, and explain why the low value of the critical angle accounts for the sparkle of diamond when cut as a gemstone.

13.18 From the data of Problem 13.7 find the critical angle for light passing from glass into water.

13.19 What will an underwater swimmer see when he looks upwards toward the surface of the water?

13.20 A point source of light is located on the bottom of a steel tank and a circular card of radius 5.65 cm is placed above it. Water is added to the tank, the card floating on the surface with its center directly above the light-source. No light is seen by an experimenter until the water has reached a depth of 5 cm. Why is this so and what is the refractive index of water?

13.21 A straight rod of circular cross-section is made of glass of refractive index 1.52. The end faces of the rod are at right angles to its axis. Show that a ray of light striking the center of an end face and refracted into the glass is totally internally reflected when it strikes a curved face. What is the relevance of this result to the transmission of light along glass fibers?

13.22 A prism is a cylinder of glass of triangular cross-section. If a prism with all its angles 60° is made of glass of refractive index 1.50, determine the least angle of incidence of a ray of light on any of the surfaces, in a plane at right angles to the axis of the prism, if it is to pass through the prism without suffering total internal reflection.

CHAPTER 14

MIRRORS AND LENSES

14.1 INTRODUCTION

Reflecting and refracting surfaces need not be plane. Surfaces of any shape may be used. But the ease of manufacture of spherical surfaces is so much greater than that of any other shape that, except in highly specialized instruments, reflecting and refracting curved surfaces are formed from parts of spheres. The center of the sphere of which the surface is a portion is known as the center of curvature of the surface.

Spherical mirrors do not form perfect images as plane mirrors do unless used under particular conditions which will be discussed later. But they have the advantage over plane mirrors that whereas the latter always produce images of the same size as the object, spherical mirrors can produce magnified or diminished images.

Single spherical refracting surfaces are seldom encountered, with the exception of the corneal surface of the eye. But a combination of two spherical, refracting surfaces, the lens, is in everyday use. Lenses, like spherical mirrors, can produce magnified or diminished images and are very much employed in consequence.

14.2 REFLECTION AT SPHERICAL SURFACES

Figure 14.1 shows a *concave* spherical mirror, the reflecting material being on the inside of the spherical surface. The center-point of the surface O is the pole of the mirror and the line at right angles to the surface at the pole is called the axis of the system. Obviously the center of curvature C must lie on this axis. Consider any ray of light BA coming into the system parallel to the axis and striking the mirror surface at A. The lengths of AC and OC are both equal to r, the radius of curvature of the mirror. The infinitesimal portion of the mirror around O is to a first approximation flat, with CA the normal at A. The laws of reflection therefore apply to the light ray at A. If the angle of incidence is θ, the ray after striking A must go off in the direction AD such that angle CAD is θ also. But the incoming ray of light BA is parallel to CD. Hence, angles BAC and ACD are equal. Hence, CAD is an isosceles triangle. It follows that $AD = CD$. In the triangle ACD

$$AD^2 = AC^2 + CD^2 - 2AC \cdot CD \cos \theta.$$

therefore

$$CD^2 = r^2 + CD^2 - 2r \cos \theta \cdot CD.$$

Therefore

$$CD = \frac{r}{2 \cos \theta}.$$

The rays in a beam of light parallel to the axis do not all cross the axis at the same point unless θ is small for all of them. This can be achieved by making the mirror of small aperture or by covering a large mirror with a screen which has a hole of small aperture in it. When this happens, all values of θ are small and for all rays $\cos \theta$ is approximately equal to 1. Thus all rays converge to a single point F on the axis called the focus of the mirror, half-way between the center of curvature and the pole. The distance from the pole to the focal point is called the focal length of the mirror and is denoted by the symbol f. It follows that $f = \frac{1}{2}r$.

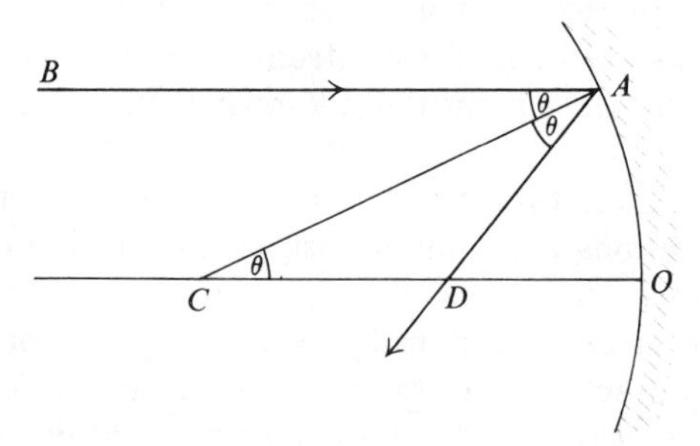

Fig. 14.1. Reflection at a concave spherical mirror.

In all subsequent work it will be assumed that the mirror employed is of small aperture, that all rays are consequently *paraxial*, and that all incoming rays parallel to the axis converge to the focal point of the mirror, which is a unique point. With this approximation, it is found that point objects near the axis produce point images, small line objects line images, and small plane objects plane images without distortion. If an object is placed at any point on the axis of the mirror, it is now possible to derive a simple formula which allows the position of the image to be calculated.

It should be noted that, throughout the chapters on optics, capital letters will normally be used to denote points and lower-case letters to denote distances. Thus F_o denotes the position of the focal point of an objective lens and f_o is its focal length. In diagrams objects and images, for reasons of clarity, will be drawn much larger than would be justifiable in practice while using the paraxial approximation.

14.3 THE MIRROR FORMULA

In Fig. 14.2 an object OO' has been placed on the axis of a concave spherical mirror whose pole is A and whose center of curvature is C. Let $AC = r$ and $AO = u$. The angle of incidence of a ray of light from O' which strikes the mirror at

A is $O'AO = \theta$. This ray must be reflected from the mirror in the direction AB, where angle BAC is θ also. Further, the ray from O' passing through C strikes the mirror at right angles and is reflected back along its original path. These two rays meet at I' and, from the definition of an image, I' must be the image of O'. If a perpendicular is dropped from I' to the axis cutting it in I, I is likewise the image of O, since as O' approaches O, I' will approach I and in the limit the statement will be proved true.

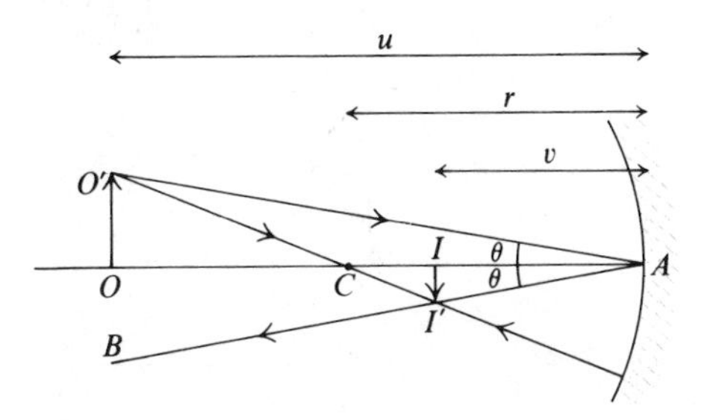

Fig. 14.2. Image location in a concave spherical mirror by ray tracing.

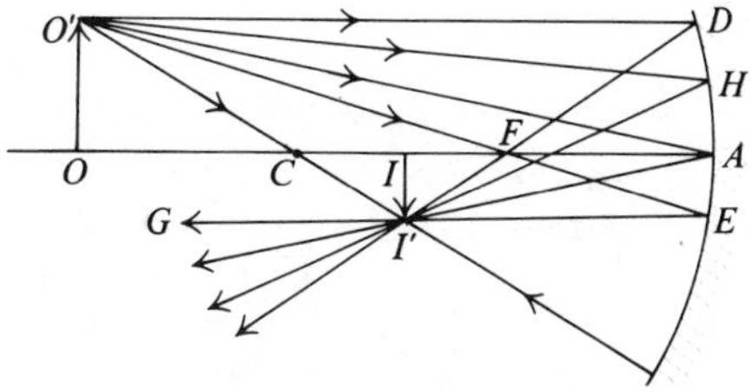

Fig. 14.3. Figure 14.2 with further rays added.

If F, the focal point of the mirror, is inserted midway between A and C, the paths of two further rays can be deduced to confirm the position of I'. The ray from O' parallel to the axis striking the mirror at D must pass through the focal point after reflection. DF is thus the path of the ray after reflection, and it also passes through I'. Finally, since any ray parallel to the axis passes through the focus after reflection, by the principle of reversibility of light rays any ray through the focus must emerge parallel to the axis after reflection at the mirror. Thus the ray $O'FE$ follows the path EG after reflection, and this ray also passes through I'.

Having found I', we can trace any other ray from O'. The general ray $O'H$ must pass through the image of O' after reflection. Thus the path of the ray after reflection is HI'.

Let $AI = v$. Triangles $O'AO$ and $I'AI$ each contain an angle θ and a right angle. Therefore the third angle in each is also equal. The two triangles are thus similar. Hence,

$$\frac{OO'}{II'} = \frac{OA}{IA} = \frac{u}{v}.$$

Similar reasoning shows that triangles $O'OC$ and $I'IC$ are similar. Thus

$$\frac{OO'}{II'} = \frac{OC}{IC} = \frac{u - r}{r - v}.$$

Therefore

$$\frac{u}{v} = \frac{u - r}{r - v}.$$

Therefore

$$ur - uv = uv - vr.$$

Therefore

$$2uv = ur + vr.$$

Therefore

$$\frac{2}{r} = \frac{1}{f} = \frac{1}{u} + \frac{1}{v}. \tag{14.1}$$

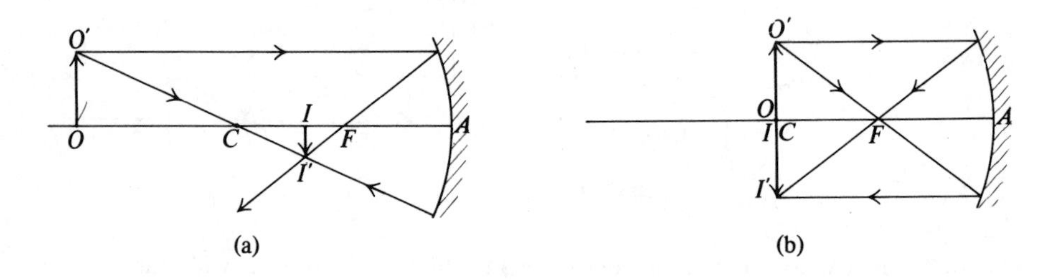

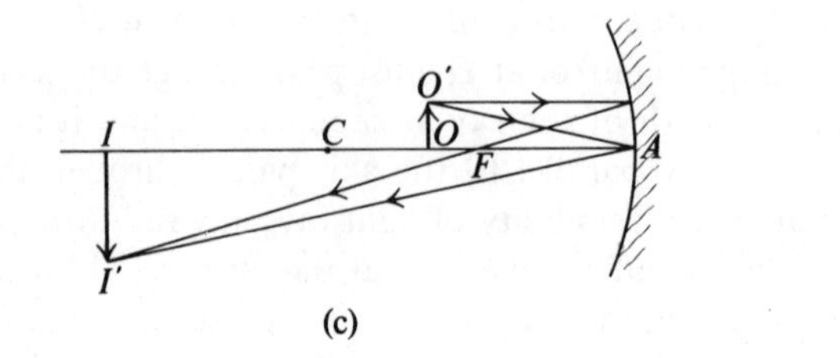

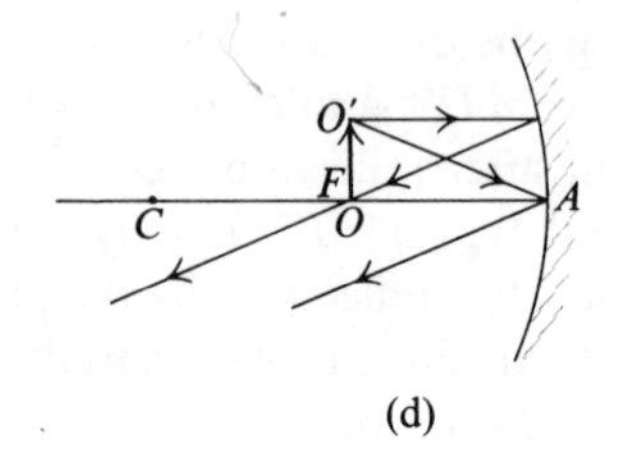

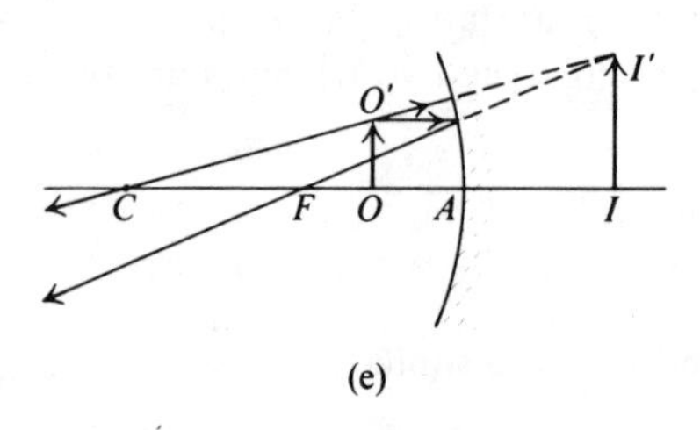

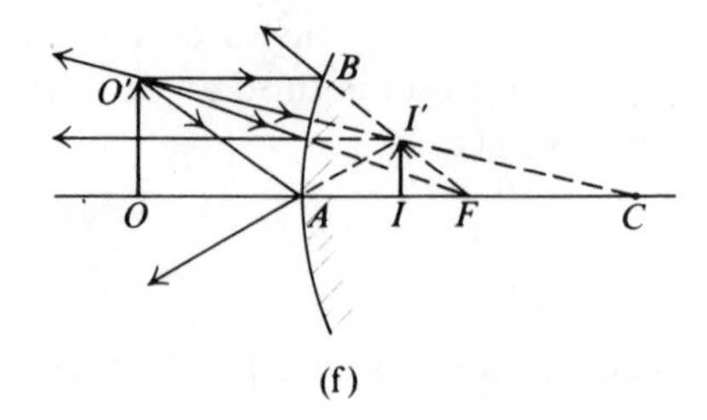

Fig. 14.4. Image formation in a spherical mirror. The image in a concave mirror when the object is (a) beyond the center of curvature, (b) at the center of curvature, (c) between the center of curvature and the focal point, (d) at the focal point, and (e) between the focal point and the pole; (f) the image in a convex mirror.

Note that the magnitude of the magnification of the image is

$$|m| = \frac{II'}{OO'} = \frac{v}{u}.$$

From a consideration of Eq. 14.1, or from the series of diagrams shown in Fig. 14.4(a)–(e), it will be seen that when $u > r$, i.e. when O lies further from A than C, the image is real, inverted, and diminished in size. If $u = r$, the image is real, inverted, and equal in size to the object. If $\frac{1}{2}r < u < r$, the image is real, inverted, and magnified. If $u = \frac{1}{2}r = f$, the image is at infinity. If $u < \frac{1}{2}r$, rays from O' after reflection do not meet but appear to come from a point behind the mirror. The image is therefore virtual and erect, and it is also magnified. Note that giving u a value less than $\frac{1}{2}r$ in the mirror formula produces a value of v which must be negative. If we adopt as a sign convention that distances measured to real objects and images (which lie in front of the mirror) are positive and that distances measured to virtual images (which lie behind the mirror) are negative, we have a consistent system of signs which makes the mirror equation we have derived always true. We also define the magnification m to be

$$m = \frac{II'}{OO'} = -\frac{v}{u}. \tag{14.2}$$

This therefore means that all inverted images turn out to have a negative sign and all erect images to have a positive sign.

Some books do not have a minus sign in the magnification formula. This is a great mistake. In such a system, inverted images have a positive magnification and erect images a negative magnification. If rays from an object are allowed successively to strike two mirrors (or lenses), the total magnification produced is the product of the magnifications produced at the individual mirrors (or lenses). The convention adopted in this book for the magnification produces the correct sign for the overall magnification in such a case, whereas the other does not. For example, suppose that the first mirror produces an inverted image, the second one then inverting this again. The final image will therefore be erect. In the system adopted here each mirror produces a negative magnification, the overall magnification (the product of these) therefore being positive. The final image is correctly deduced as erect. In the other system both magnifications are positive, as is therefore the overall magnification. But, in that system, a positive magnification implies that the image is inverted, which is obviously untrue.

The system of signs adopted applies equally well to a *convex* mirror, if we add one further convention. If the center of curvature lies in front of the mirror, r and f are considered positive, but if it lies behind the mirror, r and f, like image distances behind the mirror, are taken as negative. Figure 14.4(f) clearly shows that there is only one case that need be considered for a convex mirror. C is behind the mirror, and it is easy to show, by reasoning similar to that used in the concave case, that incoming rays parallel to the axis (for paraxial rays) all appear to diverge from the focal point F situated half-way between A and C. It is therefore possible to deduce the paths followed by certain rays from O'. The ray parallel to the axis must appear to diverge from F after striking the mirror. Because of the reversibility of light rays, the ray going toward F must travel parallel to the axis after reflection. The ray directed toward C strikes the mirror normally and returns by the same path. After reflection the ray striking the mirror at A makes the same angle with the axis but is on the opposite side of it. All these reflected rays appear to have come from the same point I'. $I'I$ is therefore the image of $O'O$.

Since I' always lies on $O'C$ and BF, it is clear that the image of a real object in a convex mirror is always virtual, erect, and diminished wherever OO' is located. This can also be seen if negative values for r and f are inserted in the mirror formula.

Example. An object 1 cm in height is placed 10 cm from a spherical mirror and produces a virtual image 2.5 cm in height. Where is the image located, and what type of mirror is being used?

Solution. It should be clear that, since the image is virtual and magnified, a concave mirror is being used. Convex mirrors can only produce diminished images. We can also deduce that the focal length of the mirror is greater than 10 cm, since u must be less than f.

If we use the formulae given earlier,

$$m = +\frac{2.5 \text{ cm}}{1 \text{ cm}} = 2.5.$$

Therefore

$$-\frac{v}{u} = 2.5.$$

Therefore

$$v = -2.5u = -25 \text{ cm}.$$

Since v is negative, the image is located 25 cm behind the mirror. Also,

$$\frac{1}{f} = \frac{1}{u} + \frac{1}{v} = \frac{1}{10 \text{ cm}} + \frac{1}{-25 \text{ cm}} = \frac{5-2}{50 \text{ cm}}.$$

Therefore

$$f = 16.7 \text{ cm}.$$

Since f is positive, the mirror is concave, and it has a radius of curvature of 33.3 cm.

14.4 REFRACTION AT SPHERICAL SURFACES—THE THIN LENS

Refraction can take place at curved surfaces as well as plane ones. Locally, any curved surface can be considered plane in the infinitesimal area around a point on it. Any ray striking at that point will be refracted into the second medium according to the laws of refraction which were developed for a plane surface. It is

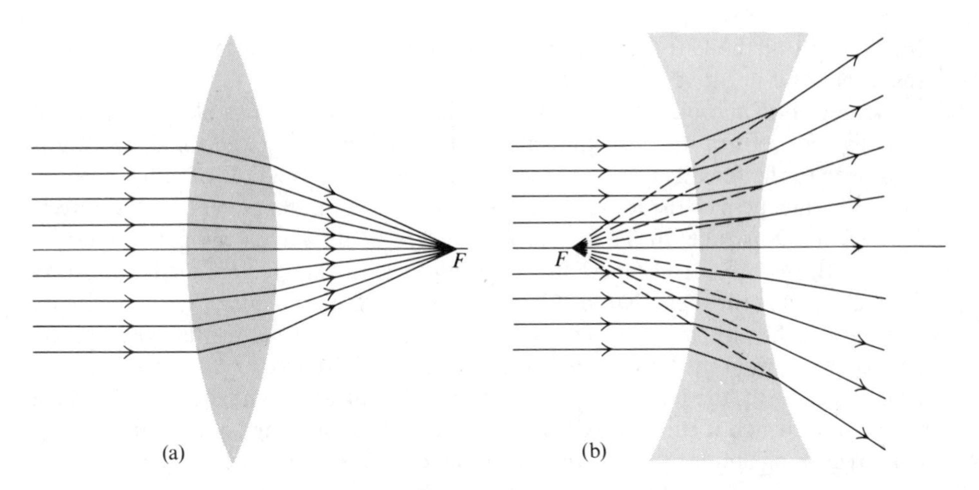

Fig. 14.5. Rays parallel to the axis of a thin lens (a) converging to the focal point of a converging lens and (b) diverging from the focal point of a diverging lens.

easy to show, as was done for the case of the spherical mirror, that, for a single, spherical, refracting surface separating two media, incoming rays parallel to the axis of the surface will be brought to a point focus only if we confine the rays to the paraxial region. If the rays are so confined, point, line, and plane objects near the axis will produce point, line, and plane images, respectively, without distortion, if the objects are small and only rays passing through the surface close to the pole are considered. However, a single, spherical, refracting surface is of little interest. What is of great importance is a combination of two spherical surfaces close together forming a thin lens. A thin lens is defined as two spherical refracting surfaces with a common axis, containing between them a refractive medium, normally glass, the distance between the poles of the surfaces being small enough to be ignored. Commonly, the outer medium on either side is air, but occasionally, as in the oil immersion microscope objective (cf. Section 19.3), the medium on one side is different from that on the other. Unless otherwise specified, it will be assumed that all lenses dealt with have air on both sides.

The remarks applying to a single spherical surface apply equally to a thin lens. So long as the objects used are not large and are located on or near the axis, and the rays from the objects are not permitted to strike anything but the central region of the lens, images are formed which faithfully reproduce the object without distortion. A beam of incoming rays parallel to the axis of the lens either converges to a point focus on the other side of the lens (when the lens is said to be a converging one) or appears to diverge from a point on the same side of the lens (when the lens is said to be a diverging one) (Fig. 14.5).

If light parallel to the axis is sent in from the other side of the lens, the lens converges or diverges the rays as before, and the focal points on the two sides of the lens are equidistant from it. The distance from the lens to either focus is called the focal length of the lens f, f being taken as positive if the distance is measured to a real focus and negative if to a virtual focus.

In Fig. 14.6 an object OO' is placed perpendicular to the axis of a thin lens and at a distance u from it. For simplicity the lens is converging and the distance u is greater than the focal length of the lens. The point A is the pole of either surface. Since the thickness of the lens is ignored, this is the common pole of both surfaces. A ray from O' parallel to the axis strikes the lens at B. Such a ray passes after refraction through F'. Hence, BF' is the path of the ray in the final medium. A ray from O' passing through F strikes the lens at D. By the principle of the reversibility of rays of light, this ray must emerge after refraction parallel to the axis. Hence, the ray in the final medium follows the path DE. These two rays meet at a point, which must be I', the image of O'. If a perpendicular is drawn from I' to the axis, the point at which it cuts the axis is the position of I, the image of O. This follows from reasoning similar to that applied in the case of the mirror.

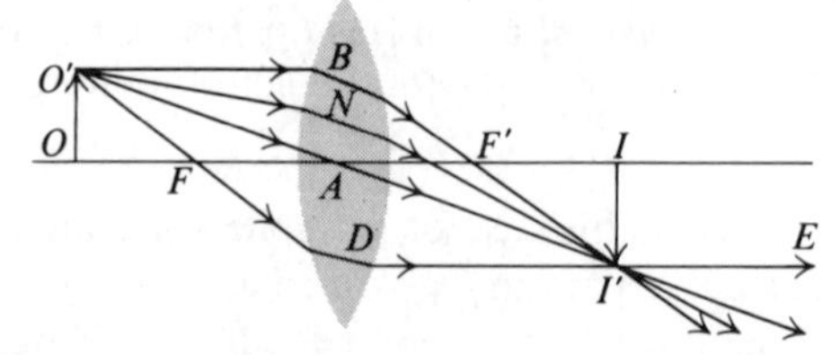

Fig. 14.6. Image location in a thin lens by ray tracing.

It will be seen that $O'AI'$ is a straight line, so that the ray from O' to A passes through the center of the lens undeviated. This is a consequence of the fact that, as was proved in Section 13.5, a parallel-sided block of glass does not deviate a ray of light but only displaces it by an amount depending on the thickness of the block. Since the portion of the lens at A is locally like a parallel-sided block, no deviation occurs in the ray $O'A$, and since the lens is of negligible thickness, the displacement is negligible also. Any general ray $O'N$ may now be plotted. After refraction by the lens, the ray must pass through I' and its path is thus NI' in the final medium.

14.5 THE LENS FORMULA

In Fig. 14.6 triangles $O'OF$ and DAF are similar, the angles at F being equal and the angles at A and O being right angles. The remaining angles must be equal also. Therefore

$$\frac{AD}{OO'} = \frac{AF}{OF} = \frac{f}{u-f}.$$

Similarly, in triangles BAF' and $I'IF'$

$$\frac{II'}{BA} = \frac{IF'}{AF'} = \frac{v-f}{f}.$$

But $BA = O'O$, and $DA = I'I$.

Therefore

$$\frac{AD}{OO'} = \frac{II'}{OO'} = \frac{f}{u - f},$$

and

$$\frac{II'}{BA} = \frac{II'}{OO'} = \frac{v - f}{f}.$$

Therefore

$$\frac{f}{u - f} = \frac{v - f}{f}.$$

Therefore

$$uv - vf - uf + f^2 = f^2.$$

Therefore

$$uv = vf + uf.$$

Therefore

$$\frac{1}{f} = \frac{1}{u} + \frac{1}{v}. \tag{14.3}$$

This is exactly the same relation as was obtained in the case of the mirror. Further, the magnitude of m, the magnification, is

$$|m| = \frac{II'}{OO'} = \frac{f}{u - f}.$$

But

$$\frac{1}{v} = \frac{1}{f} - \frac{1}{u} = \frac{u - f}{uf}.$$

Therefore

$$\frac{II'}{OO'} = \frac{v}{u}.$$

As in the case of the mirror, the magnification is defined to be

$$m = -\frac{v}{u}, \tag{14.4}$$

where a negative value for m will occur when the image is inverted.

If v is given a positive sign when the image is real and a negative sign when the image is virtual, the same convention as was used in the case of the mirror, the lens equation, like the mirror equation, will apply to all situations and to either converging or diverging systems. The series of figures showing the image formation when u has various values (Fig. 14.7) is completely analogous to that for mirrors and all the same conclusions may be drawn.

A plane through the focus perpendicular to the axis of the lens is called the focal plane. In Fig. 14.7(d), which shows the formation of the image when $u = f$, the object lies in the focal plane. Rays from any point on the object, such as O',

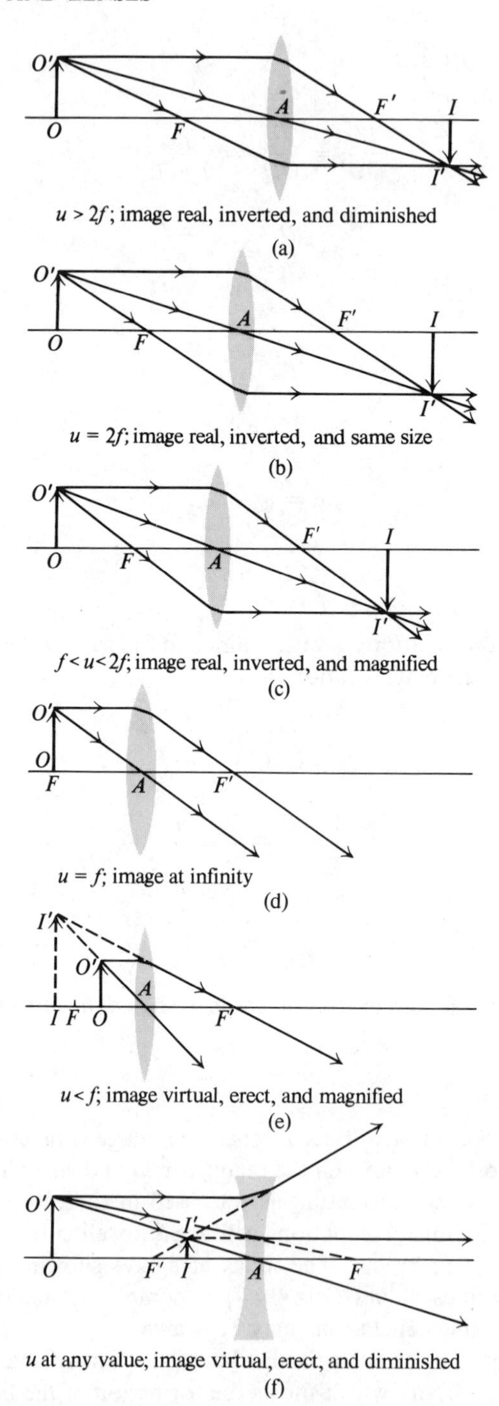

$u > 2f$; image real, inverted, and diminished

(a)

$u = 2f$; image real, inverted, and same size

(b)

$f < u < 2f$; image real, inverted, and magnified

(c)

$u = f$; image at infinity

(d)

$u < f$; image virtual, erect, and magnified

(e)

u at any value; image virtual, erect, and diminished

(f)

Fig. 14.7. Image formation by a thin lens: (a)–(e) a converging lens with the object at various positions; (f) a diverging lens.

emerge after striking the lens in a beam of rays parallel to one another but not parallel to the axis. If we reverse the light, it follows that beams of parallel rays meet after refraction by a lens in its focal plane, those parallel to the axis meeting at the focal point and those not parallel to the axis meeting elsewhere in the focal plane.

Example 1. A mirror of focal length -15 cm is used as the outside rear-view mirror on an automobile. Where is the image of a second automobile behind the first and 9.0 m from the mirror? What is the lateral magnification produced?

Solution. The mirror equation is

$$\frac{1}{v} = \frac{1}{f} - \frac{1}{u}$$

and in this case $u = +900$ cm, $f = -15$ cm. Therefore

$$\frac{1}{v} = \frac{1}{-15\text{ cm}} - \frac{1}{900\text{ cm}} = \frac{-(60+1)}{900\text{ cm}}$$

so

$$v = -900\text{ cm}/61 = -15\text{ cm}.$$

The image is virtual, since v has a negative sign and is located 15 cm behind the mirror. Also

$$m = -\frac{v}{u} = -\frac{-900\text{ cm}/61}{900\text{ cm}} = +\frac{1}{61}.$$

The image is thus erect and diminished by a factor of 61.

Example 2. A family shows its color slides of dimensions 35 mm by 35 mm by throwing an image on to a screen 10 m from the lens of a projector. The image is 1.00 m by 1.00 m in size. What distance must the slide be from the lens and what is the focal length of the lens?

Solution. The lateral magnification required has a magnitude of $1\text{ m}/35 \times 10^{-3}\text{ m} = 200/7$. Further, the image formed must be real since it is projected on to a screen. This means that the image is inverted since only inverted images formed by a single lens are real. Hence $m = -200/7$. Therefore

$$-v/u = -200/7.$$

But $v = 10$ m. Therefore

$$u = \frac{10\text{ m}}{200/7} = 35\text{ cm}.$$

Therefore

$$\frac{1}{f} = \frac{1}{u} + \frac{1}{v} = \frac{200}{7v} + \frac{1}{v} = \frac{207}{7v},$$

therefore

$$f = 7 \times 10\text{ m}/207 = 34\text{ cm}.$$

Example 3. When an object is moved along the axis of a thin lens images three times the size of the object are obtained when the object is at 16 cm from the lens and at 8 cm from the lens. Find the focal length of the lens.

Solution. Any particular value of magnification is given for one and only one value of the object distance. This can be seen quite clearly from a consideration of the nature of the image in a ray diagram as the object distance is varied from infinity to zero. It follows that if two object positions are obtained in which the image is three times the size of the object, in one case the magnification must be -3 and in the other $+3$. A magnification of -3 implies a real image and a magnification of $+3$ a virtual image. A diverging lens can produce only virtual images from real objects, so the lens being used is converging. Further, the smaller object distance must produce the virtual image since this is only possible when the object is closer to the lens than the focal point. We therefore have several checks on the answer we obtain. It must be between 8 cm and 16 cm and be positive. When $u_1 = 16$ cm,

$$m_1 = -\frac{v_1}{u_1} = -3; \qquad \text{therefore} \qquad v_1 = 3u_1 = 48 \text{ cm}.$$

When $u_2 = 8$ cm,

$$m_2 = -\frac{v_2}{u_2} = +3; \qquad \text{therefore} \qquad v_2 = -3u_2 = -24 \text{ cm}.$$

But

$$\frac{1}{u_1} + \frac{1}{v_1} = \frac{1}{f} = \frac{1}{u_2} + \frac{1}{v_2}.$$

Therefore

$$\frac{1}{16 \text{ cm}} + \frac{1}{48 \text{ cm}} = \frac{1}{f} = \frac{1}{8 \text{ cm}} - \frac{1}{24 \text{ cm}}.$$

Therefore

$$\frac{3+1}{48 \text{ cm}} = \frac{1}{f} = \frac{3-1}{24 \text{ cm}}.$$

Therefore

$$f = 12 \text{ cm}.$$

It should be noted that $f = \frac{1}{2}(u_1 + u_2)$. This is a general result and clearly the value of the magnification is not required. Assume it has not been given and that we take the magnitude of it as x. Then

$$v_1 = xu_1$$

$$v_2 = -xu_2.$$

Therefore

$$\frac{1}{u_1} + \frac{1}{v_1} = \frac{1}{f} = \frac{1}{u_2} + \frac{1}{v_2}.$$

Therefore

$$\frac{1}{u_1} + \frac{1}{xu_1} = \frac{1}{f} = \frac{1}{u_2} + \frac{1}{-xu_2}.$$

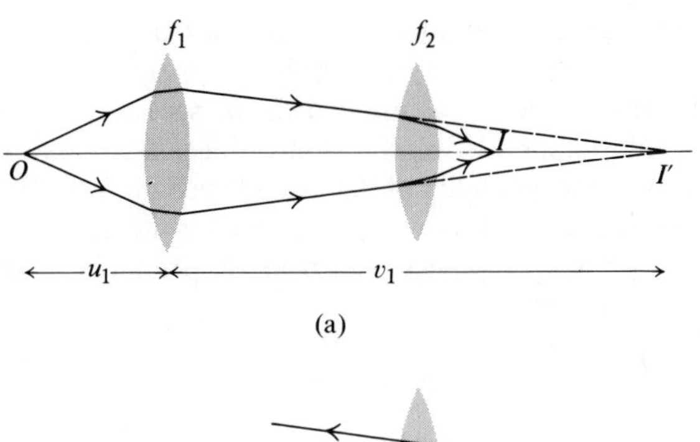

(a)

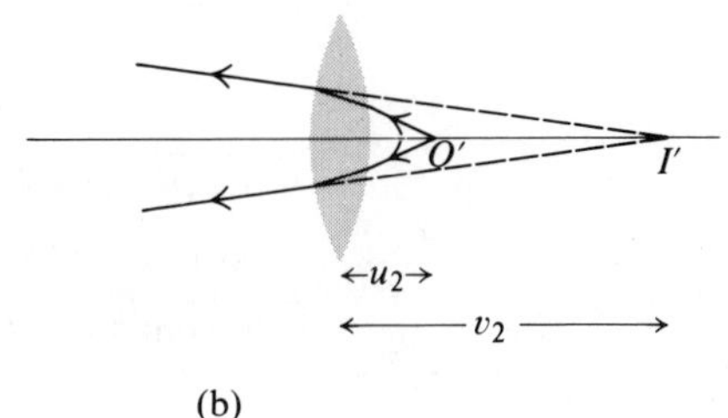

(b)

Fig. 14.8.

Therefore
$$\frac{x+1}{xu_1} = \frac{1}{f} = \frac{x-1}{xu_2}.$$

Therefore
$$u_1 = \frac{x+1}{x} f$$

$$u_2 = \frac{x-1}{x} f.$$

Therefore
$$u_1 + u_2 = 2f.$$

Example 4. A lens system consists of two converging lenses of focal lengths 10 cm and 12 cm a distance of 15 cm apart. If an object is placed 15 cm from the stronger lens, where is the final image located?

Solution. Applying the lens formula to the first lens gives

$$\frac{1}{v_1} = \frac{1}{f_1} - \frac{1}{u_1}$$

$$= \frac{1}{10 \text{ cm}} - \frac{1}{15 \text{ cm}}$$

$$= \frac{3-2}{30 \text{ cm}}.$$

Therefore
$$v_1 = 30 \text{ cm}.$$

The distance from the second lens to I' in Fig. 14.8 is thus 15 cm.

A converging beam of light strikes the second lens and is brought to a focus more quickly after refraction. We now are liable to run into trouble with virtual objects and the signs of u and v as applied to the second lens. To avoid this we consider what would happen if we placed an object O' at the point I as in Fig. 14.8(b). Because of the reversibility of light rays, the light from O' follows the same paths through the lens system as were followed by the rays from O, but in the opposite direction. Thus an object placed at O' produces in the second lens a virtual image at I'. Hence

$$\frac{1}{u_2} = \frac{1}{f_2} - \frac{1}{v_2}$$

$$= \frac{1}{12 \text{ cm}} - \frac{1}{-15 \text{ cm}}$$

$$= \frac{5 + 4}{60 \text{ cm}}.$$

Therefore

$$u_2 = 60 \text{ cm}/9 = 6.7 \text{ cm}.$$

PROBLEMS

14.1 In a Tom and Jerry cartoon, Jerry is shown as terrified when he sees his enlarged image in a silver ball hung as a Christmas tree decoration. Explain the cartoonist's mistake.

14.2 A concave spherical mirror having a diameter of 10 cm and a radius of curvature of 20 cm is illuminated by a beam of light consisting of rays parallel to the axis. Calculate the distance between the focal point and the point at which rays incident on the outer edge of the mirror cross the axis.

14.3 A small object is placed 9 cm from a concave spherical mirror of radius of curvature 12 cm. Find the position, nature, and magnification of the image.

14.4 An object 6 cm from a spherical mirror produces an image 12 cm behind the mirror. What is the radius of curvature of the mirror?

14.5 An object is situated 40 cm from a convex mirror. When a plane mirror is inserted between object and convex mirror at a distance of 32 cm from the object, the images in the two mirrors coincide. What is the radius of curvature of the spherical mirror?

14.6 What will be the diameter of the image of the moon formed by a spherical concave telescope mirror of focal length 2 m? The moon's diameter is 3500 km and the distance of the moon from the earth is 3.84×10^5 km.

14.7 When an object is moved along the axis of a spherical mirror, it is found to form images four times the size of the object when it is 9 cm and 15 cm from the pole of the mirror. What is the radius of curvature of the mirror?

14.8 A small object is placed in front of a spherical mirror and an image after reflection twice the size of the object is formed on a screen. Both object and screen are moved until the image is three times the size of the object. If the screen is moved 12 cm farther from the mirror, how far has the object to be shifted and what is the focal length of the mirror?

14.9 A child looking into a small, polished, hemispherical, aluminum kitchen-bowl sees an erect image of himself 20 cm from the bowl. He turns the bowl over and sees a further erect image of himself 6.67 cm from the bowl. What is the radius of curvature of the bowl?

14.10 Two spherical mirrors of radii of curvature -20 cm and 24 cm are placed coaxially 41.33 cm apart. An object is placed 20 cm from the convex mirror. Find the nature, position, and magnification of the final image formed after reflection in the convex mirror and then the concave mirror.

14.11 Two mirrors are placed coaxially 32 cm apart with their reflecting surfaces facing one another. A small object is placed midway between them and its image after reflection of light from the mirrors in turn is also midway between them. If one mirror is concave and of radius of curvature 24 cm, what kind of mirror is the other and what is the magnification produced by the double reflection?

14.12 An object is placed 12 cm from a diverging lens of focal length -8 cm. Find the position, nature, and magnification of the image.

14.13 Show that, if an object is placed on the axis of a lens at a distance x from the first focal point and forms an image a distance x' from the second focal point, $xx' = f^2$, where f is the focal length of the lens. (This is known as Newton's formula.)

14.14 The frames in a home movie require to be magnified 143 times before the picture formed on a screen 576 cm from the projection lens is large enough to please the family watching. What distance must the film be from the lens and what is the focal length of the lens?

14.15 A photographer employs a camera with a lens of focal length 5 cm to photograph a man 1.98 m tall. How far from the man should the camera lens be if the image on the photographic plate is to be 2 cm in height? What is the correct lens-to-plate distance?

14.16 If the distance between an object and its image in a converging lens is 50 cm and the image is four times the size of the object and inverted, what is the focal length of the lens?

14.17 A lantern slide with the dimensions 2 in. by 2 in. is to be projected on to a screen 25 ft from a projection lens in such a way as to produce an image 4 ft by 4 ft. What is the focal length of the lens and what is its distance from the slide?

14.18 An object and a screen are 48 cm apart. A converging lens of focal length 9 cm is inserted between them. For what positions of the lens will clear images of the object be seen on the screen? Compare the magnifications produced in the two cases.

14.19 Show that two thin lenses of focal lengths f_1 and f_2 in contact are equivalent in their action to a single thin lens of focal length $f_1 f_2/(f_1 + f_2)$.

14.20 Two thin converging lenses of focal lengths 5 cm and 15 cm are placed 30 cm apart. An object is placed 10 cm from the stronger lens on the opposite side from the second lens. Where is the final image located after the passage of light through both lenses, and what is the total magnification produced?

14.21 A telephoto lens combination consists of a converging lens of focal length 30 cm and a diverging lens of focal length -10 cm, the separation between the lenses being 27.5 cm. Where should a photographic plate be placed in order to photograph an object 100 m in front of the first lens?

14.22 When a luminous object is placed 50 cm from a spherical mirror, object and image coincide. If a lens is placed 10 cm from the object between object and mirror and the mirror is moved forward until it is 30 cm from the lens, object and image again coincide. What is the focal length of the lens?

CHAPTER 15

OPTICAL INSTRUMENTS

15.1 INTRODUCTION

If an object is illuminated and the rays from it are converged by a lens, under favorable conditions a magnified object can be obtained on a screen placed behind the lens. The eye (cf. Section 16.6) is limited in the amount of detail it can see in an object, and examining the magnified image may allow much more detail to be seen than examining the object directly. The improvement has been achieved by the lateral magnification of the image in comparison with the object.

It is not normally convenient to study the image of an object in this way. Since small objects are studied routinely in medicine, biology, and many other sciences, it is desirable to have available compact magnifying instruments requiring only simple adjustment, so that they can be used by operatives with little training. The rest of the chapter will therefore deal with simple instruments which satisfy these criteria. Chapters 19 and 20 will deal with more complicated instruments, or more complicated variations of the basic instruments, which perform more specialized functions.

15.2 THE SIMPLE MAGNIFIER—ANGULAR MAGNIFICATION

A single lens may be used as a magnifying glass, but without producing a real image. If the lens is a converging one and is placed so close to the object that the latter lies within the focal length of the lens, an enlarged virtual image of the object is seen through it. It is now necessary to think rather carefully about the improvement achieved when the magnified image is viewed. The image has certainly been magnified but, since the eye must be on the other side of the lens from the image and object, the image is farther away also. A close small image is often seen in more detail than a large very distant one. Therefore lateral magnification is no longer a criterion which tells anything about the ability to see the object more clearly.

What is important is the angles which the image and object subtend at the eye. If the image covers a wider angle at the eye than the object, detail will be seen more clearly. It is therefore angular magnification and not lateral magnification which is important.

At first sight this simple magnifier appears to have achieved no angular magnification at all. It is usually used with the eye close to A, and the eye looking backwards along the ray AO' sees the object and the image at the same angle θ' to the axis. But a person with normal eyesight cannot focus on an object which is

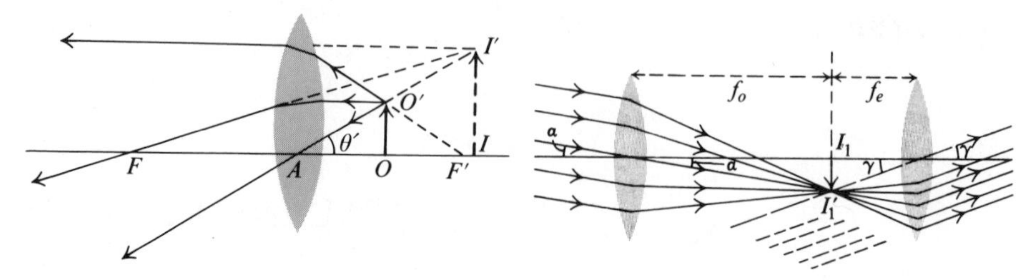

Fig. 15.1. Image formation in a magnifying glass.

Fig. 15.2. Image formation in a telescope.

brought closer to his eye than a distance of around 25 cm, which is called the least distance of distinct vision; and the simple magnifying glass is usually a lens of short focal length. The object is therefore viewed so close to the lens that if the lens were removed, the viewer could not focus on the object. The best angular viewing achieved by the unaided eye would result when the object was moved back to 25 cm from the eye. The valid comparison is therefore between the image as seen through the magnifier and the object as viewed unaided in its best position. The angular magnification is in consequence defined as

$$\frac{\text{the angle } \theta' \text{ subtended by the image}}{\text{the angle } \theta \text{ subtended by the object when at the least distance of distinct vision}}.$$

Since the angles used are small, the rays being confined to the paraxial region, the angles may be replaced by their tangents. Thus the angular magnification is β, where

$$\beta = \frac{\tan\theta'}{\tan\theta} = \frac{II'/AI}{OO'/25\text{ cm}} = \frac{25\text{ cm}}{AI} \times \frac{II'}{OO'}.$$

But $II'/OO' = m = -v/u$, and the distance $AI = -v$, since v is a negative quantity. Therefore

$$\beta = \frac{25\text{ cm}}{-v} \times -\frac{v}{u} = \frac{25\text{ cm}}{u}. \tag{15.1}$$

15.3 THE TELESCOPE

If is not necessary to confine the system to one lens in attempting to produce a laterally, or angularly, magnified image of an object. A simple refracting telescope consists of two lenses: the one nearer the object is called the objective lens, and is a converging lens of long focal length; the one nearer the eye is called the eye lens and is normally also convergent but of small focal length. The telescope views distant objects, and the light reaching it from any point of the object therefore consists of a beam of parallel rays. These will be brought to a focus in the focal plane of the objective lens, as in Fig. 15.2. If the eye lens is so placed that the focal planes of the two lenses are coincident, the rays from the point I_1' on the intermediate image, after striking the eye lens, emerge parallel, and an eye receiving these rays sees the final image at infinity. The telescope is then said to be in normal adjustment. Alternatively, if the instrument is so arranged that the focal plane of the objective falls just nearer the eye lens than its own focal plane, the image is seen as virtual and at a finite distance from the instrument.

Here the object viewed is distant, and it is not relevant to compare the angle subtended by the image with the angle subtended by the object when viewed at the normal distance of distinct vision. The latter possibility is not usually available. The magnifying power of the telescope is therefore defined as the angle (or rather its tangent, since the angle is small) subtended by the image, divided by the angle (or its tangent) subtended by the object. The angle subtended by the final image at the eye is γ. The angle subtended by the object at the telescope is α, which is also the angle subtended by I_1I_1' at the objective. Thus the magnifying power of the telescope when in normal adjustment is

$$M = \frac{\tan \gamma}{\tan \alpha} = \frac{I_1I_1'/f_e}{I_1I_1'/f_o} = \frac{f_o}{f_e}. \tag{15.2}$$

15.4 THE COMPOUND MICROSCOPE

The microscope appears very similar in construction to the telescope, but in this case the objective lens is of very short focal length. The object is placed just outside the focal plane of the objective lens, and an intermediate, real, magnified, and inverted image is produced between the two lenses (Fig. 15.3). If it is formed inside the focal plane of the eye lens, the virtual image is viewed at some finite distance from the eye lens, being again magnified in the process.

The magnifying power of the instrument is defined as the lateral magnification produced by the objective lens multiplied by the angular magnification produced

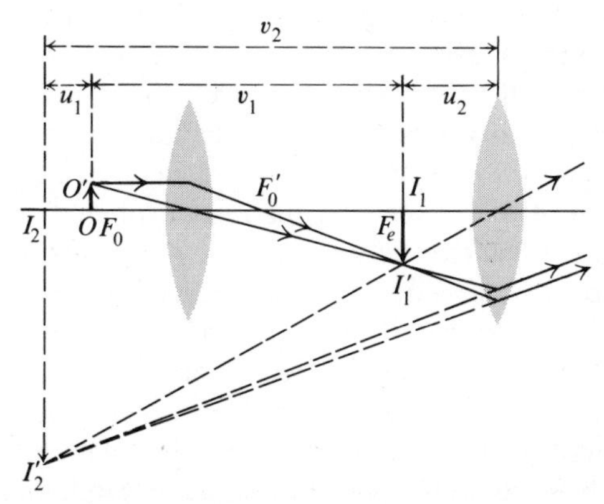

Fig. 15.3. Image formation in a microscope.

by the eye lens. Thus

$$M = -\frac{v_1}{u_1} \times \frac{25\text{ cm}}{u_2}. \tag{15.3}$$

Note that this is also the lateral magnification if the final virtual image formed is at the least distance of distinct vision. But

$$\frac{1}{u_1} + \frac{1}{v_1} = \frac{1}{f_o}.$$

Therefore

$$\frac{1}{u_1} = \frac{1}{f_o} - \frac{1}{v_1} = \frac{v_1 - f_o}{v_1 f_o}.$$

Therefore

$$M = -\frac{v_1 - f_o}{f_o} \times \frac{25\text{ cm}}{u_2} \approx -\frac{F_o' F_e}{f_o} \times \frac{25\text{ cm}}{u_2}. \tag{15.4}$$

Also, if the final image is formed at infinity, $u_2 = f_e$, and if it is formed at some other finite distance, u_2 is very close to f_e. Thus, approximately,

$$M = -\frac{F_o' F_e \times 25\text{ cm}}{f_o f_e}.$$

$F_o' F_e$ is called the tube length L of the microscope. Thus

$$M = -\frac{25\text{ cm} \times L}{f_o f_e}. \tag{15.5}$$

15.5 ABERRATIONS

In an actual telescope or microscope the only way in which rays can be confined to the paraxial region is by blocking off most of the area of both lenses by suitable apertures. The amount of light reaching the eye is then severely limited; but there is little use in producing a perfect image of a rather dim object if the image is so badly illuminated that one cannot see it. In practice the light is not confined only to the central region of the lenses, and the image suffers from a defect known as spherical aberration. This can be minimized by choosing each of the lenses so that the spread of the rays produced by one surface is to some extent compensated by the spread of the rays in the opposite direction from the second surface. It will be seen in Fig. 15.4 that a refracting surface which converges incoming rays of light converges the outer rays more strongly than the paraxial rays, whereas a refracting surface which diverges incoming rays diverges the outer rays more strongly than the paraxial rays. It is therefore apparent that a combination of these two types of surface tends to minimize the effect. For instance, the effect is almost eliminated for

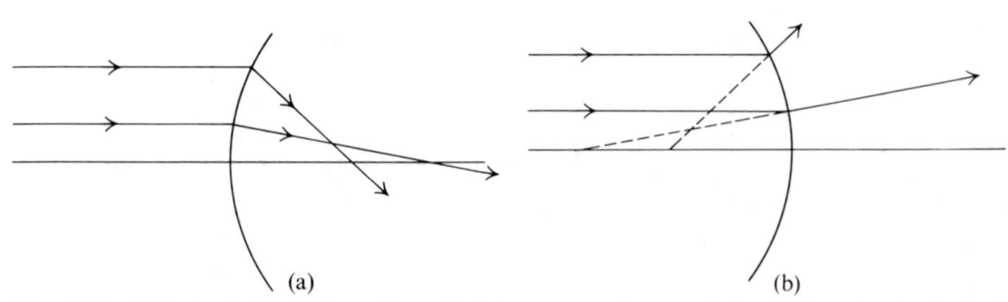

Fig. 15.4. Spherical aberration effects in (a) a converging, and (b) a diverging, spherical refracting surface.

incoming light parallel to the axis if the lens is designed as shown in Fig. 15.5, with $|r_1| = \frac{1}{7}|r_2|$. Alternatively, combinations of two or more lenses can be used for the objective and the eyepiece, each combination so chosen as to minimize the aberration. It should be noted that the aberration is minimized only for one position of the object.

Spherical aberration is not the only defect from which an image may suffer. The most important additional ones are astigmatism and chromatic aberration.

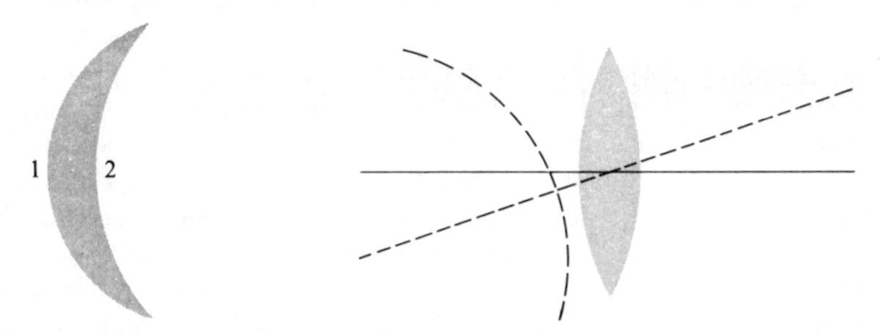

Fig. 15.5. A meniscus lens minimizes spherical aberration for an incoming beam of rays parallel to the axis.

Fig. 15.6. A spherical wave-front from an off-axis object striking a thin lens.

If a point source of light is placed off the axis of a lens, the light spreads from the source in the shape of an expanding sphere, part of which is shown in Fig. 15.6. Even if the light passing through the lens is only that portion of the sphere of light which strikes the center, the image formed by the lens suffers from the defect known as astigmatism and it is impossible to form a point image of the point object. This is because the lens possesses symmetry about its axis (shown as a full line in Fig. 15.6), whereas the incoming spherical wave-front of light which strikes the lens has symmetry about the dashed line in the same diagram. The effect of passing a wave-front with circular symmetry through a lens with circular symmetry about a different direction is to destroy the symmetry of the wave-front coming from the lens, and it is not now symmetrical about either the axis or the radius of the incoming sphere. The light cannot therefore converge to, or diverge from, a point. The resulting image is obviously more affected the farther the

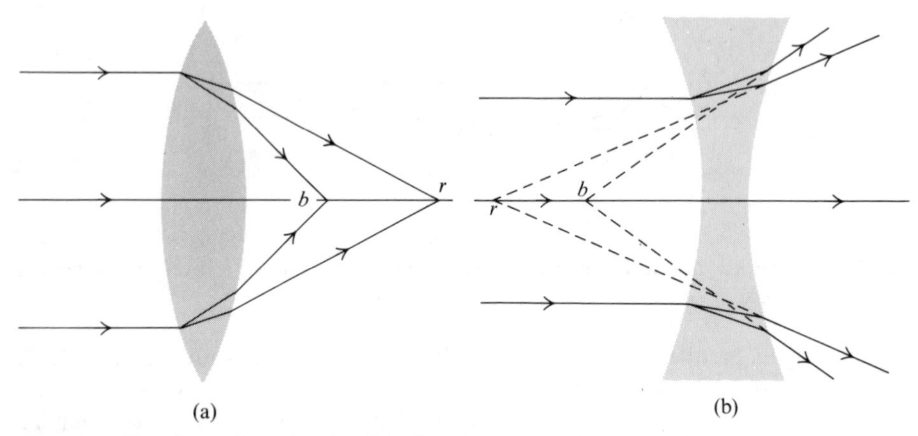

Fig. 15.7. The dispersion of white light by (a) a converging, and (b) a diverging, thin lens.

object is from the axis, and the placing of apertures at various points can ensure that only light from near-axis points is received. It is also possible by using a suitable combination of lenses instead of a single lens at objective and eyepiece to cut this defect to a minimum for a given position of the object.

Chromatic aberration is rather different. When Snell's law was quoted, it was carefully stated that the relative, or the absolute, refractive index was only a constant for a particular color of light. If the color, i.e. the wavelength, of the light is altered, it is found that the absolute refractive index for any particular substance varies also. The variation is quite small and is not simply related to wavelength. For ordinary crown glass the refractive index at the blue end of the visible spectrum is 1.523, and at the red end it is 1.517. For a large number of purposes the variation in the refractive index is relatively unimportant and can be ignored; but in forming magnified images of objects by the use of such instruments as telescopes or microscopes, when white light is employed, this is not so. White light is a mixture of all colors of visible light. Because of the variation of refractive index with wavelength, at each refraction light of a particular wavelength is deviated to a slightly different extent from light of all other wavelengths. After a number of such refractions the final image formed may consist of a number of colored images all displaced from one another by an appreciable amount. The eye, focusing on one plane, therefore sees a multicolored image which is, in consequence, blurred.

This defect cannot be eliminated from a single lens. It is, however, possible to use a combination of two lenses to minimize the effect. Different materials disperse light into its separate colors differently, and, as can be seen from Fig. 15.7, converging and diverging lenses disperse the colors in opposite directions. A combination can therefore be obtained of two lenses of two different materials such that, for incoming parallel light, the dispersion produced by each component is equal and opposite, while the deviation is not, the final emergent beam being made to converge or diverge as required. All good optical instruments use such achromatic combinations.

It should not be thought that dispersion of light into its individual colors to form a spectrum is always undesirable. In Chapter 21 spectra, particularly absorption spectra, will be discussed, and it will be found that valuable information about structural details of biological specimens can be obtained from their study.

15.6 LIMITATIONS OF TELESCOPES AND MICROSCOPES

If in a telescope or a microscope the objective lens and eye lens are both replaced by suitably chosen combinations of lenses, aberrations can be reduced to negligible proportions and a clear, magnified image of an object can be obtained. It may well be asked why one should stop at a two-lens system. Why should one not insert between objective and eyepiece one or more combinations of lenses in such a manner as to produce a further magnified, real image to be viewed through the eyepiece? Is there any reason why this process should not be continued indefinitely until a viewable image of even the tiniest object has been produced? There are, in fact, several good reasons why this is not possible.

In the first place, it has already been mentioned that at every interface between two media both reflection and refraction take place. If an incoming beam of light parallel to the axis strikes a glass surface, roughly 4% of the light is reflected back. In passing through the two lenses of a simple microscope, roughly 16% of the light never gets through to the eye to form the final image. Since the microscope, because of aberrations, has to consist not of two lenses but of at least four, one can see that the greater the number of lenses put in the less becomes the chance of the final image being seen at all clearly because of lack of illumination.

Further, some of the light scattered back from one surface may be scattered forwards again by a previous one, producing an undesirable background of light against which it is even more difficult to see the faint final image. Even worse than this is the fact that some of the scattered light may be concentrated into a particular direction and become focused in the plane of the intermediate image. It then shows up as a bright spot or area in the final picture, suggesting spurious detail in the image. For all these reasons it is undesirable to have too many components in an optical instrument.

But far more important than all that is the fact that all optical instruments are limited in their ability to resolve fine detail. It is well known from everyday experience that the eye cannot see fine detail beyond a certain limit. This is a defect inherent also in all other optical instruments. If a hundred-stage microscope, instead of a good two-stage one, were constructed and the image could still be seen clearly, it would be found that no more detail could be seen in the larger instrument. All that would be obtained would be a larger picture of the same slightly blurred image. The reason for this lies in the wave nature of light. Further discussion of this point will therefore be postponed until something is known of the subject matter of physical optics, but the resolving ability of optical instruments is such a crucial topic that the subject will be fully explored in Sections 18.3 and 19.3.

15.7 THE OPHTHALMOSCOPE

The ophthalmoscope is an instrument for viewing the retina of an eye. In its usual form it consists (cf. Fig. 15.8) of a plane mirror M attached to a handle. The handle contains a small electric light bulb B powered from a dry cell, also in the handle, and a lens L adjustable in position. The light from the bulb is suitably converged by the lens so that it focuses the light in front of the retina inside the eye to be viewed. The retina is thus illuminated by an unfocused diverging beam of light. In a normal, completely relaxed eye the light scattered from a point on the retina emerges as a parallel beam and is viewed by an observer through a hole in the center of the plane mirror. The observer is effectively using the optical system of the eye he is examining as a simple magnifying glass in order to obtain an angularly magnified image of the retina.

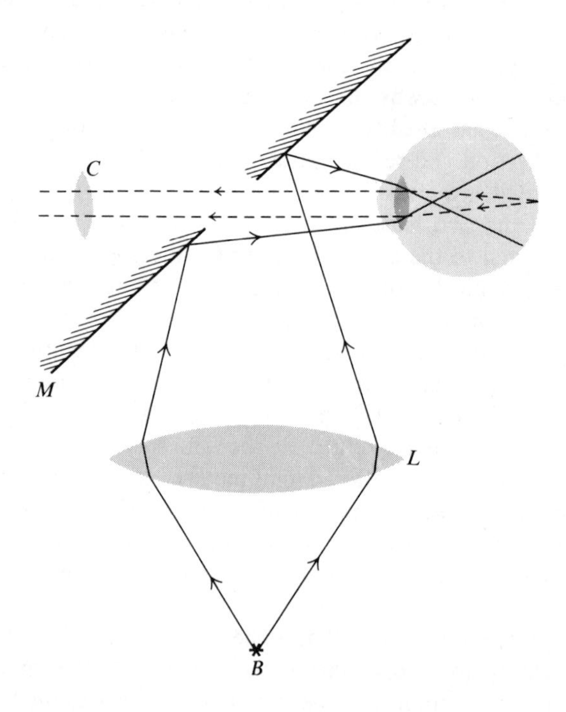

Fig. 15.8. The mode of operation of an ophthalmoscope.

To cope with the situation where the eye (or the observer's eye) suffers from an optical defect, a collection of converging and diverging lenses C is available on the circumference of a wheel attached to the mirror. In any particular case the appropriate correcting lens can be swung into position behind the hole in the mirror so that a clear view of the retina can be obtained by the observer.

Example 1. Compare the angular magnifications produced when an object is viewed through a magnifying glass of focal length 5.0 cm and the image is formed (a) at 25 cm from the lens, (b) at 100 cm and (c) at infinity.

Solution. The image in each case is virtual, the values of v in the three cases therefore being (a) -25 cm, (b) -100 cm and (c) $-\infty$. Using the lens formula gives

a) $$\frac{1}{u} = \frac{1}{f} - \frac{1}{v} = \frac{1}{5\text{ cm}} - \frac{1}{-25\text{ cm}} = \frac{6}{25}\text{ cm}$$

b) $$\frac{1}{u} = \frac{1}{5.0\text{ cm}} - \frac{1}{-100\text{ cm}} = \frac{21}{100\text{ cm}}$$

c) $$\frac{1}{u} = \frac{1}{5.0\text{ cm}} - \frac{1}{-\infty} = \frac{1}{5\text{ cm}}$$

The angular magnification is 25 cm$/u$. Thus

a) $$\beta = 25\text{ cm} \times \frac{6}{25\text{ cm}} = 6.0$$

b) $$\beta = 25\text{ cm} \times \frac{21}{100\text{ cm}} = 5.25$$

c) $$\beta = 25\text{ cm} \times \frac{1}{5\text{ cm}} = 5.0.$$

Example 2. Some old people use a magnifying glass for reading when their eyes are failing, but with the glass held just above the page rather than close to the eye. Compare the angular magnifications achieved in viewing an object 5 cm from a lens of focal length 6 cm when the eye is placed (a) at the lens surface and (b) 20 cm from the lens.

Solution. The object is viewed through the lens as in Fig. 15.9 and an image II' is seen. The angles subtended by the image at the eye are θ' when the eye is close to the lens and θ'' when the eye is 20 cm from it. The value of $\tan\theta'$ is II'/AI and the value of $\tan\theta''$ is II'/BI. The two angular magnifications achieved are $\beta' = \tan\theta'/\tan\theta$ and $\beta'' = \tan\theta''/\tan\theta$, θ being the angle subtended at the eye by the object at the near point. Therefore

$$\beta' = 25\text{ cm}/u = 25\text{ cm}/5\text{ cm} = 5$$

and

$$\beta'' = \frac{II'}{BI} \times \frac{25\text{ cm}}{OO'} = \frac{AI}{AO} \times \frac{25\text{ cm}}{BI}.$$

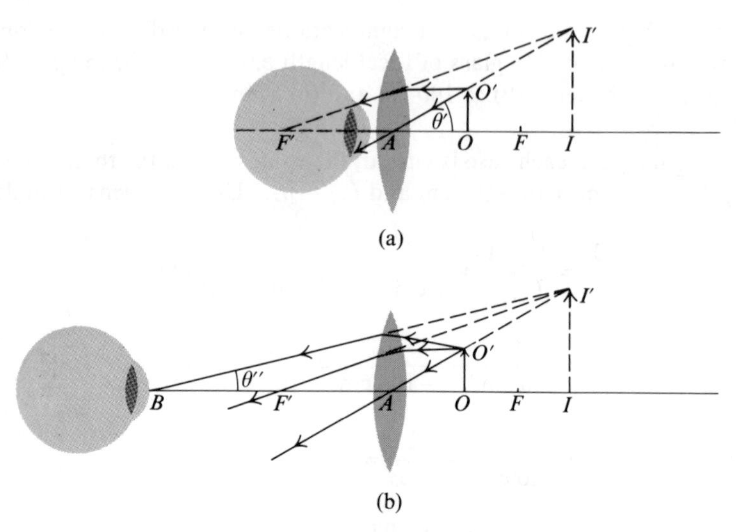

Fig. 15.9

But

$$\frac{1}{v} = \frac{1}{f} - \frac{1}{u} = \frac{1}{6\text{ cm}} - \frac{1}{5\text{ cm}} = -\frac{1}{30}\text{ cm}.$$

Therefore

$$v = -30\text{ cm} \qquad \text{and} \qquad AI = 30\text{ cm}.$$

Therefore

$$\beta'' = \frac{30\text{ cm}}{5\text{ cm}} \times \frac{25\text{ cm}}{50\text{ cm}} = 3.0.$$

The angular magnification when the lens is held far from the eye is therefore much less than that achieved if the eye is brought close to the lens, the object distance remaining the same. This is a particular example of a general result. One can always achieve better results if the lens is close to the eye, as is the case with jewelers' glasses.

Example 3. A compound microscope consists of objective and eye lenses of focal length 0.40 cm and 1.00 cm respectively. An object is placed 0.41 cm from the objective lens and the final image is formed at infinity. Calculate the magnifying power of the instrument and its tube length.

Solution. For the objective lens

$$\frac{1}{v_1} = \frac{1}{f_0} - \frac{1}{u_1} = \frac{1}{0.40\text{ cm}} - \frac{1}{0.41\text{ cm}} = \frac{0.41 - 0.40}{0.40\text{ cm} \times 0.41\text{ cm}}.$$

Therefore

$$v_1 = \frac{0.40\text{ cm} \times 0.41\text{ cm}}{0.01} = 16.4\text{ cm}.$$

Also since the final image is at infinity $u_2 = f_e = 1.00$ cm. The magnifying power is

$$M = -\frac{v_1}{u_1} \times \frac{25 \text{ cm}}{u_2} = -\frac{16.4 \text{ cm}}{0.41 \text{ cm}} \times \frac{25 \text{ cm}}{1 \text{ cm}}$$

$$= -1000.$$

The tube length is obtained from either

$$L = v_1 + u_2 - f_0 - f_e = 16.4 \text{ cm} + 1 \text{ cm} - 0.4 \text{ cm} - 1 \text{ cm} = 16 \text{ cm}$$

or

$$M = \frac{-25 \text{ cm} \times L}{f_0 f_e},$$

whence

$$L = \frac{-M f_0 f_e}{25 \text{ cm}} = +\frac{1000 \times 0.4 \text{ cm} \times 1 \text{ cm}}{25 \text{ cm}} = 16 \text{ cm}.$$

PROBLEMS

15.1 A page of print is viewed through a magnifying glass of focal length 5 cm, the final image being formed at a distance of 25 cm. What are the lateral and angular magnifications produced?

15.2 If the position of the simple magnifier in Problem 15.1 is varied, what is the minimum angular magnification that can be obtained?

15.3 An insect on a leaf is examined through a magnifying glass of focal length 10 cm, the glass being held in such a position that the final image is formed 30 cm from the lens. What is the angular magnification achieved?

15.4 If in Problem 15.3 the glass is held at 8 cm from the insect, calculate the changes in lateral and angular magnification that result.

15.5 A relaxed far-sighted eye is acting as if it were a thin converging lens of focal length 2.5 cm placed 2.3 cm in front of the retina. A doctor examines the retina through the lens of the eye. Where will the image of the retina be located, and what will be the lateral and angular magnifications achieved?

15.6 What are the angular magnifications produced when an object is viewed through a magnifying glass of focal length 4.8 cm placed 4 cm from it and the eye is (a) at the lens, (b) 10 cm from the lens, and (c) 20 cm from the lens?

15.7 A surveyor uses a telescope which has an objective lens of focal length 25 cm and a graticule consisting of two horizontal lines 3.0 mm apart in the focal plane of the eyepiece. The surveyor sights on a rod held by his assistant 10 m from the objective. What length of rod will be viewed between the graticule lines?

15.8 An astronomical telescope consists of two thin converging lenses and gives an angular magnification of 40 when used in normal adjustment with the lenses 82 cm apart. If an object 100 m from the objective is now viewed through the telescope, how far should the eye lens be moved in order that the final image may be viewed at infinity?

15.9 In one of Edgar Allan Poe's stories, the author describes his terrifying experience of focusing a telescope on a distant hill and observing a "dragon" crawling up it. The punch-line comes when he realises that the "dragon" is an ant crawling up the windowpane through which he is observing the hill. Although this makes a good story, explain why it is bad optics.

15.10 A pair of prism binoculars consists essentially of two telescopes, one for each eye, each folded into three parts to reduce the overall length. A birdwatcher wishes to see as much detail when studying ospreys through his binoculars at a distance of 100 m as he would with his unaided eyes at a distance of 50 cm. What angular magnification does he require and what is the focal length of each objective lens if each eyepiece has a focal length of 1 cm?

15.11 An astronomical telescope which has an objective lens of focal length 50 cm and an eyepiece of focal length 1 cm is used to view the moon when in normal adjustment. If the moon has a diameter of 3.50×10^6 m and the earth-moon distance is 3.84×10^8 m, what angle does the image of the moon subtend at an astronomer's eye?

15.12 A compound microscope consists of objective and eye lenses of focal lengths 0.5 cm and 1 cm, respectively. An object placed 0.52 cm from the objective produces a final virtual image 25 cm from the eye lens. What is the distance of separation of the lenses and the magnifying power of the instrument?

15.13 A compound microscope has an eyepiece of focal length 2 cm at a distance of 16 cm from an objective lens of focal length 1 cm. Where must an object be placed in order that the final image be formed at infinity and what is then the magnifying power of the instrument?

15.14 The objective and eye lenses of a compound microscope have focal lengths of 3 mm and 15 mm, respectively. The tube length of the instrument is 14.2 cm. The virtual image of an object viewed is formed at 25 cm from the eye lens. Where is the object located? Calculate from first principles the magnifying power of the instrument and compare the result with the formulae given in Section 15.4.

15.15 Repeat Problem 15.14 for the case where the final image is formed at infinity.

15.16 A scientist is observing a specimen through a microscope of tube length 16 cm, the objective lens having a focal length of 0.5 cm. If the final image is to be viewed at infinity, what should be the focal length of the eyepiece employed, if the overall magnifying power is to be -600?

15.17 The focal length of a lens for any particular color is given by the formula

$$1/f = (n - 1)(1/r_1 + 1/r_2),$$

where r_1 and r_2 are the radii of curvature of the spherical surfaces of the lens, which is made of material of refractive index n for that particular color. Crown glass has refractive indices for blue and red light of 1.523 and 1.517, respectively. What is the distance between the blue and red focal points for a biconvex lens of crown glass whose radii of curvature are both 10 cm?

CHAPTER 16

THE EYE AND VISION

16.1 INTRODUCTION

The eye is perhaps the most fascinating of all optical instruments because of its perfection both as a sensory mechanism and as an optical system. It can detect one single quantum of light, but it will register the fact only if there are sufficient quanta present to make each quantum appear part of a pattern; and it will still operate satisfactorily with intensities of light up to 10^9 times that amount. It can receive and interpret information over roughly 180° in both the vertical and horizontal directions, and yet can concentrate attention on a small area and suppress stimuli from the surroundings. It automatically adjusts the amount of light entering it in order to prevent overloading the eye, and can alter its focal length almost instantaneously so that it can view in rapid succession an object a few centimeters away and one at infinity. No manufactured instrument could hope to come up to these specifications.

16.2 THE CONSTRUCTION OF THE EYE

The eye acts in a very similar manner to a camera, consisting essentially of a lens system which forms inverted images of objects viewed on the sensitive back surface. The complete eyeball shown in Fig. 16.1 is a sphere of diameter 2.3 cm, with

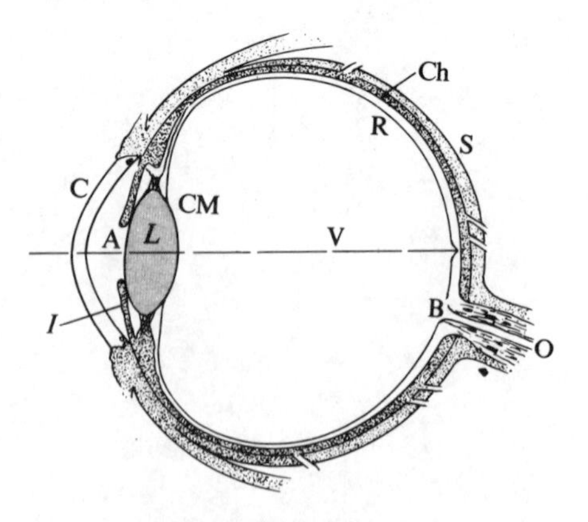

Fig. 16.1. The human eye.

at the front a transparent bulge (C) called the cornea, through which light enters. The cornea is about 12 mm in diameter, with a radius of curvature of approximately 8 mm. Most of the refractive power of the lens is due to the curved surface of the cornea. The eyeball is held in a well-lubricated socket by 6 sets of muscles, which are also used for rotation of the eye. The muscles are controlled by three pairs of nerves.

The outer covering of almost opaque fibrous tissue, the sclera (S), covers everything except the cornea. The inside of the sclera is covered with a dark, pigmented membrane, the choroid (Ch), which acts as does the blackened inside of a camera, preventing stray light from being scattered around. The colored portion of the eye is the iris (I), having a central opening, the pupil, through which light enters the inner portion of the eye. Like an automatic camera shutter, the pupil adjusts its size according to the intensity of the light falling on it, preventing the overloading of the retinal system. The crystalline lens of the eye (L), which, in spite of its name, has a cellular structure, is adjustable in focal length, its shape being altered by the ciliary muscles (CM). Between cornea and lens is a watery liquid called the aqueous humor (A), and on the other side of the lens is the jelly-like vitreous humor (V). The direction in which the eye sees most clearly is the visual axis, shown dashed in Fig. 16.1, and this is inclined to the optical axis of the system. The inside of the eye is coated with a membrane rich in blood-vessels and nerve fibers, the retina (R), which is sensitive to light. The most sensitive portion is where the visual axis strikes the retina, a small depression in the retina known as the yellow spot or fovea centralis. The nerve fibers terminate in rods and cones which are transducers, converting light energy to electrical impulses which travel along the nerve fibers. The yellow spot contains only close-packed cones, being the region most sensitive to color and to detail.

The optic nerve (O) transmits the signals from the retina to the brain, and the region where it enters the eye is not sensitive to light and is called the blind spot (B).

The ciliary muscles form a ring round the lens, and are believed under normal conditions with the eye relaxed to keep the front face of the lens fairly flat. The lens then focuses parallel light on to the retina. If near objects are to be viewed, the focal length of the lens is decreased by the ciliary muscles, which contract, move forwards, and permit the front surface of the lens to relax to a more curved shape. This ability of the eye to focus on objects over a wide range is called accommodation. The focal length cannot, however, be decreased indefinitely, and the eye cannot focus on the retina images of objects closer than a certain distance, the least distance of distinct vision, d, normally about 25 cm for a young adult. The point at which this inability to focus occurs is called the near-point of the eye.

The power of a lens is defined as $1/f$, the units being diopters when f is measured in meters. The power of a normal eye is thus $1/D$, D being the diameter of the eyeball, when relaxed, and $1/f = 1/D + 1/0.25$ m when focused on the near-point. The power of accommodation of the eye is thus $1/f - 1/D = 4$ diopters

16.3 OPTICAL DEFECTS OF THE EYE

1. Myopia (short sight). The eyeball is too long, and parallel rays are focused by the relaxed eye to a position in front of the retina. Only near objects can therefore be seen clearly, and the eye has a far-point as well as a near-point. This defect can be corrected by the use of spectacles which employ diverging lenses. If the spectacle lens is chosen to have a focal length equal in magnitude to the distance to the far-point (F), then parallel rays striking the spectacles appear to the eye to diverge from the far-point. The relaxed eye can therefore focus the rays on to the retina. Note that the least distance of distinct vision for the bespectacled eye is no longer d but is increased to x, where $1/x - 1/d = 1/-F$, or $1/x = 1/d - 1/F$. This is because an object at distance x must produce a virtual image at d in the spectacle lens in order just to be brought to a focus by the eye.

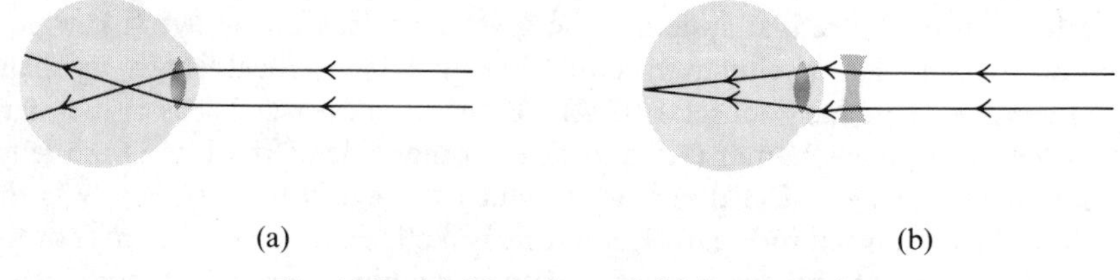

(a) (b)

Fig. 16.2. (a) Myopia, and (b) its correction.

2. Hypermetropia (long sight). This is the opposite effect. The eyeball is too short, and parallel rays are focused to a point behind the retina. In order to focus distant objects, the accommodating power of the eye must be employed. Since only four dioptres of accommodation are available, this means that the near-point is much farther from the eye than normal, and in extreme cases it is not even possible to read a book.

This defect is corrected by using converging spectacle lenses. If the near-point is at d', then an object at d requires the lens to produce a virtual image of it at d' which will then be visible to the fully accommodated eye. In other words, the focal length of the spectacle lenses must be f, where $1/f = 1/d - 1/d'$.

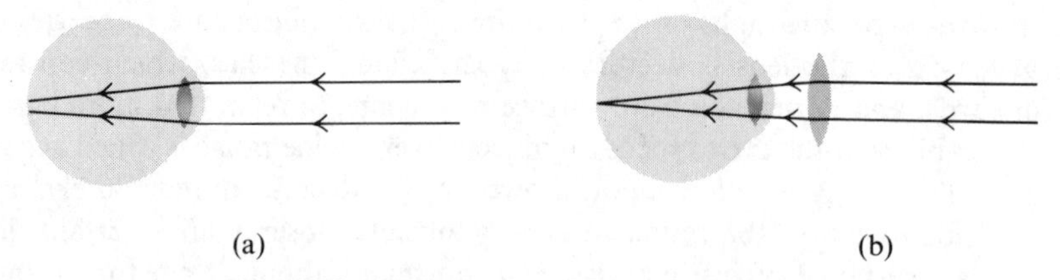

(a) (b)

Fig. 16.3. (a) Hypermetropia, and (b) its correction.

3. Presbyopia (old sight). As people get older, the ciliary muscles weaken and the lens loses some of its elasticity. The power of accommodation therefore diminishes with age. In order to compensate, converging spectacle lenses are employed as in the case of hypermetropia, but the patient only requires these for reading or similar close work, since his vision for distant objects is unimpaired.

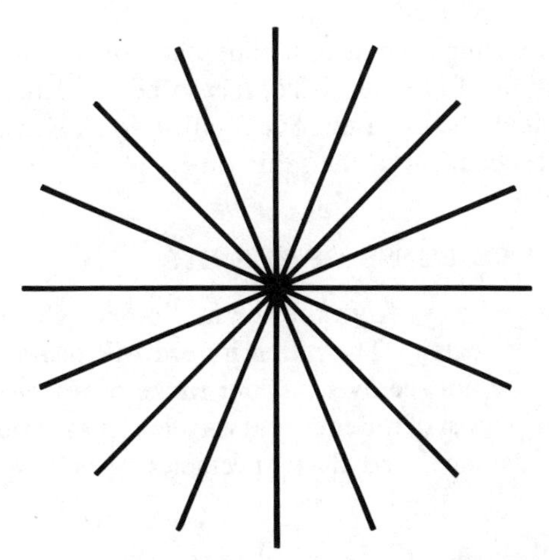

Fig. 16.4. A diagram for the detection of astigmatism.

A short-sighted person on aging will normally be prescribed bifocal spectacles. The upper half of each lens is diverging and corrects the myopia when the wearer is looking ahead at distant objects. The lower half corrects the presbyopia with a suitable converging lens, and the wearer looks through this part when reading.

4. Astigmatism. When astigmatism is present, point objects do not form point images on the retina. This is normally due to the cornea's having unequal curvature in different directions. If the curvature is greater in a horizontal section than in a vertical section, rays are brought to a focus more quickly in the horizontal than in the vertical plane. A collection of horizontal and vertical lines cannot all be brought into focus simultaneously. A normal eye will see all the lines in Fig. 16.4 equally black. An astigmatic eye will see variations in the intensity of the lines. From the distribution of the intensity variation an optometrist can deduce the variations in curvature of the cornea. The defect is corrected by the use of cylindrical spectacle lenses. For an otherwise normal eye one surface of each lens will be flat and the other a section of a circular cylinder (Fig. 16.5). For the particular case of astigmatism mentioned above, the lens will be set in the spectacle-frame with the axis of the cylinder horizontal. Rays passing through in a horizontal plane are unaffected. Those in a vertical plane are converged by the circular surface,

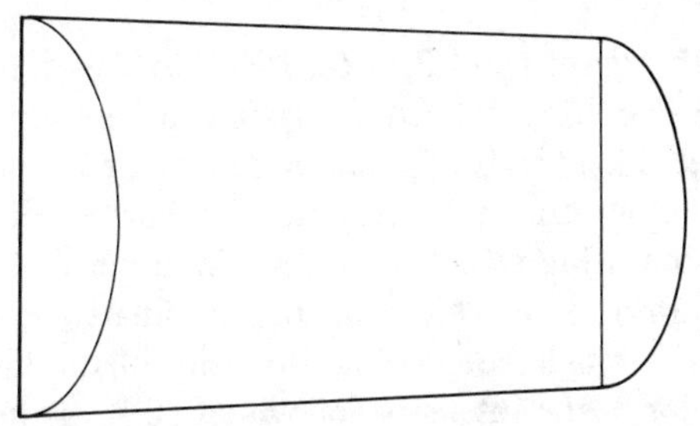

Fig. 16.5. A cylindrical lens as used in the correction of astigmatism.

the extra convergence being such as to compensate for the lack of curvature of the cornea in that direction. If another defect has to be corrected for as well, the lens surface is made toroidal, the two radii of curvature so chosen as to compensate for that defect and the asigmatism at the same time.

16.4 WAVELENGTH RESPONSE OF THE EYE

The radiation reaching the eye from the sun and other light-sources often extends well beyond the visible region. The retina indeed will respond far into the ultra-violet. Why then do we not see over a wider range of wavelengths?

The cornea absorbs most of the energy at wavelengths shorter than 3.0×10^{-7} m. This causes corneal damage, and dark spectacles, which will absorb ultraviolet

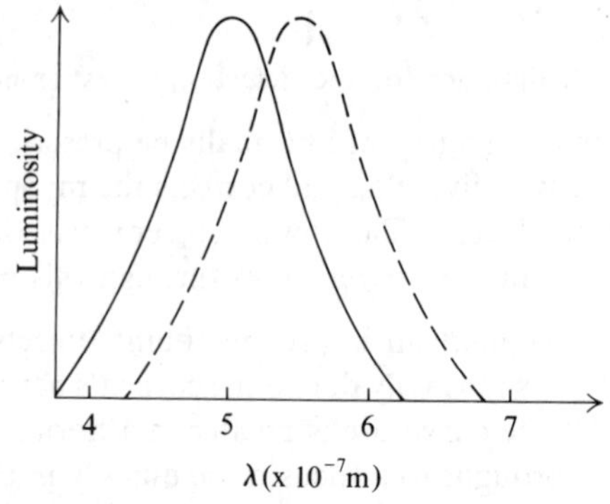

Fig. 16.6. Luminosity curves for the human eye; the full line is for scotopic vision and the dashed line for photopic vision.

light, should always be worn when an ultraviolet lamp is used, or even during sunbathing in strong sunlight. The crystalline lens strongly absorbs radiation of wavelength below 3.8×10^{-7} m, and this is the minimum wavelength which can reach the retina.

On the long wavelength side the water molecules in the cornea and aqueous humor absorb most of the energy in wavelengths above 12×10^{-7} m. However, the eye pigments are unresponsive to light above 8.0×10^{-7} m and are not very sensitive above 7.0×10^{-7} m.

In the wavelength region from 3.8×10^{-7} m to 7.0×10^{-7} m the eye is not equally responsive to all colors. Indeed it responds differently in bright light and in dim light. If a subject is kept in bright surroundings, he is said to be light-adapted and to be using photopic vision; if kept in dark surroundings, he is said to be dark-adapted and to be using scotopic vision. In either condition he may be sat in front of a screen, an area of which can be illuminated with a flash of light of small duration and of a single color. The intensity of the light is varied and the threshold for that color is the intensity for which 50% of the flashes are seen by the subject. The reciprocal of this quantity is called the luminosity. Luminosity curves for a typical eye are shown in Fig. 16.6, the luminosity being to an arbitrary scale. The full line represents the results for scotopic vision and the dashed line the results for photopic vision. The curves have been reproduced to the same maximum height, although in fact much greater intensities are needed for photopic vision. This is a result with which everyone is familiar. A light-adapted eye cannot see anything in a dark room when the lights are first switched off. After an interval of adaptation to the dark, however, the objects in the room become visible.

The two curves are displaced one from the other by about 0.5×10^{-7} m, and this would indicate that two different types of receptor are responsible for vision. Since photopic response is greatest at the fovea, where there are only cones, and scotopic response greatest at the periphery of the eye (it is easiest to see a faint star "out of the corner of the eye"), where there are more rods, it was natural for early experimenters to assume that these receptors operated under different conditions of intensity. This now appears to be incorrect. Cones do not appear to play an appreciable part, if any, in scotopic vision, but rods appear to be active in both scotopic and photopic vision.

16.5 QUANTUM RESPONSE OF THE EYE

When a dark-adapted eye sees a flash of light of wavelength 5.05×10^{-7} m at the threshold intensity, this being the wavelength to which the eye is most sensitive, how many quanta are affecting each receptor? Such a question is extremely difficult to answer, since for every 100 quanta from the source which strike the cornea only a few will reach the retina, the rest being absorbed by the intermediate materials. A measurement of the intensity of light I reaching the cornea does not therefore allow a direct calculation of the average number of photons a received by a single receptor. Nonetheless, the two must be linearly related, and thus $a = kI$; k is a constant depending on so many factors that it would be almost impossible to calculate it theoretically.

If I is not too large, a will be a small number, and thus any individual receptor has the possibility of receiving 0, 1, 2, 3, . . . , a, $a + 1$, . . . photons, although on average a single receptor will receive a photons. The probability $P_m(a)$ that any receptor will receive m photons follows a Poisson distribution (cf. Section 2.13). Thus

$$P_m(a) = \frac{e^{-a}a^m}{m!} \tag{16.1}$$

If n is the number of photons which a receptor must receive in order to stimulate vision, the probability P that n or more photons will be absorbed during one flash is

$$P = P_n(a) + P_{n+1}(a) + P_{n+2}(a) + \cdots + P_\infty(a).$$

But the receptor must certainly receive some number of photons between 0 and ∞. Thus

$$1 = P_0(a) + P_1(a) + P_2(a) + \cdots + P_\infty(a).$$

Therefore

$$P = 1 - [P_0(a) + P_1(a) + P_2(a) + \cdots + P_{n-1}(a)]$$
$$= 1 - \sum_{m=0}^{n-1} P_m(a). \tag{16.2}$$

The value of P can be calculated from Eq. (16.2) for any values of n and a. Curves may be drawn of P against a, but it is generally more convenient to draw curves of P against $\log a$. It will be seen from Fig. 16.7 that the curves obtained for different values of n are quite different in shape. Since $a = kI$, curves of P against $\log I$ will have corresponding shapes. The fraction of times a subject sees the light of a given intensity I (which is a measure of the probability of his receptors responding) may therefore be plotted against $\log I$. The shape of the obtained curve tells the experimenter which value of n is applicable.

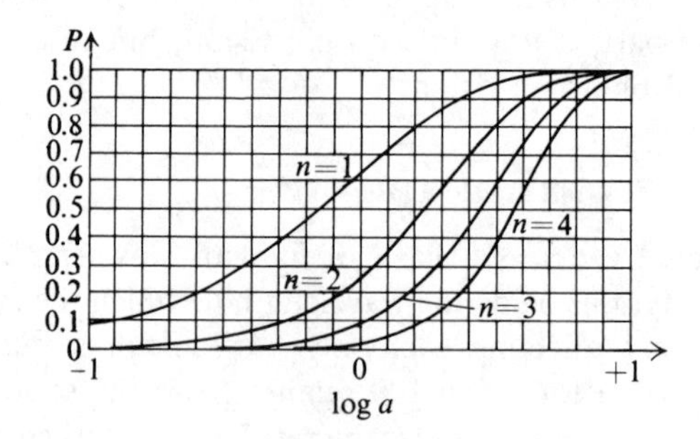

Fig. 16.7. The frequency of visual response as a function of the intensity of the light reaching the eye, when 1, 2, 3, and 4 photons are necessary to stimulate a receptor.

In human subjects n may be as high as 8 or as low as 1. A single photon is thus all that is necessary in any individual receptor in the most favorable subjects. Experiments of this type have been performed in the main on dark-adapted eyes, and the result can thus only be said to be established for rods. There is no reason to believe, however, that cone response is any different.

It would be quite misleading to give the impression that one quantum absorbed by a single receptor produces a sensation of seeing. The rods and cones are connected to neurons between which are innumerable interconnections. No electric impulses are transmitted to the optic nerve from the retinal cells unless several receptors are activated. This is similar to the situation which applies with many pieces of scientific equipment such as coincidence counters, where several events must be recorded before a signal will be emitted. Such a mechanism is clearly necessary for the eye. The statistical fluctuations in the thermal energy available in the retinal cells will ensure that randomly, at intervals of around 1 s, individual receptors will receive sufficient energy to become activated. If all signals produced in this way were transmitted, the brain would constantly be alerted quite needlessly.

16.6 VISUAL ACUITY

The acuity of the eye is defined as the minimum angular separation of two equidistant points of light which can just be resolved by the eye. If they are closer than this minimum value, the eye will merely see the light as coming from a single point source. The acuity of scotopic vision is highest about 20° from the fovea, where the density of rods is greatest; for photopic vision it is highest on the fovea, where the cones are most dense. Photopic acuity is always greater than scotopic. These observations confirm the belief that cones are only concerned in the seeing process for light-adapted eyes.

In Sections 18.3 and 19.3 we shall see that the wave nature of light places a lower limit on the resolving ability or acuity of the eye. With the formula given there and using green light of wavelength 5×10^{-7} m, an iris diameter of 5 mm would imply an acuity of 10^{-4} rad. Experiments have, however, shown that for most people acuity is of the order of 5×10^{-4} rad, although one or two exceptional people may be able to resolve down to 2×10^{-4} rad. In the centre of the fovea, cones are separated by distances of about 2×10^{-6} m. An acuity of 2×10^{-4} rad implies that the images of the two point sources are separated on the retina by a distance of about 5×10^{-6} m. It is clear that at the very least the two images have to be separated by one unexcited cone before the images will be resolved. The limiting factor in the acuity therefore does not lie in the wave nature of the light but in the detail of the structure of the retina.

16.7 PHOTOSENSITIVE PIGMENTS IN THE EYE

It has been known for many years that there existed in the eye pigments which played some part in the vision process. The most fully studied of these is rhodopsin (also called visual purple because of its characteristic color), which is a complex protein with a molecular weight of about 40,000. Rhodopsin is found in rods. The action of light on it is to split from the protein a small hydrocarbon group called retinine. Retinine is very similar in structure to vitamin A, and ingestion of the latter is the source of rhodopsin in the eye. Anyone suffering from vitamin A deficiency cannot resynthesize the rhodopsin which has been broken up by the action of light, and suffers from "night-blindness". There is insufficient rhodopsin in his rods to permit him to see objects in the dark.

All this evidence that rhodopsin is responsible for rod vision is confirmed by examining the absorption spectrum of rhodopsin. This is found to be almost identical with the luminosity curve for scotopic vision shown in Fig. 16.6.

If a section from a rod cell is examined under the electron microscope, it is found to contain a collection of parallel planes. It is in these planes that the rhodopsin is found. Why they are set down in planes is not understood, although this layered arrangement is found in all cells which use light, such as chloroplasts which are responsible for photosynthesis in plants. It may be that the rhodopsin molecules are arranged in this neat fashion so that an electron produced in the disruption of one protein can be transferred easily to the end of the structure to start the electrical response.

Iodopsin, found in cones, is also a protein which contains a retinine group. It has an absorption spectrum different from rhodopsin and also, unfortunately, different from any known luminosity or response curve. Its action is therefore still not understood. No other pigments have been isolated in the human eye, although others have been found in animals and fish.

16.8 COLOR VISION

The ability of the eye to see colors is a very complex one. With low intensities of light, scotopic vision being used, color response is absent. Thus many faintly seen objects are much more beautiful than we would imagine. Nebulae which are seen as black and white through telescopes are often highly colored, and a completely new world was opened up to skin-divers when they began to take powerful lights with them down to the sea-bed. Even with photopic vision, color is not distinguished in objects far from the center of the field of view. Since the images of these objects will be formed on the retina far from the fovea, where only rods are present, this strengthens the view that cones are responsible for color discrimination.

A physicist understands perfectly what he means by yellow light. It is electromagnetic radiation in a band with wavelengths around 6×10^{-7} m. But the eye sees as yellow not only this light but also a mixture of red light and green light, with the physicist's yellow light entirely absent. Physiological color is therefore something quite different from physical color. Indeed, this might have been expected. The eye is unlikely to consist of an infinity of different receptors each of which would respond to one of the infinity of wavelengths present in white light.

Many experiments have been performed to investigate the color response of the eye. It has been conclusively shown that the effect of all colors can be produced by mixing in varying proportions three primary colors only, normally chosen as red, green, and blue. It is a simple step from this discovery to the hypothesis that three pigments in the eye must be responsible for color vision. The original exponents of a three-color theory hoped that three pigments would be discovered in the human eye. Only two have been found, and their absorption spectra do not agree with the theoretically predicted response curves of the three-color theory.

It might seem surprising that it is possible to obtain theoretical response curves. The existence of color-blind humans allows this to be done.* If there are three types of pigment in the eye, one might have anticipated that there would be among the various types of color-blindness three corresponding to eyes lacking

---

* About 8% of the male population and rather less than 1% of the female population are color-blind. The factor is sex-linked and transmitted by the female. (See Problem 2.10.)

one of the necessary pigments. Such a color-blind person is called a dichromat, since all his sensations of color can be reproduced by mixing only two primary colors. There are indeed three types of dichromats, two very common and the third somewhat rare. By comparing the response to color of a normal person and each type of dichromat, it is possible to deduce the response curves of the three different pigments assumed to be present in the normal eye. These theoretical curves are shown in Fig. 16.8. None of the curves is similar to the absorption curves of either rhodopsin or iodopsin. Presumably these pigments are merely concerned with the gross detection of light and therefore require to be present in relatively large quantity, particularly the rhodopsin. The sensation of color, which is only present if the light intensity is great enough, would then be produced by pigments which may be present in quite small amounts. A simple technique has shown that they almost certainly do exist.

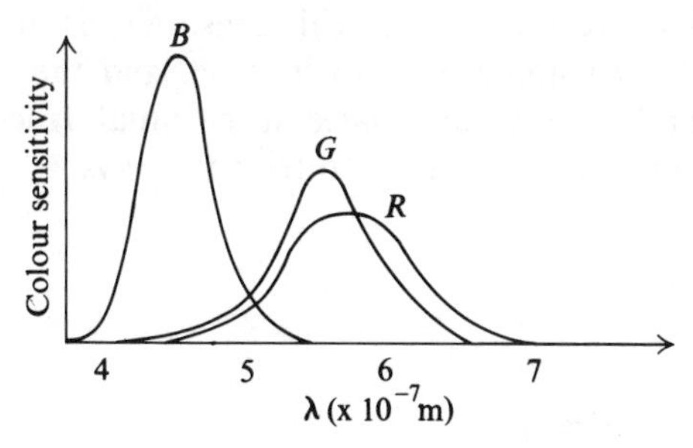

Fig. 16.8. Theoretical response curves for the three color-sensitive pigments in the human eye.

If an ophthalmoscope is used on the eye, the light returning to the instrument is reflected from the eyeball after having passed twice through the pigmented layer. This light will therefore show the absorption spectrum of all the constituents of the retina, not only the pigments present but also the blood-vessels and various types of cell material. It would be difficult from this to disentangle the spectrum of any particular pigment were it not for the fact that each pigment, when subjected to light, is broken down; its concentration therefore decreases and its absorption spectrum correspondingly changes. If the absorption spectrum of an eye is measured with a modified ophthalmoscope and then measured again while the eye is being simultaneously subjected to another intense beam of light, the difference in the two spectra is entirely due to the change in concentration of the pigments which have been affected. By performing such an experiment on a normal eye and on the eyes of the three different types of dichromat, it is possible to deduce the absorption curves of the three types of pigments assumed present in the normal eye. One of these curves agrees well with the curve marked G in Fig. 16.8. The agreement with the other two curves is not so satisfactory. Nonetheless, the experiments appear to indicate that three color-sensitive pigments are present in a normal eye.

16.9 THE INSECT EYE

The eyes of almost all vertebrates are very like the human eye, although not all vertebrates are believed to have color sense. Since color vision has been measured until recently mainly by subjective methods, it will be appreciated that it is difficult to be definite about color discrimination in animals. Invertebrates have many different types of eye, the only other highly developed type being that found in insects. Much research has gone into the insect eye, the bee being one insect which has been particularly studied. In certain respects bees have advantages over humans. Their eyes can see ultraviolet light of wavelengths down to about 3×10^{-7} m, and they would appear to have color discrimination over most of that range. Further, bees can tell where the sun is by looking in a direction in which the sun is not. Bees are very sensitive to the state of polarization of light (cf. Sections 17.6 and 17.7), and, since the sunlight scattered from air molecules is polarized, this allows bees to tell where the sun is. This sensitivity is almost certainly tied up with the bee's ability to home back to its hive, and the information on honey location passed by a bee to its hive-mates in the ritual dance appears to include a direction indication in terms of angles to the sun's rays.

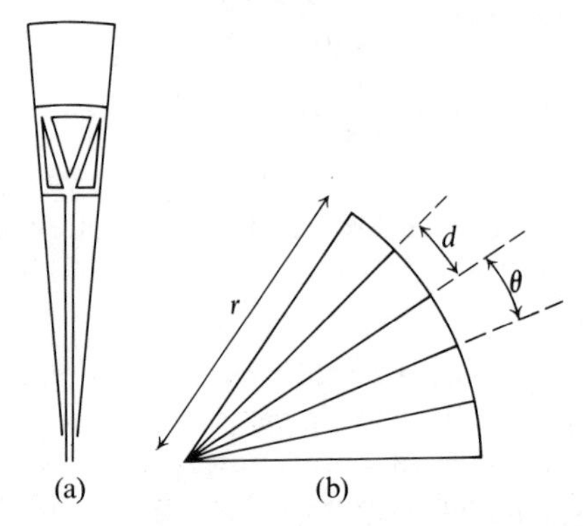

Fig. 16.9. The insect eye: (a) shows a single cell and (b) shows a portion of the eye containing several cells.

In every other way the insect eye is inferior. It consists of a number of conical-shaped cells (Fig. 16.9a) called ommatidia which are packed together in the insect's head. Their shape is conditioned by the fact that they must occupy part of the volume of a sphere. The end of each cell is transparent and acts like a light-pipe to transmit the light to the sensitive regions farther on. The cell terminates in a nerve fiber.

The insect eye lacks many of the refinements of the human eye and, in particular, its acuity is poor. Each of the cells will in general receive light from sources in a slightly different direction. Only light in the angle θ (Fig. 16.9b) can be received in a single cell, where $\theta = d/r$, d being the diameter of the end of the cell and r the radius of the eye. The acuity of the eye is thus of the order of θ. For a bee d is about 3×10^{-5} m and r about 3×10^{-3} m. The acuity is thus of the order of 10^{-2} rad, two orders of magnitude greater than that of the human eye. Why has the bee not evolved slimmer ommatidia to increase its acuity?

The answer makes good physical sense. If the value of d were made too small, diffraction effects (cf. Section 18.2) would become important. The angle of bending of light due to diffraction is of the order of λ/d, which for a bee's eye and ultraviolet light has a value of about 10^{-2} radian. This is roughly the same value as for the acuity. If d were decreased, the bending of light due to diffraction would be greater than the angle of vision covered by a single cell. Hence, the light from a single point object would be able to enter more than one cell and the ability to distinguish two point sources by the fact that they produced responses in two different cells would be lost. Evolutionary nature works on sound scientific principles!

Example 1. A man has a near point at 50 cm from his eyes and a far point at infinity. What is his useful accommodating power?

Solution. Whatever the object distance, the image distance is the distance D from the eye lens to the retina. Thus, when the eye is focused on infinity,

$$P_\infty = 1/f_\infty = 1/\infty + 1/D;$$

and, when the eye is accommodating to its maximum,

$$P_N = 1/f_N = 1/0.50 \text{ m} + 1/D.$$

Therefore

$$P_N - P_\infty = 1/0.5 \text{ m} - 1/\infty = 2 \text{ diopters}.$$

The useful accommodating power is 2 diopters.

Example 2. What spectacle lenses would be prescribed for the man of Example 1?

Solution. The man is suffering from long sight and it is necessary to provide him with a lens which will bring his near point back to the normal value. The spectacle lens must produce a virtual image of an object placed at the new near point of 25 cm which will appear to the man to lie at the old near point; for then the eye, fully accommodated, is able to focus the rays leaving the lens to form an image on the retina. Figure 16.10 illustrates this. If f is the focal length of the prescribed lens, it follows that with $u = 25$ cm and $v = -50$ cm (the image being virtual),

$$1/f = 1/25 \text{ cm} - 1/50 \text{ cm} = (2 - 1)/50 \text{ cm}.$$

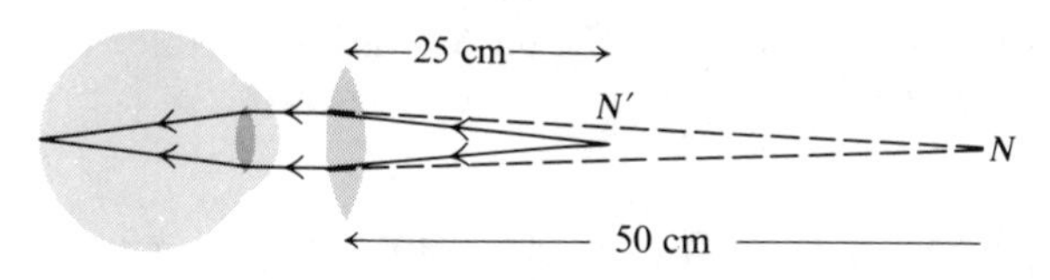

Fig. 16.10

Therefore

$$f = 50 \text{ cm}$$

or

$$P = 1/f = 1/0.5 \text{ m} = 2 \text{ diopters.}$$

Example 3. A myopic male has near and far points of 20 cm and 250 cm respectively. What spectacle lens is prescribed for his defect and where is his near point

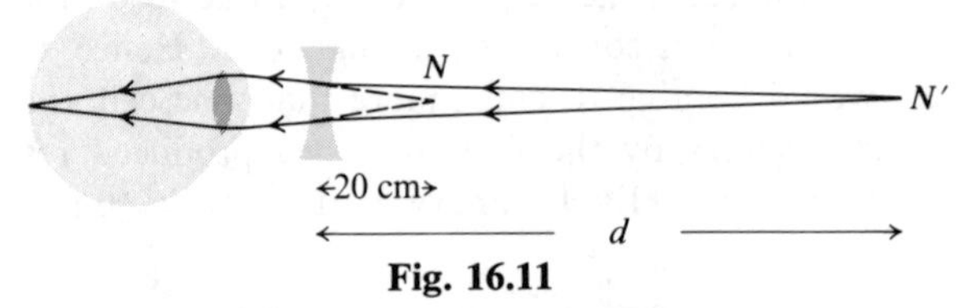

Fig. 16.11

Solution. The prescribed lens must produce a virtual image at the far point of the eye of an object at infinity so that the relaxed eye can focus on to the retina the rays coming from the lens and striking it. If f is the focal length of the lens, u is infinite and v is -250 cm, being virtual. Hence

$$1/f = 1/\infty - 1/250 \text{ cm}.$$

Therefore

$$f = -250 \text{ cm}$$

or

$$P = -1/2.5 \text{ m} = -0.4 \text{ diopter.}$$

The lens is diverging as expected with a focal length equal in magnitude to the distance to the far point

The near point when wearing the spectacles will be at distance d from the eye. The virtual image produced by the lens must be at the near point of the eye in order that the fully accommodated eye can focus the incoming rays on the retina. Hence $v = -20$ cm and

$$-1/250 \text{ cm} = 1/d - 1/20 \text{ cm}.$$

Therefore

$$1/d = \frac{1}{20 \text{ cm}} - \frac{1}{250 \text{ cm}} = (25 - 2)/500 \text{ cm}.$$

Therefore

$$d = 500 \text{ cm}/23 = 21.7 \text{ cm}.$$

Example 4. The results of an experiment in which a dark-adapted subject attempted to observe flashes of light against a dark background are shown in the table, f being the fraction of times the subject saw the flash of arbitrary intensity I. How many photons must be received by the subject's receptors before it is possible for vision to be stimulated?

f	0	0.09	0.19	0.25	0.44	0.58	0.80	0.89	0.95	0.99	1.00
I	25	49	83	100	150	200	320	400	500	700	800

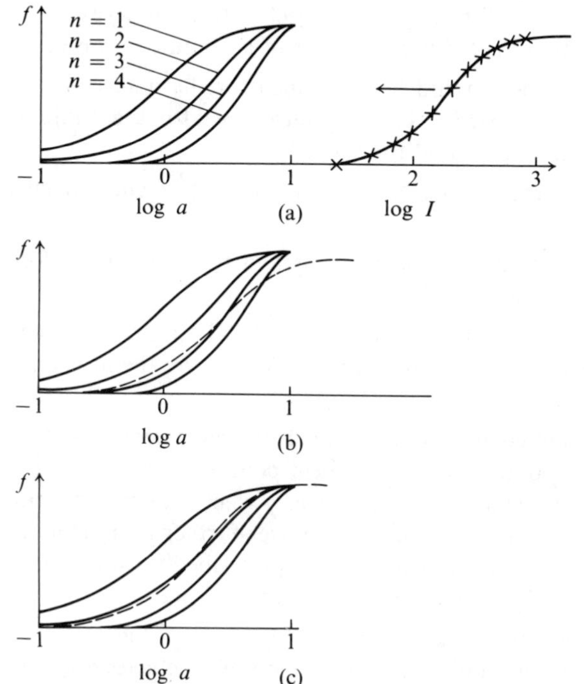

Fig. 16.12

Solution. Figure 16.12(a) shows the experimental results from the table plotted on a graph of f vs. $\log I$. It also shows Fig. 16.7 replotted to the same scale. This shows theoretical graphs of f vs. $\log a$ for various values of n, a being the average number of quanta received by a single receptor in the eye and n being the number which must be received before response is stimulated. But since we recall that $a = kI$, where k is a constant, and thus $\log a = \log k + \log I$, the experimental curve should be identical with one of the theoretical curves but displaced from it along the x-axis by an amount $\log k$. If the experimental curve is plotted on transparent paper, it can be moved sideways along the x-axis to the position of the theoretical curves, and the latter will still be seen through the paper. Figure 16.12(b) shows this and clearly indicates that the experimental curve does not fit the theoretical curves for $n = 3$ or $n = 4$. Further movement produces the situation of Fig. 16.12(c) where the experimental curve and the theoretical curve for $n = 2$ coincide almost exactly. Clearly each receptor of the subject's eyes requires to receive two photons before the vision response is stimulated.

PROBLEMS

16.1 What is the power of the lens system of an eye when the object viewed is (a) at infinity, (b) at 500 cm from the eye and (c) at 25 cm from the eye? The image distance in each case may be taken as 2.0 cm.

16.2 What is the power of accommodation of the eye of Problem 16.1?

16.3 A man can see distinctly only objects which lie between 25 cm and 400 cm from his eyes. What is the useful power of accommodation of his eyes?

16.4 What spectacles should be prescribed for the man in Problem 16.3? If he reads a book while wearing these spectacles, what is the minimum distance from his eyes at which he can hold the book?

16.5 A hypermetropic male has a near point of 60 cm. What lenses would be prescribed for him?

16.6 A man is prescribed spectacle lenses of focal length +75 cm. Where is the near point of his eye located?

16.7 A man while wearing spectacles of focal length −200 cm sees clearly all objects lying between 25 cm and infinity. Where are the near and far points of his unaided eye located?

16.8 A short-sighted man is prescribed diverging lenses of power 2.5 diopters. Where is his far point located? His near point is 10 cm from his eye. What is the closest distance at which he can see clearly when wearing the spectacles?

As he ages, the man's near point moves further away from him. What is the other component of the bifocal lens now prescribed for him when the near point is (a) at 20 cm and (b) at 30 cm from the eye?

16.9 An elderly man can see distinctly only objects which lie in the range between 75 cm and 250 cm from his eyes. What kind of spectacles does he require in order to see both distant objects and a book held at 25 cm from his eyes? Between what distances are objects not clearly seen by him when he is wearing the spectacles?

16.10 Bifocal lenses are prescribed for a patient, the components having focal lengths of 40 cm and −300 cm. What are the near- and far-points of the patient's eye?

16.11 The following table shows the results for an experiment performed with a dark-adapted subject. Row A shows the intensity in arbitrary units of the light reaching the subject's cornea when a flash is emitted, and row B gives the percentage of times the light flash was perceived.

A	25	40	100	200	300	350	400
B	0	2	25	73	95	100	100

By plotting the frequency of perception of the flash against the log of the intensity, determine how many photons the subject's receptors must receive in order to stimulate vision.

16.12 A subject requires only one photon to be received on each receptor in his retina before vision is stimulated. When 24 photons reach his cornea, he sees a light flash in 45% of cases. Estimate the percentage of times he sees the flash when (a) 80 photons and (b) 120 photons reach the cornea.

16.13 If cones in the foveal region are separated by distances of 2×10^{-6} m and the eye is a sphere of diameter 2.3 cm, estimate the minimum possible value for the visual acuity of an eye.

CHAPTER 17

INTERFERENCE AND POLARIZATION

17.1 INTRODUCTION

In nature energy is transferred from one point to another in two main ways. In the first method matter is bodily transferred from one place to another and can be made to give up its energy at any point. Thus water from a dam may be allowed to fall through tunnels, acquiring kinetic energy in the process, and if the water is allowed to hit the blades of a turbine, energy is transferred to them, causing the turbine to rotate. In a similar manner the energy carried by a moving mass of air can be transferred to a windmill.

In the other method the energy is transferred by a wave motion, the energy passing from one place to another without the transfer of the material of the medium taking place. Thus if a stone is dropped into water, the disturbance caused can rock and perhaps sink a model ship some distance away. The water between the point of impact and the ship does not move bodily from one spot to the other. The disturbance in the water molecules at one point affects neighboring molecules, which in turn affect molecules farther on. Thus the disturbance is propagated onwards without individual molecules doing more than being displaced slightly from their normal positions.

Wave motion was discussed in Chapter 11, where the transfer of energy was considered as resulting from the coupling of oscillators. It is easy to see the relevance of this approach in the case of water waves, waves on a string, and waves in air; and it is easy to see that there will be two different types of wave, longitudinal and transverse, where the displacements are, respectively, in the direction of travel and at right angles to this direction. In the case of electromagnetic waves it is much more difficult to visualize what is happening. There are no material particles involved, oscillating about mean positions. Energy is still being transferred from place to place, but at any particular location what is happening is that electric and magnetic fields (cf. Sections 22.6 and 25.2) associated with the wave are oscillating about a mean, normally zero, value. It is this disturbance which is passed on from point to point as the wave progresses. It should be made clear that in the case of light it is the electric field which is responsible for specifically optical effects.

It is not necessary to be more precise than that. The optical phenomena which are specific to the wave nature of light can be explained qualitatively and quantitatively by the assumption that light propagates as a transverse, sinusoidal wave, which, when it moves in the positive x-direction, is described by the equation below, given in several different alternative forms,

$$y = A \sin 2\pi \left(\frac{x}{\lambda} - \frac{t}{T}\right) = A \sin 2\pi \left(\frac{x}{\lambda} - ft\right) = A \sin (kx - \omega t). \quad (17.1)$$

In these expressions A is the amplitude of the wave; λ, T, f, and ω are the wavelength, period, frequency, and angular frequency of the wave respectively; and $k = 2\pi/\lambda$ is known as the wave number. The magnitude of the light-sensation at any point is measured by the intensity I, where $I \propto A^2$. The angle $2\pi[(x/\lambda) - (t/T)]$, or any of its equivalent expressions, is the phase angle, and it determines at any given distance at any time what point of the cycle of operation has been reached and what is the actual value of the displacement.

17.2 INTERFERENCE

The fact that light is a wave motion was proved by the demonstration that light can produce interference phenomena, for only by assuming a wave motion can such phenomena be explained. It is therefore necessary to go very carefully into the effects that occur when more than one wave arrives at a particular place. The resulting intensity may be a simple addition of the intensities of the individual waves; but if the conditions are correct, the effects are much more complicated.

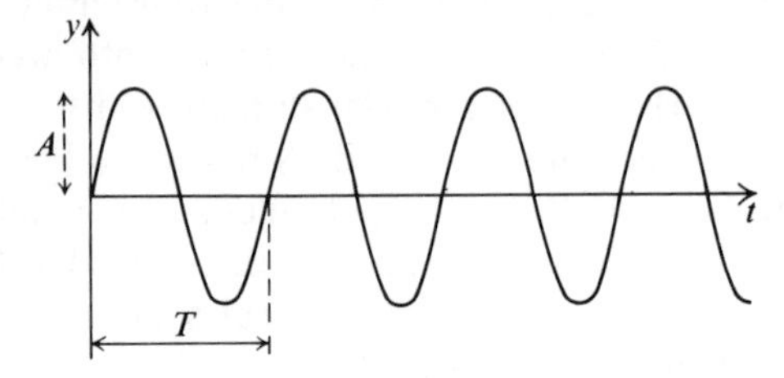

Fig. 17.1. The variation of a wave disturbance as a function of time.

If an observer at a particular location were able to plot the variation with time of the disturbance at that point, caused by the arrival of a single light wave, he would obtain a graph similar to the one shown in Fig. 17.1.

If two identical wave trains arrive simultaneously at the point by different routes, each will produce at all times a disturbance identical to that produced by the other. The two waves have the same phase at that point at all times. The resultant disturbance, as shown in Fig. 17.2, is thus always twice that of either wave alone. The fact that effects add algebraically in this way is only one example of the principle of superposition, which applies to many situations throughout all branches of physics.

This result is achieved if the two waves start off in phase at the same instant and arrive at the observation point after traveling equal paths. A similar effect arises if the other conditions stay the same but the lengths of the two paths differ by one complete wavelength. One wave will arrive while the other is still one wavelength away, but when the first train begins to produce a second oscillation at that point, the second train is starting to produce its first oscillation. The omission of the dotted wave from the second train shown in Fig. 17.2 gives a correct representation of the situation. Apart from the first oscillation produced at the point, all subsequent disturbances produced by the wave trains are identical. The waves are 2π out of phase, but this produces identical displacements at all times after the first period. Since a wave train contains millions of waves (a pulse of light lasting for only 1 s contains about five hundred million million waves), the effect of the first oscillation is unnoticeable.

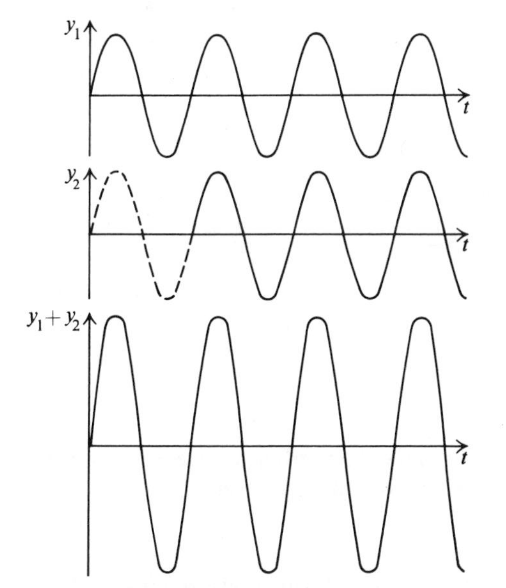

Fig. 17.2. The combined effect due to two identical monochromatic wave trains in phase or 2π out of phase.

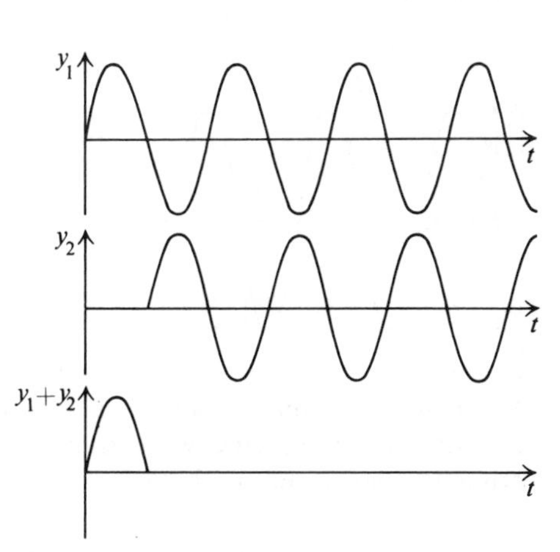

Fig. 17.3. The combined effect due to two identical monochromatic wave trains π out of phase.

The same is obviously true when the difference in the length of the path traveled by the two trains is any exact number of wavelengths. Apart from the first few waves, which are ignored, at all subsequent times the two waves are producing identical displacements at the point and double the displacement results. The two waves are equivalent at that point to a single wave of twice the amplitude and thus produced four times the intensity of either.

If the path difference between the waves is half a wavelength, the phase difference therefore being π, the situation is quite different, as can be seen from Fig. 17.3. Whenever the first wave train is attempting to produce a displacement in one direction, the second is trying to produce an equal displacement in the opposite direction, and the effects therefore cancel. At all times after the first half-cycle the net effect is zero and no disturbance results. The two wave trains cancel each other out and produce no effect at all. The same is obviously true, if we ignore the first few waves, whenever the path difference is an odd number of half-wavelengths or the phase difference is $(2n + 1)\pi$, where n is an integer.

When the path difference between the two wave trains is some intermediate value between an odd number of half-wavelengths and an exact number of complete wavelengths, an intermediate effect is obtained. Figure 17.4 illustrates the resultant wave for a general path or phase difference.

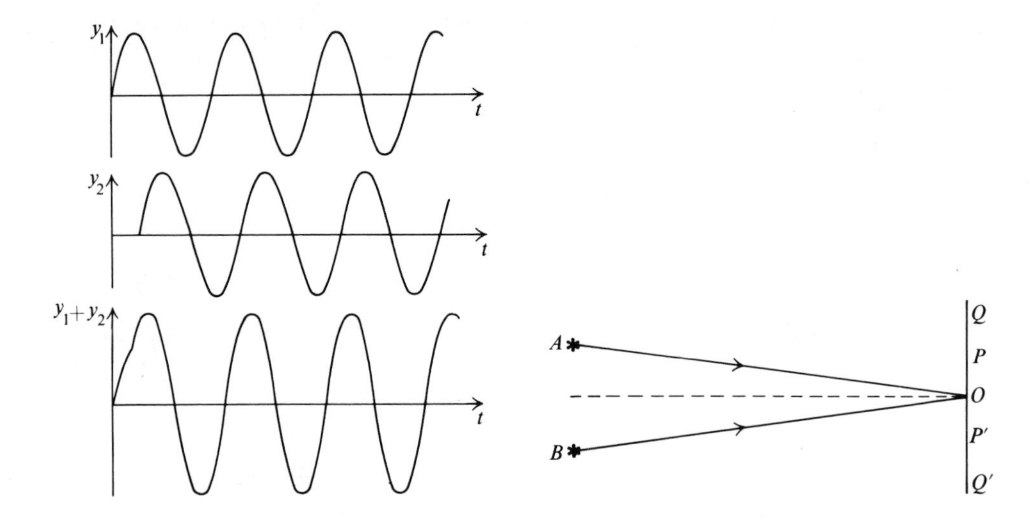

Fig. 17.4. The combined effect due to two identical monochromatic wave trains an arbitrary amount out of phase.

Fig. 17.5. Light from two identical sources being received on a screen.

Figure 17.5 represents two sources of light A and B, which are sending out at every instant identical wave trains. These strike a screen placed some distance away. It is obvious that the path difference between the waves varies continuously from one point of the screen to another. Thus since the point O is at an equal distance from the two sources, there will be zero path difference there between the wave-trains. At O four times the intensity due to one source alone is obtained. As one moves from O in either direction along the screen, a path difference is introduced between the waves from the two sources, and eventually at some point P (and P') a half-wavelength path difference exists. The two wave trains cancel each other out and no disturbance occurs at P. P is therefore a point of complete darkness. Farther on, at Q (and Q'), one complete wavelength path difference exists and the amplitude of vibration is once again twice what it would be if only one source were present.

Thus as the point of observation is moved along the screen from O, the light intensity varies gradually from a maximum at O (four times the intensity produced by a single source) to complete darkness at P and P', gradually increases again to maximum brightness at Q and Q', and so on. The screen therefore displays a series of dark and light bands called interference fringes.

17.3 THE MATHEMATICS OF INTERFERENCE

The screen (Fig. 17.6) is located parallel to the plane containing the sources S_1 and S_2 and distance D from it. The sources are d apart and the wavelength of the waves emitted by the sources is λ. The point on the screen on the perpendicular bisector of the line S_1S_2 joining the sources is O and the distance of the point P on the screen from O is x. PS_1 and PS_2 are joined and the perpendicular from S_1 to PS_2 strikes the line at X. The phenomenon of interference is only seen if P is close to O and if d is small in comparison with D. Hence, to a first approximation $S_1P = XP$. Let the angle OSP be θ. Then, since S_1S_2 is perpendicular to SO and S_1X is to a first approximation perpendicular to SP, angle S_2S_1X is θ also. Thus S_2X is $d \sin \theta$. But this is the path difference $S_2P - S_1P$.

Since θ is a very small angle, $\sin \theta$ is approximately equal to $\tan \theta$, which has the value x/D.

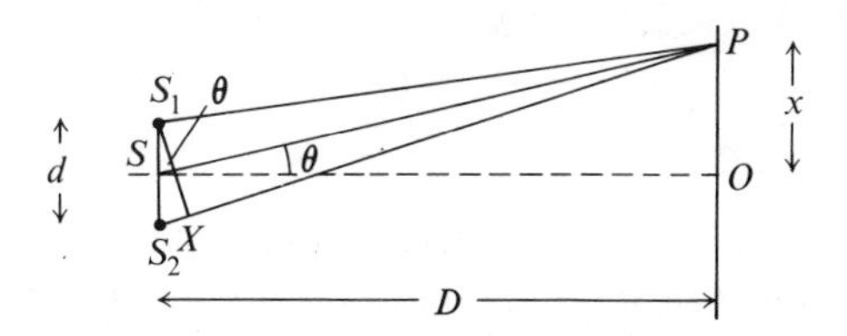

Fig. 17.6. Interference at a point on a screen of the light from two identical sources.

Therefore

$$\text{the path difference} = \frac{dx}{D} = m\lambda,$$

where m is an integer, if P is a point of maximum intensity; and

$$\frac{dx}{D} = (m + \tfrac{1}{2})\lambda,$$

if P is a point of darkness.

Thus if x_m is the distance from O to the mth bright fringe, and x_{m+1} the distance from O to the $(m + 1)$st bright fringe,

$$\frac{dx_m}{D} = m\lambda,$$

$$\frac{dx_{m+1}}{D} = (m + 1)\lambda,$$

and, therefore

$$x_{m+1} - x_m = \frac{D\lambda}{d}. \tag{17.2}$$

The separation of the bright fringes is thus independent of m. Similarly,

$$x_{m+3/2} - x_{m+1/2} = \frac{D\lambda}{d}. \tag{17.3}$$

Thus dark fringes are equidistant, as are bright fringes, and the separation of the dark fringes is the same as the separation of the bright fringes.

Note that if the fringes lie too close together the eye cannot distinguish them (cf. Sections 16.6 and 18.3) and sees only a general illumination. For $x_{m+1} - x_m$ to be appreciable so that the fringes are visible, D/d must be large, since λ is small. If D has a value around 100 cm, d must therefore be of the order of 1 mm or smaller in order to make the fringes visible.

The distance apart of the fringes depends on λ. It has been assumed that a simple sinusoidal wave train is being emitted by each source; in other words, each source is emitting monochromatic light (light of only one wavelength). If monochromatic light is not being used, each wavelength produces its own interference pattern, differing in spacing from all the others. The fringe patterns therefore overlap and cancel out. The net effect is to produce a system of colored fringes around O (which is the point of zero path difference for all wavelengths), which gradually tail off into a general background of light.

17.4 OPTICAL PATH LENGTH

It is appropriate at this point to be precise about what is meant by path length. If both wave trains travel in the same medium for the whole of their paths, no problem arises. This will not be true in all applications dealt with in subsequent chapters and it must therefore be made clear that geometrical distance is not all that is involved.

It has already been stated that the velocity of light is less in a dense medium than in air. The greater the refractive index of a medium, the smaller the velocity of light in that medium. Since the frequency of a wave train cannot vary, it follows that the wavelength varies from medium to medium. In any medium $f = v/\lambda$. In particular, for vacuum, $f = c/\lambda_0$. Thus

$$\lambda = \lambda_0 v/c = \lambda_0/n. \tag{17.4}$$

If two waves travel the same geometrical distance D, but one wave travels in a medium of refractive index n_1 and the other in a medium of refractive index n_2, the number of wavelengths traveled is different in the two cases. One wave has traveled D/λ_1 wavelengths and the other D/λ_2 wavelengths. But

$$\frac{D}{\lambda_1} = \frac{Dn_1}{\lambda_0}$$

and

$$\frac{D}{\lambda_2} = \frac{Dn_2}{\lambda_0}.$$

The relevant distance for interference and other similar phenomena is not the geometrical path distance D but the *optical path distance*, Dn_1 or Dn_2. Whenever path length is quoted, it is always optical path length that is implied, since it is this quantity which is important.

In the previous section a certain number of approximations were made in order to achieve the final expression. The necessity for this would have been avoided if the screen had been placed in the focal plane of a lens as in Fig. 17.7. The light from S_1 and S_2 reaching P must have traveled parallel paths S_1A and S_2B before striking the lens (cf. Section 14.5). The light at B has traveled an extra path, strictly equal to $d \sin \theta$, as compared with the light at A. But from A to P and from B to P the path is more complicated because the lens is in the way. What then is the relation between the path lengths of the wave trains when they arrive at P?

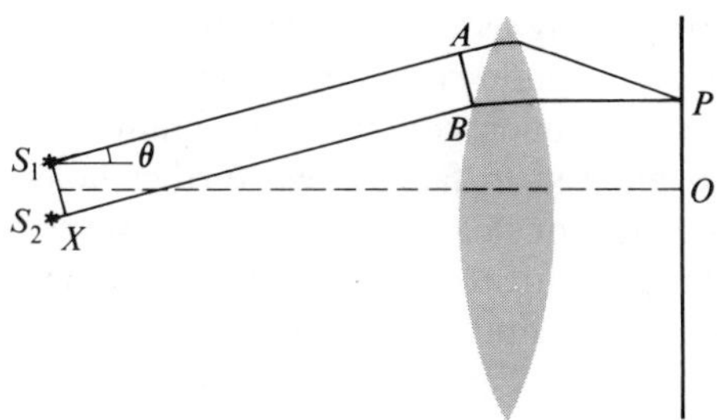

Fig. 17.7. Interference at a point in the focal plane of a lens.

The answer is very simple. The wave trains have a path difference between them, still of amount $d \sin \theta$, when they reach P. Although one wave, after leaving A, travels geometrically farther in getting to P, it travels through less glass because of the narrowing of the lens. Its optical path length is exactly the same in getting from A to P as is that of the other wave in going from B to P. This applies to all lens situations. The light from all parts of any plane wave-front arrives in phase in the focal plane of a lens. Further, any wave train from a point on an object arrives at the image in a lens exactly in phase with all other wave trains from the same point. If lenses are interposed in any interference (or diffraction) experiment, the effect produced on path and phase differences is therefore always zero.

17.5 COHERENCE

In all the previous discussion of wave motion and interference, the waves were said to be identical and the sources were assumed to emit identical wave trains. In practice this places a severe limitation on the sources which can be used. If two sources of a particular monochromatic light are set up as in Fig. 17.5, each source having been manufactured as far as possible in an identical manner, it will be found to be impossible to produce interference fringes from the combined action of the light from the two sources.

The reason for this lies in the realms of atomic physics, a subject which is not discussed until a later chapter. It is, however, possible to give a simple explanation now, the detailed reasons for which should be apparent in due course. The light from any source other than a laser (cf. Section 30.13) does not consist of a continuous train of waves oscillating at right angles to the direction of motion. Because of the fact that the light is emitted independently, except in the case of the laser, from a large number of different atoms over any short period of time, the wave train emitted from the source consists of short pulses between which the phase changes abruptly and discontinuously. Since the rapid and random changes occur quite differently from one source to the next, the wave trains from different sources do not stay in the same phase relationship to one another at any point for any appreciable length of time. Under these circumstances it is obviously impossible to obtain a stable interference pattern. This point was made from a different viewpoint in Sections 10.4 and 10.5. To avoid this trouble it is necessary that the two interfering wave trains originate from the same source. Thus in the simple Young's interference experiment described in Sections 17.2 and 17.3, it is necessary to have a single source of monochromatic light illuminating a hole, or slit, S, the light from which is allowed to fall on two further holes, or slits, S_1 and S_2, which act as the interference sources. In this way the waves from the two sources have the same phase characteristics and the sources are said to be coherent. Figure 17.8(a) shows such an experimental arrangement.

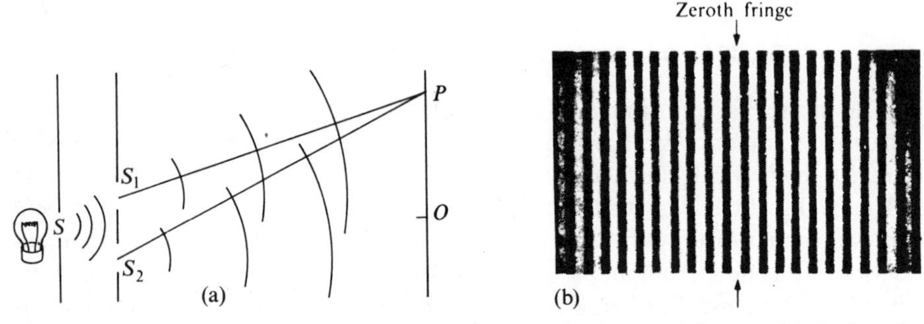

Fig. 17.8. (a) The slit arrangement in a Young's interference experiment; (b) the interference fringes produced. (Reproduced from *University Physics* by F. W. Sears and M. W. Zemansky, Addison-Wesley, by courtesy of the authors.)

Note that, whether S, S_1 and S_2 are all holes or all slits, the same interference pattern will be seen on the screen—a collection of equidistant light and dark bands. The use of slits allows more light to pass through and improves the illumination of the fringes. Figure 17.8(b) shows a typical interference pattern.

17.6 POLARIZATION

Interference is not the only phenomenon which demonstrates that light is a wave motion. Polarization is another, which in addition proves that light is a transverse wave motion. It must not be thought that normal light waves are such that the disturbance propagated oscillates in only one direction at right angles to the direction of motion. This is not the case. The vibration takes place in all possible directions at right angles to the propagation direction. But it is possible to obtain light which is vibrating in only one direction, and such light is said to be plane-polarized. It will be appreciated that only transverse, and not longitudinal, waves can be polarized. There are several methods by which light can be rendered plane-polarized, but the most important method is the use of certain types of crystal.

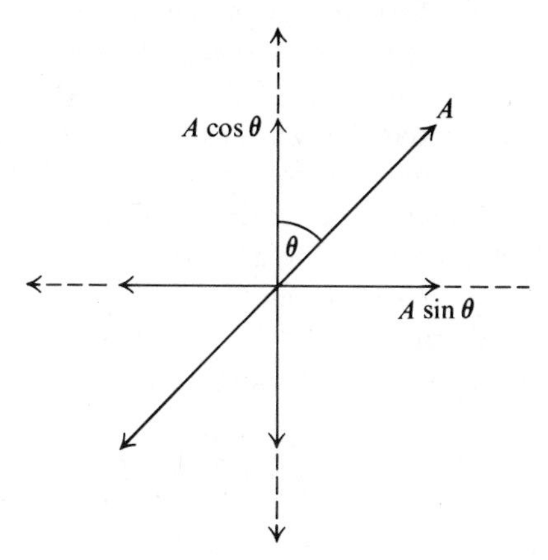

Fig. 17.9. The splitting of polarized light into components by a birefringent crystal.

A crystal such as calcite has the property that a beam of light falling upon it is split up in general into two beams, each of which contains light that is plane-polarized. In the two beams the planes of vibration of the light (the plane containing the directions of propagation and vibration) are at right angles to one another. A light wave vibrating in a plane other than those two, on striking the crystal is split into component waves along the two permitted directions. Thus in Fig. 17.9 the vibration of amplitude A inclined at an angle θ to one of the permitted vibration directions is split up on striking the crystal into two components, one of amplitude $A \cos \theta$ in one direction and one of amplitude $A \sin \theta$ in the direction at right angles. This property of a crystal of splitting up incoming light into two beams in which the vibrations are at right angles is called double refraction or birefringence. If the crystals are large enough and correctly cut and the incoming beam is fairly narrow, it is possible to separate out the two beams and use one or both of them as beams of plane-polarized light for experimental purposes. If the incoming light is unpolarized and of intensity I, the two beams emerging must each by symmetry considerations have intensity $I/2$.

In some crystals in addition, tourmaline being an example, both beams are absorbed on passing through the crystal but one much more strongly than the other. If the crystal is thick enough, only one beam emerges from the other side. This phenomenon is known as dichroism. The width of the incoming beam may therefore be as large as the cross-section of the crystal. Unfortunately, dichroic crystals seldom grow to very large size. Recently it has become possible to manufacture large sheets of Polaroid, which are in fact sheets of plastic packed with tiny crystals of a dichroic substance all carefully orientated parallel to one another. This collection of tightly packed crystals all orientated in the same direction has therefore the same effect as a large, single dichroic crystal. One may therefore use sheets of Polaroid to produce a beam of plane-polarized light of any width.

If a beam of light is sent through a sheet of Polaroid, plane-polarized light emerges. If a second Polaroid sheet is placed in the path of the light and orientated in exactly the same way as the first sheet, light slightly reduced in intensity because of absorption emerges from the second sheet. If the second Polaroid is now swung through 90° about the direction of the light beam, the light entering it is vibrating in the direction such that heavy absorption takes place. No light therefore emerges from the second Polaroid. Such a pair of Polaroids is called a polarizer-analyzer pair and in this relative orientation are said to be crossed. Passage through, or reflection from the surface of, a substance may alter the state of polarization of a beam of plane-polarized light, and valuable information about the composition of, or the molecular constitution of, the substance may be obtained from an investigation of the nature of the change which has taken place.

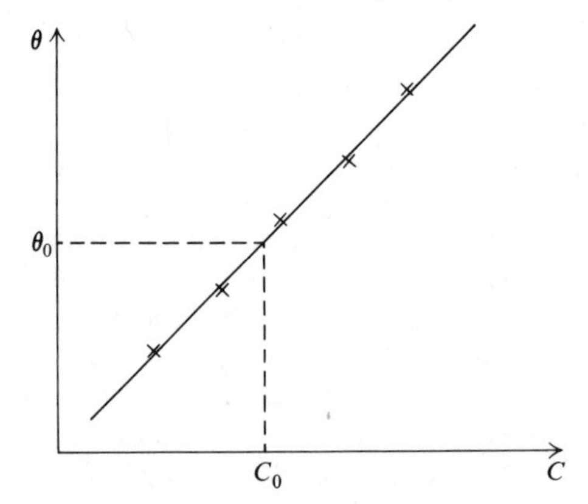

Fig. 17.10. The rotation of the plane of polarization as a function of the concentration of an optically active solution.

To take one simple example, there are a number of substances which consist of corkscrew-like molecules. In certain cases both right-handed and left-handed molecules exist, but very often only one type occurs naturally. Some of these substances have the power of rotating about the transmission direction the plane of vibration of plane-polarized light, right-handed substances rotating it in one direction, left-handed substances in the other. The substances are said to be optically active. The amount of rotation that takes place depends on the length of the path in the substance and, if a solution, on the concentration.

Sugars are optically active substances. If a tube of sugar solution is placed between crossed Polaroids, it will be found necessary to turn the second Polaroid about the direction of the light beam before darkness can once more be obtained. The light entering the sugar has a vibration direction at right angles to the transmission direction of the second Polaroid. Passage through the sugar rotates the vibration plane through an angle θ. The second Polaroid must be rotated in the same direction through an angle θ to make its transmission direction once more at right angles to the plane of vibration.

If, for the same length of tube, sugar solutions of varying concentration are inserted, it will be found that θ is directly proportional to the concentration C. One may therefore draw a linear calibration graph as in Fig. 17.10 relating any θ to the corresponding concentration. If a solution containing an unknown concentration of sugar is now obtained in an analysis or an experiment, the concentration can be simply and quickly found by finding the value of θ_0 of the rotation produced under standard conditions. The corresponding concentration may be read off from the graph.

17.7 POLARIZATION BY REFLECTION AND BY SCATTERING

If the light reflected from a glass surface is examined with a Polaroid sheet, it will be found that rotation of the Polaroid in the beam produces a variation in the intensity of the light transmitted. The light is partially plane-polarized. For one particular angle of incidence of the light on the surface, known as the Brewster angle, the reflected light is completely plane-polarized, the light vector vibrating parallel to the surface only. If the transmitted light is examined, it is found to be partially plane-polarized, having been robbed of some light vibrating parallel to the surface. This phenomenon is found to take place at the surfaces of all media, although more complex effects take place in certain cases.

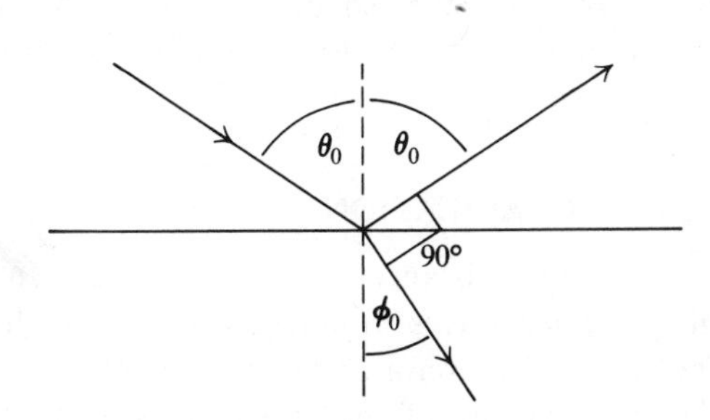

Fig. 17.11. Polarization on reflection at the Brewster angle.

If θ_0 is the Brewster angle, it is found that

$$\tan \theta_0 = n, \tag{17.5}$$

where n is the refractive index of the glass. It is not difficult to see why the light reflected is plane-polarized. From Snell's law

$$\sin \theta_0 / \sin \phi_0 = n.$$

Combining this equation with Brewster's equation gives

$$\sin \phi_0 = \cos \theta_0.$$

Therefore the angle between the reflected and transmitted rays is 90°.

The light falling on the glass surface causes electric vibrations there. These vibrations are in directions at right angles to the direction of the transmitted beam, and are equivalent (cf. Section 17.6) to vibrations only parallel to the surface and at right angles to this direction. Vibrations in both these directions can be carried by the transmitted beam, but since the second of these directions is the direction of propagation of the reflected light, this light would not be a transverse wave if it carried this vibration. It can therefore be vibrating only in the plane of the surface and is therefore plane-polarized.

It is for the same reason that light scattered at right angles to its direction of travel is always plane-polarized, light scattered in any other direction being partially plane-polarized. As was mentioned in Section 16.9, sunlight scattered by air molecules is plane-polarized in a direction at right angles to the direction of the direct rays from the sun and partially plane-polarized in other directions. Bees and other insects appear able to detect the state of polarization of light entering their eyes and in consequence have an innate method of determining direction.

The polarization effect on reflection gives a method of determining the refractive index of substances which, because they scatter or absorb light passing through them, cannot be used in more conventional methods. The angle of incidence of a beam of light is varied until the reflected beam is found to be plane-polarized. At this point the angle of incidence is the Brewster angle θ_0, and n is $\tan \theta_0$.

17.8 OTHER TYPES OF POLARIZATION

Many biologically interesting substances are birefringent. In birefringent substances the two plane-polarized waves produced travel through the material with different speeds. The material therefore has two different refractive indices for these light waves. If the thickness of the specimen is D, the two waves in the material have optical path lengths Dn_1 and Dn_2. A path difference of $D(n_1 - n_2)/\lambda_0$, and therefore a phase difference of $2\pi D(n_1 - n_2)/\lambda_0$, exists between the two beams when they emerge from the birefringent substance. The disturbance produced by the waves is therefore that due to two sinusoidal vibrations of different phase oscillating at right angles to each other. From the discussion in Section 10.4 it will be realized that either a circular or elliptical disturbance is produced, and the light is said to be circularly or elliptically polarized.

Use can be made of this type of polarization, whose characteristics may be investigated by apparatus more complicated than anything we have so far described. It is, however, more usual to make use of the phase difference introduced in another manner. In Section 20.2, such uses will be described.

Example 1. Two monochromatic wavetrains of the same frequency and with equal amplitudes A arrive at an observation point 45° out of phase. What is the amplitude of the resulting waveform?

Solution. We shall do the example in three different ways.

1) It is always possible to use graphical means to solve a problem of this sort. In Fig. 17.12 two wavetrains of equal amplitude and 45° out of phase have been carefully drawn. At every point along the t-axis the algebraic sum of the displacements due to the two waves has been plotted and the points joined by a curve. This is the wavetrain which is the superposition of the individual wavetrains. It is of the same frequency as the other two, has amplitude 1.85 A and is 22.5° out of phase with each of the other waves.

2) The two wavetrains can be represented by the equations

$$y_1 = A \sin \omega t$$

$$y_2 = A \sin \left(\omega t - \frac{\pi}{4}\right).$$

Therefore

$$\begin{aligned} y = y_1 + y_2 &= A \sin \omega t + A \sin \left(\omega t - \frac{\pi}{4}\right) \\ &= A \sin \omega t + A\left(\sin \omega t \cos \frac{\pi}{4} - \cos \omega t \sin \frac{\pi}{4}\right) \\ &= \sin \omega t\left(A + \frac{A}{\sqrt{2}}\right) - \cos \omega t \frac{A}{\sqrt{2}}. \end{aligned}$$

But y must be expressible as

$$\begin{aligned} y &= B \sin (\omega t - \phi) \\ &= B \sin \omega t \cos \phi - B \cos \omega t \sin \phi. \end{aligned}$$

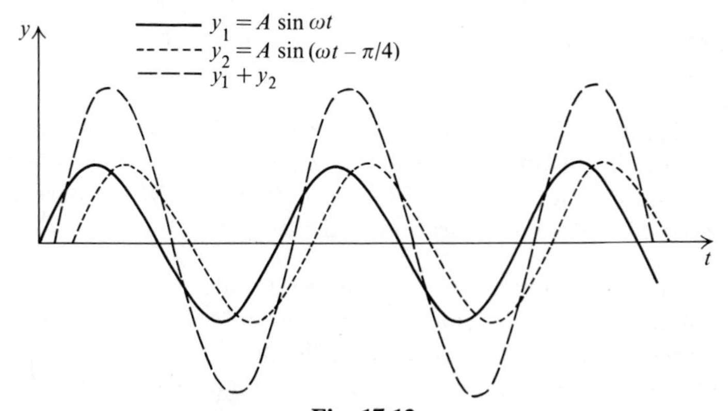

Fig. 17.12

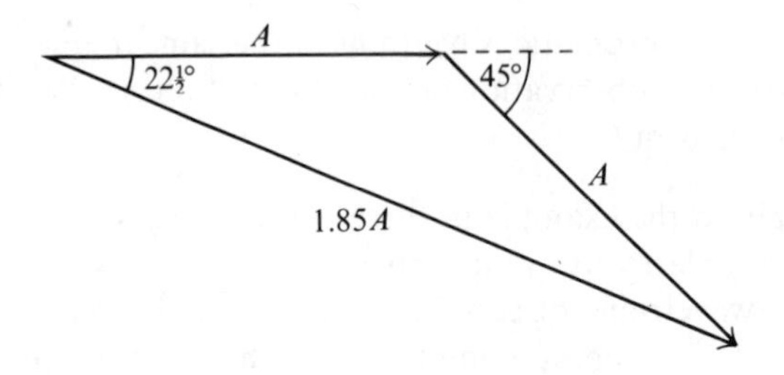

Fig. 17.13

Comparing the two expressions for y

$$B\cos\phi = A\left(1+\frac{1}{\sqrt{2}}\right); \qquad B\sin\phi = \left(\frac{A}{\sqrt{2}}\right).$$

Therefore

$$B^2 = B^2\cos^2\phi + B^2\sin^2\phi = A^2\left(1+\frac{1}{\sqrt{2}}\right)^2 + \frac{A^2}{2}$$

$$= A^2\left(1+\frac{1}{2}+\frac{2}{\sqrt{2}}\right) + \frac{A^2}{2} = A^2(2+\sqrt{2})$$

$$\tan\phi = \frac{B\sin\phi}{B\cos\phi} = \frac{A/\sqrt{2}}{A(1+1/\sqrt{2})} = \frac{1}{1+\sqrt{2}} = 0.414.$$

Therefore

$$B = 1.85\,A, \qquad \phi = 22.5°.$$

3) As was explained in Sections 10.4 and 10.5, each wave may be represented by a suitable vector. In Fig. 17.13, the first wave is represented by a vector in the positive x-direction of length A. The second wave is represented by a vector also of length A but inclined at an angle of $-45°$ to the x-axis. If the two vectors are added, then by measurement or by calculation the resultant is a vector of length $1.85\,A$ inclined at $-22.5°$ to the x-axis. It follows that the resultant wave has an amplitude of $1.85\,A$ and is out of phase with each of the other waves by an angle of 22.5°.

Example 2. Light of wavelength 550 nm is emitted by two slits in a Young's interference experiment and the resultant interference pattern is observed on a screen in the focal plane of a lens of focal length 25 cm. If the dark fringes are 0.500 mm apart, what is the separation of the two slits?

Solution. Light waves emitted from the sources parallel to the axis (shown dashed in Fig. 17.14) meet and interfere at the focal point of the lens. Light waves emitted from the sources in the direction θ meet at the point P in the focal plane. If an auxiliary wave (shown dotted in Fig. 17.14) also at angle θ to the axis passes through the center of the lens it is undeviated and also passes through P. It is clear that the dashed and dotted lines intersect at angle θ. Hence $x/f = \tan\theta$. But the angle involved ($\theta \approx 0.05/25$) is small. Hence $x/f \simeq \sin\theta$.

A maximum occurs at P if $d\sin\theta = m\lambda$, that is if

$$\frac{dx}{f} = m\lambda.$$

It follows that the separation of the fringes is

$$\Delta x = x_{m+1} - x_m = \frac{(m+1)\lambda f}{d} - \frac{m\lambda f}{d} = \lambda f/d.$$

Hence

$$d = \frac{\lambda f}{\Delta x} = \frac{550 \times 10^{-9}\ \text{m} \times 25 \times 10^{-2}\ \text{m}}{0.5 \times 10^{-3}\ \text{m}}$$

$$= 2.75 \times 10^{-4}\ \text{m} = 0.275\ \text{mm}.$$

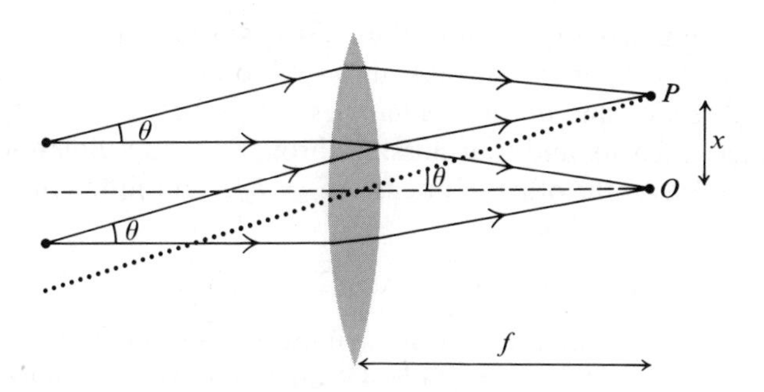

Fig. 17.14

Example 3. Light of wavelength 500 nm in air passes normally through a glass block of thickness 0.5 cm and refractive index 1.50. How many more wavelengths are there in the block than in 0.5 cm of air?

Solution. The wavelength in air is equal to the wavelength λ_0 in vacuum to the accuracy to which we are working and, in the glass block of refractive index n, it is $\lambda = \lambda_0/n$. The number of wavelengths in a length d in air is d/λ_0 and in glass it is $d/\lambda = nd/\lambda_0$. The extra number of wavelengths is

$$\frac{nd}{\lambda_0} - \frac{d}{\lambda_0} = \frac{d(n-1)}{\lambda_0}$$

$$= \frac{0.5 \times 10^{-2}\ \text{m} \times 0.50}{5.0 \times 10^{-7}\ \text{m}}$$

$$= 5.0 \times 10^3.$$

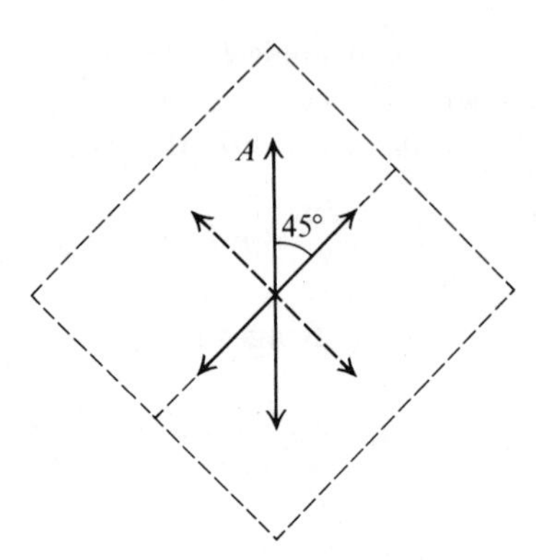

Fig. 17.15

Example 4. Two sheets of Polaroid are placed in contact with an angle of 45° between their transmission directions. If unpolarized light is passed through the combination, what fraction of the original intensity emerges?

Solution. The light striking the first Polaroid has amplitude A_0 and intensity I_0, where $I_0 \propto A_0^2$. The light is resolvable into two component vibrations, one in the transmission direction and one at right angles to this direction. By symmetry these will be of equal intensity and transmission through the first Polaroid removes one component and leaves the other of intensity $I = I_0/2$. It follows that the amplitude is

$$A \propto \sqrt{I} = \sqrt{I_0/2} \propto A_0/\sqrt{2}.$$

This vibration of amplitude A on striking the second Polaroid can be split as in Fig. 17.15 into a component $A \cos 45°$ along the transmission direction and a component $A \sin 45°$ at right angle to this direction. This latter component is absorbed in passage through the Polaroid. The amplitude emerging from the second Polaroid is thus A', where

$$A' = A \cos 45° = \frac{A_0}{\sqrt{2}} \times \frac{1}{\sqrt{2}} = \frac{A_0}{2}.$$

Therefore

$$I' \propto A'^2 = A_0^2/4 \propto I_0/4.$$

Only one-quarter of the intensity is transmitted by the pair of Polaroids. Note that we are ignoring any absorption in the Polaroids of the transmitted components of the beam.

PROBLEMS

17.1 What is the amplitude of the resultant wave at a point where two waves of the same frequency are arriving, one of amplitude A, the other of amplitude $2A$, the latter being 60° out of phase with the former?

17.2 Two waves of the same frequency and of equal amplitude A arrive at a point 90° out of phase. What is the amplitude of the resulting waveform?

17.3 Two slits 0.5 mm apart are illuminated normally by a parallel beam of light of wavelength 5893 Å and the interference pattern is observed on a screen 100 cm beyond the slits. How far from the central fringe on the screen is the sixth dark fringe?

17.4 What is the separation of the bright fringes on the screen in Problem 17.3?

17.5 White light from a slit source falls on two further slits close together. One of these slits has a yellow filter over it and the other a red filter. What sort of interference pattern will be seen on a screen placed behind the slits?

17.6 A parallel beam of light containing only the wavelengths 4.90×10^{-7} m and 5.88×10^{-7} m falls normally on a pair of slits 0.3 mm apart. At what distance from the central bright fringe on a screen 300 cm from the slits does a maximum of one wavelength first coincide with a maximum of the other?

17.7 If the apparatus described in Problem 17.3 is immersed in water, will the interference pattern alter in any way?

17.8 A slit is placed 1 mm above a mirror and a screen is placed 200 cm from the slit in a plane at right angles to the plane of the mirror. If the light coming from the slit has a wavelength of 6.28×10^{-7} m, what is the separation of the dark fringes on the screen?

17.9 A radio telescope is sited on the edge of a cliff overlooking the sea and operates on a wavelength of 10 m. A radio star rises above the horizon and is tracked by the telescope. The first minimum of the received signal occurs when the star is 30° above the horizontal. Explain why the minimum occurs and determine the height of the cliff, assuming that radio waves suffer a phase change of π on reflection at a water surface.

17.10 Light of frequency 5.20×10^{14} c/s passes normally through a parallel-sided block of glass of thickness 5 mm and refractive index 1.50. How many wavelengths are there in the glass?

17.11 Two identical wave trains of wavelength 5.5×10^{-5} cm in vacuum travel identical paths in reaching a point, except that one has traveled through 5 cm of vacuum while the other was traveling through 5 cm of air of refractive index 1.0003. What is the phase difference between the two wave trains when they arrive at the final point?

17.12 The planes of transmission of three Polaroids are at angles measured with respect to an arbitrary direction of 0°, 45°, and 105°, respectively. What fraction of the intensity of a beam of unpolarized light incident on the first Polaroid emerges from the third?

17.13 If a polarizer-analyzer pair is crossed, no light will emerge from the combination when unpolarized light enters the other side. If a third Polaroid is inserted between the first two, can this alter the situation? Justify completely the answer you give.

17.14 A sugar solution of concentration 1 g cm^{-3} is found to rotate the plane of polarization of plane-polarized light passing through it by 6.64° per centimeter of length. The same light passed through a tube of 10 cm length containing a sugar solution of unknown concentration undergoes a rotation of its plane of polarization of 8.30°. What is the concentration of the solution?

17.15 What is the angle of the sun above the horizon when the light reflected from a still lake is completely plane-polarized?

17.16 The critical angle for sodium light in a particular material is 30°. At what angle of incidence must sodium light strike the material in order that the reflected light will be completely plane-polarized?

17.17 When blue light of wavelength 4.26×10^{-7} m in vacuum passes through a particular biological specimen, the light splits into two beams plane-polarized at right angles, the refractive indices for these two beams being 1.316 and 1.332, respectively. With what velocities and wavelengths do the two beams travel through the specimen?

17.18 If the specimen in Problem 17.17 is 0.001 mm thick, what phase difference is introduced between the two beams of light in traveling through it?

17.19 When sodium light of wavelength 5893 Å in vacuum falls on quartz, it splits into two plane-polarized beams whose indices of refraction are 1.544 and 1.553. What is the minimum length of quartz which will introduce a phase difference of π between the two beams?

CHAPTER 18

DIFFRACTION

18.1 INTRODUCTION

In the simple Young's interference experiment described in the previous chapter, a source of light was placed behind a hole or slit, the light then being allowed to fall on two further holes or slits. If light traveled in straight lines, the light arriving at a screen placed beyond the two holes would consist of two quite separate images.

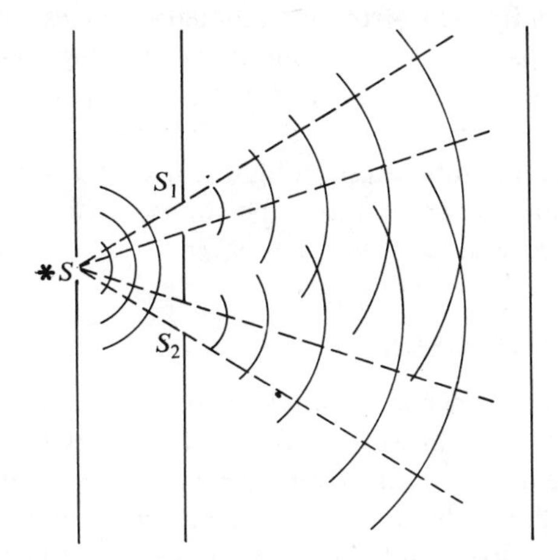

Fig. 18.1. Light diffracted by the slits interferes in a Young's experiment.

Indeed, if the holes are large, this is what happens. It is because light is a wave motion and a phenomenon known as diffraction takes place at each hole that interference can take place at all. The initial hole or slit, if sufficiently narrow, acts as a source of light which is sent out, not necessarily with equal intensity, in all directions. Similarly, the two further holes or slits send out light in all directions and interfering wave trains are therefore received at all points of the screen (Fig. 18.1).

If further holes or slits are punched between the two which produce the final interfering beams, the interference pattern will of course alter, but variations of light and darkness will still occur across the viewing screen. The positions of complete darkness and maximum brightness would now, of course, be a little more complicated to calculate. Indeed, if the material which separates the holes or slits is dispensed with altogether, the resulting larger hole or slit may be regarded as a collection of very small holes or slits placed next to one another over the area. Variations of light and shade over the viewing screen are therefore still to be expected, the resultant pattern being known as the diffraction pattern of the hole or slit.

The situation may be looked at from a different point of view, as was originally done by Huygens. Light from a point source will be emitted equally in all directions, forming at any given time a wave-front which is spherical. This wave-front may be considered as a source of light waves which will spread out from every point of the wave-front to form future wave-fronts at any subsequent time. If the wave-front is regarded as the seat of an infinite number of secondary sources of light, it is possible to show that the interference of the beam from these secondary sources produces at future times a spherical wave-front of increased dimensions, the light intensity being equal at all points on the new wave-front. This model for the spreading of the light is therefore such as to give a satisfactory description of future events.

It is clear that if a full wave-front of secondary sources is necessary to produce a future wave-front, part of a wave-front cannot produce the same effect. Thus, restricting the wave-front by an obstacle, such as a screen with a hole or slit in it, must produce at future times a wave-front which is no longer spherical and is not of equal intensity at all points. Restricting the light in a wave-front must produce variations of light and shade at future positions. Diffraction is defined as the phenomenon observed when a wave-front is limited.

18.2 DIFFRACTION AT A SLIT

The simplest case to consider in an attempt to work out the effect of diffraction is that of a single slit. The slit will be assumed to be very long and any end effects produced will be ignored. Away from the end the system is identical in any plane perpendicular to the slit, and the problem can therefore be reduced to one in two dimensions, the three-dimensional effect merely being that produced by a set of similar planes stacked one on top of the other. To reduce the complexity of the problem, it is assumed that the light falling on the slit is monochromatic and in the form of a parallel beam at right angles to the plane of the slit. All wave trains of light striking the slit are therefore at the same phase at all points on the slit.

The slit of width d shown in Fig. 18.2 can be treated as a collection of narrow slits stacked one beside the other, each acting as a secondary source of light sending waves in all directions. Suppose that there are n of these narrow slits numbered from 1 to n from edge A to edge B. Consider the first slit and the $(\frac{1}{2}n + 1)$st slit, i.e. the slit just below A and the slit just below the center X of the large slit. The light from these two slits interfere. The separation of the two slits is $\frac{1}{2}d$, and from the discussion of Young's interference experiment (cf. Sections 17.2 and 17.3), wave trains from these slits sent in a direction θ to the original beam, which meet at infinity, or in the focal plane of a lens, will produce maximum intensity of light there if $\frac{1}{2}d \sin \theta = m\lambda$, m being an integer, and darkness if $\frac{1}{2}d \sin \theta = (m + \frac{1}{2})\lambda$. The same is true of the second and $(\frac{1}{2}n + 2)$nd slit, the third and $(\frac{1}{2}n + 3)$rd slit, and so on, right down to the pair consisting of the $\frac{1}{2}n$th slit and the nth slit. Thus the light from each member of a pair of slits cancels out the light from the other member of the pair, and therefore there is darkness for the complete beam of light from the large slit in the direction θ, when

$$\tfrac{1}{2}d \sin \theta = (m + \tfrac{1}{2})\lambda, \qquad \text{or} \qquad d \sin \theta = (2m + 1)\lambda. \tag{18.1}$$

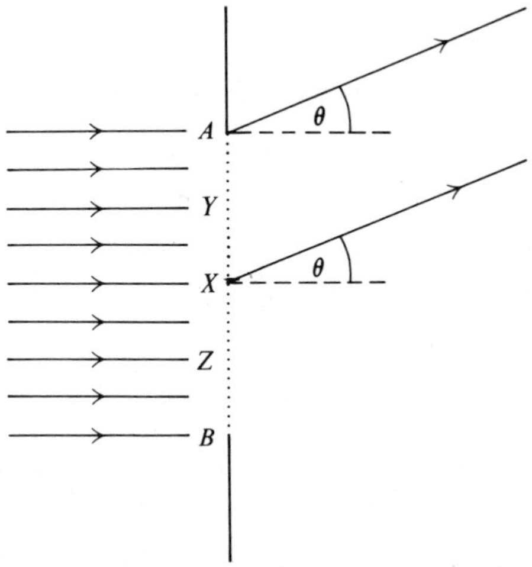

Fig. 18.2. Diffraction at a slit.

It may not be clear at first sight that the same is not true of the direction of maximum brightness from the whole slit as compared with the direction of maximum brightness from the pairs of slits. It is certainly true that the straight through beam ($m = 0$, $\theta = 0$) produces a maximum, because all points on this wave-front are in phase after it has passed through the slit. But it is not true that the condition $\frac{1}{2}d \sin \theta = m\lambda$ produces a maximum in general. In fact it also produces darkness.

Although each pair of slits produces maximum brightness in the direction θ given by the equation $\frac{1}{2}d \sin \theta = m\lambda$, the light from the first slit, {and the $(\frac{1}{2}n + 1)$st}, in that direction is slightly out of phase with light from the second slit, {and the $(\frac{1}{2}n + 2)$nd}, because the latter light has a slightly longer path to travel. Similarly, light from the second slit is out of phase with light from the third slit, and so on. The light from each pair of slits is thus not in phase with the light from any other pair of slits and interference effects take place.

It is, however, possible to deduce that darkness occurs in that direction by pairing the narrow slits in a different way. This time the first and the $(\frac{1}{4}n + 1)$st are paired, these being the slits just below A and just below Y, the point one-quarter of the way across the slit. Similarly, the second and $(\frac{1}{4}n + 2)$nd, the third and $(\frac{1}{4}n + 3)$rd, etc., are paired right down to the $\frac{1}{4}n$th and $\frac{1}{2}n$th. Similarly, all slits below X and all below Z are paired, i.e. the $(\frac{1}{2}n + 1)$st with the $(\frac{3}{4}n + 1)$st, the $(\frac{1}{2}n + 2)$nd with the $(\frac{3}{4}n + 2)$nd, and so on, down to the $\frac{3}{4}n$th and the nth. In each pair of slits the distance of separation is $d/4$. Hence, darkness results from each pair of slits, and thus from the complete large slit, in directions for which

$$\tfrac{1}{4}d \sin \theta = (p + \tfrac{1}{2})\lambda, \qquad \text{or} \qquad d \sin \theta = (4p + 2)\lambda. \tag{18.2}$$

If the same procedure is adopted for pairs of slits $d/8$, $d/16$, $d/32$, etc., apart, the relations for the directions in which darkness will occur will be, respectively,

$$d \sin \theta = (8p + 4)\lambda, \tag{18.3}$$

$$d \sin \theta = (16p + 8)\lambda, \tag{18.4}$$

$$d \sin \theta = (32p + 16)\lambda, \text{ etc.} \tag{18.5}$$

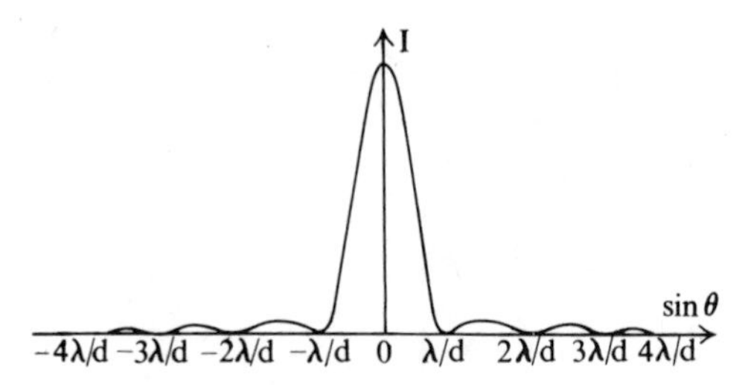

Fig. 18.3. The intensity of light diffracted by a slit as a function of the angle of diffraction.

The first relation obtained, $d \sin \theta = (2m + 1)\lambda$, indicates that whenever $d \sin \theta$ is an odd number of wavelengths, darkness occurs; the other relations merely show the same for values of $d \sin \theta$ which are various types of even number of wavelengths. The final conclusion is therefore that, for a slit, light is sent out in various directions, darkness occurring whenever

$$d \sin \theta = n\lambda, \tag{18.6}$$

n being a whole number not equal to zero. There are no other directions in which darkness occurs.

It would be tedious to work out the detailed distribution of intensity in all directions and the result is, for the most part, of no great interest. Figure 18.3 shows the intensity of the light sent in various directions plotted against $\sin \theta$. It will be seen that the intensity drops off fairly rapidly with increasing θ.

Between each two neighboring directions where darkness occurs, there must be a position of maximum brightness. Except for $\theta = 0$, where the maximum is centrally located, this position is almost, but not exactly, midway between the nearest positions of darkness. The intensities of the successive maxima are roughly in the ratio $1 : \frac{1}{20} : \frac{1}{60} : \frac{1}{120}$: and so on. The central maximum is thus the only one of great importance.

Note that if light were linearly propagated, the light beam received on a screen would have a width d. Since light is a wave motion and diffraction takes place, the angular spread of the light measured by the half-width of the central maximum is arc sin (λ/d), or

$$\theta \approx \sin\theta = \lambda/d, \tag{18.7}$$

since $d > \lambda$ in general and the angle is in consequence small. Diffraction is therefore more pronounced as d gets smaller. If $d \gg \lambda$, the angular spread of the diffraction effect tends to zero and pure geometrical effects are obtained.

If the light does not come in as a normal, parallel beam, the diffraction pattern obtained will be slightly different, but the general shape will remain the same.

If the screen contains not a slit but a circular hole of diameter d, the mathematics is somewhat more complicated, but it is possible to show that the diffraction pattern is a central bright circle of light surrounded by concentric alternate dark and light rings. The intensity of the successive bright rings falls off rapidly and all but the central maximum are fairly faint as in the case of the slit. A one-dimensional section through the two-dimensional pattern looks very similar to Fig. 18.3. The lines joining the hole to the first dark ring form a cone, and it is found that the semi-angle of this cone is θ, where

$$\theta \approx \sin\theta = \frac{1.22\lambda}{d} \tag{18.8}$$

18.3 RESOLVING POWER

We are now in a position to see why it is not possible to construct a telescope in which fine detail of any given smallness can be resolved. If a point source of light is being viewed by a telescope, light spreads out in a spherical wave-front from the source but only a small portion of this wave-front is allowed to pass through the objective lens and proceed onwards to form the image. The wave-front inside the telescope has been severely limited and diffraction effects are therefore present. Even if all the components of the telescope are perfect and produce no aberrations whatsoever, the final image of the point source is that of the diffraction pattern due to the light being limited by a circular aperture. Instead of seeing a point image one therefore obtains a circular spot of finite extent surrounded by concentric dark and light rings.

If a second point source is brought up, its image will be similar. As the two sources are brought closer and closer together, the diffraction images draw closer too and eventually overlap. A stage will be reached at which it is impossible to tell whether two merged images are present or only one. This is perhaps more easily seen if we consider Fig. 18.4, where the case of diffraction at a slit is illustrated. As the two diffraction images get closer and closer together, the dotted curve showing the combined effect of the two eventually becomes indistinguishable from a single diffraction curve. Figure 18.5 shows the opposite effect. Here the sources are at a fixed distance apart at all times but the size of the circular aperture is being varied.

It is almost impossible to give a practical test for the ability to distinguish or resolve a diffraction pattern as being that due to two combined ones. With modern photometric equipment one can distinguish two diffraction patterns even when very close together, if the intensities are about the same. But if one intensity is very much smaller than the other, the diffraction patterns must be much farther apart before they can be resolved.

A theoretical criterion was proposed by Rayleigh and is a good working

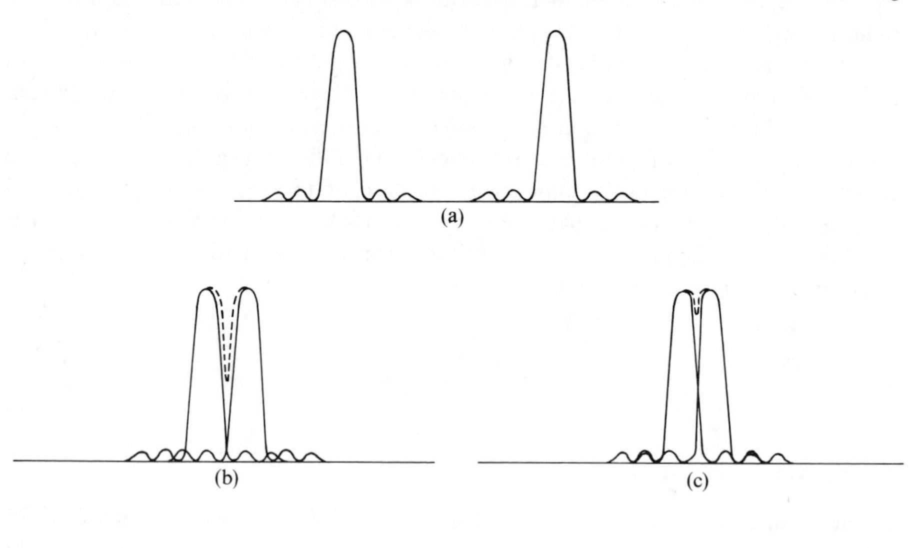

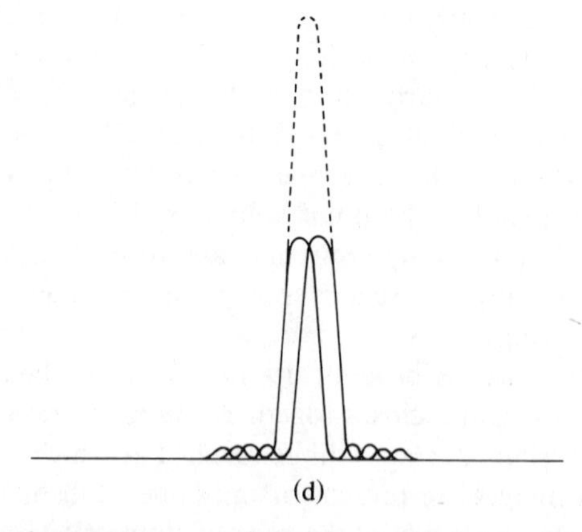

Fig. 18.4. The diffraction patterns produced by a slit due to two point sources (a) far apart, (b) close together, (c) so close that the patterns are just resolvable, and (d) so close that the patterns are unresolved.

criterion. It is the following: "Two diffraction patterns will be resolvable as long as the central maximum of one is no closer to the other than the position of the first minimum." This is the position illustrated in Fig. 18.4(c), and it will be seen to be about the limit of resolution to be expected.

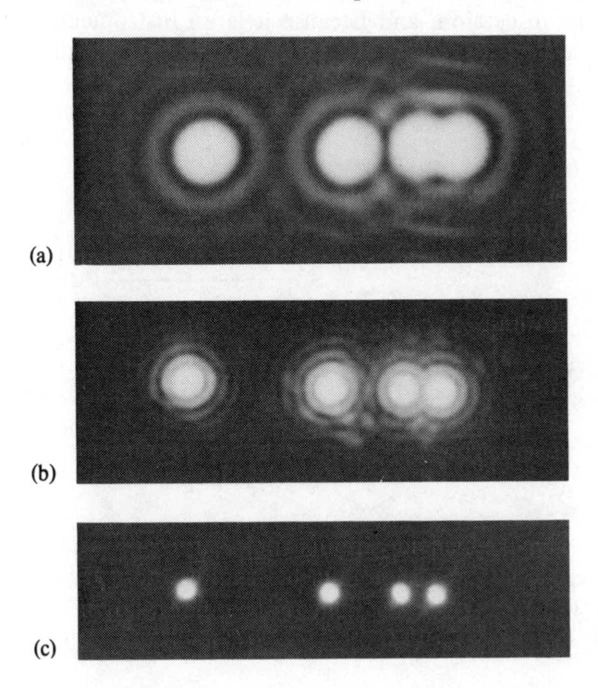

Fig. 18.5. Diffraction patterns of four "point" sources, with a circular opening in front of the lens. In (a) the opening is so small that the patterns at the right are just resolved by Rayleigh's criterion. Increasing the aperture decreases the size of the diffraction patterns, as in (b) and (c). (Reproduced from *University Physics* by F. W. Sears and M. W. Zemansky, Addison-Wesley, by courtesy of the authors.)

In Fig. 18.6 rays of light from two sources are seen passing through the objective lens of a telescope, which has a diameter d. The rays which pass undeviated through the center of the objective lens, go to the centers of the diffraction images formed by the lens. So long as the angle subtended by the images at the center of the lens, i.e. the angle between the central rays, is larger than θ, where $\theta = 1.22\lambda/d$, the two images can be seen resolved. But this is the angle subtended by the objects at the objective, and therefore the telescope can resolve two point objects only if they subtend an angle greater than $1.22\lambda/d$ at the objective. Note that increasing indefinitely the magnifying power of later stages is useless. All that will be seen is a bigger and better picture of two diffraction patterns, and if these are unresolved after the objective, they cannot be resolved by magnification thereafter.

If one is dealing with self-luminous objects, this is the end of the matter.

Diffraction caused by limitation of the wave-fronts from the objects produces finite images of point objects and limits the detail that can be seen. The case of non-self-luminous objects, which is normally what one views in a microscope, is more complicated, and must be looked at more closely in the next chapter. To prepare for this discussion, and because it is an instrument of considerable use in the investigation of spectra, we conclude this chapter with sections on the diffraction grating.

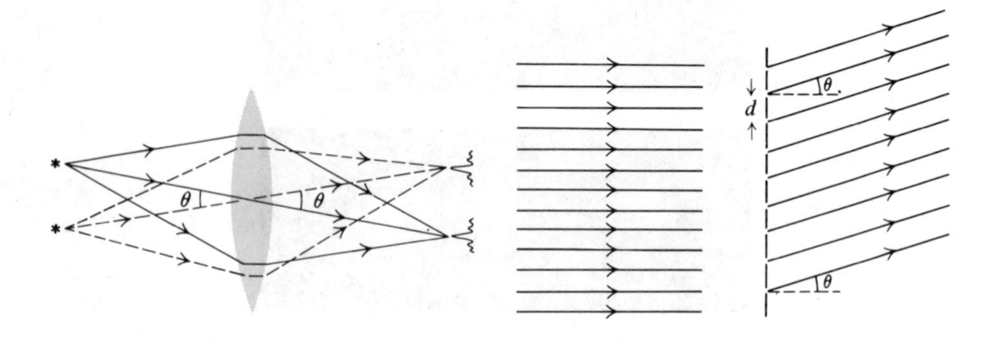

Fig. 18.6. The objective lens of a telescope produces two diffraction patterns as the images of two point sources.

Fig. 18.7. Diffraction by a one-dimensional grating.

18.4 THE DIFFRACTION GRATING

The diffraction grating consists in theory of a number of slits and stops placed side by side. In practice, gratings have generally been produced by ruling lines on a glass or metal surface with a diamond point. The untouched surface acts as a slit. The grooves scatter light in all directions, forming a general background of illumination. They therefore act somewhat in the manner of stops. It is possible by this method to produce gratings with perhaps 10,000 or 20,000 lines per inch, a feat which could not be equalled until recently by more conventional methods. The improvements in reduction photography, which have led to the use of microdots in espionage, have also made possible the production of diffraction gratings by more straightforward means. A collection of black and white lines can now be photographically reduced to give gratings of very small grating spacing.

In Fig. 18.7 monochromatic light is shown as falling normally on a diffraction grating whose grating spacing, the distance from the center of one slit to the center of the next slit, is d. If the wavelength of the light is λ, the light scattered in a direction θ to the incident direction from any two adjacent slits will be a maximum if $d \sin \theta = m\lambda$, (cf. Section 17.3). But if this is true of any two adjacent slits, it must be true for all slits, since they are all an equal distance apart, and thus the light from the whole grating adds up to give maximum intensity in the directions for which

$$d \sin \theta = m\lambda. \tag{18.9}$$

In the case of the two slits of Young's experiment the intensity in the various directions θ varied gradually from maximum brightness to darkness, as shown in Fig. 18.8(a). With more than two slits the maximum brightness occurs in directions given by the same formula, but the effect of adding more and more slits is to cause greater and greater destructive interference between the incoming wave-trains in all other directions. The net effect is therefore as in Fig. 18.8(b). The more slits the grating possesses (and therefore the smaller d) the narrower becomes the angle over which light is actually diffracted. With a large number of slits, 10,000 lines per inch or greater, the angular width of each diffraction maximum is negligible and the light may be considered to be scattered in a few discrete directions only.

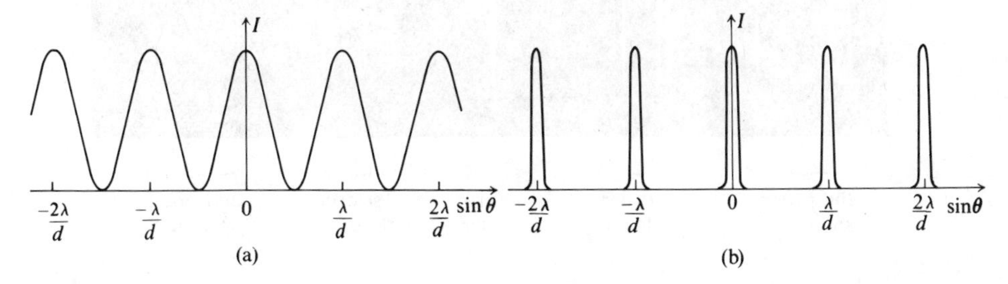

Fig. 18.8. The intensity distribution in the focal plane of a lens due to (a) two slits, and (b) a diffraction grating of the same spacing.

If the grating has 100 lines per centimeter, d therefore having the large value of 10^{-2} cm, diffraction maxima of light of wavelength 5×10^{-5} cm will occur at angles θ such that $\sin \theta = m \times 5 \times 10^{-5}$ cm $/10^{-2}$ cm. The values of $\sin \theta$ for the various possible values of m are 5×10^{-3}, $2 \times 5 \times 10^{-3}$, $3 \times 5 \times 10^{-3}$, and so on. There will be 200 maxima in the range θ from 0° to 90°. If, on the other hand, there are 4000 lines per cm, maxima occur for $\sin \theta = m \times 5 \times 10^{-5} \times 4 \times 10^{3} = 2m \times 10^{-1}$. There are now only four maxima in the range from 0° to 90°. These are well separated and each is very narrow.

Since θ depends on λ, if light other than monochromatic light is used, a spectrum of colors will be produced at all maxima except the zero order one. The diffraction grating is in fact used to examine spectra as is a prism, having the advantage over the prism that a good grating, one with a large number of lines per centimeter, spreads the spectrum over a wider angle than the prism and therefore gives better chromatic resolution.

18.5 INTENSITY DISTRIBUTION IN A GRATING

The formula for the positions of the maxima, for a given wavelength, depends on d. Two gratings of the same spacing, one with large slits and small stops, the other with the exact reverse, produce their maxima in identical directions. The amount of light scattered into each maximum will, however, be different in the two cases. The one with large slits will have more light in the zero-order maximum than the one with small slits. Indeed slits and stops are not necessary to produce diffraction maxima in specific directions. So long as there is a variation in scattering ability which repeats itself periodically with period d, diffraction maxima must occur in directions for which $d \sin \theta = m\lambda$. The striated muscle from a rabbit can be used as a diffraction grating, for example, since the density varies regularly and periodically because of the structure.

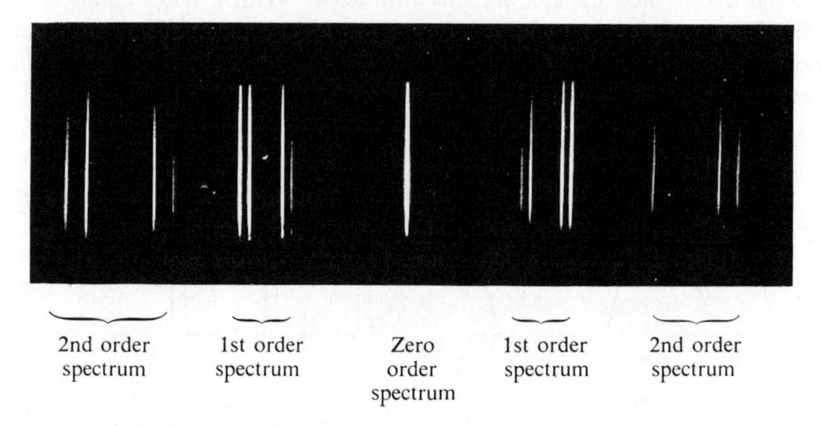

Fig. 18.9. The diffraction pattern due to a one-dimensional grating illuminated by mercury light. The zero-order spectrum is in the centre and the two diffraction orders visible on each side contain lines due to four distinct wavelengths.

18.6 TWO-DIMENSIONAL GRATING

If a grating has lines drawn across it in two directions at right angles, a two-dimensional grating results. Figures 18.9 and 18.10 show the diffraction patterns from a one-dimensional and from a two-dimensional grating. It will be seen that the pattern in the latter case consists of discrete dots. It is easy to see why this is so. One of the sets of lines, that in the x-direction, gives maxima only where

$$d_1 \sin \theta_1 = h\lambda, \tag{18.10}$$

d_1 being the grating spacing in that direction, θ_1 being the angle measured from the incident beam in the x, z-plane, and h being an integer. The lines in the y-direction give maxima only where

$$d_2 \sin \theta_2 = k\lambda, \tag{18.11}$$

d_2 being the spacing in the y-direction, θ_2 being the deviation in the y, z-plane, and k being an integer. The effect of the two sets of lines is to produce maxima only where both equations are satisfied simultaneously. This occurs only in the directions where the two sets of line spectra, which would be produced by the sets of lines acting individually, cross. Diffracted beams occur in discrete directions only and strike a screen placed in the focal plane of a lens in discrete points. Each light spot on the screen can be labeled (h, k), denoting that it is the result of the hth-order spectrum from one of the one-dimensional gratings and the kth-order spectrum from the other coinciding and producing a maximum.

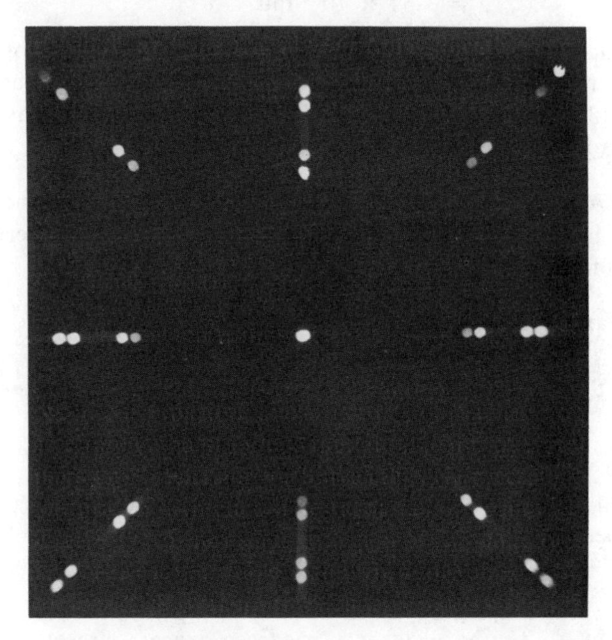

Fig. 18.10. The diffraction pattern due to a two-dimensional grating. The incident light contained four discrete wavelengths and each diffraction order except the zeroth shows four discrete spots.

It should again be emphasized that so long as the grating repeats periodically with spacing d_1 in one direction and d_2 in the other, diffraction maxima will occur in the same directions whatever the distribution of scattering matter. The latter affects only the variation of the intensity among the maxima. This point must be grasped if X-ray crystallography (cf. Section 20.8) is to be understood. There three-dimensional diffraction is involved, but the same principles apply.

Example 1. The visual acuity of an observer was shown in Chapter 16 to be dependent on the structure of the retina. Work out from diffraction theory the limit of resolution of an eye with an iris diameter of 5 mm and compare the result obtained with the visual acuity quoted earlier.

Solution. A mean wavelength for light is 5.5×10^{-7} m. By the same reasoning as was applied to the objective lens of a telescope, the limit of resolution of the eye will be

$$\theta = 1.22\lambda/d,$$

where d is the diameter of the iris aperture. Therefore

$$\theta = \frac{1.22\lambda}{d} = \frac{1.22 \times 5.5 \times 10^{-7}\ \text{m}}{5 \times 10^{-3}\ \text{m}}$$

$$= 1.34 \times 10^{-4}\ \text{rad}.$$

This compares with a value of 5×10^{-4} rad normally found, or a minimum of 1.74×10^{-4} rad in the most exceptionally gifted individuals as calculated in Problem 16.13. Clearly the structure of the retina and not the wave nature of the light places a lower limit on the visual acuity.

Example 2. A diffraction grating has 4000 lines per centimeter and is illuminated normally with light of wavelength 500 nm. Determine the directions in which maxima occur.

Solution. Maxima can only occur in directions for which

$$\sin\theta = m\lambda/d = m \times 5.00 \times 10^{-7}\ \text{m} \times \frac{4.00 \times 10^{3}}{1\ \text{cm}} \times \frac{100\ \text{cm}}{1\ \text{m}}$$

$$= m \times 0.200.$$

But $\sin\theta$ can only have values lying in the range $-1 < \sin\theta < 1$. The only possible values for $\sin\theta$ are thus

$$\sin\theta = 0,\ \pm 0.200,\ \pm 0.400,\ \pm 0.600,\ \pm 0.800$$

or

$$\theta = 0,\ \pm 11°32',\ \pm 23°35',\ \pm 36°52',\ \pm 53°8'.$$

Example 3. White light falls normally on a diffraction grating of spacing 2.4 μm. What is the angular spread in the highest order fully visible? What is the range of overlap with the spectrum of one order less?

Solution. For the red end of the spectrum, $\lambda = 7.0 \times 10^{-7}$ m. Hence, for $\sin\theta = 1$, the grating equation gives

$$m = \frac{d \times 1}{\lambda} = \frac{2.4 \times 10^{-6}\ \text{m}}{7.0 \times 10^{-7}\ \text{m}} = 3.4.$$

Thus the highest order spectrum fully visible is the third and the two ends of the spectrum occur at angles where

$$\sin\theta = \frac{3 \times 7.0 \times 10^{-7}\ \text{m}}{2.4 \times 10^{-6}\ \text{m}} = 0.875$$

and

$$\sin \theta' = \frac{3 \times 3.8 \times 10^{-7}\ \text{m}}{2.4 \times 10^{-6}\ \text{m}} = 0.475$$

Thus

$$\theta - \theta' = 61°3' - 28°21' = 32°42'.$$

For the red end of the second order spectrum

$$\sin \theta'' = \frac{2 \times 7.0 \times 10^{-7}\ \text{m}}{2.4 \times 10^{-6}\ \text{m}} = 0.583.$$

This occurs in the third order spectrum at the position occupied by wavelength λ, where

$$\lambda = \frac{d \sin \theta''}{m} = \frac{2.4 \times 10^{-6}\ \text{m} \times 0.583}{3}$$

$$= 466\ \text{nm}.$$

Thus the range of the third order spectrum which overlaps with the second is from 466 nm to 380 nm.

Also, since $\theta'' = 35°40'$,

$$\theta'' - \theta' = 35°40' - 28°21' = 7°19'.$$

The angular overlap is thus 7°.

PROBLEMS

18.1 Since light waves do not diffract round buildings, why do radio waves do so?

18.2 A slit of width 0.1 mm is illuminated normally with light of wavelength 5.5×10^{-7} m. At what angle to the straight-through beam does the first minimum occur?

18.3 If the slit in Problem 18.2 is replaced by a circular hole of diameter 0.1 mm, what is now the value obtained?

18.4 A slit is placed in front of a lens of focal length 50 cm and is illuminated with light of wavelength 5.5×10^{-7} m. The diffraction pattern is observed in the focal plane of the lens, and it is found that the separation of the first minimum on either side of the center of the pattern is 2.2 mm. What is the width of the slit?

18.5 The 40-in refracting telescope at Yerkes Observatory has an objective lens of focal length 65 ft. Two stars are just resolved by the instrument. How far apart are their images in the focal plane of the objective? Take the effective wavelength of the light as 5.5×10^{-5} cm.

18.6 A car is approaching on a straight road on a pitch-black night. The lights of the car are 100 cm apart. How far from an observer will the car be when he can just be sure that he is seeing two lights and not one? Take the aperture of the iris of the eye as 0.5 cm and the effective wavelength of the light as 5500 Å.

18.7 Would the fact that the light emitted by the automobile in Problem 18.6 is not monochromatic make the resolution by the observer easier or more difficult?

18.8 Two stars are just resolvable in the 200-in Mount Palomar telescope. If the stars are 5 light-years away, how far are they apart? Take the mean wavelength of the light as 5500 Å.

18.9 What is the maximum separation of the two slits in a Young's interference experiment when illuminated by light of 5893 Å, if the interference pattern formed on a screen 150 cm from the slits is to be observable by an eye with a resolving power of 1 minute of arc and a least distance of distinct vision of 25 cm?

18.10 A telescope has an objective lens of diameter 12 cm and focal length 50 cm. What is the maximum useful magnification of the eyepiece if an average eye may be considered to have an iris diaphragm of 3 mm diameter and the mean wavelength of the light entering the telescope is 5.00×10^{-7} m?

18.11 A diffraction grating with 4000 lines per centimeter is illuminated normally with white light. What is the maximum wavelength observable in the fifth-order spectrum?

18.12 The visible spectrum contains wavelengths from 3.8×10^{-7} m to 7.0×10^{-7} m. What is the spacing of a diffraction grating which will spread this spectrum over 15° in the second order?

18.13 A diffraction grating with 2000 lines per centimeter is illuminated normally with light which contains the two wavelengths 4.84×10^{-7} m and 6.05×10^{-7} m. At what angles to the incident beam will maxima due to the two wavelengths coincide?

18.14 A diffraction grating with 2000 lines per centimeter produces a strong diffraction maximum at 30° when illuminated normally with a monochromatic beam of visible light. What are the possible wavelengths of the incident light?

If there is also a strong maximum at 22° 1′, which of these wavelengths is the correct one?

18.15 Light from a sodium lamp contains two very close wavelengths, the mean value being 5893 Å. When a diffraction grating with a spacing of 2.5×10^{-4} cm is illuminated normally with sodium light, the angular separation of the components in the third-order diffracted beam is 10^{-3} rad. What is the difference in wavelength of the two components?

18.16 What is the angular spread of the visible spectrum in the first-order diffraction pattern from a grating of spacing 2.5×10^{-4} cm illuminated normally with white light. If the spectrum is focused on to a photographic plate by a lens of focal length 25 cm, what is the length occupied by the spectrum on the plate?

CHAPTER 19

THE OPTICAL MICROSCOPE

19.1 INTRODUCTION

In Chapter 18 the discussion of resolving power made it possible to understand the limitations placed on the use of a telescope for viewing self-luminous objects. In a microscope the objects investigated are viewed by light from a separate source and they are seen because they scatter or diffract the light which falls on them. It is much more difficult in consequence to obtain an expression for the limit of resolution of a microscope, and it will not be attempted in this book. The mathematical sophistication required would be beyond most readers. It is, however, possible by a quite simple discussion to see on what factors the limit of resolution will depend and to show that under certain conditions a microscope, wrongly used, can give insufficient or even wrong detail in the image.

Although it is unlikely that one would, in practice, attempt to view a diffraction grating through a microscope, the diffraction pattern obtained from a grating is so simple that it makes discussion of image formation easier. We will therefore deal with this situation in the following section.

19.2 IMAGE FORMATION BY DIFFRACTION

Figure 19.1 shows a diffraction grating of spacing d placed just outside the first focal plane of the objective lens of a microscope and illuminated normally by a parallel beam of monochromatic light of wavelength λ. The lens collects the light diffracted by the grating and focuses it to form an image beyond its second focal plane. The only light scattered by the grating is in the specific directions given

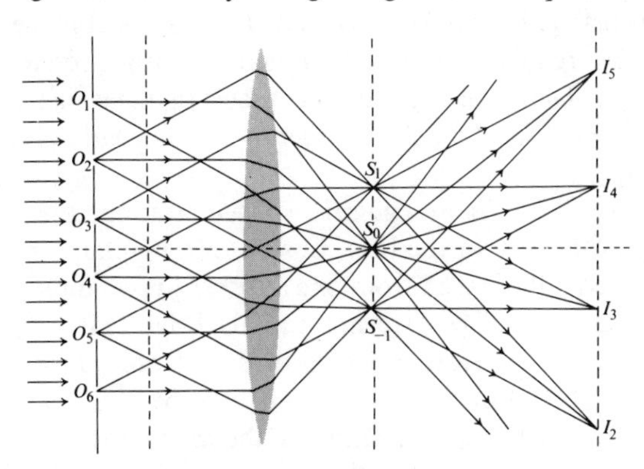

Fig. 19.1. The formation of the image of a diffraction grating in the objective lens of a microscope.

by the formula $\sin \theta = m \times (\lambda/d)$. As m takes the values 0, 1, 2, . . . , −1, −2, −3, . . . , we speak of the zero-order, first-order, second-order, . . . , minus first-order, minus second-order, . . . , etc., diffraction maxima. The zero-order diffraction maximum consists of beams of light from the individual slits, the beams being parallel to the axis of the system. These are brought to a focus in the focal plane of the lens at S_0. Similarly, the parallel beams of light from the slits forming the first-order, second-order, etc., maxima are focused in the focal plane at S_1, S_2, etc. Thus a screen placed in the focal plane shows a number of fine lines which are the images produced by the various diffraction orders from the grating.

As Fig. 19.1 shows, the beams focusing at S_0, S_1, etc., diverge again to form the images I_1, I_2, I_3, etc., of the original object slits O_1, O_2, O_3, etc. Unless the lens is almost infinite in extent it will not collect the light from diffraction maxima of high order, which are diffracted at large angles to the incident beam. In the focal plane of a small lens not all diffraction orders are present and the image finally formed is therefore not a perfect image. It is the perfect image of an object which only produces the number of diffraction maxima which have focused in the focal plane.

In particular, if the lens is stopped down so that only the zero-order diffraction maximum can pass through, only the straight-through undeviated beam contributes to the image, and this is what is obtained in any case if no grating is in the way. Absolutely no detail at all would therefore be seen in the image plane. If the first-order maximum on either side is allowed through, detail is now seen, although in fact the detail is little related to the true object. One would in fact see a sinusoidal variation of intensity over the image. As more and more diffraction orders are permitted through the lens, the detail becomes more and more clear, until, when a reasonable number of orders are let through, the image obtained is very similar to the object. But one must beware of the danger that the image may not, if too few orders are getting through, bear much relation to the object.

This is strikingly demonstrated in the following experiment. An intense beam of parallel light is allowed to strike a lens and is brought to a point focus by the lens in the second focal plane. A wire gauze, which is essentially a two-dimensional grating, is inserted in the beam outside the first focal plane of the lens. Light is diffracted by the regularly spaced arrangement into discrete diffraction orders (cf. Section 18.6). These all individually focus in the same plane as the zero-order beam, the second focal plane of the lens. At some later plane the image of the gauze is in focus.

Since the wire gauze is a regularly spaced grating, the diffraction pattern obtained in the focal plane of the lens consists of a regular two-dimensional array of bright spots all equally spaced from one another in two directions at right angles and very close together. These are not equally bright. If the gauze has one set of wires horizontal and the other set vertical, the only diffraction spots of any intensity are those shown in Fig. 19.2. Spots will be seen quite clearly in a horizontal line and in a vertical line through the zero-order maximum, and more faintly in directions at 45° to these. Any other spots are hardly visible.

If a slit is inserted in the second focal plane of the lens, the plane in which the diffraction pattern is in focus, and left wide open, a good magnified image of the gauze can be obtained on a screen some distance beyond the slit. If the slit is placed vertical and narrowed down until only the light forming the vertical line of spots AB passes through, the image is found to consist of horizontal lines only. Half of the detail has been completely lost. This is because the line AB consists of spots which are portions of the diffraction lines that would be obtained from a one-dimensional grating of horizontal slits and stops. This is all the information that the diffraction spots along AB carry, and the image produced from them in the image plane can therefore contain no other detail.

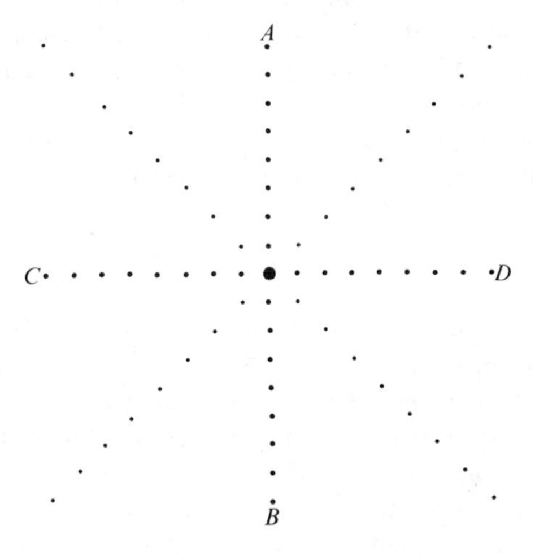

Fig. 19.2. The diffraction pattern seen in the focal plane of a lens when a parallel beam of light falls on a wire gauze.

Similarly, if the slit is rotated through 90°, so that only the spots on the horizontal line CD fall on it, the image contains only vertical lines. Once again, for the same reason as before, only half the detail is obtained in the image.

Much more striking is the image obtained when the slit is now rotated through 45°. Spots on one of the diagonal lines are striking it, and the image consists of a collection of parallel lines at right angles to the direction of the slit, i.e. also at 45° to both the horizontal and the vertical directions. The reason for this is as before. Diagonal slits and stops would produce diagonal lines as a diffraction pattern and portions of such a pattern are falling on the slit. The light from these spots therefore forms this spurious image which bears no relation to the original object.

If the slit is opened slightly, an incorrect but different image will be obtained, the correct image crossed by the diagonal lines seen before. In other words, a correct image has been obtained but with quite spurious detail added.

This experiment strikingly illustrates the dangers of limiting the light scattered from an object and the possibility of superimposing completely false detail on the image, or even producing an image which bears no relation to the object. One must always ensure in using a microscope that sufficient diffraction orders are admitted to produce a reasonably true image of the object and that there takes place no selection of the diffraction orders admitted which might produce spurious detail. This is why it is necessary to employ an objective lens of small focal length, to ensure that the angle subtended by the aperture of the instrument at the object is as nearly half the complete solid angle around the object as possible.

19.3 NUMERICAL APERTURE

It is obviously somewhat more complicated to work out the limit of resolution of a microscope than it was to do the same thing for a telescope focused on distant self-luminous objects. The crucial factor is how many diffraction orders take place in image formation, and it is a difficult task to relate this to the limit of resolution. We merely quote the result. Two point objects distance d apart examined in a microscope will be resolved and seen as separate point images if

$$d = \frac{\lambda}{2n \sin i}, \tag{19.1}$$

where λ is the wavelength of the light used, n is the refractive index of the medium between object and objective lens, and i is the semi-angle of the cone formed by joining the objects to the perimeter of the objective lens. This is not an exact expression, since it obviously depends on the method of illumination of the objects, but it is a good working criterion. The quantity $n \sin i$ is called the numerical aperture of the microscope and is even more important than the magnifying power of the instrument in predicting what the microscope can do.

The eye is an optical instrument which has an aperture of a few millimeters which is variable, depending on the intensity of the light falling on it. If we use for the eye the expression $\theta = 1.22\lambda/a$, which was derived for the telescope, it will be seen that objects at an angular separation of less than about 1 minute of arc will not be resolvable (see Section 16.6 for a more complete description of the resolving power of the eye). This corresponds roughly to objects separated by 0.1 mm at the least distance of distinct vision. The useful magnification of the microscope is thus that magnification which will bring an object separation of $d = \lambda/(2n \sin i)$ up to a separation of 0.1 mm in the final image. Assuming $\sin i$ to be 1 and n to be 1.5, which is achieved with an oil-immersion objective, d has a value of about 2×10^{-5} cm. The useful magnification is thus

$$\frac{10^{-2}\text{ cm}}{2 \times 10^{-5}\text{ cm}} = 500.$$

More than this merely gives a better view of the diffraction patterns. It does not help the resolution. Magnifications of 1000 or 1500 are often used in practice to minimize the strain on the eye in looking at very small image detail.

So long as light is used, detail below a value of about 2×10^{-5} cm cannot be seen. To achieve even this, the objective lens combination must have a focal point just outside the first glass surface so that the object can be as near the glass as possible and thus make sin i have a value as near 1 as possible. It must of course lie outside the focal point in order to obtain a real image for the eyepiece to view. It is normal practice to have on good microscopes a turret of three objectives which can be swung in turn to the bottom of the microscope tube. The object is first viewed at low magnification to survey a whole region, and then a higher-powered objective is switched in to study part of the image in greater detail. Similarly, the microscope has alternative eyepieces which can be dropped into the top of the tube to alter the magnifying power. But it is useless to use a combination which gives a much greater total magnification than 500. The eyepieces are all stamped with the magnification they produce, and the objectives normally with magnification and numerical aperture.

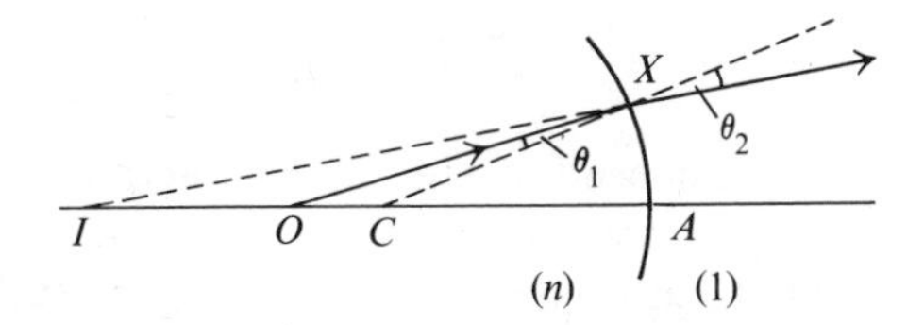

Fig. 19.3. Refraction at a spherical refracting surface.

In normal working with a microscope, the medium in which the object lies is air and the numerical aperture is therefore less than, or equal to, 1. To improve the resolution, one may use an oil-immersion objective in which the region between the object and the first lens is filled with oil of refractive index around 1.5. This reduces the limit of resolution by a factor of two-thirds. If the oil is of the same refractive index as the glass of the first lens, then the whole region from the object to the final surface of the lens is effectively continuous and no refraction of the rays from the object entering the instrument takes place until they hit the back surface of this lens. Further, if the refraction at the second surface of the lens is examined in detail, it will be seen that if the surface has a radius of curvature r and the object is at a distance $r + (r/n)$ from the pole of the surface, spherical aberration is no longer present. The situation is depicted in Fig. 19.3. Employing the sine law in the triangle XOC,

$$\frac{\sin XOC}{\sin \theta_1} = \frac{XC}{OC} = \frac{r}{r/n} = n.$$

But, by Snell's law,

$$\frac{\sin \theta_2}{\sin \theta_1} = n.$$

Therefore

$$\sin XOC = \sin \theta_2.$$

Also, angle $IXC = \theta_2$. Therefore in triangles OXC and IXC

$$\theta_1 + \theta_2 + XCO = 180° = \theta_2 + XIC + XCO.$$

Therefore

$$XIC = \theta_1.$$

Applying the sine law to the triangle IXC,

$$\frac{\sin\theta_1}{\sin\theta_2} = \frac{XC}{IC}.$$

But

$$\frac{\sin\theta_1}{\sin\theta_2} = \frac{1}{n}.$$

Therefore

$$IC = nr,$$
$$IA = r + nr.$$

This expression does not depend on the angle θ_1, and no approximations have been used. It is therefore true for all rays leaving O. They all appear to diverge from I. O and I are said to be aplanatic points of the curved surface, and the image at I suffers from no spherical aberration. It is also not difficult to show that the image has been magnified n^2 times. The oil-immersion objective was used to increase the numerical aperture and thus improve the resolution. We find that, in addition, the proper location of the object frees the resulting image from spherical aberration and produces a magnification of n^2 times.

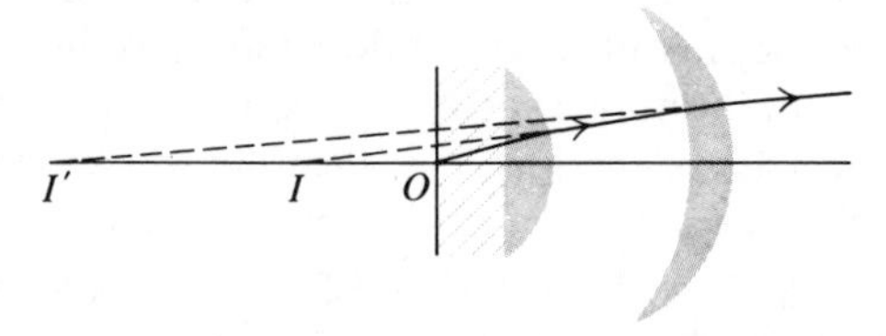

Fig. 19.4. Successive image formation in the first two components of an oil-immersion microscope objective.

This process can be continued. If one introduces, as in Fig. 19.4, a second lens, with its first face concave and its second face convex, the radii of curvature being correctly chosen, the same process is repeated. If I is located at the center of curvature of the first face of this second lens, the rays emerging from the first lens strike this surface at right angles and pass through undeviated. If, in addition, I is at the first aplanatic point of the second surface, all rays after refraction will appear to have come from the second aplanatic point I'. Once again spherical aberration is not present and the image is magnified a further n^2 times.

One cannot, as might seem possible, go on doing this indefinitely. Other defects, particularly chromatic aberration, are still present. It is therefore only usual in high-power microscope objectives to use the two stages shown and then employ one or two further, conventional, converging lenses in the system. The further lenses are so constructed as to minimize their own aberrations and also compensate for the aberrations present after passage through the first two lenses, although the compensation for chromatic aberration is sometimes left until the eyepiece stage. In the latter case the eyepiece is over-compensated for its own color defect in order to correct also for residual chromatic effects from the objective.

The eyepiece normally consists of two lenses separated by a small distance. The four radii of curvature and the distance of separation are so chosen that the correct focal length is given to the combination and defects of the image are minimized.

Example 1. What are the limit of resolution and the useful magnification of a microscope viewing objects in air with light of wavelength 600 nm if the angle i in Equation 19.1 is 80°?

Solution. The limit of resolution is d, where

$$d = \frac{\lambda}{2n \sin i} = \frac{6 \times 10^{-7}\ \text{m}}{2 \times 1 \times \sin 80°}$$

$$= \frac{6 \times 10^{-7}\ \text{m}}{2 \times 1 \times 0.985}$$

$$= 3.05 \times 10^{-7}\ \text{m}.$$

The microscope is usefully employed only in magnifying from the limit of resolution of the microscope up to the limit of resolution of the eye, which is 0.1 mm = 10^{-4} m at the least distance of distinct vision. Further magnification permits no more detail to be seen. Thus the useful magnification is

$$M = \frac{10^{-4}\ \text{m}}{3.05 \times 10^{-7}\ \text{m}} = 328.$$

Example 2. A paper weight is made of a sphere of glass of radius 5 cm and refractive index 1.50 with a small portion lopped off to provide a flat surface on which it will stand. A design is to be stuck to the flat bottom surface. What is the best height for the paper weight so that the center of the image seen when viewing vertically will be free from spherical aberration?

Solution. If the center of the image is to be free from spherical aberration, the center of the object must be placed at the first aplanatic point, i.e. at a distance from the top curved surface equal to

$$r + \frac{r}{n} = 5\ \text{cm}\left(1 + \frac{1}{1.5}\right)$$

$$= \frac{25\ \text{cm}}{3} = 8.33\ \text{cm}.$$

The flat surface is a plane at right angles to the vertical diameter 3.33 cm from the center.

PROBLEMS

19.1 A diffraction grating with 2500 lines per centimeter is placed 0.5 cm from the objective lens of a microscope which has a diameter of 5 cm and is illuminated normally with light of wavelength 5.5×10^{-7} m. How many diffraction orders enter the instrument?

19.2 The objective lens of a microscope has a focal length of 0.5 cm and an aperture diameter of 2.55 cm. If an object is viewed at a distance of 0.51 cm from the objective lens in air, what is the numerical aperture involved?

19.3 What is the numerical aperture of a microscope with an oil-immersion objective which has a limit of resolution of 2.5×10^{-7} m, when objects are viewed in light of wavelength 5.5×10^{-7} m?

19.4 What is the limit of resolution of point objects in air when viewed in a microscope which employs light of mean wavelength 5.5×10^{-7} m, if the angle i is (a) 10° and (b) 85°?

19.5 What would be the useful magnification of the microscope in Problem 19.4 in each of the cases quoted?

19.6 A microscope has an objective lens of focal length 0.3 cm and diameter 0.8 cm. Estimate the limit of resolution of the microscope when objects are viewed in air.

19.7 A small object is embedded in glass 20 cm from the pole of a spherical surface separating the glass and air. A virtual image free from spherical aberration is formed due to refraction in the surface 30 cm from its pole. Find the radius of curvature of the surface and the refractive index of the glass.

CHAPTER 20

SPECIALIST MICROSCOPY

20.1 INTRODUCTION

For many medical and biological uses the limit of resolution of a microscope is not the only important parameter. *Contrast* is even more important in many cases. As has been brought out in previous sections, the ability to see an object depends on the fact that, where a discontinuity occurs, light is diffracted. The purpose of most optical instruments is to collect the diffracted light and refocus it into an image. In much medical and biological work the object examined consists of cells in solution, and cells have a refractive index little different from the water surrounding them. Thus little scattering of light takes place at the cell boundaries and an image of the cell is difficult to produce.

The zero-order diffracted beam gives no detail at all, since this would be produced if no cell was present. It is the first-, second-, third-order, etc., beams which, joining with the zero-order beam, form a visible image in the final plane. If the amount of light sent into these orders is minute, any detail that would be produced is swamped by the light passing straight through and is extremely difficult to see. The same end-result will occur where the refractive index does not change abruptly but gradually varies over an appreciable distance. This is a situation also encountered in biological work.

The intensity difference between the image of an object and the image of its surroundings is called the contrast. It is clear that, in the situations described above, the contrast is poor and that interesting objects in the solution will be barely, if at all, visible. Increasing the resolving power of a microscope is useless if, owing to lack of contrast, the final image achieved is indistinguishable from its background.

If parts of an object will selectively take up a dye, the absorption of light by the dye will cause the intensity of the image to vary from point to point. This staining technique has been employed widely, since it is possible to use dyes which are selectively taken up by different cellular components and thus to render these components visible in the image. Unfortunately, most staining techniques kill the cells and it is never possible to be sure that all of the particular component takes up the dye.

It is better to employ techniques for making the cells visible which do not have the disadvantages inherent in staining. In the next few sections we shall describe specialist instruments which are in wide use for the purpose in biology and medicine as well as in many other branches of science.

20.2 THE POLARIZING MICROSCOPE

As was mentioned in Sections 17.6 and 17.8, many objects of biological interest contain asymmetrical regions or are made up of oriented molecular structures. These may well appear transparent and almost invisible when viewed in ordinary light, but if plane-polarized light is used to illuminate them, intensity differences can be produced between the cellular portions which are oriented and those which are randomly arranged. This may be due to the oriented material either being birefringent or rotating the plane of polarization.

In the polarizing microscope this fact is employed. The light used to illuminate the object is passed through a polarizer so that the vibrations are confined to one plane. In the microscope an analyzer, crossed with the polarizer,

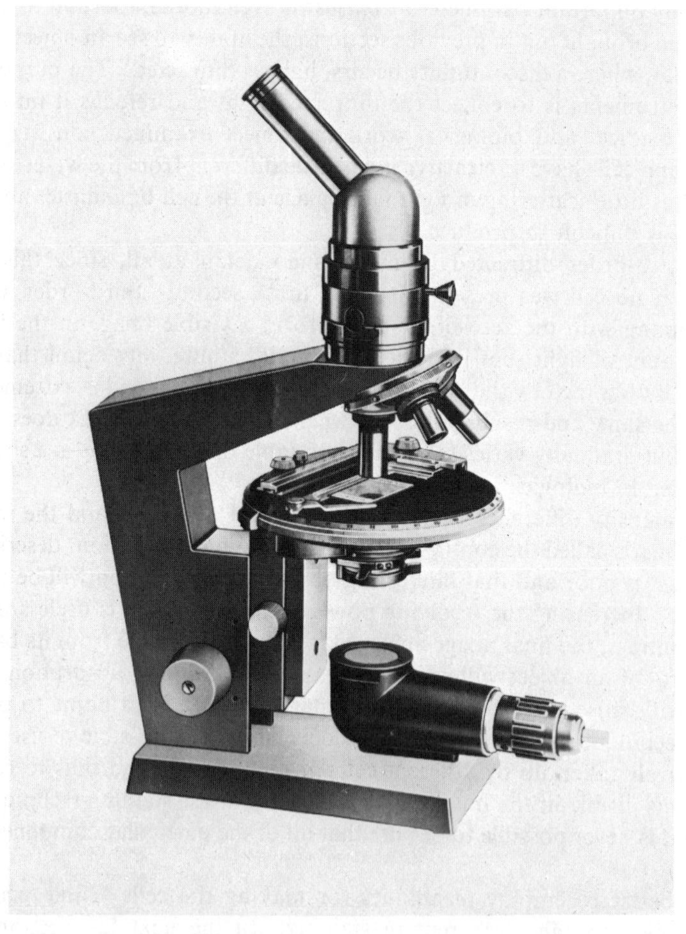

Fig. 20.1. The Zeiss Jena Laboval polarization microscope. (Reproduced by courtesy of Carl Zeiss of Jena.)

is inserted between objective and eyepiece. If no object is viewed or if the object has the same properties throughout, the field of view is quite dark. Oriented regions in an object will, however, appear bright, the intensity of any part depending on a number of factors.

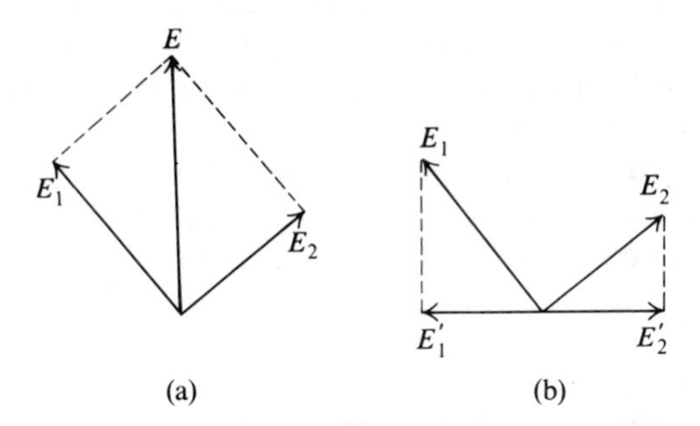

Fig. 20.2. Plane-polarized light from a polarizer is split up in (a) into two components by a birefringent material. Each of these waves, on arrival at the analyzer, has, as shown in (b), a component in the permitted vibration direction. These components, being in the same direction and out of phase, interfere.

What happens when the plane of polarization is rotated we have already dealt with in Section 17.6. Proteins, nucleic acids, and other important cell constituents are birefringent, and it is worth while considering what happens to light passing through a medium of this type. A light wave striking the birefringent material is split up into two components E_1 and E_2 (see Fig. 20.2a) vibrating in directions at right angles to each other. For reasons connected with its structure, the material has different refractive indices n_1 and n_2 with respect to these two light waves. If the thickness of the material is t, the optical path difference between the two waves in passing through the material is $(n_1 - n_2)t$. The phase difference between them is thus $\theta = (2\pi/\lambda)(n_1 - n_2)t$. Each wave on arriving at the analyzer will have in general a component along the permitted vibration direction (E_1' and E_2' in Fig. 20.2b). These components add in the image plane, the amplitude and phase of the resultant wave being dependent on the value of θ, which in turn is dependent on the thickness t. If $\theta = 0$ or 2π, E_1' and E_2' are vibrating in opposite directions and cancel as far as possible. For this situation the amplitude, and thus the intensity, of the resultant wave is a minimum. If $\theta = \pi$, by similar reasoning it will be seen that the amplitude is a maximum. For any other value of θ the amplitude will have an intermediate value.

The actual value of t may be deduced by inserting into the beam from the object a wedge of a known birefringent material such as quartz. By orientating it correctly and sliding in the wedge to a suitable thickness one may cancel the phase difference θ developed in the biological material by a phase difference $-\theta$ developed in the quartz, and thereby make the intensity of the image as small as possible From the known properties of quartz one may deduce t.

Even if a determination of t is not required, a quartz wedge is often used with a polarizing microscope. Many biological samples are very thin and θ is correspondingly small. The image of the biological material viewed may therefore not be well illuminated. The quartz wedge is used in this case to improve the contrast. It is so oriented in the microscope that any phase difference it introduces between E_1' and E_2' enhances that due to the biological material. From our earlier discussion it will be seen that this improves the amplitude, and thus the intensity, of the resultant light wave in the image plane.

Use of the polarizing microscope has confirmed many of the facts obtained by staining and the use of an ordinary microscope, and corrected others. Since the cells are not killed, as they are by staining, it has also been possible to view changes while they occur in living cells, such as the process of cell division.

20.3 THE ULTRAVIOLET MICROSCOPE

Since the limit of resolution of the microscope is given by the formula

$$d = \frac{\lambda}{2n \sin i}, \tag{19.1}$$

one method of decreasing d is to decrease λ. Using ultraviolet rather than visible light should therefore theoretically reduce the limit by a factor dependent on the wavelength used. In practice this does not happen. The final image must be caught on a photographic plate since the eye is insensitive to ultraviolet light; and since glass absorbs strongly in this wavelength region, special materials have to be used for the lenses. The first of these factors means that focusing is very much more difficult; and, unless it is perfect, resolution is lost. The second factor means that it is much more difficult and expensive to compensate for aberrations. Indeed mirrors are often used instead of lenses, when spherical aberration may be very difficult to eliminate. For these reasons the resolution of an ultraviolet microscope is seldom as good as that of an optical one.

One may well ask why such a microscope is used at all. The reason lies in the fact that cellular components, such as protein and nucleic acid, are strong absorbers of ultraviolet radiation. They therefore show strong image contrast in ultraviolet light without the need for staining. The roles of DNA and RNA in living cells before, during, and after cell division are examples of the mysteries which have been unraveled by the use of the ultraviolet microscope.

20.4 THE INTERFERENCE MICROSCOPE

If light from a point source is split up into two parallel beams of equal intensity and one of these beams is passed through a solution containing an opaque object, then on a screen placed beyond the object a dark patch will be seen of the same

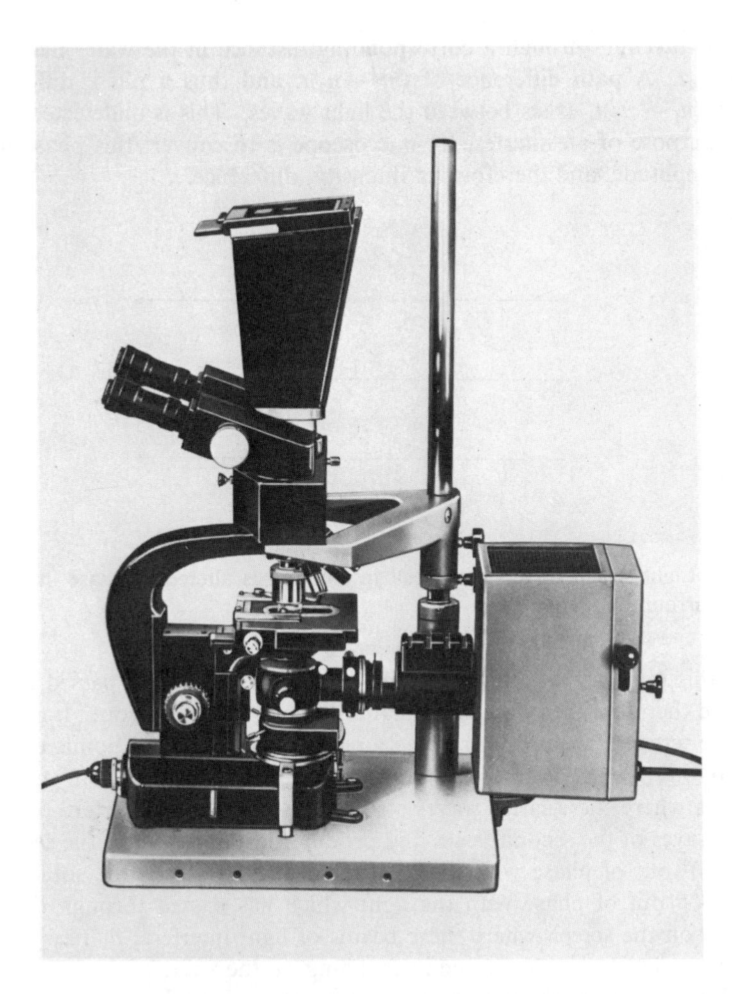

Fig. 20.3. The Labex 67 ultraviolet microcsope incorporating a beam changer which allows observations to be made in visible light, ultraviolet radiation, or a mixture of ultraviolet radiation and visible light. (Reproduced by courtesy of Carl Zeiss of Jena.)

size as the object. However, if the solution contains a cell, the screen is likely to appear equally illuminated all over. The light waves passing through the cell and those passing through the water have equal amplitudes, and therefore intensities, on the screen. The eye is only able to detect differences in intensity (no question of color being involved in this case), and the fact that the two regions on the screen are different goes undetected. For they are different: they differ in the phase of the light reaching them.

The cell material has a different refractive index n_1 from that of the surrounding water, n_2. Light traversing the cell of thickness t travels an optical distance $n_1 t$,

and light traveling through a corresponding distance in the water has an optical path of n_2t. A path difference of $(n_1 - n_2)t$, and thus a phase difference of $\theta = (2\pi/\lambda)(n_1 - n_2)t$, exists between the light waves. This is undetected by the eye but the purpose of an interference microscope is to convert this phase difference into an amplitude, and therefore an intensity, difference.

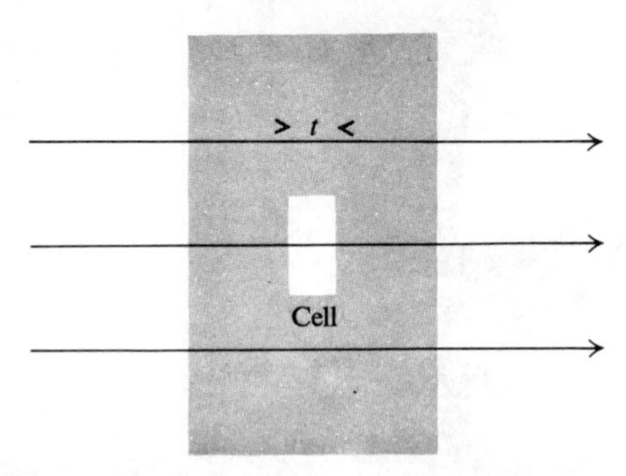

Fig. 20.4. Light which traverses the cell in solution is altered in phase in comparison with the surrounding light.

For this purpose the second parallel beam is used. It has passed through no object and therefore the phase is constant all along the wave-front. It is allowed to fall on the screen after having traveled a path which makes all points on its wave-front 180° out of phase with those light waves that have traversed only the solution. At regions where these latter waves strike the screen, they interfere destructively with the waves of the second beam and produce darkness. But if the second beam arrives 180° out of phase with the light which has traversed the solution, it will be $180° + \theta$ out of phase with the light which has passed through the cell. At the points on the screen where these beams of light interfere, there is therefore a resultant amplitude, the magnitude depending on the value of θ. For example, if θ were 180°, complete enhancement would result and maximum brightness would be obtained.

Cells do not generally have the same thickness at all points. The variation in thickness will therefore be shown on the screen as a variation in brightness of different regions of the image. The thickness of the cell at any point can be deduced by the use of a quartz wedge used in precisely the same way as was described in connection with the polarizing microscope.

It will be appreciated that a practical interference microscope is much more complicated than the foregoing analysis would suggest. The image of the cell has to be magnified by an objective and an eyepiece in the normal manner to make it large enough to view, at the same time as the interference technique is being applied. The system cannot therefore be as simple as just producing two parallel beams. Nonetheless, however complicated the actual mechanism, the treatment given above explains basically what is happening.

20.5 THE PHASE-CONTRAST MICROSCOPE

The phase-contrast microscope is a special type of interference microscope, the construction and adjustment of which are very much simpler than in the basic instrument. Here the interference is produced between the diffracted light and the undiffracted light rather than between two independent beams. Unfortunately, images in this instrument are often surrounded by halos, produced by diffraction at the phase plate. These can sometimes be troublesome.

The method of illumination of the object in a phase-contrast microscope is very important. The light passing through the object on the microscope stage is in the form of a hollow cone of parallel light whose apex coincides with the object.

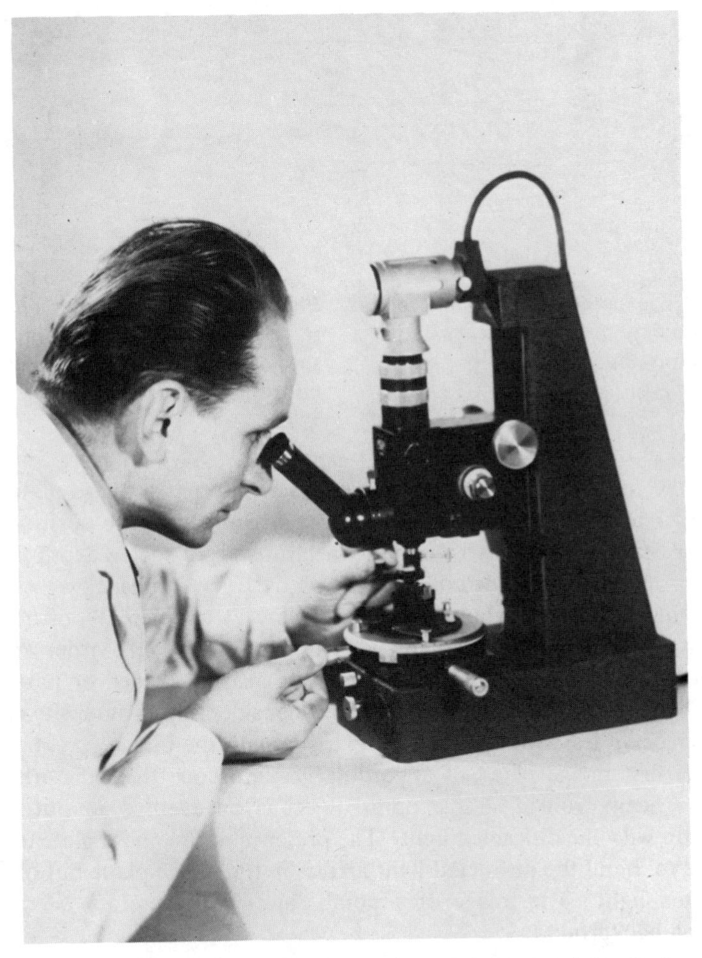

Fig. 20.5. The Zeiss Jena large interference microscope being used for the inspection of the surface of a gage block. The instrument uses multiple-beam interference techniques. (Reproduced by courtesy of Carl Zeiss of Jena.)

Assuming that there is no structure in the object, the light passes through the objective lens of the microscope and forms an annular zero-order diffraction image in the focal plane of that lens. If an object is present, the zero-order diffracted beam still forms an annular image in the focal plane as before (shown by the full-line rays in Fig. 20.6). A diffracted beam of any other order is also brought to a focus in the focal plane of the objective lens but at a different position. (One such beam is shown by the dotted-line rays in Fig. 20.6.) All the light, after being brought to a focus in the focal plane, then passes on to interact and form an enlarged image of the object at a later stage.

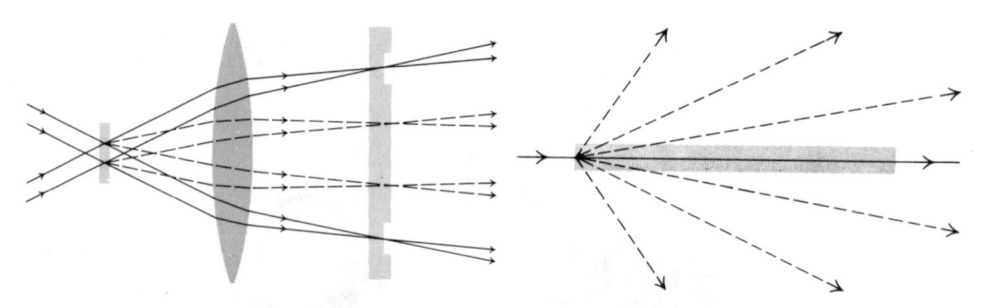

Fig. 20.6. The method of illumination and the positioning of the phase plate in a phase-contrast microscope.

Fig. 20.7. The zero-order diffracted light travels a greater distance in a particular specimen than does the light diffracted into other orders.

Into the focal plane of the objective lens is inserted a phase plate. This is a circular disk of glass with an annular groove cut into it at the position where the zero-order beam is brought to a focus. The groove is of such a depth that the light passing through it is advanced in optical path by $\frac{1}{4}\lambda$, or in phase by $\frac{1}{2}\pi$, in comparison with light passing through the phase plate at a different position. By this method a $\frac{1}{2}\pi$-phase difference is introduced between the zero-order and all other diffracted beams. The inside of the groove is also blackened or part-silvered so that the light passing through it is reduced considerably in intensity, making the zero-order beam less swamping in comparison with the others.

In the first image plane of an ordinary microscope the light arriving in the zero-order beam from a simple point object is on average $\frac{1}{2}\pi$ out of phase in comparison with the diffracted light. The presence of the phase plate increases this difference to π and the diffracted light arrives in the image plane out of phase with undiffracted light. The image of a point object will therefore be seen as dark against the background.

In addition, a complex object introduces phase changes between beams passing through it at different points and in different directions. To take a simple example, imagine light striking an object which is long and narrow like that shown in Fig.

20.7. At the boundary of the object, diffraction effects take place. The zero-order beam continues along the object, but the various diffracted beams are sent in different directions and pass through very little of the object. If the object and its surroundings have different refractive indices, a phase difference is introduced between the zero-order beam and the diffracted light; and when they meet in the image plane, they will not combine to produce an image as dark as that of a simple point object. If the object is in fact fairly complex, the resulting image will show variations of light and shade across it. These will be due to differing phase changes, introduced between the zero-order and diffracted beams at various parts of the object, having been converted to differing amplitude, and thus intensity, variations at the corresponding points of the image. Even quite transparent objects in solution will therefore show up in the image, and the variations in intensity in the image will correspond to variations in depth and internal structure in the object as well as to variations of surface detail. The phase-contrast microscope is now an instrument in routine use in medical and biological laboratories.

20.6 THE TRANSMISSION ELECTRON MICROSCOPE

All the specialist microscopes so far discussed have limits of resolution greater than that of the ordinary optical microscope, and this latter instrument can only resolve objects which are at least 2×10^{-7} m apart. Many organisms of interest, such as viruses, may be smaller than this and cellular material has structural detail below this limit. Indeed if anything of molecular dimensions is to be viewed, the limit of resolution must be reduced drastically. The only way to make this possible is to use a microscope which employs radiation of wavelengths many orders of magnitude smaller than are found in the optical region. The electron microscope fulfils these conditions.

It may seem odd that electrons can be used in a microscope; but we have already found out, and it will be discussed more fully in a later chapter, that electromagnetic radiation, when taking part in atomic processes, appears to act not like waves spreading out from the source as ripples spread over the surface of a pond but rather like discrete wave packets with particle-like properties. In 1926 de Broglie suggested that atomic particles might, by analogy, be expected to exhibit wave-like properties and show the phenomena of diffraction and interference under suitable conditions. This, as is described in Section 30.17, was confirmed experimentally.

If electrons are capable of being diffracted, they must have a wavelength associated with them. De Broglie inferred that the wavelength would be given by the relation

$$\lambda = h/p, \tag{20.1}$$

where p is the momentum of an individual electron (mv if the particle's speed is not near that of light, when relativistic effects are present); h is a very important atomic constant known as Planck's constant. Its value is 6.63×10^{-34} J s. If an electron, which has a mass of 9.11×10^{-31} kg, is traveling with a speed of 0.5

$\times\ 10^7$ m s^{-1}, which speed it would acquire in falling through a potential difference of 100 V, its associated wavelength is thus

$$\lambda = \frac{6.63 \times 10^{-34}\ \mathrm{J\ s}}{9.11 \times 10^{-31}\ \mathrm{kg} \times 0.5 \times 10^{-7}\ \mathrm{m}} = 1.3 \times 10^{-10}\ \mathrm{m} = 1.3\ \text{Å},$$

which is of the same order as for X-rays. After falling through a potential difference of 50 kV the operating potential of many electron microscopes, the wavelength would be 0.08 Å.

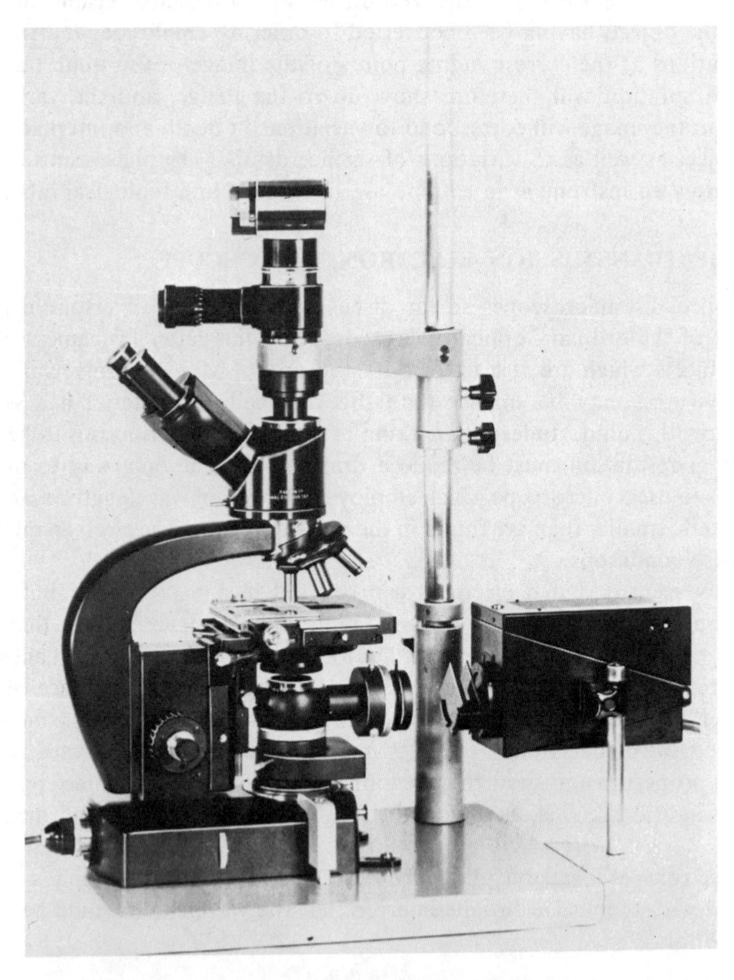

Fig. 20.8. The Zeiss Jena phase-contrast-fluorescence microscope. In biological and medical investigations, phase-contrast illumination is often used in conjunction with fluorescence to determine the morphology and chemical constitution of samples. The instrument permits this to be done simultaneously without changing microscopes. (Reproduced by courtesy of Carl Zeiss of Jena.)

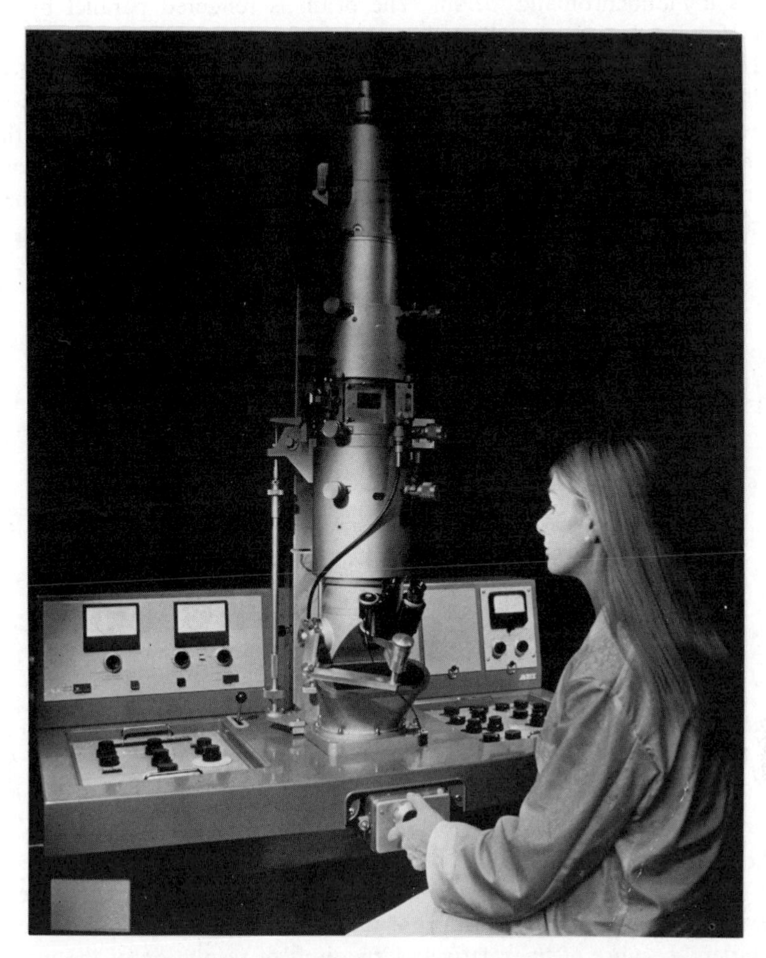

Fig. 20.9. The A.E.I. EM.802 electron microscope. (Reproduced by courtesy of GEC-AEI (Electronics) Ltd.)

It would therefore appear as if the limit of resolution of electron microscopes should be of the order of 0.08 Å. Unfortunately, it has proved impossible to design an electron microscope objective with a numerical aperture of better than 2×10^{-2}. This would theoretically put the limit of resolution at 2 Å. Aberrations are also present and, even with microscopes operating at higher voltage than 50 kV, it has not been possible to get beyond a limit of resolution of about 4 Å.

Being charged particles, electrons are affected by electric and by magnetic fields. Suitably shaped fields of either type act as the lenses in an electron microscope. The electrons are emitted from a heated filament (cf. Section 33.2) and accelerated through a high potential difference, producing a monoenergetic, and thus a monochromatic, beam. The beam is rendered parallel by suitably shaped electron lenses and allowed to strike the specimen. Electrons are diffracted, in the same way as light waves are in an optical microscope, and the diffracted beams are collected and brought to a focus by an objective lens system. The intermediate magnified image is further magnified by the eyepiece, in this case called a projector lens, because it projects a real image on to a photographic plate or fluorescent screen. Since electrons are quickly absorbed by air, the whole instrument must be completely evacuated.

Fig. 20.10. Parts of some bacterial flagella showing their stringlike structure. Magnification × 50,000 reduced to 72%. (From L. W. Labaw and V. M. Mosley, *Biochim. Biophys. Acta*, **15**, 325 (1954).)

The overall magnifying power of a good electron microscope may be around 10^4 or greater. A further factor of 10 can be obtained by enlarging the photographic image. Magnifications of more than this amount could be achieved but are pointless, because of the relatively large limit of resolution of the instrument. This is almost entirely due to aberrations in the lenses, particularly spherical aberration. Electrons must therefore stay close to the axis of the system and this limits the numerical aperture to the low figure already quoted.

The specimens used in an electron microscope must be very thin, to ensure that there is no appreciable absorption of the electrons passing through. If this happened, there would be differing speeds, and therefore wavelengths, in the electron beam passing into the microscope. The image would then suffer from chromatic aberration. The specimens are supported on collodion or Formvar films, of thickness around 150 Å, or even thinner films of carbon. These in turn are supported on meshes made of fine wire, having around 200 wires per inch. The areas between the wires are large in comparison with the field of view of the instrument.

Suspensions containing bacteria or viruses are dropped on to the supporting

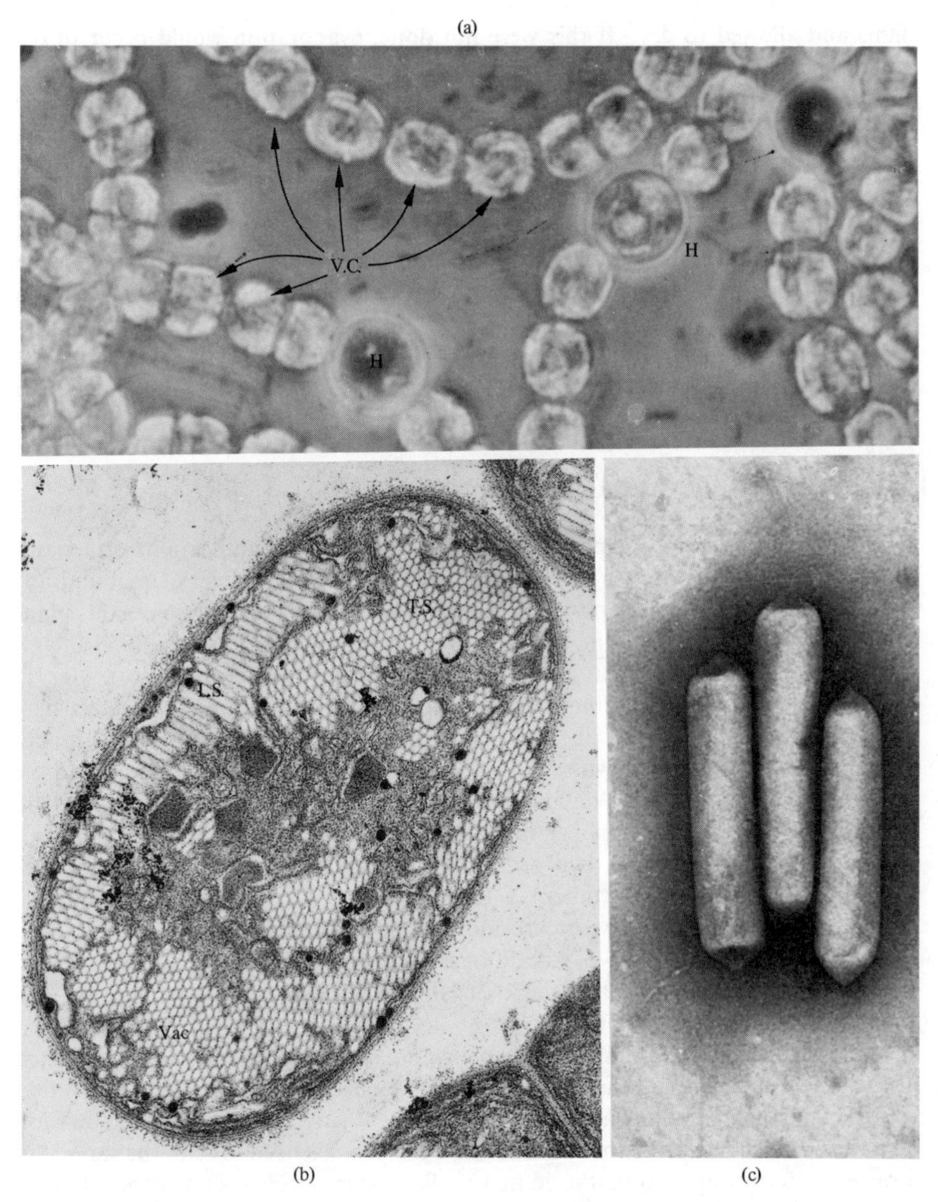

Fig. 20.11. Photographs of *Anabaena Circinalis*, a blue–green alga that forms thick growths on the surface of lakes. (a) A phase-contrast picture of several filaments ($\times 2{,}000$). The three large cells are heterocysts, the sites of nitrogen fixation, and the others are vegetative cells. (b) A transmission electron microscope picture of one of the vegetative cells ($\times 20{,}000$). It is packed with gas vacuoles, some seen end on and some side on, which provide the buoyancy necessary to keep the alga on the surface. If the vacuoles are ruptured, the alga sinks to the bottom. (c) A transmission electron microscope picture of three isolated gas vacuoles ($\times 106{,}000$). (Photographs by courtesy of W. D. P. Stewart and M. F. J. Daft.)

films and allowed to dry. If this were not done, evaporation would occur in the microscope and the vacuum would be affected. Unfortunately, normal drying may distort or flatten the specimens and freeze drying is often adopted. Tissue specimens are embedded in plastic and sectioned with a good microtome. Contrast can still be a problem here and the specimens are often coated with metallic atoms. These are sprayed on from the side in vacuum. Electrons are scattered by the metal atoms and the metal-covered parts show up as light against a darker background. The resultant electron micrograph looks as if the objects in the specimen were casting shadows from a light-source located where the source of metal atoms was placed.

The advent of electron microscopy has allowed detailed study of viruses, chloroplasts, cell membranes, subcellular structure, and many other items too small to be resolved in an optical microscope.

20.7 THE SCANNING ELECTRON MICROSCOPE

With either a light microscope at anything but very low magnifications or a transmission electron microscope, the depth of the field of view is very small and any picture is in sharp focus in only one plane. Any material above or below this plane can interact with light passing through in such a way as to blur the final image. In the transmission electron microscope in addition, differing absorptions in different parts of a thick specimen can produce varying speeds among the electrons, and the resulting chromatic aberration can be serious. Best results are thus obtained with thin samples or with flat samples viewed by reflected radiation, and the pictures obtained are very definitely two-dimensional. The scanning electron microscope gives a three-dimensional quality to its pictures, although its limit of resolution is around 100 Å, two orders of magnitude worse than for the corresponding transmission instrument. But, whereas in transmission the sample thickness has to be kept down to around 50 nm, a distance several hundred times smaller than the depth of a typical cell, in scanning whole organisms can be viewed at a time, and specimens such as the insect shown in Fig. 20.14 can be inserted into the instrument while still alive.

In a scanning electron microscope the whole of the specimen is not illuminated at the same time. Instead the electron beam is focused to needle sharpness so that it strikes an area of only about 10 nm radius at any instant. Several million electrons per second strike the specimen and the beam is used to scan the sample in a raster pattern similar to that used in TV. Secondary electrons are knocked from the specimen by the energetic beam and these are collected by a detector maintained at an electric potential of around 200 V. The signal received by the detector is used to modify the intensity of an electron beam which is tracing a raster on a TV tube in unison with the scanning raster. The resulting picture has therefore similar properties to the one obtained on a normal TV set. The magnification can be simply varied by limiting the extent of the scan. One can thus zoom in for a close-up of any part of a specimen, but in doing so the depth of focus remains

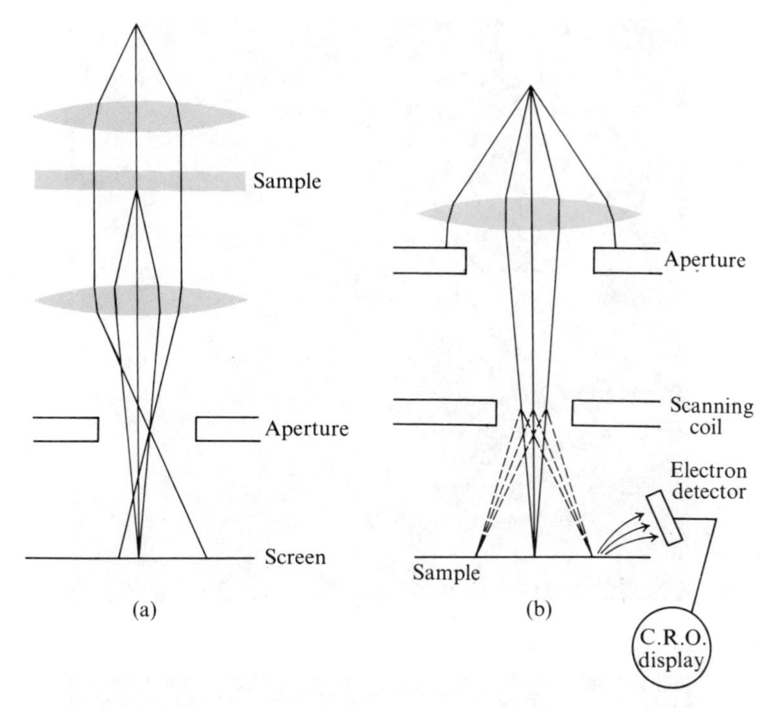

Fig. 20.12. Comparison of the schematic ray diagrams of (a) a transmission and (b) a scanning electron microscope.

unaltered, a distinct advantage when compared with the flatness of focus that results when zooming in for a close-up on TV.

The secondary electrons come from a very narrow surface layer of the specimen of the order of 5 nm thick. If the scanning beam strikes a sloping or sharply contoured surface, it produces more secondaries than if a similar surface were placed at right angles to the beam. This produces in the final picture variations in light and shade over a contoured surface which give it a three-dimensional effect, as is seen in Figs. 20.13 and 20.14. It also means that a flat surface can be tilted in the instrument to improve the intensity of the final picture.

In order that collection of the secondaries shall be done efficiently and in order that the efficiency shall be the same for all points of the specimen, the latter should be grounded. This means that the specimen should be joined electrically to the sample holder and should be conducting. If it is living, no problem arises, but any insulating sample must have a thin film of gold or other suitable metal evaporated over its surface.

The preparation of specimens is in general much easier in this instrument than in the transmission electron microscope; for many, indeed, no preparation is required. Figures 20.13 and 20.14 give some idea of the power and beauty of the results achieved.

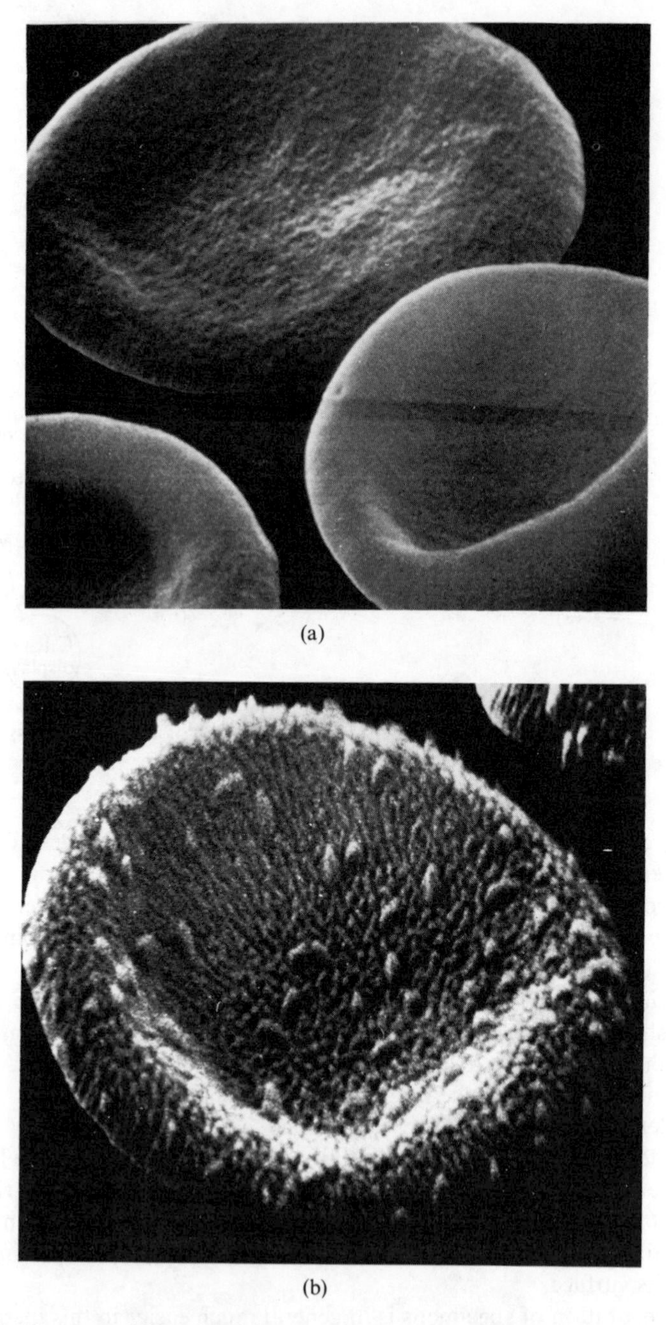

(a)

(b)

Fig. 20.13 (a) Human red blood cells as seen in the scanning electron microscope after ion-etching in hydrogen. Without ion-etching it would not be possible to distinguish between young cells, which are roughened by etching, and old cells, which remain smooth (× 10,000). (b) Human red blood cells with the hemoglobin abnormality Hb-Köln as seen in the scanning electron microscope after ion-etching in hydrogen. The hemoglobin aggregates known as "Heinz Bodies," are clearly visible (× 10,000). (Photographs by courtesy of The National Physical Laboratory, London. Crown copyright reserved.)

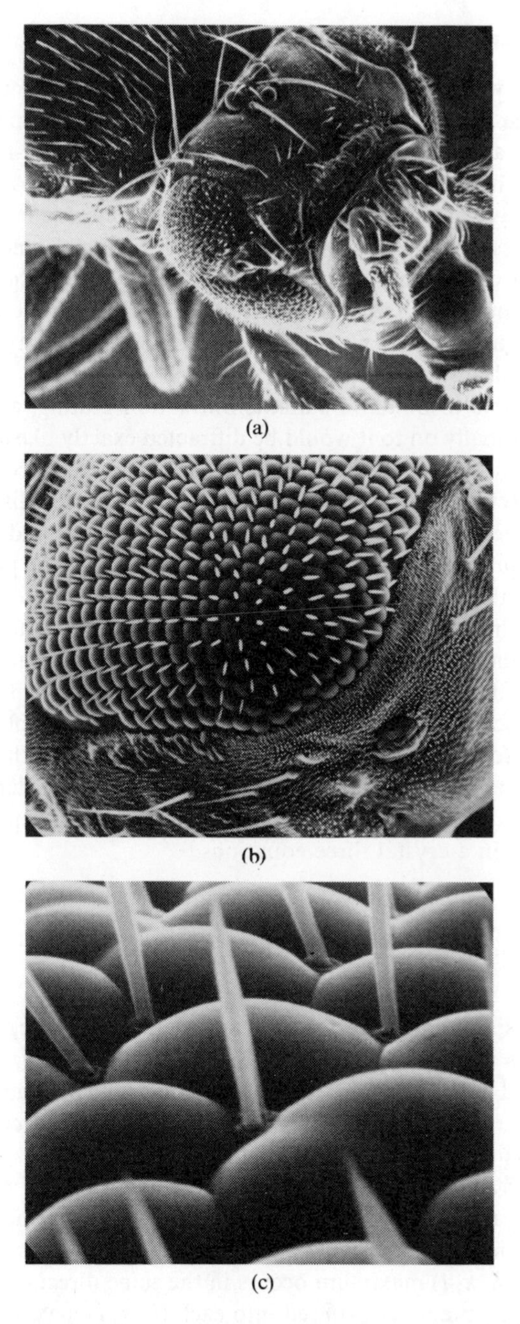

(a)

(b)

(c)

Fig. 20.14. The eye of a fruit fly at various magnifications as seen in the scanning electron microscope. The first micrograph shows the entire head, the others the eye alone in greater detail. The successive magnifications are $\times 180$, $\times 450$ and $\times 4500$. The fly can be viewed in the microscope while still alive so that no metal coating or other preparative technique is needed. (Reproduced by permission from H. Hartman and T. L. Hayes, *Journal of Heredity*, **62,** 41, 1971, and T. E. Everhart and T. L. Hayes, *Scientific American*, January, 1972.)

20.8 X-RAY CRYSTALLOGRAPHY

Since X-rays have wavelengths of roughly the same order of magnitude as those associated with the electrons in an electron microscope, it is natural to ask why X-ray microscopes are not used. It is an unfortunate fact that X-rays are negligibly refracted by any known substance, and it is not therefore possible to construct X-ray lenses. X-rays are certainly diffracted by objects (their wave nature was proved by the classical diffraction experiment of Friedrich and Knipping), but it is impossible to collect the diffracted beams and to bend them back together again to form an image of the scattering object. Although we cannot therefore construct an X-ray microscope, we can none the less employ X-ray diffraction in the investigation of molecular structure.

If it were possible to obtain a suitable diffraction grating, a parallel beam of X-rays directed normally on to it would be diffracted exactly like any other electromagnetic radiation. Thus X-rays would be detected only in directions obeying the law $d \sin \theta = m\lambda$ (cf. Section 18.4). To get reasonably spaced diffraction maxima, d would have to be of the same order of magnitude as λ, and thus the grating spacing would require to be only a few ångstroms. It is not possible to manufacture gratings with such small spacings, but nature provides rather more complicated gratings, three-dimensional ones, in the form of crystals. Here the molecules of the substance are spaced in all directions in a regular manner, effectively producing a series of one-dimensional diffraction gratings in the three directions of space. Radiation scattered from such an array is only detectable in certain quite definite directions (compare the two-dimensional case dealt with in Section 18.6).

In the one-dimensional case the positions of the maxima depend on d in the form given by Eq. (18.9). In the two-dimensional case they depend on d_1 and d_2 in similar fashion. In a crystal three equations,

$$\left.\begin{aligned} d_1 \sin \theta_1 &= h\lambda, \\ d_2 \sin \theta_2 &= k\lambda, \\ d_3 \sin \theta_3 &= l\lambda, \end{aligned}\right\} \qquad (20.2)$$

must be satisfied simultaneously: d_1, d_2, and d_3 being the repeat distances, or grating spacings, in the x-, y-, and z-directions in the crystal; θ_1, θ_2, and θ_3 being the angles measured in the y, z-, z, x-, and x, y-planes; and h, k and l being integers. From measuring the directions in which diffraction maxima occur, one may thus determine the spacing distances in the crystal.

Since each diffraction maximum obeys three equations of the form $d \sin \theta = (h, k \text{ or } l)\lambda$, we can call a particular maximum the (h, k, l) one. If a number of crystals have the same repeat distances, no matter how different are the molecules they contain, each (h, k, l) maximum occurs in the same direction for each crystal. But the amplitude or intensity scattered into each (h, k, l) maximum is completely dependent on the underlying structure detail. Thus different substances produce different intensities in fixed diffraction directions if they have the same repeat spacings when in a crystalline form. If in any of these cases the diffracted beams could be focused by a lens, an image of the crystal would be obtained when the beams combined, in exactly the same manner as in an optical microscope.

Since there are no lenses to focus the scattered radiation, one must measure the intensities of all the diffracted beams and attempt to combine them mathematically to form the image. Unfortunately, the intensity information is not enough. We require to know the relative phases of the beams either when they leave the object or when they combine to form the image. There is no way at present of recording this information, and stratagems have to be adopted either to try to deduce these or to work out the structure without them.

If one forms a small cuboid in the crystal made up of one repeat distance in each of the three principal directions, the resulting volume is called the unit cell. This unit cell is almost endlessly repeated throughout the crystal in all directions, and it is the purpose of X-ray crystallography to determine its exact contents. If the unit cell contains only a small number of atoms, it is normally a fairly easy matter to find their positions in the volume and their scattering power from a deduction based on the pattern of intensities obtained in the diffraction directions. Since there are a large number of diffraction maxima, one has a large amount of information and has to determine only a small number of unknown parameters, the x-, y-, and z-coordinates of all the atoms. The deduction is not too difficult. In theory, one places the atoms at all possible combinations of the positions they may occupy, works out the distribution of scattered intensity that would result in each case, and finds by inspection which result fits the experimental data. Since in practice this would be a tedious and very lengthy task, all available physical and chemical information is first carefully studied. This normally limits the possible locations to a more manageable number. In addition, there are available many mathematical techniques which allow an approximation to the correct positions to be reached fairly rapidly. As the number of atoms in the unit cell becomes larger, the problem becomes more and more complicated and soon impossible without additional aid.

The scattering ability of an atom for X-rays depends on its electrons, and atoms of high atomic number with large numbers of electrons thus scatter very much more than light atoms. If a heavy atom, such as iodine or bromine, can therefore be combined with a structure whose atomic composition is to be determined, the relative phases and the intensities of the diffracted beams depend very largely on the relative phases and intensities of the radiation scattered by the heavy atoms. One may therefore initially ignore all atoms but the heavy ones and, using the methods detailed in the last paragraph, deduce their approximate positions. Having done so, one can calculate the intensities and relative phases of the diffraction maxima that would be obtained from a crystal that contained only the heavy atoms. Using the experimentally obtained intensities and the relative phases so calculated, one may do mathematically what a lens does automatically: produce the image obtained when all the diffracted waves are recombined in an image plane. Since the phases are not the correct ones, the image is rather blurred. The heavy atoms show up clearly but the others are rather shapeless and indistinct.

None the less, one obtains their approximate locations and can recalculate the intensities and relative phases of the diffraction maxima with the scattering from these atoms included. The relative phases so obtained are a better approximation to the correct values than those initially calculated. Using these and the experimentally determined intensities, one redraws the image. The atoms show up more clearly. The relative phases are recalculated, and so on. By a process of successive approximation one eventually obtains a reasonably accurate image. This method is known as the heavy atom technique.

With large molecules, such as those of biological interest, which may contain more than 1000 atoms, even this technique is of little use. Although the heavy atom scatters much more radiation than any of the other atoms, there are so many of the latter that the contribution of the heavy atoms is no longer decisive. What one may be able to do in this case, however, is to use the method of isomorphous replacement. One attempts to form two compounds identical in every respect and in crystal formation, except that in the two cases one heavy atom is replaced by another atom at some particular site in the molecule whose structure is to be determined. The amount of X-radiation diffracted in the diffraction directions differs in the two cases only in the contribution from the replaceable atom. By noting the difference in intensity diffracted, one may, by similar means to those described in the preceding paragraph, deduce the position of the replaceable atom. Further, from the difference caused by the replacement one may deduce what are the relative phases of a number of the diffracted beams. In this way one may hope to determine the approximate structure of the rest of the molecule and then to refine it by successive approximations.

This is a very condensed and superficial account of the process of solving structures from X-ray diffraction data. In practice it is a good deal more difficult than the foregoing account might suggest, and there are many more mathematical techniques which may be required before a structure is successfully solved. The final information is given in the form of electron density maps which locate the position of atoms. These do not appear as small point structures, because the electrons are not localized, and in addition thermal vibration of the atoms in the structure tends to smear the electron density out. It will be seen from Fig. 20.15, where the resolution is better than 1 Å, that X-ray crystallography is a powerful tool for examining fine detail, even although X-ray lenses are not available. Man can painstakingly produce X-ray images without the aid of an X-ray microscope.

The triumphs of X-ray crystallography have been so numerous that it is only possible in the space available to mention a few highlights. The determination of the structures of penicillin and vitamin B_{12}, when chemical methods had failed, allowed these compounds to be synthesized for widespread use in medicine. In the biological field after more than twenty years of painstaking work, the structure of the first protein, myoglobin, was elucidated a few years ago. The X-ray work on fibrous proteins, on purines and pyrimidines, and on DNA led to the double-helix model of Crick and Watson. At the present time many X-ray investigations are

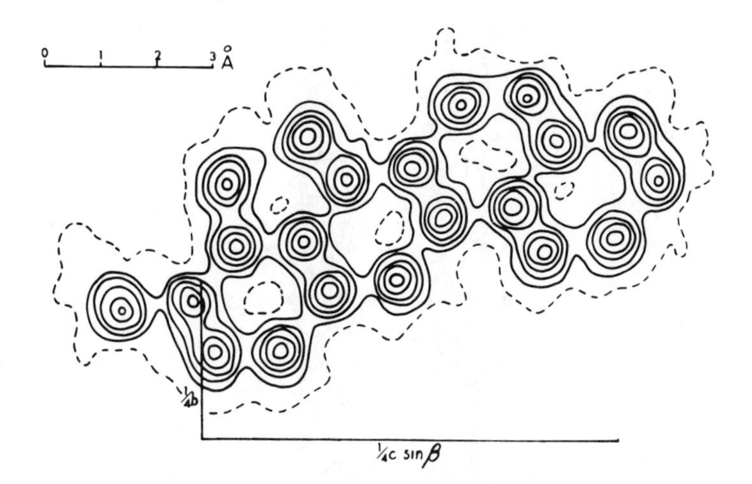

Fig. 20.15. Electron density map of 20-methylcholanthrene projected down the *a*-axis, as obtained by X-ray diffraction analysis. Contours are at intervals of one electron per Å^2. (From J. Iball and S. G. G. MacDonald, *Z. Krist.*, **114**, 439 (1960).)

proceeding on biological materials. In particular, the gross structures of RNA and of various types of viruses have been deduced and more detailed investigations are proceeding. It is possible to predict with confidence that a much greater insight into the detailed structure of cellular materials will be available in a very short time as a result of the large number of X-ray investigations in this field which are now taking place.

20.9 HOLOGRAPHY

In Section 20.4, in the description of the interference microscope, it was pointed out that information not only about the amplitude of light scattered from an object but also about its phase characteristics could be obtained by allowing the light to interfere with a reference beam. If this is done and the resulting interference pattern is photographed and turned into a transparency, the resulting two-dimensional picture is called a hologram. If the hologram has been constructed correctly, passing a beam of light through the hologram produces a reconstructed image of the original object.

Holography was first suggested in 1948 as a means of improving the then rather poor resolution of an electron microscope. This, as we have seen in Section 20.6, is almost entirely due to the spherical aberration of the electron lenses. A wide-angle beam of electrons allowed to enter the instrument produces a very poor image in the final plane. The suggestion was made that a hologram, and not a final image, should be obtained by introducing a reference beam into the system. It was hoped that the image could be reconstructed with visible light, the beam sent

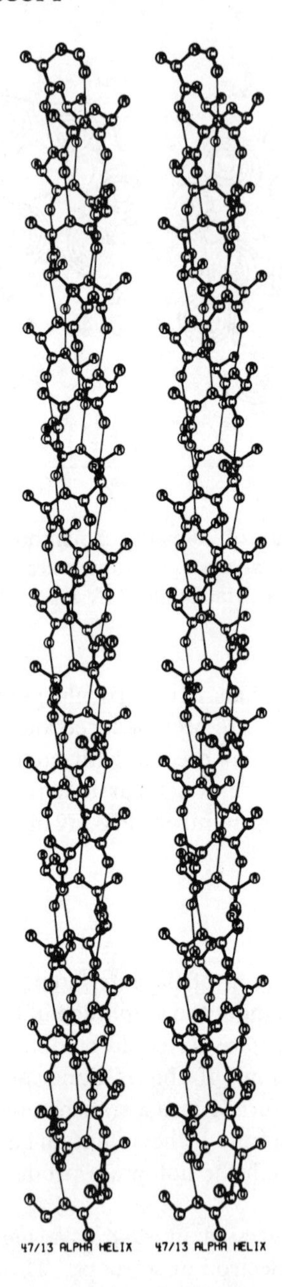

Fig. 20.16. A stereoscopic pair of perspective projections showing the α-helix which is present in poly-L-alinine. The data for computing these projections were obtained by X-ray analysis. (Reproduced from "OR TEP: A Fortran Thermal-Ellipsoid Plot Program for Crystal Structure Illustrations" by Carroll K. Johnson by permission of the author.) Stereo viewing is accomplished by placing a sheet of cardboard between the helices.

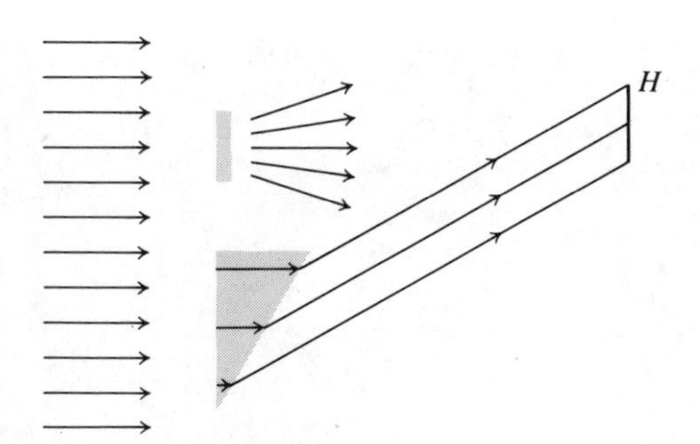

Fig. 20.17. One method of producing a hologram.

in faithfully reproducing the spherical aberration of the original electron lens and thus producing a true enlarged image of the original object.

The suggestion attracted considerable attention and optical holograms were produced and investigated, but the lack of sources capable of producing a beam coherent over a large wave-front and over reasonable periods of time appeared to place holography in the position of an interesting novelty without any serious practical application. The discovery of the laser (cf. Section 30.13) revived interest in the subject and an enormous amount of work is now being undertaken in this field.

In the production of a hologram an intense beam of coherent monochromatic radiation is allowed to pass through an object (or be reflected from it if it is not transparent) and fall on a photographic plate. At the same time a reference beam from the same source is taken past the object and allowed to fall on the plate. A very complicated interference pattern is produced on the emulsion: in general, the

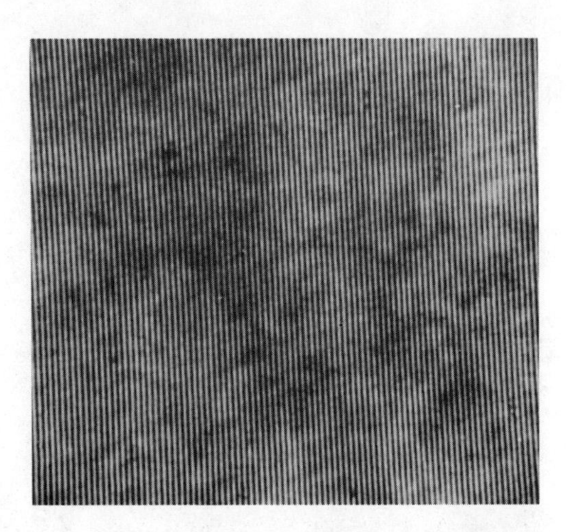

Fig. 20.18. A small portion of a hologram ×500. (From R. J. Collier, C. B. Burckhardt and L. H. Lin, *Optical Holography*, Academic Press.)

Fig. 20.19. Demonstration of the three-dimensional properties of a holographic image. The individual pictures are photographs of the same reconstructed image shot from different directions. (From R. J. Collier, C. B. Burckhardt and L. H. Lin, *Optical Holography*, Academic Press.)

more detail there is in the object, the more complex the resulting pattern. When the photographic plate has been developed and turned into a suitable transparency it is called a hologram. Figure 20.18 shows an enlarged picture of a small portion of a hologram.

If the original reference beam is directed at the hologram in the same direction as it originally struck the photographic plate, the beam is diffracted by the hologram in such a way as to recreate the light wavefront originally sent out from the object. An observer looking back through the hologram sees this wavefront diverging from a virtual image of the original object located at the position it formerly occupied. The image displays all the depth properties associated with real three-dimensional objects. If the reference beam is sent in from the other side of the hologram, a real image of the original object in the correct position is obtained. In practice, for technical reasons which need not concern us, both the real image and the virtual image will normally be seen when a hologram is illuminated from either side by a suitable reference beam.

It should be clear why a source coherent over a considerable period of time, as is a laser, must be used in holography. Since the path difference between light beams scattered from different portions of a large object to the hologram can be appreciable, there must be no discontinuous phase changes in the source during the times of arrival of such beams. The use of pulsed lasers which produce beams of high energy and of very short duration allow holograms to be obtained of rapidly moving objects.

When a transparency is viewed by placing it between a light-source and the eye, only that part of the picture directly between the light and the eye can be seen. If, however, a diffuse screen is placed between source and transparency, the whole picture can be seen. In the same manner, if a diffusing screen is placed between source and object when the hologram is made, the whole image can be seen in the reconstruction. Since the image is three-dimensional, viewing it from different angles will give different perspectives. It should be noted that even a portion of a hologram will reconstruct the full image, although with less resolution. Figure 20.19 shows reconstructed images of the same object taken from different angles.

The above is obviously a rather simplified account of a highly technical subject. With more complicated apparatus and with the use of a microscope holography has been used in a two-step process for producing enlarged images with a much greater depth of focus than is possible in normal microscopy. This has great application in biological work. Figure 20.20 shows an example of the large depths of focus possible.

Holograms taken at X-ray wavelengths have been reconstructed with visible light. This raises the possibility of getting around the difficulty created by the absence of X-ray lenses. Unfortunately, good holograms cannot be obtained because of the lack of space-coherent and time-coherent sources of X-rays, and until the equivalent of an X-ray laser is available, there appears to be little future in this direction. Much more interesting is the possibility of marrying ultrasonics and light through holography. If two coherent beams of ultrasound, one a reference beam, the other scattered from an object in water, are allowed to interfere at the water surface, a system of stationary, standing waves is produced. This has all the characteristics of a hologram; but the latter cannot be produced directly, because of the impossibility of getting a photographic record of ultrasound. None the less, the standing-wave pattern may be photographed with a small-aperture lens, and the photographic plate obtained will be the equivalent of a hologram. Light from a laser may be passed through it and a reconstructed image of the original object will be produced. At the moment the quality of the image is not very good.

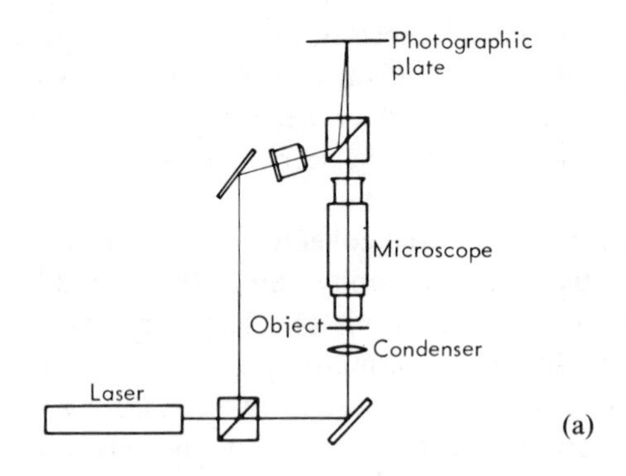

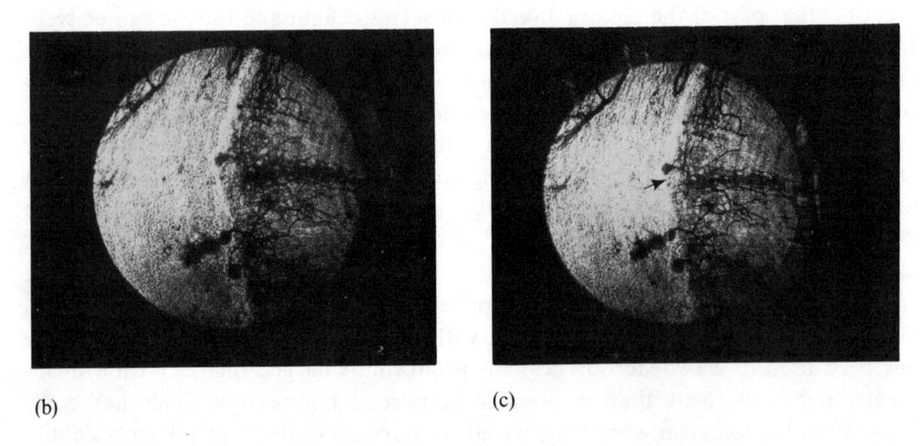

Fig. 20.20. Example of an optical holographic microscope: (a) schematic diagram of the experimental arrangement; (b) and (c) reconstructed images of a stained specimen of neurons. The reconstructions were made from the same hologram but in planes at different depths. The difference in depth corresponds to a defocusing of 40 μm in the object. The arrow in (c) indicates detail of approximately 1 μm. The hologram was made on Kodak 649-F spectroscopic plate with an exposure time of 20 s. A 4-mW He-Ne gas laser was used. (From R. F. Van Ligten and H. Osterberg, "Holographic Microscope," *Nature*, **211**, 282 (1966).)

Ultrasound (cf. Section 11.3) is now being used widely in medicine for many purposes. One of these applications involves the scattering of ultrasound by the internal organs. If the resultant pressure disturbances, due to the interference of scattered waves and a reference beam, could be transferred to an optical hologram, a reconstructed three-dimensional image of the inside of the body could be obtained. The value of this in diagnosis does not need to be stressed. Investigations on these lines are proceeding.

In many aspects of biological and medical work the limitation of recording three-dimensional information on two-dimensional film is well known. Pictures have to be taken from several angles and the total effect is painfully built up. For these reasons, and for many others, the possibility of 3-D hologram television is attracting great attention.

One way of doing this would be to transmit a hologram of the scene being studied every one-thirtieth of a second. At the receiving end a corresponding hologram would be produced by scanning a suitable photosensitive material which would then be illuminated to produce a three-dimensional image. The hologram would be erased from the material and the process repeated. Owing to persistence of vision the scene would appear continuous as films and TV do now. At the present time, unfortunately, it would take a good deal more than one-thirtieth of a second to transmit sufficient information to reproduce a hologram with enough detail to get good resolution in the image, nor do we possess a suitable material on which to record the received hologram. Research is proceeding into these problems.

Holograms have recently been produced which enable colored images to be reconstructed. These holograms are made from thick photographic emulsions, information being recorded in planes throughout the thickness. With the emulsion in position, a scene is illuminated first with the red light from a helium-neon laser and then with the light from an argon ion laser which emits green light and two wavelengths in the blue region. If white light is reflected from the resulting thick hologram, a colored three-dimensional image of the original object is obtained. The information for any color is recorded in the emulsion in parallel planes, the separation of which depends on the wavelength of the color. Emulsions tend to shrink when they are developed and this would result in shifts in wavelength in the reconstructed images so that colors would not be faithfully reproduced. Thick holograms are thus artificially swelled to compensate for the shrinkage of the emulsion. The ability to produce colored images increases the potential uses of holography even further.

Example 1. A biological sample contains birefringent material of thickness 5.00 μm. Light of wavelength 690 nm in vacuum splits into beams which traverse the material with velocities of 2.257×10^8 m s^{-1} and 2.262×10^8 m s^{-1}. What phase difference is introduced between the two beams in passage through the birefringent material?

Solution. The refractive indices of the birefringent material for the two beams traversing them are $n_1 = c/v_1$ and $n_2 = c/v_2$, where v_1 and v_2 are the respective velocities. The thickness D of the material represents D/λ_1 and D/λ_2 wavelengths respectively for the two beams and the phase difference introduced is thus

$$2\pi\left(\frac{D}{\lambda_1} - \frac{D}{\lambda_2}\right) = 2\pi D\left(\frac{n_1}{\lambda_0} - \frac{n_2}{\lambda_0}\right)$$

$$= \frac{2\pi D}{\lambda_0}(n_1 - n_2) = \frac{2\pi Dc}{\lambda_0}\left(\frac{1}{v_1} - \frac{1}{v_2}\right)$$

$$= \frac{2\pi \times 5.00 \times 10^{-6}\ \text{m} \times 3.00 \times 10^{8}\ \text{m s}^{-1}}{690 \times 10^{-9}\ \text{m}}$$

$$\times\left(\frac{1}{2.257 \times 10^{8}\ \text{m s}^{-1}} - \frac{1}{2.262 \times 10^{8}\ \text{m s}^{-1}}\right)$$

$$= \frac{2\pi \times 5.00 \times 10^{-6}\ \text{m} \times 3.00 \times 10^{8}\ \text{m s}^{-1}}{690 \times 10^{-9}\ \text{m} \times 10^{8}\ \text{m s}^{-1}}\left(\frac{2.262 - 2.257}{2.257 \times 2.262}\right)$$

$$= 0.134\ \text{radian.}$$

Example 2. Electrons which have been accelerated to a speed of 1.87×10^{7} m s^{-1} are used in a transmission electron microscope. What numerical aperture must the microscope have if the limit of resolution is 5.00 nm?

Solution. At a speed of 1.87×10^{7} m s^{-1}, relativistic effects are negligible to the accuracy with which we are working. Thus the associated wavelength is

$$\lambda = \frac{h}{mv} = \frac{6.63 \times 10^{-34}\ \text{J s}}{9.11 \times 10^{-31}\ \text{kg} \times 1.87 \times 10^{7}\ \text{m s}^{-1}}$$

$$= 0.0389\ \text{nm.}$$

If the numerical aperture is N and the limit of resolution d, then

$$d = \lambda/2N.$$

Therefore

$$N = \lambda/2d = \frac{0.0389\ \text{nm}}{2 \times 5.00\ \text{nm}} = 3.89 \times 10^{-3}.$$

PROBLEMS

20.1 Biological material of thickness 0.01 mm is birefringent, the two waves having refractive indices of 1.328 and 1.333 for light of wavelength 5.4×10^{-7} m. What phase difference is introduced between the two waves in passage through the material?

20.2 A quartz wedge varies in thickness from 3 mm to 3.1 mm. When light of wavelength 5.4×10^{-7} m passes through it, it splits up into two beams with refractive indices of 1.544 and 1.553. Determine the change in phase difference between the beams when the wedge is moved through the beams.

20.3 For what thicknesses of the quartz wedge of Problem 20.2 would the phase difference calculated in Problem 20.1 be canceled out if the wedge is inserted in the beams after the biological material with the phase difference in the opposite sense?

20.4 Two identical beams of light of wavelength 5.5×10^{-7} m pass through a solution. One traverses a cell of thickness 0.1 mm of refractive index 1.332 while the other is traversing an equal distance in water of refractive index 1.330. Determine the phase difference introduced between the two beams.

20.5 What are the wavelengths associated with (a) a rifle bullet of mass 5.0 g moving with a speed of 500 m s^{-1} and (b) a hydrogen molecule of mass 3.34×10^{-27} kg moving with a speed of 2.5×10^{5} cm s^{-1}?

20.6 What is the wavelength associated with an electron which has a velocity of 3.00×10^{6} m s^{-1}? If these electrons are employed in an electron microscope with a numerical aperture of 5×10^{-3}, what is the theoretical limit of resolution of the instrument?

20.7 A parallel beam of X-rays of wavelength 1.5 Å falls on a crystal. If the rows of atoms parallel to the beam are (a) 30 Å, (b) 5 Å, and (c) 2.5 Å apart, at what minimum angle are the X-rays diffracted by these rows in each case?

CHAPTER 21

SPECTROPHOTOMETRY

21.1 INTRODUCTION

Dispersion is the splitting up of light into its constituent colors. In Section 15.5 it was mentioned that this effect, undesirable in telescopes and microscopes and limiting the resolution obtainable, was of great importance in biological and medical work. The pattern seen when light is split up into its individual colors is called a spectrum. The spectra emitted by the atoms of different elements are quite distinct and extend into the ultraviolet region. The presence of any element in a sample can be detected very simply by examining the spectrum emitted by the sample and identifying the known spectra of the elements of which it is composed. The same technique can be applied to compounds. Identification is possible in circumstances where the concentration is too small for chemical analysis or where the latter is for some other reason impossible.

Much more important in biological work than emission spectra are absorption spectra. If white light, or bands of ultraviolet or infrared radiation, are passed through substances, normally in solution, these absorb radiation just as characteristically as they emit it. Samples where the conditions necessary to produce emission spectra would alter the constitution of the constituents, or in a form in which the production of emission spectra is not feasible, can be identified with ease from the absorption spectra produced. In particular, certain chemical groupings which occur frequently in biology have highly characteristic absorption spectra. In biological systems there are often present similar molecules which are difficult, if not impossible, to separate by chemical means. One such group is the heme proteins; but each member of the group may be identified by its absorption spectrum, since each one differs from all the others.

The use of spectra for identification purposes is now routine. In addition, many spectral instruments are employed to determine the concentration of important constituents in medical and biological samples. The extent of the absorption at some characteristic wavelength permits the calculation of the amount of a constituent present. It is therefore important to discuss how spectra are produced and how the information which they yield can most simply and accurately be obtained.

It is necessary to define a few terms current in this field. A spectroscope is an instrument in which the spectrum produced is examined by eye. If, in addition, the wavelengths at which various features appear can be measured, the instrument is called a spectrometer. An instrument which can measure and compare the intensities emitted from, or passing through, two samples is called a photometer. A spectrometer combined with a photometer is called a spectrophotometer.

21.2 PRODUCTION OF A PURE SPECTRUM

Optical spectra are most easily produced by use of a prism, which is a cylinder, commonly of glass, of triangular cross-section. Light from the source to be investigated is focused on to a slit which is placed at the focal point of a lens. The light from the slit emerges from the lens in a beam of rays parallel to its axis and is allowed to strike the prism. After refraction at the two surfaces of the prism, the light has been split up into its individual colors. All the blue light of a particular wavelength emerges in a parallel beam which is not traveling in the same direction as the parallel beam of light of any other particular wavelength. A parallel beam of light (cf. Section 14.5) converges to a point in the focal plane of a lens. If a lens is interposed in the beams from the prism, all rays of one color form an image in the focal plane at a different place from the image formed by the rays of any other color. In the focal plane of this lens a pure spectrum of the light from the source has therefore been produced.

Had the two lenses not been employed to render parallel the light striking the prism and properly converge the outgoing light, the emerging rays of light of any particular color would not have formed a parallel beam and the resulting image for any wavelength would have spread over a finite area. The overlap of the images for different wavelengths would thus have produced a blurred or impure spectrum.

A special instrument is in common use for making the procedure mentioned quite routine. It is a form of spectrometer. The prism is mounted on a rotatable table which carries a circular scale. Attached to this is a fixed arm, called a collimator, whose purpose is to render parallel the incoming light. It consists of a hollow metal tube, adjustable in length, with a variable slit at one end and a corrected lens at the other. In use the length of the tube is adjusted to be the focal length of the lens.

Also attached to the table, but rotatable about a vertical line through its center, is a telescope. It carries a vernier scale which moves over the scale attached to the table so that the angular position of the telescope at any time may be noted accurately. The position of the telescope is related to the wavelength of the light located in the center of the field of view. The purpose of the telescope is to collect the light emerging from the prism. The objective lens forms an image of the spectrum which is viewed by the eyepiece. In more complex instruments the telescope is replaced by an objective lens and photocell.

The basic arrangement is shown in Fig. 21.1. It should be noted that the red light has a greater geometrical path to travel from A to C than from B to D, but has an identical optical path in both cases. Thus the plane wave-front AB emerges as a plane wave-front after passing through the prism, all waves along CD having the same phase at all times. The same applies to the light of any other color.

21.3 GRATING SPECTRA

As was mentioned in Section 18.4, a diffraction grating will produce a spectrum of the light falling on it in each of the orders visible. The spectrum from a grating is reversed with respect to that from a prism, since in the former red light is deviated more than blue; but in the latter, less. A good grating will spread a spectrum over a much wider angle than will a prism, allowing better resolution of the color components. The angular spread is greater in a higher diffraction order than in a lower one but, unfortunately, the second and all higher-order spectra overlap if the full range of optical colors is present in the incoming light.

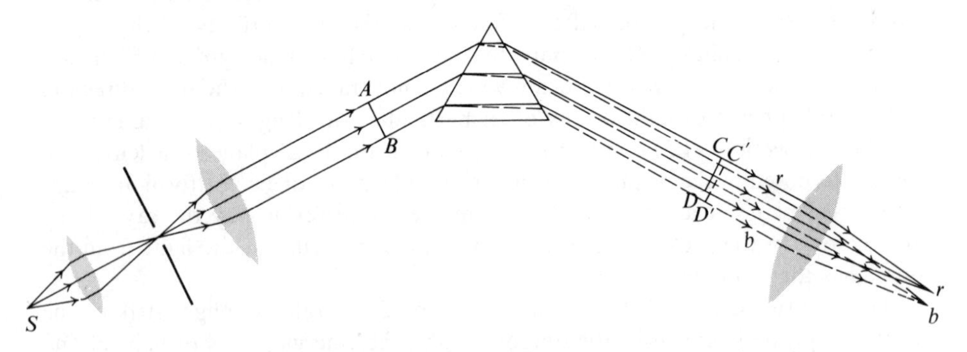

Fig. 21.1. The production of a pure spectrum by means of a prism.

In use, a grating is mounted on a spectrometer table in place of the prism, the incoming light being rendered parallel and the resultant spectra being viewed in the telescope.

21.4 ULTRAVIOLET AND INFRARED SPECTRA

There is little difficulty in producing spectra in the visible region. An incandescent lamp which produces white light, or a gas-discharge tube which emits only one or more discrete wavelengths, can be used as sources of light, and the lenses, grating, and prism may be made of glass.

In the ultraviolet region glass cannot be used, since it absorbs radiation in this range strongly; all prisms, transmission gratings, and lenses must be made of quartz. Gas-discharge tubes emit ultraviolet light but this is normally absorbed by the glass envelope. For use as ultraviolet sources such tubes must be made of quartz and filters must be incorporated to remove the visible light. Below 2000 Å ultraviolet spectra should be examined in vacuum.

In the study of infrared spectra similar problems arise. Prisms must be made of rock salt, but it is almost impossible to make lenses transparent to infrared radiation and mirrors are normally employed for converging, or rendering parallel, infrared beams. Hot, glowing objects are used as sources, the visible radiation being filtered off. Nernst glowers of rare earth oxides and Globars of carborundum are often used as infrared sources.

It should be clear that, since in spectrophotometers light intensities are to be compared, the intensity of a source must not vary during an experiment. Practically all sources are run from electrical power supplies and the intensity of the radiation emitted is proportional to a high power of the applied voltage. The latter must therefore be strictly regulated and, if necessary, feedback techniques (cf. Section 33.11) must be employed to do this. In technical language, the lamp source must be powered by a strictly regulated power supply.

21.5 THE ABSORPTION OF LIGHT

If monochromatic radiation of intensity I is allowed to fall on an absorbing medium of small thickness Δx, it is found that the loss of intensity in passing through the medium is proportional to I and to Δx. Thus

$$-\Delta I = \mu I \Delta x, \tag{21.1}$$

where μ is called the absorption coefficient of the material. If Δx becomes smaller and smaller, in the limit the equation may be written as

$$\frac{dI}{dx} = -\mu I, \tag{21.2}$$

whence integration leads to

$$I = I_0 e^{-\mu x}. \tag{21.3}$$

If both absorption and scattering take place, μ may be written as $\mu = \mu_a + \mu_s$. For most substances, assuming that the absorbing thickness is small and that, if it is a solution, the concentration is not great, scattering can be ignored. Alternatively, the combined effect of scattering and absorption may be considered to be the relevant quantity to be measured. For low concentrations $\mu = \beta c$, where β is known as the extinction coefficient and c is the concentration. Concentrations may therefore be determined from measurements of μ.

The measured quantity in spectrophotometry could be μ, where

$$\mu = \frac{1}{x} \ln \frac{I_0}{I}. \tag{21.4}$$

In practice it is customary to measure the optical density D, defined as

$$D = \log \frac{I_0}{I}. \tag{21.5}$$

Since $\ln y = 2.303 \log y$,

$$D = \mu x / 2.303 \tag{21.6}$$

and

$$c = 2.303 D / \beta x. \tag{21.7}$$

Not only do μ, β, and D vary from absorber to absorber but they also vary for one absorber with wavelength. It is therefore necessary to pass through the absorber only one wavelength at a time. This is virtually achieved by placing a narrow slit in the focal plane of the second lens shown in Fig. 21.1, i.e. in the plane where the spectrum is first in focus. Only a small range of wavelengths passes on to the absorbing and measuring stages of the instrument. Rotation of the prism, or the grating in the case of the other arrangement described, brings different parts of the spectrum on to the slit. If the rotation of the table is coupled to a recording instrument, the intensity received as the wavelength of the radiation falling on the absorber changes can be automatically charted. Typical absorption spectra are shown in Fig. 21.2.

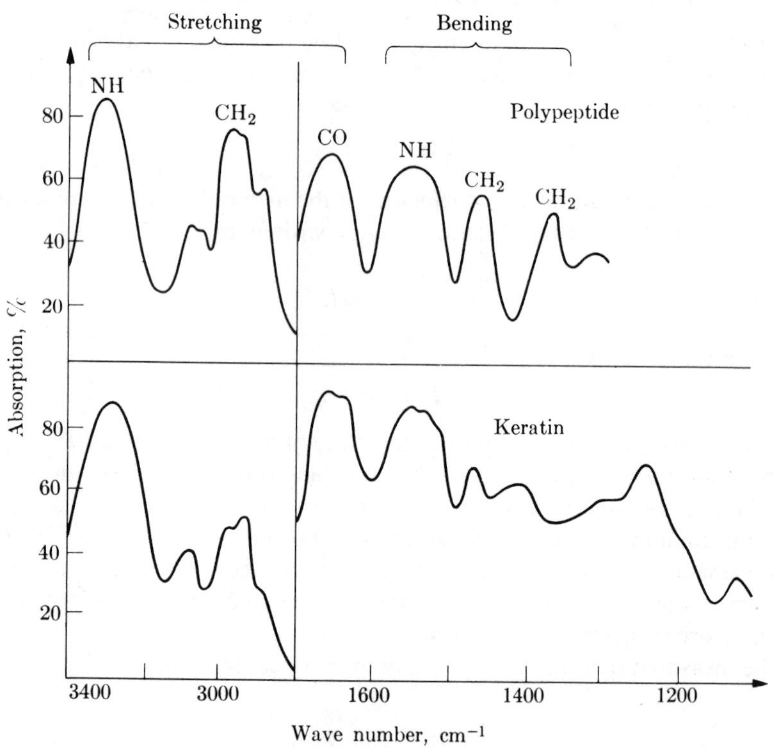

Fig. 21.2. Absorption spectra of polyphenylalanine-leucine and keratin, showing identification of the individual bands. (From S. E. Dalmon and G. B. B. M. Sutherland, *J. Am. Chem. Soc.*, **69**, 2074 (1947).)

21.6 THE ABSORBER

The absorber is almost invariably a solution which is contained in a cuvette. This is made of glass, or quartz in the ultraviolet region, and has parallel sides which are placed at right angles to the beam emerging from the slit mentioned in the preceding section. This beam is rendered parallel by a lens placed between slit and cuvette. The beam, after passage through the cuvette, strikes a photocell, photomultiplier, or some similar recording device, the signal from which is normally amplified electrically and applied to the pen of a chart recorder.

If it is desired to study the change in absorption at some significant wavelength during a rapid reaction, a cuvette is not suitable as the container, and flow systems are often employed. The two reactants are allowed to flow continuously into a tube, part of which intercepts the beam of the spectrophotometer. The two reactants mix just before reaching the beam and the rate of flow determines how far the reaction has proceeded before the light absorption takes place. Thus a recording of the optical density, made as the flow rate is decreased, effectively shows the increase of absorption with time during the reaction. The flow may even be stopped to permit the reaction to be completed in the path of the light beam.

21.7 THE SPLIT-BEAM SPECTROPHOTOMETER

As was mentioned in Section 21.4, control of the intensity of the source is vital in normal spectrophotometers. Since the solvent may absorb as well as the solute, it is normal to plot the absorption curves for pure solvent and for solution, and to interpret the difference as the absorption curve of the sample. If the intensity of the source varies between the two determinations, the results will be relatively meaningless and certainly inaccurate. For this reason, very accurate instruments often employ split beams so that the difference recording is made immediately.

In such instruments light-choppers are employed. A simple type is shown in Fig. 21.3. The monochromatic light falls on the light-chopper S, which consists of a mirror rotating about an axis OO'. The mirror consists of a complete circle from which two 90°-segments have been removed. As it rotates, it reflects the incoming light through A for a quarter-cycle, allows the light through to the fixed mirror M_1 for the next quarter cycle, and so on. For half of the time, light goes through cuvette A, which contains the sample, and for the rest of the time it goes through cuvette B, which contains only solvent. The light beams alternately reach photocell P via the subsidiary mirrors M_2 and M_3, and ancillary equipment records only the difference in the readings of the photocell.

21.8 ROUTINE MEASUREMENTS

It will be clear from what has been said in the previous sections that spectrophotometry has innumerable uses in the biochemical field. For identification, for the studying of the products formed during complex reactions, and for the unraveling of structural details through the recognition of the characteristic absorption spectra of various known chemical groupings, this type of instrument is invaluable. In cruder form the spectrophotometer is used routinely in medicine.

The state of working of a human organ can very often be diagnosed from the quantity of a particular chemical present in some part of the body. If a sample is

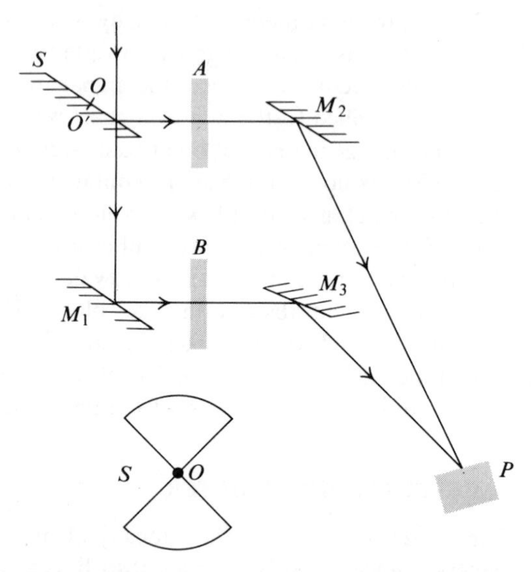

Fig. 21.3. The light-chopper system in a split-beam spectrophotometer.

taken from a patient, one adds a fixed amount of another chemical which produces a specific reaction with the important constituent. The reaction is so chosen that the product either is colored or absorbs strongly at a specific wavelength. The extent of the absorption of the light of a chosen wavelength is directly related to the amount of the absorbing product produced. This in turn is related to the quantity of the initial chemical present in the sample. Important information on the patient's state of health has been given by a routine analysis which can be performed by a semiskilled assistant. The determination of blood sugar levels, for example, is performed by this method.

21.9 ELECTRON SPIN RESONANCE AND NUCLEAR MAGNETIC RESONANCE

Neither electrons (cf. Section 28.2) nor nuclei (cf. Section 30.2) are simple point structures. The former always, and the later sometimes, appear as if they were spinning about an axis like a top. The orientation of the axis can only be in certain specific directions in a magnetic field (cf. Section 25.2), and if radiation carrying the correct amount of energy passes through a substance, the electrons and the nuclei in the substance can flip into a different orientation, absorbing energy in the process. The correct radiation to do this is in the microwave region (wavelength around 1 cm) for electrons, and in the high-frequency radio region (wavelength around 30 m) for nuclei.

Although the instruments used for ESR and NMR absorption experiments are very different from spectrophotometers and are too complicated to detail here, it will be clear that absorption experiments in the wavelength regions mentioned can also give valuable information about structure and can be used for identification purposes just as infrared, visible, and ultraviolet experiments can.

Example 1. If white light is diffracted by a grating, show that there is always overlap between the second- and third-order spectra.

Solution. The red end of the second-order spectrum occurs at an angle θ_1, where

$$d \sin \theta_1 = n\lambda = 2 \times 7.0 \times 10^{-7}\ \mathrm{m} = 14 \times 10^{-7}\ \mathrm{m}$$

and the blue end of the third order spectrum at an angle θ_2, where

$$d \sin \theta_2 = 3 \times 3.8 \times 10^{-7}\ \mathrm{m} = 11.4 \times 10^{-7}\ \mathrm{m}.$$

It follows that

$$\sin \theta_1 > \sin \theta_2,$$

and, therefore,

$$\theta_1 > \theta_2.$$

Part of the second-order spectrum occurs at greater angles of deviation than portions of the third-order spectrum. These spectra therefore overlap.

Example 2. A diffraction grating with 10,000 lines per centimeter is illuminated normally with white light. Find the angular extent of the spectrum in the first order.

Solution. The spectrum will extend from angle θ_1 to angle θ_2, where

$$\sin \theta_1 = \frac{\lambda_1}{d} = 3.8 \times 10^{-7}\ \mathrm{m} \times \frac{10^6}{1\ \mathrm{m}} = 0.38,$$

$$\sin \theta_2 = \frac{\lambda_2}{d} = 7.0 \times 10^{-7}\ \mathrm{m} \times \frac{10^6}{1\ \mathrm{m}} = 0.70.$$

Thus

$$\theta_1 = 22°20' \quad \text{and} \quad \theta_2 = 44°26'.$$

Therefore

$$\theta_2 - \theta_1 = 22°.$$

Example 3. When light of wavelength 575 nm is passed through a cuvette of thickness 5 cm containing vegetable oil, 40% of the incident radiation is absorbed. Calculate the absorption coefficient of vegetable oil at that wavelength.

Solution. The optical density of the vegetable oil at the given wavelength is

$$D = \log (I_0/I) = \log (I_0/0.6I_0) = \log (1.67) = 0.223.$$

The relation between the optical density and the absorption coefficient is given by Eq. (21.6),

$$D = \mu x/2.303.$$

Therefore

$$\mu = 2.303D/x$$

$$= \frac{2.303 \times 0.223}{5.00 \times 10^{-2}\ \text{m}}$$

$$= 10.3\ \text{m}^{-1}.$$

PROBLEMS

21.1 A prism has a cross-section in the form of an isosceles triangle, its refracting angle having a value of A. A ray of light traveling in a plane at right angles to the axis of the prism strikes one face, at such an angle that, after refraction into the prism, it travels parallel to the base before refraction through the second face. Passage through the prism causes the beam to deviate through an angle D. Show that the refractive index of the prism is given by the formula

$$n = \sin \tfrac{1}{2}(D + A)/\sin \tfrac{1}{2}A.$$

21.2 The prism in Problem 21.1 has a refracting angle of 60° and refractive indices for red and blue light of 1.604 and 1.620, respectively. A parallel beam of white light falls on the prism at the same angle as the ray mentioned in Problem 21.1. Calculate the deviations of the blue and red light caused by refraction through the prism.

21.3 The light emerging from the prism in Problem 21.2 is focused on to a screen by a lens of mean focal length 30 cm. Calculate the length of the spectrum seen on the screen.

21.4 A parallel beam of white light falls normally on a diffraction grating with 5000 lines per centimeter. Over what angles are the first- and second-order spectra diffracted?

21.5 The light emerging from the grating in Problem 21.4 is focused on to a screen by a lens of mean focal length 30 cm. Calculate the length on the screen of the first- and second-order spectra.

21.6 Show that, if white light strikes a diffraction grating normally, all adjacent spectra except the first- and second-order spectra overlap.

21.7 If the third-order and fourth-order spectra are not to overlap when a grating is illuminated normally, determine what low-wavelength portion of a white-light spectrum must be removed.

21.8 When light falls on an absorbing material, one per cent is absorbed in the first millimeter of length. What is the absorption coefficient of the material?

21.9 A photograph of a spectrum shows the angular position and the wavelength of each line appearing. How may one tell whether the spectrum was produced by a diffraction grating or a prism?

21.10 Light is passed through a cuvette of thickness 1 cm and the optical density is measured as 0.6. What is the absorption coefficient of the solution in the cuvette?

CHAPTER 22

ELECTROSTATICS

22.1 INTRODUCTION

From early times it has been known that if amber is rubbed with fur, the amber acquires the ability to pick up small pieces of dust or paper and show other unusual phenomena. The same is true if glass is rubbed with silk, or if hair is combed smartly with a dry plastic comb. Further, if the amber so electrified, as it is called, is touched against a light metallic ball suspended from a silk fiber and a similar electrified ball on a silk thread is brought up, the two electrified balls repel each other. However, if a ball touched by electrified amber is brought near a ball touched by electrified glass, the balls no longer repel but attract. It would appear therefore that there are two types of electrification, which we shall call positive and negative electrification.

A more detailed investigation shows that although the amber carries a negative *electric charge*, the fur with which it was rubbed shows a positive charge; and that rubbed glass shows a positive charge, although the silk with which it was rubbed shows a negative charge. Very detailed investigation would show that in each case the positive and negative charges produced are equal in magnitude, which is a particular case of the principle of conservation of electric charge. Charge can neither be created nor destroyed, although it may be neutralized.

If this phenomenon is further investigated, it is found that rubbing will produce electric charges in almost any pair of substances used, although in certain cases it is necessary, for reasons which will appear later, to be very careful about the conditions or the phenomenon may not always be observed.

It would therefore appear that all substances normally contain positive and negative charges, in equal amounts since all substances are normally electrically neutral; one or other of these charges, in fact the negative charges, can be removed fairly easily by frictional effects. In that case, the substance which has acquired the negative charge from the other becomes negatively electrified and the second substance, having lost some of its negative charge, is now left with a surplus of positive electricity and shows positive electrification.

As will be seen in the chapters on modern physics, the Bohr-Rutherford theory of the atom envisages it as consisting of a central nucleus of small size (about 10^{-13} cm in diameter), which contains most of the mass of the atom and all the positive charge, surrounded by a cloud of electrons which travel round the nucleus in circles and ellipses, of approximate linear dimension 10^{-8} cm. These electrons are very light; each has a single negative charge, and there are as many electrons as there are positive charges on the nucleus, so that the atom as a whole is electrically neutral. The electrons nearest to the nucleus are very tightly bound to it in closed shells, as they are called. An atom with from 2 to 10 electrons has the first 2 tightly bound in a closed shell. Atoms with more than 10 electrons have the first 2 and the next 8 bound tightly in closed shells; and so on. The electrons outside the closed shells are fairly loosely bound to the atom and it is these which can be rubbed off. It is also these which produce the optical spectra of the atoms and take part in chemical binding. The charge of each electron is found to be, as we shall see later, -1.6×10^{-19} coulomb.

22.2 CONDUCTORS AND INSULATORS

If a rod of glass is held in the hand, rubbed with silk, and then brought up to an electrified ball on a silk thread, the ball is repelled or attracted. If the same is done with a metal rod, the ball is unaffected. On the other hand, if the metal rod is held in a glass handle and the experiment is repeated, movement of the ball results. One is led to believe that the electric charge can move about on the metal rod and can flow into or through a human body but that this is not so in the case of the glass. Substances are therefore divided into two classes: conductors, such as the metal rod, along which charge can flow or be conducted; and insulators, along which charge cannot flow. In conductors some of the electrons are *conduction electrons*, which are not bound to any atom but can move about relatively freely in the structure. In insulators the electrons tend to be bound to individual atoms and do not have this ability to move about in the substance as a whole. The dividing line between conductors and insulators is not a sharp one. In this region are the substances known as semiconductors which have become increasingly important in recent years. Semiconductors alter their ability to conduct markedly with temperature, impurity, etc. This change in conducting properties is what makes semiconductors so important (cf. Sections 33.2, 33.3, and 33.4).

22.3 CHARGING BY INDUCTION

If a positively charged rod is brought close to a metal object which is uncharged, the electrons in the object are attracted toward the rod and tend to accumulate in the portion of the object nearest to the rod. The opposite side of the object is then positively charged, since there is a deficiency of electrons at that place. This phenomenon of induction can be used to charge the object. For if it is attached to the ground (or to a large conductor), electrons will flow to the object, attracted by the surplus positive charge. An alternative way of looking at this is to say that the object now consists of itself and the earth. Electrons are attracted to the part of the object nearest to the charged rod, leaving an excess of positive electricity at the other side of the composite body, i.e. on the earth. If the connection is now broken and the charged rod is then removed, the object retains the electrons which it has acquired from the ground and shows negative electrification. The opposite effect is produced if a negatively charged rod is brought up.

If the object is not connected to ground, the positive and negative electric charges redistribute themselves over the object, when the rod is removed, to produce electrical neutrality.

Note that while the rod is close to the object, the two will be attracted to each other. The force of attraction between the positively charged rod and the negative charge at one end of the object is greater than the force of repulsion between the rod and the positive charge on the other end of the object, since the latter is farther away.

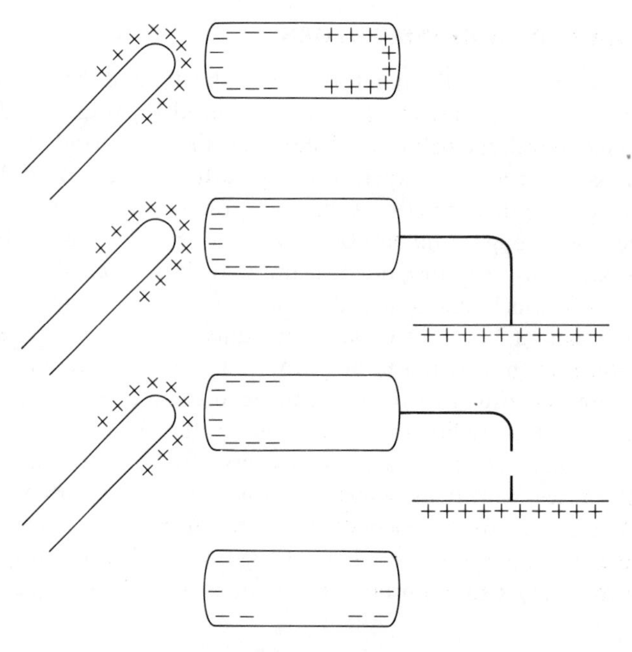

Fig. 22.1. Charging by induction.

It should also be noted that electric charge does not in general distribute itself evenly over the surface of a conductor. It is most dense where the curvature is greatest, and therefore tends to be concentrated at any pointed portion of the surface. If indeed there is a sharp point on the surface, the density of charge may become so great that air molecules striking it acquire charge from the point and are instantly and violently repelled. A noticeable wind can be felt near a charged point.

The action of a lightning conductor depends on this fact. A lightning conductor has a collection of points at its topmost end, the other end being grounded. Clouds, particularly those with turbulent currents in their interior, are often charged, usually negatively on the base and positively on the top. There is a danger that they will discharge by a lightning stroke to earth. When such a cloud passes over a lightning conductor, induction takes place, positive charge being induced on the points on the conductor. Positive charges spray upwards because of the wind effect mentioned earlier. This tends to neutralize the negative charge on the base of the cloud and prevent a lightning discharge. Even if this desirable effect is not achieved, any discharge that takes place is likely to do so along the track of the positive charges. Buildings in the vicinity of the conductors are thus protected.

22.4 COULOMB'S LAW

It is obviously necessary to introduce some quantitative measurement into the subject. To do so it is necessary to have available a collection of electrified bodies, the ratios of the charges on which are known. This can be achieved by acquiring a collection of identical metal balls on insulating handles and charging one of them. This ball will then possess an electric charge of unknown magnitude q. If the ball is touched to another, from symmetry considerations it will be realized that the charge will be shared equally and that each ball will now possess a charge $q/2$. If one of these balls is touched to an uncharged ball, each of them will possess a charge $q/4$. Hence, one can in like manner obtain charges of magnitude $q/8$, $q/16$, etc. Further, if one touches balls containing charges $q/2$ and $q/4$, each will acquire a charge $3q/8$. If one touches balls containing charges $q/2$ and $q/8$, one can obtain charges of $5q/16$, and so on. It is thus possible to prepare a series of charged balls containing charges in almost any ratio desired.

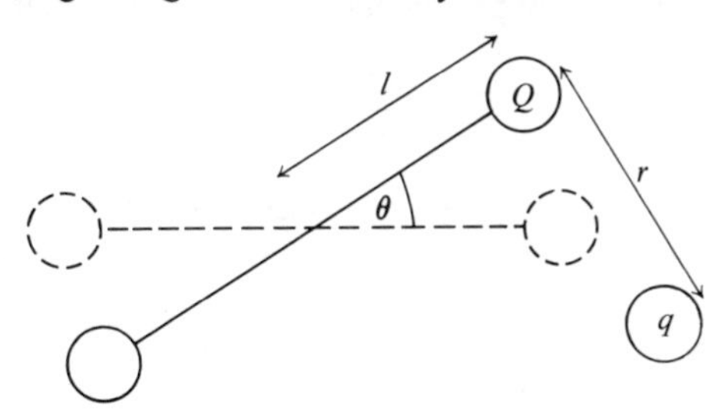

Fig. 22.2. The experimental verification of Coulomb's law of force between charged spheres.

In Fig. 22.2 an insulating rod which has equal, uncharged, metal balls at either end is suspended horizontally by a torsion fiber. One of the balls is given a charge Q. Another ball, on an insulating handle and carrying a charge q of the same sign, is brought up in the same horizontal plane. The hanging ball will be repelled and the suspended rod will rotate through an angle θ until the torque produced by the force of repulsion is balanced by the torsional torque in the twisted fiber of the suspension. The ball with charge q must always be brought up in such a way that at equilibrium the line joining the centers of the charged balls is at right angles to the length of the rod. If this is done, if l is the length shown in Fig. 22.2, and if F is the force of repulsion between the charged balls when their centers are at a distance r apart,

$$Fl = \tau\theta, \tag{22.1}$$

where τ is the torsion constant of the fiber. Since τ and l are constants of the instrument, $F \propto \theta$, and the dependence of the force of repulsion on various factors may be investigated by studying the consequent variation of θ.

Three experiments are, in fact, performed. In the first place, with the charges q and Q unchanged, the distance r between the centers of the balls is altered and the value of θ is measured for each value of r. It is found that θ varies as $1/r^2$. Hence, $F \propto 1/r^2$.

As a second experiment, various balls with different charges are brought up, the ratios of the charges being found in the manner described above. The charge Q remains unchanged at all times and the balls are brought up in such a manner that at equilibrium the distance of separation of the centers of the charged balls is always the same. In this case θ varies directly as q. Hence, $F \propto q$.

As a final experiment, the ball with charge q is used at all times but the value of Q is varied by touching the hanging ball with identical uncharged balls. Throughout the experiment the hanging ball thus holds charges Q, $Q/2$, $Q/4$, etc., for the reasons given above. With each value of charge, the other charged ball is brought up in such a way that at equilibrium the distance apart of the centers of the charged balls has a fixed value, and θ is measured. It is found that θ varies directly as the charge on the hanging ball. Hence, $F \propto Q$.

From the whole series of experiments it is concluded that

$$F \propto \frac{qQ}{r^2} = k\frac{qQ}{r^2}, \tag{22.2}$$

where k is a constant. Since the units of distance and force are known, this equation, known as Coulomb's law, can be used to define the unit of electric charge.

22.5 THE MKSA UNIT OF CHARGE

In the early days of the study of electrostatics, it seemed natural to define the unit of charge in such a way that the constant k would have a value of unity. This is a practice adopted wherever possible in physics. In those days the system of units in general use was the CGS (centimeter-gram-second) system, in which distances were measured in centimeters and forces in dynes. The unit of charge, defined in terms of these units and with $k = 1$, was known as an electrostatic unit of charge or statcoulomb. Unfortunately, magnetism started off from different experimental facts, although it was eventually found that the two subjects were linked. The charge worked out from previously defined magnetic units, the electromagnetic unit of charge, turned out to be different in value from the electrostatic one. Both systems of units were inconvenient for practical work, and a third set of units, the practical units, was also in use. This confusion resulted in the development of the MKSA system, which produces a set of units identical with the practical ones.

The unit of charge in the MKSA system is the coulomb, 1 coulomb (1 C) of charge being that quantity of charge which when placed 1 m from an identical charge in vacuum repels it with a force of 8.99×10^9 N. The constant k is therefore neither unity nor a dimensionless quantity. It has the value 8.99×10^9 N m^2 C^{-2}. It is normally written as $1/4\pi\varepsilon_0$, where ε_0 is called the permittivity of free space, this form being used to simplify equations which occur in the higher branches of the subject.

If the medium in which the charges are placed is not vacuum but a material medium, then it is found that

$$F = \frac{1}{4\pi\varepsilon_0\varepsilon_r} \times \frac{qQ}{r^2} = \frac{1}{4\pi\varepsilon} \times \frac{qQ}{r^2}. \tag{22.3}$$

The constants ε_r and ε are called the relative and abolute permittivities of the medium: ε_r is dimensionless and has no units, whereas ε has the same units as ε_0. For gases ε_r is around 1 (for air ε_r is 1.0006), and for most substances ε_r has a value in the range 1–10, although a few liquids and crystals have much higher values. The quantity ε_r is often called the dielectric constant of the medium.

Note that ε_0 has the value 8.85×10^{-12} C^2 m^{-2} N^{-1}.

22.6 THE ELECTRIC FIELD

Since a charged body exerts a force on another charged body in its vicinity, a charged body can be regarded as surrounded by a field of force, i.e. a region in which charged bodies experience forces and in which uncharged bodies may suffer induction or other effects. If a small positive test charge is introduced at a point in the field, the effects produced by the field at any point can be investigated. The test charge must be small both in the magnitude of its charge and in its dimensions so that it will produce as small an effect on the field being investigated as possible. If the test charge is placed at any point, it will experience a vector force **F**. The electric intensity (often loosely called the field strength) at that point is defined as

$$\mathbf{E} = \mathbf{F}/q, \tag{22.4}$$

where q is the magnitude of the test charge. Note that **E** is also a vector, having the magnitude F/q and the direction of **F**. It will be measured in the unit of newtons per coulomb. It would therefore be possible to map out the field by attaching at each point of it a vector **E**. Alternatively, and more usually, the field is mapped out by drawing lines in it so that the tangent to the line at any point is the direction in which a positive test charge would move if placed at that point. Typical mappings of fields are shown in Fig. 22.3.

Note that in Fig. 22.3(c) there is a null point, midway between the two charges

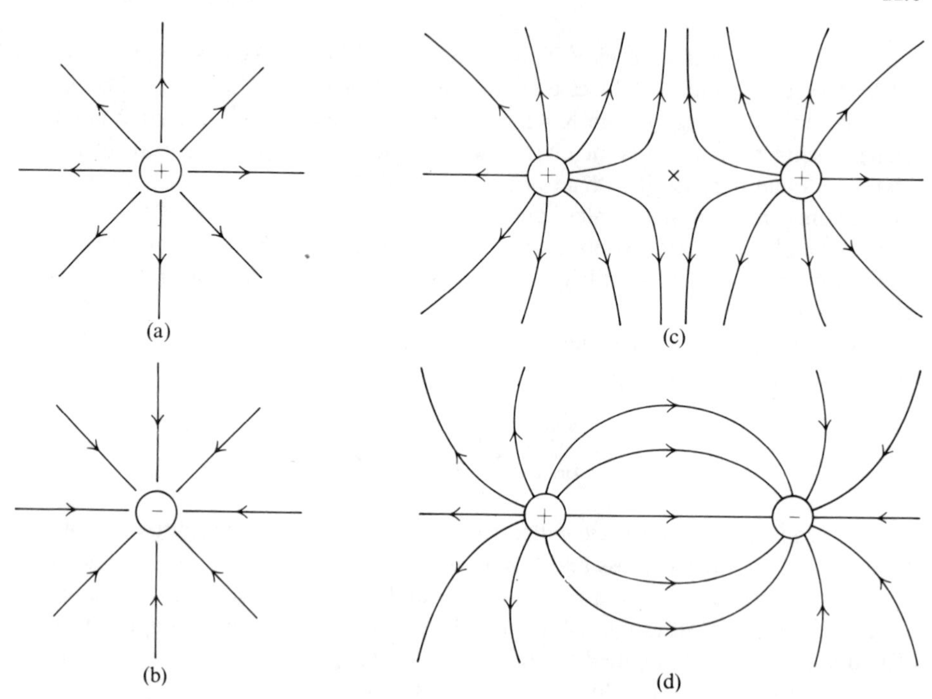

Fig. 22.3. The electric fields due to (a) a positively charged sphere, (b) a negatively charged sphere, (c) two spheres containing equal quantities of positive charge, and (d) two spheres containing equal but opposite quantities of charge.

if they are of equal magnitude. A test charge placed there would suffer no net force. Lines of force obviously start on positive charges, and end on negative charges or at infinity.

If an isolated point charge of magnitude Q is considered, the force on a positive test charge is always away from the charge, as therefore is the direction of the electric intensity (cf. Fig. 22.3a). Further,

$$F = \frac{qQ}{4\pi\varepsilon_0 r^2}$$

and thus

$$E = \frac{Q}{4\pi\varepsilon_0 r^2}. \tag{22.5}$$

In order to make the field diagrams meaningful, they should be drawn with large numbers of lines where E is large, and with small numbers of lines where E is small. To achieve this, the number of lines N passing through any small area A placed at right angles to the direction of $\mathbf{E}$ at that place is chosen to be proportional to A and proportional to E. For convenience the number of lines is chosen so that $N = \varepsilon_0 EA$.

If a sphere is drawn around, and concentric with, a point charge Q, on any small area A of the surface of the sphere $N = \varepsilon_0 EA$. But at all points of the sphere **E** has the same value and is everywhere normal to the surface. Thus the number of lines of force N_0 passing from the sphere is given by $N_0 = \varepsilon_0 EA_0$, where A_0 is the surface area of the sphere. Thus

$$N_0 = \varepsilon_0 EA_0 = \varepsilon_0 \times \frac{Q}{4\pi\varepsilon_0 r^2} \times 4\pi r^2 = Q. \tag{22.6}$$

The number of lines leaving the sphere is equal to the magnitude of the charge contained inside it.

This result is obviously true whether Q is at the center of the sphere or not. It is still true if the surface is ellipsoidal or has any simple shape, for the same number of lines of force must pass through the surface. If the closed surface, as in Fig. 22.4, comes back on itself, any line that emerges from and then reenters the surface must come out again. If a line is counted as negative when it goes in and as positive when it comes out, the same result as before is seen to be true. If the surface does not surround a charge, as many lines of force must leave the surface as enter, since lines of force can only end on a negative charge (Fig. 22.5). Thus $N = 0$ and $Q = 0$ also. If the idea is extended to the case in which any number of charges are inside the surface, by similar reasoning it can be shown that the net number of lines of force coming from a closed surface is always equal to the total quantity of charge contained inside the surface.

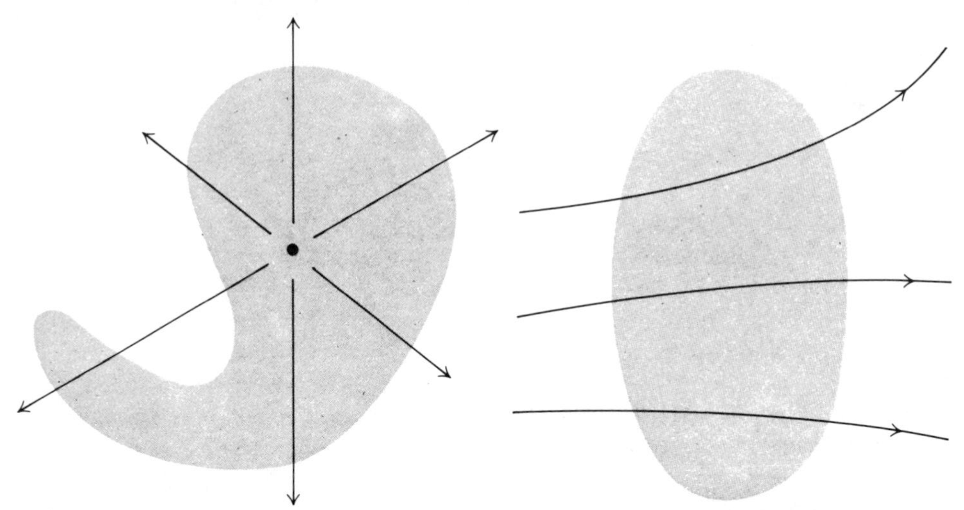

Fig. 22.4. Lines of force leaving a closed surface which contains a charge.

Fig. 22.5. Lines of force passing through a closed surface which contains no charge,

This is one form of *Gauss's theorem*, which is more usually given in another form. Since N is proportional to E and the lines run in the same direction as **E**, there must be a relation between E evaluated and summed all over the surface and the charge included inside the surface. In general, since **E** varies over the surface and is not necessarily perpendicular to the surface, an integration is required to obtain the summation for **E**. The general situation is thus too complicated for us to deal with. But when **E** is everywhere perpendicular to the surface and has the same magnitude at all points of the surface, from Eq. (22.6) it follows that

$$E = \frac{1}{\varepsilon_0 A_0} \sum q, \tag{22.7}$$

where A_0 is the total area of the closed surface and $\sum q$ denotes the sum of all the charges inside the surface.

22.7 THE POINT CHARGE AND THE CHARGED SPHERICAL CONDUCTOR

It has been tacitly assumed in the preceding section that Coulomb's law, which was verified for charged, spherical conductors, also applies to point charges. It is necessary to prove this. The use of Gauss's law in the form given in Eq. (22.7) allows this to be done.

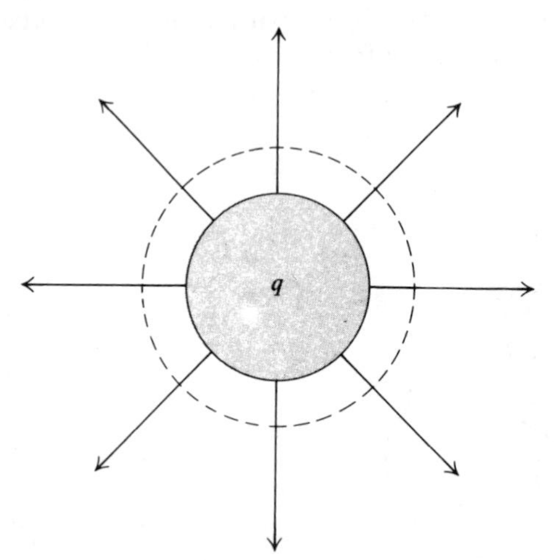

Fig. 22.6. Lines of force leaving a closed surface concentric with, and containing, a charged sphere.

Figure 22.6 shows a spherical conductor of radius a carrying a charge q. By symmetry the lines of force must be at right angles to any spherical surface of radius r concentric with, and surrounding, the conductor. Further, **E** must have the same magnitude at all points on the surface. Therefore

$$E = \frac{1}{\varepsilon_0 \times 4\pi r^2} \times q = \frac{q}{4\pi\varepsilon_0 r^2}.$$

This result does not involve the radius of the conductor: it is true whether the radius is large or negligible. A conducting sphere thus acts, for all points outside it, as though all the charges were concentrated at the center. It therefore acts like a point charge.

At the surface of the sphere, since the charge will be distributed uniformly,

$$E = \frac{q}{4\pi\varepsilon_0 a^2} = \frac{q/4\pi a^2}{\varepsilon_0} = \frac{\sigma}{\varepsilon_0}, \tag{22.8}$$

where σ is the surface density of charge. This is true whether the charged sphere is surrounded by another spherical conductor or not. If it is not, and a is made larger and larger, the relation $E = \sigma/\varepsilon_0$ holds whatever the value of a. In particular, when a is infinite, the relation holds outside an infinite plane conductor. If there is a further conductor of radius $a + d$ concentric with the one considered, increasing a and keeping d fixed leads eventually to the result that the relation $E = \sigma/\varepsilon_0$ holds in the region between two infinite parallel conducting plates. If the plates are close together, so that d is small, the relation will still be true except at the edges of the system, when the two plates are finite in extent. This is a most useful and important result.

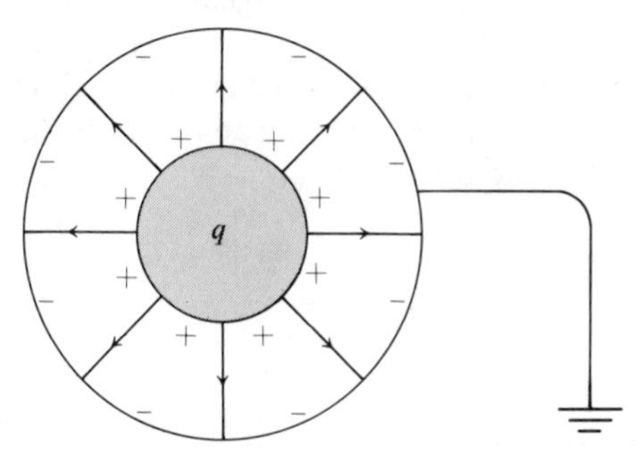

Fig. 22.7. Electrostatic screening.

If we revert to the case of two spheres of finite radius, it will be seen that the inner charged sphere is the source of lines of force, all of which terminate on the inner surface of the outer sphere. Since the inner sphere carries charge q, the inner surface of the outer sphere must carry an induced charge $-q$. An insulated sphere would therefore have a charge $+q$ induced on its outer surface and lines of force from this charge would produce effects on objects in the vicinity. If, however, the outer sphere is grounded, the charge $+q$ is not present on the outer surface (cf. Section 22.3) and objects near the sphere suffer no electrical effects. This is the basic idea of electrical screening (Fig. 22.7). Any piece of apparatus which would produce an electrical field of force is enclosed in a grounded metal container. All lines of force are confined to the inside of the container and none begin on its outside. The apparatus cannot therefore interfere with the working of other pieces of apparatus in its neighborhood.

By similar reasoning it will be seen that any sensitive apparatus may be shielded from fields which might affect its operation by being contained in a grounded metal box. In radio and television sets, and in most electronic equipment, many components are shielded in this way. If the container is made of a ferromagnetic material (cf. Section 25.2), the apparatus is also shielded from magnetic fields.

22.8 ELECTRIC POTENTIAL

In the preceding section, a formula was obtained which allowed the calculation of the value of **E** for a simple case, that of an isolated point charge. If it is required to find the value of **E** at the origin due to a collection of point charges (or to any system which may be regarded as such), as shown for the simple case of three charges in Fig. 22.8, it is easy to calculate $\mathbf{E}_1$, $\mathbf{E}_2$, and $\mathbf{E}_3$ for each of the point charges. Unfortunately, all of these quantities are vectors, and to get the combined effect of the three, it is necessary to add them vectorially. This is normally highly inconvenient. Fortunately, there is a more convenient method of mapping the field. This is to map it in terms of the potential. The potential at a point in a field is defined as the energy per unit charge required to bring a small positive charge to that point in the field from infinity against the electrical forces, the potential at infinity being taken conventionally as zero. The potential is not defined as the energy necessary to bring up 1 C, since a charge of 1 C would seriously disrupt the whole field. The energy required to bring an infinitesimal charge q to the point from infinity is calculated and divided by q.

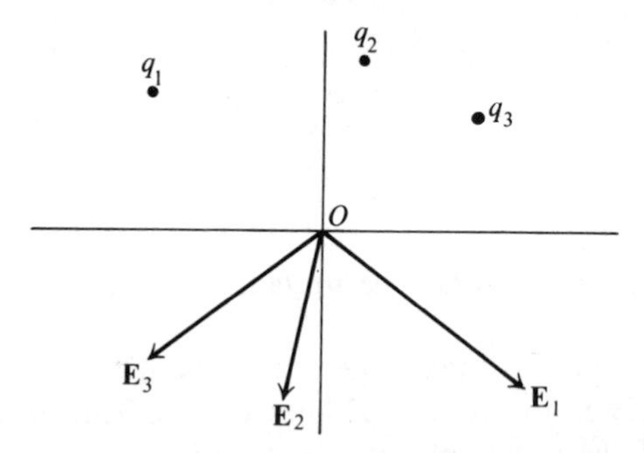

Fig. 22.8. The electric intensity at the origin due to three point charges.

Consider the case of a point charge Q. The electrical force on a charge q at distance r from it is

$$F = \frac{qQ}{4\pi\varepsilon_0 r^2}.$$

If the charge moves a distance dr under the action of the electric force, the work done *by* the electrical force is

$$F dr = \frac{qQ}{4\pi\varepsilon_0 r^2}\, dr.$$

The work done *by* the force in moving the charge q from ∞ to a distance R from the point charge Q is then (cf. Chapter 3)

$$\int_{\infty}^{R} F dr = \int_{\infty}^{R} \frac{qQ}{4\pi\varepsilon_0 r^2}\,dr = \left[-\frac{qQ}{4\pi\varepsilon_0 r}\right]_{\infty}^{R} = -\frac{qQ}{4\pi\varepsilon_0 R}.$$

Since the answer is negative, work must be done *against* the electric force of amount $qQ/4\pi\varepsilon_0 R$, and the energy per unit charge required is

$$V = \frac{Q}{4\pi\varepsilon_0 R}. \tag{22.9}$$

This is the potential at the point a distance R from the point charge Q. The energy supplied per unit charge in bringing a charge up to that point is stored as potential energy and can be released if the charge is permitted to return to infinity.

Potential is a scalar quantity. It is independent of the means of bringing the charge up to the point. Using any route, we would obtain the same result. The potential due to a collection of point charges can thus be calculated easily, since the effects of the individual charges merely add algebraically. The unit of potential is the joule per coulomb, which is given the special name of the volt (V). The volt is defined as the difference of potential between two points when it takes 1 J of work per coulomb to transport a charge from one point to the other against the electrical forces.

Equipotential surfaces are surfaces joining points of equal potential. Any conductor must be an equipotential surface. Charge can flow freely over a conductor and will always attempt to go to the place of lowest potential energy. If a potential difference existed between two points on a conductor, charge would continue to flow until equipotential conditions had been obtained. In an equilibrium state with no charge flowing a conductor must therefore be at constant potential throughout. Note that equipotential surfaces and lines of force must meet at right angles. If they did not do so, **E** would have a component along the surface, and thus charge would flow along the surface under the action of the electrical force, a condition which contradicts the statement that equilibrium has been established. The electric intensity **E** must always point in the direction of decreasing potential. It also follows that E has the value of zero inside a conductor, since V is constant everywhere and therefore E cannot have any value but zero throughout the body of the conductor.

Between two parallel conducting plates, distance d apart, one with charge q and the other with charge $-q$, the intensity is uniform and of value $E = \sigma/\varepsilon_0$ (cf. Eq. 22.8). If charge q' is moved from the negative plate to the positive plate, the work done on the charge is $\mathscr{W} = q'Ed$. Thus the difference in potential between the plates is $V = q'Ed/q' = Ed$. Between the plates $E = -V/d$ and is called the potential gradient, the minus sign being incorporated to show that **E** points in the direction of decreasing potential. The units of E can be given as volts per meter as well as newtons per coulomb.

The potential due to a point charge and that due to a charged metal sphere are the same at all points outside the sphere. This follows from the definition of V and the earlier proof that $\mathbf{E}$ is the same for a spherical conductor and a point charge. In particular, at the surface of a sphere of radius a the potential is $Q/4\pi\varepsilon_0 a$.

22.9 MILLIKAN'S OIL-DROP EXPERIMENT

The unit of charge has been defined from Coulomb's law, but the basic quantity of charge is that carried by the electron in the atom. How are these two charges related? This was discovered by Millikan in his famous experiment (Fig. 22.9).

Two parallel plates are kept at all points accurately a small distance d apart by the use of an optically flat spacer. A potential difference V can be maintained between them in either direction. Water-free oil, which does not evaporate, is sprayed through a small hole in the top of the upper plate and enters the space between the plates. Through a window in the spacer, light from a lamp (having been passed through a water cell to remove heat rays) is focused on the center of the system, where the oil drops are falling. The light scattered from the drops is picked up in a long-focus microscope directed through a further window at 120° to the first, and the oil drops are therefore seen as brilliant points of light against a dark background. Some of the drops may be charged by friction in the spray nozzle, but a third window in the spacer at 120° to the other two permits γ-rays to enter from a radioactive source to ionize air molecules so that the drops may acquire charges.

If the drops are watched and a suitable one of mass m is chosen and allowed to fall under gravity, it will rapidly reach the state in which its weight is balanced

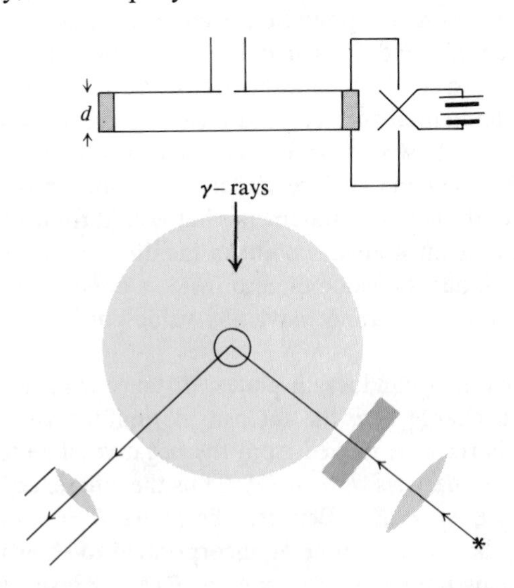

Fig. 22.9. Millikan's oil-drop apparatus in elevation and in plan.

by the air-resistance forces, which are proportional to its speed. After that it drops with a constant terminal velocity. The time t_1 which it takes to fall through a distance s while traveling with its terminal velocity v is noted. A potential difference is put across the plates such as will cause the drop to move upwards. It will quickly acquire a terminal velocity u, and its time over the same distance s is measured as t_2. In the first case, when the drop has reached its terminal velocity,

$$mg = bv.$$

In the second case, under the same conditions,

$$Eq = \frac{qV}{d} = mg + bu = bv + bu.$$

Therefore

$$q = \frac{d}{V} b(v + u) = \frac{d}{V} bs \left(\frac{1}{t_1} + \frac{1}{t_2}\right). \tag{22.10}$$

If the same drop is watched for hours, it picks up and loses charges by collision with charged air molecules and q takes different values. The time t_1 stays constant but t_2 varies. The quantity $1/t_1 + 1/t_2$ is always found to be a multiple of a basic unit. It follows that the charge on the drop is always a multiple of a basic quantity of charge.

The quantities d, V, and s are all measurable. It is known what form b has from other experiments (cf. Appendix B). It is thus possible to work out values for q, and this is repeated for other drops. From these experiments it is found that q is always a multiple of -1.6021×10^{-19} C, which is the charge on a single electron. The magnitude of this quantity, $+1.6021 \times 10^{-19}$ C, is known as the elementary charge and is denoted by the symbol e.

22.10 THE ELECTRON-VOLT

In studies in atomic physics the joule is far too large a unit to describe the energies possessed by atomic particles. It is normal to use instead the electron-volt (eV) as a unit of energy in these situations. This is the energy acquired by an electron if it is permitted to fall through a potential difference of 1 V. The extra potential energy possessed by the electron at the first place over that possessed at the second place appears in the form of kinetic energy. Since 1 J is the energy employed in taking 1 C through the potential difference of 1 V, 1.6021×10^{-19} J is the energy used up in taking one electron through a potential difference of 1 V. Thus the electron in falling back through the 1-V potential difference acquires an energy of 1.6021×10^{-19} J, and this is 1 electron-volt.

22.11 CAPACITANCE

If electric charge is continually added to an insulated metal sphere, it is found that, beyond a certain level, the process is self-defeating. No more charge is actually retained by the sphere. This is because charge can be neutralized by ions attracted from the air and can leak away through dirt and grease on the insulating support; but even if all leakage and similar losses could be removed, there would eventually occur an electric discharge through the air or the support. Any material can only support an electric field of a given strength before a discharge of electricity takes place through it. Every material, in other words, has a dielectric strength, which is the maximum potential difference per unit length which it can withstand. Dry air has a dielectric strength of 30 kV per cm, glass a dielectric strength of about 400 kV per cm. But if spheres of different radii are used, it will be found that breakdown takes place on some before others for the same amount of charge added. Thus the same amount of added charge does not produce the same potential on all conductors. The capacitance of a conductor is defined as the charge acquired by the conductor when it has unit potential. Thus in the MKSA system the unit of capacitance is the coulomb per volt, which has the special name of the farad (F). A conductor has a capacitance of 1 F if a charge of 1 C given to it produces a potential on it of 1 V. The capacitance of a spherical conductor is thus

$$C = Q \Big/ \frac{Q}{4\pi\varepsilon_0 a} = 4\pi\varepsilon_0 a; \tag{22.11}$$

the larger the sphere the more charge it holds for a given potential, and thus the more charge it can acquire before breakdown.

An isolated conductor is seldom used for storing charge. The capacitance can be increased by employing two conductors close together, one of them normally grounded. If a positive charge is given to one plate, this induces a negative charge on the grounded plate, the resultant positive charge appearing on the ground. The positive and negative charges on the plates are mutually attracted to the inside of each plate, which helps to prevent leakage, and all the lines of force from the positive charges tend to end on the negative charges on the other plate. The plates therefore contain equal and opposite charges. The positive charge, if on an isolated plate, would produce a certain potential on it. The presence of the nearby negative charge must produce a further negative potential on the plate, the net value being smaller than that which it would have had in isolation. The grounded plate has raised the capacitance of the charged plate. Such a system of two plates with equal and opposite charges is called a capacitor. Its capacitance is defined in the same way as was that of the conductor. It is the charge on the positive plate for unit potential difference between the plates, and is measured in units of farads.

The simplest type of capacitor is the parallel-plate capacitor (Fig. 22.10). It has already been seen that between charged plates the potential difference is V, where $V = Ed = (\sigma/\varepsilon_0) \times d$. Hence,

$$C = \frac{Q}{V} = \frac{A\sigma}{\dfrac{\sigma}{\varepsilon_0}d} = \frac{\varepsilon_0 A}{d}, \tag{22.12}$$

where A is the area of each of the plates.

By performing experiments, or by repeating the investigation of E and V in the case of a dielectric medium, a more general formula for the capacitance is

$$C = \frac{\varepsilon_0 \varepsilon_r A}{d} = \frac{\varepsilon A}{d}, \tag{22.13}$$

when there is a substance of permittivity ε between the plates. It is more usual to determine the value of ε with a capacitor than by means of Coulomb's law. It

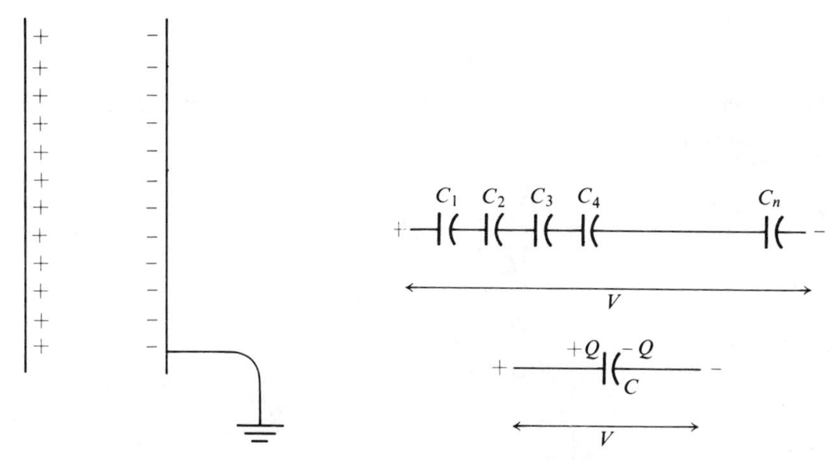

Fig. 22.10. A parallel-plate capacitor.

Fig. 22.11. Capacitors in series, and the equivalent capacitor.

would be almost impossible to determine ε by the latter means, and the value obtained would certainly be inaccurate. The capacitances of a capacitor with and without a dielectric, C and C_0, respectively, are simply and accurately measurable. The ratio C/C_0 is obviously equal to ε_r.

22.12 CAPACITORS IN SERIES AND IN PARALLEL

If a number of capacitors are joined in series and charged, a potential difference V will be established from beginning to end of the combination, as shown in Fig. 22.11. Let the capacitors have capacitances $C_1, C_2, C_3, \ldots, C_n$. The combination will be equivalent to a single capacitor of capacitance C which has charges $\pm Q$ on its plates. If charge Q has flowed into the system, the first plate of C_1 must also possess a charge Q. This induces a charge $-Q$ on the other plate, and thus a charge $+Q$ on the first plate of C_2, since these joined plates had zero charge initially and must have zero net charge at all times, because of the principle of conservation of charge. Similarly, the second plate of C_2 has a charge $-Q$, the first plate of C_3 a charge $+Q$, and so on, finishing finally with $-Q$ on the second plate of C_n, as required.

If the potential differences between the plates of $C_1, C_2, \ldots, C_n$ are $V_1, V_2, \ldots, V_n$, respectively, then

$$V = V_1 + V_2 \ldots + V_n = \frac{Q}{C_1} + \frac{Q}{C_2} + \ldots + \frac{Q}{C_n}.$$

But $$V = Q/C.$$

Therefore

$$\frac{1}{C} = \frac{1}{C_1} + \frac{1}{C_2} + \frac{1}{C_3} + \ldots + \frac{1}{C_n}. \tag{22.14}$$

The reciprocal of the equivalent capacitance is the sum of the reciprocals of the individual capacitances.

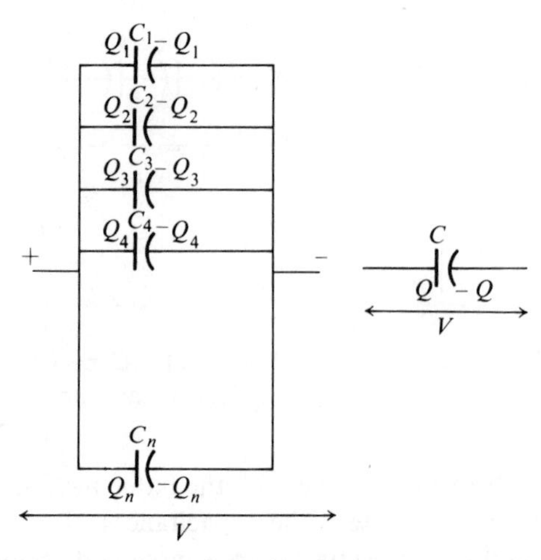

Fig. 22.12. Capacitors in parallel, and the equivalent capacitor.

If the capacitors are connected in parallel (Fig. 22.12), the combination will again be equivalent to some single capacitor of capacitance C with charges $\pm Q$ on its plates and a potential difference of V across it. Each individual capacitor has a potential difference of V between its plates, and the charges on the positive plates of $C_1, C_2, \ldots, C_n$ will be $Q_1, Q_2, \ldots, Q_n$, respectively. Since the two systems must be equivalent,

$$Q = Q_1 + Q_2 + Q_3 + \ldots Q_n = C_1V + C_2V + C_3V + \ldots + C_nV.$$

But $$Q = CV.$$

Therefore

$$C = C_1 + C_2 + C_3 + \ldots C_n. \tag{22.15}$$

The equivalent capacitance is the sum of the individual capacitances.

22.13 ENERGY IN A CHARGED CAPACITOR

Energy is required to charge a capacitor, this energy being stored as potential energy, which may be released to do work by joining the plates by a conductor. When a capacitor of capacitance C is charged so that a potential difference V exists between its plates, the capacitor must possess a unique quantity of energy however the charge was brought up. If this were not so, it would be possible to charge the capacitor by the method which required least energy and then return the charge to its original position by the reverse of the method which required most energy. In the process more work would be obtained than had originally been used on the system. Part of the work could be used to charge the capacitor again by the minimum-energy method, and part would be left over for additional work. This process could be continued indefinitely, work being obtained from the system as a product. Such a perpetual-motion machine violates the laws of physics. A charged capacitor must contain the same amount of energy however it was charged.

That being so, it is only sensible to choose a method of charging which makes simple the calculation of the energy stored. If the final charge on the positive plate is Q, this charge will be brought up from the negative plate in a large number n of small units of charge q. To transport the first charge q from the initially uncharged negative plate to the initially uncharged positive plate requires no work, since no electrical force is present. As soon as this charge is on the positive plate, however, a potential difference $v = q/C$ exists and an amount vq of work is required to bring up the second charge. Similarly, to bring the third charge up requires $2v \times q$ of work, and so on. In particular, when fQ of charge, f a fraction, is on the positive plate and therefore $-fQ$ on the negative plate, the potential difference between the plates is $fQ/C = fV$. The work done to bring the next charge up is fqV. Thus the work required to bring up charge q to the positive from the negative plate varies linearly from 0 to qV as f varies from 0 to 1, as Fig. 22.13 shows.

Because the relation is a linear one, the average amount of work done in bringing up a small charge is thus $w = \frac{1}{2}qV$. But n charges were brought up.

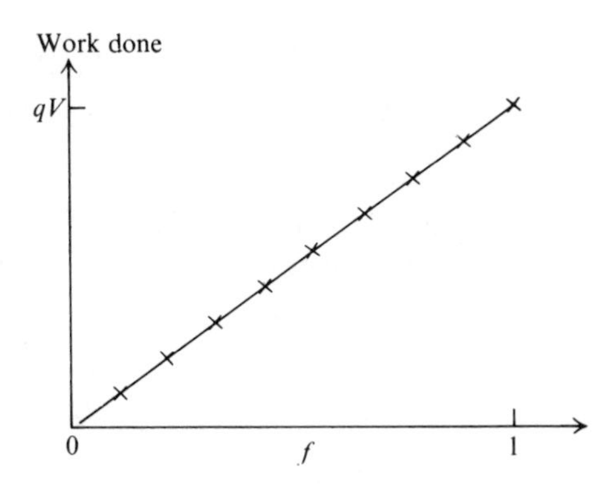

Fig. 22.13. The work done in transferring a small charge q from the negative to the positive plate of a capacitor as a function of the fraction of the total charge transferred.

Thus the total work done is

$$\mathscr{W} = nw = n \times \tfrac{1}{2}qV = \frac{Q}{q} \times \tfrac{1}{2}qV = \tfrac{1}{2}QV. \qquad (22.16)$$

This can alternatively be written as

$$\mathscr{W} = \tfrac{1}{2}CV^2 = \tfrac{1}{2}\frac{Q^2}{C}, \qquad (22.17)$$

since

$$C = Q/V. \qquad (22.18)$$

Example 1. Charges $Q_1 = +1.00 \times 10^{-12}$ C and $Q_2 = -4.00 \times 10^{-12}$ C are situated 5.00 m apart in air. A third charge $Q_3 = +1.00 \times 10^{-14}$ C is placed midway between them. Find the magnitude and direction of the resultant force on the third charge.

Solution. The third charge Q_3 is acted upon by two forces, one due to charge Q_1 of magnitude $Q_1Q_3/4\pi\varepsilon_0 r^2$, and one due to Q_2 of magnitude $Q_2Q_3/4\pi\varepsilon_0 r^2$. The net force in the direction from Q_1 to Q_2 is thus

$$F = \frac{Q_1Q_3}{4\pi\varepsilon_0 r^2} - \frac{Q_2Q_3}{4\pi\varepsilon_0 r^2}$$

$$= \frac{Q_3}{4\pi\varepsilon_0 r^2}(Q_1 - Q_2)$$

$$= \frac{8.99 \times 10^9 \text{ N m}^2 \text{ C}^{-2} \times 1.00 \times 10^{-14} \text{ C } [1.00 \times 10^{-12} \text{ C} - (-4.00 \times 10^{-12} \text{ C})]}{(2.5 \text{ m})^2}$$

$$= \frac{8.99 \times 10^9 \times 1.00 \times 10^{-14} \times 5.00 \times 10^{-12}}{(2.5)^2} \text{ N}$$

$$= 7.19 \times 10^{-17} \text{ N}.$$

Example 2. Determine the electric intensity and the electric potential midway between charges of $+1.00 \times 10^{-12}$ C and -4.00×10^{-12} C which are separated by 5.00 m in air.

Solution. The situation is similar to that illustrated in Fig. 22.14. From the solution to Example 1 it is clear that the electric intensity at the position of Q_3 is

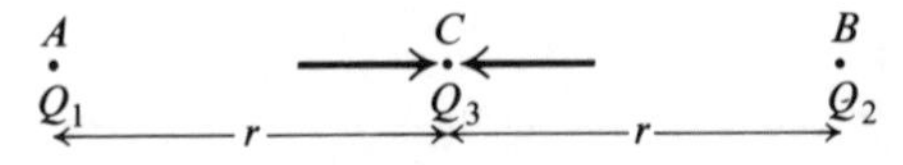

A C B
Q_1 Q_3 Q_2
r r

Figure 22.14

$(Q_1 - Q_2)/4\pi\varepsilon_0 r^2$ in the direction from Q_1 to Q_2 and that the magnitude has the value 7.19×10^{-3} N C^{-1}.

Work has to be done on Q_3 to take it from infinity to its position in the diagram. If only the charge Q_1 were present, this would be $Q_1Q_3/4\pi\varepsilon_0 r$, and so the electric potential, which is defined as the work done per unit charge, would be $Q_1/4\pi\varepsilon_0 r$. Similarly, if only the charge Q_2 were present, the potential would be $Q_2/4\pi\varepsilon_0 r$. Clearly the total potential is the sum of these two quantities. Thus

$$V = \frac{Q_1}{4\pi\varepsilon_0 r} + \frac{Q_2}{4\pi\varepsilon_0 r} = \frac{Q_1 + Q_2}{4\pi\varepsilon_0 r}$$

$$= \frac{8.99 \times 10^9 \text{ N m}^2 \text{ C}^{-2} [1.00 \times 10^{-12} \text{ C} + (-4.00 \times 10^{-12} \text{ C})]}{(2.5 \text{ m})}$$

$$= \frac{-8.99 \times 10^9 \times 3.00 \times 10^{-12}}{2.5} \text{ N m C}^{-1}$$

$$= -1.08 \times 10^{-2} \text{ V}.$$

The negative sign indicates that work is done *by* the positive charge Q_3 as it moves from infinity to the midpoint, the attractive force exerted by Q_2 being greater than the repulsive force exerted by Q_1.

Example 3. A water droplet of radius 1 μm remains stationary in a Millikan oil-drop apparatus under the influence of an electric field of intensity 51 kV m^{-1}. How many electronic charges does it carry?

Solution. The droplet is stationary under the influence of two forces, its weight acting downwards and the electric force on it acting upwards, and these forces must be equal in magnitude. Thus

$$Eq = mg = \tfrac{4}{3}\pi\rho r^3 g.$$

Therefore

$$q = \frac{\frac{4}{3}\pi\rho r^3 g}{E}$$

$$= \frac{\frac{4}{3}\pi \times 10^3 \text{ kg m}^{-3} \times 10^{-18} \text{ m}^3 \times 9.8 \text{ m s}^{-2}}{5.1 \times 10^4 \text{ V m}^{-1}}$$

$$= \frac{41.05 \times 10^{-15} \text{ N}}{5.1 \times 10^4 \text{ V m}^{-1}} = \frac{41.05 \times 10^{-15} \text{ N}}{5.1 \times 10^4 \text{ N C}^{-1}}$$

$$= 8.0 \times 10^{-19} \text{ C}.$$

Since $e = 1.6 \times 10^{-19}$ C, it follows that the drop carries

$$\frac{8.0 \times 10^{-19} \text{ C}}{1.6 \times 10^{-19} \text{ C}} = 5 \text{ electronic charges.}$$

Example 4. A hydrogen nucleus (a proton) of mass 1.67×10^{-27} kg is accelerated from rest through a potential difference and acquires a velocity of 4.00×10^6 m s^{-1} in the process. What is the magnitude of the potential difference?

Solution. The kinetic energy acquired by the proton is $\frac{1}{2}mv^2$, m being its mass and v its final speed. The energy therefore has the value

$$\tfrac{1}{2} \times 1.67 \times 10^{-27}\ \text{kg} \times (4.00 \times 10^6\ \text{m s}^{-1})^2$$

$$= \frac{\tfrac{1}{2} \times 1.67 \times 10^{-27} \times 16.00 \times 10^{12}\ \text{J}}{1.60 \times 10^{-19}\ \text{J eV}^{-1}}$$

$$= 8.35 \times 10^4\ \text{eV}.$$

The proton carries the elementary electronic charge e. When it is accelerated through a potential difference of 1 V, it acquires an energy of 1 eV. In this case it must be accelerated through 83.5 kV.

Example 5. One early form of X-ray set obtained its energy by charging capacitors to a high voltage and then discharging them. If a 0.25-μF capacitor is charged to 50 kV and discharged every 0.1 s, find the charge stored in each charging operation, and the average rate at which charge flows through the circuit.

Solution. The charge acquired will be

$$Q = CV = 0.25 \times 10^{-6}\ \text{F} \times 5 \times 10^4\ \text{V}$$

$$= 1.25 \times 10^{-2}\ \text{C}$$

and this is stored in the capacitor. This charge flows through the circuit on discharge and the average rate of flow is obtained by dividing the charge delivered to the circuit by the time for which it is delivered. Hence

$$Q/t = 1.25 \times 10^{-2}\ \text{C}/0.1\ \text{s}$$

$$= 125\ \text{mC s}^{-1}.$$

Example 6. A 1.0-μF capacitor is charged to 100 V and a 2.0-μF capacitor is charged to 200 V. If they are now joined positive plate to positive plate, what is the common potential difference acquired?

Solution. Before joining, $C_1 = 1\ \mu$F, $V_1 = 100$ V and

$$Q_1 = C_1V_1 = 100\ \mu\text{C}.$$

Also $C_2 = 2\ \mu$F, $V_2 = 200$ V and thus

$$Q_2 = C_2V_2 = 400\ \mu\text{C}.$$

Joined conductors must come to a common potential. Thus each capacitor finishes up with the same potential difference V across its plates when joining takes place. In coming to a common potential difference, the charges redistribute themselves and the capacitors finish up with charges $\pm Q_1'$ and $\pm Q_2'$ on their positive and negative plates respectively. But no charge can be lost in the process. Therefore

$$Q_1' + Q_2' = Q_1 + Q_2 = 500\ \mu\text{C}.$$

Also

$$V = \frac{Q_1'}{C_1} = \frac{Q_2'}{C_2}.$$

Therefore

$$Q_1'/Q_2' = C_1/C_2.$$

Therefore

$$(C_1/C_2)Q_2' + Q_2' = 500\ \mu\text{C}.$$

Therefore

$$Q_2' = \frac{500\ \mu\text{C}}{\frac{1}{2} + 1} = 333\ \mu\text{C}.$$

Therefore

$$V = \frac{333\ \mu\text{C}}{2\ \mu\text{F}} = 167\ \text{V}.$$

PROBLEMS

22.1 If a charged rod is brought up to dry pieces of dust, the latter are often attracted to the rod. Once they touch the rod, the pieces of dust are often violently repelled. Why does this happen?

22.2 An electron has a mass of 9.1×10^{-31} kg and an electric charge of -1.6×10^{-19} C. The gravitational force between two bodies of masses m and M a distance d apart is

$$F = GMm/d^2,$$

where $G = 6.6 \times 10^{-11}\ \text{Nm}^2\ \text{kg}^{-2}$. Compare the gravitational and electrical forces acting between two electrons.

22.3 When a negatively charged rod is placed near one part of an uncharged insulated metallic sphere, electrons are repelled from that part. Why does this flow of electrons cease, since there is a very large quantity of electrons in the sphere?

22.4 A charged rod is placed near two touching metal spheres which are initially uncharged, and the spheres are separated while the rod is still in position. The spheres are found to attract each other with a force of 8.99×10^{-5} N when 10 cm apart. How many electrons moved from one sphere to the other during the process of induction?

22.5 When a negatively charged sphere is brought near a suspended body, the latter is attracted to it. Is it correct to assume that the body is positively charged?

22.6 The center of a conducting sphere carrying a charge of 1.0×10^{-8} C is situated 3 cm above the center of a small, light, metalized, ball which carries a negative charge and has a mass of 0.05 g. If the ball does not fall to the earth, what is the magnitude of the charge it carries?

22.7 Two equal conducting spheres of negligible size are charged with 5.0×10^{-14} C and -6.0×10^{-14} C, respectively, and placed with their centers 20 cm apart. If they are moved to positions 50 cm apart, compare the forces between them in the two positions.

The spheres are connected by a thin wire. What force does each now exert on the other?

22.8 Two light conducting balls each of mass 1 g are suspended by silk threads 1 m long from a common point. When equal charges are placed on the balls, the system comes to equilibrium with each string making an angle of 5° with the vertical. What is the charge on each ball? Would the threads make equal angles with the vertical if the charges were unequal?

22.9 Four equal charges of 5×10^{-10} C are located at the four corners of a square of side 10 cm. Calculate the magnitude and direction of the force on each charge.

22.10 Three equal charges of 5×10^{-10} C are located at three of the four corners of a square of side 10 cm. Calculate the electric intensity and the potential at the fourth corner.

22.11 Two point charges of magnitude 4×10^{-8} C and -9×10^{-8} C are 50 cm apart in vacuum. At what points are the electric intensity and the potential zero?

22.12 At a point on the earth's surface the potential gradient is 100 V m^{-1}. Calculate the charge density on the earth's surface at that point. If the earth and the moon were each charged with a surface density of this magnitude, calculate the electrostatic repulsion between them and compare this with the gravitational attraction.

22.13 The electron in a hydrogen atom in its normal state occupies an orbit of radius 0.52×10^{-10} m. Calculate the potential at any point on the orbit due to the nuclear charge.

22.14 Calculate the work done against electrostatic forces in moving a charge of -100 pC from a position 10 cm below a point charge of 10 μC to one 1 m below it. In the final position the negatively charged body remains suspended, the electrostatic and gravitational forces on it being equal and opposite. What is the mass of the body?

22.15 In a sodium chloride crystal the unit cell is cubic with length of side 5.64 Å. The Na^+ ions occupy positions at the center of the cube and at the centers of each side. The Cl^- ions occupy positions at the corners and at the centers of each face. Calculate the potential energy of the Na^+ ion in the center of the cube in such an arrangement.

22.16 The potential difference causing a lightning-flash may be as high as 1 GV and the charge passing may go up to around 40 C. How much energy is transformed during the flash and to what temperature would this amount of energy raise 10^6 kg of water from an initial temperature of 20°C?

22.17 If a pocket comb is rubbed with fur, it is possible to raise its potential to 10^4 V. Why is this large voltage not dangerous but the much smaller voltage of a domestic electric supply is?

22.18 Calculate the radius of a water droplet which, while carrying 50 electronic charges, floats near the earth's surface where the electric field is vertical and has magnitude 100 Vm^{-1}.

22.19 An oil droplet of radius 3.41×10^{-4} cm and density 0.96 g cm^{-3} is falling between vertical plates a distance of 1 cm apart with a constant terminal velocity of 0.253 cm s^{-1}. When a potential difference of 6000 V is applied between the plates, the droplet acquires in addition a constant horizontal velocity of 0.207 cm s^{-1}. How many excess electrons does the drop contain?

22.20 What are the velocities of (a) electrons with energies of 100 eV, (b) protons of mass 1.67×10^{-27} kg with energies of 0.5 MeV, and (c) oxygen atoms of mass 2.68×10^{-26} kg with energies of 5 MeV?

22.21 The dielectric strength of air is 3 MV m^{-1}. What is the maximum charge that can be placed on a conducting sphere of radius 5 cm without breakdown occurring? What is then the potential of the sphere?

22.22 Three capacitors have capacitances of 3 μF, 10 μF, and 15 μF. How may they be connected to produce capacitances of (a) 2 μF, (b) 9 μF, and (c) 12.5 μF?

22.23 The three capacitors of Problem 22.22 are connected in series with a 100-V cell. What is the charge and potential difference on each capacitor?

22.24 The three capacitors of Problem 22.22 are connected in parallel and the combination is joined in series with a 100-V cell. What is the charge and potential difference on each capacitor?

22.25 A 2-μF capacitor is charged to 500 V and a 5-μF capacitor to 100 V. The capacitors are then joined positive plate to positive and negative plate to negative. What is the charge and potential difference on each capacitor? What loss of energy has taken place and where has this energy gone?

22.26 Repeat Problem 22.25 but this time join the positive and negative plates of the charged capacitors.

22.27 A capacitor with air between its plates is connected to a 100-V cell. The capacitor is then immersed in non-conducting oil of dielectric constant 2.5, and the charge on the positive plate increases by 200 μC. What was the original capacitance of the capacitor?

22.28 A tuning capacitor has a capacitance which can be changed from 0.1 μF to 0.005 μF by rotation of the plates. It is charged to a potential difference of 50 V at the maximum value of capacitance and disconnected from the electrical source.

If the plates are now turned to produce minimum capacitance, what is the potential difference between the plates and what work had to be done during the rotation of the plates?

22.29 Calculate the equivalent capacitance of the arrangement in Fig. 22.15.

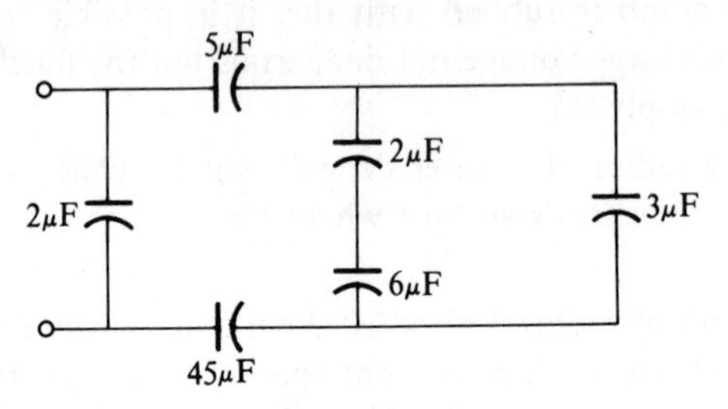

Figure 22.15

CHAPTER 23

CONDUCTION IN SOLIDS

23.1 INTRODUCTION

If the plates of a charged capacitor are joined by a conducting wire, charge will travel from one plate to the other under the action of the potential difference. Although what happens in most cases is that electrons flow through the wire (cf. Section 25.7), it is conventional to indicate the direction of the flow of charge as though positive charge were flowing from the place of high potential to the place of lower potential. This convention was adopted before the mechanism of charge transfer was understood and is so deep-rooted that it has never been changed.

The flow of charge is called a current, and in the example considered the current will vary from a high value to zero as the potential difference between the plates of the capacitor diminishes. The average current over a period of time is defined as Q/t, the total charge passing any point in the circuit divided by the time for which it passed. The instantaneous current will be measured in the same way, the time of passage being taken as small as possible. Thus $i = dQ/dt$. The units of current are coulombs per second, which are given the special name of amperes. If the current is a steady one, keeping the same value at all times, $i = dQ/dt = Q/t$, and the instantaneous current is the same as the average current. In order to get a steady current, the potential difference between the two ends of the wire through which the charge is flowing must be kept constant. Cells and generators which can do this will be discussed later.

23.2 OHM'S LAW

We will assume that there are available a number of identical sources of electric power of some kind—cells, for example—each of which will maintain a constant potential difference between two points of a conductor. There are various devices dependent only on the phenomena which have so far been discussed which would allow the measurement of the current passing through the wire. Since these are now obsolete except for highly specialized purposes, they will not be described; but let us assume that there is available one such device which registers the amount of current flowing by displacing a needle on a dial. With these pieces of apparatus a circuit as shown in Fig. 23.1 is set up. AB is a conductor, G is the current-measuring device, and D represents a collection of cells.

The experiment is performed several times, first with one cell in the circuit, then with two, three, and so on. The potential difference across AB is assumed to take the values V, then $2V$, $3V$, etc. This assumption is not quite correct but is almost correct, as will be seen later. It is found that, when the potential difference across AB is $V, 2V, 3V, \ldots, nV$, the current passing through AB is $I, 2I, 3I, \ldots, nI$. The conductor has been carefully chosen so that even with nI flowing through it the temperature does not rise appreciably. From the results of the experiment Ohm's law has been verified. This law states that if a conductor is kept at constant temperature, the current flowing through it is directly proportional to the potential difference between its ends.

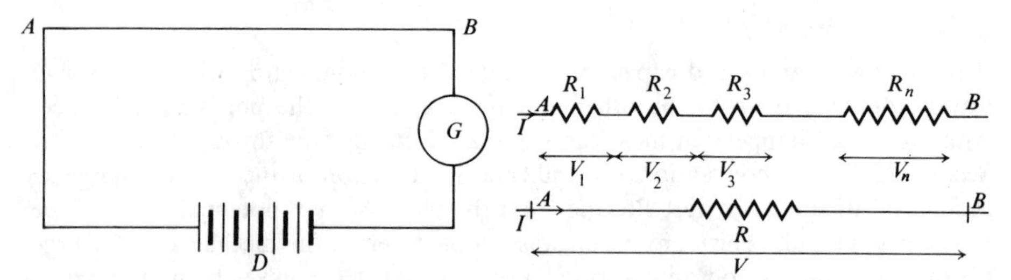

Fig. 23.1. An experimental arrangement for verifying Ohm's law.

Fig. 23.2. Resistors in series, and the equivalent resistor.

If the law is put in mathematical form,

$$\frac{V}{I} = \text{constant} = R. \tag{23.1}$$

R is called the resistance of the conductor and is therefore measured in volts per ampere, or ohms (Ω). One ohm is the resistance of a conductor if a current of 1 ampere flows through it when there is a potential difference of 1 volt across its ends.

A component which possesses resistance is known as a resistor.

23.3 RESISTANCES IN SERIES AND IN PARALLEL

It is necessary to know the combined effect of joining resistors in series and in parallel. In Fig. 23.2 there is shown a steady-state situation in which a current I is flowing through a number of resistors joined in series whose resistance are R_1, $R_2, \ldots, R_n$. It must be the same current through each of the resistors or charge would accumulate at some point, altering the potential there, and thus the current, and it would not be a steady-state situation. If there is a potential difference of V across the ends of the whole network, there must be an equivalent single resistor of resistance R which could replace the chain of resistors without altering the potential difference V or the current I. Then, by Ohm's law,

$$IR = V = V_1 + V_2 + V_3 + \ldots + V_n = IR_1 + IR_2 + IR_3 + \ldots + IR_n.$$

Therefore

$$R = R_1 + R_2 + R_3 + \ldots + R_n. \tag{23.2}$$

It is therefore seen that the equivalent resistance has a value equal to the sum of the component resistances.

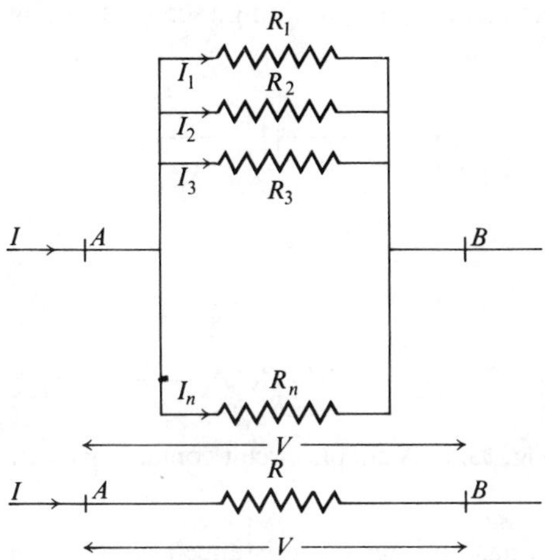

Fig. 23.3. Resistors in parallel, and the equivalent resistor.

Figure 23.3 shows the resistances connected in parallel. Once again, there must be a single resistor which can replace the collection of resistors without altering the circuit conditions. Since there can be no accumulation of charge at the points A and B,

$$I = I_1 + I_2 + I_3 + \ldots + I_n.$$

The potential difference across each of the single resistors, and also across the equivalent resistor of resistance R, must be V. Therefore, by Ohm's law,

$$\frac{V}{R} = I = I_1 + I_2 + I_3 + \ldots + I_n = \frac{V}{R_1} + \frac{V}{R_2} + \frac{V}{R_3} + \ldots + \frac{V}{R_n}.$$

Therefore

$$\frac{1}{R} = \frac{1}{R_1} + \frac{1}{R_2} + \frac{1}{R_3} + \ldots + \frac{1}{R_n}. \tag{23.3}$$

Thus the sum of the reciprocals of the resistances equals the reciprocal of the equivalent resistance.

Note that these results are the opposite of those obtained for combinations of capacitors.

23.4 ELECTROMOTIVE FORCE

In dealing with resistors in series, we obtained the potential difference across the equivalent resistance by adding up the potential differences across the individual resistors. In other words, $V = V_1 + V_2 + \ldots + V_n$. This is obviously true, since the work done in taking a unit positive charge from one end of the network to the other is merely the sum of the amounts of work done in taking the charge across each individual resistor. But this is only true if no cell or generator is in the way. Consider a complete circuit containing a cell, as in Fig. 23.4.

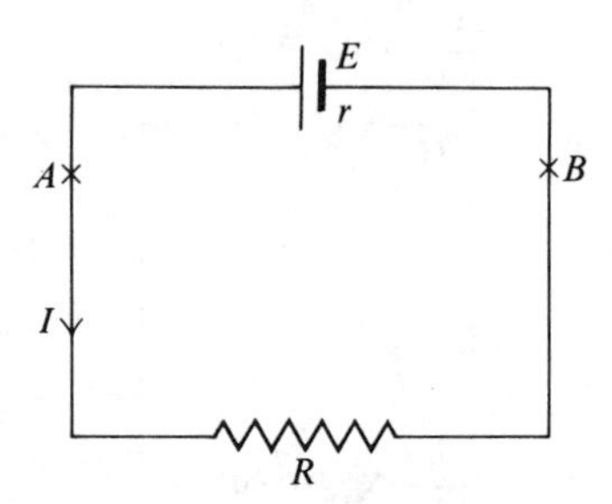

Fig. 23.4. A simple circuit containing a cell.

The cell in general has an internal resistance r which is small. This is why one does not quite get nV when n cells are inserted in series in a circuit, as in the experimental verification of Ohm's law. In going from A to B in Fig. 23.4 through the resistance R, $I \times R = V$, the potential difference from A to B. In going from B to A through the battery, there is an additional potential drop $I \times r$. But in arriving back at A again, the potential must be the same as at the start of the journey. In going through the cell from negative plate to positive plate, the charge is being raised to a higher potential and work must be being done on it. In fact, to put the charge in motion and keep it going round the complete circuit, the charge gaining kinetic energy in so doing, must require the conversion of some sort of energy (chemical in a cell, mechanical in a generator) into electrical energy. If the quantity of energy necessary to pass unit charge round the complete circuit is E, this is called the electromotive force (emf) of the source. E thus has the units volts in the MKSA system and

$$E = IR + Ir. \tag{23.4}$$

Thus

$$V = IR \qquad \text{and} \qquad E - V = Ir.$$

The potential difference from A to B through the external resistor is $V = IR$ and through the cell is $V = E - Ir$. The potential difference from B to A through the cell is thus $-E + Ir$. The potential difference between any two points, if no cell is in the path, is merely the sum of all the products IR along the path, I being taken as positive if one moves with the current and negative if one moves against it. If cells occur along the route, the potential difference is obtained from the sum of all the products IR for every part of the path, I being signed as before, plus the sum of all the emf's included, E being taken as positive if the route goes from positive plate to negative plate, and vice versa.

Consider the following example to see how this works in practice.

Example. Two cells, one of emf 1.5 V and 0.2 Ω internal resistance, the other of emf 1.0 V and 0.3 Ω internal resistance, are connected in opposition, and joined by a conductor of 5-Ω resistance (Fig. 23.5). What is the current in the circuit?

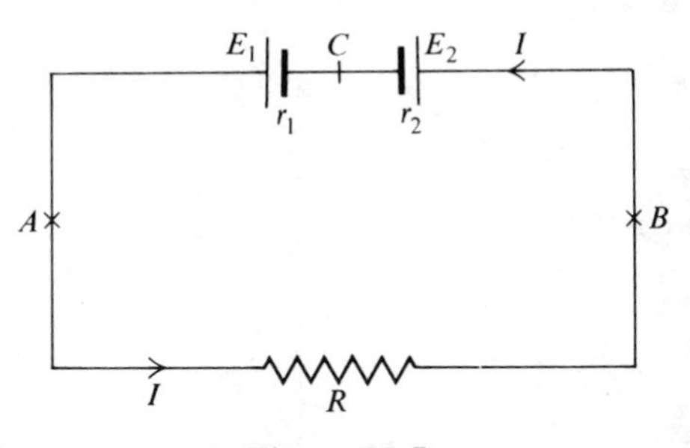

Figure 23.5

Solution. The same current I flows through all three resistors in series. Then, by the rules detailed above,

$$V_{AB} = IR,$$
$$V_{BC} = E_2 + Ir_2,$$
$$V_{CA} = -E_1 + Ir_1.$$

But

$$V_{AB} + V_{BC} + V_{CA} = 0.$$

Therefore

$$0 = E_2 - E_1 + I(r_1 + r_2 + R).$$

Therefore

$$E_1 - E_2 = I(r_1 + r_2 + R)$$

or

$$0.5\text{ V} = I(0.2\ \Omega + 0.3\ \Omega + 5\ \Omega).$$

Therefore

$$I = \frac{0.5\text{ V}}{5.5\ \Omega} = \frac{1}{11}\text{ A}.$$

It will be seen that an expression analogous to Ohm's law for a conductor applies to the full circuit. The algebraic sum of the emf's in the circuit is equal to the current flowing times the total resistance in the circuit.

Note that the emf is the total energy required to send a unit positive charge round the full circuit, while the terminal voltage V across the plates is the energy required to send unit positive charge round the external circuit. E is only equal to V if no current is flowing, because a quantity of energy Ir is necessary to send unit positive charge through the internal resistance of the cell.

23.5 RESISTIVITY

If a number of wires are selected, the wires identical except that each has a different length, the same potential difference V can be established across each one in turn. It is found that if the current I flowing through the wire is plotted against $1/l$, where l is the length of the wire, a straight-line graph is obtained, as in Fig. 23.6. For fixed V, however, I is proportional to $1/R$, where R is the resistance of the wire. Therefore $R \propto l$.

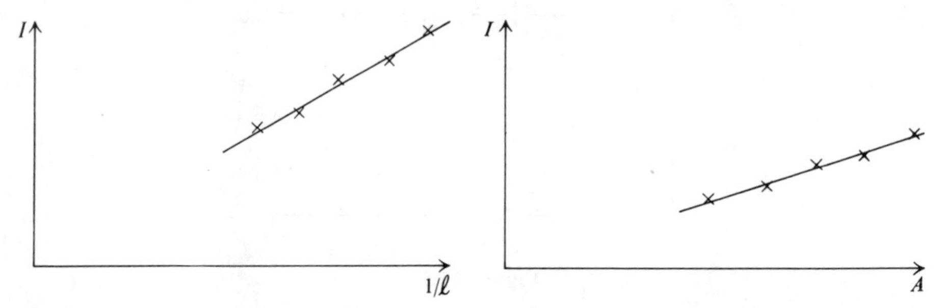

Fig. 23.6. The current through a wire which has a constant potential difference maintained between its ends as a function of the reciprocal of the length.

Fig. 23.7. The current through a wire which has a constant potential difference maintained between its ends as a function of the cross-sectional area.

Similarly, if the experiment is repeated with wires of the same material and length but of different cross-sectional areas, a plot of I against A, where A is the cross-sectional area, for fixed V, gives a straight-line graph once more (Fig. 23.7). Thus $R \propto 1/A$. Combining these two results, one may write

$$R \propto l/A = \sigma l/A, \tag{23.5}$$

where σ is defined as the resistivity of the material. Resistivity thus has the units ohm-meters and the resistivity is defined as equal to the resistance of a cube of the material of side 1 meter when the current flows perpendicular to opposite faces.

23.6 TEMPERATURE COEFFICIENT OF RESISTANCE

In Section 23.2, it was carefully stated that the current through a conductor was proportional to the potential difference between its ends only if the temperature remained constant. The resistivity of most substances increases with temperature, although a few, such as carbon and many semiconductors, have resistivities which decrease with temperature, and some alloys have resistivities at room temperature which vary little over a fair range of temperatures. It is possible to investigate the variation in resistance of a resistor by immersing a coil of wire in insulating oil in a heating tube and immersing the latter in a water-bath. Then, if the same potential difference is applied across the ends of the coil at all temperatures, it is

Table 23.1. The resistivity and the temperature coefficient of resistance of some common substances at room temperature

Substance	$\sigma(10^{-8}\ \Omega\ \text{m})$	$\alpha(10^{-3}\ \text{deg}^{-1})$
Aluminum	2.64	3.9
Copper	1.72	4.0
Silver	1.53	3.8
Tungsten	5.50	4.6
Constantan	49	0.002
Nichrome	106	0.3
Carbon	3,500	−0.5

found that the current flowing through the coil varies with temperature. With most substances the decrease in current is found to be roughly proportional to the temperature difference, which shows that the increase in resistance varies directly with temperature difference over a small temperature range, such as from 0°C to 100°C. Over this range most conductors obey the equation

$$R_\theta = R_0[1 + \alpha(\theta - \theta_0)]. \tag{23.6}$$

Thus

$$\alpha = (R_\theta - R_0)/R_0(\theta - \theta_0),$$

where $\theta - \theta_0$ is the temperature difference, and α is called the temperature coefficient of resistance and is the change in resistance per unit resistance per unit rise in temperature. Its units will be deg^{-1}. Over a wider temperature range it is found that

$$R_\theta = R_0[1 + \alpha(\theta - \theta_0) + \beta(\theta - \theta_0)^2]. \tag{23.7}$$

This effect is not linked to the increase in the dimensions of a resistor due to thermal expansion. The coefficient of linear expansion of a conductor is of the order of $10^{-5}\ \text{deg}^{-1}$, and the temperature coefficient of resistance is of the order of $10^{-3}\ \text{deg}^{-1}$.

23.7 THE WHEATSTONE BRIDGE

In practice the value of a resistance is seldom determined by measuring the current passing through it and the potential difference across its ends, and dividing one by the other. This is not accurate for reasons which will become fairly obvious. A voltmeter is a conventional instrument for measuring voltage and an ammeter one for measuring current. Both of these instruments, which will be described later (cf. Section 25.5), have a certain resistance themselves. A voltmeter should have as high a resistance as possible and an ammeter as low a resistance as possible, in order that they shall disturb as little as can be arranged the circuit in which they are making measurements. If it is decided to measure a resistance by using a voltmeter which has a resistance R' and which is put in parallel with the resistor of resistance R, and by using an ammeter which has a resistance r and is put in series with R, the two possible methods of connection are shown in Fig. 23.8.

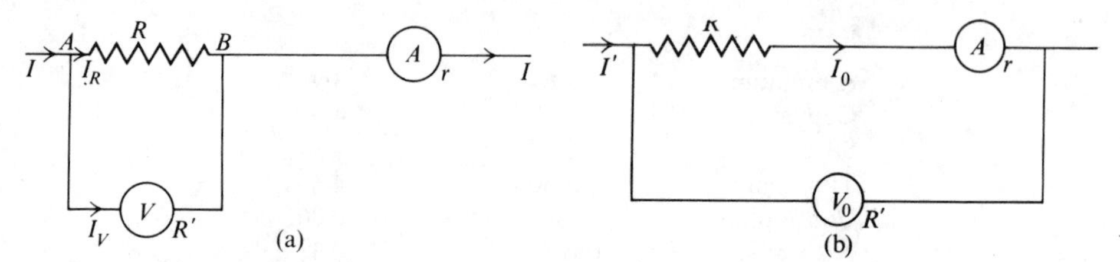

Fig. 23.8. The two methods of connection of a circuit which attempts to measure resistance by the use of a voltmeter and an ammeter.

In the first situation I divides into I_R and I_V, where $I = I_R + I_V$, and since V_{AB} is the same across either route from A to B,

$$I_R R = I_V R' \qquad \text{or} \qquad \frac{I_R}{I_V} = \frac{R'}{R}.$$

Therefore

$$\frac{I_R}{I_R + I_V} = \frac{I_R}{I} = \frac{R'}{R + R'}.$$

Therefore

$$V = I_R R = \frac{RR'}{R + R'} I.$$

Therefore

$$\frac{V}{I} = \frac{RR'}{R + R'} = R\left(\frac{1}{1 + \dfrac{R}{R'}}\right).$$

As R' tends to infinity, V/I tends to R; but if R and R' are comparable, then V/I may be considerably different from R.

In the second situation

$$V_0 = I_0(R + r).$$

Therefore

$$\frac{V_0}{I_0} = R + r.$$

If r tends to zero, V_0/I_0 tends to R; but if r is comparable with R, the answer may differ widely from R.

It would be desirable for accurate work to have a simple and accurate method of determining resistance, and this is provided by the Wheatstone bridge.

It consists of a network of four resistors, as in Fig. 23.9, X being the unknown, R_1, at least, being variable and of a known magnitude, and R_2 and R_3, or at least their ratio, being of known magnitude. It helps the sensitivity of the bridge if all four resistors have approximately the same resistance.

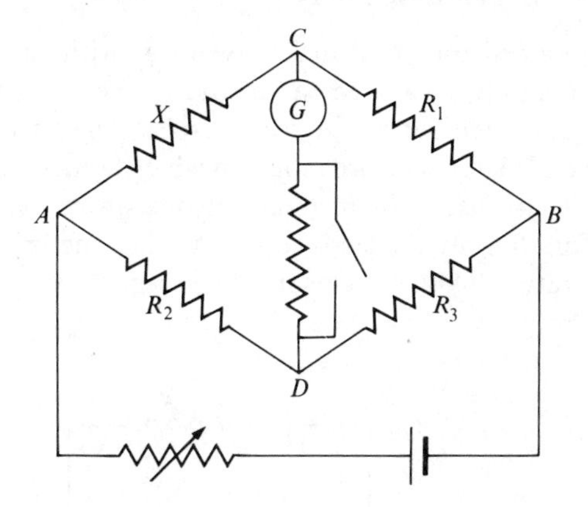

Fig. 23.9. The basic Wheatstone bridge network.

Points A and B are linked through a cell and a variable resistor, the latter to control the current from the cell; and C and D are linked by a current-detecting device, known as a galvanometer and to be described later (cf. Section 25.4), with a protective resistance in series to limit the current which passes through the galvanometer. This protective resistance may be shorted out to give greater accuracy near the balance point. One or more of the known resistances is adjusted until no current flows through the galvanometer. Then, if current I_1 flows through AC, it must also at balance flow through CB; similarly, I_2 flows through AD and DB at balance. Since no current flows through the galvanometer, C and D must be at the same potential:

$$V_{AC} = V_{AD} \quad \text{or} \quad I_1 X = I_2 R_2$$

and

$$V_{CB} = V_{DB} \quad \text{or} \quad I_1 R_1 = I_2 R_3.$$

Therefore

$$X = \frac{R_2}{R_3} \times R_1. \tag{23.8}$$

In practice the network does not look like the simple arrangement in the diagram. In the meter-bridge form, which is often used in the laboratory, R_2 and R_3 together form the resistance of a long uniform wire. Balance is obtained by adjusting a sliding contact D on the wire until no current flows through the galvanometer. The ratio R_2/R_3 is then merely the ratio of lengths AD/DB. In more complicated arrangements the whole network is often enclosed in a box, the ratio R_2/R_3 being adjusted on a dial, X being clipped on to a pair of sockets, and R_1

being adjustable on a dial. The cell and galvanometer may either be joined to terminals on the box or even be included inside the box as part of the arrangement.

23.8 THE RESISTANCE THERMOMETER

Since the resistance of a resistor varies in a known way with temperature and since resistance can be accurately measured, a resistance thermometer can obviously be constructed to measure temperature by the variation in resistance of a resistor. This is particularly useful in temperature ranges in which conventional thermometers cannot work, and it is also useful in enclosed regions where the thermometer may be placed once and for all, only the leads to the Wheatstone bridge requiring to be left free. The temperature of the enclosure may then be measured without disturbing it.

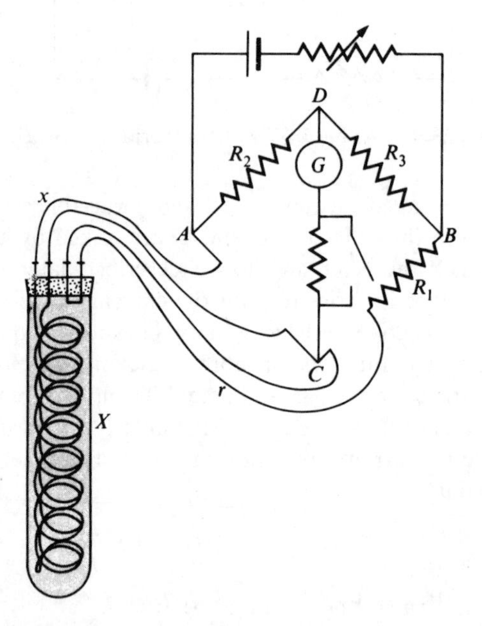

Fig. 23.10. A platinum resistance thermometer and the associated bridge network.

The thermometer, often of platinum because it has a high melting point and does not take part easily in chemical reactions, consists of a coil of fine wire wound on an insulating quartz rod. The leads from the two ends of the resistor tend to heat up also and change their resistance in consequence, and it is normal to have a dummy pair of identical leads, joined together in the thermometer as shown in Fig. 23.10, running close beside the real leads. The unknown resistance in the Wheatstone bridge then consists of the thermometer of resistance X plus its leads of resistance x. The variable resistance consists of the standard resistance

R_1 plus the resistance r of the dummy leads which are placed in series with it. The ratio of R_2/R_3 is arranged to be 1. On balance,

$$\frac{X + x}{R_1 + r} = 1.$$

Therefore

$$X + x = R_1 + r.$$

But since the real and dummy leads are identical and in a similar environment,

$$x = r,$$

therefore

$$X = R_1,$$

and the standard resistance is, at balance, equal to the resistance of the thermometer at all temperatures.

At the high temperatures normally measured, the approximate relation given before is not accurate enough and one must use the full relation

$$R_\theta = R_0[1 + \alpha(\theta - 20°\text{C}) + \beta(\theta - 20°\text{C})^2], \tag{23.7}$$

α and β being constant, θ being the Celsius temperature and R_0 normally being the resistance at room temperature, taken to be 20°C. If the resistance is measured at the ice-point, the steam-point and the boiling point of pure sulfur, the three constants α, β, and R_0 can be determined and the thermometer can be calibrated. Alternatively, a graph can be drawn of R against θ for a number of known temperatures and any other temperature obtained from the graph. If temperatures above 660°C, or well below 0°C, are to be measured, even the equation given above is no longer accurate and it is necessary to employ the relation

$$R_\theta = R_0[1 + \alpha(\theta - 20°\text{C}) + \beta(\theta - 20°\text{C})^2 + \gamma(\theta - 20°\text{C})^3(\theta - 100°\text{C})]. \tag{23.9}$$

One of the disadvantages of a conventional resistance thermometer is its large size. It cannot be used in very small enclosures and has an appreciable response time when an attempt is made to follow varying temperatures. Because of this it is becoming more and more usual to employ *thermistors* for this purpose. Initially, these were made of metallic oxides compressed and fired at elevated temperatures and produced in the form of small beads, disks, etc. Nowadays a variety of semiconductor devices are also used.

Thermistors have the advantage of very fast response times. They have appreciable resistances, and negative temperature coefficients of resistance which are, in general, much larger in magnitude than those of metals or alloys. At 20°C they may vary in resistance by around 5% for every degree change of temperature. With an accurate resistance-measuring device it is possible to measure temperature differences of the order of 10^{-3} deg. The average thermistor works over a range from around -50°C to 300°C.

More recently, man-made diamond crystals have been employed. These have very high stability and fast response, and are completely corrosion-resistant. A single diamond thermistor can record temperatures from around $-200°C$ to 650°C and is the only device which can measure temperature over such a wide range without any discontinuity whatsoever.

23.9 THE STRAIN-GAGE

The resistance thermometer is one type of *transducer*, the latter being defined as a device which converts a signal of one type into a signal of another type. Since the ease of measurement and accuracy obtainable differ markedly from one physical quantity to another, transducers are much employed to convert physical signals to a more manageable form. Electrical measurements can, in general, be made with ease and accuracy, and there are many devices for conversion of signals to an electrical output. In the case of the resistance thermometer, temperature readings are converted to electrical readings, which are easier to handle and give greater accuracy.

Another quantity difficult to measure but often required is pressure. A convenient transducer which converts pressure effects to electrical effects is the strain-gage. Since the resistance of a wire is given by Eq. (23.5) as

$$R = \frac{\sigma l}{A},$$

any stress applied to the wire which changes its length (or, indeed, its area) must produce a corresponding change in the resistance of the wire. The quantity $(\Delta R/R)/(\Delta l/l)$ is called the strain factor of a particular device and is denoted by the symbol K. It follows that

$$\Delta R = K\frac{\Delta l}{l}R = K\varepsilon R, \tag{23.10}$$

where ε is the strain produced.

A typical wire strain-gage is shown in Fig. 23.11. The wire is formed in a grid pattern and bonded to some backing material. It is usual to employ materials

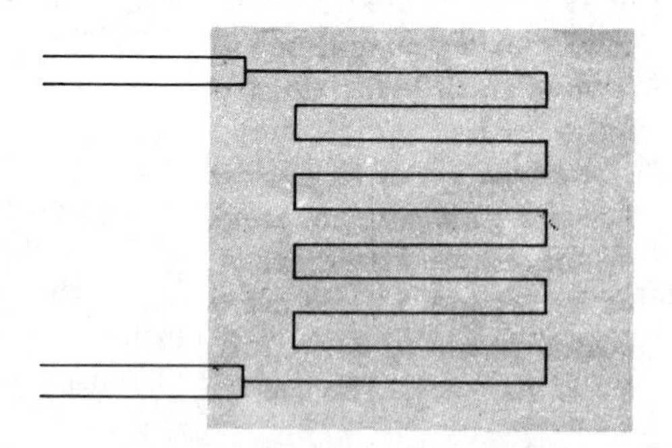

Fig. 23.11. A typical wire strain-gage.

such as constantan which have negligible temperature coefficients of resistance at normal temperatures. The size of the gage can be made as small as 2 or 3 mm across. Since the wire must not be taken beyond its elastic limit, the maximum permissible value of ε is around 0.1%. K has a value of around 2 and this means that $\Delta R/R$ is of the order of 0.2%. A very sensitive Wheatstone bridge arrangement is therefore required.

A strain-gage of this type is often used attached to a diaphragm, the deflection of which under changes of pressure produces strain in the gage element. Such gages are often swallowed by patients, the gage and the leads to it being encased in surgical rubber. This permits pressures in human tracts to be accurately measured.

Since strain-gages are often used under conditions of varying temperature, it is often necessary for accuracy to include in the measuring bridge circuit a compensating gage which is subjected to the same changes in temperature but no changes of pressure. This is very important in the case of semi-conductor strain-gages. These types have values of K perhaps 30 times that for wire strain-gages, but (cf. Section 33.2) are much more sensitive to temperature changes.

Strain-gages can be used to measure pressure variations of several atmospheres. They are routinely employed in blood-pressure measurements. Attached to a diaphragm they may be inserted, using a catheter, into veins or arteries. Rather less successfully and less accurately they have been used to measure the blood flow from the strains produced in a gage strapped by cuffs tightly round the arm over an artery. If the latter method could be improved, it would offer a simple and accurate method of measurement which did not involve discomfort and unpleasantness for the patient, which gave a continuous record, and which did not require the patient to remain relatively inactive. Strain-gages are also used to measure respiratory and urinary flow rates.

More recently, transistors have been employed as pressure and force transducers. If a tiny diaphragm is coupled to the emitter-base junction of a transistor (cf. Section 33.4), large reversible changes take place in the transistor characteristics for movements of the diaphragm. A force of around 2.5×10^{-3} N can alter the output of a transistor amplifier by around 1 V, the linearity of the change being better than 1% over the range. The small size of transistors makes them very useful for implanting into surfaces in contact with liquid or gas flow.

23.10 PIEZOELECTRICITY

Certain mineral crystals such as quartz may also be used as force or pressure transducers. When subjected to squeezing or stretching, an emf is set up in them proportional to the force applied. Conversely, if they have electricity passed through them, they expand or contract in length by an amount proportional to the voltage across them. It is believed that parts of the body, in particular skin, show piezoelectric properties and that this may explain why tactile sensations are converted to nerve impulses. Piezoelectric crystals are used in radiogram pick-ups to convert pressure effects from the needle running in the groove of the record to electrical impulses to be fed to the electronic circuits. It should be clear that a piezoelectric crystal can be used instead of a strain-gage as a pressure transducer and may therefore be employed in measuring blood flow, urinary flow, and respiratory rates.

23.11 THE POTENTIOMETER

The potentiometer is a device for the accurate measurement of potential differences and emf's. If a voltmeter is placed across the terminals of a cell, a small current flows and the voltage measured on the voltmeter is less than E by an amount Ir. Similarly, if a voltmeter is put in parallel with a resistor to measure the voltage across it, the voltage drops to $I_R R$ or $I_V R'$ (cf. Section 23.7) instead of IR. A precision instrument for the measurement of potential differences and emf's is therefore required. The potentiometer is such an instrument.

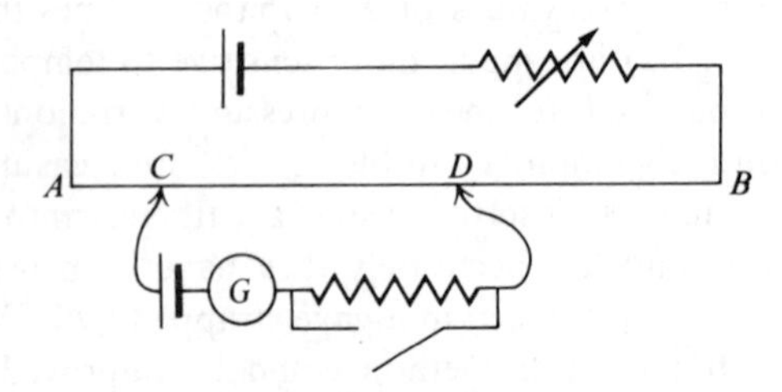

Fig. 23.12. The basic potentiometer network.

The instrument, drawn in basic form in Fig. 23.12, consists in essence of a long uniform wire joined in series with a battery and a variable resistor, the latter to adjust the drop in potential across the wire to a suitable value. If the potential difference across AB is V and the wire is uniform, the potential drops uniformly along the wire since voltage is proportional to resistance for constant current. Thus across any length CD the potential drop is

$$V_{CD} = V \times \frac{CD}{AB}.$$

To C is joined the positive terminal of a cell of unknown emf, the negative terminal being connected to a galvanometer, the other end of which is connected through a shortable safety resistance to a sliding contact which can move along AB. If D is moved along the wire until no current flows through the galvanometer, then V_{CD} equals the potential drop across the inserted cell. But no current is flowing through the cell, and when this happens, the terminal voltage of the battery equals the emf, E. Therefore $E = (CD/AB) \times V$. The potential difference V is not known but a standard cell can be inserted instead of the cell of unknown emf. A standard cell is a type of cell which has an accurately known emf at a fixed temperature, so long as no current is being drawn from it. A balance point is found when the standard cell, whose emf E' is known, is connected across length $C'D'$ of the wire AB. Therefore

$$E' = \frac{C'D'}{AB} \times V.$$

Hence,

$$\frac{E}{E'} = \frac{CD}{C'D'} \quad \text{or} \quad E = \frac{CD}{C'D'} \times E'. \tag{23.11}$$

In practice the potentiometer is all enclosed and may even contain a driving cell, a standard cell, and a galvanometer inside it. If not, terminals are provided on the case for joining in any or all of these components. Terminals will always be provided to join in the unknown emf or potential difference. The potentiometer wire is no longer in the form of a wire in such an instrument, but is as shown in Fig. 23.13. It takes the form of several resistors of equal resistance, normally 17, plus a slide wire whose resistance is equal to that of each of the resistors. The slide wire is connected from the point A and is joined by a thick strip of metal of negligible resistance to the resistors, which then run round to the point B. The unknown emf is joined to a sliding contact which can run over the wire. The other terminal is joined through a galvanometer and safety resistance to a wiper which runs over the studs to which the resistors are joined. It is thus possible to get the potential difference across an exact number of resistors plus part of the slide wire equal to that of the unknown potential difference. The studs joining the resistors are normally marked 0.1 V, 0.2 V, . . . up to 1.7 V, and the slide wire is divided into 100 equal divisions each of which registers a difference of 0.001 V. In order to read directly an unknown potential difference, the potentiometer must therefore be calibrated. This is done by switching in to the circuit a standard cell, setting

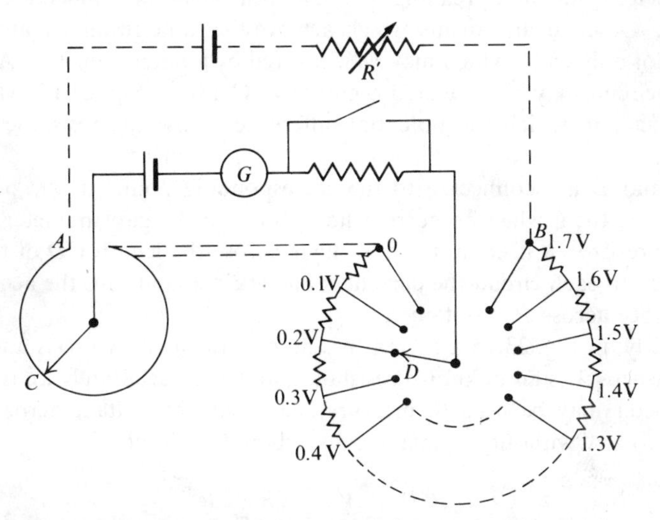

Fig. 23.13. The circuit diagram of a typical commercial potentiometer.

the dials on the potentiometer to, say, 10 of the unit resistors and 18.6 of the divisions on the wire for a Weston standard cell of emf 1.0186 V. The variable resistor R' is adjusted until no reading appears on the galvanometer. When this is done each resistor, and the whole of the slide wire, has a potential drop of 0.1 V across it. Note that in the more advanced potentiometers it may not be necessary to adjust the dials on the potentiometer as the standard cell may already be connected inside the box across the correct part of the circuit AB. After calibration the potentiometer will read directly in volts. The studs of the resistors are normally mounted in a circle so that the wiper can pass easily from one to another, and the slide wire is normally also mounted circularly for ease of moving the sliding contact over it. This instrument is called a direct-reading potentiometer.

If it is desired to measure very large voltages, one must cut the voltage down by use of a potential divider to manageable size. This is done by putting the potential difference across a resistor and tapping off a known fraction of it. If, on the other hand, one wishes to measure very small voltages, such as thermoelectric emf's (cf. Section 24.6), or piezoelectric emf's (cf. Section 23.10), which are of the order of millivolts, sufficient accuracy will not be obtainable from the normal direct-reading potentiometer. It is therefore necessary to modify the potentiometer set-up. Essentially the same arrangement is still required, but in addition, in series with the equal resistors and the slide wire, is placed a further resistor whose resistance is, say, accurately 999 times the resistance of slide wire and equal-resistor network. If V is the voltage dropped across AB, then only $10^{-3}\,V$ is dropped across the direct-reading portion. Thus the readings will now be in millivolts instead of volts. The standard cell is inserted by the manufacturer across a suitable portion of the large added resistor plus the direct-reading network so that, when it is switched into the circuit and zero reading is obtained on the galvanometer, the direct-reading portion will give an immediate reading in millivolts. Other arrangements which achieve the same result are also employed.

It is not only emf's which may be measured by a potentiometer. Any potential drop in a circuit may be measured accurately. Consider Fig. 23.14, which shows a typical circuit in which the potential difference across the resistance R is to be measured.

If C and D are connected to the corresponding points in the potentiometer of Fig. 23.13, then, when no current flows through the galvanometer, the voltage across the resistor R is equal to the voltage across the length CD of the potentiometer wire. In each circuit the current is undisturbed and thus the accurate potential difference across R is obtained.

Similarly, in Fig. 23.15 a current I is flowing through two resistances R and X in series, where X is an unknown resistance, and R an accurately known one. The potentiometer may be used to measure accurately the voltage across R and the voltage across X without, at balance, disturbing the circuit. Then,

$$\frac{CD}{C'D'} = \frac{V_R}{V_X} = \frac{IR}{IX} = \frac{R}{X}.$$

Therefore

$$X = \frac{C'D'}{CD} R.$$

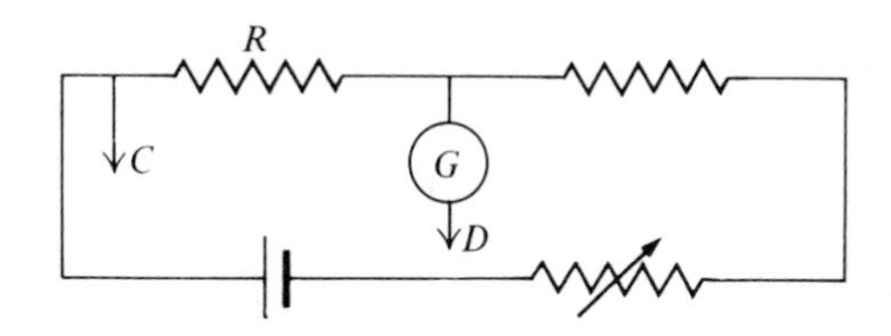

Fig. 23.14. The measurement of potential difference with a potentiometer.

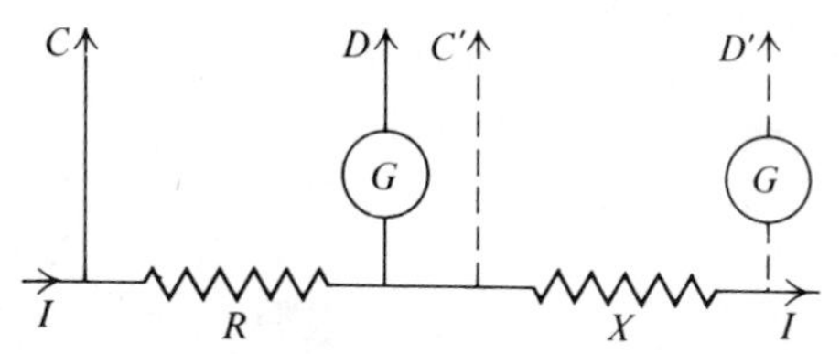

Fig. 23.15. The comparison of resistances with a potentiometer.

The Wheatstone bridge is not an accurate method of measuring small resistances, where $R \leqslant 1\ \Omega$, because resistances of contacts and leads become of importance. Under these circumstances the potentiometer method gives greater accuracy.

23.12 NON-OHMIC RESISTANCES

It should not be thought that all devices obey Ohm's law. This statement can be tested by varying the voltage across any device and measuring the current through it. If the device is a simple conductor, a straight-line graph is obtained; and when V is reversed, I reverses also (Fig. 23.16). But if one does the same with a rectifier, which consists of layers of copper and copper oxide, one finds that the graph is quite different. Practically no current flows if the voltage is put across in one direction; whereas if the voltage is reversed, a current passes which increases non-linearly (Fig. 23.17). The same will be true if the device is a semiconductor rectifier (cf. Section 33.3). Such devices may thus be used to turn alternating current into unidirectional current.

Note that a normal conductor, if it takes too much current and heats up, loses a great amount of energy by radiation; and as this is equivalent to an increase in resistance, its measured resistance increases non-linearly with voltage after a certain point.

23.13 THE HEATING EFFECT OF A CURRENT

If a potential difference exists between two points, a unit positive charge has higher potential energy at the point of higher potential than at the point of lower potential by an amount V. A charge Q has an extra quantity QV of potential energy. Thus if the charge Q flows from the place of higher potential to the point of lower potential, QV of work must be liberated. If the charge flows from the positive plate to the negative plate in a cell (cf. Section 24.4), work is used mainly in changing the chemical composition; if it flows through an electric motor (cf. Section 25.4), a large part of the work is changed to mechanical energy; but if the charge flows through a resistor, it has collisions with the atoms in the conductor, increasing their vibrational kinetic energy. The work therefore finally appears in the form of heat.

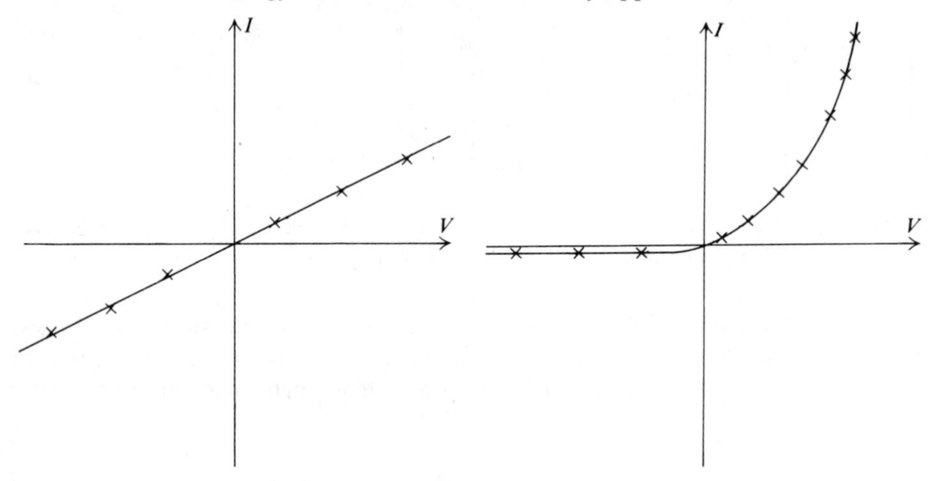

Fig. 23.16. A typical voltage-current characteristic curve for an ohmic resistor.

Fig. 23.17. A typical voltage-current characteristic curve for a rectifier.

Thus if charge Q flows through a conductor with a potential difference V maintained between its ends and the current I is constant, the heat produced is

$$H = QV = IVt = I^2Rt = V^2t/R. \tag{23.12}$$

If we use MKSA units for the quantities involved, the heat will be given in joules. The power developed in the circuit, this being the rate of doing work, is thus

$$\frac{H}{t} = IV = I^2R = \frac{V^2}{R}. \tag{23.13}$$

With MKSA units used, the power will be measured in watts.

Note that if the whole circuit is considered, R being the external resistance and r the internal resistance involved, the total heating is

$$I^2Rt + I^2rt = EIt, \tag{23.14}$$

which is the work done by the cell in the time t. The rate of working of the cell, i.e. the rate of production of electrical energy from some other form of energy, is therefore EI.

The way in which energy is measured, and sold, commercially is in kilowatt-hours, the normal electrical unit. Obviously, if a power of 1 W is developed for 1 s, 1 J of work is done. If 1 kW of power is developed for 1 h, the energy supplied is $10^3 \times 60 \times 60 = 36 \times 10^5$ J. It follows that 1 kWh is equal to 3.6×10^6 J.

23.14 THE CHARGING OF A CAPACITOR THROUGH A RESISTOR

In Fig. 23.18 is shown a circuit containing a resistor, a capacitor, and a switch in series with a cell of negligible internal resistance and emf E. If the switch is closed

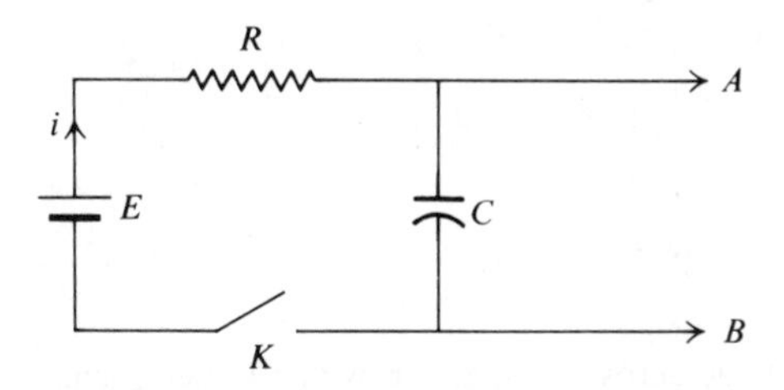

Fig. 23.18. A circuit for the charging of a capacitor through a resistor.

at time $t = 0$, the capacitor charges up, a current i, which varies with time, flowing through the circuit. At any instant the sum of the potential differences across the resistor and capacitor must equal the applied emf. Thus

$$E = iR + q/C,$$

where q is the charge on the positive plate of the capacitor at that instant. But

$$i = \frac{dq}{dt}.$$

Therefore

$$E = R\frac{dq}{dt} + q/C. \tag{23.15}$$

It would be possible to integrate this equation and obtain an expression for q at any time; but we will merely quote the result and verify that it satisfies the equation. In fact

$$q = CE(1 - e^{-t/RC}). \tag{23.16}$$

Therefore

$$\frac{dq}{dt} = CE\left(\frac{1}{RC}\right)e^{-t/RC} = \frac{E}{R}e^{-t/RC}.$$

Substituting back into Eq. (23.15) gives

$$E = R\left[\frac{E}{R}e^{-t/RC}\right] + E[1 - e^{-t/RC}] = E.$$

The solution therefore satisfies the equation. The voltage across the capacitor at any time is thus

$$v = q/C = E[1 - e^{-t/RC}]. \tag{23.17}$$

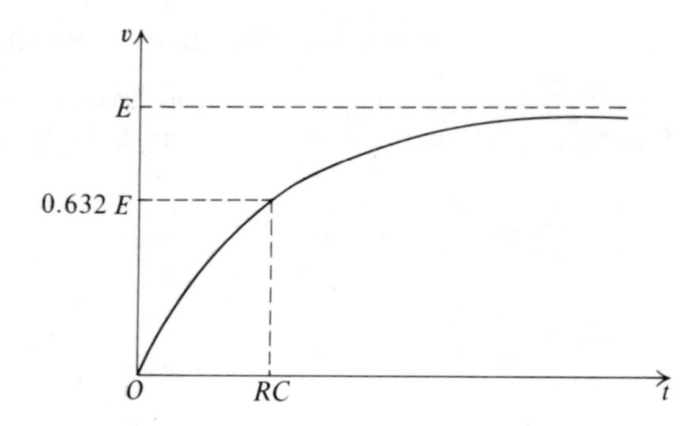

Fig. 23.19. The voltage across a capacitor which is being charged through a resistor as a function of time.

A graph of v against time is shown in Fig. 23.19. The quantity RC, which has the dimensions of a time, is called the time constant of the circuit. At time $t = RC$ the voltage has the value $E(1 - e^{-1}) = E(1 - 0.368) = 0.632\,E$. The quantity RC therefore determines how long it takes the voltage across the capacitor to approach its final, steady-state value. If R has the value 1 $M\Omega$ and C the value 5 μF, the time constant of the circuit is 5 s. For times small in comparison with this value $e^{-t/RC}$ is approximately $1 - t/RC$. Thus, with this approximation,

$$v = E[1 - (1 - t/Rc)] = Et/RC. \tag{23.18}$$

For circuits with a large time constant the voltage across the capacitor varies linearly with time over short time periods. This is very useful in a large number of applications (cf., for example, Sections 24.15 and 33.6). Across AB, in parallel with the capacitor, is placed a neon tube or a thyratron, devices which will only conduct if the voltage across them reaches a prescribed value V. The capacitor charges up linearly; and when the potential difference reaches V, a discharge takes place through the neon tube or thyratron. Almost instantaneously the potential difference across the capacitor drops to zero and it begins to charge once more. The voltage variation across the capacitor with time is as shown in Fig. 23.20. Such a sawtooth waveform is required in many types of apparatus.

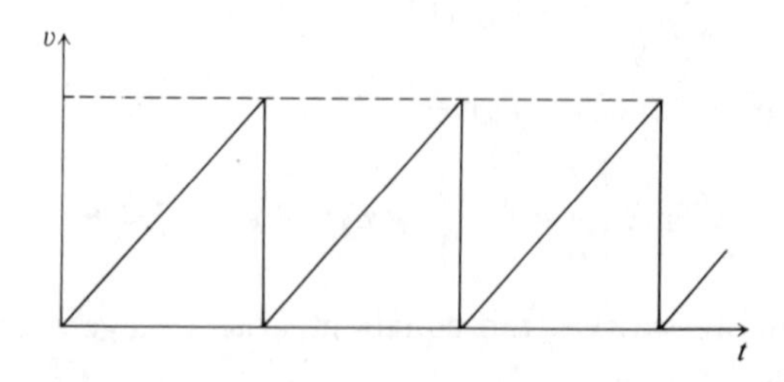

Fig. 23.20. The sawtooth wave-form across a charging and discharging capacitor.

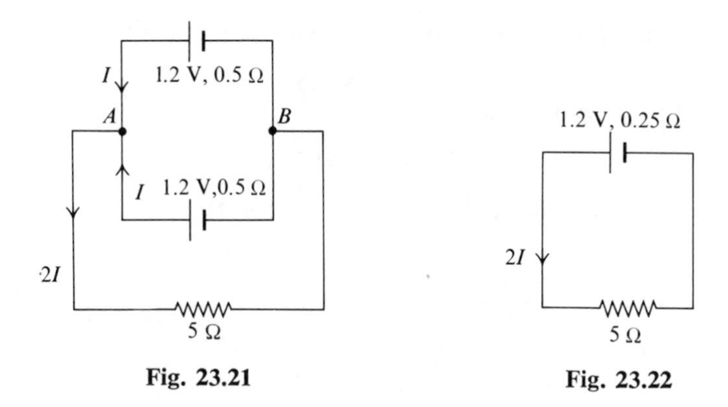

Fig. 23.21 **Fig. 23.22**

Example 1. Two cells each of emf 1.20 V and internal resistance 0.50 Ω are joined in parallel and connected to a resistance of 5.0 Ω. Determine the current in each branch of the circuit.

Solution. By symmetry the same current I must pass through each cell. Since no charge can accumulate anywhere in the circuit, it follows that the current through the resistor is $2I$.

The potential drop from A to B through the resistor is $2I \times 5\ \Omega$ and through a cell it is $1.20\ \text{V} - I \times 0.5\ \Omega$

$$\therefore \quad 2I \times 5\ \Omega = 1.20\ \text{V} - I \times 0.5\ \Omega$$

$$\therefore \quad I = \frac{1.20\ \text{V}}{10.5\ \Omega} = 114\ \text{mA}$$

$$\therefore \quad 2I = 228\ \text{mA}.$$

Alternatively, one may say that the effect of putting the cells in parallel is to leave the emf of the combination as 1.2 V. However, the cells being in parallel, the combined internal resistance is now r, where

$$\frac{1}{r} = \frac{1}{0.5\ \Omega} + \frac{1}{0.5\ \Omega}$$

$$\therefore \quad r = 0.25\ \Omega.$$

The circuit is thus equivalent to that shown in Fig. 23.22 and it is clear that, as before

$$2I = \frac{1.20\ \text{V}}{5.25\ \Omega} = 228\ \text{mA}.$$

Example 2. A 10-Ω resistor is to be made from a copper wire of circular cross section and diameter 1 mm. If the resistivity of copper is $1.7 \times 10^{-8}\ \Omega\ \text{m}$, what volume of copper will be required?

Solution. The length of copper required is given by the equation

$$l = AR/\sigma,$$

and the volume required is thus

$$V = Al = A^2R/\sigma.$$

But $A = \pi d^2/4$, where d is the diameter of the wire. Hence

$$V = \frac{\pi^2 d^4 R}{16\,\sigma}$$

$$= \frac{\pi^2 \times (10^{-3}\text{ m})^4 \times 10\,\Omega}{16 \times 1.7 \times 10^{-8}\,\Omega\text{ m}}$$

$$= 3.63 \times 10^{-4}\text{ m}^3.$$

Example 3. A bank of cells of total emf 100 V and negligible resistance is placed in series with resistances of 2000 Ω and 3000 Ω. A voltmeter of resistance 6000 Ω is placed across each resistance in turn. What readings are obtained? Explain why the sum of these does not add up to 100 V.

Solution. In Fig. 23.23(a) the voltmeter is in parallel with the 2000-Ω resistor. Hence, if R_1 is the combined resistance,

$$\frac{1}{R_1} = \frac{1}{2000\,\Omega} + \frac{1}{6000\,\Omega} = \frac{3+1}{6000\,\Omega}$$

$$= \frac{1}{1500\,\Omega}.$$

The circuit is thus equivalent to a cell and two resistance of 1500 Ω and 3000 Ω in series. By Ohm's law, the current flowing is

$$I = E/R = 100\text{ V}/(1500\,\Omega + 3000\,\Omega)$$

$$= (1/45)\text{ A}.$$

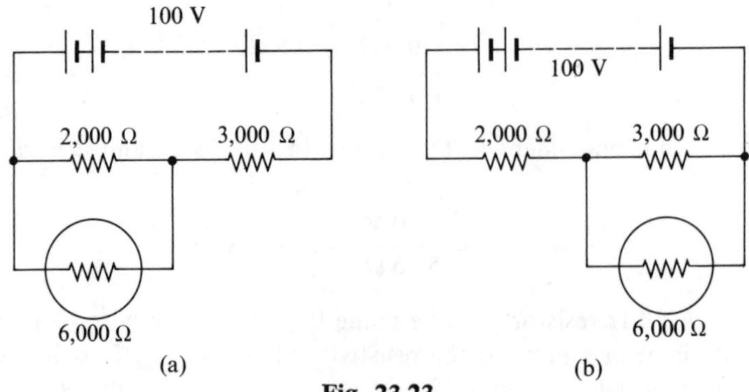

Fig. 23.23

The potential difference across voltmeter and 2000-Ω resistor is the potential across the 1500-Ω resistor in the equivalent circuit, i.e.

$$V_1 = I_1 R_1 = 1500\ \Omega \times (1/45)\ \text{A} = 33.3\ \text{V}.$$

Similarly, since

$$\frac{1}{R_2} = \frac{1}{6000\ \Omega} + \frac{1}{3000\ \Omega} = \frac{1}{2000\ \Omega},$$

the circuit in Fig. 23.23(b) is equivalent to a cell and two resistances of 2000 Ω in series. Thus

$$I_2 = 100\ \text{V}/4000\ \Omega = (1/40)\ \text{A}.$$

The potential difference across voltmeter and 3000-Ω resistor is

$$V_2 = I_2 R_2 = 50\ \text{V}.$$

The sum of V_1 and V_2 is only 83.3 V which is markedly different from 100 V. The reason is that the voltmeter has too low a resistance and it takes an appreciable quantity of current in both cases. The circuit is very badly disturbed by the insertion of the voltmeter. Only if the voltmeter has a negligible effect will the readings add up to the emf of the cell. In this problem the effect of the voltmeter is to change the voltage division from 40 V across the 2000-Ω resistor and 60 V across the 3000-Ω resistor to a 33.3 V – 66.7 V split in the first case and a 50 V – 50 V split in the second. One should clearly beware of using a voltmeter that has too small a resistance.

Example 4. The resistance of a platinum resistance thermometer is found to be 10.10 Ω at 20°C and 24.24 Ω when placed in a furnace which is known to be at a temperature of less than 660°C. What is the temperature of the furnace if α and β for platinum are $3.6 \times 10^{-3}\ \text{deg}^{-1}$ and $-16.0 \times 10^{-7}\ \text{deg}^{-2}$ respectively?

Solution. We know that

$$R_\theta = R_0[1 + \alpha(\theta - \theta_0) + \beta(\theta - \theta_0)^2].$$

Let us replace $\theta - \theta_0$ by x deg. Then

$$\frac{24.24}{10.10} = 2.4 = 1 + 3.6 \times 10^{-3}\, x - 16.0 \times 10^{-7}\, x^2.$$

Therefore

$$8x^2 - 18 \times 10^3 x + 7 \times 10^6 = 0.$$

Therefore

$$x = \frac{18 \times 10^3 \pm \sqrt{18^2 \times 10^6 - 4 \times 8 \times 7 \times 10^6}}{16}$$

$$= \frac{18 \times 10^3 \pm 10 \times 10^3}{16}$$

$$= 1.75 \times 10^3 \quad \text{or} \quad 0.5 \times 10^3.$$

It follows that $\theta - \theta_0 = 500$ deg. Therefore

$$\theta = 520°\text{C}.$$

Example 5. The plates of a cell are connected to a 4-Ω resistor and the potential difference measured across them by a potentiometer is found to be 1.6 V. If the 4-Ω resistor is replaced by a 3-Ω resistor, the potential difference is now measured to be 1.5 V. Find the emf and internal resistance of the cell.

Solution. The potentiometer is measuring the terminal voltage of the cell in the two cases. Hence

$$V_1 = I_1 \times 4\,\Omega$$

$$V_2 = I_2 \times 3\,\Omega.$$

But

$$I_1 = E/(r + 4\,\Omega)$$

$$I_2 = E/(r + 3\,\Omega),$$

where E and r are the emf and internal resistance of the cell respectively. Hence

$$\frac{1.6\text{ V}}{4\,\Omega} = \frac{E}{(r + 4\,\Omega)}$$

$$\frac{1.5\text{ V}}{3\,\Omega} = \frac{E}{(r + 3\,\Omega)}$$

Therefore

$$\frac{1.6}{1.5} = \frac{4\,\Omega(r + 3\,\Omega)}{3\,\Omega(r + 4\,\Omega)}$$

Therefore

$$r \times 4.8\,\Omega + 19.2\,\Omega^2 = r \times 6.0\,\Omega + 18.0\,\Omega^2.$$

Therefore

$$r \times 1.2\,\Omega = 1.2\,\Omega^2.$$

Therefore

$$r = 1\,\Omega.$$

Therefore

$$E = \frac{1.6\text{ V} \times 5\,\Omega}{4\,\Omega} = 2.0\text{ V}.$$

Example 6. A manufacturer's calatog states that the 2-kW kettle they sell will boil 2 liters of water in 5 min for a cost of less than $\frac{1}{2}$ cent. If electricity costs 3 cents per kW h, check the accuracy of each of these claims.

Solution. Assuming the temperature of the domestic cold water supply to be 20°C, the heat used in raising 2 liters to 100°C is thus

$$Q = 4.18\text{ J g}^{-1}\text{ deg}^{-1} \times 2 \times 10^3\text{ cm} \times 1\text{ g cm}^{-3} \times (100°\text{C} - 20°\text{C})$$
$$= 6.69 \times 10^5\text{ J}.$$

The heat produced by the element in the kettle in 5 min is

$$Q' = 2 \times 10^3 \text{ J s}^{-1} \times 5 \times 60 \text{ s}$$
$$= 6.00 \times 10^5 \text{ J}.$$

The first part of the claim is clearly not true even although we have overestimated the starting temperature of the water and taken no account of heat used in raising the temperature of the kettle or lost to the surroundings.

The cost of boiling the water is

$$\frac{6.69 \times 10^5 \text{ J}}{3.60 \times 10^6 \text{ J(kW h)}^{-1}} \times 3 \text{ cents (kW h)}^{-1} = 0.56 \text{ cents}.$$

The cost claim is not true either.

Example 7. Three resistances of 5 Ω, 10 Ω, and 15 Ω are joined in parallel. If each is rated at $\frac{1}{2}$ W, what is the maximum voltage that may be applied to the combination and what current then flows in each resistor?

Solution. Since the same voltage V is applied to each resistor, the power dissipated in the resistors is V^2/R_1, V^2/R_2 and V^2/R_3.

Clearly most power is dissipated in the smallest resistance. Hence, if V_0 is the maximum voltage possible,

$$\frac{V_0^2}{5\ \Omega} = \tfrac{1}{2}\text{W}$$

$$\therefore \quad V_0 = \sqrt{\frac{5}{2}}\ \text{V} = 1.58 \text{ V}.$$

The currents flowing are

$$I_1 = \frac{V_0}{5\ \Omega} = 0.32 \text{ A}, \qquad I_2 = \frac{V_0}{10\ \Omega} = 0.16 \text{ A}, \qquad \text{and} \qquad I_3 = \frac{V_0}{15\ \Omega} = 0.11 \text{ A}.$$

PROBLEMS

23.1 A cell has an emf of 1 V and an internal resistance of 0.5 Ω. How many cells are necessary to pass a current of 1 A through an external resistance of 6 Ω?

23.2 If the cells in Problem 23.1 have internal resistances of 6 Ω and the external resistance is 0.5 Ω, how many cells are now required and how should they be connected?

23.3 If the cells in Fig. 23.24 have negligible internal resistance, what is the potential of point A with the switch open and closed?

23.4 Determine the current in each resistor and the voltage across AB in Fig. 23.25.

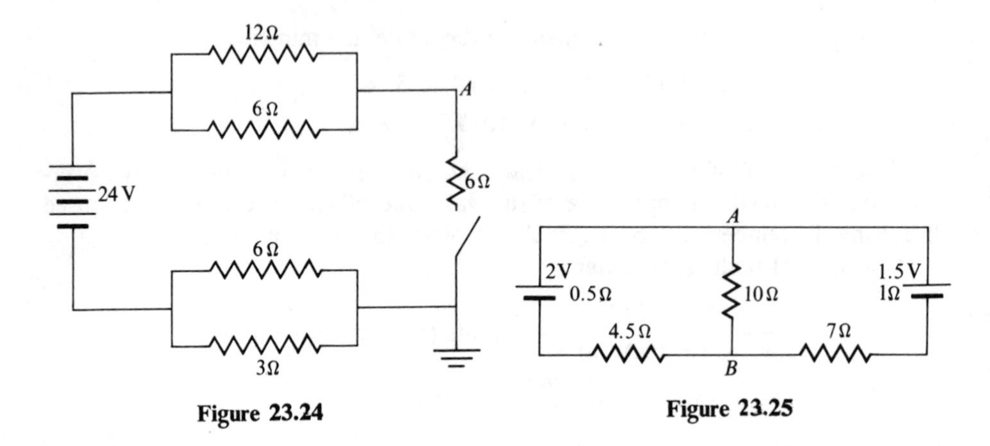

Figure 23.24 **Figure 23.25**

23.5 A cell is joined in series with a resistor and the current flowing is found to be I_1. When a second, identical, cell is added to the circuit, the current increases to I_2. If the two cells are now joined in parallel and connected in series with the resistor, the current becomes I_3. Show that

$$3I_2I_3 = 2I_1(I_2 + I_3).$$

23.6 Two cells, one of emf 2.4 V and internal resistance 1 Ω, the other of emf 4 V and internal resistance 0.2 Ω, are connected in parallel and the combination is connected in series with an external resistance of 10 Ω. What current passes through the external resistance?

23.7 A resistor is in series with an ammeter in a electrical circuit. The reading on the ammeter is 0.1 A when the potential difference across the resistor is 3.5 V. A further resistor is joined in parallel with the first, the current rising to 0.2 A and the potential difference dropping to 3.15 V. What are the resistances of the resistors?

23.8 What are the greatest and smallest resistances obtainable from combinations of 10-Ω, 20-Ω, 25-Ω, and 50-Ω resistors?

23.9 The potential difference between the terminals of a cell is 50 V when 10 A are being drawn and 54 V when 6 A are being drawn. What are the emf and internal resistance of the cell?

23.10 A copper wire of circular cross-section 1 mm in diameter carries a current of 1 A. If there are 8×10^{28} free electrons per m^3 of copper, estimate the average velocity with which the electrons move through the wire.

23.11 A 100-m length of wire is made of silver of resistivity 1.5×10^{-8} Ωm, and has a circular cross-section of radius 1 mm. A second wire is made from the same mass of silver but has double the radius. What is the resistance of each wire?

23.12 A 2-Ω resistor is to be made from 100 cm^3 of copper of resistivity 1.7×10^{-8} Ωm. If the copper is drawn into a wire of circular cross-section, what is its diameter?

23.13 Calculate the current passing through and the voltage across each resistor in Fig. 23.26.

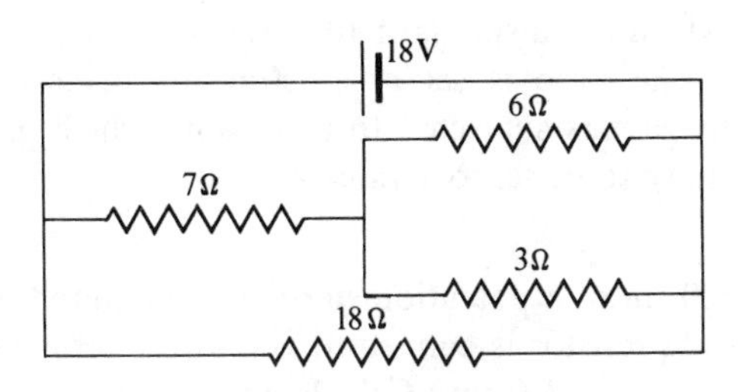

Figure 23.26

23.14 Platinum has values of α and β of 3.6×10^{-3} deg^{-1} and -16.0×10^{-7} deg^{-2} respectively. If a platinum resistance thermometer has a resistance of 10.50 Ω at 20°C, what is the temperature of a furnace in which the thermometer has a resistance of 23.10 Ω?

23.15 A bobbin of wire is connected in one arm of a Wheatstone bridge, the second arm containing a standard 1-Ω resistor. At balance the other arms contain resistances of 50 Ω and 4 Ω, respectively. A portion of wire 120 cm long is now cut from the bobbin and inserted in the first arm of the bridge. To obtain balance, the resistances in the ratio arms have to be altered to 2 Ω and 1 Ω, respectively. What length of wire was originally on the bobbin?

23.16 A cell of emf 10 V and negligible internal resistance is connected in series with a 100-Ω resistor and a 500-Ω resistor. A voltmeter of resistance 1000 Ω is connected in turn across each resistor and across the terminals of the cell. What are the three readings obtained?

23.17 A cell of emf 12 V and negligible internal resistance is connected in series with two resistors. A voltmeter of resistance 5000 Ω is connected across each resistor in turn and measures 3 V and 6 V, respectively. What are the resistances of the two resistors?

23.18 The tungsten filament of a 100-W lamp which operates from a 110-V line has a working temperature of 2020°C. If the values of α and β for tungsten are 4.7×10^{-3} deg^{-1} and -1.3×10^{-7} deg^{-2}, calculate the resistance of the filament at 20°C.

23.19 When a 60-W, 240-V lamp is connected to a 120-V line, the current flowing is found to be 0.2 A. Using the data of Problem 23.18, estimate the difference in temperature of the filament of the lamp when working on the 240-V and 120-V lines. The temperature of the filament when working on the 240-V line is 2020°C.

23.20 In a Wheatstone bridge X is an unknown resistance, R_1 is a 50-Ω standard resistor, and R_2 and R_3 are in the ratio arms. To get exact balance, R_1 is shunted by a 225-Ω resistor. When R_2 and R_3 are interchanged, exact balance is achieved by shunting R_1 with a 4950-Ω resistor. What are the values of X and R_2/R_3?

23.21 The element of a semiconductor strain-gage is joined to one arm of a Wheatstone bridge, which has a standard 10-Ω resistor in the second arm and the ratio arms variable. The ratio is set at 1.050:1 for balance when the gage is not in use. When the gage is subjected to a pressure which produces a strain of 10^{-3}, the ratio arms have to be set to a value of 1.102:1. What is the strain factor of the gage?

23.22 A length of 300 cm of a potentiometer wire is required to balance the emf of a cell. When a 10-Ω resistor is connected across the terminals of the cell, the length required for balance is 250 cm. Calculate the internal resistance of the cell.

23.23 The equation $P = I^2R$ suggests that the rate of joule heating in a resistor is reduced if the resistance decreases in value, whereas the equation $P = V^2/R$ seems to imply precisely the opposite. Reconcile the different viewpoints.

23.24 The heating element in the plate of an electric cooker consumes 2 kW when connected to the 120-V electrical line. If it consists of a ribbon of metal of length 2 m, breadth 0.1 cm, and thickness 0.0025 cm, calculate the resistivity of the metal.

23.25 An electric cooker run from a 240-V electrical line possesses a hotplate which is powered by two coils of resistance 30 Ω which may be used separately, or in series, or in parallel. Calculate the possible power consumptions.

23.26 Two resistors when connected in parallel across a cell of negligible internal resistance use four times the power that they would use when connected in series across the same cell. If one of the resistors has a resistance of 20 Ω, what is the resistance of the other?

23.27 A small factory uses 100 kW of power, which is delivered through lines of resistance 5 Ω. How much less power is lost if the power is delivered at 50,000 V instead of 10,000 V?

23.28 A 100-Ω resistor and a 300-Ω resistor are connected in parallel and the combination is joined in series with a 50-Ω resistor. If each resistor is rated at $\frac{1}{4}$ W, what is the maximum safe potential difference which can be applied across the whole arrangement? What is then the rate of heat production in each resistor?

23.29 An electromagnet runs from a 240-V line and takes 100 A. It is water-cooled, the water entering at a temperature of 15°C. If the output temperature of the water is not to exceed 50°C, what is the minimum flow rate necessary?

23.30 An electric light instalation consists of 25 lamps in parallel between the ends of leads which have a resistance of 1 Ω and bring current from a source of emf 120 V and 9 Ω internal resistance. If each lamp takes 0.4 A, find its resistance and the number of watts dissipated in each part of the circuit.

23.31 A 25-μF capacitor is attached to a 60-V cell and reaches a potential difference of 50 V in 10 ms. What is the total resistance in the circuit?

23.32 A 1-μF capacitor is in series with a resistance of $5 \times 10^3\ \Omega$ and a 50-V cell. What is the time constant of the circuit and the initial current flowing when the circuit is made? How long would it take fully to charge the capacitor if the current continued at its initial value?

CHAPTER 24

IONIC CONDUCTION

24.1 INTRODUCTION

When an acid, base, or salt is dissolved in water, the molecules of the substance tend to split up into ions, atoms or groups of atoms carrying positive or negative charges. Thus NaCl splits up into Na^+ and Cl^- ions, $CuSO_4$ into Cu^{++} and SO_4^{--} ions, H_2SO_4 into H_2^{++} and SO_4^{--} ions, and so on. The solution so formed is called an electrolyte. If such an electrolyte has introduced into it conducting terminals, known as electrodes, and a potential difference is established between them, the positive ions tend to flow to the negative electrode and vice versa. These ions, on arriving at their respective electrodes, give up their charges and either deposit on the electrode or react with the electrode or the electrolyte. Thus if copper electrodes are immersed in copper sulfate, copper is deposited on the cathode, or negative electrode, and sulfate ions react with the copper anode, or positive electrode, producing copper sulfate. Thus effectively copper is removed from the anode and deposited on the cathode. Any similar reaction, such as occurs in silver nitrate with silver electrodes, can be used as a means of plating an object with the appropriate metal. The object is used as the cathode and a lump of the metal as the anode. The object therefore receives an even layer of plating if the current passed through the electrolyte is not too high.

The behavior of the ions on arrival at the electrodes is determined by the nature of the electrodes, and may differ from one type to another. Thus if dilute sulfuric acid has a current passed into it through platinum electrodes, hydrogen is given off at the cathode, and since the platinum and sulfate groups do not react, at the anode the reaction

$$2H_2O + 2SO_4 = 2H_2SO_4 + O_2$$

takes place. But with copper electrodes, copper sulfate would form at the anode.

24.2 FARADAY'S LAWS OF ELECTROLYSIS

If a collection of electrolytic cells, each containing a solution and appropriate electrodes, are joined in series to a source of electric current, it will be noted that the larger the current and the longer the time for which it flows the greater is the deposition at any electrode. Further, the mass deposited or liberated at each electrode bears a constant relation to the similar quantity at any other electrode.

These observations are made precise in Faraday's two laws of electrolysis. They are as follows.

1) The mass of any substance liberated or deposited at an electrode in electrolysis is directly proportional to the quantity of charge which has passed through the electrolyte. Therefore $M \propto Q$.

2) The masses of the elements liberated or deposited at electrodes in electrolysis by the same quantity of electricity passing through different electrolytes are directly proportional to their chemical equivalent weights. Therefore $M \propto W_c$.

Note that the chemical equivalent weight of an element is its atomic weight divided by its valency.

Combining the results of Faraday's two laws gives

$$M \propto W_c Q \propto W_c It = zIt, \tag{24.1}$$

where z is known as the electrochemical equivalent weight. For copper, for example, z has the value 3.3×10^{-7} kg C^{-1}.

24.3 CONDUCTIVITY MEASUREMENTS

If a small potential difference is set up between the electrodes immersed in an electrolyte, it will be found that the current flowing gradually decreases. This is because the electrolytic action at one or both of the electrodes produces there a layer, normally of gas, which impairs the contact of the electrode with the solution. The resistance of the electrolyte thus appears to increase. Further, an electrolytic cell working in the reverse direction to the applied voltage is effectively set up, producing a back-emf. The general effect is called polarization. It can be overcome by applying a.c (cf. Chapter 27) rather than d.c. to the cell, the frequency of the a.c. being such that current flow takes place in any one direction for too short a time to cause deposition on the electrodes. Alternatively, a large enough d.c. voltage can be applied to overcome the back-emf and force a current through the electrolyte. In the latter case one must also note that a concentration polarization effect may be set up with certain electrodes. If metal is deposited on the cathode and removed from the anode, the concentration of the electrolyte decreases near the cathode and increases near the anode. Using platinum electrodes reduces the effect, which can be eliminated by continuously agitating the electrolyte.

If the concentration of the electrolyte and the temperature are kept fixed, it is found that Ohm's law is obeyed. It is not normal to talk about the resistance of the electrolyte, but rather its reciprocal, the conductance C. If the area A of the electrodes and their distance apart l are varied, it is found that

$$C = \frac{kA}{l}, \tag{24.2}$$

where k is the conductivity of the electrolyte. The conductivity is found to depend on the temperature and the concentration of the electrolyte. For small concentrations k is roughly proportional to the concentration. The concentration of a particular constituent in a biological or medical sample is often determined by measuring the conductivity of the sample in a standard electrolytic tank at fixed temperature.

24.4 THE VOLTAIC CELL

When a metal plate is immersed in an electrolyte, there is a tendency for the metal to go into solution. If zinc is immersed in dilute hydrochloric acid, for instance, doubly ionized zinc ions tend to go into solution, leaving an excess of two electrons on the electrode. This effect is very marked for zinc. If copper is also inserted, the tendency is very small for copper to go into solution, and the zinc electrode tends to become negatively charged with respect to the copper electrode. If the two electrodes are joined externally by a wire, positive charge flows from the copper to the zinc. Since the zinc keeps going into solution, the potential difference is maintained and the current persists. The cell therefore acts as a source of electrical energy, transforming chemical energy to provide it. Such a cell has an emf of around 1 V, and has a very limited current capacity. Putting cells in series increases the potential difference they can provide; putting them in parallel increases the current.

All simple voltaic cells of this type suffer from the polarization effects mentioned in the last section and various ingenious methods have been adopted to produce depolarization. This normally means that the cells are more complicated in their construction than the above account of a simple Zn-Cu cell would indicate. Further, depolarization is usually only accomplished slowly, and cells which work perfectly correctly when only required to provide current occasionally will rapidly polarize if required to produce continuous current for lengthy periods.

Simple voltaic cells are only used nowadays when no other source can be conveniently employed. Dry cells, the most commonly used form, are dry only in the sense that the electrolyte has been mopped up by sawdust or some similar material, which is then packed between the outer case, the negative terminal, and a central rod, the positive terminal. The sawdust is sealed in with pitch.

For many applications—hearing aids and cardiac pacemakers (cf. Section 24.15), for example—mercury cells are now employed. These have many more hours of useful life than a conventional voltaic cell of comparable size and weight, last longer in store, and are little affected by climate. In miniature form they have a very large energy/volume ratio. In a normal cell the polarizer is often hydrogen; and if this is not dealt with efficiently, the output voltage may fluctuate because of increased internal resistance in the cell, and will certainly drop towards the end of its useful life. In a mercury cell the polarizing material is mercury, which, being a conductor, does not alter the resistance. The cell thus has a very stable output over its whole lifetime. It is because of their reliability and their long lifetime that mercury cells are used when power sources are to be implanted in the body. They have a useful lifetime of perhaps 24 months in cardiac pacemakers.

24.5 SECONDARY CELLS

Cells which obtain their energy directly from the chemical energy of the materials used are called primary cells. Secondary cells are much more useful. Here the chemical nature of the two electrodes is changed by passing a current through the cell. At a later stage this energy of chemical change can be obtained by running the cell in the reverse direction, and this can be done again and again.

The simplest secondary storage cell is the lead-acid battery. After use, the electrodes consist of grids packed with lead sulfate mixed with dilute sulfuric acid, this in order to maximize the electrode surface. If current is passed through this from what, in use, will be the positive plate through the electrolyte of dilute sulfuric acid at a specific gravity of 1.18, and out of the negative plate, then at the positive plate the reaction is

$$PbSO_4 + SO_4 + 2H_2O = PbO_2 + 2H_2SO_4.$$

At the negative plate

$$PbSO_4 + H_2 = Pb + H_2SO_4.$$

The specific gravity rises to 1.21 during the charging process, since water is removed from the solution and sulfuric acid is added.

If the cell is now used on its own, the positive plate is at a higher potential than the negative, and current flows from one to the other in the external circuit and the other way internally. At the positive plate

$$PbO_2 + H_2 = PbO + H_2O,$$
$$PbO + H_2SO_4 = PbSO_4 + H_2O.$$

At the negative plate

$$Pb + SO_4 = PbSO_4.$$

The plates return to their former condition, the specific gravity, because of the added water and the loss of sulfuric acid, dropping to 1.18.

The emf is of the order of 2 V and is very constant over a long period until the specific gravity drops to almost 1.18. One gets various sizes of cell giving 30, 40, 60, etc., ampere-hours of charge.

Too high a charging current makes the deposit flake off and fall to the bottom of the container. Leaving a battery too long discharged makes the plates form a white insoluble layer which prevents recharging. If water is added to such a cell it must always be distilled water. Tap-water contains impurities such as zinc which form small cells inside the battery and these remove energy from the external circuit.

24.6 THERMOELECTRICITY

In a voltaic cell an emf is set up when two dissimilar metals are inserted into an electrolyte. A very similar effect occurs when the metals are joined directly, as in Fig. 24.1. If junction B is kept at constant temperature and the temperature of A is raised, a small current, of the order of milliamperes, is found to be flowing in the circuit. This is known as the Seebeck effect, after its discoverer.

As will be explained in Section 33.2, some of the electrons in a metal occupy energy levels in a conduction band, the levels not being localized but extending over the whole metal lattice. In two separate metals the occupied levels will have different energies. When the metals are joined, electrons from one metal can drop into unoccupied levels of lower energy in the other metal. This leaves the first metal with an excess positive charge and the second with an excess negative charge. A contact potential has therefore been set up across the junction (Fig. 24.2).

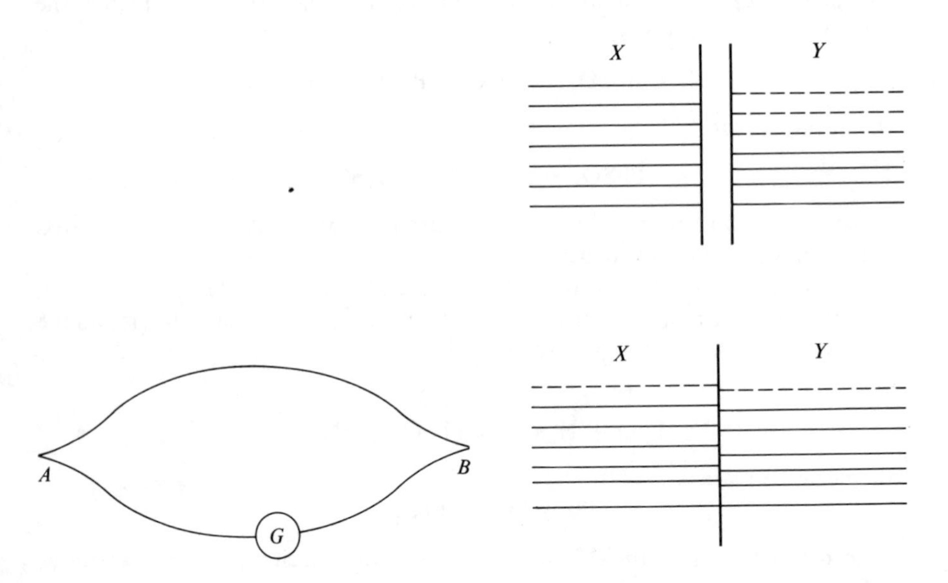

Fig. 24.1. A simple circuit to observe thermoelectricity.

Fig. 24.2. The change in occupancy of electron levels in two different metals on joining them.

A similar potential will of course occur at both junctions A and B. Since the effects are equal and opposite, no electric current flows when A and B are at the same temperature; but the value of the contact potential is markedly dependent on temperature. If the temperature of A is raised, the contact potentials are no longer equal and opposite and there is a net emf in the circuit.

Seebeck investigated this effect for a number of metals and found that he could make up a thermoelectric series, Bi, Ni, Co, Pd, Pt, Cu, Mn, Ti, Hg, Pb, Sn, Cr, Mo, Rh, Ir, Au, Ag, Zn, W, Cd, Fe, Sb, such that the direction of the current at the hot junction is from the metal coming earlier in the series to the one coming later. In general, the emf is greater the farther apart are the two metals in the series.

A few years later Peltier noted the inverse effect. If a current is passed through the junction of two dissimilar metals, heat is either absorbed or liberated at the junction. If the current is reversed, the heat is now liberated or absorbed, i.e. the opposite of what happened before. The quantity of heat absorbed or liberated is directly proportional to the charge passed through. The direction of the current to produce heating at the junction must be the reverse of the current through the hot junction produced by the Seebeck effect. It is obvious that this must be so from considerations of conservation of energy; otherwise the heating produced would augment the effect, which would produce more heating, and so on, giving rise to a perpetual-motion machine.

The Seebeck effect can obviously be used as a very sensitive method of measuring temperature, since the emf produced depends on the temperature difference between the two junctions, and this emf can be measured very accurately by using a potentiometer. A pair of dissimilar metals used in this way is known as a thermocouple. It is normal practice to keep one of the junctions at a fixed temperature, often the ice-point, and to measure the emf as the temperature of the other junction is altered.

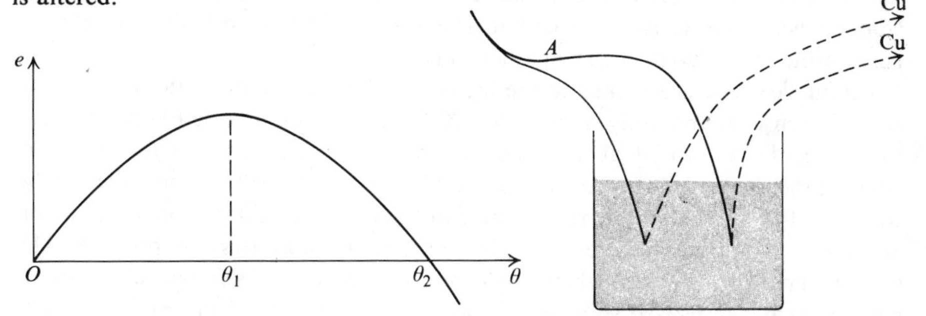

Fig. 24.3. The variation of the emf of a thermocouple as the temperature of the hot junction is varied.

Fig. 24.4. The correct method of joining a thermocouple to the rest of the measuring apparatus.

A typical graph is as shown in Fig. 24.3. The maximum emf is obtained at θ_1, after which the emf falls off and reverses at θ_2. It is obviously ridiculous to use a thermocouple which has its peak around θ_1 to measure temperatures in the region of θ_1, since the results will be ambiguous and have two possible solutions. One therefore uses a thermocouple which has its peak nowhere near the region to be measured.

Note that the thermocouple must be joined to galvanometers or potentiometers which have terminals of different metals from those of the thermocouple. It must be arranged that there is no difference in temperature between any of these junctions; otherwise emf's may be introduced by the contact potentials which would alter the response of the thermocouple. It is usual to join copper leads to the thermocouple at the cold junction as in Fig. 24.4. This tends to eliminate any thermoelectric emf's in other parts of the circuit, and the introduction of the copper has no effect on the performance of the thermocouple. One should always be on one's guard in experimental work against junctions of dissimilar metals in any circuit, for these will produce thermoelectric emf's if the junctions alter in temperature.

The great advantage of the thermocouple is its accuracy, allied with the smallness of the junction which needs to be inserted in the region in which the temperature is to be measured. It is therefore widely used to measure temperatures in very small enclosures. Variations in temperature in various parts of the body, in animals and in insects, are often measured in this way.

24.7 RESISTANCE IN THE HUMAN BODY

The tissues and fluids beneath the skin in the human body conduct electricity almost as well as metals, the conduction being principally ionic. However, if two electrodes are placed at separate locations on the body, the resistance between them may appear large because of the poor contact between the electrodes and the skin. Special conducting gel is often placed between the electrodes and the skin in consequence. Skin resistance varies markedly over the human body, the response being principally due to the sympathetic nervous system.

If an electrode is attached to the leg of a patient and another in the form of a roller is run over the body, a 50-V to 100-V cell being connected between them, the value of the current at any point is an inverse indication of the resistance between the electrodes. In a normal subject the resistance will vary in a predictable way, but if the nerves in any region are damaged, or if carcinoma or other tumors are compressing the nerves, resistance increases appreciably near that position. This is very marked in the chest when lung cancer is present, and this type of resistance measurement has proved very effective as a simple and rapid means of detecting the disease. The intensity of the effect is also an indication of the state of advancement of the cancer.

In addition to its use in the detection of carcinoma, the method has been useful in the diagnosis and location of peripheral vascular disease and regional nerve lesions.

24.8 ELECTRICAL TRANSMISSION IN NEURONS

One of the distinguishing features of animals is that they respond much more rapidly to external stimuli than do plants, and this is a function of the fact that they possess nervous systems. The speed of the response suggests that the process involved is an electrical one. If nerves are investigated, it is found that they consist of bundles of nerve fibers or axons. An axon is part of a single cell called a neuron, and is often much greater in length than the dimensions of other types of cell. Neurons which control the action of the fingers run the whole length of the arm, the cell body and nucleus being located in the spinal column. Axons are separated from one another in a nerve bundle by insulated sheaths.

Electrical impulses are sent along axons, but it is clear that the conduction process is not the same as for the simple processes already described in this book. Whereas normal electrical circuits set up fields at a speed of 3×10^8 m s^{-1}, so that a current starts virtually simultaneously at all parts, it is not difficult to show by experiments on neurons that electrical impulses are transmitted along them at the much slower speed of around 30 m s^{-1}. The transmission of an electrical impulse can proceed in either direction along a neuron, but can pass in one direction only across the synapse which joins one neuron to another. There are sensory axons which conduct toward the central nervous system and motor axons which conduct away from it.

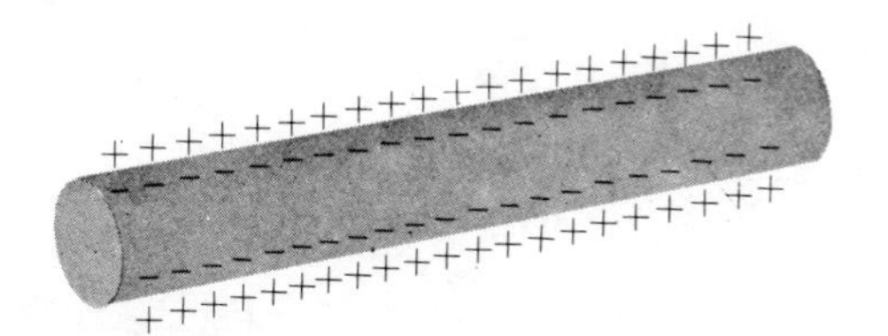

Fig. 24.5. The charge distribution in a resting nerve cell.

All nerve fibers, indeed possibly all cell membranes, are charged electrically (cf. Fig. 24.5). An axon consists of a hollow insulating membrane with one conducting fluid, cytoplasm, inside, and another, extracellular fluid, outside. If a needle electrode is inserted into the interior of the axon and another is held in the extracellular fluid, it is found that in the resting state there is a difference of potential of around 90 mV between them, the interior being negative with respect to the outside. It should be noted that contact potentials and polarization potentials may be set up in such an arrangement and must be eliminated. It is also worthy of note that the dielectric strength of cell membranes must be of the order of 10^8 V m^{-1}, about two orders of magnitude higher than for dry air, in order to withstand the difference of potential.

The difference of potential is maintained at the expense of the metabolic energy of the cell. In the resting state the membrane has an unequal concentration of sodium and potassium ions on its two sides. The membrane is somewhat permeable to potassium ions, but much less permeable, by a factor of around 75, to sodium ions. There are other ions present but these appear to play little part in the electrical behavior of nerves. There are more potassium ions inside than outside the membrane; if they diffuse outwards through the membrane owing to the concentration gradient, they do so against the electrical field set up in the membrane. In the case of sodium there are more ions outside than inside and one would therefore expect a net inward diffusion under the action of both the concentration gradient and the field. The ions indeed attempt to do so, but the cell contains a device known as a *sodium pump* which actively ejects sodium ions across the membrane to the outside. How this pump operates is not known. Evidence for its existence will be given later.

The maintenance of the ion separation across the membrane produces the potential difference across it, but it is not clear on exactly which factors this depends. There is evidence that increasing the rate of sodium ejection increases the potential difference across the membrane. On the other hand, inhibiting the action of the sodium pump does not appear to alter the resting potential except by a very gradual decline explicable in terms of the slow changes that take place in the concentration gradients. Further, the rate at which sodium ions are transported appears to be unaffected if the magnitude of the resting potential is increased; what it does depend on markedly is any alteration of the sodium concentration inside the membrane from the normal low level.

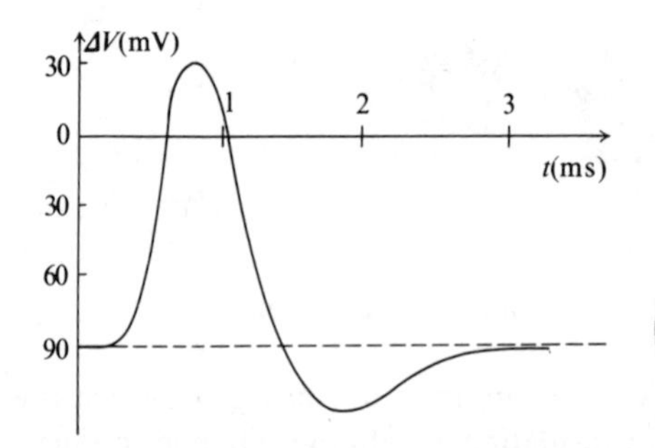

Fig. 24.6. The difference in potential across a nerve cell membrane as a function of time as an electrical impulse passes.

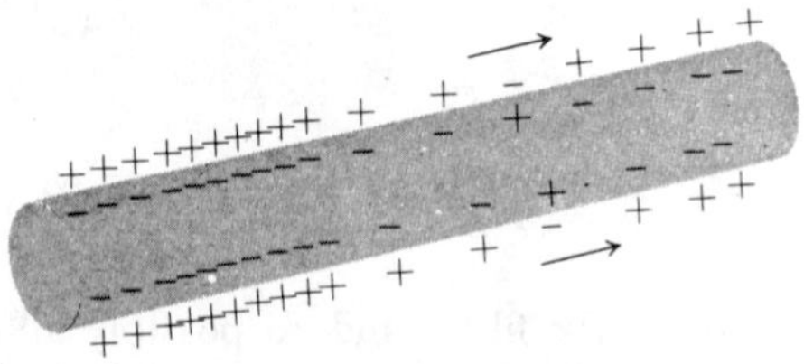

Fig. 24.7. The charge distribution along a nerve cell as an electrical impulse passes.

All that can be said definitely is that the concentration gradients of sodium and potassium ions are maintained at the expense of the metabolic energy of the cell. As a consequence of these gradients and the differing rates of diffusion of potassium and sodium ions through the membrane, a potential difference is established across it.

If the axon is slightly stimulated at some point, a small electrical pulse is sent along the axon in both directions but is attenuated rapidly. If, however, the stimulation is great enough to lower the potential across the membrane by around 20 mV or more, a pulse of standard form is transmitted without attenuation. An *action potential* or *spike potential* moves along the neuron in both directions. If the difference of potential between inside and outside of the membrane is plotted at any point as the spike potential passes along, the variation is as shown in Fig. 24.6. The potential difference reverses in sign as the pulse passes, and then returns to normal after a certain amount of oscillation. The distribution of charge along the axon is shown in Fig. 24.7. The stimulation may be electrical, mechanical, chemical, or thermal in nature.

By using tracer techniques (cf. Section 29.5) on the sodium and potassium ions, it is possible to deduce what happens as the pulse moves along. The membrane becomes permeable first to sodium ions and then to potassium ions. This causes a rapid flow of sodium ions inwards, followed quickly by a flow of potassium ions out. This overshoots the equilibrium position, causing a reversal of the potential. Very rapidly the impermeability is reestablished, and the concentrations of sodium and potassium ions return to the resting values. What causes the great increase in permeability of the membrane is not known.

Stimulation of greater than the threshold value thus produces changes in the permeability of, and transient reversal of the potential difference across, the cell membrane in the region of the stimulation. This affects neighboring regions, and the transient reversal of the potential and the permeability changes spread at a steady state along the axon in both directions. During this process the axon cannot conduct a second pulse, so that there is a finite repetition time for signals along a neuron.

Why does the difference of potential across the membrane re-establish itself? Sodium and potassium ions must return to their former positions against a concentration gradient, so that the mechanism involved cannot be simple diffusion. Here the sodium pump must once more be involved and it is during this operation that its presence can most easily be established. This can be shown by removing the energy-producing mechanism from the cell, when it is found that the ions do not re-establish the resting potential across the membrane.

The normal cell, and neurons are no exception, obtains its metabolic energy from ATP (adenosine triphosphate), the energy being liberated when one or two of the phosphate groups are split off. Inhibitors are known which, if injected into a cell, block the production of ATP. If this is done in the case of a stimulated neuron, the return of the sodium ions to the outside of the membrane ceases, showing that energy is necessary for the transport. If the inhibitor is washed out, the transport of the ions takes place once more. When the sodium ions are outside, they cannot easily return because of the impermeability of the membrane.

There are other axons besides the type mentioned above. Medullated axons have segmented insulating sleeves round the axon membrane. Each segment is about 2 mm long and segments are separated by exposed membrane surfaces at the nodes of Ranvier. Since the mechanism of impulse transmission described above cannot occur where the insulating sleeve exists, a signal passes along the axon inside the sheath between the nodes much in the manner that a signal will pass along a submarine cable. It travels much faster in consequence, but, as we have indicated earlier, the signal is severely attenuated. However, the sleeve is only 2 mm long and at the next node the disturbance is great enough to cause the changes in permeability and transient reversal of potential mentioned above at the exposed membrane. This produces a spike potential of standard size and the disturbance is passed into the next insulated section.

The nodes clearly act as amplifying stations along the axon, ensuring that the signal continues to be transmitted. The insulated sections provide faster transmission of the signal. Indeed medullated axons are used for high-speed signaling in the nervous system.

24.9 CONDUCTION ACROSS A SYNAPSE

Conduction will only take place one way from one neuron to another, or from a nerve terminal to a muscle fiber. The junction is known as a synapse (Fig. 24.8), and it is clear that conduction across the synapse cannot be the same as that along the neuron, in which the spike potential traveled in both directions from the point of stimulation. It has indeed been shown only for the crayfish that synaptic conduction is purely electrical. In all other synaptic conduction studied, the transmission is chemical. It is by no means clear that the same chemical is involved at every synapse, but the only chemical transmitter positively identified is ACL (acetylcholine).

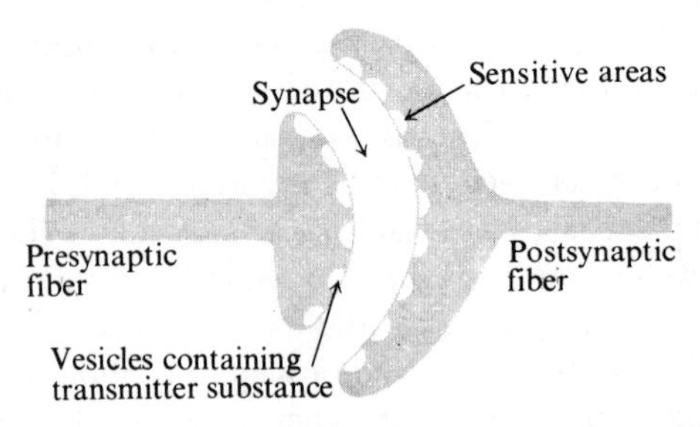

Fig. 24.8. A schematic representation of a synapse.

When the nerve impulse arrives at the neuron terminal at the synapse, it causes ACL, or some similar substance, to be released from vesicles which secrete it. Perhaps 10^7 molecules are released and these cross the gap, of the order of 1 μm, to the end of the receptor fiber. This takes of the order of a millisecond. The molecules are absorbed there, and change the ionic permeability of the membrane in the second fiber, causing a spike potential to start along it. The absorbed molecules are destroyed by an enzyme; in the case of ACL this is acetylcholinesterase.

Note that a synapse may act in the simple manner above, or it may fulfil the action which takes place in electronic computers and add, subtract, divide, multiply, or even differentiate or integrate. The first of these processes will result if the synapse only transmits sub-threshold responses, several therefore being necessary to produce a spike potential. Although ACL produces a spike potential at motor end-plates, it inhibits heart muscle and can therefore be considered to be subtracting. At other synapses several arriving potentials may be necessary to produces a single response, and the synapse "divides". If several terminals from one neuron synapse with differing time delays at a single other fiber, the original spike potential may be considered to be "multiplied".

Since the response at a synapse may show a very complex time dependence, processes analogous to differentiation and integration may be considered to take place. Indeed the processes at synapses are essentially non-linear and highly complex, and the response may, in general, be more complicated than it is in a simple electronic circuit.

24.10 CONDUCTION IN MUSCLES

If the synapse is between a nerve fiber and a muscle fiber, it is often called a myoneural junction. The transmission across such a junction produces a spike potential in the muscle fiber very similar to that in a neuron, except that the pulse at any point lasts much longer in a muscle. Also, the muscle fibers are not insulated from one another, and it is believed that the spike potential spreads to other fibers, ensuring that a number of them act in unison.

When a spike potential passes along a fiber, it produces contraction a short time later. Once the peak of the spike has passed and the membrane potential is returning to normal, heat production in the muscle increases and, a fraction of a millisecond later, there is a slight relaxation and then a mechanical contraction. How the spike triggers the production of the necessary chemicals is not known.

In all skeletal muscles are organs known as proprioceptors which sense what is happening and send back reports to the central nervous system, so that the state of working is known at all times and appropriate signals to change the action if necessary can be transmitted. The whole system is thus highly complex, using negative feedback (cf. Section 33.11) to control what is happening.

Muscle fatigue appears to be due to fatigue setting in in the nervous system either at the myoneural junctions or at synapses in the central nervous system, and would therefore appear to be connected with the chemical transmitter there. Similarly, the effect of drugs such as caffeine and nicotine appears to be similar to that of the transmitter substance, causing muscle contraction and general jitteriness. On the other hand, curare inhibits the action of the transmitter and prevents muscle action. It would therefore appear that the chemical transmitter plays a fundamental role in the nervous system.

24.11 ELECTRICAL ACTIVITY IN THE HEART

The mammalian heart consists of four chambers, right and left auricles and right and left ventricles (Fig. 24.9). Blood from all parts of the body except the lungs enters the right auricle, being pumped from there into the right ventricle. The blood is forced from there into the lungs and back to the left auricle. From there it is pumped to the left ventricle and thence through the aorta to the arteries except those to the lungs.

The purpose of keeping blood circulating is manyfold. Carbon dioxide from the tissues is transported to the lungs and removed. Similarly, oxygen from the lungs is returned to the tissues. Food and metabolic products are transported to and from various sites, and endocrine secretions are transported from the ductless glands to the organs they control. Finally, the blood carries agents to fight any invading organisms. Because the blood plays such an essential function in the

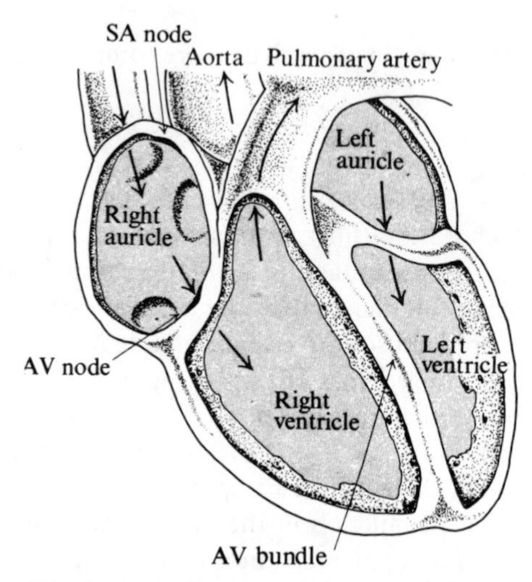

Fig. 24.9. A diagram of the human heart.

organization of the organism, the heart occupies a key role. If it stops beating, the organism cannot long survive.

The heart maintains a rhythmic pulsation which is initiated at the SA (sino-auricular) node, which puts out about 76 pulses per minute. The spike potentials produced travel over the auricles in all directions, canceling out on the opposite side when two spikes meet. The auricles then contract, forcing blood through valves into the ventricles.

The pulse from the SA node also stimulates the AV (auriculo-ventricular) node. After a time rather shorter than one-tenth of a second, the AV node puts out a pulse which passes along the group of fibers called the AV bundle. From the end of this bundle the pulse spreads as spike potentials over the walls of the ventricles, causing these in due course to contract. This closes the valves to the auricles and opens those to the arteries, forcing blood into them. The entire process repeats.

The spike potentials are entirely similar to those in neurons and muscles, except that the recovery time is very much greater than it is in muscle fibers. The transmission of the potentials is once more due to removing the impermeability of the membranes to sodium and potassium ions.

If the SA node fails, the AV node takes control of the rhythmic process. The auricular contraction is no longer properly synchronized to the ventricular and the repetition rate falls to perhaps 50 per minute. Neither of these occurrences is fatal and the heart continues to function, although with considerably impaired efficiency.

If the AV node fails, the walls of the heart continue to contract but the repetition rate falls to around 30 per minute. Further, the auricular and ventricular effects are not correlated and occur independently. The heart continues to function, but with difficulty.

24.12 ELECTROCARDIOGRAPHY

The potentials produced by the heart spread to the surface of the body and may be picked up by electrodes placed there. The process of recording these potentials is known as electrocardiography (e.c.g.). The potentials are so small that they require to be electronically amplified before they are recorded.

Since nerves and muscles may also be producing action potentials, it is necessary to eliminate these as far as possible. It is therefore normal to record e.c.g. traces with the patient lying down and completely relaxed. There are also various d.c. potentials existing on the surface of the body of the order of perhaps 0.1 mV. It is not difficult to discriminate against steady d.c. potentials with electronic apparatus.

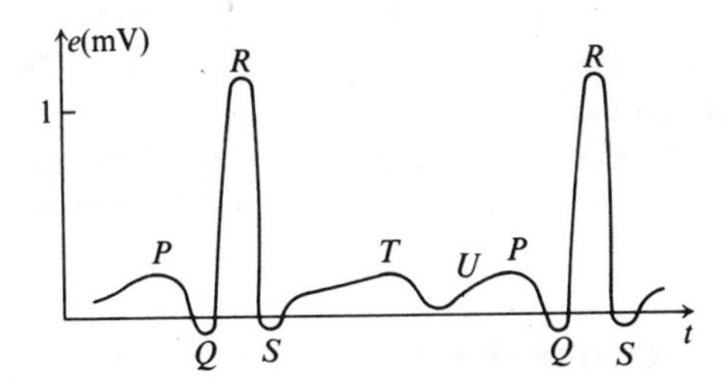

Fig. 24.10. A typical e.c.g. trace showing the variation of the potential with time.

The e.c.g. potentials may be observed between an electrode fixed to almost any part of the body and a neutral electrode, but it is traditional to attach three electrodes, one to each arm and one to the left foot. The potential of a single electrode, or that between any two of the three electrodes, always has the same form; that shown in Fig. 24.10. The *P*-wave precedes auricular contraction; the *QRS*-complex occurs at the start of ventricular contraction and the *T*-wave at its end.

Studies of e.c.g. recordings permit early and simple diagnosis of heart disease. It has now become standard practice to attach 12 rather than three leads to the patient, the new method being called vector, as opposed to the old scalar, e.c.g. It is claimed that more precise information about faults in the heart can be gained in this way.

24.13 AUDITORY HEART WARNING SYSTEM

After an accident, in an ambulance on the way to hospital, and during an operation, it is often essential to have immediate warning of the cessation of the electrical activity of the heart so that remedial measures can be taken at once. The pulses emitted by the heart are therefore often picked up by electrodes on the chest, amplified, and applied to a loudspeaker system, so that pulses of sound are heard so long as the heart is working. This gives immediate and noticeable warning the moment anything goes wrong.

24.14 ELECTROMYOGRAPHY

It is appropriate to deal with electromyography here, since what it studies was mentioned in a previous section. Electromyography is the process of registering the electrical activity of muscle. It is possible to do this by placing an electrode above the particular muscle to be studied and measuring the potential difference between it and a neutral electrode. However, the activity of a single muscle fiber is too small to produce a significant response, and larger potentials from groups of muscles may be attenuated and distorted in passing through tissue. It is therefore normal practice to use needle electrodes which are inserted into the muscle to be investigated, the voltage fluctuations therefore being detected at source.

The potential variation, which only occurs when the muscle is conducting, is displayed on a CRO (cf. Section 33.6) and is also rendered into an audible variation in sound on a loudspeaker. The particular response obtained will vary from place to place and from muscle to muscle. Essentially, what is being measured is the spike potential described earlier.

It is claimed that the use of this technique leads to efficient diagnosis of diseases which affect the lower motor neurons, myoneural junctions, and skeletal muscle fibers.

24.15 ARTIFICIAL PACEMAKERS

If the heart stops, external electrical stimulation can start it again. If electrodes are fixed to the chest on the long axis of the heart, pulses may be passed through the chest at regular intervals corresponding to the repetition time of the normal heartbeat. This may be achieved by charging a capacitor through a resistor up to 60 V, at which point it discharges through the electrodes. The artificial stimulation is very painful to a conscious patient and must be regarded as purely emergency treatment to restart the heart.

With certain heart conditions this emergency treatment would have to be continued indefinitely or at irregular intervals, the heart being either incapable of resuming its electrical activity or suffering from an intermittent fault. In these circumstances pacemakers are now implanted routinely in patients. Electrodes, usually of platinum to prevent chemical reaction taking place, are fixed to the heart or surrounding tissue, and pulses are sent through them from an electrical circuit which consists in essence of a capacitor charging up to a fixed voltage, at which point it discharges. The values of the capacitance and the resistance through which it charges determine the repetition rate (cf. Section 23.14). The power is provided by mercury cells. The patient must be operated on at intervals of around $1\frac{1}{2}$–2 years for renewal of the cells.

In certain cases of heart block a temporary return to normal rhythm may occur, although this is unlikely to be permanent. If the pulses of the pacemaker fall into a vulnerable period of the normal heart-cycle, there is a possibility of repetitive firing and ventricular fibrillation taking place. Demand pacemakers have therefore been introduced which only emit a pulse if the heart has not already done so within a fixed period. The capacitor charges up and, if it reaches the prescribed potential, triggers a discharge through the implanted electrodes. If, however, a pulse is emitted by the heart before the capacitor is fully charged, the latter discharges without emitting a pulse and begins to charge again. The repetition time of the pacemaker is adjusted to be a little greater than that of the heart, so that it never triggers a pulse so long as the heart keeps beating normally.

A pacemaker is enclosed in a substance, often epoxy resin, which does not react with body fluids. The encapsulation process often compresses the components, setting up mechanical stresses which may cause damage to both components and resin. It has been known for electrolyte from the cells to track along connecting wires and corrode resistor and capacitor, causing dangerous variations in pulse rate. Further, body fluids have sometimes penetrated the resin and reacted with pacemaker contents. Apart from any other effects, cells may fail long before the expected time.

A newer form of pacemaker therefore has no cells. The power is provided by the heart's own contractions acting on a piezoelectric crystal (cf. Section 23.10). This produces an alternating current (cf. Chapter 27), which is converted to d.c. by a subminiature rectifier (cf. Section 33.3), the power being fed to, and stored in, a capacitor. At regular intervals a transistor (cf. Section 33.4) switches the charge to a pulse generator which powers the heart. The whole apparatus is in subminiature form, and takes little space. It should be noted that in the case of failure external chest massage will operate the crystal and restore the current.

An external pacemaker which avoids a large number of the difficulties associated with implanted pacemakers will be described in Section 26.9.

24.16 DIRECT STIMULATION OF NERVES

Stimuli experienced by the human body are finally converted to electrical impulses which are transported along the nerves to the brain for interpretation. The brain has no built-in method of knowing what all these stimuli mean. During childhood a painful process of learning takes place which results in the interpretation of the nerve impulses and the correlation of their signals with the stimuli received from a number of senses. The process also takes place in adults who have restored to them a sense which has been missing since birth.

Physics has increasingly throughout the years discovered effects which bear some relation to the processes which take place in the conversion of stimuli to electrical messages. Thus the photoelectric effect (cf. Section 30.5) converts light to electrical energy as the eye does, and microphones perform the same function for sound. The possibility therefore exists that it may be possible, particularly now that miniaturization is with us, to use physical apparatus to record stimuli and feed the electrical impulses obtained directly to the appropriate nerves. Experimentation on these lines is in active progress.

The human body produces its effects with marvellous economy in a very small space. It is likely that the artificial eyes and ears will be, in the initial stages at least, much bulkier and very much less efficient than the real things. The results achieved may well be very crude in comparison with what nature provides. Nonetheless, the benefits to a deaf or blind person would be enormous.

Example 1. A cylinder of length 2.0 cm and diameter 0.40 cm is to be used by an orthopedic surgeon as a pin. Before he can insert it in the patient's body, the pin has to be silver-plated since the material of which it is made would tend to react with body fluids. The pin is placed in an electrolytic tank through which is passed a current of 0.2 A for 10 minutes. If the density of silver and its electrochemical equivalent are 1.05×10^4 kg m^{-3} and 1.118×10^{-6} kg C^{-1} respectively, what thickness of silver is deposited on the pin?

Solution. The mass of silver deposited will be

$$m = zIt,$$

where z is the electrochemical equivalent of silver and I is the current which flows for time t. The volume deposited is thus $v = m/\rho$, where ρ is the density of silver. The cylinder of length l and radius r has an area of $2\pi r \times l$ on its curved surface and πr^2 on each of the two ends. Thus the thickness of silver deposited is

$$\begin{aligned} d &= \frac{v}{2\pi r l + 2\pi r^2} = \frac{zIt}{2\pi r(l + r)\rho} \\ &= \frac{1.118 \times 10^{-6}\ \text{kg C}^{-1} \times 0.2\ \text{A} \times 10 \times 60\ \text{s}}{2\pi \times 0.2 \times 10^{-2}\ \text{m} \times 2.2 \times 10^{-2}\ \text{m} \times 1.05 \times 10^4\ \text{kg m}^{-3}} \\ &= 4.62 \times 10^{-5}\ \text{m}. \end{aligned}$$

Example 2. Electrical cells are often rated by the number of ampere-hours of electricity they can supply. Find the conversion factor between this unit and the corresponding standard MKSA unit.

A mercury cell used in a particular type of pacemaker can supply 100 μW of electrical power for 2 years. Give the ampere-hour rating of the cell, and determine the average current drawn from the cell if its emf is 1.35 V.

Solution. One ampere-hour is 3600 ampere-seconds. An ampere flowing for one second represents one coulomb of charge passing any point in the circuit. The rating of the cell is thus in terms of the amount of charge stored which can be passed through an external load and

$$1\ \text{A h} = 3.6 \times 10^3\ \text{C}.$$

The mercury cell can supply 100 μW of power for 2 years. The total amount of energy supplied is thus

$$10^{-4}\ \text{W} \times 2 \times 3.156 \times 10^7\ \text{s} = 6.312 \times 10^3\ \text{J}.$$

But this energy is also the quantity Q of charge that has passed any point in the circuit times the emf of the cell. Hence

$$Q = \frac{6.312 \times 10^3 \text{ J}}{1.35 \text{ V}} = 4.68 \times 10^3 \text{ C}.$$

The rating of the cell is thus

$$\frac{4.68 \times 10^3 \text{ C}}{3.6 \times 10^3 \text{ C (A h)}^{-1}} = 1.30 \text{ A h}.$$

The power supplied by the cell is the average current I drawn from the cell times its emf. Thus

$$I = \frac{100 \ \mu\text{W}}{1.35 \text{ V}} = 74.1 \ \mu\text{A}.$$

Example 3. Part of the table linking the emf and the temperature of the hot junction of a particular platinum–platinum rhodium thermocouple, the other junction being maintained at 0°C, is the following:

emf in millivolts

Temp (°C)	0	10	20	30	40	50	60	70	80	90
400	3.25	3.35	3.44	3.54	3.64	3.73	3.83	3.93	4.02	4.12
500	4.22	4.32	4.42	4.52	4.62	4.72	4.82	4.92	5.02	5.12

The thermocouple is inserted in an enclosure and the emf measured on a potentiometer is 4.58 mV. What is the temperature of the enclosure?

Solution. The temperature lies between 530°C and 540°C since the emf lies between 4.52 mV and 4.62 mV which differ by 0.1 mV. The actual emf is (4.52 + 0.06) mV. The temperature is therefore

$$530°\text{C} + \frac{0.06}{0.10} \times (540°\text{C} - 530°\text{C})$$

$$= 530°\text{C} + 6°\text{C}$$

$$= 536°\text{C}.$$

Example 4. In an experiment of von Helmholtz, a length of nerve attached to muscle fiber was electrically stimulated at the free end and then at 40 mm from that end. Muscle contraction occurred at times of 0.0244 s and 0.0231 s respectively after the stimulations. Calculate the speed of travel of electrical impulses along the nerve.

Solution. The second pulse traveled a distance d and produced a muscle contraction a time of 0.0231 s after stimulation. The first pulse travelled a distance (d + 40 mm) and produced contraction 0.0244 s after stimulation. It follows that for a pulse to travel 40 mm along the nerve requires a time of (0.0244 − 0.0231)s = 0.0013 s. The constant speed of travel is thus

$$v = 40 \text{ mm}/0.0013 \text{ s} = 4 \times 10^{-2} \text{ m}/1.3 \times 10^{-3} \text{ s}$$

$$= 30.8 \text{ m s}^{-1}.$$

PROBLEMS

24.1 Why are dentists always careful to ensure that the silver amalgam of a tooth filling is never allowed to touch the gold of a capped tooth?

24.2 A wire of resistance 2.7 Ω is immersed in 220 g of water in a calorimeter of negligible heat capacity. It is connected in series with a copper voltameter consisting of two copper electrodes immersed in copper sulfate solution, and a current is passed through the system. If 0.88 g of copper is deposited at the cathode in half an hour, what will be the rise in temperature of the water in the same time? The electrochemical equivalent of copper is 3.3×10^{-7} kg C^{-1}.

24.3 A sphere of radius 1.00 cm is silver-plated in an electrolytic tank through which a current of 0.4 A is passed for 15 min. If the density of silver is 1.05×10^4 kg m^{-3} and its electrochemical equivalent is 1.118×10^{-6} kg C^{-1}, calculate the thickness of the deposited layer.

24.4 A steady current is passed through two voltameters in series, one containing a dilute solution of sulfuric acid and platinum electrodes, the other containing copper sulfate and copper electrodes. If 0.050 g of hydrogen is liberated at one electrode, calculate the mass of copper deposited and the number of molecules of oxygen liberated. One molecule of oxygen has a mass of 5.30×10^{-26} kg.

24.5 One method of measuring the rate of heatbeat of an athlete is to strap a miniature electrolytic cell to his chest and pass the electrical impulses from the heart through it. In a particular case each heartbeat deposits a layer of silver 1.2×10^{-8} m thick on the cathode. The athlete runs a 3-mile race in 13 min 10 s and the deposited layer is 1.44×10^{-5} m thick. Calculate his average rate of heartbeat during the run.

24.6 A space satellite in the form of a sphere of radius 20 cm is being gold-plated in an electrolytic tank which passes a current of 0.5 A for 24 hours. If the electrochemical equivalent and density of gold are 2.042×10^{-6} kg C^{-1} and 1.930×10^4 kg m^{-3} respectively, what is the thickness of the deposited layer?

24.7 A piece of costume jewelry with a surface area of 75 cm^2 is to be covered with a layer of silver 10^{-5} m thick. How long should a current of 0.25 A be passed through an electrolytic tank containing silver nitrate using the piece of jewelry as cathode?

24.8 When plates of area 5 cm^2 are immersed in KCl solution 2.5 cm apart and a 50-V potential difference is established between them, a current of 1.2 mA flows. Calculate the conductivity of the electrolyte.

24.9 As the concentration of a particular biological solution varies from 1.00 to 10.0 millimoles per liter, the conductivity of the solution varies from $0.600 \times 10^{-3}\ \Omega^{-1}\ m^{-1}$ to $6.00 \times 10^{-3}\ \Omega^{-1}\ m^{-1}$. A sample is obtained from a research worker and its measured conductivity is found to be $2.34 \times 10^{-3}\ \Omega^{-1}\ m^{-1}$. What is the concentration of the sample?

24.10 Twenty-five secondary cells of total internal resistance 0.5 Ω and each of emf 2 V are to be charged from a 120-V source of internal resistance 10 Ω. If the charging current is not to exceed 1 A, what extra resistance should be placed in series?

24.11 A particular size of mercury cell has an emf of 1.35 V and an internal resistance of 0.03 Ω whereas a comparable dry cell has an emf of 1.5 V and an internal resistance of 0.5 Ω. If a hearing aid, which takes 2 W and requires 4 V for its operation, is to be powered by cells, show that three mercury cells can act as the supply but not three dry cells.

24.12 Sets of (a) three mercury cells in parallel and (b) three dry cells in parallel are connected in turn to a prosthetic device which has a resistance of 0.4 Ω. If the cells have the characteristics of those in Problem 24.11, show that set (a) can provide the 3 A necessary for correct functioning but not set (b)

24.13 A thermocouple, used with one junction at the ice-point, has an emf given by the relation $E = a\theta + \frac{1}{2}b\theta^2$, where θ is the Celsius temperature of the other junction. If the emf has the values 4.05 mV and 9.85 mV for $\theta = 100°C$ and $\theta = 300°C$, respectively, what is the emf when $\theta = 500°C$?

24.14 Electrical impulses are applied to a cathode ray oscilloscope (CRO) and to the end of a nerve fiber of length 16 cm simultaneously. On arrival of the impulse at the other end of the fiber, a second pulse is applied to the CRO. The time measured between the arrival of the two impulses at the CRO is 2.5 ms. Calculate the speed of transmission of the electrical impulse along the nerve fiber.

24.15 The time of transmission of a signal across a synapse 18 nm wide is timed to be 0.8 ms. What is the speed of diffusion of ACL molecules?

24.16 The membrane of a particular axon is 50 Å thick. Calculate the dielectric strength of the membrane.

24.17 For a heart which operates at 76 pulses per minute intermittently, a demand pacemaker operating at 70 pulses per minute is implanted. If the capacitor which is charged up to provide the pulse has a capacitance of 10 μF, and each charging period is one-third of the time constant of the arrangement, what is the value of the resistance through which the capacitor charges?

CHAPTER 25

MAGNETISM

25.1 INTRODUCTION

It has been known since ancient times that certain naturally occurring substances called magnets have the property of attracting to them iron placed in their proximity. This is very reminiscent of induction (cf. Section 22.3) in electrostatics. Furthermore, such magnets, if freely suspended, always take up a position pointing north-south.

These magnets act as though they contained two magnetic poles, one at the north-seeking end called the north pole, and the other at the south-seeking end called the south pole. These poles are fictitious, since poles are never found separately but only in equal and opposite pairs. A magnet which is cut in half immediately shows two poles in each half. But these poles act as though like poles repel and unlike poles attract. This can be shown by obtaining very long magnets in which the second pole of each magnet is too far away to influence the effect of the first one, and by performing an experiment similar to the one which proved Coulomb's law for charges. One then finds that the equation giving the law of force between poles is

$$F \propto P_1 P_2 / r^2, \tag{25.1}$$

where P_1 and P_2 are the pole strengths of the poles of the magnets and r is the distance between the poles. Poles are therefore analogous quantities in magnetism to electric charges in electrostatics.

It is possible, and at one time this was the normal method of approach, to develop magnetostatics by exact analogy with electrostatics, going through magnetic field, magnetic potential, etc. This offers little of interest to biologists or doctors, and we shall ignore magnetostatics completely. What is important is that current flow produces magnetic effects, and this is the point at which we take up the study of magnetism.

25.2 MAGNETIC EFFECTS OF CURRENTS

A flowing current produces effects on suspended magnets in its vicinity. Therefore the region around a current-carrying conductor is a region of magnetic field. In this field suspended magnets are deflected from their free north-south directions and a free north pole, if one could be found, would move in the field in the same

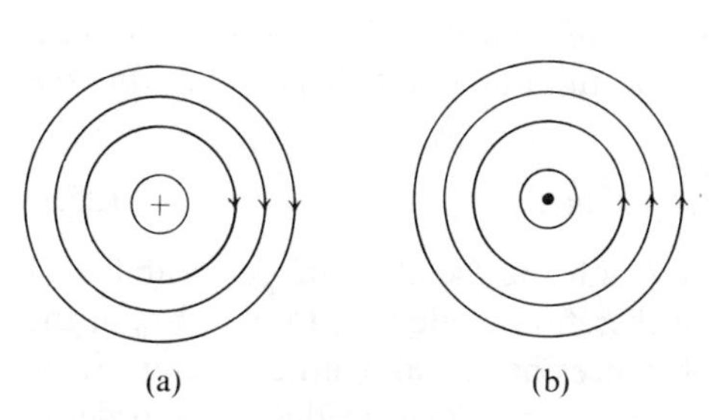

Fig. 25.1. The magnetic field due to a long, straight, current-carrying conductor in a plane at right angles to the conductor (a) with the current going into the plane, and (b) with the current coming out of the plane.

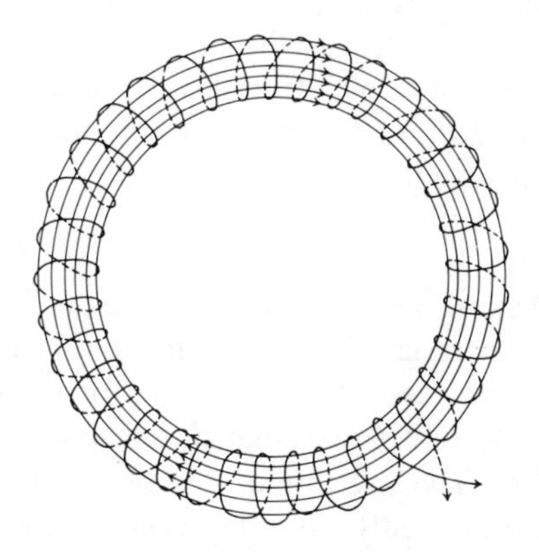

Fig. 25.2. The magnetic field of a toroidal coil.

way as a positive charge does in an electric field. The force per unit pole at a point in the field is called the magnetic induction, is denoted by the symbol **B**, and has the units of webers per square meter (Wb m^{-2}) or tesla (T). A more satisfactory definition of **B** will be given later. It is obviously the magnetic analog of the electrical intensity.

For a straight conductor the magnetic field, plotted by lines of force in exactly the same way as in the electric case, must be the same in any plane perpendicular to the conductor, and is as shown in Fig. 25.1. It consists of a set of clockwise concentric circles viewed from above if the current is going into the paper. The rule is that if the thumb of the right hand is thrust in the direction of the current, the fingers wrap round in the direction of the lines of force. Note that there is a convention for indicating which way a current is flowing when the conductor is passing at right angles to the plane of the paper in a diagram. The current is likened to the arrow fired by an Indian in a Western movie. If it is approaching us out of the paper, we see the point only. If it is going away from us into the paper, we see only the feathers. In Fig. 25.1(a), we see the tail feathers. The current is therefore going into the plane of the paper. In Fig. 25.1(b), we see only the point. The current is therefore coming out of the plane of the paper.

One may in general produce more powerful magnetic fields by using currents than by using natural magnets. For instance, a toroidal coil, a wire wound closely as if round a solid but invisible doughnut (cf. Fig. 25.2), produces a magnetic field inside the windings which is uniform and can be made very large. The magnitude of the magnetic induction is given by the equation

$$B = \mu_0 nI, \tag{25.2}$$

where μ_0 is called the permeability of free space and has the value $4\pi \times 10^{-7}$ Wb A^{-1} m^{-1}, n is the number of turns per unit length of the coil, and I is the current flowing. If a closed ring of iron or some other *ferromagnetic* material is inserted into the toroidal coil so that it fills the doughnut-shaped portion, the magnetic induction increases to

$$B = \mu_r \mu_0 nI = \mu nI, \tag{25.3}$$

where $\mu_r = \mu/\mu_0$ is the relative permeability, μ being the absolute permeability of the material. The quantities μ and μ_0 are therefore analogous to ε and ε_0 in the electric case; but, whereas ε_r for most substances has a value little different from unity, μ_r for ferromagnetic substances such as iron may have a value of the order of 5000. The insertion of iron or steel therefore increases the strength of the magnetic field markedly. If a gap is cut in the ring, the magnetic field in the gap is not as strong as in the iron but is much stronger than it would be if no iron were present at all. Such an arrangement is called an electromagnet (Fig. 25.3).

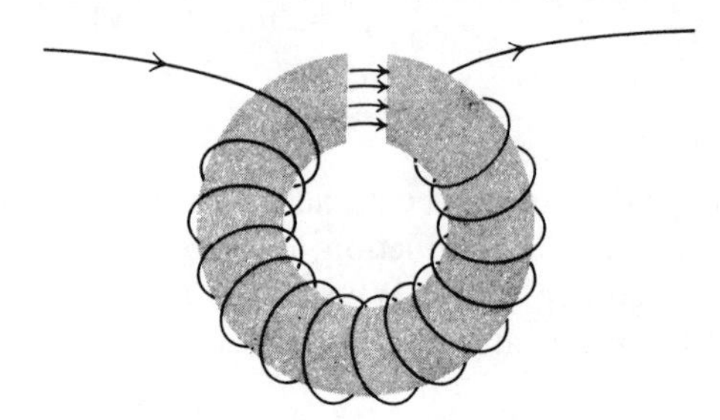

Fig. 25.3. An electromagnet.

It is possible to obtain formulas for calculating the field strengths produced by various shapes of current-carrying circuits with various materials inserted inside, but this is of interest to experts only. The only effect with which we shall be concerned for the moment is the effect of passing a current through a wire which is in an already established magnetic field.

25.3 THE FORCES ON A CHARGED PARTICLE AND ON A CURRENT ELEMENT IN A MAGNETIC FIELD

If a particle of charge q is moving with velocity $\mathbf{v}$ through a region in which is established a uniform magnetic field of induction $\mathbf{B}$, and $\mathbf{v}$ is at right angles to $\mathbf{B}$, then the particle is found to follow a path in the field which is the arc of a circle of radius R. Figure 25.4 shows the tracks of electrons and positrons in a cloud chamber, and it will be seen that, owing to the magnetic field established at right angles to the plane of the chamber, the particles follow circular paths. In Fig. 25.5 is shown the path of a particle passing between the poles of an electromagnet in a plane parallel to the pole faces, the direction of the magnetic field being shown by the arrow convention already described for currents. The deflection which occurs is that which a positively charged particle would experience.

Since the curved portion of the path is found to be accurately circular, it follows that the acceleration, and therefore the force acting on the particle, are radial and indeed centripetal. The speed of the particle therefore stays constant during the deflection, although the velocity changes its direction. If the particle is that described in the first sentence of this section, the force **F** acting on the particle at right angles to the path during the period when it is in the magnetic field is found to have a magnitude given by the equation

$$F = Bqv. \tag{25.4}$$

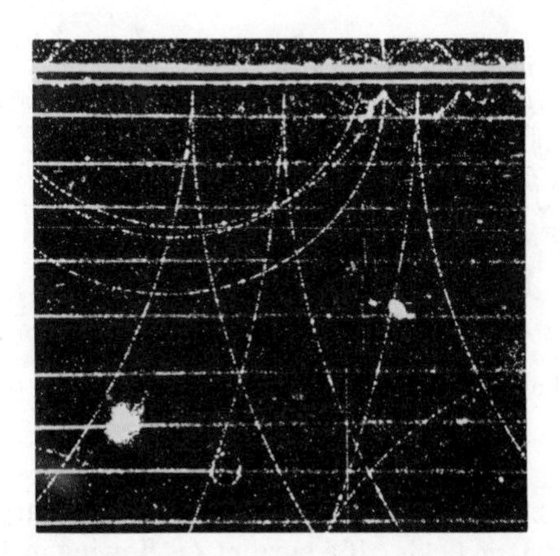

Fig. 25.4. Cloud-chamber tracks of three electron-positron pairs in a magnetic field. Three gamma-ray photons entering at the top materialize into pairs within a lead sheet. The coiled tracks are due to low-energy photoelectrons ejected from the lead. (Courtesy of Lawrence Radiation Laboratory, University of California.)

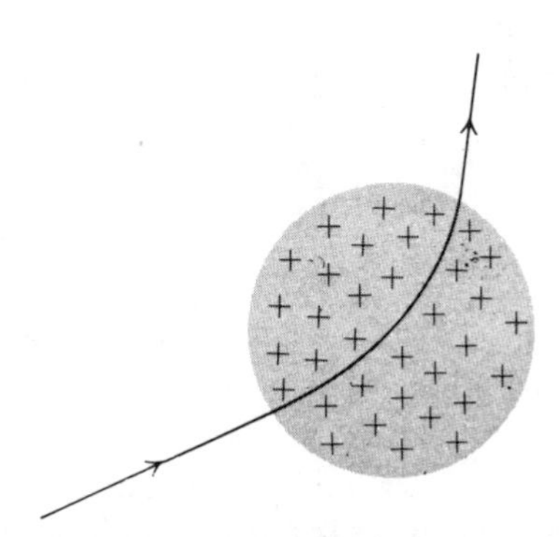

Fig. 25.5. The path of a moving charged particle in a magnetic field.

The direction of the force is found to be always at right angles to both **B** and **v**. The actual direction is given by a mnemonic. Since the positive charge in motion acts like a tiny current and since the effect is experienced by currents as well, the mnemonic runs: "Swim with the current, look along the lines of the magnetic field, and the motion is to the left". This is illustrated in Fig. 25.6.

If the particle has a mass m, the centripetal acceleration must be v^2/R (cf. Section 4.8), and therefore

$$F = Bqv = \frac{mv^2}{R}.$$

Therefore

$$R = \frac{mv}{Bq}. \tag{25.5}$$

The time of travel round a complete circle will be

$$T = \frac{2\pi R}{v} = \frac{2\pi m}{Bq}, \tag{25.6}$$

which is independent of the speed of the particle. This remarkable result is used in the design of machines which accelerate particles to high energy for the atom-smashing experiments mentioned in Section 32.1.

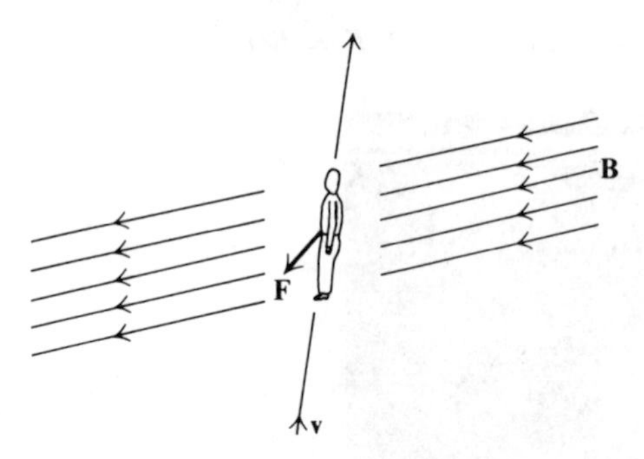

Fig. 25.6. Diagrammatic representation of the rule giving the direction of movement of a current in a magnetic field.

Of greater interest than isolated charges traveling through magnetic fields are currents in wires in already established magnetic fields. If a current I is flowing in a conductor of length l at right angles to a field of strength **B**, then there will be n charge-carriers per unit length of the conductor, each traveling with some velocity **v**. The force on each charge is qvB, on unit length of the wire is $nqvB$, and thus on length l of the wire is

$$F = nqvlB.$$

But after 1 s all the charges originally in a length v adjacent to any particular cross-section of the wire have passed through that cross-section. It follows that

$$I = nqv. \tag{25.7}$$

Therefore

$$F = IlB. \tag{25.8}$$

This relation can be used to give a more satisfactory definition of the units of B. One Wb m^{-2} (1 tesla) is defined as that magnetic field which will exert a force of 1 N on a 1 m-length of wire carrying a current of 1 A and lying in a direction at right angles to the magnetic field. The unit of 1 Wb m^{-2} can therefore be alternatively written as 1 N A^{-1} m^{-1}.

25.4 THE ELECTRIC MOTOR AND THE GALVANOMETER

A rectangular coil of height l and breadth b is suspended by the center of its upper side of length b in such a way that this side lies horizontally in a uniform magnetic field whose lines of force are also horizontal. The torque acting on the coil when a current I passes through it can be calculated in the following manner.

Any forces on the currents in the horizontal arms are vertical and are therefore balanced by the weight of the coil or by tension in the suspending fiber. The force F exerted on each vertical current in the coil is $F = IBl$ and is horizontal, as shown in Fig. 25.7. Each of the two opposite forces F has a turning effect about the line of the suspending fiber and the total torque is $N = Fb \cos \theta = IBlb \cos \theta = IBA \cos \theta$, where A is the area of the coil. If the coil has n turns,

$$N = nABI \cos \theta. \tag{25.9}$$

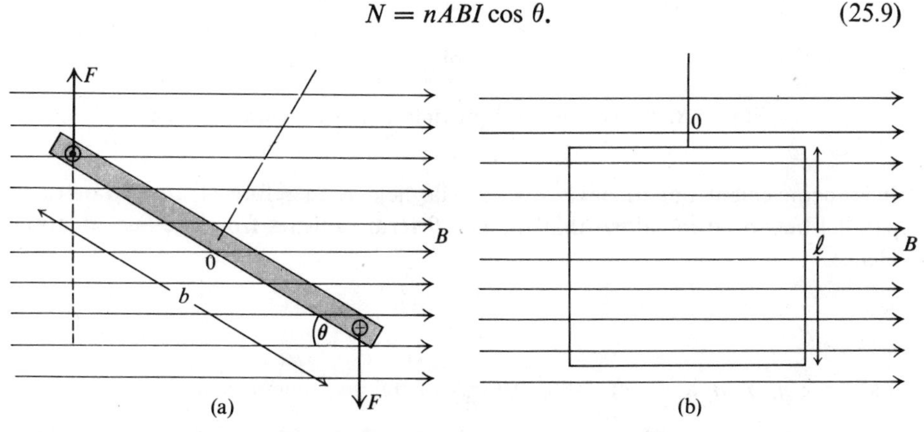

Fig. 25.7. A current-carrying coil in a magnetic field (a) in plan and (b) in elevation.

If no restraint were put on the coil, it would rotate until the coil was at right angles to the magnetic field. Because of the inertial effect of its motion it would carry past the position. However, the torque remains such as to restore it to an equilibrium position at right angles to the field, which it will achieve after some oscillation and damping. But if the current is reversed after the position at right angles to the field is reached, the directions of the forces on the coil reverse also and the torque is such as to continue the motion. If, indeed, the current is arranged to reverse every time the coil is at right angles to the field, the rotation continues indefinitely. This is in essence the principle of operation of the electric motor.

It is not difficult to arrange the reversal, if the current is led in and out by a split-ring, as shown in Fig. 25.8. As the coil rotates, the current enters the top wire and leaves by the bottom wire. When the coil is at right angles to the field, the leads P and Q have reached the gaps in the split-ring. As this position is passed, Q contacts the top half of the ring and P the bottom half. The current therefore reverses as required and will continue to do so each half-cycle.

On the other hand, if the coil is suspended by a torsion fiber (or is mounted in pivots with restraining springs), the rotation only continues until the torque is balanced by that due to the restraining mechanism. If a pointer is attached to the coil, the amount of sweep of this pointer over a scale is a measure of the current passing through. If τ is the torsion constant of the suspending fiber and if the coil comes to rest in equilibrium after rotating through an angle θ,

$$\tau\theta = nABI\cos\theta.$$

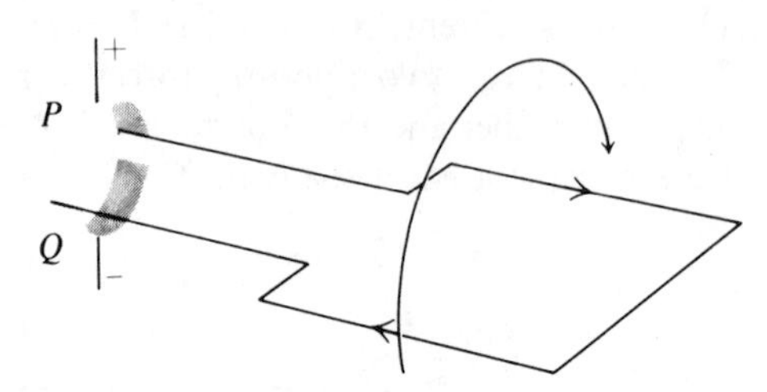

Fig. 25.8. Diagrammatic representation of the electric motor.

It is inconvenient not to have $\theta \propto I$. The field is therefore arranged not to be parallel but to be radial, so that the cos θ term vanishes from the last equation. Hence,

$$\theta = \frac{nAB}{\tau} I, \tag{25.10}$$

and since n, A, B, and τ are all constants of the instrument, $\theta \propto I$.

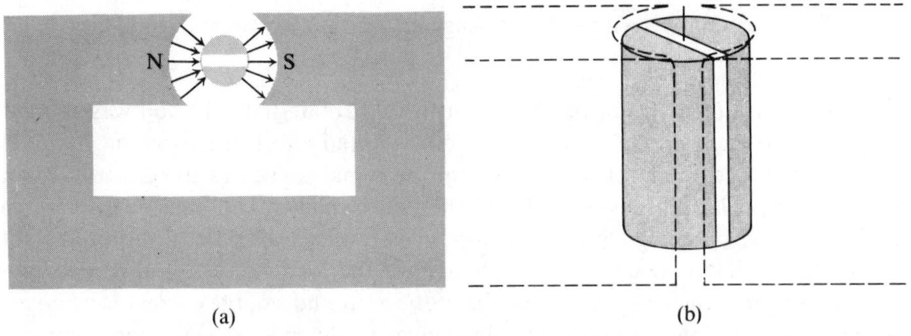

Fig. 25.9. The construction of a moving-coil galvanometer: (a) gives the general plan and (b) shows the suspended coil in more detail.

The actual constructional arrangement is as shown in Fig. 25.9. The coil is suspended between the curved poles of a permanent magnet and is mounted on a circular iron cylinder. This helps to concentrate the field and ensure its radial character.

This is the basic current-measuring device which is used for practically all routine current and voltage measurement. In order to be used in all such situations, it must be adapted to suit any particular function. We shall consider how it requires to be adapted in the next section.

25.5 THE AMMETER AND VOLTMETER

The galvanometer is a very sensitive instrument. By attaching a mirror to the coil and following its movement by the corresponding movement of a light beam reflected from the mirror, one may measure currents of the order of a micro-ampere. More generally, currents of the order of milliamperes may be measured on the basic instrument. This is because the coil of the galvanometer is in general quite small, for constructional reasons and because the resistance should be kept as low as possible in order not to disturb a circuit into which the galvanometer may be inserted. Hence, the instrument in its simple form is only suitable for measuring small currents. Large currents would in fact burn out the coil or cause much too large a deflection. But it is possible to adapt the instrument for use in measuring large currents, and it can also be adapted to measure voltages.

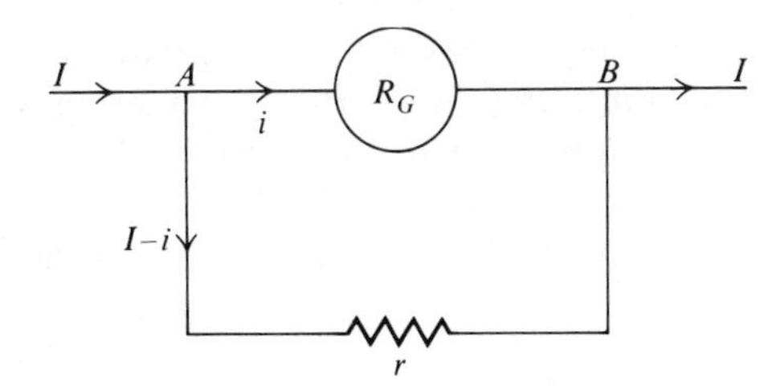

Fig. 25.10. Adapting a moving-coil galvanometer for use as an ammeter.

Suppose that the resistance of the galvanometer is R_G, and that it gives full-scale deflection for a small current i. Suppose that it is desired to use it in a circuit which is carrying a large current I. It is clear that a resistance r must be placed in parallel with the galvanometer, r being so chosen that current i passes through the galvanometer and the remaining $I - i$ passes through the added resistor, as in Fig. 25.10. Then, since the potential difference between A and B is the same by any route,

$$iR_G = (I - i)r.$$

Therefore

$$r = R_G \frac{i}{I - i}. \tag{25.11}$$

This allows the calculation of the value of r in any particular case. Also,

$$\frac{r}{R_G} = \frac{i}{I - i} \quad \text{or} \quad \frac{r}{R_G + r} = \frac{i}{I}.$$

Therefore

$$I = i\left(\frac{R_G + r}{r}\right). \tag{25.12}$$

A galvanometer may therefore be converted to read any current in a circuit. The galvanometer is shunted, as it is called, with a suitably chosen resistance r, given by Eq. (25.11). When the galvanometer gives full-scale deflection, a current i is passing through it, but I is the current in the main circuit, and I is related to i by Eq. (25.12). Each reading on the scale must be multiplied by a constant factor $(R_G + r)/r$ in order to obtain the current in the main circuit. This is normally already done by the manufacturer so that the meter reads directly without further calculation.

When a current i is flowing through the galvanometer, the potential difference across it is iR_G. The current readings on the galvanometer can therefore be treated as equivalent to readings of potential difference across the ends and can be converted to these by multiplication by a factor R_G. But R_G is very small, as is i, so that the potential difference measured is very small also. The resistance R_G is in any case too small for the galvanometer to be used as a voltmeter, since we have already seen that a voltmeter should have as high a resistance as possible so as not to disturb the circuit in which it is employed. But the situation is quite different if a high resistance R is placed in series with the galvanometer. Then, if the current flowing through the instrument is i, the potential difference across the galvanometer is $V_G = iR_G$ and the potential difference across the resistance R is $V_R = iR$. Thus the total potential difference is

$$V = V_G + V_R = i(R + R_G), \tag{25.13}$$

and

$$R = \frac{V}{i} - R_G. \tag{25.14}$$

Thus by suitable choice of R it is possible to measure the potential difference across a circuit element by putting the galvanometer and series resistor in parallel with it and calculating V by multiplying the current reading by $R + R_G$, which is a constant factor. Once again, the manufacturer normally does this before the voltmeter leaves the factory.

An example may help to illustrate what have been established as general rules.

Example. A galvanometer has a resistance of 2 Ω and gives full-scale deflection when a current of 1 mA passes through it. How can it be converted for use as (a) an ammeter reading up to 10 A, and (b) a voltmeter reading up to 50 V?

Solution. The potential difference across the galvanometer at full scale is $V_G = 10^{-3}$ A × 2 Ω = 2 mV.

a) To convert into an ammeter reading up to 10 A requires the addition of a shunt resistor of resistance r, where

$$r = \frac{i}{I - i} R_G = \frac{10^{-3}\ \text{A}}{(10 - 10^{-3})\ \text{A}} \times 2\ \Omega = \frac{10^{-4}}{1 - 10^{-4}} \times 2\ \Omega$$
$$= 2 \times 10^{-4}\,(1 + 10^{-4})\ \Omega = 2.0001 \times 10^{-4}\ \Omega.$$

b) To convert into a voltmeter reading up to 50 V requires the addition of a series resistor of resistance R, where

$$R = \frac{V}{i} - R_G = \frac{50\text{ V}}{10^{-3}\text{ A}} - R_G = 5 \times 10^4\ \Omega - 2\ \Omega$$
$$= 49{,}998\ \Omega.$$

25.6 THE OHMMETER

Although the ohmmeter does not work on the same principle as an ammeter or voltmeter, it seems most suitable to complete this discussion by describing a routine instrument for measuring resistance, even if not very accurately.

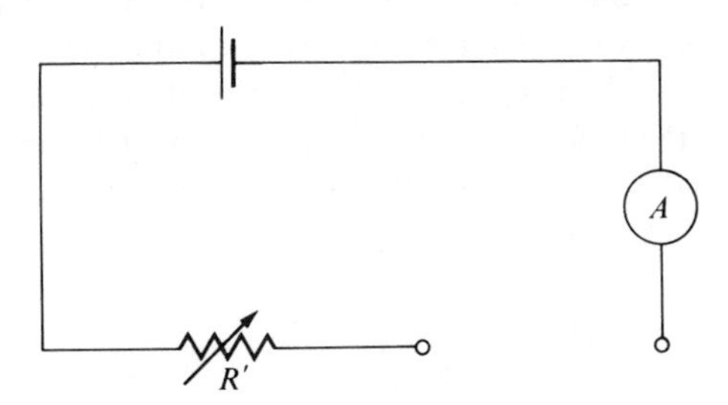

Fig. 25.11. A simple ohmmeter circuit.

Figure 25.11 shows a typical series circuit which consists of a cell, a variable resistor R', an ammeter, and a couple of terminals between which an unknown resistance can be inserted. The terminals are shorted and the variable resistor is adjusted to give full-scale deflection on the ammeter. This point on the scale is labeled zero ohms. Known resistances of increasing value are inserted between the terminals one after the other. Since the resistance in the circuit is increasing, the current drops and the various positions at which the pointer settles are labeled with the values of the appropriate known resistances. The scale will therefore read from zero at the full-scale deflection position to infinity at the zero deflection end. If I is the full-scale current and R' the value of the variable resistance which gives this, then $R' = V/I$.

If insertion of R in the circuit gives a current of fI, where f is a fraction, then

$$R + R' = \frac{V}{fI}, \qquad \text{or} \qquad R + R' = \frac{R'}{f}.$$

Therefore

$$R = R'\left(\frac{1-f}{f}\right). \tag{25.15}$$

For $f = \frac{1}{2}$, $R = R'$; for $f = 0.1$, $R = 9R'$, and so on. If an unknown resistance is inserted in the circuit, its value can be read off on the scale now constructed.

The readings on the scale will only be reasonably accurate over a restricted range of values. This range may be preselected by suitable choice of R' and V.

25.7 THE HALL EFFECT

Another interesting magnetic effect which may be mentioned is that discovered by Hall in 1879. Figure 25.12 shows a straight conductor of rectangular cross-section, in which a current I is flowing, located in a uniform magnetic field of strength $\mathbf{B}$ perpendicular to one of the faces of the conductor. The charges which are flowing along the conductor will experience a force which is at right angles to both $\mathbf{B}$ and the direction of I, and, hence, tend to drift sideways. A potential difference is therefore set up between the side faces of the conductor F and G. This potential difference will increase, but only up to a certain point. The potential difference sets up an electric field, so that the charge carriers are subjected to an additional force due to this field. When the forces due to the electric and magnetic fields are equal and opposite, there will be no further tendency for sideways drift of the charges. The potential difference which is established between the faces when this occurs is called the Hall emf.

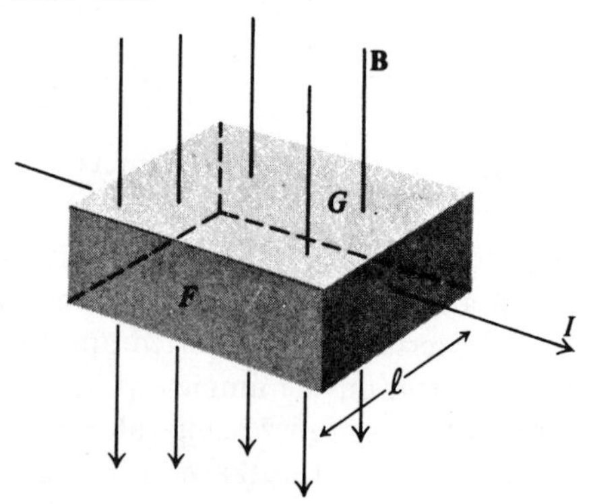

Fig. 25.12. The production of the Hall effect.

Whether the charge carriers are positive or negative, the drift is always towards G; but if the carriers are negative, F becomes of higher potential than G; and vice versa if the carriers are positive. This is therefore a means of determining the nature of the charge carriers in a conductor. In most cases the direction of the emf is such as to indicate that these are negative, i.e. the movement of electrons is causing the current. In some semiconductors, however, the Hall emf is in the opposite direction, apparently showing that the charge carriers are positive. This does not in fact mean that positive charges are carrying the current in these cases. These semiconductors have holes or vacancies where electrons are missing (cf. Section 33.2), and, as electrons move into vacancies under the action of an applied emf, the holes appear to drift in the opposite direction. The current therefore is seen as a movement of absences of negative charge and appears as a flow of positive charge.

If the electric intensity due to the Hall emf is denoted by E_H, then, since F and G act like the charged plates of a parallel plate capacitor, l apart, $E_H = V/l$, where V is the Hall emf. Therefore

$$qE_H = qvB$$

or

$$\frac{V}{l} = E_H = vB = \frac{I}{nq} \times B,$$

from Eq. (25.7). Therefore

$$n = IBl/qV. \tag{25.16}$$

For a known magnetic field and current, the Hall emf may be measured and the number of charge carriers per unit length of the conductor may be determined. It would be difficult to determine n simply by any other means.

Once n has been determined for a specific conductor with a known B, by use of the last equation one may measure the strengths of other magnetic fields by measuring V for a fixed current. Such a device is called a Hall probe, and the conductor or semiconductor used can be made very small indeed. Thus one may measure the field almost at a point and plot the variation of a variable magnetic field with relative ease. This is the most accurate simple method of measuring magnetic induction.

25.8 ELECTROMAGNETIC PUMPING

If a metallic fluid is contained in a closed system, part of which is shown in Fig. 25.13, the fluid may be kept in motion by means of electromagnetic pumping. Over a portion of the system a magnetic field is established, and a current is passed through the fluid at right angles to the direction of the magnetic field. The portion of the fluid carrying the current experiences a force to the left and therefore moves through the tube. Each fluid element brought to the location of the pumping section experiences this force, and fluid therefore moves through the channel so

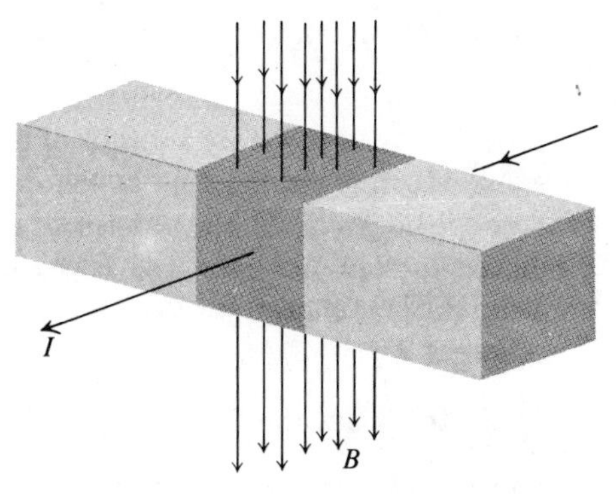

Fig. 25.13. A simple electromagnetic pumping system.

long as the magnetic field and current persist. Note that the pumping action of the electromagnetic pump is achieved with no moving parts.

Electromagnetic pumps have until recently only been used in nuclear reactors using liquid metals such as sodium as coolant, or in foundries where molten metals were hard to control. Recently, however, the idea has been transferred to the movement of blood. In artificial kidney machines the heart is assisted by auxiliary pumping, and in artificial heart machines the pumping must be completely accomplished by the machine. Mechanical pumps, since they contain moving parts, damage blood cells passing through. Electromagnetic pumps remove this danger completely.

The establishment of a magnetic field presents no problem and the ions in the blood will carry an electric current through it. The pumping action is easily achieved, the blood cells are not harmed, and a closed system with few leakage problems can be achieved.

Example 1. Protons in a cyclotron pick up energy each revolution and thus increase the radius of their orbit continually until they reach the outside of the instrument. In a cyclotron with a radius of 2 m and a magnetic field of strength 0.05 Wb m^{-2}, what is the energy of the protons which reach the side of the cyclotron?

Solution. In any orbit, by Eq. (25.5),

$$R = mv/Bq.$$

Therefore

$$v = BqR/m.$$

Therefore

$$\begin{aligned} E &= \tfrac{1}{2}mv^2 = B^2q^2R^2/2m \\ &= \frac{(0.05\ \text{Wb m}^{-2})^2 \times (1.6 \times 10^{-19}\ \text{C})^2 \times (2\ \text{m})^2}{2 \times 1.67 \times 10^{-27}\ \text{kg}} \\ &= 7.7 \times 10^{-14}\ \text{Wb}^2\ \text{m}^{-2}\ \text{C}^2\ \text{kg}^{-1} \\ &= 7.7 \times 10^{-14}\ \text{N}^2\text{A}^{-2}\text{A}^2\text{s}^2\ \text{kg}^{-1} \\ &= 7.7 \times 10^{-14}\ \text{J}. \end{aligned}$$

Example 2. Copper has 8.0×10^{28} conduction electrons per cubic meter. A copper wire of length 1 m and cross-sectional area 8.0×10^{-6} m^2 carrying a current and lying at right angles to a magnetic field of strength 5.0×10^{-3} Wb m^{-2} experiences a force of 8.0×10^{-2} N. What is the drift velocity of the free electrons in the wire?

Solution. The conduction electrons are the charge carriers in the wire and their linear density is

$$\begin{aligned} n &= 8.0 \times 10^{28}\ \text{m}^{-3} \times 8.0 \times 10^{-6}\ \text{m}^2 \\ &= 6.4 \times 10^{23}\ \text{m}^{-1}. \end{aligned}$$

The force experienced by the wire in the magnetic field is

$$F = nqvlB$$

$$\therefore \quad v = F/nqlB$$

$$= \frac{8.0 \times 10^{-2}\ \text{N}}{6.4 \times 10^{23}\ \text{m}^{-1} \times 1.6 \times 10^{-19}\ \text{C} \times 1\ \text{m} \times 5.0 \times 10^{-3}\ \text{Wb m}^{-2}}$$

$$= 1.6 \times 10^{-4}\ \text{N m}^2\ \text{C}^{-1}\ \text{Wb}^{-1}$$

$$= 1.6 \times 10^{-4}\ \text{N C}^{-1}\ \text{A m N}^{-1}$$

$$= 1.6 \times 10^{-4}\ \text{m s}^{-1}.$$

Note how low this value is. Although the electric field set up follows any changes almost instantaneously, the actual movement of electrons in the wire is very slow.

Example 3. A galvanometer has a resistance of 5 Ω and gives full-scale deflection when a current of 5 mA passes through it. If the galvanometer, shunted by a resistance of 5×10^{-4} Ω, is inserted into a circuit, what is the current in the circuit when the galvanometer shows full-scale deflection?

The galvanometer has the shunt removed and a resistance of 1995 Ω joined in series with it. What is the maximum voltage it can measure?

Solution. From Eq. (25.12), the current in the circuit I and the current shown on the galvanometer are related by

$$I = i\left(\frac{R_G + r}{r}\right)$$

$$= 5\ \text{mA}\left(\frac{5\ \Omega + 5 \times 10^{-4}\ \Omega}{5 \times 10^{-4}\ \Omega}\right)$$

$$= 5\ \text{mA}(5.0005/5 \times 10^{-4})$$

$$= 50\ \text{A}$$

to the accuracy justified.

From Eq. (25.14), the voltage between the terminals of the voltmeter and the current shown on the galvanometer are related by

$$V = i(R + R_G)$$

$$= 5\ \text{mA}(1995\ \Omega + 5\ \Omega)$$

$$= 5\ \text{mA} \times 2000\ \Omega$$

$$= 10\ \text{V}.$$

Example 4. In an ohmmeter, the value of the variable resistance for full-scale deflection of the instrument, which occurs for 1 mA, is 1000 Ω. Compare the ranges of resistance values covered by the second and third quarters of the scale of the instrument.

Solution. At full scale

$$V = R'I = 1000\ \Omega \times 1\ \text{mA} = 1\ \text{V}.$$

At one-quarter of the full-scale reading

$$R_1 + R' = 1\ \text{V}/0.25\ \text{mA} = 4000\ \Omega.$$

At one-half of the full scale reading

$$R_2 + R' = 1\ \text{V}/0.5\ \text{mA} = 2000\ \Omega.$$

At three-quarters of the full-scale reading

$$R_3 + R' = 1\ \text{V}/0.75\ \text{mA} = 1333\ \Omega.$$

Thus

$$R_2 - R_3 = 667\ \Omega,$$

but

$$R_1 - R_2 = 2000\ \Omega.$$

PROBLEMS

25.1 Calculate the magnetic induction at the center of a long, air-cored solenoid which has 5×10^3 turns per meter and carries a current of 5 A. An iron core of relative permeability 3000 is now inserted in the solenoid. What is the new value for the magnetic induction?

25.2 Electrons of 1 keV energy move in a N-S direction across 20 cm in a TV tube. If the vertical component of the earth's magnetic field is 5.5×10^{-5} Wb m^{-2}, by how much are the electrons deflected in the passage across the tube?

25.3 Protons, which have a single positive charge and a mass of 1.67×10^{-27} kg, are accelerated through a potential difference of 1 MV and enter a region where they are subjected to a field of magnetic induction of 1 Wb m^{-2} at right angles to their direction of travel. What is the radius of the circular path they travel?

25.4 Protons in a cyclotron are being acted upon by a field of 0.1 Wb m^{-2}. What is their time of revolution in the orbit?

25.5 A wire which has a linear density of 0.025 kg m^{-1} is lying in a horizontal plane which contains magnetic field lines perpendicular to the wire. If the magnetic induction is 2.5×10^{-2} Wb m^{-2} and the wire is prevented from falling by the magnetic force, calculate the current it is carrying.

25.6 A galvanometer of resistance 20 Ω gives full-scale deflection when a current of 5 mA passes through it. What modification must be made to it so that it will give full-scale deflection for (a) a current of 1 A and (b) a potential difference of 100 V?

25.7 The galvanometer shown in Fig. 25.14 has a resistance of 20 Ω and gives full-scale deflection when a current of 5 mA passes through it. It has been converted into a multi-range instrument by the addition of three resistors. It gives full-scale

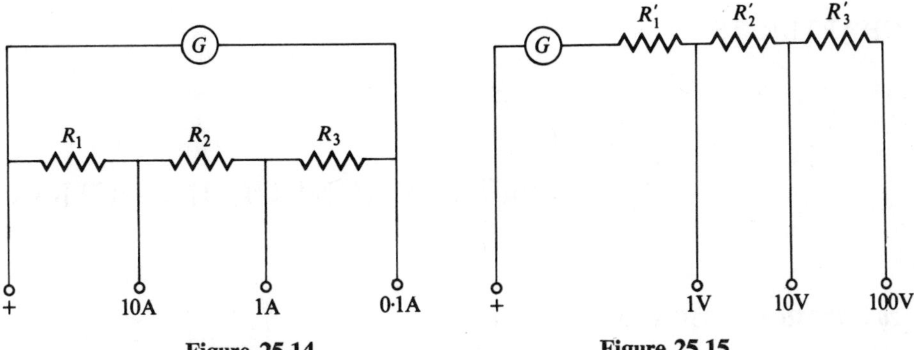

Figure 25.14.

Figure 25.15.

deflections of 0.1 A, 1 A, or 10 A when the current enters at the terminal marked + and leaves by the appropriately marked one of the other three terminals. Calculate the magnitudes of R_1, R_2, and R_3.

25.8 The galvanometer of Problem 25.7 has been converted in Fig. 25.15 for use as a multirange voltmeter, the connections being as for the previous problem. Determine the magnitudes of R_1', R_2', R_3'.

25.9 An ohmmeter is constructed from a 2-V cell, a variable resistance, and an ammeter with a full-scale reading of 5 mA. The calibration is performed but the emf of the cell subsequently falls by 10%. The variable resistance is adjusted so that full-scale deflection still corresponds to zero ohms. What is the true value of a resistance which is measured on the instrument as 2000 Ω?

25.10 If in Fig. 25.12 the breadth and depth of the conductor are each 1 mm, the magnetic induction is 0.5 Wb m^{-2}, and the material is copper which has 8.0×10^{28} free electrons per m^3, calculate the Hall emf when the drift velocity of the electrons is 0.30 mm s^{-1}, and the magnitude of the current flowing.

25.11 The Hall emf set up between the faces of a conductor 0.2 cm apart, when a current of 10 A is flowing through it, is 1 μV when the conductor is at right angles to a field of magnetic induction of 0.2 Wb m^{-2}. Calculate the number of free electrons per meter length of the wire.

CHAPTER 26

ELECTROMAGNETIC INDUCTION

26.1 INTRODUCTION

In physics inverse effects are often very common. Many are mentioned throughout this book. It is very simple, once the reasons for various phenomena are known, to see that the opposite effect must happen as well; but in the early stages of an investigation, this is seldom clear. Effects such as the Seebeck and Peltier effects (cf. Section 24.6) were discovered quite independently.

The last chapter was concerned with the movement of conductors when currents pass through them in magnetic fields. This chapter is concerned with the inverse effect. When a conductor moves through a magnetic field, a current is induced in it. In fact the effect is more complex than that. This effect was investigated quite independently of the effects mentioned in the last chapter.

It should be clear that electricity and magnetism are intimately related. It is one of the fundamentals of physics that the laws of physics must appear the same to two observers in uniform motion relative to each other. An observer holding a charged insulated sphere knows that the region round him contains an electric field. A second observer, moving uniformly past the first, sees the charge moving. To him it therefore represents a current and he knows that the region contains both an electric field and a magnetic field. It is clear that since the two descriptions are equivalent, magnetism and electricity are two sides of the same coin. The next sections describe some very useful results of the intimate relation between the two quantities.

26.2 INDUCED EMF'S

If a magnet is brought rapidly up to a loop of wire in series with a galvanometer, the galvanometer gives a brief deflection. If the magnet is as rapidly withdrawn, a deflection in the opposite direction occurs.

The same phenomena are seen if the magnet is kept fixed and the loop is moved first towards and then away from it. It is natural to conclude that relative motion between a magnetic field and a circuit induces an emf in the circuit. This phenomenon is known as electromagnetic induction (Fig. 26.1). If the effect is investigated more fully, it is found that the magnitude of the deflection, i.e. the magnitude of the induced current, depends directly on the speed of relative movement between the field and the circuit. The faster the magnet is brought up or taken away the bigger is the induced effect. Also, using a more powerful magnet produces a greater effect for the same speed of withdrawal; and a larger loop, or a greater number of turns on the loop, also increases the effect.

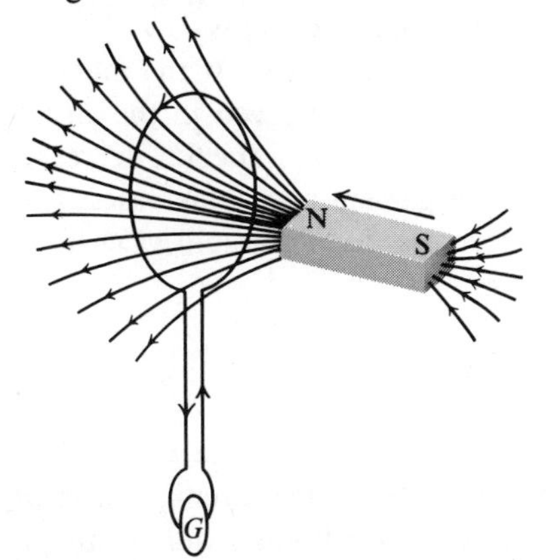

Fig. 26.1. The induction of a current in a loop when a magnet is rapidly brought up to it.

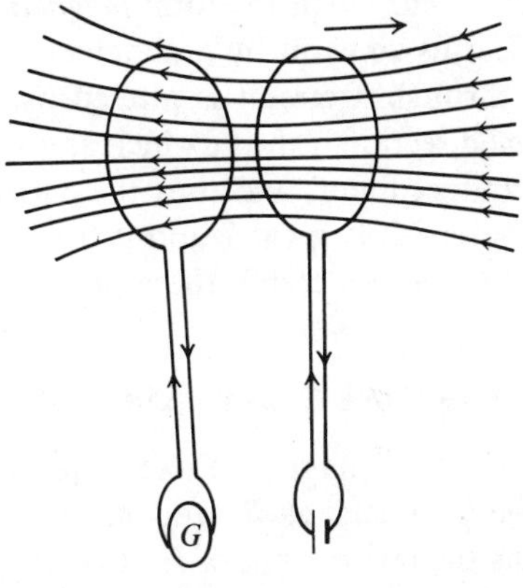

Fig. 26.2. The induction of a current in a loop when a current-carrying circuit is rapidly moved away from it.

The same effect is seen (Fig. 26.2) if one loop is moved relative to another loop which is carrying a steady current. This should not surprise us, since we know that there is a magnetic field associated with a flowing current, and we would expect, and find, that the variation with speed, with field strength, i.e. with the magnitude of the current flowing, and with the size of loop is as before. But we also find that, if the two loops are stationary and side by side and the current in the main (or primary) circuit is switched on or off, a current momentarily flows in the other (or secondary) circuit. In either case the magnetic field has to grow from (or fall to) zero to (or from) its full, steady value, and the field through the secondary loop is thus changing.

Finally, if a loop is constructed as shown in Fig. 26.3, the circuit being

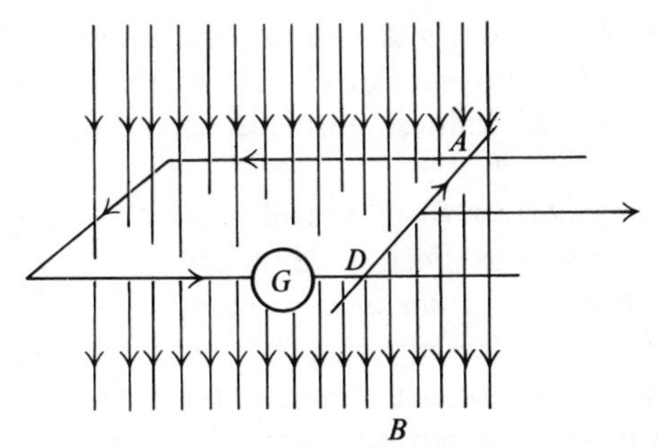

Fig. 26.3. The induction of a current in a circuit when the magnetic flux through it is increased.

completed by the wire AD, and placed in a steady magnetic field, movement of AD causes a deflection of the galvanometer. In this case the total number of lines of force passing through the loop is clearly being altered. It would appear that induced effects occur when the total *magnetic flux* through a circuit changes.

Let us sum up by a statement which covers all three cases. If the magnetic flux through a circuit is altered by any means, an emf is induced in the circuit, in one direction if the flux increases and in the reverse direction if the flux decreases. The induced emf, and thus the current shown on a galvanometer, increases if the field strength is greater and if the increase (or decrease) is more rapid. Let us see whether we can codify these phenomena into laws.

26.3 FARADAY'S LAW AND LENZ'S LAW

We define the magnetic flux through a circuit in air as the sum over the whole circuit of the products of B_n and dA for each small area dA of the circuit, where B_n means the component of $\mathbf{B}$ at right angles to dA. In general, this requires integration, since the magnitude and direction of $\mathbf{B}$ vary over the area. We shall always deal with cases where $\mathbf{B}$ is constant at any instant over the whole circuit. Then, the instantaneous flux is

$$\Phi = BA \cos \theta, \tag{26.1}$$

A being the total area of the circuit and $\mathbf{B}$ being at angle θ to the normal to the area. Φ is thus measured in webers.

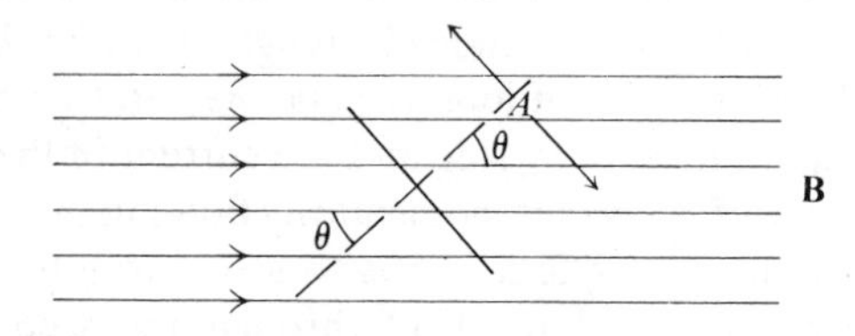

Fig. 26.4. The magnetic flux through a coil.

With this definition, Faraday's law states that the magnitude of the induced emf in a circuit is equal to the rate of change of the magnetic flux through it.

Lenz's law states that the induced emf is in such a direction as to oppose the change causing it.

Lenz's law must be correct from considerations of conservation of energy. If the induced emf enhanced the effect, it would produce an increase in the induced effect, which would produce further enhancement, and so on. Thus if the emf opposes the effect causing it, the current produced in the secondary circuit must always cause a magnetic effect which decreases the magnetic field if the magnetic flux in the circuit is increasing and enhances the magnetic field if the magnetic flux is decreasing. Because of this, one can always deduce the direction of the induced current, and one should check that the direction of the induced current is correct in each of Figs. 26.1, 26.2, and 26.3. Remembering Lenz's law, Faraday's law is therefore written as

$$e = -\frac{d\Phi}{dt} \tag{26.2}$$

26.4 THE EMF INDUCED IN A ROTATING COIL

In Fig. 26.5 a coil is being rotated in a uniform magnetic field **B** about an axis OO', the emf produced being transferred to an external circuit by means of the slip rings S_1 and S_2. If the coil is rotating with constant angular speed ω about its center-line OO', the flux through the coil at the position shown is $BA \cos \theta$. But $\theta = \omega t$; therefore $\Phi = BA \cos \omega t$, or $d\Phi/dt = -BA\,\omega \sin \omega t$. From Faraday's law, therefore, the instantaneous induced emf is $e = BA\omega \sin \omega t = E_0 \sin \omega t$. If the coil consists not of a single turn but of n turns, then

$$e = BAn\omega \sin \omega t. \tag{26.3}$$

Note that the expression holds for a coil of any shape.

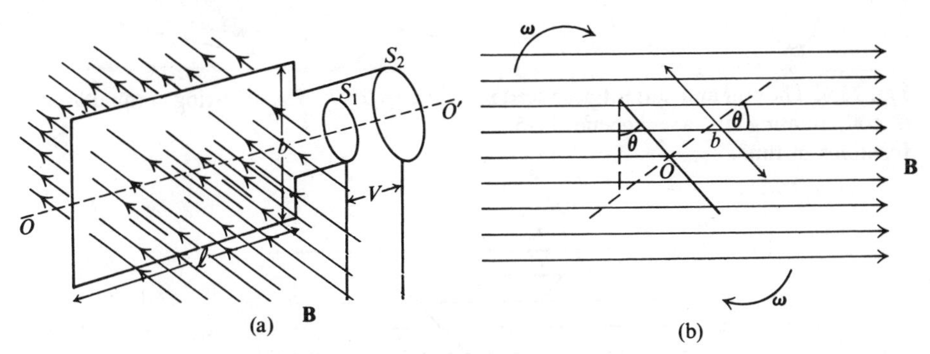

Fig. 26.5. The production of an alternating emf in a coil rotated in a magnetic field: (a) gives a general view; (b) defines the relation of the coil to the direction of the magnetic field.

The emf induced rises to a maximum when $\omega t = \theta = \frac{1}{2}\pi$, i.e. when the coil is parallel to the field, is zero when the plane of the coil is at right angles to the field, and changes in direction as it passes through the latter position. If the value of e is plotted against time, the form will be as shown in Fig. 26.6. This is said to be an alternating emf. If a resistor is connected between the slip rings, a current i will flow through the circuit. The form of the current is shown also in Fig. 26.6 and the current produced is called an alternating current.

This is the type of current and voltage provided by the normal electrical supply, a rather more complicated system of coils being rotated in a magnetic field by mechanical means. Heat from oil, coal, or a nuclear reactor may be used to turn water into steam to drive a turbine; or falling water from a hydroelectric scheme may be made to pass its kinetic energy to turbine blades. In either case other forms of energy are being turned into electrical energy, which is a form more easily transported and used.

In commercial generators the rotating portion is called the armature and consists of a large number of coils wound on iron cores to increase the magnetic flux. The magnetic field may be produced by static permanent magnets, but nowadays is usually supplied by field coils which are wound round the static frame of the generator. When a current is passed through these, the magnetic field is produced. The circuit of such a generator is fairly complicated and we shall not discuss it in further detail.

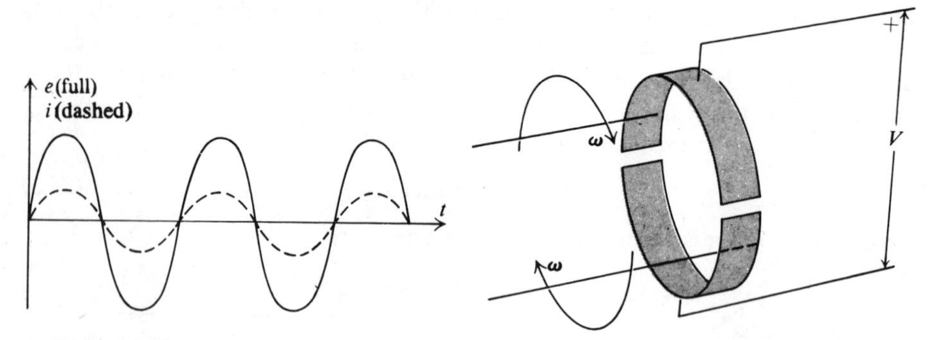

Fig. 26.6. The emf and current produced in a coil rotating in a magnetic field as functions of time.

Fig. 26.7. The split-ring commutator.

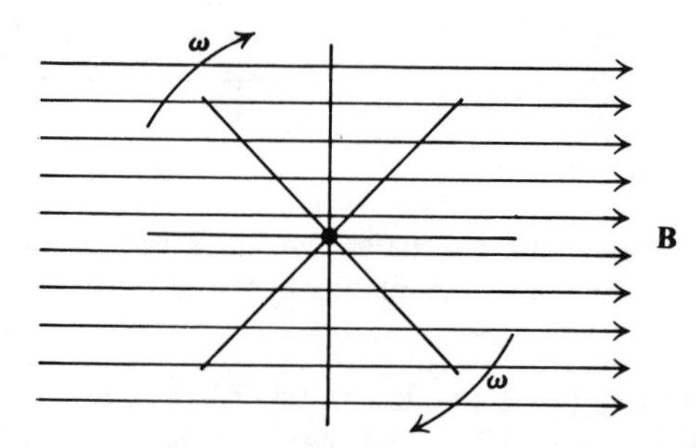

Fig. 26.8. An arrangement of coils to produce a reasonably constant emf when rotated in a magnetic field.

One can produce a current which does not change direction by using a split-ring commutator, as in Fig. 26.7. The ends of the coils go to a split ring. While the coil is rotating, the top lead runs over the top half of the ring and the bottom lead over the bottom half. The emf induced is such that, say, the upper half of the ring is at higher potential than the lower half. When the leads are halfway along the ring sections, the emf has its maximum value. When the leads come to the gap, the coil is at right angles to the field. At this point the emf reverses. But so, too, do the leads on the split ring. Hence, the upper half always remains positive with respect to the lower half. This produces a unidirectional emf but a varying one. However, if one has several coils, as in Fig. 26.8, and not just one, all connected in series, the emf in each coil is at a different position with respect to its maximum and the combined effect is an almost constant unidirectional emf.

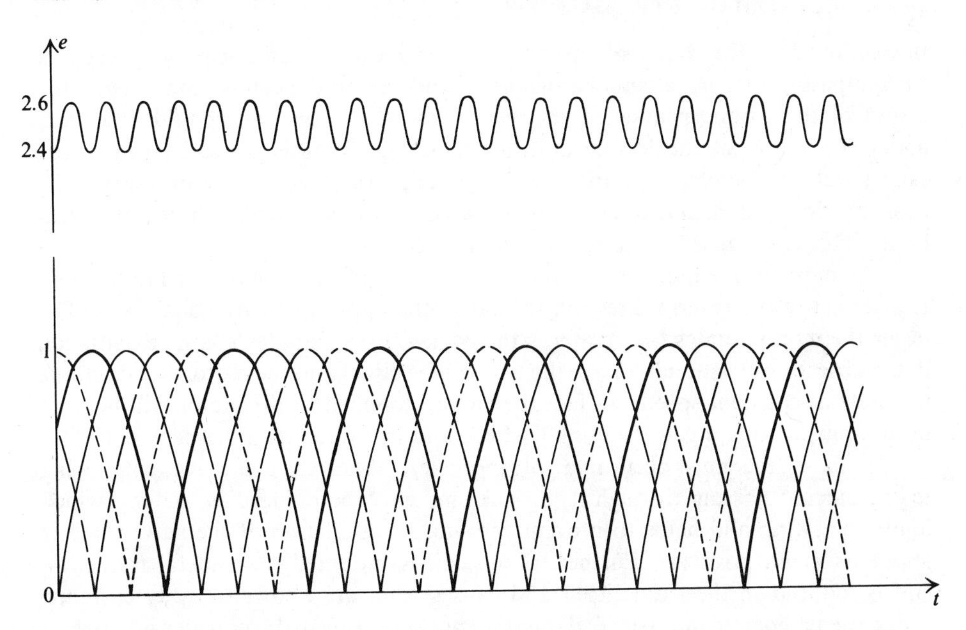

Fig. 26.9. The emf's induced in each of the coils of the arrangement of Fig. 26.8 as functions of time, and the combined effect of all four coils.

Figure 26.9 illustrates the results achieved. The four coils of the generator of Fig. 26.8 each produce an emf shifted in phase with respect to the others. The combined effect of the four is shown by the full line, which is unidirectional and has only a small ripple. More coils would smooth the final output even more.

26.5 THE SEARCH COIL

Another method of measuring a magnetic field can be provided by the use of electromagnetic induction. If a small coil of area A and number of turns n is placed at a spot where the magnetic field is B, then the magnetic flux through the coil is $\Phi = nAB$. If the coil is snatched from this position to one where the field is zero, Φ changes to zero in a very short time. If the coil is connected in series with a galvanometer, specially constructed to have a fair amount of damping, the emf produced by the electromagnetic induction in the search coil sends a short burst of charge through the galvanometer. The consequent swing of the galvanometer is a measure of the change of flux. Since the only thing that is different at the end of the measurement from the beginning is B, one may use this deflection as a measure of B. Standardization is achieved by finding the deflection in known fields.

26.6 ELECTROMAGNETIC DAMPING

In Section 25.4 the mode of operation of an electric motor was described. A current passed through a coil in a magnetic field caused the coil to rotate under the action of the torque produced. In theory, therefore, the coil should accelerate under the action of the torque until air damping, friction in the bearings, etc., cancel out the accelerating torque. In practice the electric motor reaches its constant final speed much earlier than would be expected from these considerations. This is because of electromagnetic induction.

The moment the motor starts turning, a back-emf must be induced in the coil. It must act in the reverse direction to that of the applied emf by Lenz's law. The faster the motor rotates the greater is the back-emf by Faraday's law. Eventually, in the absence of frictional forces, the two will be equal and the motor will continue to rotate at constant speed. In fact the energy supplied to the motor will be used up in joule heating, in overcoming friction, and in overcoming the back-emf.

Similar effects must occur in a galvanometer. When the coil is swinging owing to the current passing through it, a back-emf must be induced in the coil which limits the swing and helps to prevent too violent oscillation of the galvanometer about its mean position. Further, if the coil is mounted on a metal frame, an emf is induced in the frame itself and causes what are known as eddy currents. These use up energy and, hence, damp the motion. If the galvanometer is properly constructed, oscillations can be prevented completely and the coil of the galvanometer swings to, and stays at, its final position. This damping effect is known as electromagnetic damping.

26.7 MUTUAL INDUCTANCE AND SELF-INDUCTANCE

If two coils are placed near to one another and one is made to carry a current, a magnetic field is produced which passes through the other. As has already been seen, the changing of the current in the primary circuit causes an induced emf in the other. If the instantaneous current in the primary circuit is i_1, the magnetic induction B at any part of the secondary circuit is directly proportional to i_1, as is therefore the magnetic flux Φ_2 through the secondary circuit. Thus the induced emf e_2 in the secondary circuit is given by the relation

$$e_2 = -\frac{d\Phi_2}{dt} \propto -\frac{di_1}{dt}.$$

The actual constant of proportionality between e_2 and di_1/dt will depend on the geometrical arrangement of the two circuits and their separation, and could be calculated. In practice it is easier to find the constant of proportionality by experiment. We therefore define

$$e_2 = -M\frac{di_1}{dt}, \tag{26.4}$$

where M is called the mutual inductance between the circuits. It is defined as the emf induced in the secondary circuit for unit rate of change of current in the primary. M has thus the unit of Vs A^{-1}, which is the same as Ω s, and we give this the special name of henry (H). The mutual inductance between two circuits is 1 H if a rate of change of current of 1 A s^{-1} in the primary produces an induced emf of 1 V in the secondary.

Mutual inductance is of much less interest than the similar quantity for a single circuit, self-inductance. If a single circuit consists of several loops of wire, the changing of the current in the circuit causes the magnetic field through each of these loops to vary. This induces an emf in each of the loops tending to oppose the change causing it. Thus when a current is switched on, as it grows it produces a magnetic field which gradually increases up to the steady-state value. The increasing field induces an emf which produces a current opposed to the one which is growing. This limits the rate of rise of the field and current. On the other hand, when the current is switched off and the field decreases, an induced emf opposing the effect tries to increase the field and current, thus augmenting the current in the circuit. If the fall-off is very rapid, the induced emf and current may be quite large. This is why, on breaking a circuit, sparks often jump across the switch, and why electrical apparatus is most liable to break down at switch-off under the sudden surge of current. Very high momentary magnetic fields may, for instance, be produced by short-circuiting an electromagnet.

By analogy with the two-coil case, we define for any circuit a self-inductance L. If a current changes in a circuit, it induces a back-emf in that same circuit. This emf must be proportional to the rate of change of current. Thus

$$e \propto -\frac{di}{dt} = -L\frac{di}{dt}, \tag{26.5}$$

where L is the self-inductance of the circuit. All circuits thus possess self-inductance. Few circuit elements, however, possess appreciable self-inductance, and those which do are called inductors. When inductance is mentioned in subsequent chapters, it is always self-inductance which is meant. L has the unit of henry.

Resistors made of wire are often wound as shown in Fig. 26.10, i.e. back on themselves. This is called non-inductive winding, because any effects induced in one half of the wire will be equal but opposite to the effects induced in the other half. This is because the current follows the same path, but in reverse directions, in the two portions of the wire.

26.8 THE ENERGY STORED IN AN INDUCTOR

Consider a current being established in a circuit in which the final steady value will be I. A magnetic field is set up in an inductor in such a circuit in this process and energy is stored in the magnetic field of the inductor. Since

$$e = -L\frac{di}{dt}, \qquad eidt = -Lidi.$$

But idt is the charge dq passing through the inductor in the time dt. Therefore

$$edq = -Lidi.$$

But the charge dq is passed through the inductor against the back-emf, and thus $-edq$ is the work done in passing the charge through against the induced electrical forces. Therefore

$$d\mathscr{W} = Lidi.$$

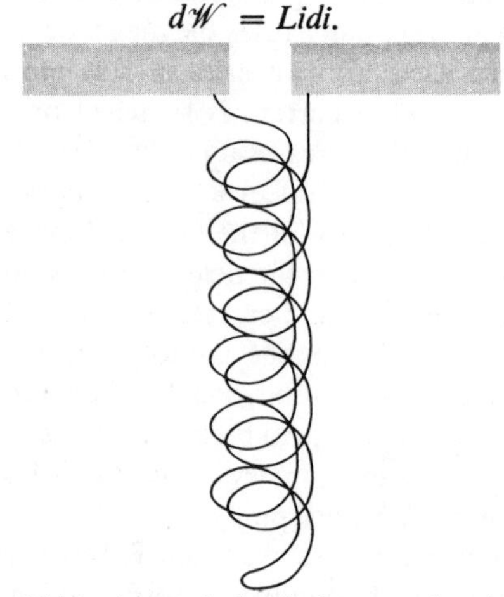

Fig. 26.10. A non-inductively wound resistor.

As in Section 22.13, where the charging of a capacitor was considered, it cannot matter how the final situation comes about, otherwise a perpetual-motion machine could be achieved. The simplest method of establishing the final steady current I in in n equal steps di, where $n = I/di$. The work necessary to establish each increment of current thus varies linearly from 0 to $LIdi$.

Because the work varies linearly, and only because of this, the average work done $\overline{d\mathscr{W}}$ is half the sum of the initial and final amounts of work. Thus

$$\overline{d\mathscr{W}} = \tfrac{1}{2}(0 + LIdi).$$

The total amount of work necessary to establish the current is therefore

$$\mathscr{W} = n \times \overline{d\mathscr{W}} = \frac{I}{di} \times \tfrac{1}{2}LIdi = \tfrac{1}{2}LI^2. \tag{26.6}$$

This work is stored as energy in the magnetic field of the inductor while the current is maintained at its steady value, and is then returned to the circuit, to appear finally as heat, when the current is switched off, during the process where the back-emf attempts to keep the current going.

The inductance L may be regarded as the electrical inertia of the circuit, being the property of the inductor which resists a change of current, just as a mass is the inertial property of a body which resists changes in the state of its motion.

26.9 ARTIFICIAL EXTERNAL PACEMAKERS

Pacemakers to assist an ailing heart in pumping blood through the body were discussed in Section 24.15. When these are implanted inside the body, the cells which provide the motive power fail eventually and a surgical operation is necessary to replace them. Indeed if the pacemaker source fails suddenly, an emergency operation has to be undertaken immediately. Nor is it uncommon for such failure to occur. The cells are encased in a substance such as epoxy resin, but the fluids in the body sometimes find a way inside, and on occasion react with the substances which compose the cells themselves. This does neither the cells nor the patient any good.

Several pacemakers have therefore been devised in which the power supply is external to the body and strapped to it. The power is supplied to a coil taped to the chest and a current is induced in a corresponding coil implanted inside the chest and attached to the heart. Although more power needs to be supplied to this arrangement than is necessary when the whole pacemaker is implanted, this is of little consequence as compared with the ease of the changing of the cells and the lack of complications, since the cells cannot be attacked by body chemicals.

26.10 ELECTROMAGNETIC MEASUREMENT OF BLOOD FLOW

The discussion of electromagnetic induction in this chapter has centered around the movement of closed conducting loops in a magnetic field. When this movement takes place, a current flows in the circuit. If the circuit is not closed, an emf is still induced between the ends of the conductor, a current only flowing when the circuit is completed. In particular, the conducting bar of length l shown in Fig. 26.11 will have an emf induced in it, due to its motion at right angles to the magnetic field of induction of strength B. If terminals are attached to the ends of the bar, the emf set up may be measured on a potentiometer.

The magnitude and direction of the induced emf are easily calculated. If the speed of the bar is v, positive and negative charges of magnitude $\pm q$ in the bar experience forces qvB in opposite directions given by the rule of Section 25.3. The direction of the emf set up will therefore be as shown in Fig. 26.11(a) and (b). The tendency of charges to drift to the ends of the bar will be opposed by the electric field E that will be set up by the separation of the charges (cf. Section 25.7). When equilibrium is achieved, V will be the difference of potential between the ends of the bar, and the electric and magnetic forces on any charges will be equal and opposite.

Hence,

$$Eq = Vq/l = qvB.$$

Therefore

$$V = vlB. \tag{26.7}$$

This result will be true irrespective of the shape of the conductor; and it is clear that if the dimensions of the conductor and the strength of the magnetic field are known, measurement of the potential difference across the conductor permits calculation of the speed v.

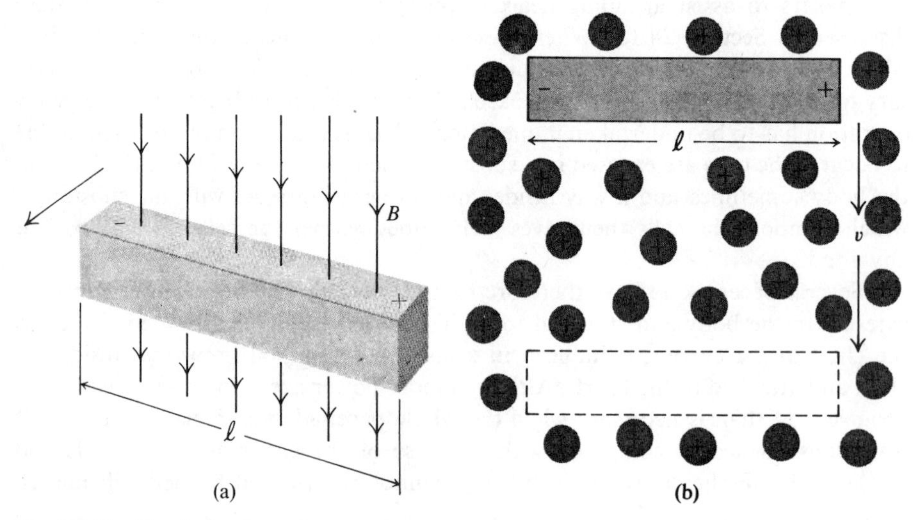

Fig. 26.11. The induction of an emf between the ends of a conductor moving at right angles to a magnetic field: (a) is a general view and (b) shows the movement through the field.

If the moving conductor is not a solid bar but a circulating conducting liquid, any disk of liquid moving through the region containing the magnetic field will have an emf induced across it for the same reasons as are given above. If the pipe containing the liquid is non-conducting, electrodes must be inserted through the walls to be in contact with the moving fluid. This system has been used in measuring the flow rates of molten metals.

In the human body the arteries which carry the blood are themselves conducting, and the use of electromagnetic induction therefore allows the estimation of blood flow rates from the potential difference set up between the intact, external walls of the arteries. This has enormous advantages over conventional methods in which the measuring device must be inserted into the artery by surgical operation.

Initially, a constant magnetic field was employed in measurements of this type. Unfortunately, polarization effects (cf. Section 24.3) occur at the electrodes, and elimination of these presented difficulties and made the apparatus employed very cumbersome. Further, the constant, induced, d.c. emf, of the order of 0.1 mV, may be masked by cardiac, and other, action potentials. The passage of alternating current (cf. Chapter 27) through an electromagnet produces an alternating magnetic field, the emf set up by induction being then alternating also. This eliminates (cf. Section 24.3) the polarization effects and makes it easier to discriminate against other body voltages.

This use of an alternating magnetic field introduces other problems. Even at zero flow rate of the fluid, a signal will still be obtained because of the electromagnetic induction in the loop formed by the lead wires joining the electrodes to the amplifying equipment. This can be eliminated by joining in series with the circuit a compensating loop in which an equal but opposite emf is induced, although this is not always easy to do.

An alternative method of tackling this problem is to vary the form of the alternating magnetic field. If this is sinusoidal, then the induced emf due to the flow, and the emf induced in the leads, are both sinusoidal also. This is because the flow-emf is of the same form as the magnetic field, and the lead-emf of the form of the rate of change of the magnetic field; and the rate of change of a sinusoidal variation is also sinusoidal. If, however, the magnetic field has a square-wave or sawtooth-wave form, the rate of change of the field has no longer the same form as has the field, and the two emf's can be separated out from one another. This is the normal method of dealing with the problem.

The artery to be used in such measurements has to be laid bare. This presents no great difficulties if the flow rate is to be determined in static situations such as occur during surgical operation. For the recording of blood flow in cases where the patient is still to be left active, implantation has many advantages. Miniaturized versions of the apparatus have therefore been developed and have been implanted, encased in suitable materials, in animals. The technique will no doubt be sufficiently refined to be used on human subjects in due course.

Example 1. A uniform magnetic induction of 0.1 Wb m^{-2} extends over a plane circuit of area 1 m^2 and is normal to it. How quickly must the field be reduced to zero if an emf of 100 V is to be induced in the circuit?

Solution. The flux through the circuit is

$$\Phi = BA$$

and

$$e = -\frac{0 - BA}{t}$$

The negative sign merely indicates that the induced emf is in opposition to the change causing it but has no other significance. The time must clearly be positive. Therefore

$$t = \frac{BA}{|e|} = \frac{0.1 \text{ Wb m}^{-2} \times 1 \text{ m}^2}{100 \text{ V}} = \frac{0.1 \text{ V s}}{100 \text{ V}}$$
$$= 1 \text{ ms}.$$

Example 2. A flat coil of 500 turns, each of area 50 cm^2, rotates about a diameter in a uniform field of intensity 0.14 Wb m^{-2}, the axis of rotation being perpendicular to the field and the angular velocity of rotation being 150 rad s^{-1}. The coil has a resistance of 5 Ω and the induced emf is connected via slip rings and brushes to an external resistance of 10 Ω. Calculate the peak current flowing.

Solution. The emf generated by the motion is given in standard notation by the equation

$$e = BAn\omega \sin \omega t.$$

The current flowing is thus

$$i = \frac{BAn\omega}{R} \sin \omega t,$$

and its peak value is

$$\begin{aligned} I_{\text{max}} &= \frac{BAn\omega}{R} \\ &= \frac{0.14 \text{ Wb m}^{-2} \times 50 \times 10^{-4} \text{ m}^2 \times 500 \times 150 \text{ s}^{-1}}{15\ \Omega} \\ &= 3.5 \text{ A.} \end{aligned}$$

Example 3. A metal bar of length 1.0 m falls from rest under the action of gravity while remaining horizontal with its ends pointing towards the magnetic east and west. What is the potential difference between its ends when it has fallen 10 m? The horizontal component of the earth's magnetic field is 1.7×10^{-5} Wb m^{-2}.

Solution. The direction of motion is vertical and thus the vertical component of the earth's magnetic field produces no effect. An induced emf will be produced due to motion relative to the horizontal component of the earth's magnetic field.

The equation of motion of the bar is

$$v^2 = v_0^2 + 2g(y - y_0),$$

where v_0 is the initial velocity of zero. When it has fallen 10 m, $(y - y_0) = 10$ m and

$$v^2 = 2 \times 9.8 \text{ m s}^{-2} \times 10 \text{ m.}$$

Therefore

$$v = 14 \text{ m s}^{-1}.$$

The magnitude of the potential difference between the ends of the bar is

$$\begin{aligned} V &= vlB \\ &= 14 \text{ m s}^{-1} \times 1 \text{ m} \times 1.7 \times 10^{-5} \text{ Wb m}^{-2} \\ &= 0.24 \text{ mV.} \end{aligned}$$

PROBLEMS

26.1 A coil of area 10 cm^2 is in a uniform magnetic field of induction of magnitude 0.1 Wb m^{-2}. The field is reduced to zero in 1 ms. What is the value of the induced emf?

26.2 A rectangular coil which has 50 turns each of area 50 cm^2 is rotatable about an axis joining the mid-points of two opposite sides. When a current of 10 A is passed through it while its plane is at right angles to a uniform magnetic field, it experiences a torque of 1.5 N m. What is the magnitude of the magnetic induction?

26.3 A long straight conductor carrying a current lies parallel to a side of, and in the same plane as, a rectangular coil. If the current is suddenly reversed, in what direction does an induced current flow in the coil?

26.4 A flat coil of area 10 cm^2 and with 200 turns rotates about an axis in the plane of the coil which is at right angles to a uniform field of magnetic induction of 0.05 Wb m^{-2}. If the speed of rotation of the coil is 25 rad s^{-1}, what is the maximum emf induced in the coil, and the instantaneous value of the emf when the coil is at 45° to the field?

26.5 A flat coil, consisting of 500 turns each of area 100 cm^2, rotates with a constant angular velocity of 100 rad s^{-1} about a diameter in a uniform field of magnetic induction 0.05 Wb m^{-2}, the axis of rotation being perpendicular to the field. The coil has a resistance of 10 Ω and is connected to an external resistance of 20 Ω. What is the peak current flowing, and the average power consumed in the external circuit?

26.6 A search coil has 50 turns each of area 5 cm^2. It is snatched in 10 ms from a place where the magnetic induction is 0.25 Wb m^{-2} to one where there is no magnetic field. Calculate the emf induced in the coil.

26.7 The back-emf induced in a coil when the current changes from 1 A to zero in 1 ms is 4 V. What is the self-inductance of the coil?

26.8 An airplane with a wingspan of 50 m is flying in a northerly direction at a speed of 480 m.p.h. in a region where the vertical component of the earth's magnetic field is 5.05×10^{-5} Wb m^{-2}. What is the emf induced between the wing-tips?

26.9 A conducting liquid flowing through a tube of diameter 3 cm is subjected to a magnetic field of induction of magnitude 0.05 Wb m^{-2} at right angles to its direction of travel. The maximum emf set up across the diameter of the tube is 1.5 mV. What is the rate of flow of the liquid?

CHAPTER 27

ALTERNATING CURRENT

27.1 INTRODUCTION

The electricity which is run into our homes, our offices, and our laboratories is not direct current and voltage but alternating current and voltage. It will be shown later that there is good reason for preferring this type of electrical supply. It is marginally easier to produce but very much simpler to transmit without appreciable loss. Our task for the present, however, is to examine how circuits behave when alternating current (a.c.) and not direct current (d.c.) is passed through them.

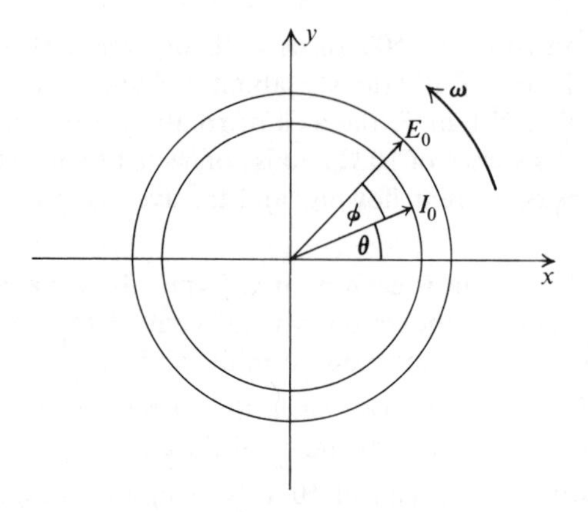

Fig. 27.1. The rotor diagram.

A convenient diagrammatic method of dealing with a.c. is the use of a rotor diagram. This method has already been used (cf. Section 10.4), being an aid to handling problems in any situation in which sinusoidal variation is involved. An explanation will once more be given of the method, to emphasize the main features of the treatment and to point out matters particularly relevant to a.c.

In Section 26.4 it was shown that a.c. can be produced by rotating a coil in a magnetic field and that the instantaneous emf and current are given by the equations $e = E_0 \sin \omega t$ and $i = I_0 \sin \omega t$, respectively. Throughout the chapter small letters are used to denote instantaneous values of current or voltage; capital letters denote effective or r.m.s. values (see below); and capital letters with subscript zero denote peak values. The angular frequency of rotation of the coil is ω, and it is often preferable to write ω as $2\pi f$, where f is the frequency of the a.c. in cycles per second or hertz. Suppose that I_0 is represented by a vector which is allowed to rotate as in Fig. 27.1 about a fixed origin point with constant angular velocity ω or constant cyclic frequency f. The projection on the y-axis of this vector at any time, starting with the vector lying along the x-axis at $t = 0$, is $i = I_0 \sin \theta$, where θ is the angle made by the vector with the x-axis. But $\omega = \theta/t$; therefore $i = I_0 \sin \omega t$. Thus the projection on the y-axis represents the instantaneous value of the current at any time.

Similarly, if the applied emf, which will also be of sinusoidal form, is represented by a rotating vector E_0, $e = E_0 \sin (\omega t + \phi)$ will be the emf at any time and will be the projection of E_0 on the y-axis. E_0 and I_0 will not necessarily be coincident. If they are, they are said to be in phase and ϕ, the phase constant, is zero. If they are separated by an angle ϕ in the rotor diagram, they are said to be out of phase by ϕ. If ϕ is positive, the voltage is said to lead the current; and if ϕ is negative, it is said to lag behind the current. Since $2\pi + \phi$ is the same as ϕ, ϕ and $\phi - 2\pi$ are the same angle. It is therefore usual, if ϕ has a value between $0°$ and $180°$, to say it leads; and if it has a value between $180°$ and $360°$, to say it lags.

27.2 EFFECTIVE OR R.M.S. VALUES OF CURRENT AND VOLTAGE

Since the current and the voltage in an a.c. circuit vary periodically with time, all effects produced by either vary with time also. What therefore is meant by saying that an a.c. current of 1 A is flowing in a circuit? The peak value is not meant, since this is not a very significant measure. The effective value of the current is intended, where the effective value is the magnitude of the d.c. current which would produce the same heating effect. Let us calculate what the effective value is in terms of the peak value.

At any instant the current has the value i. Suppose that it is passing through a resistor of resistance R. The instantaneous heating power of the current is $i^2R = I_0^2R \sin^2 \omega t = \frac{1}{2}I_0^2R(1 - \cos 2\omega t)$. It is now necessary to average $\frac{1}{2}I_0^2R(1 - \cos 2\omega t)$ over all values of t for a number of complete cycles of operation. But $\cos 2\omega t$ is a curve as shown in Fig. 27.2. Over one cycle it has a number of positive values and an exactly equal number of negative values. Thus the average value of $\cos 2\omega t$ over a complete cycle is zero. The average heating effect per second of the alternating current is thus $\frac{1}{2}I_0^2R$.

But this must be equal to I^2R, where I is the appropriate d.c. value and is thus the effective value of the a.c. current. Therefore

$$I^2 = \tfrac{1}{2}I_0^2 \quad \text{or} \quad I = \frac{I_0}{\sqrt{2}}. \tag{27.1}$$

Since the effective value is also the square root of the mean, or average, value of the square of the current over the cycle, it is more usually called the root-mean-square, or simply r.m.s., value of the current.

Applying a similar argument to e^2/R or v^2/R, we obtain the information that $E = E_0/\sqrt{2}$ or $V = V_0/\sqrt{2}$. The r.m.s. values of current, emf, and voltage are thus the peak values divided by $\sqrt{2}$.

When values of currents and voltages in a.c. circuits are quoted, it is always the r.m.s. values which are meant. Thus the normal electrical supply in Britain is 230 V and in the U.S.A. is 110 V. This means that the incoming sinusoidal voltages have peak values of $\pm 230\sqrt{2} = \pm 326$ V, and $\pm 110\sqrt{2} = \pm 156$ V, respectively.

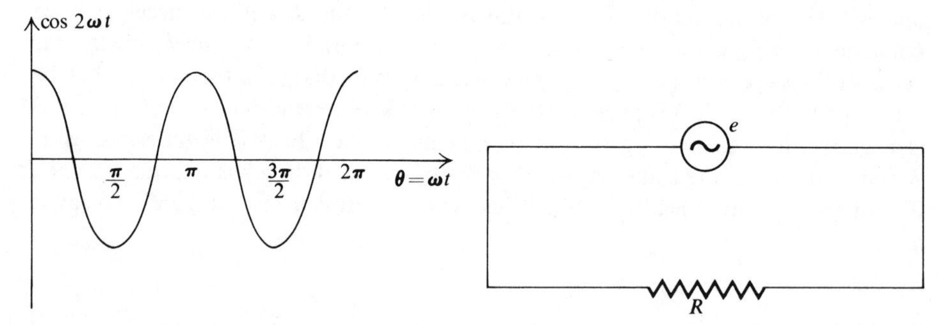

Fig. 27.2. The variation of the function cos $2\omega t$ with time.

Fig. 27.3. An a.c. circuit containing resistance only.

27.3 A.C. CIRCUIT CONTAINING RESISTANCE ONLY

An a.c. source, as in Fig. 27.3, is in series with a resistor of resistance R. By Ohm's law, $i = e/R$ and thus if

$$i = I_0 \sin \omega t, \tag{27.2}$$

$$e = I_0 R \sin \omega t. \tag{27.3}$$

The current and the voltage are in phase, $E_0 = I_0 R$, and, obviously, $E = IR$. The instantaneous power, since e and i are in phase, is

$$p = ei = E_0 I_0 \sin^2 \omega t. \tag{27.4}$$

Therefore (cf. Section 27.2)

$$P = EI = \tfrac{1}{2} E_0 I_0. \tag{27.5}$$

The rotor diagram applicable to this case is shown in Fig. 27.4. No essential difference has been brought in with the introduction of a.c. current.

27.4 A.C. CIRCUIT CONTAINING AN INDUCTOR

It is impossible in practice to produce a pure inductance, since an inductor always has a certain resistance; but it is useful to see what the effects of a pure inductance would be.

Since the inductor has no resistance, there is no potential difference across it, but a back-emf is induced in the inductor because of the continually changing magnetic flux through it caused by the continually changing current.

This back-emf is given by (cf. Section 26.7)

$$e_b = -L\frac{di}{dt} = -L\frac{d}{dt}(I_0 \sin \omega t)$$
$$= -LI_0 \frac{d(\sin \omega t)}{dt}.$$

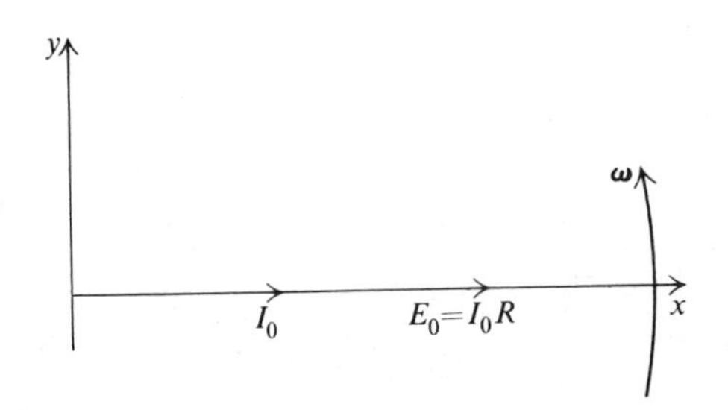

Fig. 27.4. The rotor diagram applicable to the circuit of Fig. 27.3.

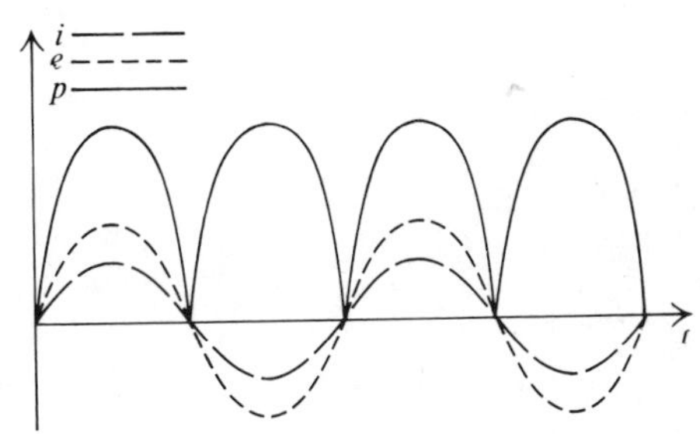

Fig. 27.5. The variations with time of applied emf, current, and power for the circuit of Fig. 27.3.

From the list of derivatives given in Table 3.2 it follows that

$$e_b = -\omega LI_0 \cos \omega t.$$

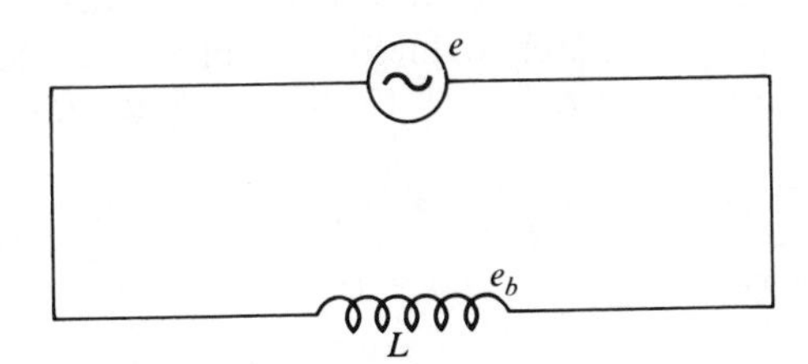

Fig. 27.6. An a.c. circuit containing inductance only.

The circuit contains no resistance. Thus if unit charge is moved once round the circuit, the work done is only against the back-emf and is $-e_b$; but it is also, by definition, the applied emf. Therefore $e = -e_b$. Hence,

$$e = \omega LI_0 \cos \omega t$$
$$= \omega LI_0 \sin (\omega t + \tfrac{1}{2}\pi). \tag{27.6}$$

Thus the current and voltage are 90° out of phase, the voltage reaching a maximum while the current is zero, and dropping as the current increases. Hence, the pure inductance causes the current to lag 90° out of phase with the applied voltage. This is shown in Fig. 27.7.

Note that $E_0 = \omega L I_0$, and therefore $E = \omega L I$. Thus ωL plays the role in this case that R does in the resistive case. The quantity

$$\omega L = X_L = 2\pi f L \tag{27.7}$$

is called the reactance of the circuit and will have the unit of ohm. Thus an inductor impedes the current due to its reactance, which is higher the higher the frequency of the a.c. employed.

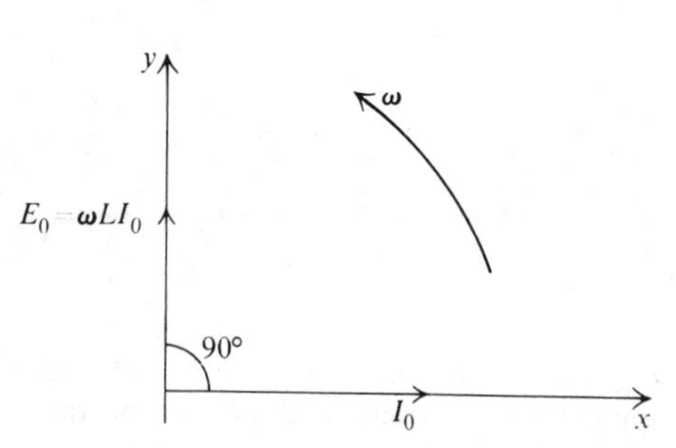

Fig. 27.7. The rotor diagram applicable to the circuit of Fig. 27.6.

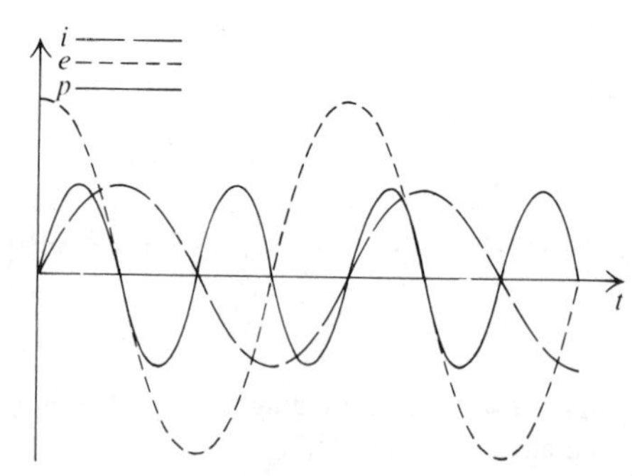

Fig. 27.8. The variations with time of applied emf, current, and power for the circuit of Fig. 27.6.

The instantaneous power is $p = ei = E_0 I_0 \sin \omega t \cos \omega t = \frac{1}{2} E_0 I_0 \sin 2\omega t$. Like $\cos 2\omega t$, $\sin 2\omega t$ averages to zero over a complete cycle (cf. Fig. 27.10), and thus the power dissipated in a purely inductive circuit is zero. As will be seen from Fig. 27.8, the power varies over the cycle, being sometimes positive and sometimes

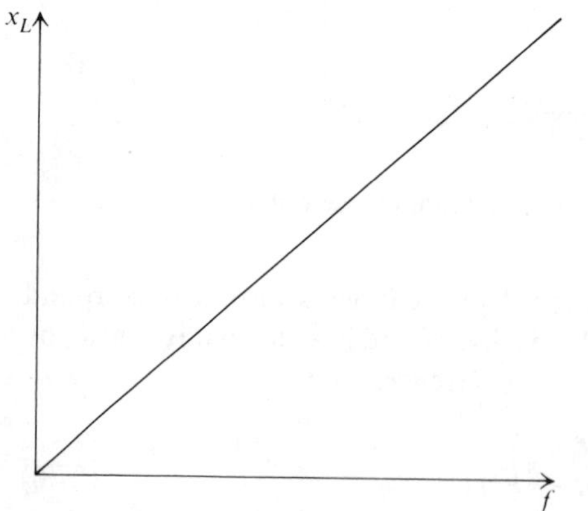

Fig. 27.9. The variation of inductive reactance with frequency.

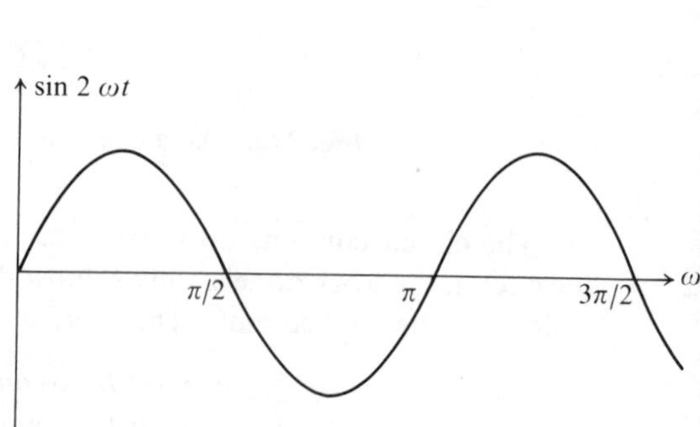

Fig. 27.10. The variation of the function sin $2\omega t$ with time.

negative. The energy supplied builds up and is stored in the magnetic field of the inductor; then it is given back to the circuit, and built up into a magnetic field in the opposite direction. The changing of the magnetic field's direction means that the energy stored changes in sign during each half-cycle, but no energy is dissipated at any time in the circuit, since it contains no resistance.

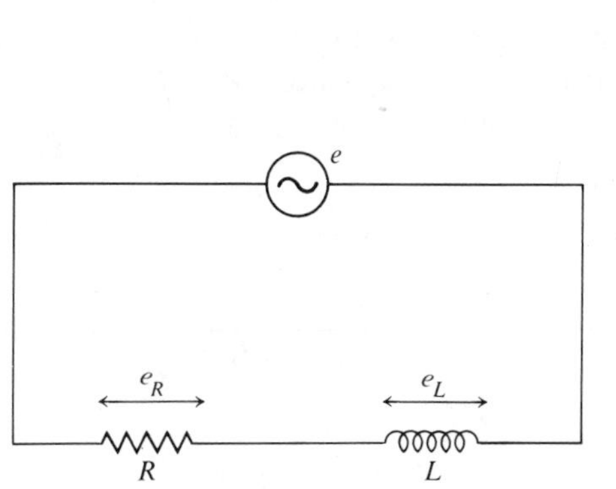

Fig. 27.11. An a.c. circuit containing inductance and resistance.

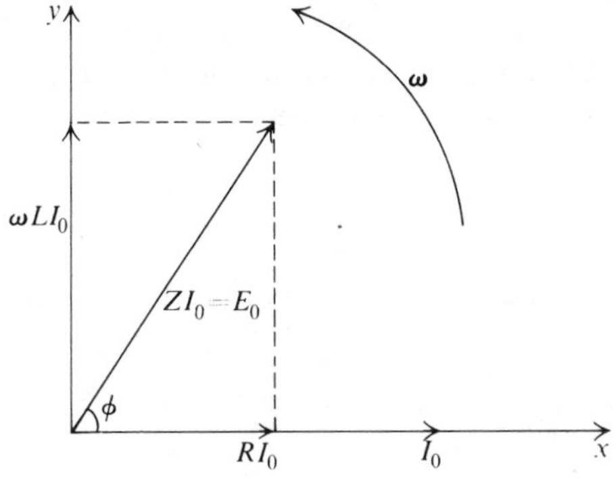

Fig. 27.12. The rotor diagram applicable to the circuit of Fig. 27.11.

If the inductor is a real one, then it has a certain amount of resistance. For convenience, the resistor is regarded as a separate component from the inductor, as shown in Fig. 27.11. The same current must be flowing through both components. It is obvious from Fig. 27.11 that

$$e = e_R + e_L.$$

But, from the results obtained in this and the previous sections,

$$e_R = I_0 R \sin \omega t \qquad \text{and} \qquad e_L = \omega L I_0 \cos \omega t.$$

Therefore

$$e = I_0 R \sin \omega t + \omega L I_0 \cos \omega t.$$

Let $R = Z \cos \phi$, and $\omega L = Z \sin \phi$. Therefore

$$e = I_0 Z[\sin \omega t \cos \phi + \cos \omega t \sin \phi] = I_0 Z \sin (\omega t + \phi). \tag{27.8}$$

Therefore the current lags behind the voltage by a phase constant ϕ, where

$$\tan \phi = \frac{\omega L}{R}, \tag{27.9}$$

and $E_0 = I_0 Z$, where Z is called the impedance of the circuit and

$$Z = \sqrt{R^2 + \omega^2 L^2}. \tag{27.10}$$

It is clear that $E = IZ$ also.

The inductive part of the circuit is non-dissipative, as has already been demonstrated, only the resistive part using up power. Thus

$$P = I^2R. \tag{27.11}$$

27.5 A.C. CIRCUIT CONTAINING A CAPACITOR

In a circuit, such as that in Fig. 27.13, containing a capacitor, the current flowing at any time is transferring charge to the plates of the capacitor; thus $i = dq/dt$. This produces a potential difference between the plates of amount v, where $v = q/C$.

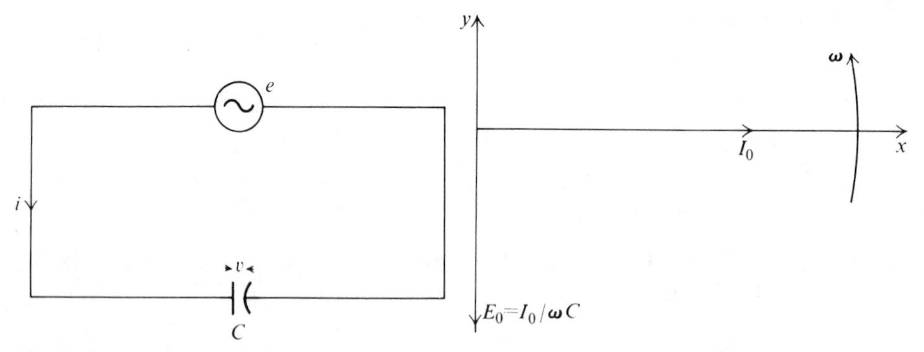

Fig. 27.13. An a.c. circuit containing capacitance only.

Fig. 27.14. The rotor diagram applicable to the circuit of Fig. 27.13.

But at all times the applied emf e must equal v. It follows from all these relations that

$$i = \frac{dq}{dt} = \frac{d(Cv)}{dt} = C\frac{de}{dt}.$$

If $i = I_0 \sin \omega t$, then

$$\frac{de}{dt} = \frac{I_0}{C} \sin \omega t, \qquad \text{or} \qquad e = \int \frac{I_0}{C} \sin \omega t \, dt.$$

Consulting the list of integrals in Table 3.2, it will be seen that

$$e = -\frac{I_0}{\omega C} \cos \omega t = \frac{I_0}{\omega C} \sin (\omega t - \tfrac{1}{2}\pi). \tag{27.12}$$

Thus the current and voltage are 90° out of phase, the voltage being zero when the current is a maximum, the latter then dropping in value as the voltage increases. Hence, a pure capacitance causes the current to lead the applied voltage by 90°. This is shown in Fig. 27.14. Note that

$$E_0 = \frac{I_0}{\omega C} \qquad \text{or} \qquad I_0 = \frac{E_0}{1/\omega C}.$$

Therefore

$$I = \frac{E}{1/\omega C}. \tag{27.13}$$

In this case $1/\omega C$ plays the role that R does in the resistive case.

The quantity $1/\omega C$ is called the reactance of the circuit. Where X_L is the inductive reactance,

$$X_C = \frac{1}{\omega C} = \frac{1}{2\pi f C} \tag{27.14}$$

is the capacitive reactance. Thus a capacitor impedes the current due to its reactance, which is lower the higher the frequency of the a.c. The units of capacitive reactance are also ohms. Note that a capacitor blocks d.c. completely but permits a.c. to flow through the circuit.

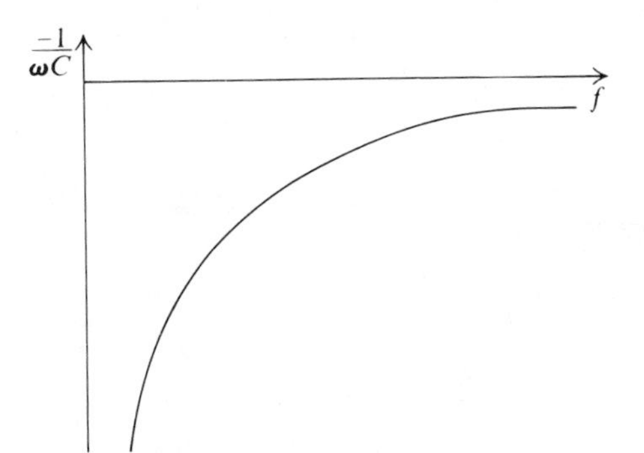

Fig. 27.15. The variation of capacitive reactance with frequency.

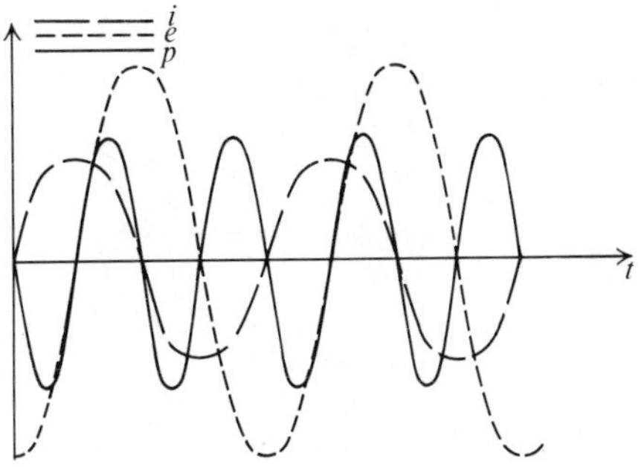

Fig. 27.16. The variations with time of applied emf, current, and power for the circuit of Fig. 27.13.

Note also that, as in the inductive case, the instantaneous power is $p = ei = -\frac{1}{2}E_0 I_0 \sin 2\omega t$, and that this averages to zero over a complete cycle. Thus a capacitor charges up during part of the cycle and returns the charge to the circuit when it discharges in another part of the cycle. It then charges up in the opposite direction, and so on. A single, loss-free capacitor thus does not dissipate power in an a.c. circuit.

27.6 A.C. CIRCUIT WITH RESISTANCE, INDUCTANCE, AND CAPACITANCE IN SERIES

With the three components in series, R may be considered to consist of a resistor and the resistive part of the inductor lumped together. An instantaneous current through this circuit produces an instantaneous potential difference across each of the three components. But the instantaneous applied emf must be equal to the sum of the instantaneous potential differences across R, L, and C.

If the instantaneous current is $i = I_0 \sin \omega t$, then, from the results of the last few sections,

$$\begin{aligned} e &= e_R + e_L + e_C \\ &= I_0 R \sin \omega t + I_0 \omega L \cos \omega t - \frac{I_0}{\omega C} \cos \omega t \\ &= I_0 \left[R \sin \omega t + \left(\omega L - \frac{1}{\omega C} \right) \cos \omega t \right]. \end{aligned}$$

Let

$$R = Z \cos \phi \qquad \text{and} \qquad \left(\omega L - \frac{1}{\omega C} \right) = Z \sin \phi.$$

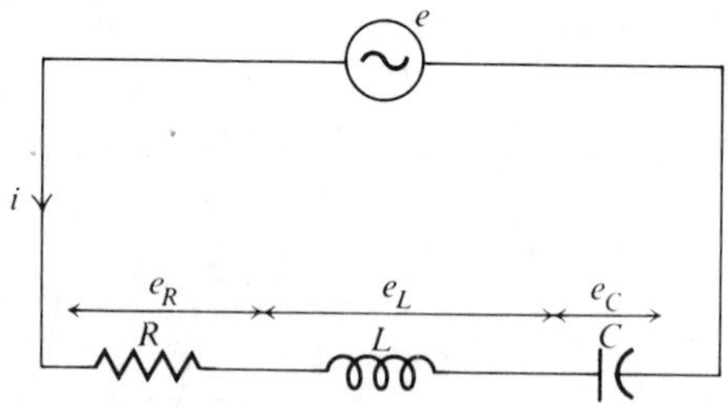

Fig. 27.17. The general series a.c. circuit.

Therefore

$$e = I_0 Z(\sin \omega t \cos \phi + \cos \omega t \sin \phi) = I_0 Z \sin (\omega t + \phi), \tag{27.15}$$

where

$$Z = \sqrt{R^2 + \left(\omega L - \frac{1}{\omega C} \right)^2} \tag{27.16}$$

and

$$\tan \phi = \frac{\omega L - \dfrac{1}{\omega C}}{R}. \tag{27.17}$$

Therefore

$$E_0 = I_0 Z \qquad \text{and} \qquad E = IZ.$$

Therefore

$$I = \frac{E}{\sqrt{\left\{ R^2 + \left(\omega L - \dfrac{1}{\omega C} \right)^2 \right\}}}. \tag{27.18}$$

The current and voltage are therefore out of phase by an angle ϕ given by Eq. (27.17). The current lags the voltage if ωL is greater than $1/\omega C$, i.e. if the inductive reactance is greater than the capacitive reactance, and leads otherwise.

The impedance Z plays the same part as R does in the d.c. case, and the only part of the circuit which is dissipating power is the resistive part.

The rotor diagram shows how the total circuit effect is built up from the effects produced by the separate components. It is customary to draw I_0 along the x-axis. Thus I_0R is along this direction also. I_0X_L is in the positive y-direction, since the potential difference across the inductor leads the current by 90°. I_0X_C is thus in the negative y-direction. $I_0(X_L - X_C)$ is then obtained by subtracting the last two effects, and may be either positive or negative. $E_0 = I_0Z$ is obtained by vectorially compounding I_0R and $I_0(X_L - X_C)$. The angle E_0 makes with the x-axis is then the angle of lag or lead. If ϕ is above the x-axis, then ϕ is positive, and the voltage leads the current; if ϕ is below the x-axis, the opposite is true. The rotor diagram then represents the position at $t = 0$. The total effect or the effect for any component may be simply obtained by allowing the whole rotor diagram to rotate.

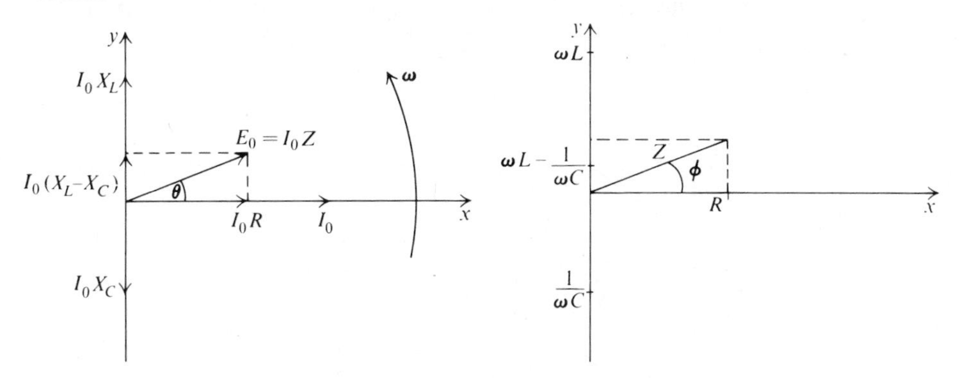

Fig. 27.18. The rotor diagram applicable to the circuit of Fig. 27.17.

Fig. 27.19. Vector diagram for the calculation of impedance.

Note that the rotor diagram allows the calculation of Z from R, L, and C. By dividing all vectors by I_0, one obtains the diagram of Fig. 27.19, which shows the vector diagram for calculation of Z from its constituent resistances and reactances.

Although the instantaneous values of e_R, e_L, and e_C add algebraically to give e, the impressed emf, an a.c. voltmeter placed across R, L, and C in turn will give values, the sum of which is not equal to the applied emf. This is because a voltmeter reads effective or r.m.s. values, and these have to be added vectorially, since they are not in phase with one another.

The power dissipated in the circuit is dissipated in the resistor only. Therefore $P = E_RI$, where E_R is the r.m.s. potential difference across the resistance. Therefore

$$P = \frac{I_0R}{\sqrt{2}} \times \frac{I_0}{\sqrt{2}} = \tfrac{1}{2}I_0^2R = I^2R. \tag{27.19}$$

But, from the rotor diagram,

$$I_0R = E_0 \cos \phi.$$

Therefore

$$P = \tfrac{1}{2}E_0 I_0 \cos\phi = EI\cos\phi. \tag{27.20}$$

The factor $\cos\phi$ is known as the power factor of the circuit, and will be zero when the circuit is purely reactive. Its value can increase from zero up to 1, reaching the latter stage when the circuit is purely resistive. A low power factor means that for a given power, I is large, producing large heating in resistors in the circuit. If ϕ is too large because of a large value of X_L (or X_C), one can always add a capacitor (or inductor) to make the circuit less reactive.

If E and I are measured separately, the apparent power EI is not the power dissipated in the circuit because the $\cos\phi$ term is not measured. A dynamometer is required to measure the true power. The power factor may therefore be considered as the ratio of the true power to the apparent power.

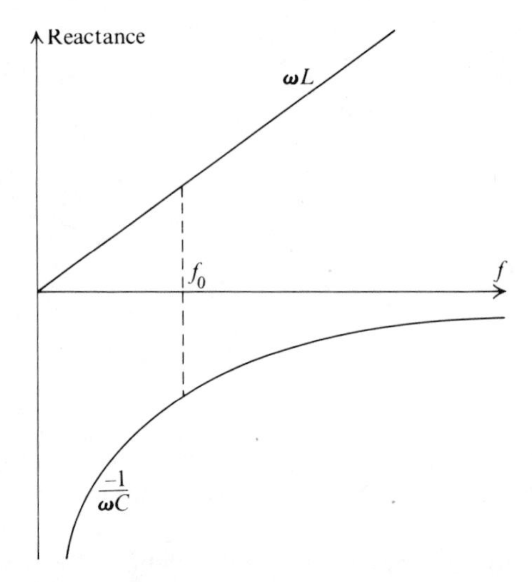

Fig. 27.20. The variation of capacitive and of inductive reactance with time, showing the position of the resonant frequency.

27.7 RESONANCE

If, in the circuit shown in Fig. 27.17, it is possible to alter the frequency of the a.c. generator, a frequency can be found (cf. Fig. 27.20) for which $\omega_0 L = 1/\omega_0 C$. Thus

$$\omega_0 = \frac{1}{\sqrt{LC}}, \qquad \text{or} \qquad f_0 = \frac{1}{2\pi\sqrt{LC}}. \tag{27.21}$$

The circuit is then said to be at resonance and f_0 is the resonant frequency.

The inductive and capacitive reactances are equal, so that the circuit acts as though it were purely resistive. $I = E/R$, and thus I has a maximum value at resonance. If R is fairly small, the resonance effect is very sharp and the current at, and very close to, resonance is markedly greater than it is at any other frequency. This is the basis of tuning in radio, TV, and other similar circuits.

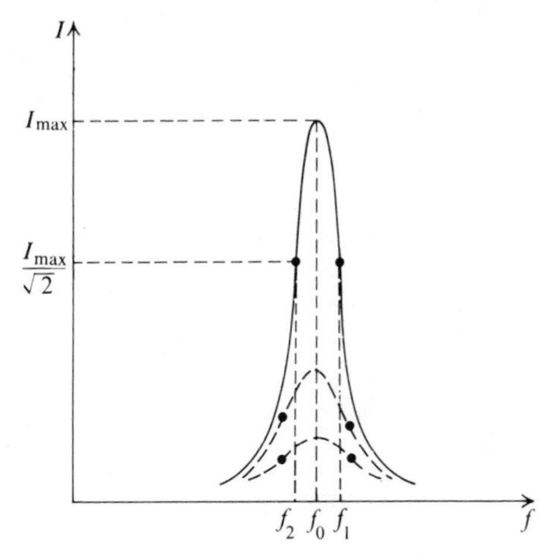

Fig. 27.21. Resonance curves showing the half-power points and the decreasing sharpness of the maximum as the resistance in the circuit is increased.

At resonance the current and impressed voltage will be in phase, since the impedance is a minimum and is purely resistive. Also,

$$\tan \phi = \frac{\omega_0 L - \dfrac{1}{\omega_0 C}}{R} = 0.$$

The impressed emf is therefore fully dropped across the resistance. But very large voltages may well be across the inductance and the capacitance, although they must be equal and opposite and therefore cancel out. The ratio of the voltage developed across L (or C) at resonance to the impressed voltage, which is the voltage dropped across R, is

$$\frac{V}{E} = \frac{\omega_0 LI}{RI} = \frac{\omega_0 L}{R} = \frac{1}{R}\sqrt{\frac{L}{C}}. \tag{27.22}$$

This can be a very large ratio and is known as the voltage multiplication of the circuit. It is more usually called the Q-factor of the circuit and defines the sharpness of tuning at resonance. It can be easily shown that $Q = \omega_0 L/R$ is also 2π times the maximum energy stored in the inductor divided by the energy dissipated per cycle, and is thus also a measure of the dissipative features of the circuit.

If the value of R is increased, the sharpness of the maximum is decreased, as is shown by the different curves in Fig. 27.21. Another method of evaluating the sharpness of the resonance response is the width of the curve at half-power, i.e. when the current has dropped to $(1/\sqrt{2}) \times I_{max}$. If this occurs at f_1 and f_2 on either side of f_0, it is not difficult to show that $(f_1 - f_2)/f_0 = 1/Q$. Thus the greater Q the narrower and sharper the resonance. Q is often defined from the half-power width.

Note that in most circuits C rather than f is varied to obtain resonance. Also note that at resonance $\phi = 0$, $\cos \phi = 1$, and thus the power in the circuit is maximal and the real and apparent powers are equal.

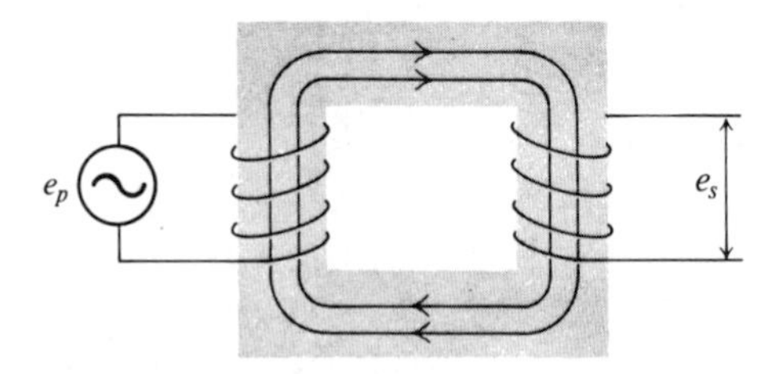

Fig. 27.22. A diagrammatic representation of a transformer.

27.8 THE TRANSFORMER

In the transmission of electrical power from one place to another, economy of operation is the first essential. It is inevitable that if electric current is carried along wires from generator to point of distribution, a certain amount of heating loss is unavoidable, but this must be reduced to the minimum.

The resistance of the transmission wires has a constant value R, and thus the losses in passing a current I through them is I^2R. It is possible by using a transformer to raise a.c. to a high voltage and a low current. As we shall see, the quantity VI is to a first approximation constant. Thus if we start with V and I and, before transmitting the electricity, raise V to V' and correspondingly lower I to I', the heat losses will be substantially reduced. It is quite normal in the United States of America to generate at 2400 V, to step this up to 120,000 V for transmission, reducing the current by a factor of 50, stepping down the voltage again to 2400 V at the substation, and distributing it to housing areas where it is further stepped down to 120 V. Thus the current transmitted is about 1/1000 of the current used in the house or factory, and the transmission losses are correspondingly reduced by a factor of 10^{-6}. Since the transformer is an a.c. instrument only, and a very simple one, the ease of transmission of a.c. with little loss makes it much more suitable than d.c. as an electrical supply.

One cannot apply the above argument to the voltage, since one has no way of knowing *a priori* how much of the voltage will be dropped along the transmission wire.

A transformer consists of two mutually coupled coils (Fig. 27.22). In most transformers the two coils are linked by a magnetic core of iron which leads all the magnetic flux from the primary coil through the secondary. Where the linkage does not require to be so great, as in some radio receiver coils, the core may be of air, and the primary and secondary are interleaved. An alternating current is passed through the primary coil, causing an alternating magnetic field which passes through the secondary coil also. The changing magnetic flux does two things. It induces an emf in the secondary, but it also produces a back-emf in the primary. Since the primary winding has little resistance, the back-emf in the primary will be almost equal to the emf applied.

Let us first consider the situation when no load is attached to the secondary and the winding there is left open. If N_p and N_s are the number of turns on the primary and secondary, respectively, e_b, the instantaneous back-emf induced in the primary, equals $-N_p(d\Phi/dt)$, where Φ is the flux linkage per turn at any instant. Similarly, $e_s = -N_s(d\Phi/dt)$, since there will be negligible loss of flux from the core. Therefore

$$\frac{|e_s|}{|e_p|} = \frac{N_s}{N_p}, \qquad \text{or} \qquad \frac{E_s}{E_p} = \frac{N_s}{N_p}. \tag{27.23}$$

The ratio of the voltage in the secondary to the voltage in the primary is thus the same as the ratio of the number of turns on the secondary and on the primary. If N_s is greater than N_p, the transformer is said to be a step-up transformer; and if N_s is less than N_p, it is said to be a step-down transformer.

Suppose that a load is now attached to the secondary. Since the secondary circuit is closed, e_s produces a current, which itself produces a magnetic flux in the core. This flux by Lenz's law must reduce the flux due to the primary current, and thus reduces the back-emf in the primary. But the source supplies a constant emf. Further, because of the lack of resistance in the primary, this emf must be balanced by the total back-emf. To achieve this the current in the primary increases in order to raise the back-emf to equality with the applied emf.

By the necessity for conservation of energy, assuming no losses in the iron core, the power delivered to the secondary circuit must equal the power supplied by the primary circuit. Therefore

$$E_s I_s = E_p I_p, \qquad \text{or} \qquad \frac{I_s}{I_p} = \frac{E_p}{E_s} = \frac{N_p}{N_s}. \tag{27.24}$$

It is seen that the current in the secondary reduces in the same ratio as that by which the voltage increases, i.e. in the ratio of the turns on the secondary and the primary.

In order that the above analysis should be correct, the losses in the transformer must be negligible. This is achieved by using as material for the core a substance which suffers little hysteresis loss, and by constructing it of thin laminations, well insulated from one another, so that any currents induced in the material have no easy path to follow and are consequently negligibly small. The efficiency of transformers can be arranged to be as high as 99%.

27.9 A.C. METERS

It is obvious that a normal moving-coil galvanometer cannot be used for measuring a.c. currents. The coil would be attracted first one way and then the other, and the alterations in the current would be so fast that the inertial properties of the coil would be far too great for it to follow such rapid changes. The meter would therefore fail to read at all. It is of course possible to use in the circuit a rectifier, such as the copper-copper oxide one mentioned earlier, which makes the current unidirectional, and in certain multi-range instruments this is done.

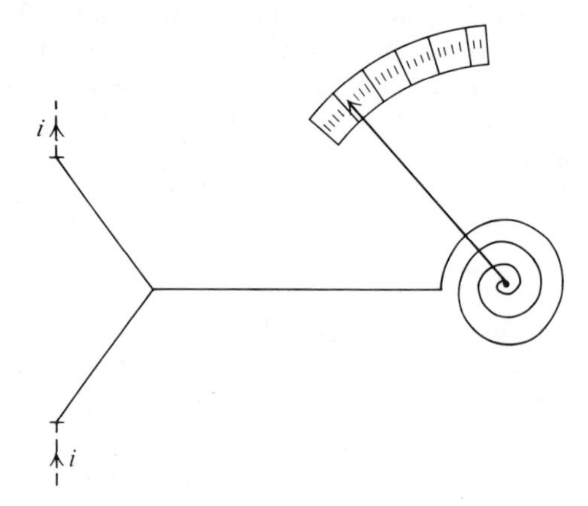

Fig. 27.23. The essential features of a hot-wire a.c. ammeter.

It is, however, also possible to use a system which does not depend on the direction of the current, and heating is one effect which is often employed. For instance, an instrument incorporating a heating wire may be used. When a current is allowed to pass through it, the wire heats up and will eventually reach a temperature at which the heat produced exactly equals the heat given out by radiation, conduction, and convection. The higher the current the greater this final equilibrium temperature. If one end of a thermocouple is joined to the wire, a unidirectional current is produced in the thermocouple circuit, the magnitude of the current depending on the temperature and thus on the current in the original circuit. If this thermoelectric current is passed through a conventional moving-coil galvanometer, the deflection produced is therefore dependent on the current in the original circuit. The scale over which the pointer moves is no longer linear and must be calibrated.

More simply (Fig. 27.23), the heating wire may have its mid-point attached to a taut wire which controls the contraction of a spring, to which is attached a pointer. When the wire heats up, it expands. The taut wire moves and the spring is allowed to uncoil to a degree which is dependent on the expansion of the heating wire, which is itself dependent on the current flowing through it. Again, the scale is not linear and requires to be calibrated.

27.10 IMPEDANCE MATCHING

Most readers, for the rest of their careers, will very probably be using electrical instruments in black boxes, where the output from one is applied to the input of the next one, and so on. The very important, but little appreciated, question of matching one instrument to the next immediately arises. Let us illustrate what is in general a problem of impedance-matching by a very simple case of resistance-matching.

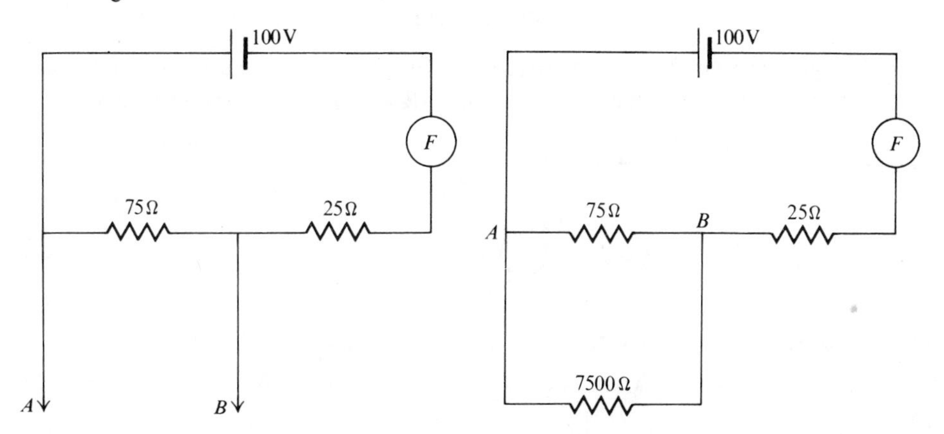

Fig. 27.24. A first attempt at a power-pack circuit.

Fig. 27.25. The power pack of Fig. 27.24 with a 7500-Ω load.

Suppose that we were asked to design and construct a simple d.c. circuit which would act as a power supply for a second piece of apparatus, which requires 75 V for its operation. Having available a bank of cells of negligible internal resistance, and emf 100 V, we construct the circuit shown in Fig. 27.24, the leads to the next piece of apparatus being taken off at A and B. The cells are such that not more than 2 A must flow through them; a 2-A fuse F has therefore been incorporated in the circuit. The second piece of apparatus may be considered for all practical purposes as a resistor. It may contain a complex of resistors; but by combining these one after the other by the laws of resistors in series and in parallel, we shall eventually arrive at a single resistor equivalent to the complete network. If the equivalent resistance of the second circuit turns out to be very high, nothing much will happen when A and B are joined to it. If, for example, the equivalent resistance is 7500 Ω, then the effective resistance between the points A and B of Fig. 27.25 is given by the relation

$$\frac{1}{R} = \frac{1}{75\ \Omega} + \frac{1}{7500\ \Omega} = \frac{101}{7500\ \Omega},$$

and thus R is approximately 75 Ω. Joining up the two circuits has no serious effect on either.

However, if the second circuit has a resistance equivalent to 5 Ω (Fig. 27.26), then

$$\frac{1}{R} = \frac{1}{5\,\Omega} + \frac{1}{75\,\Omega} = \frac{16}{75\,\Omega},$$

and thus R is 4.7 Ω. The current in the power-pack circuit is then

$$\frac{100\text{ V}}{(25 + 4.7)\,\Omega}$$

which is greater than 2 A, and therefore the fuse blows. Furthermore, the voltage supplied to the second circuit is in any case all wrong. It is no longer 75 V.

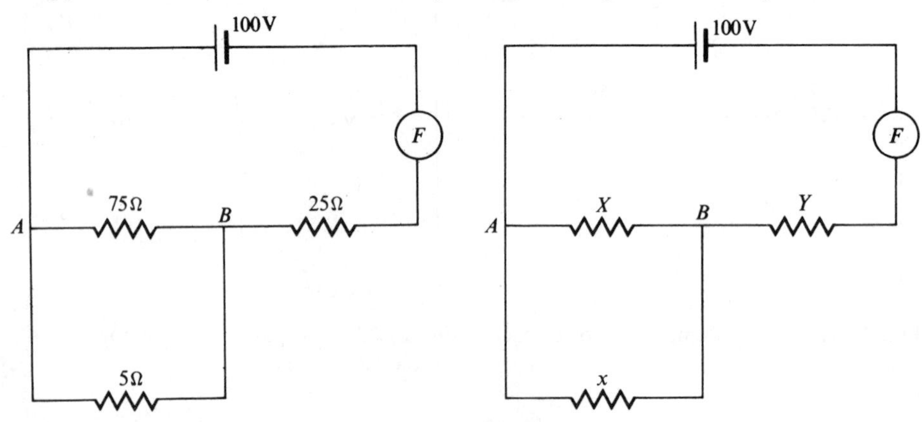

Fig. 27.26. The power-pack of Fig. 27.24 with a 5-Ω load.

Fig. 27.27. A circuit to determine the correct resistance values for power-pack and load.

Even when the equivalent resistance is 7500 Ω and no trouble arises, the current through the cells is 1 A and the power being delivered by them is 100 W. The current through the 7500-Ω resistor is approximately 1/100 A delivered at 75 V. The power supplied to the second circuit is thus only $\frac{3}{4}$ W. The power-pack is therefore dissipating practically all the power; and if the purpose is to supply power to the second circuit it is not doing its job efficiently.

This should show clearly that circuits cannot be joined together indiscriminately. They must be designed carefully to ensure that they will work in unison and with maximum efficiency. In the example the first circuit should consist of two resistances X and Y in series with the cells. The second circuit will act like an equivalent resistance x. The resistance between the points A and B when the circuits are joined will be given by $1/R = (1/X) + (1/x)$. Furthermore, $R = Xx/(X + x) = 3Y$, if the voltage across AB is to be 75 V. Further, $100\text{ V}/4Y = 2$ A for maximum current. Therefore

$$Y = 12\tfrac{1}{2}\,\Omega.$$

Therefore

$$\frac{Xx}{X+x} = 37\tfrac{1}{2}\ \Omega.$$

Therefore

$$X = \frac{37\tfrac{1}{2}\,\Omega \times x}{x - 37\tfrac{1}{2}\ \Omega}.$$

In order that X shall be positive, x must be at least $37\tfrac{1}{2}\ \Omega$. If x is not much greater than $37\tfrac{1}{2}\ \Omega$, X is very large. It is obviously best to have x as small as possible and X large, so that maximum current will flow from the power-pack into the second instrument. Since the value of x decreases as X increases, the minimum value of x is $37\tfrac{1}{2}\ \Omega$ when X is very large. A power of 200 W is being delivered by the cells in the power-pack of which 150 W is being passed to the second circuit. Maximum power transfer is thus achieved for the circuit shown in Fig. 27.28.

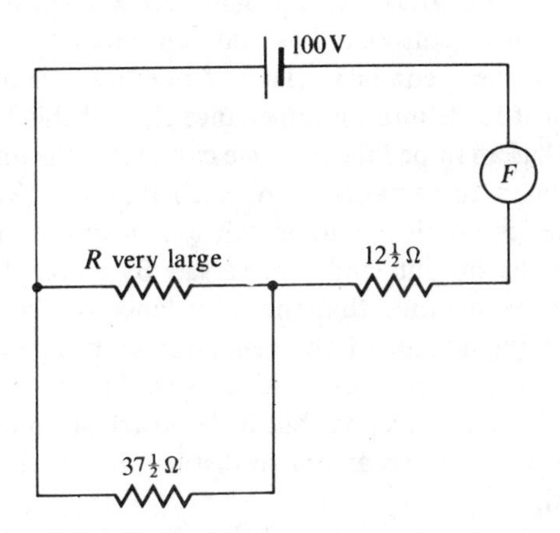

Fig. 27.28. The circuit of Fig. 27.27 with the correct values for maximum power transfer.

This simple example should demonstrate how important it is to match, in this case the resistances, in the general case the impedances, of circuits which are to be joined together. With impedances the situation may be even more crucial, because the effect of adding a further circuit may upset the main circuit completely. It may take the main circuit from resonance, or establish resonance in it; it may put the current and voltage out of phase, or do many other similarly disturbing things. The matching of apparatus in this way is thus of crucial importance.

We have already talked about a simple method of impedance-matching, although we have not described it as such. This is the use of a transformer. Let us look at the transformer from the viewpoint that it is an impedance-matching device.

Suppose that the secondary is connected to a load with impedance Z_s. Then,

$$I_s = \frac{E_s}{Z_s},$$

and the current in the primary is

$$I_p = \frac{N_s}{N_p} \times I_s,$$

the voltage being

$$E_p = \frac{N_p}{N_s} \times E_s.$$

Hence, the effect of the load Z_s is to produce an effective impedance Z_p in the primary, where

$$Z_p = \frac{E_p}{I_p} = \frac{E_s}{I_s} \times \left(\frac{N_p}{N_s}\right)^2 = Z_s \times \left(\frac{N_p}{N_s}\right)^2. \tag{27.25}$$

Thus Z_p can be considered as the impedance reflected into the primary circuit, and the transformer can be considered as matching the impedance in the primary with the impedance in the secondary. Thus if one has a main circuit which will only work properly if it feeds into an impedance Z_p and this has to be joined to a second circuit which has an impedance Z_s, one can match the impedances through a transformer which has a turns ratio N_p/N_s such that Eq. (27.25) is true. Thus, a step-up transformer is employed to match a low impedance at the input to a high impedance at the output, and vice versa. By employing a suitable transformer, one may always arrange that the impedance of the load is matched as far as possible to the impedance of the generator or primary circuit in order to ensure maximum power transfer to the load, as we did in the simple resistance case.

Instruments which accept a signal but little power are often arranged to have a high impedance at the input so as not to disturb the previous circuits to which they may be attached.

27.11 IMPEDANCE MATCHING IN GENERAL

In electrical measurements we have a driving electrical pressure, the emf, which drives a flow of electricity through a circuit, the magnitude of the flow being determined by the circuit impedance. This idea of impedance allows us to determine the flow characteristics if we know the applied pressure. This concept of impedance is one reason why the branch of physics known as electricity has been such a successful one. The concept is very simple and allows complicated problems to be broken down into component parts. Each component part exhibits a characteristic impedance. The separate impedances may be joined together by simple rules for impedances in series and in parallel (just as for the simple resistance case). The complicated problem reduces to a series of simpler ones and is handled with ease.

The success of the concept has in fact been responsible for its being taken over into other branches of physics. For instance, a pressure difference produces a flow of fluid through a channel, and the amount of flow resulting can be considered as being determined by the impedance of the channel. Similarly, sound waves are produced by pressure disturbances and sound may flow through various tubes and channels because of this disturbance. The amount of sound reaching any further point of the circuit depends on the impedance of the circuit arrangement. In any instance in which we have a driving force, or pressure, and a corresponding flow, we can introduce the notion of a circuit impedance which gives us a constant of proportionality between the driving quantity and the resultant flow. Branches of the circuit which are in series or in parallel can be handled by formulas which link impedances in series or in parallel and produce an equivalent impedance effect of the whole. In this way the handling of problems in other branches of physics has been much simplified; but it should be immediately obvious that if two general circuits are joined together, unless the impedances are matched, the presence of the second circuit may seriously affect the impedance, and therefore the response, of the first. In fact, all the difficulties we have mentioned in the previous section as applicable to the joining of electrical circuits are equally applicable to more general circuits in which the idea of impedance is possible.

If one designs a loudspeaker badly and mismatches it to the previous electrical circuit, the sound issuing from the loudspeaker may be very small or completely distorted. Similarly, in the case of sound, one may design a megaphone very badly. In its simplest form it consists of a tube plus a horn; and if the impedances of tube and horn are mismatched, little sound proceeds into the horn (it would be reflected back into the tube), or the sound issuing may be very badly distorted.

Take the simple case of a stethoscope. Here one has a system consisting essentially of a small horn joined to a tube which branches into two further tubes, each of which finishes in a small horn. If the stethoscope is badly designed and the impedances are mismatched at the several junctions, it is by no means clear that the sound finally heard by the doctor need bear much resemblance to the sound originating in the patient's chest. Over the years stethoscopes have obviously been designed by trial and error to a condition of reasonable efficiency. Nonetheless, doctors do not like to swop stethoscopes, because they claim that sounds often appear different from one stethoscope to another. It is not really surprising.

In a case, by no means isolated, which came to the attention of the authors some time ago, a catheter was being inserted through the heart valves of patients in order to measure the changes in pressure as the catheter moved through the valve. The results of such experiments are very useful in diagnosing heart-valve trouble. Unfortunately, the apparatus joined to the catheter in order to measure and record the effects taking place was so badly mismatched that the readings taken were utterly meaningless. The recording apparatus was so seriously affecting the pressure measured by the catheter that the output bore no relation to the actual variations at the heart valve. It is unnecessary to stress the unfortunate consequences that could arise from wrong diagnosis with such an instrument.

This is merely one illustration of many examples that could be given of the great care which is necessary in joining pieces of apparatus together. If mismatching occurs, distortion or complete falsification of the results may be produced in the final instrument. It is therefore always worth while checking on the characteristics of each piece of apparatus used and making sure that no instrument affects the results of the instruments previous or subsequent to it. It should be made clear that when a complete set of pieces of apparatus is bought from a manufacturer, a design engineer has already solved all these problems.

27.12 ELECTROENCEPHALOGRAPHY

The electrical potential between the surface of the cortex of a normal adult and an electrically neutral area of the body, such as an ear lobe, may be as high as 10 mV. It is more usual to measure the potential not on the surface of the cortex but on the surface of the scalp, where the value drops to 100 μV because of the high resistance of bone. This electrical activity can be measured all over the scalp, the form being similar at all points although displaced slightly in time from one site to another. Electroencephalography (e.e.g.) is the study of the electrical potentials on the surface of the head.

It is normal in such study to hook electrodes to the scalp at up to 24 locations and record on chart recorders the potential differences experienced between pairs of these electrodes. It should be noted that other electrical disturbances in the body must be filtered out, and that the signals must be separated from background noise and amplified before they can be recorded. Certain types of feature appear more strongly in some areas of the scalp than others, and one trace may prove more valuable than another for a particular purpose. The potential difference between any two points is found to be alternating, but not in any simple fashion.

The most outstanding normal feature is the α-rhythm, which consists of waves 8–13 c/s in frequency, and very sinusoidal and regular in shape. There are also β-waves, occurring in the range 14–50 c/s, and found in all adults, which are smaller in amplitude than α-waves and often spindle-shaped. There are other types of regular waves which occur, but these are normally associated with abnormality.

The α-rhythm is stronger near the occipital region, associated with vision. If the eye focuses on a bright image, the α-waves disappear, although they return if the concentration is continued for long. When the eyes are closed, the α-rhythm is slower than when they are open. Sleeping alters the wave patterns completely. As a subject drifts off to sleep, the α-waves disappear. β-waves take over during the dream stage; but with full sleep only very slow waves of around 1 c/s are very evident. Any stimulus during sleep produces a burst of α-waves on top of these slow waves.

These wave activities do not appear to correspond to any simple body process, and the origin and function of e.e.g. waves is obscure. Two suggestions, neither having any evidence to commend it, are (1) that the waves represent a scanning mechanism by which the brain acquires incoming information from sensory neurons, and (2) that the waves represent feedback loops for information storage.

All that *can* be said is that e.e.g. waves represent an activity of the central nervous system as yet unidentified.

The great usefulness of e.e.g. patterns comes from their employment in diagnosis, and occasionally in treatment. Epilepsy produces typical e.e.g. patterns, and tumors near the surface of the brain can be located from their characteristic effect on the e.e.g. patterns in their vicinity. Likewise, brain damage in accidents can be assessed from alterations in e.e.g. waves. Without this information, difficult, painful, and dangerous brain operations would be required to make such diagnoses.

27.13 IMPEDANCE PLETHYSMOGRAPHY

Volume changes in biological conducting regions can be followed by the resulting changes in the electrical impedance of that region. This fact has been principally applied to change in shape and content of the lungs on breathing and to changes brought about by the pulsatile flow of blood through arteries.

One may only be interested in making sure that these changes take place. Thus in patient-monitoring, continual measurement of the changing current caused by the changing impedance across the chest may be used as a means of sounding an alarm if the current remains constant. This might also be achieved by placing thermistors (cf. Section 23.8) or pressure transducers (cf. Section 23.9) in the nostrils or otherwise directly in the air flow, but, in general, this is not so convenient. There is always the chance of accidental blockage, particularly in infants; and if the insert becomes covered with mucus or vomit, stoppage of the air flow may go unnoticed. Attaching leads to the face is also likely to be uncomfortable and the danger of the leads being pulled off is greater . The attachment of leads to the chest in a jacket is likely to be less restrictive.

If estimates of volume flow are to be made as well, greater care may have to be taken. It is normal in this case to use four electrodes, attached to the legs or arms for blood-flow measurements, attached in transthoracic mid-axillary configuration for air-flow measurements. The frequency of the a.c. supply is in the region of 15–20 kc/s, since the impedance change, which is small, tends to be maximal in this range. The current passed must be small enough to be unnoticed by the patient and amplifiers must therefore be used to make the current change of reasonable magnitude.

27.14 DIATHERMY

If two electrodes of large area are attached to the body of a patient and a high-frequency a.c. generator is connected between them, an alternating current passes between the electrodes and heat is produced in the tissues. Since the current does not pass in a straight line from electrode to electrode, but spreads out through various low-impedance paths, heating is largely confined to tissues near the electrodes. The heating effect is very useful in the treatment of many diseases. It is employed in treatment of rheumatic and arthritic conditions of various types; or of lesions in the vagina, urethra, and cervix of women; or of the epididymis, prostate, and vesicles of men.

If v.h.f. currents are employed, the penetration of the heating effect is greater, and more deep-seated sites may be treated. This fact has been used in bone and joint diseases, in sinusitis, and in axillary and cervical adenitis. The rapid healing of wounds may also be achieved by such short-wave medical diathermy.

Diathermy may also be used surgically. If one of the electrodes is reduced in size to button or needle form, the intensity of the current at that electrode may be made so great that cells are coagulated and killed by the concentrated heating effect. This fact has been used in the treatment of both benign and malignant tumors.

If, in addition, the frequency and power are increased and the small electrode is made in the form of a needle or a loop, an intense arc can be produced at that point. Soft tissue can be cut instantly and cleanly by the moving electrode, the small blood-vessels and lymphatics being sealed at the same time. This method can be used for the removal of malignant tissue and has been also employed in extensive dissection, such as the removal of the breast. It has been especially successful in the operation of transurethral prostatic resection.

Example 1. A sinusoidal a.c. voltage of peak value 180 V is applied to a resistor and produces a current of peak value 12 A. What are the r.m.s. values of voltage and current, and the power dissipated in the resistor?

Solution. The r.m.s. values are

$$E = \frac{180\text{ V}}{\sqrt{2}} = 127.3\text{ V},$$

$$I = \frac{12\text{ A}}{\sqrt{2}} = 8.48\text{ A}.$$

The power dissipated is

$$P = EI = \frac{180\text{ V} \times 12\text{ A}}{\sqrt{2} \times \sqrt{2}}$$
$$= 1080\text{ W}.$$

Example 2. An inductor is in series with a 100-V, 50-c/s a.c. generator. The current is 10 A and is found to be lagging behind the voltage by 60°. What are the resistance and inductance of the inductor?

Solution. Clearly the inductor has an appreciable resistance since the current and voltage are not 90° out of phase. Also

$$Z = \sqrt{R^2 + \omega^2 L^2} = 100\text{ V}/10\text{ A} = 10\,\Omega$$

and

$$\tan\phi = \omega L/R = \tan 60° = \sqrt{3},$$

therefore

$$\omega^2 L^2 = 3R^2,$$

therefore

$$\sqrt{R^2 + 3R^2} = 2R = 10\ \Omega,$$

therefore

$$R = 5\ \Omega,$$

therefore

$$L = \frac{\sqrt{3}\,R}{\omega} = \frac{\sqrt{3} \times 5\ \Omega}{2\pi \times 50\ \text{c/s}},$$

$$= 28\ \text{mH}.$$

Example 3. At what frequency will a capacitor of capacitance 5.0 μF have a reactance of 1000 Ω?

Solution. The reactance is 1000 Ω. Therefore

$$\frac{1}{\omega C} = 1000\ \Omega,$$

therefore

$$\omega = \frac{1}{1000\ \Omega \times C},$$

therefore

$$f = \frac{1}{2\pi \times 10^3\ \Omega \times 5 \times 10^{-6}\ \text{F}},$$

$$= \frac{100}{\pi}\ \text{c/s} = 32\ \text{c/s}.$$

Example 4. A 110-V, 4000-rad s^{-1} a.c. generator is in series with a 15-Ω resistor, a 10-μF capacitor and an inductor of resistance 5.0 Ω and inductance 0.010 H. Calculate the potential difference across each component, and the phase angle. Check by use of a rotor diagram.

Solution. The important quantities are

$$R = R_0 + R_L = 15\ \Omega + 5\ \Omega = 20\ \Omega$$

$$\omega L = 4000\ \text{s}^{-1} \times 0.01\ \text{H} = 40\ \Omega$$

$$1/\omega C = 1/(4000\ \text{s}^{-1} \times 10^{-5}\ \text{F}) = 25\ \Omega,$$

therefore

$$Z = \sqrt{(20\ \Omega)^2 + (40\ \Omega - 25\ \Omega)^2}$$

$$\doteq \sqrt{(20\ \Omega)^2 + (15\ \Omega)^2} = 25\ \Omega.$$

The current in the circuit is

$$I = E/Z = 110\ \text{V}/25\ \Omega$$

$$= 4.4\ \text{A}.$$

The resistor has a resistance of 15 Ω, the capacitor a reactance of 25 Ω, and the inductor an impedance Z', where

$$Z' = \sqrt{R_L^2 + \omega^2 L^2} = \sqrt{(5\ \Omega)^2 + (40\ \Omega)^2}$$
$$= 40.3\ \Omega.$$

The voltages across the components are therefore

$$V_{R_0} = 4.4\ \text{A} \times 15\ \Omega = 66\ \text{V}$$
$$V_C = 4.4\ \text{A} \times 25\ \Omega = 110\ \text{V}$$
$$V_L = 4.4\ \text{A} \times 40.3\ \Omega = 177\ \text{V}$$

The phase angle is such that

$$\tan\phi = \frac{\omega L - \dfrac{1}{\omega C}}{R} = \frac{15\ \Omega}{20\ \Omega} = 0.75,$$

therefore

$$\phi = 36°52'.$$

The applied voltage leads the current by 37°.

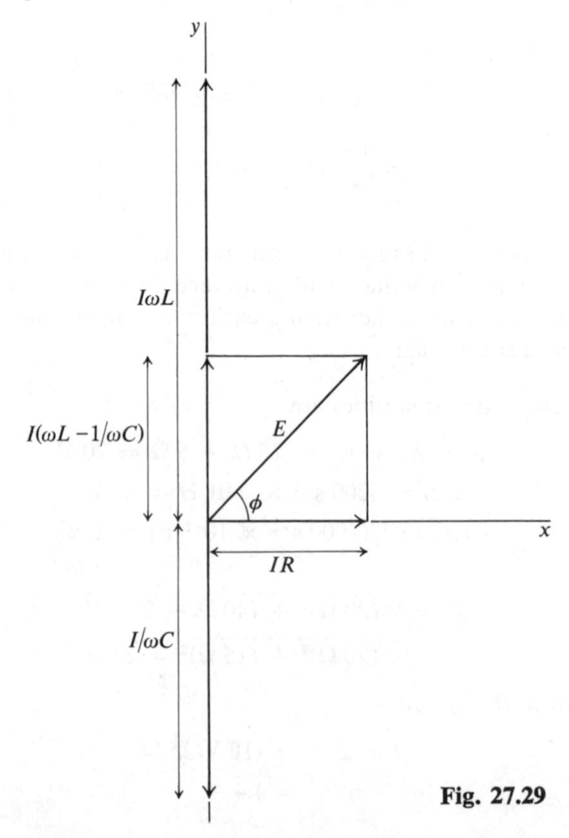

Fig. 27.29

The quantities IR, $I\omega L$, and $I/\omega C$ are plotted on the rotor diagram and combined in the usual manner as shown. E has a length corresponding to 110 V as required and ϕ is measured by a protractor as 37°.

Example 5. Calculate the resonant frequency, the current at resonance and the Q-factor for the circuit of Example 4.

Solution. The resonant frequency is given by the equation

$$f_0 = \frac{1}{2\pi\sqrt{LC}} = \frac{1}{2\pi\sqrt{10^{-2}\text{ H} \times 10^{-5}\text{ F}}}$$
$$= 503 \text{ c/s}.$$

Since the circuit is purely resistive at resonance, the current at that time is

$$I = \frac{E}{R} = \frac{110\text{ V}}{20\ \Omega}$$
$$= 5.5 \text{ A}.$$

The Q-factor is given by the equation

$$Q = \frac{1}{R}\sqrt{\frac{L}{C}} = \frac{1}{20\ \Omega}\sqrt{\frac{10^{-2}\text{ H}}{10^{-5}\text{ F}}}$$
$$= 1.58.$$

The low Q-factor is attributable to the high resistance of the circuit.

Example 6. An a.c. source of internal resistance 9000 Ω is to supply current to a load of resistance 10 Ω. How should the source be matched to the load and what is then the ratio of the currents passing through load and source?

Solution. The matching may be done by means of a transformer. If the numbers of turns on primary and secondary windings are N_1 and N_2, then the internal resistance of the source R_1 and the load resistance R_2 are related by the equation

$$R_1 = (N_1/N_2)^2 R_2.$$

Therefore

$$\left(\frac{N_1}{N_2}\right)^2 = \frac{R_1}{R_2} = \frac{9000\ \Omega}{10\ \Omega} = 900,$$

therefore

$$\frac{N_1}{N_2} = 30.$$

A transformer with a turns ratio of 30:1 must be employed. The transformer lowers the voltage in the ratio 30:1, but correspondingly increases the current in the same ratio.

PROBLEMS

27.1 When a cathode-ray tube has its deflector plates connected across a 100-V battery, the spot on the fluorescent screen is deflected 12 cm. If the plates are now connected across a resistor which is in parallel with an a.c. voltmeter and in series with an a.c. generator, the length of the trace on the screen is 17 cm and the voltmeter reads 50 V. How can these apparently contradictory figures be explained?

27.2 A 120-Ω resistor is connected across the 240-V a.c. supply. What are the current through the resistor, the power dissipated in it, and the maximum instantaneous potential difference across it?

27.3 An inductor is in series with a 100-Ω resistor and a 110-V, 60-c/s a.c. generator. The voltage across the resistor is 50 V and the voltage is found to be leading the current by 45°. What are the resistance and inductance of the inductor?

27.4 At what frequencies would a 0.1-H inductor and a 10-μF capacitor have reactances of 500 Ω?

27.5 An a.c. circuit consists of a resistance of 100 Ω, a capacitance of 25 μF, and an inductance of 0.1 H, all in series with a 240-V, 300-rad s^{-1} generator. Find (a) the current flowing, (b) the voltage across each circuit element, (c) the power factor and phase angle, (d) the power dissipated, (e) the resonant frequency of the circuit, and (f) the Q-factor.

27.6 A capacitor is in series with a non-inductive resistor and a 130-V, 350-rad s^{-1} a.c. supply. A potential difference of 50 V is set up across the resistor, and the current flowing is 1.5 A. What are the values of capacitance and resistance involved?

27.7 A series circuit consists of a capacitor, a 20-mH inductor of negligible resistance, a 125-Ω resistor, and an a.c. generator operating at 300 rad s^{-1}. The current is found to lead the voltage by 52°. What is the capacitance of the capacitor?

27.8 When a 12-V cell is connected in series with two electrical elements, the current is 1.2 A. If a 12-V, 50-c/s a.c. generator replaces the cell, the current is 1.0 A. What are the values of the circuit elements?

27.9 Four black boxes are identical, each having two terminals on top. Between the terminals internally are connected either a resistor, a capacitor, an inductor, or a Cu–CuO rectifier, one of these elements to each box. The boxes are unlabeled. If a cell, an a.c. generator, and a.c. and d.c. ammeters are available, how may one discover which box is which?

27.10 A 5-μF capacitor, a 25-mH inductor of negligible resistance, and a 10-Ω resistor are connected in series with a 100-V a.c. generator of variable frequency. Calculate the current flowing when the angular frequency of the a.c. is 50 rad s^{-1}, 100 rad s^{-1}, and 1000 rad s^{-1}.

27.11 An inductor of resistance 5 Ω and inductance 10 mH is in series with a capacitor and a 10-V, 1000-c/s a.c. source. The capacitor is adjusted to give resonance in the circuit. Calculate the capacitance of the capacitor and the voltages across capacitor and coil.

27.12 What are the maximum and minimum values of the capacitance of the tuning capacitor in the first stage of a radio which has an inductance of 10 mH, if it is to receive all broadcasts on the band from 550 kc/s to 1.6 Mc/s?

27.13 A series circuit consists of an inductor of inductance 40 mH and resistance 5 Ω, a 4-μF capacitor, and a variable-frequency, 10-V a.c. generator of negligible internal impedance. What is the resonant frequency of the circuit, and the frequencies for which the current is $1/\sqrt{2}$ of its maximum value? Hence find the Q-value of the circuit, and check the answer by using a different formula.

27.14 A 220-V a.c. generator is connected to a load of 100 Ω through a transmission line of resistance 10 Ω. Calculate the power dissipated in the line. If a transformer is used to step the voltage up to 11,000 V before transmission and step it down correspondingly at the load, how much power is saved?

27.15 An a.c. voltage of 110 V is applied to the primary of a transformer whose efficiency is 99%, the currents in the primary and secondary circuits being 1.60 A and 132 mA, respectively. What is the voltage on the secondary and the turns ratio of the transformer?

27.16 A power-pack acts as though it were a battery of emf E and internal resistance r. Find the equivalent resistance of any load attached to it to ensure maximum power transfer.

If two loads connected in parallel are attached to the power-pack, one load to consume four times the power of the other, find the equivalent resistances of the two loads.

27.17 Maximum power is transferred from an a.c. generator when it feeds into an impedance of 2500 Ω. If the generator is to deliver power to a load of resistance 100 Ω, what is the turns ratio of the transformer which should be employed?

CHAPTER 28

THE ATOMIC NATURE OF MATTER

28.1 INTRODUCTION

The earliest evidence for the atomic and molecular nature of matter, which is at the heart of the study which constitutes modern physics, was obtained from the field of chemistry. Early chemical studies showed that all substances were built up from 92 basic substances called *elements*. The smallest portion of an element which would still show the chemical properties of that element was called an atom and the smallest portion of a more complicated substance which would show the properties of that substance was called a molecule. Thus a molecule is a collection of atoms bound together by some kind of attractive force.

Three laws, (a) Dalton's law, (b) Gay-Lussac's law, and (c) Avogadro's hypothesis, gave some information about these atoms and molecules. Dalton's law states that elements combine in fixed proportion by weight. Thus it is always found that 16 g of oxygen combine with 2 g of hydrogen to form water without leaving any oxygen or hydrogen uncombined. Similarly, 16 g of oxygen always combine with 12 g of carbon to form carbon monoxide. It is evident that 16 g of oxygen must contain the same number of atoms as 12 g of carbon and 1 g of hydrogen. Thus, although nothing may be known about the masses of atoms in terms of grams, an atomic mass scale can be set up on which hydrogen will be approximately 1 a.m.u. (atomic mass unit), carbon will be 12 a.m.u., oxygen 16 a.m.u., and so on. The elements are then found to have atomic masses ranging from around 1 for hydrogen to about 238 for uranium on the atomic mass unit scale, although the atomic masses on this scale are not in general integral.

Gay-Lussac's law states that chemical combination of gases involves integral proportion by volume; and Avogadro's hypothesis, which is a deduction from this, states that equal volumes of all gases under the same conditions of temperature and pressure will contain the same number of molecules. This assumes that Gay-Lussac's law can be extended to gases which do not combine, such as the rare gases, and this extension is found to be valid.

The mass in grams of an element equal in magnitude to its atomic mass on the a.m.u. scale, is called the gram-atomic mass or gram-atom; similarly, the mass in grams of a substance equal to its molecular mass in a.m.u. is called the gram-molecular mass or mole. It follows from Avogadro's hypothesis that all gram-atoms and all moles will have the same number of atoms or molecules, respectively, whether they are in gaseous, liquid or solid form. The number of molecules in a mole, or atoms in a gram-atom, is called Avogadro's constant N_0, and its value is found to be $6.0249 \times 10^{23}\ \text{mol}^{-1}$. Modern practice is to use mole whether the gram-atom or the gram-molecular mass is meant.

Further evidence about atomicity is provided by Faraday's laws of electrolysis. If we refer back to Section 24.2, it becomes obvious that the valency rules in chemistry are related to atomic structure and have an electrical origin, But, in addition, it is possible to obtain some quantitative results. If it is assumed that a salt in solution splits up into positive and negative ions, which are attracted to opposite electrodes, then, in the simple case where the ion is simply a charged atom, the atoms appear (in general liberated or deposited) at the electrodes and the charges they give up constitute the current through the electrolyte. Therefore

$$\frac{E}{m} = \frac{\text{the charge of an ion}}{\text{the mass of an ion}} = \frac{\text{the total quantity of electricity passed}}{\text{the total quantity of mass deposited}}.$$

But 1 C of electric charge is that quantity of electricity which in electrolysis deposits 11.177×10^{-7} kg of silver. Therefore

$$\left(\frac{E}{m}\right)_{\text{Ag}} = \frac{1\ \text{C}}{11.177 \times 10^{-7}\ \text{kg}} = 8.9471 \times 10^5\ \text{C kg}^{-1}.$$

Therefore

$$\left(\frac{N_0 E}{N_0 m}\right)_{\text{Ag}} = 8.9471 \times 10^5\ \text{C kg}^{-1}.$$

But $N_0 m$ for silver is 1 mol of silver, which is known to have a mass of 0.10788 kg. Therefore

$$N_0 E = 96{,}522\ \text{C mol}^{-1}.$$

All monovalent elements yield the same value for $N_0 E$; divalent elements yield twice this value. If it is assumed that all monovalent ions carry a single elementary charge e, and divalent ions twice this charge, it follows that

$$N_0 e = 96{,}522\ \text{C mol}^{-1}. \tag{28.1}$$

This very important physical quantity, known with high accuracy, is called the faraday. It is the quantity of charge necessary to liberate 1 mol of a monovalent element in electrolysis. If either N_0 or e can be determined accurately, the other follows from a knowledge of the faraday. It may be surprising to learn that the most accurate method of determining e, the elementary charge, is by measuring the faraday from electrolytic experiments and Avogadro's constant from X-ray experiments.

From the above analysis it follows that if M' is the mass of a silver atom and M the mass of a hydrogen atom,

$$\frac{e}{M'} = 8.9471 \times 10^5 \text{ C kg}^{-1}$$

and

$$\frac{e}{M} = 8.9471 \times 10^5 \text{ C kg}^{-1} \times \frac{107.88 \text{ a.m.u.}}{1.0076 \text{ a.m.u.}} = 9.5794 \times 10^7 \text{ C kg}^{-1}.$$

28.2 THE DISCOVERY OF THE ELECTRON

The real starting point of modern physics lies in the investigation of the passage of electricity through gases. It was found that if a glass tube containing a gas at low pressure had two electrodes inserted into it and a high potential difference was maintained between them, then electricity flowed through the gas. When this happened, the gas emitted visible light. But it was also found that if the gas was gradually pumped out, the emitted light changed in form and eventually disappeared but that electricity still passed across the tube. If the anode was put into a side tube, the glass opposite the cathode glowed with fluorescent light, showing that the particles emitted from the cathode and carrying the electricity through the tube traveled in straight lines. This was verified by putting obstacles in the path of the particles emitted from the cathode, when sharp geometrical shadows appeared in the fluorescent glow.

J. J. Thomson investigated the effect, showed that the electricity carriers were in fact charged particles by deflecting them in electric and magnetic fields, proved that they were negatively charged particles from the sense of the deflection, and called them electrons. He showed that their properties were independent of the substance used for the electrodes, and therefore correctly assumed that they were constituents of all matter. From the results of the deflection experiments discussed in Sections 5.5 and 5.6, he was able to show that their charge to mass ratio e/m was 1.7589×10^{11} C kg^{-1}.

Millikan's famous experiment to determine the value of e has already been described, and the value obtained for e was 1.6021×10^{-19} C. It follows that the mass of the electron is 9.1083×10^{-31} kg. Note that the (e/m) value for the electron is about 2000 times greater than the value of the same ratio for the hydrogen ion. If one assumes, as is correct, that the hydrogen ion is a hydrogen atom with a single positive charge, it follows that the mass of the hydrogen atom is 1836 times the mass of the electron. The hydrogen atom thus has a mass of 1.672×10^{-27} kg.

28.3 POSITIVE RAYS

If a discharge tube is made by putting two electrodes in a gas chamber with a high potential difference between them and a little residual gas is left after pumping, the electrons traveling across the tube strike some of the gas atoms, knocking electrons from them. The gas atoms will be left as positive ions, and will be attracted toward the cathode. If a hole is bored through the cathode, some of these ions will pass through and may be made to enter an evacuated chamber, constantly pumped out to keep gas from trickling through the hole. One therefore has a fast-moving stream of positive ions, or positive rays, fairly finely collimated, passing along the evacuated vessel, and this stream can be deflected in electric and magnetic fields.

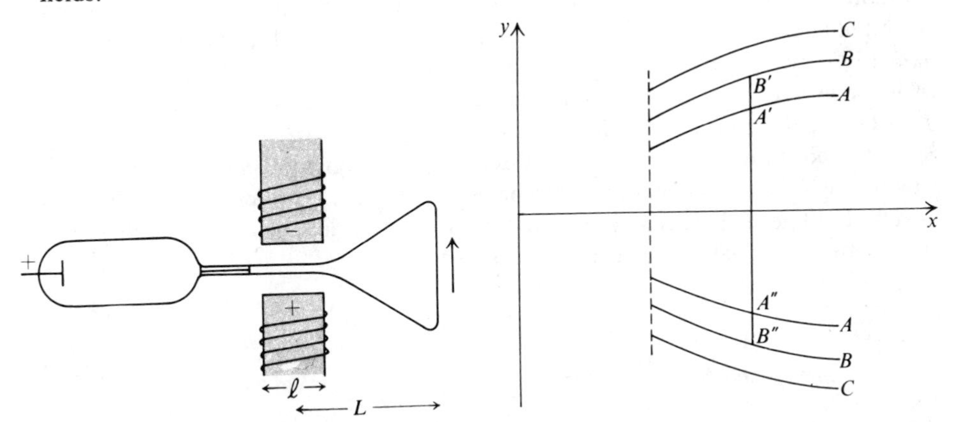

Fig. 28.1. Thomson's apparatus for investigating positive rays.

Fig. 28.2. A set of positive-ray parabolas.

The cathode in Fig. 28.1 is water-cooled because it heats up appreciably. This effect is due to positive ions which do not pass through the hole striking it. It is also covered with an iron shield to keep stray electric and/or magnetic fields from affecting the ions passing through, and deflecting them from a straight path through the fine hole. Note that the beam arriving in the evacuated chamber may consist of ions with a single positive charge, ions with more than one positive charge owing to multiple collisions, and uncharged gas atoms. The latter may be present either because ions have recombined with electrons during their movement through the tube or because they are neutral gas atoms which have been struck by fast-moving ions and knocked on. Further, these atoms and ions can have a wide range of velocities, depending on where they were formed in the discharge tube or whether they have suffered collisions which have slowed them down.

In the original experiment by J. J. Thomson an electric and a magnetic field were established in the same direction. The effect of this was to produce a deflection in the x-direction due to the electric field E and a deflection at right angles, in the y-direction, due to the magnetic field B. If the velocity of a particular ion entering the fields is v_0, then, from Eqs. (5.16) and (5.19),

$$x = Ll\left(\frac{e}{m}\right)\left(\frac{1}{v_0^2}\right)E,$$

$$y = Ll\left(\frac{e}{m}\right)\left(\frac{1}{v_0}\right) B.$$

Therefore

$$\frac{y^2}{x} = Ll\frac{B^2}{E}\left(\frac{e}{m}\right) = b\left(\frac{e}{m}\right), \tag{28.2}$$

where b is a constant for the apparatus. For a fixed value of (e/m), y^2/x is a constant, which is the equation of a parabola. Thus all ions with the same (e/m) value fall at some point of a parabolic trace. In Fig. 28.2 all particles on trace A have the same (e/m) values. So do all particles forming trace B, although their (e/m) value is different from that for the particles on trace A. Similarly, the ions on trace C have the same (e/m) value. The uncharged atoms are of course undeflected. Since the position of the x-axis is not defined, the magnetic field is reversed halfway through the experiment and the corresponding traces are produced below the x-axis. If a line such as $B'A'A''B''$ is drawn parallel to the y-axis at any value of x, the distances $A'A''$ and $B'B''$ can be measured:

$$\frac{(A'A'')^2}{(B'B'')^2} = \frac{(2y_A)^2}{(2y_B)^2} = \frac{bx\left(\frac{e}{m}\right)_A}{bx\left(\frac{e}{m}\right)_B} = \left(\frac{e}{m}\right)_A \Big/ \left(\frac{e}{m}\right)_B. \tag{28.3}$$

The ratio of (e/m) values may therefore be determined for the various constituents of a known gas; and if one value of (e/m) is known, the others can be obtained. When the (e/m) values for a large number of atoms and molecules have been obtained, this may be used as a method of determining the constituents of an unknown gas. The interesting feature of this apparatus is that it separates out according to charge and mass, and, for singly charged particles, which in most cases is the overwhelming majority, according to mass. If hydrogen is used, one may, from the geometry of the apparatus and the known values of the electric and magnetic fields, calculate the value of the ratio (e/m) for hydrogen. If we know the value for e—from Millikan's experiment, for example—the mass of the hydrogen atom may be determined in grams. When the apparatus has been calibrated for hydrogen, the masses of any other atomic or molecular constituents may be determined.

In Thomson's original experiment neon was used in the tube. He expected to obtain a single trace corresponding to a mass of 20.2 a.m.u., this being the known atomic mass of neon. Instead he obtained a trace corresponding to mass 20 and a fainter line corresponding to a mass of 22 a.m.u.. It was clear that neon did not consist of identical atoms all of mass 20.2 a.m.u. Instead it consisted of two different species. one of mass 20 a.m.u. and the other of mass 22 a.m.u., in such a proportion that the combined average mass was 20.2 a.m.u. Both of these species must have identical chemical properties but they obviously differ in physical properties. Such species are called isotopes, and further investigation showed that all elements with markedly non-integral atomic mass consisted of mixtures of isotopes. All elements, if separated into their constituent isotopes, are found to have masses which are almost but not quite integral on the a.m.u. scale.

28.4 THE MASS SPECTROGRAPH

If one wishes to separate the atoms and molecules of a sample into their separate isotopic constituents in order to measure isotopic masses, one would wish for a more convenient apparatus than Thomson's. Since mass spectrographs, as they are called, are used more or less as routine pieces of apparatus in many fields for identification, separation, and other purposes, we shall describe a convenient form of such an instrument in a little detail. In addition to knowing the masses of the constituents, it is often necessary to know the percentage concentration of each constituent. If an apparatus is used in which each constituent has its atoms falling along a parabolic trace, it becomes difficult to compare the concentration of one constituent with another. It is usual nowadays to use a form of apparatus in which each single constituent is concentrated either to a point or, more usually, to a thin line. Such an apparatus is Bainbridge's mass spectrograph (Fig. 28.3).

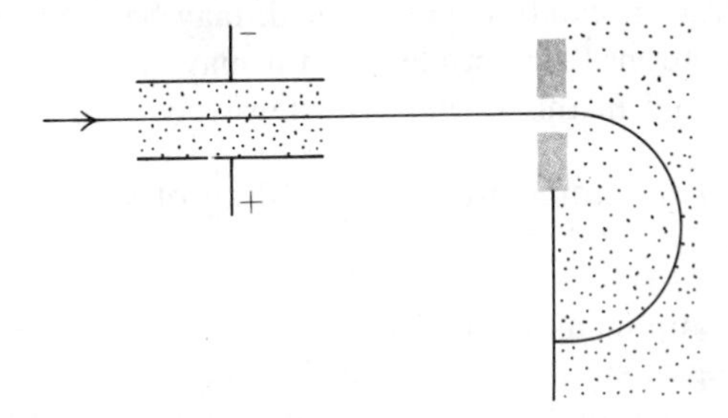

Fig. 28.3. The essential features of a Bainbridge mass spectograph.

The ion source is a high-voltage, low-current discharge tube which may contain the substance for investigation in vapor form. More normally the substance is used to coat a filament so that, when the filament is heated, atoms of the substance will be emitted into the body of the discharge tube. These atoms are bombarded by the electrons which pass across the discharge tube, and ions are produced which emerge through a fine slit in the cathode and enter a region which is highly evacuated. In this region is a velocity selector, in which there are an electric and a magnetic field at right angles. The ions emerging into this region have a variety of energies and velocities, depending on where they were produced in the discharge tube. But on each the electric field produces a force Ee and the magnetic field a force Bev. Particles which are undeflected in passing through the velocity selector go through a further slit into the next section of the apparatus. All other particles impinge on the surroundings of the slit and are absorbed. But particles which are undeflected must have suffered equal but opposite forces from the electric and magnetic fields. If this happens, $Ee = Bev$, or $v = E/B$. Ions of fixed velocity therefore pass through the second slit, and different velocities may be selected by varying either E or B.

The selected ions emerge from the second slit into a region of uniform magnetic field B'. They follow a path which is the arc of a circle, and therefore the magnetic force on the particle must be providing the necessary centripetal force. Thus

$$B'ev = \frac{mv^2}{R}, \qquad \text{or} \qquad R = \frac{mv}{B'e}. \tag{28.4}$$

For all singly charged ions v, B', and e are the same. Thus R is proportional to m. If the ions are doubly charged, they act as though they were singly charged ions of $\frac{1}{2}m$ mass. In any case the great majority of ions are singly charged and each species of ion will follow a path of fixed radius. If the particles are allowed to complete a semicircle, the distance from the slit at which they strike a photographic plate or enter a counter is $2R$. Ions of different mass collect at points along the collecting plate, the distances from the slit at which they deposit being proportional to their mass. Thus the mass may be measured and the relative concentration of each constituent may be obtained from the number of ions arriving at each point.

Such an apparatus, if carefully constructed, may be used to measure accurate masses and concentrations. In a cruder form it may be used for simple identification purposes only, and in this form it is extensively used in many laboratories.

Example 1. How many atoms are there in 10 kg of silver which has an atomic mass of 108 a.m.u.?

Solution. Since the atomic mass of silver is 108 a.m.u., 1 mole of silver has a mass of 108 g. Thus 10 kg of silver contains $10^4/108$ moles. Each mole contains $N_0 = 6.02 \times 10^{23}$ atoms. The total number of atoms in 10 kg of silver is thus

$$\frac{10^4}{108} \times 6.02 \times 10^{23} = 5.57 \times 10^{25}.$$

Example 2. In one type of mass spectrometer the charged particles pass through a velocity selector before entering the magnetic field. In another the particles pass through a strong electric field before entering the magnetic field. Compare the ratio of the radii of singly charged lithium ions of mass 6 a.m.u. and 7 a.m.u. in the two cases.

Solution. In the magnetic field an ion moves in a circle, the centripetal force necessary being provided by the magnetic force on it. Thus

$$qvB = mv^2/R,$$

therefore

$$v = qBR/m.$$

When the ions have passed through a velocity selector, both ions have the same velocity. In addition they have the same charge and are in the same magnetic field. Hence

$$\frac{R_6}{m_6} = \frac{R_7}{m_7},$$

therefore

$$R_6/R_7 = 6/7 = 0.857.$$

If the ions have passed through a strong electric field, they have both acquired the same energy. But

$$\tfrac{1}{2}mv^2 = (m/2) \times (qBr/m)^2$$
$$= q^2B^2r^2/2m.$$

Therefore

$$\frac{r_6^2}{m_6} = \frac{r_7^2}{m_7},$$

therefore

$$r_6/r_7 = \sqrt{6/7} = 0.926.$$

PROBLEMS

28.1 How many atoms are there in 1 kg of gold, which has an atomic mass of 197 a.m.u.?

28.2 Silver has a density of 1.05×10^4 kg m^{-3} and an atomic mass of 108 a.m.u. Estimate the diameter of a silver atom.

28.3 Calculate the number of molecules in every cubic meter of a gas at STP. The density and electrochemical equivalent of hydrogen are 9.0×10^{-2} kg m^{-3} and 1.0×10^{-8} kg C^{-1}, respectively.

28.4 If one faraday deposits 197.1 g of gold in electrolysis, what is the charge-to-mass ratio of a gold ion?

28.5 Two positive ray parabolas are obtained in a particular experiment and one is known to be due to hydrogen ions. At a particular x-value the y-values of the parabolas are in the ratio 4:1. Determine the (e/m) value for the other constituent. If the second constituent is assumed to be singly charged, what element is it?

28.6 Singly charged ions of isotopes of nitrogen of mass 13.006 and 14.003 a.m.u. are accelerated through the same potential difference and enter a magnetic field of induction of 0.1 Wb m^{-2} at right angles to the velocity direction. If the heavier isotope traverses a circular path of radius 20 cm, what is the radius of the path of the other isotope? Through what potential difference were the ions accelerated? The mass of a hydrogen atom is 1.67×10^{-27} kg.

28.7 Hydrogen ions of mass 1.67×10^{-27} kg are passed through a velocity selector in which the magnetic induction is 0.1 Wb m^{-2} and the electric field is provided by maintaining a potential difference of 100 V between two plates 1 cm apart. They then enter a region of uniform magnetic induction of value 0.01 Wb m^{-2} at right angles to the direction of travel of the ions. What is the radius of the circular path they follow?

28.8 A doubly charged α-particle of 1 MeV energy moves in a circular path in a uniform magnetic field. What energies must (a) a proton and (b) a deuteron possess if they are to traverse the same path? The masses of a proton, a deuteron, and an α-particle are 1.008, 2.014, and 4.003 a.m.u., respectively.

CHAPTER 29

RADIOACTIVITY

29.1 INTRODUCTION

At the end of the last century a phenomenon was noted which provided considerable knowledge about the constitution of atoms. Becquerel discovered while working with uranium salts that they emitted particles or radiation which could affect photographic plates. Investigation showed that this was true of all uranium salts of whatever type, and it was therefore clear that this was a property of the uranium atoms and not of another constituent of the salt. Rutherford and others began an extensive investigation of the phenomenon, the main lines of research being into (a) what sort of particles were being emitted, and (b) what sort of laws the emission mechanism obeyed.

29.2 THE NATURE OF RADIOACTIVE EMISSIONS

As regards the first line of investigation, it was found that the emanations must consist of three different types, the first being easily absorbed, the second rather less easily absorbed, and the third scarcely absorbed at all. If electric and magnetic fields were applied to the emissions, the easily absorbed constituent deflected in such a way as to show that it consisted of positively charged particles. The slightly less easily absorbed emanation deflected in such a way as to show that it consisted of negatively charged particles, and the remaining emanation was unaffected by such fields. The positively charged particles were called α-particles and the negative ones β-particles; and the undeflected emanation, probably radiation, was called γ-rays. The negatively charged particles were obviously very light on the evidence of the ease with which they were scattered and the extent of their deflection in an electric field. Their (e/m) value was found to be identical with that of the electron, and it was therefore clear that β-particles and electrons were the same thing. The α-particles acted as though their (e/m) ratio was $\frac{1}{2}(e/M)$. They were therefore either singly charged particles of atomic mass 2 (and no such particle was then known) or doubly charged particles of atomic mass 4 (i.e. doubly charged helium atoms).

In two famous experiments Rutherford established beyond doubt the identity of these particles. In the first he allowed a number of α-particles to strike, and give up their charge to, a metal container which was then attached to an electroscope in order to measure the acquired charge. From this the charge carried by each particle was worked out to be $+2e$. In the second experiment α-particles were emitted into a chamber containing gas molecules in collision with which the α-particles were stopped, lost their charges, and therefore finished up as gas atoms. After a few days the gas was forced into a discharge tube and an electric current was passed through it. The light emitted by the discharge tube was examined and, although previously no trace of the helium spectrum had been observed, now the lines of the helium spectrum appeared. These two experiments showed that the α-particles must be doubly charged helium atoms.

Particles and radiation are quite differently absorbed in passage through matter. A particle can lose its energy gradually, but a single γ-ray, X-ray, or photon must normally lose all of its energy in one event. The laws of absorption of particles and of radiation are therefore quite different. The way in which γ-rays were absorbed was characteristic of radiation. When it was also shown that γ-rays could be diffracted, the fact that it was radiation was definitely confirmed. The frequency of the γ-rays is in general found to be higher than that for X-rays, although the two types of radiation slightly overlap in frequency. The difference between X-rays and γ-rays is that the former are emitted from the electronic parts of an atom, the latter from the nucleus.

29.3 THE LAW OF RADIOACTIVE DECAY

On the second line of investigation mentioned in the introductory section, the law obeyed by the emitted radiation was not at first easy to discover. At first sight the energy emitted appeared to be inexhaustible. But it was soon found that this was because of the presence of several radioactive elements and not just one. Once it had been discovered what types of particle were being emitted from radioactive elements, the picture was easier to see. An atom of element X, which has an atomic mass number A, the nearest whole number to its mass in a.m.u., and an atomic number Z, the number of positive charges on its nucleus or the number of electrons it possesses, is denoted by the symbol ${}^{A}_{Z}X$. If it emits an α-particle, it turns into an atom ${}^{A-4}_{Z-2}Y$. Similarly, an atom ${}^{A'}_{Z'}P$ on emitting a β-particle will become an atom of ${}^{A'}_{Z'+1}Q$. Thus if ${}^{238}_{92}\mathrm{U}$ emits an α-particle, it turns into an atom of atomic mass 234 and atomic number 90. If this atom is also radioactive, as it is, and emits a β-particle, it becomes element 91. This particular isotope of element 91, which has a mass of 234, is also radioactive and emits a β-particle. The element now formed is an isotope of uranium of mass 234. It also is radioactive emitting an α-particle, and so on. When the radioactivity from natural uranium is being investigated, what is in fact being examined is a complex radioactive emission from a whole chain of radioactive elements. This obviously makes the task of finding the law of radioactive emission almost impossible, since, while some elements are decaying and their radioactivity is consequently decreasing, other elements are being formed and this radioactivity is increasing. However, if the uranium isotope of mass 234 is separated chemically from a uranium mixture and its products are continually removed, it is found to emit α-particles only and the α-activity is

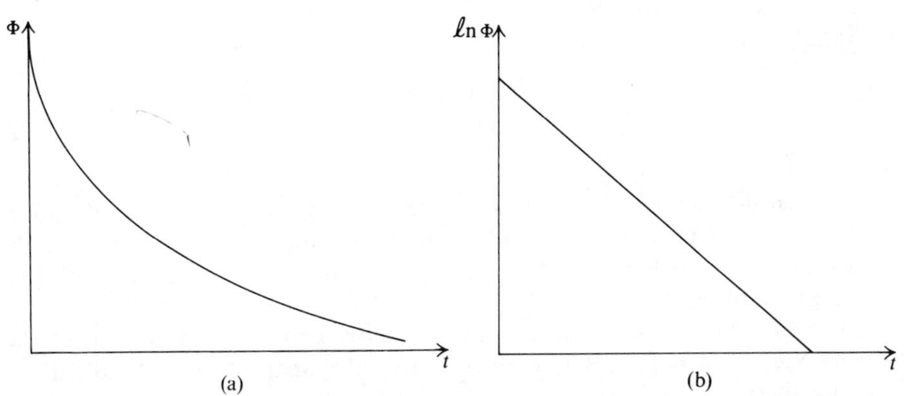

Fig. 29.1. The activity of a radioactive isotope as a function of time: (a) shows the exponential decay of Φ, and (b) the linear decay of $\ln \Phi$.

found to decay exponentially. In fact if any radioactive element is chemically separated out and the products formed by the action of its radioactivity are continually removed, it is always found that the activity decays exponentially.

If the activity Φ of a radioactive element is defined as the number of particles emitted per second, then the decay curve of any radioactive substance is as shown in Fig. 29.1. The number of atoms remaining of the element at any time is

$$N = N_0 e^{-\lambda t}, \tag{29.1}$$

where N_0 is the number of atoms originally present and λ is the decay constant. Alternatively, one may say that the number of atoms decaying per second, which is the same as the activity, is proportional to the number of atoms present at any time. For, from Eq. (29.1),

$$\Phi = -\frac{dN}{dt} = N_0 \lambda e^{-\lambda t} = \lambda N, \tag{29.2}$$

the constant of proportionality being the decay constant. Since the number of atoms present decreases exponentially with time, so does the activity.

Since the number of atoms decaying at any time depends only on the number present at that time, being unaffected by pressure, temperature, or any other physical property, it is clear that the atoms of the radioactive element are unstable and decay spontaneously owing to this instability. The process is therefore a purely statistical one. One cannot predict which radioactive atom will decay at any given time, but over any period one can predict with considerable accuracy the number of atoms that will decay.

It is often preferable to work not with the decay constant but with the half-life of the radioactive substance, this being the time it takes for the number of atoms, or the activity, to reduce to half of the initial value. Thus

$$\tfrac{1}{2}N_0 = N_0 e^{-\lambda \tau},$$

where τ is the half-life period. Thus

$$2 = e^{\lambda \tau}, \qquad \text{or} \qquad \ln 2 = \lambda \tau.$$

It follows that

$$\tau = \frac{\ln 2}{\lambda}. \tag{29.3}$$

The half-life period is independent of the initial number of atoms, or the initial activity. If a sample contains N_0 atoms, it takes a time τ for this number to be reduced to $\frac{1}{2}N_0$ and then a further time τ for a further reduction to $\frac{1}{2}$ of $\frac{1}{2}N_0$, i.e. to $\frac{1}{4}N_0$, and so on.

The constant τ is also the time for the activity to reduce by one-half. For $\Phi_0 = \lambda N_0$, and $\Phi = \lambda N$. But, since $N = \frac{1}{2}N_0$, the final activity is also half the initial activity.

Any radioelement obeys the above laws and emits one or other but not generally both of the possible particles. It may, in addition, emit radiation. The instability of the atom will determine which of the particles it emits. Normally, when a particle is emitted from a parent atom, the daughter atom formed is in its minimum possible energy state. It is, however, possible for the particle to be emitted with less than the maximum possible available kinetic energy, the rest of the energy being retained by the daughter, which is then said to be in an excited state. Since atoms prefer to exist in the state of lowest possible energy, the daughter atom gets rid of its excess energy fairly quickly by emitting it in the form of γ-radiation. Experiment has shown that this is the true explanation of the origin of γ-rays, because γ-radiation is always emitted after a particle and always from the daughter atom.

The phenomenon of radioactivity gives the first direct evidence that all atoms are almost certainly made from the same building bricks, since the emission of fairly simple particles can transform one type of atom into another. It is therefore a reasonable inference that one type of atom differs from another only in the number of basic building bricks which it contains. Further investigation will be necessary in order to find out what these building bricks may be and how these constituents are joined together.

It is now possible to produce radioactive substances artificially. By bombarding elements with fast charged particles from accelerating machines or with neutrons in a nuclear reactor, radioactive isotopes of any known element can be produced. These artificial radioisotopes are of much greater use in medicine, biology, and industry than the naturally occurring ones, as we shall see in some of the following sections.

29.4 CARBON AND URANIUM DATING

A simple direct use of radioactivity is in dating. It is often necessary in biology, geology, and other fields to determine when a particular fossil was alive, when an artifact was made, or when a geological stratum was laid down. The use of radioactivity, where this is possible, is almost always the most accurate method of dating and is sometimes the only one available.

In living plants and animals the carbon is mainly ^{12}C, the normal isotope, but a small but detectable quantity of ^{14}C atoms are also present. These result from the bombardment of atmospheric nitrogen by cosmic rays. The ^{14}C isotope is radioactive, with a half-life of 5.76×10^3 years. In living organisms the ^{14}C atoms will be decaying but, since the atoms are continually renewed by uptake from the environment, the ratio of ^{14}C to ^{12}C atoms is found to remain constant at all times. It is further assumed that this ratio has not altered significantly over a considerable period of time.

A quite different situation arises as soon as the organism dies. No further renewal of the radioactive carbon takes place and the number of ^{14}C atoms decays exponentially with time, as therefore does the radioactivity exhibited. If the activities of a quantity of carbon from a recently alive organism and of the same quantity of carbon from a subject to be dated are Φ_0 and Φ, then

$$\Phi = \Phi_0 e^{-\lambda t}.$$

Therefore

$$t = \frac{1}{\lambda} \ln \frac{\Phi_0}{\Phi} = \tau \frac{\ln (\Phi_0/\Phi)}{\ln 2}. \tag{29.4}$$

The time t so calculated is the time for which the subject has been dead.

Since the half-life of ^{14}C is 5.76×10^3 years, and the activity of carbon found in organisms is in any case small, the limit to the carbon-dating process is around 50,000 years. For geological purposes this is a negligibly small period, and uranium dating is used for events more distant in time. In many rocks, uranium, which consists mainly of the isotope 238, is found in conjunction with the lead isotope 206. This is not surprising, since ^{238}U is the first member of one of the naturally occurring radioactive series, the end-product of which is stable ^{206}Pb. It is believed that when the rocks were formed, only ^{238}U was present, the ^{206}Pb now found representing the product of the radioactivity which has taken place since that time. The half-life of ^{238}U is 4.5×10^9 years, and all radioactive elements in the series have half-lives very much smaller than this. When a radioactive series is in equilibrium, the numbers of atoms of each element present at any time are in the direct ratio of their half-lives. The numbers of atoms present of elements other than ^{238}U and ^{206}Pb are therefore negligibly small and may be ignored in any calculation.

In any sample the number of atoms n and n_1 of ^{238}U and ^{206}Pb respectively, are calculated from the respective masses and Avogadro's number. It is believed that the number of atoms of ^{238}U present at the formation of the rocks was $n_0 = n_1 + n$. Therefore, if t is the time since the rocks were laid down,

$$n = n_0 e^{-\lambda t}.$$

Therefore

$$t = \tau \frac{\ln (n_0/n)}{\ln 2} = \tau \frac{\ln [(n_1 + n)/n]}{\ln 2}. \tag{29.5}$$

29.5 RADIOACTIVE ISOTOPES AS TRACERS

Many elements are taken up by biological organisms, but their role in the functioning of the organism is not always clear. Chemical and biochemical techniques are not always able to provide the answers to the problems raised, and the use of radioactive isotopes as tracers has allowed many obscure points to be cleared up. Although the normal isotope of an element cannot be identified after it has entered a biological system, the absorption, metabolic uptake, movement to particular sites, etc., of a radioactive isotope can be followed by the activity it exhibits. This can be done without operating on the system or in any way interfering with its functioning.

The unit of radioactivity is the curie, originally chosen to be the activity of 1 g of radium, but now internationally defined as 3.700×10^{10} distintegrations per second. For biological and medical purposes activities of this order are far too high and would kill the host material into which such radioactivity was introduced. Activities of millicuries and microcuries are much more usual.

The particles and radiation emitted by the radioactive isotopes can be detected by a number of instruments, and a great amount of work has been undertaken to improve the detectors, to make them suitable for specific purposes, and to provide associated electronic equipment by means of which the output can be suitably recorded. The most usual instruments employed in biological and medical fields are geiger and scintillation counters. These are described in Sections 33.7 and 33.8, to provide the reader with some knowledge of how counters work; but it seems unnecessary to say more than this about counting equipment, since these are now purchased from a manufacturer as standard packages with full instructions for use.

It should, however, be made clear that any counter, even when far removed from a radioactive sample, will still record counts. This is because of the cosmic radiation, natural radioactivity in the air and the ground, and various types of man-made radioactivity around us. A background count should therefore always be recorded before, after, and often during, a series of tests. This is subtracted from the count made during an investigation to give the count due to the sample alone. If the sample activity is of the same order as, or less than, the background count, the counter may require to be shielded to reduce the background effect and increase the accuracy of the result. Attention is also drawn to Section 2.13, where the question of reducing the error in a radioactive count was dealt with.

A discussion of the use of tracer techniques could fill a book several times as large as this one. The next few sections will therefore be concerned with dealing with a few examples of the use of tracer techniques to indicate the type of problem that can be tackled and to show the simplicity and usefulness of the method.

29.6 STUDIES OF METABOLIC UPTAKE

In patients suffering from pernicious anemia, the absorption of vitamin B_{12} is low. It is difficult, however, by normal methods to determine just how much of the vitamin is being absorbed by the body. Since vitamin B_{12} contains a cobalt atom, it is possible to synthesize the vitamin with the radioactive isotope ^{58}Co and to feed a known dose to the patient. Comparison of the activity of the dose given with the subsequent activity of the feces allows the calculation of the fraction of the dose which has been absorbed by the gut.

Practically all iodine ingested and absorbed by the gut is taken up by the thyroid. This stored iodine is used to form the hormones thyroxine and di-iodotyrosine, which are then circulated through the body. A measured dose of the radioactive isotope ^{131}I can be given by mouth to normal subjects and the activity can be measured with a standard configuration at the neck in the region of the thyroid gland. A fixed blood sample taken after 48 h will also show activity because of the labeled iodine in the hormones.

A patient suffering from hyperthyroidism has an over-active thyroid which absorbs too much iodine too quickly and thus produces an excessive amount of hormone. By feeding such a patient radioactive iodine and comparing the activity of the thyroid and of blood samples with the data obtained from normal subjects, one can easily and quickly diagnose the complaint.

In the two examples dealt with, how the ingested substance was used by the system was of no great interest. It is often important in biochemical work to know which portion of an ingested molecule is used in the synthesis of another compound, and tracer techniques can answer this type of question. For example, glycine is used in the formation of protoporphyrin. Its structural chemical formula is

```
      NH₂  O
       |  //
    H—C₁—C₂
       |   \
       H    OH
```

the two carbon atoms being numbered for identification purposes. It is possible to synthesize glycine with either C_1 or C_2 the radioactive ^{14}C isotope. If C_1 is radioactive, the radioactivity appears in the porphyrin ring; it does not if C_2 is radioactive. This shows that glycine is used in the synthesis of protoporphyrin but that the carboxyl carbon is removed during the synthesis.

29.7 TRANSPORT STUDIES

Carbon dioxide is absorbed by the leaves of plants and takes part in the process of photosynthesis, in which carbohydrates are produced. A simple method of determining where the carbohydrates go in the plant after they are produced is provided by radioactive-tracer techniques. Carbon dioxide is made from radioactive carbon ^{14}C, and a leaf of a growing plant is enclosed in a transparent container through which the prepared gas is passed. After several hours a geiger counter placed at various parts of the plant will indicate where the labeled carbon atoms are now located.

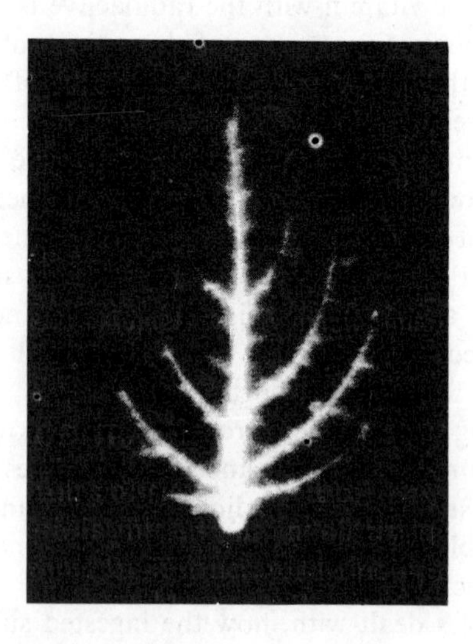

Fig. 29.2. Autoradiograph of a leaf which has been maintained in an atmosphere of radioactive carbon dioxide, showing the uptake of the labeled carbohydrates. (Photo by courtesy of Brookhaven National Laboratory.)

More usually a technique called autoradiography is employed. Leaves and portions of the stem can be removed from the plant and placed between photographic plates. The β-particles from the ^{14}C atoms strike the emulsion of the plates, each one producing a developable spot. The more radioactive atoms there are at a particular part of a leaf the more spots will be produced at that point of a film. If the carbohydrate is spread throughout a leaf, a picture of the whole leaf will be produced on the film; but the regions where the concentration is highest will appear darker than the others. It is found that the concentration is greater in the vascular system of the leaf than elsewhere (cf. Fig. 29.2).

It is further found that if a young leaf is used in the experiment, all the radioactive atoms remain in that leaf, the hydrocarbon being needed for the growth of the leaf itself. On the other hand, if the leaf is a mature one, some of the radioactive carbohydrate is passed on to young leaves which are unable to supply all their own needs. It would have been very difficult to obtain such information by any means other than the use of radioactive tracers.

29.8 CHROMOSOME DIVISION

The use of radioactive tracers allied to autoradiography has been extremely useful in the study of replication. As an example of this, let us consider chromosome division in broad-beam root tips, which are often used for this type of purpose, since the chromosomes, or their autoradiographs, can be made clearly visible in a microscope. If thymidine labeled with ^{3}H is used in growing the specimens, the ^{3}H is incorporated only into the chromosomes and autoradiographs of cells will therefore show the chromosomes, or portions of them, clearly.

Cell division may be inhibited with colchicine, which does not affect the division of chromosomes. If cells containing labeled chromosomes are treated with colchicine, chromosome division will proceed to different stages in different cells. After the process is completed, some cells will contain two sets of chromosomes, some four sets, and so on. If an autoradiography film is placed over the preparation, pictures of these cells will be obtained and the distribution of ^{3}H after one, two, three, etc., divisions can be followed. In this way one obtains information about the manner in which the actual material of the chromosome divides and how it replicates from the medium.

29.9 ISOTOPIC DILUTION

Radioactive techniques may also be used to determine the total volume occupied by a fluid. If the activity of a known quantity of suitably labeled fluid is measured and this fluid is then injected into the volume, when the radioactive tracer has dispersed throughout the whole region, the activity of an extracted sample indicates by how much the tracer has been diluted. The total volume may therefore be calculated.

In one such experiment 5×10^{-6} m^3 of water, with ^{3}H replacing the normal hydrogen isotope, was injected into the antecubital vein of a human subject. The activity of 10^{-6} m^3 of the injected sample was 30,200 counts per second. After $\frac{1}{2}$, 1, 2, and 3 h blood samples were withdrawn from the same vein and the plasma was separated out. Samples of volume 10^{-6} m^3 were taken, the activities being 3.02, 3.46, 3.28, and 3.40 counts per second, respectively. It was clear that equilibrium had been achieved before 1 h and it was assumed that complete mixing had taken place in that time. The average of the last three activities was 3.38

counts per second. The activity had decreased by a factor of 3.38/30,200. The total volume finally occupied by the labeled water was therefore

$$5 \times 10^{-6}\ \text{m}^3 \times \frac{30{,}200}{3.38} = 4.47 \times 10^{-2}\ \text{m}^3.$$

This is the total volume of the subject's body occupied by water.

29.10 LOCATION OF HEMORRHAGE

Recently radioactive-tracer techniques have been applied to the location of hemorrhage. It is often difficult to tell whether a hemorrhage is taking place or, if it is, its location. The isotope ^{51}Cr has been used in various studies of blood because it is taken up by the red cells. In this new application ^{51}Cr-labeled blood is injected into the patient. With normal blood circulation the radioactivity is distributed throughout the circulatory system. If hemorrhage is occurring, the radioactivity will markedly increase at some region of the body, and the rate at which the activity increases is an indication of the volume of blood being lost. This is a simple and very effective way of dealing with a difficult problem.

29.11 RADIOCARDIOGRAPHY

A routine method of investigating heart conditions is the insertion of a catheter into the bloodstream, from where it is worked into the heart. This is a skilled operation which is time-consuming and is not without its attendant risks. Radiocardiography is a much simpler procedure which gives the same information almost routinely and with no risk whatsoever. It also gives information about pulmonary conditions.

A tracer element ^{137}Ba with a half-life of 127 s is used. Ten millicuries are injected rapidly into the sub-clavian vein and enter the right ventricle almost immediately. A counter directed from above at the heart detects the presence of the tracer. There is a dip in the recording as the tracer is pumped out to the lungs and a rise when it returns to the heart. A counter at the back, collimated to pick up radiation from the aorta, follows the flow from the heart. If the flow through other parts of the body is of interest, further counters may be located in these regions. An e.c.g. trace is normally taken at the same time.

If the heart and lung functions are normal, the recording obtained from the counters will have a typical form. Blockage or malfunction will lead to a nonstandard pattern, its features being interpretable in terms of various known conditions. The diagnosis of atrial-septal defect, of ventricular-septal defect, and of congenital, pathological, cardio-pulmonary conditions can be rapidly and easily established.

29.12 NON-RADIOACTIVE TRACERS

It would be quite misleading to give the impression that tracer techniques cannot be employed without using radioactive atoms. So long as the tracer atoms can be identified, this is all that is required. If a non-radioactive but unusual isotope is used, its presence can be detected either by use of the mass spectrograph (cf. Section 28.4) or, more usually, by centrifugation (cf. Section 9.5). This is not as convenient as the use of geiger counters or autoradiography but may have to be employed in certain cases.

For example, one may use a non-radioactive isotope of nitrogen ^{15}N in determining the lifetime of red blood cells in humans. The isotope ^{15}N is present in normal samples of nitrogen to the extent of 0.36%. Abundances greater than this amount are indications of the presence of the added tracer. The amino acid glycine can be synthesized using this isotope and fed to a subject. It is incorporated into hemoglobin and the presence or absence of ^{15}N in samples of blood can be investigated at intervals after the start of the experiment. It is found from such studies that the average lifetime of red cells is about four months (cf. Fig. 29.3).

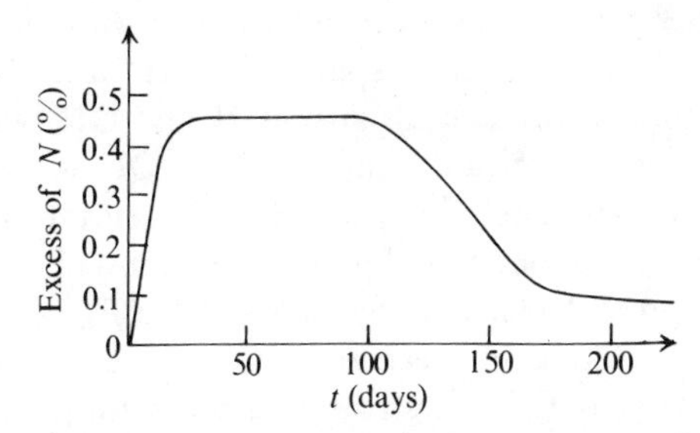

Fig. 29.3. The excess of ^{15}N in heme of human erythrocytes after feeding ^{15}N-labeled glycine for three days. (From D. Shemin and D. Rittenberg, *J. Biol. Chem.*, **166,** 627 (1946).)

The classic experiment of Meselson and Stahl, which confirms the Watson-Crick model of DNA replication, also employed this technique. Bacteria were grown in a medium in which almost all the nitrogen atoms were of the isotope ^{15}N. They were allowed to grow for 14 generations, when the DNA was extracted and purified and placed in a cesium chloride solution in an ultracentrifuge spun at 14,000 g (cf. Section 9.8). The DNA stabilized at a different level from that of DNA obtained from bacteria grown in a normal ^{14}N medium, because of the different mass characteristics.

The experiment was repeated with ^{15}N-labeled bacteria allowed to grow in a ^{14}N medium. In the ultracentrifuge two stabilized layers now appeared. After one reproduction there is only one stabilized layer intermediate between those for heavy and for normal DNA. After several reproductions this intermediate layer is still present but a layer corresponding to normal DNA is also found. The latter becomes denser in comparison with the hybrid layer with each successive reproduction.

These results indicate that in replication the two strands of the DNA come apart, each one then duplicating from the surrounding medium the strand it has lost. Once the original DNA has divided, the only things that can be produced are hybrids or completely unlabeled cells. The number of hybrids can never increase while the unlabelled cells continue to increase in number.

29.13 RADIOTHERAPY

Radioactive isotopes are also used to destroy unwanted cells. The particles and radiation emitted produce two main effects on the cells. They can supply energy to a molecule, resulting in the ejection of an electron. Since the outermost electrons are the most loosely bound, these are the ones most frequently ejected. These are also the electrons which take part in chemical bonding, and the loss of one or more electrons can result in the breaking-up of the molecule into its component parts. Cells may consequently be destroyed if these molecules play a vital role in their function.

Just as important is the fact that all biological material contains water, and irradiation of the water produces products which react with the biological material. Although a water molecule is normally split up into H^+ and OH^- ions, it can also be split up into the electrically neutral groups H and OH, which are called free radicals. These are very chemically reactive. The OH is a powerful oxidizing agent, attracting electrons strongly in order to turn itself into the stable OH^- ion. In doing so it breaks chemical bonds and produces in consequence biological effects.

Radiotherapy is possible only if the unwanted cells can be destroyed by a dose of radiation which does not permanently damage the surrounding healthy tissue. In general, the cells to be destroyed, such as tumor cells, must be more sensitive to radiation than the healthy tissue.

Tubes containing radium, or some similar isotope, can be implanted around the offending volume and left there for a carefully calculated time. The surrounding tissue will generally receive an intense dose of radiation, and radiation sickness may be produced. Since radium is a long-lived isotope, it must be removed after the prescribed time. To avoid an unnecessary operation, radon or a gold isotope may be used instead. The half-lives are only of the order of three days and the tubes are not removed.

Even better: if a specific radioactive isotope is concentrated by the organ to be treated, it can be ingested or similarly introduced into the patient. Thus a suitable dose of ^{131}I is given for the treatment of hyperthyroidism, and ^{32}P, which is taken up by bone marrow, is used for the treatment of polycythemia.

Example 1. Six α-decays and a number of β-decays are necessary before $^{232}_{90}Th$ achieves stability, the final product in the chain being $^{208}_{82}Pb$. Deduce the number of β-decays that take place.

Solution. In the emission of an α-particle the mass number drops by 4 and the atomic number by 2. Hence six α-decays alone would change $^{232}_{90}Th$ to an atom with mass number 208 and atomic number 78. In the emission of a β-particle only the atomic number alters, increasing by one. To obtain ${}^{208}_{82}Pb$, it is clear that 4 β-decays are necessary in addition to the α-decays.

Example 2. Atoms of ${}^{238}Pu$ emit 5.10-MeV α-particles, the half-life of the activity being 90 years. The energy from 180 mg of ${}^{238}Pu$ is used to power one particular form of cardiac pacemaker. Calculate the initial power supplied by the device if Avogadro's constant is 6.02×10^{23} per mole and the device is assumed to be one hundred per cent efficient.

Solution. The number of atoms in 238 g of ${}^{238}Pu$ is 6.02×10^{23}. Therefore the number in 180 mg is

$$N = \frac{180 \times 10^{-3}}{238} \times 6.02 \times 10^{23}.$$

The initial activity of the ${}^{238}Pu$ is

$$\Phi = \lambda N = N \ln 2/\tau,$$

which is the number of atoms emitted per second. Each atom carries energy $E = 5.10$ MeV. The initial power for the pacemaker is

$$\begin{aligned} \Phi E &= NE \ln 2/\tau \\ &= \frac{180 \times 10^{-3} \times 6.02 \times 10^{23} \times 5.10 \text{ MeV} \times 1.60 \times 10^{-13} \text{ J(MeV)}^{-1} \times 0.693}{238 \times 90 \text{ yr} \times 3.16 \times 10^{7} \text{ s yr}^{-1}} \\ &= 90.5 \text{ mW}. \end{aligned}$$

Example 3. Fossils are found in a sandstone quarry associated with rubidium 87 and strontium 87 in the ratio of 1 to 0.0035 by weight. Estimate the age of the fossils.

Solution. ${}^{87}Rb$ decays to stable ${}^{87}Sr$ with a half-life of 4.7×10^{10} years. It is assumed that only ${}^{87}Rb$ was present when the sandstone containing the fossils was laid down. The rubidium and the strontium have the same mass number and thus equal masses of the two substances contain equal numbers of atoms. Thus in the sandstone the ratio of the number of atoms of strontium to the number of atoms of rubidium n_1/n is 0.0035/1. The age of the sandstone is thus

$$\begin{aligned} t &= \tau \frac{\ln (n_1 + n)/n}{\ln 2} = \tau \frac{\log (n_1 + n)/n}{\log 2} \\ &= \tau \frac{\log 1.0035}{\log 2} \\ &= 4.7 \times 10^{10} \text{ years} \times \frac{0.0015}{0.3010} \\ &= 2.4 \times 10^{8} \text{ years}. \end{aligned}$$

Example 4. A tube containing radon, which has a half-life of 4 days, is to be implanted into a patient and left in position for 8 days. Due to unforeseen circumstances the implanting has to be delayed. What is the maximum possible delay that can be tolerated if the dose the patient is to receive cannot be reduced?

Solution. The implant was to have been left in for two half lives. After that time only $\frac{1}{2} \times \frac{1}{2} = \frac{1}{4}$ of the original atoms would not have disintegrated. The dose required for the patient is the radioactive emission from three-quarters of the atoms. Clearly the implant becomes useless once less than three-quarters of the original atoms remain (and even before this time the implant period becomes very long). The maximum possible delay time is thus t, where

$$\tfrac{3}{4} N_0 = N_0 e^{-\lambda t}.$$

Therefore

$$\tfrac{4}{3} = e^{\lambda t},$$

therefore

$$\ln (4/3) = \lambda t = \frac{t}{\tau} \ln 2,$$

therefore

$$t = \tau \frac{\ln (4/3)}{\ln (2)} = \tau \frac{\log (4/3)}{\log 2},$$

$$= \frac{0.1249}{0.3010} \tau = 0.415\, \tau,$$

therefore

$$t = 1.62 \text{ days}.$$

PROBLEMS

29.1 Eight α-decays and 6 β-decays are necessary before an atom of $^{238}_{92}U$ achieves stability. What are the atomic number, atomic mass number, and chemical name of the final atom?

29.2 A solution containing a radioactive isotope which emits β-particles with a half-life of 12.26 days surrounds a geiger counter which records 480 counts per minute. What counting rate will be obtained 49.04 days later?

29.3 Radium 226 has a half-life of 1620 years. What is the mass of a sample which undergoes 20,000 disintegrations per second?

29.4 A fixed quantity of a radioactive isotope is delivered to a hospital at the same time every week. One day a doctor finds an unopened bottle of the isotope with no label attached. He places it in front of a geiger counter and records 4200 counts per second. When he substitutes a bottle which has just arrived, he records 47,500 counts per second. If the isotope has a half-life of 8 days, how long has the unlabeled bottle been in the hospital?

29.5 Radioactive ^{24}Na, which has a half-life of 15 h, is sent from the A.E.R.E. Laboratories at Harwell to a London hospital. What should be its activity when it leaves Harwell if the activity is to be 10 mCi (millicuries) when it is used in the hospital 3 h later?

29.6 A curie was originally defined as the activity of the amount of radon in equilibrium with 1 g of radium 226. If the half-life of the radium is 1620 years, how many disintegrations per second does this represent?

29.7 A counter detecting the β-particles from a radioactive source takes the following readings:

t (min)	0	30	60	120	240	360	480	600
Φ (counts/min)	154	148	140	128	109	92	81	72

When the source is removed, the counter records an average counting rate of 30 per minute. What is the half-life of the radioactive source?

29.8 What are the masses of 1 Ci of ^{227}Th, ^{30}P, and ^{212}Po if the respective half-lives are 1.90 years, 2.55 min, and 3.00×10^{-6} s?

29.9 A burial site yields a wooden artifact which gives 11.6 counts per minute per gram of carbon present. The corresponding count from wood from living trees is 15.3. When was the artifact made? The half-life of ^{14}C is 5600 years.

29.10 In a sample of rock the ratio of ^{238}U to ^{206}Pb is 1:0.75 by weight. Estimate the age of the rock. The half-life of ^{238}U is 4.5×10^{9} years.

29.11 A patient is injected with 5×10^{-6} m^3 of blood labeled with ^{51}Cr, the activity being 60,000 counts per minute. The activity of similar-sized samples of blood withdrawn from the patient at intervals stabilizes at a value of 82.7 counts per minute. What is the total volume of blood in the patient's body?

29.12 A patient is fed a radioactive isotope which has an activity of 1 μCi. The activity of the feces measured 36 h later is 0.25 μCi. How much of the isotope has been taken up by the patient's body if the half-life of the isotope is 15 days?

CHAPTER 30

THE RUTHERFORD-BOHR ATOM

30.1 INTRODUCTION

From the discussions in the last two chapters it will be seen that at the beginning of the century the view held of the atom was that it consisted of a nucleus and several electrons. The nucleus was positively charged, of almost integral mass on the a.m.u. scale, contained practically all the mass of the atom, and was apparently made of the same building units whatever the type of atom. In addition, each atom possessed a number of electrons, the number ranging from 1 to 92 and varying from element to element. These electrons were negatively charged, very light, and the same for all atoms. But there was still no idea of how the nucleus and the electrons were joined together. The first idea came from J. J. Thomson and has come to be known as the plum-pudding model.

If Avogadro's constant (the number of atoms in a mole) and the size of a mole of any solid are known, it is possible to work out fairly easily the volume occupied by each atom, and thus the atom's approximate dimensions. Avogadro's constant is easily obtained by dividing the faraday by the magnitude of the electronic charge. This gives a value of 6.0249×10^{23} per mole. If this value is used, it is found that although the size of an atom may vary somewhat from element to element, its linear dimension is of the order of 10^{-10} m. Thomson assumed that the nucleus occupied most of this space, the electrons, like currants in a plum-pudding, being scattered throughout the total volume. It was unclear whether they were stationary, being held in position by various attractive forces from portions of the nucleus, or whether the electrons were in motion to prevent them from being annihilated by contact with the positive charges. This view of the atom was shattered by the scattering experiments of Lord Rutherford.

30.2 THE RUTHERFORD ATOM

If α-particles are fired into foils, so thin that the α-particles emerge from the other side, it would be expected that they would be scattered slightly in their passage through the foil owing to occasional encounters with atoms. But if the atom is a diffuse plum-pudding-like structure with the electrons intermingled with the nucleus, the total quantity of negative charge being the same as the total quantity of positive charge, the electric field exerted by such a mixture cannot be very large. One would therefore expect the angle of scatter of the α-particles to be very small. Unfortunately, Rutherford's results did not show this. A substantial proportion of an incoming beam of α-particles was found to be scattered through large angles after passage through a foil. Indeed, a few of the α-particles were scattered through angles greater than 90°. It was impossible to explain such large angles of scatter if the atom was a diffuse mixture of equal quantities of positive and negative charge. The α-particles traveled with speeds of the order of 10^4 m s $^{-1}$ and should have been relatively unaffected by such weak fields.

Rutherford found it necessary to postulate that the repelling nucleus had a linear dimension very much smaller than the 10^{-10} m found for the atom. In fact the nucleus had to act like a point charge in order to produce a sufficiently concentrated field to deflect fast-moving α-particles through large angles. Far away from the atom the α-particle would experience the effects due to a nucleus at the center of the atom and a distributed cloud of electrons 10^{-10} m from the nucleus. The combined effect of these would be almost zero. As the α-particle approached nearer and nearer to the atom, the combined effect of nucleus and electrons would still remain small. But once the α-particle had penetrated the cloud of electrons and was closer to the nucleus than 10^{-10} m, the effect of the electrons would be negligible. The α-particle could be considered to be inside a uniform spherical shell of negative charge, and in Section 22.8 it was shown that the electric field inside a charged sphere was zero. The force exerted on the α-particle by the nucleus, however, would still be present and would be very large. The α-particle would therefore be deflected through large angles.

Rutherford, working on this basis, obtained a formula which would predict the proportion of α-particles scattered at any given angle if the nucleus acted like a point charge. The formula depended on the charge of the nucleus and the velocity of the incoming particles. His theory was tested by scattering α-particles from various foils over a range of velocities of the incoming α-particles. Experiments fully confirmed his predictions, so long as the velocity of the incoming particles was not too great.

30.3 THE SIZE OF THE NUCLEUS

The fact that the observed scattering of the α-particles departs from Rutherford's prediction when the velocity of the incoming particles is fairly large is very valuable. It is clear that the building bricks of which the atom is made must be held together in some fashion. Some of the bricks at least must have positive charges, and without attractive forces between the constituents the nucleus would spontaneously disrupt.

The attractive force acting between any two constituents of a nucleus must act over only a very short distance, and this distance is a measure of the size of the nucleus. If incoming α-particles have small speeds and therefore low energies, they cannot penetrate anywhere near the nucleus before they are repelled. As the speed, and therefore the energy, increases, the incoming particles penetrate closer and closer to the nucleus before being repelled and deflected. It is easy to work out how close they can possibly get. When departures from Rutherford scattering start to take place, the incoming particles are just penetrating to the edge of the nucleus, where the attractive forces are felt. This therefore gives a measure of the size of the nucleus. It is found to have a linear dimension of around 10^{-15} m, which is one-hundred-thousandth of the dimension of the atom as a whole. It is clear that the atom consists of a minute nucleus surrounded by mainly empty space with a few electrons circling in orbits which have a linear dimension of around 10^{-10} m.

Note that until this time the number from 1 to 92, which had been given to an element as its atomic number, had been assigned on the basis of an arrangement of the elements in order of increasing mass. It had long been believed, without any proof being available, that this number represented both the charge on the nucleus and the number of electrons that the atom possessed. Chadwick, assuming Rutherford's scattering formula to be true at low energies, worked out the charge on the nucleus for several atoms by bombarding foils of the elements with α-particles and noting the deflections obtained. He showed that the number from 1 to 92 previously assigned corresponded with the number of charges on the nucleus.

30.4 THE QUANTUM THEORY

The first clues about the way in which atomic processes worked, and how the pieces of the atom fitted together, came from the study of black-body radiation. A black body is one which absorbs all radiation falling on it. It must also have a spectrum of radiations which it emits at various temperatures. Figure 30.1 shows three such experimentally obtained spectra for a black body at different temperatures. One notes that the total amount of energy emitted increases with temperature (actually as T^4), the maximum amount emitted also increasing, and that the position of the maximum in the spectrum shifts (actually $\lambda_{max}T =$ constant).

Many attempts were made to find by the methods of classical physics a formula for the emission of the radiation as a function of wavelength at any temperature, but all such attempts proved failures. It was left to Planck to show how such an equation could be obtained. It had until then been assumed that the atoms in a black body could emit energy continuously of any quantity. Planck found that the only way to obtain theoretically the experimentally observed form of the curves was to throw aside these deep-rooted beliefs of classical physics. He showed that the correct form was given if it was assumed that the atoms could only emit a discrete, fixed quantity of radiation at a time. These discrete units he called quanta, although they are also known as photons. Subsequent analysis showed that the amount of energy in any quantum was hf, where h was a constant, now known as Planck's constant, of value 6.626×10^{-34} J s, and f was the frequency of the radiation. On this assumption all the difficulties disappeared. Planck accepted that the assumption led to the correct results, but had considerable doubts about its validity and took it no further. It was left to Einstein to show that the energy was also transported by the radiation in the form of quanta, and absorbed in this form as well. He demonstrated this by explaining a phenomenon known as the photoelectric effect which had been baffling scientists for some years.

30.5 THE PHOTOELECTRIC EFFECT

When light falls on a metal plate which is electrically insulated, the plate emits electrons and will continue to do so until the potential rises, on account of the acquired positive charge, to the order of 1 V, when emission stops. This effect is much more marked with some metals than others, and for some will only occur if ultraviolet light is employed. The alkali metals show the effect strongly for visible light. If the plate is not insulated but instead is grounded, the emission of electrons continues until the incident radiation is cut off, a current flowing continuously through the connecting wire to ground. This effect is only shown if the surface is pure and clean. It is often masked by occluded surface layers of gas or surface impurities, so that photoelectric surfaces should always be prepared and used in a vacuum. Contact potentials and thermoelectric emf's should also be avoided, as these tend to mask the effect.

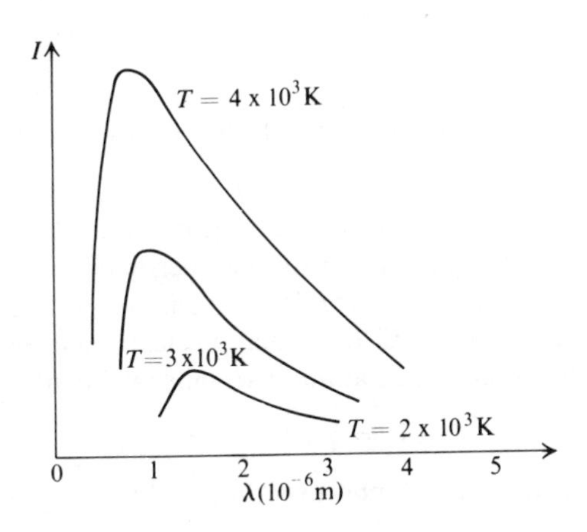

Fig. 30.1. The distribution of energy in the spectrum of a black body at different temperatures.

It is easy to check that electrons of low energy are being emitted by an observation of their deflection in electric and magnetic fields. The energies of the photoelectrons in a given case vary in value from a definite maximum downwards. The maximum energy, $\frac{1}{2}mv^2$, can be found by illuminating a plate which is insulated and allowing it to rise in potential until the emission ceases. At this point the potential V is known as the stopping potential and can be measured accurately.

When the potential has this value, the potential energy possessed by the electrons is just sufficient to counterbalance the maximum kinetic energy which is available under normal circumstances. Alternatively, a potential difference can be set up between a photo-cathode and an anode, and altered until it is just sufficient to stop the emission of the electrons. In either case $eV = \frac{1}{2} mv^2$, from which equation the maximum kinetic energy $\frac{1}{2} mv^2$ and the maximum velocity v are calculable.

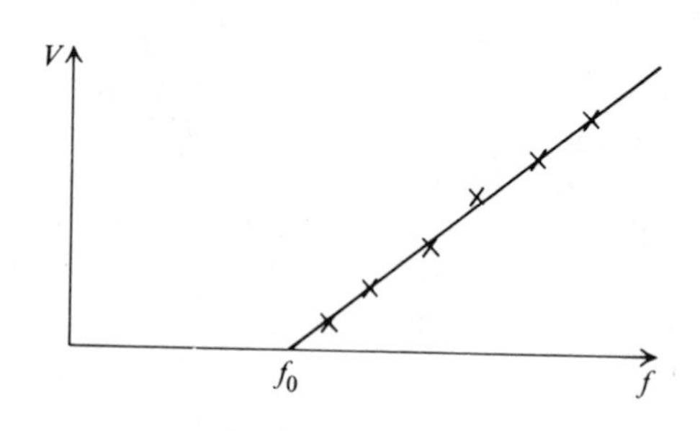

Fig. 30.2. Variation of stopping potential with frequency in the photoelectric effect.

Millikan investigated the effect in detail, in particular the dependence of the stopping potential (and thus the maximum energy of emission of the electrons), on the frequency of the incident radiation. Figure 30.2 shows his results. The effect does not occur at all until the radiation has frequency f_0, and rises linearly with frequency thereafter. Thus at any particular value of f

$$eV = \tfrac{1}{2}mv^2 = C(f - f_0), \tag{30.1}$$

where C is a constant. It is found experimentally that C is equal to h, Planck's constant.

The conclusions to be drawn from Millikan's extensive experiments were the following.

a) The intensity of the radiation affects only the number of electrons emitted, not their energy. This is inexplicable by classical theory. More intense radiation should provide more energy for the emitted electrons.

b) The effect is frequency-dependent. By classical theory the frequency should not matter, only the intensity being important; and the fact that there is a threshold frequency below which the effect will not occur is also inexplicable.

c) As the intensity of the incident radiation is reduced, it would be anticipated that the electrons would become delayed in emission. They would be expected to acquire energy a little at a time until they had sufficient to overcome the energy binding them to the plate. No such delay is observed in practice. However low the intensity, emission starts as soon as radiation falls on the plate.

Einstein explained the phenomenon by using Planck's ideas on the emission of radiation from black bodies. If the radiation is not only emitted in quanta but transmitted and absorbed in this form also, then the phenomenon is explicable in terms of quantum theory. If a photon is absorbed in the surface, the energy hf carried by it may be given to an electron. Part of this energy $\mathscr{W}$, called the work function, is used in overcoming the potential energy with which the electron is bound to the solid, the remainder being available to the electron as kinetic energy. Thus

$$\tfrac{1}{2}mv^2 = hf - \mathscr{W}.$$

If the frequency of the radiation necessary to release the electron without acquiring any kinetic energy is f_0, then it follows that $0 = hf_0 - \mathscr{W}$. Therefore

$$\tfrac{1}{2}mv^2 = hf - hf_0 = h\frac{c}{\lambda} - h\frac{c}{\lambda_0}. \tag{30.2}$$

If the electron is released from the surface, it appears outside the plate with all this energy. If the electron is freed in the body of the metal, it loses energy by collision and other processes in making its way to the surface. It therefore emerges with less energy than the maximum possible amount. The electrons taking part in this process are the free electrons in the metal, and the energy required to release them is therefore the same as in thermionic emission (cf. Section 33.2).

30.6 THE COMPTON EFFECT

Another effect, also explicable only in terms of quantum theory, further emphasizes the particle nature of radiation. Compton noticed that when X-rays were scattered by a light element such as carbon, the scattered radiation consisted of two components, one having the same wavelength as the incident beam, the other with slightly longer wavelength. The difference in wavelength of the two components $\Delta\lambda$ varied with the scattering angle. When the angle of scatter was 90°, the difference in wavelength was 2.36×10^{-12} m, a value which did not depend on the wavelength of the incident radiation or the nature of the scattering atoms.

The only way in which Compton could explain this second component of the scattered radiation was to assume that in the scattering process a photon and an electron suffered an elastic collision, behaving like two particles colliding in a dynamical problem, and conserving energy and momentum in the process.

During last century, in order to explain a phenomenon known as radiation pressure, it had been necessary, on purely classical arguments, to assume that a beam of radiation has both an energy density E and a momentum density p. The relation between these was $E = pc$. Compton therefore assigned to a colliding photon not only an energy hf but also a momentum hf/c, a value which is also justified by relativity theory for a particle of vanishingly small mass. With these assigned quantities the problem of the collision of a photon with an electron is easily soluble.

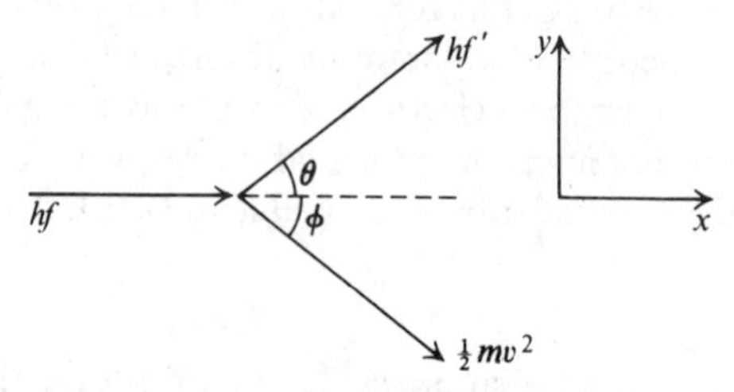

Fig. 30.3. A photon is scattered through an angle θ in a Compton collision with an electron which recoils at an angle ϕ to the incident direction.

Figure 30.3 shows the result of such a collision. A photon of energy hf collides with an electron of mass m, assumed to be free and to have negligible energy. The photon gives up some of its energy to the electron, which recoils with velocity v at an angle ϕ to the incident direction. The photon, now of energy hf', is scattered through an angle θ.

Since the collision is elastic, by the principle of conversion of energy,

$$hf = hf' + \tfrac{1}{2}mv^2. \tag{30.3}$$

Applying the principle of conservation of momentum in both the x- and y-directions,

$$\frac{hf}{c} = \frac{hf'}{c}\cos\theta + mv\cos\phi, \tag{30.4}$$

and

$$\frac{hf'}{c}\sin\theta = mv\sin\phi. \tag{30.5}$$

Dividing Eq. (30.3) by mc^2 and Eqs. (30.4) and (30.5) by mc, and writing

$$\frac{hf}{mc^2} = \alpha, \quad \frac{hf'}{mc^2} = \alpha_0, \quad \frac{v}{c} = \beta,$$

gives

$$\alpha - \alpha_0 = \tfrac{1}{2}\beta^2, \tag{30.6}$$

$$\alpha - \alpha_0\cos\theta = \beta\cos\phi, \tag{30.7}$$

$$\alpha_0\sin\theta = \beta\sin\phi. \tag{30.8}$$

Therefore

$$\beta^2 = \beta^2(\cos^2\phi + \sin^2\phi) = \alpha^2 + \alpha_0^2\cos^2\theta - 2\alpha\alpha_0\cos\theta + \alpha_0^2\sin^2\theta.$$

Therefore

$$\beta^2 = \alpha^2 + \alpha_0^2 - 2\alpha\alpha_0 \cos\theta.$$

From Eq. (30.6)

$$(\alpha - \alpha_0 + 1)^2 = (1 + \tfrac{1}{2}\beta^2)^2.$$

Therefore

$$\alpha^2 + \alpha_0^2 + 1 - 2\alpha\alpha_0 + 2\alpha - 2\alpha_0 = 1 + \beta^2,$$

ignoring the term $\frac{1}{4}\beta^4$, which will be negligible, since $v \ll c$.

Therefore

$$\alpha^2 + \alpha_0^2 - 2\alpha\alpha_0 \cos\theta = \alpha^2 + \alpha_0^2 - 2\alpha\alpha_0 + 2\alpha - 2\alpha_0.$$

Therefore

$$\alpha - \alpha_0 = \alpha\alpha_0 (1 - \cos\theta).$$

Therefore

$$\frac{hf}{mc^2} - \frac{hf'}{mc^2} = \frac{h^2 ff'}{m^2c^4}(1 - \cos\theta).$$

Therefore

$$\frac{1}{\lambda} - \frac{1}{\lambda'} = \frac{h}{mc\lambda\lambda'}(1 - \cos\theta).$$

Therefore

$$\lambda' - \lambda = \frac{h}{mc}(1 - \cos\theta). \tag{30.9}$$

Inserting the values of the atomic constants gives

$$\lambda' - \lambda = 0.0243\,(1 - \cos\theta)\text{ Å}. \tag{30.10}$$

Thus, for $\theta = 90°$,

$$\lambda' - \lambda = 0.0243\text{ Å},$$

a figure which compares favorably with the experimentally determined value. The value at other angles is also found to be in accord with experiment.

If the electron is not free but is bound to an atom, and if the energy acquired in the collision is not sufficient to free it, the recoil involves not the electron but the atom as a whole. The mass m appearing in the formula should therefore be the mass of the atom, which is thousands of times greater than the mass of the electron. The change in frequency of the scattered radiation in this case is negligible. This explains the presence of the scattered component of unchanged wavelength.

The Compton effect is one of the major ways in which a beam of X- or γ-radiation is attenuated in passage through matter (cf. Section 31.4), and an understanding of the mechanics of the effect is therefore crucial. As with all other atomic processes of this type, classical physics cannot provide an explanation. Only a quantum explanation fits the observed facts.

30.7 THE WAVE NATURE OF MATTER

Similar experiments requiring an explanation in quantum terms have been found throughout the whole reach of atomic physics. Quantum physics is now a firmly established branch of the subject. This concept of discrete bundles of radiation does not appear at first sight to agree with the earlier interpretation of radiation in terms of waves. One must merely accept that atomic particles and radiation on the atomic scale behave in a manner that cannot be represented by a model on the macroscopic scale. Radiation which acts sometimes like particles and sometimes like waves is a phenomenon impossible to visualize, but this duality exists throughout the whole of atomic physics. Individually, photons act like particles; in large numbers, they exhibit wave properties. Although it is impossible to build a visual model to represent them, their behaviour can be accurately predicted from a now firmly established mathematical model.

It was postulated in 1926 by de Broglie that the inverse effect must also take place, i.e. that particles should on occasion behave like radiation, exhibiting diffraction and interference effects as though they had a wavelength $\lambda = h/p$, where p was the momentum of the particle. Davisson and Germer in the United States and G. P. Thomson in Britain showed that electrons could indeed be diffracted by crystalline material. Since then it has been shown that these phenomena are exhibited by other fundamental particles. This type of behavior, this wave-particle duality, is thus a fundamental characteristic of all atomic species.

30.8 THE BOHR THEORY OF THE ATOM

It became obvious to Niels Bohr that an explanation of the construction of the atom must be made in terms of quantum theory. Before he applied quantum ideas to the atom, it was known that on the Rutherford model the nucleus was very tiny and was situated at the centre of the atom. The electrons were some distance away and must be circling in orbits to explain why they did not fall into the nucleus under the attraction of the positive charge. But this raised some awkward problems. By classical theory an accelerating electron must radiate energy. A circling electron is being constantly accelerated toward the nucleus, should therefore in the classical view be continually radiating, and thus in losing energy should spiral in toward the nucleus. Obviously, this did not happen.

Further, spectroscopy was a well-established subject, and it was known that atoms could be excited and would emit spectral lines as they fell back to their normal energy states. These spectral lines appeared to fall into series. For instance, the hydrogen atom emits a number of spectral series in the infrared, ultraviolet, and visible regions, all of which obey laws of the type

$$\frac{1}{\lambda} = R\left(\frac{1}{n^2} - \frac{1}{m^2}\right), \tag{30.11}$$

each series corresponding to a fixed value of n and a variation of m through integral values from $n + 1$ to infinity. The constant R is known as the Rydberg constant and has the experimentally determined value of 1.097×10^7 m $^{-1}$. Classical theory was completely unable to explain the presence of these series. Needless to say, hydrogen, with a singly charged positive nucleus and only one electron, being the simplest atom possible, its spectra had been investigated most fully, and it was to hydrogen that Bohr now attempted to apply quantum theory in order to explain the observed facts. He started off with two postulates.

1) An atom can only exist in one or other of a number of fixed energy states. It only radiates or absorbs energy if it changes from one energy state to another.

2) If the atom changes from a state of energy E_m to a state of energy E_n, then a quantity of energy $|E_m - E_n|$ is absorbed or radiated. If radiated, the frequency of the radiation emitted is

$$f = \frac{|E_m - E_n|}{h}.$$

It is clear that by this definition radiating particles are constantly changing from one energy state to another but that in an atom this only occurs if one of the electrons changes from one fixed orbit to another. It is also clear that the quantum principle is being applied in this case. Only certain fixed energies are possible: and when the energy state is altered, all the energy is emitted in one quantum of radiation.

There must also be some further condition which indicates which are the allowable fixed energy states. Bohr found, by studying the observed wavelengths of the spectral lines emitted, that the condition had to be such that an electronic orbit was only permissible if the angular momentum of the electron in its orbit was an integral number of units of $h/2\pi$. Bohr only considered circular orbits, and the condition may therefore be stated in an alternative form. The linear momentum of the electron in its circular orbit multiplied by the length of the orbit must be an integral multiple of h. Thus

$$mv \times 2\pi r = nh, \tag{30.12}$$

where n is an integer, m and v are the mass and speed of the electron, and r is the radius of the circular orbit. It is necessary to combine this equation with that for motion in a circular orbit. In this case the centripetal force is provided by the electrical force between nucleus and electron. Therefore

$$\frac{mv^2}{r} = \frac{e \times e}{4\pi\varepsilon_0 r^2}, \tag{30.13}$$

where e is the electronic charge and ε_0 the permittivity of free space.

From Eq. (30.12)

$$v^2 = \frac{n^2h^2}{4\pi^2r^2m^2}.$$

Substitution in Eq. (30.13) yields

$$\frac{n^2h^2}{4\pi^2r^3m} = \frac{e^2}{4\pi\varepsilon_0 r^2}.$$

Therefore

$$r = \frac{n^2h^2\varepsilon_0}{\pi me^2}. \tag{30.14}$$

If the values of the known constants are inserted in the last equation,

$$r = n^2 \times 0.52 \times 10^{-10}\ \text{m}.$$

The first orbit ($n = 1$), the one of lowest energy and therefore the one normally occupied by the electron, is roughly 10^{-10} m in diameter, which agrees with the earlier estimate of atomic size.

In an orbit the electron possess kinetic energy due to its motion and potential energy due to being in the field of the nucleus. The total energy it possesses is thus

$$E = \tfrac{1}{2}mv^2 - \frac{e \times e}{4\pi\varepsilon_0 r},$$

$$E = \frac{e^2}{8\pi\varepsilon_0 r} - \frac{e^2}{4\pi\varepsilon_0 r} = -\frac{me^4}{8\varepsilon_0^2 n^2h^2}. \tag{30.15}$$

The negative sign implies that the electron is bound to the nucleus and requires energy to release it. The energy will be zero when the electron is in an infinitely far orbit.

When the electron jumps from the mth to the lower nth orbit, the excess energy is radiated as one quantum. Thus

$$hf = E_m - E_n.$$

Therefore

$$\frac{1}{\lambda} = \frac{f}{c} = \frac{E_m - E_n}{hc} = \frac{me^4}{8\varepsilon_0^2ch^3}\left(\frac{1}{n^2} - \frac{1}{m^2}\right) \tag{30.16}$$

If the known atomic constants are inserted in this equation, $me^4/8\varepsilon_0^2ch^3$ is found to have the value $1.097 \times 10^7\ \text{m}^{-1}$. But experimentally it was found that

$$\frac{1}{\lambda} = R\left(\frac{1}{n^2} - \frac{1}{m^2}\right), \tag{30.11}$$

where R had exactly the value quoted above. The agreement is thus excellent on this count also.

Sommerfeld extended the ideas of Bohr to elliptic as well as to circular orbits, and showed that this extension explained the fine structure observed in the lines of the hydrogen spectrum. Subsequent refinements to the theory have explained other fine details.

Once it has been shown that electrons on occasion behave like waves, one has a physical explanation of why the angular momentum of the electron in a permitted orbit must be an integral multiple of $h/2\pi$. Waves, as was shown in Section 11.4, will only persist in any situation if they produce a resonant condition in the system, i.e. produce standing waves. One may apply this physical idea to the electron in its orbit. Unless the wavelength associated with the electron is such that the orbit the electron traverses is an exact number of wavelengths, so that standing waves can be set up, the situation is not a stable, persistent one. Thus

$$2\pi r = n\lambda = nh/mv,$$

which is merely a rewriting of Bohr's condition (cf. Eq. 30.12).

It should be noted that Bohr theory is only a first approximation to the truth. It is necessary, in order to get exact agreement with experiment and also in order to deal with atoms having more than one electron, to employ the newer, and more correct, technique of wave mechanics. However, Bohr theory gives a picture by which it is possible to visualize what is happening.

30.9 THE PERIODIC TABLE

Although the electronic structure of atoms can only be fully unraveled by the use of wave mechanics, it is possible to give a reasonable description in terms of the ideas so far developed. The fact that elliptical orbits are allowable as well as circular orbits, and the fact that orbits can occupy various locations in space, add refinements to the simple Bohr theory. For any energy level characterized by the quantum number n there are now several permissible orbits, not just one. Indeed, there are n^2 orbits for each value of n, the orbits, for reasons which we will not consider, differing slightly in total energy. Further, the electron is now known not to be a simple point but to have structural features. In particular, it acts as though it were spinning about an axis like a top, and it is possible for it to be spinning either clockwise or anticlockwise.

Pauli's exclusion principle states that any orbit can only hold two electrons and that these must have contrary spins. Any electron will attempt to occupy an orbit as close to the nucleus as possible, since such an orbit is of lower energy than one farther from the nucleus. A hydrogen atom in its normal, or ground, state will thus have its electron in the single ($n = 1$) orbit. A helium atom will have its two electrons, with contrary spins, in the same orbit, and this orbit will now hold no further electrons. This completes the ($n = 1$) shell, and a closed shell atom is always highly stable and very unreactive chemically.

The ($n = 2$) shell is closed when a further 8 electrons have been added, and the atom which has 10 electrons is neon, a rare gas like helium. For values of n of 3 and 4, closed shells should occur when the total numbers of electrons in the atom are 28 and 60. Atoms with these numbers of electrons are not rare gases. The other rare gases, argon, krypton, and xenon, occur at $Z = 18, 36$, and 54. The reasons for this apparently anomalous behavior are: first, that closed-shell characteristics now begin to appear for the completion of sub-group shells; and secondly, that because of the departure from the simple model caused by the presence of large numbers of electrons, some orbits from a particular shell have lower energies than have some of the orbits in the shell one lower in number. The addition of electrons to an unoccupied orbit of lowest energy does not therefore proceed in what might initially have appeared to be the logical sequence.

However, a rare gas always occurs on completion of a shell or a subgroup. The atom with the next highest atomic number thus always has one electron outside a completed shell or subgroup, and such an electron is very loosely bound to the rest of the atom. Such an atom shows characteristic chemical behavior, since it will donate its extra electron freely, and a characteristic emission spectrum very similar to that of hydrogen. All atoms with an atomic number one greater than a rare gas are alkali metals. Similarly, atoms with two electrons outside a closed shell or subgroup exhibit similar properties and are called alkali earths.

All atoms which lack one electron to form a closed shell or sub-group also show highly characteristic behavior. They will attempt to acquire a further electron to complete their closed shell. Such atoms are called halogens.

It should now be clear, without our continuing to multiply examples, why chemists have been able to work from a periodic table for so long. The repeat pattern of chemical properties with increasing atomic number is linked to the electronic structure of the atom, the chemical behavior being determined by the number of electrons outside of, or necessary to complete, a closed shell or sub-group.

30.10 MOLECULAR ENERGY LEVELS

The manner in which atoms join to form molecules is also determined by electronic structure. When an alkali metal atom and a halogen atom join to form a molecule, the former donates its loosely bound electron to the latter, which requires this electron in order to form a closed shell. The formation of two ions in this way requires the expenditure of a certain amount of energy, but the electrostatic energy of attraction of the two ions far exceeds this, and a stable molecule is formed. Such binding of atoms is called ionic binding and a typical example is common salt (NaCl).

The other common form of molecular binding is covalent binding, such as occurs in the hydrogen molecule. Each of two hydrogen atoms requires a further electron to complete a closed shell. Since neither can rob the other of its electron, they share the two available, each electron then occupying an orbit which encompasses both nuclei. This produces an arrangement of lower energy than that of the two separate atoms. In the formation of such a molecule, the repulsion of the two positively charged nuclei must be overcome. The two atoms must therefore approach with sufficient kinetic energy for the atoms to come close enough for the covalent orbit to be formed. Once the molecule is formed, the orbits are such that the electrons spend most of their time between the two nuclei, screening one nucleus from the other and minimizing the repulsion effect. As in the case of the atom, orbits of differing energy are permissible. The electrons will normally occupy the orbit of lowest energy but may be excited into higher orbits. They return to the normal, ground state emitting photons in the process.

Note that other energy states of the molecule are now possible. The atoms may vibrate about their equilibrium positions, shortening and lengthening the bond distance between them. Also, the molecule may rotate about axes through its centre of mass. It is found that quantum principles are applicable to these energetic processes also. Only certain, definite, fixed-energy, vibrational and rotational energy states are permissible, and transfer from any one of these states to one of lower energy is accompanied by the emission of the excess energy in the form of one quantum of radiation. The emission and absorption spectra associated with changes in vibrational and rotational energy levels are as characteristic of a particular substance as are those associated with electronic changes.

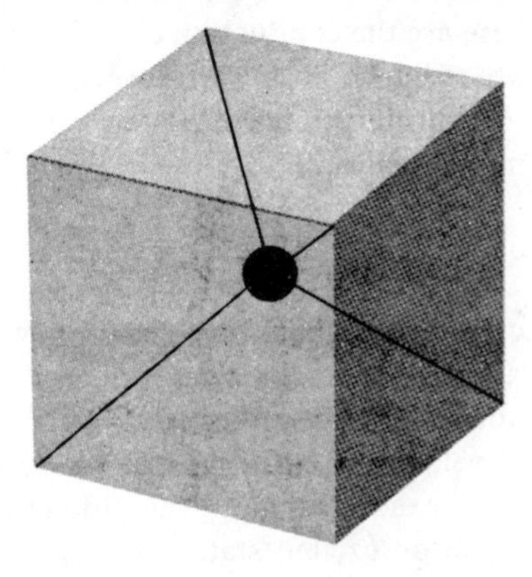

Fig. 30.4. The highly localized electron bonds in methane.

A carbon atom has four electrons outside a closed shell and tends to share these to fill up its ($n = 2$) shell. Thus methane, CH_4, is a simple covalent combination of carbon with four hydrogen atoms. Any two shared electrons tend to concentrate between the carbon atom and the appropriate hydrogen atom, forming an electron cloud there. These electron clouds are mutually repulsive and tend to separate as far as possible. Thus, if the carbon atom is considered as being at the centre of a cube, the hydrogen atoms will occupy two opposite corners of the upper face and the other corners of the lower face, the shared electrons being highly localized along the four corresponding semi-diagonals, as shown in Fig. 30.4.

Carbon atoms are very special in that they have four electrons in, and need a further four electrons to complete, the ($n = 2$) level. They can therefore join one with the other. Indeed they can form long chains of carbon atoms containing thousands of atoms, which are intimately concerned in biological processes.

30.11 METALLIC BINDING

When ionic molecules are joined together to form crystals, the electrostatic force holding the ions together is still present but the effects are modified by the presence of other ions. Thus any positive ion is now surrounded by several negative ions and it is no longer clear which negative ion is the other half of its molecule. Nonetheless, no basic change has occurred in the binding mechanism.

Similarly, for molecules covalently bound, crystallization retains the identity of each molecule which is held to others by various forces such as van der Waals' forces and hydrogen bonding. But if metal atoms are joined together to form a solid, something quite different occurs. It is now energetically possible for two atoms to get so close together that the outermost electrons cannot pass between them. These electrons therefore attempt to circle the complex of the two nuclei plus their central, closed shells. But if large numbers of atoms are present, the same phenomenon is occurring between any pair. For the loosely bound electrons the consequence is that the orbit of the electron is not localized at all but extends over the whole solid. These are the conduction electrons which are free to move to any part desired. They are still held in the framework of the solid, since it requires energy to release them, but inside the solid they can move freely. This question will be taken up again in Section 33.2.

30.12 ABSORPTION AND EMISSION OF ENERGY

In atoms (and in molecules and metals) any electron will normally occupy the level of lowest energy available; but if it acquires sufficient energy, in a collision with an atomic particle or a photon, it may be either removed from the atom altogether or raised to a higher energy level. In the latter case the atom is said to be excited and the energy absorbed must be exactly equal to the difference in energy of the levels.

The atom cannot stay in an excited state. It may return to the ground state directly, emitting the excess energy in the form of one photon of radiation, or it may cascade down via intermediate energy levels. If the latter, a number of photons are emitted, each one corresponding to a particular change from one to another energy state. The total energy of all the photons emitted must be equal to the photon energy emitted in a direct jump to the ground state.

The optical, ultraviolet, and infrared spectra emitted by atoms and molecules are due to changes in the outermost, valence electrons (or to vibrational or rotational energy changes in the case of molecules). A particularly important application of this type of emission is given in the next section. The emission of X-rays involves changes in electronic levels deeper in the atom. We shall deal with X-rays in the next chapter.

30.13 THE LASER

In Section 17.5 it was mentioned that an ordinary vapor lamp emits light because electrons in vapor atoms which have been raised to excited states drop back to lower energy states, each emitting the excess energy in the form of one photon of light. Each train of waves emitted lasts only for about 10^{-9}s, and the light from the vapor lamp thus changes its phase discontinuously on average every 10^{-9}s or less. Since the eye has a response time of around 0.1 s, no interference effects are observed between beams of light from two different sources of this type. The same would not be true if it were possible to use perfect lasers, kept under strictly controlled, vibration-free conditions, as the light-sources, since such lasers would produce wave trains in which no discontinuous changes in phase took place over times of the order of 0.1 s. In ordinary circumstances a laser does not work to this perfection. Nonetheless it produces a highly collimated beam of light which acts like a sequence of continuous sine waves each of length about 20 m or 30 m. The light is thus coherent over relatively long periods of time. The laser takes its name from the first letters of the words – Light Amplification (by) Stimulated Emission (of) Radiation.

An electron in an atom will go to a level of higher energy if it can acquire the necessary energy in collision, absorption of a photon, or some similar energetic process. It may return to the lower energy state spontaneously, emitting a photon in the process; but it can also be stimulated to do so if it collides with a photon of the same energy as it will emit. If it does so, the photon emitted is in phase with the one which provided the stimulation. The light signal represented by the incoming photon has thus been amplified.

In any substance under normal conditions, the majority of atoms will have their electrons in the lowest possible energy states. Only a small proportion of the atoms will have excited electrons. If a beam of light were sent into such a substance, stimulated emission would certainly occur, but many more of the incoming photons would be absorbed in raising electrons to higher energy levels than would be employed in producing stimulated emission. In a laser it is obviously necessary to arrange that a *population inversion* has taken place. The majority of atoms must have electrons in the correct excited levels before the laser action takes place.

This situation can be produced by various methods which supply the correct amount of energy to electrons in suitable substances. One method employed is optical pumping, which is used in ruby lasers (Fig. 30.5). Flashes of light of wavelength 5500 Å are concentrated on to a ruby rod which is doped with chromium oxide. Electrons which acquire the energy hf_1 available in a photon of the incoming light are raised to a level E_2, 2.2 eV above their normal level. A few hundred microseconds later, most of these excited electrons will have spontaneously

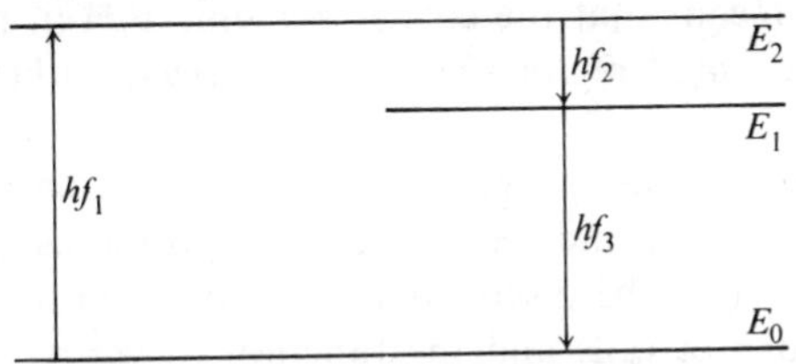

Fig. 30.5. The energy level scheme and possible transitions in a ruby crystal.

left this level, some to return to their normal level E_0 but some to drop to an intermediate level E_1 which has an excitation energy of 1.8 eV. This level is said to be metastable because electrons can remain in this level for several milliseconds before descending to the normal level E_0. If the pumping action is rapid and efficient, it is easy to arrange that a population inversion takes place and that most atoms have electrons in state E_1 rather than E_0.

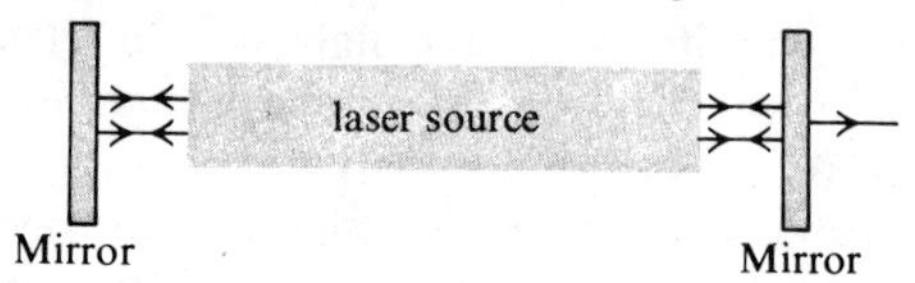

Fig. 30.6. The basic construction of a laser.

When an electron does descend from level E_1 to level E_0, it emits a photon of light of wavelength 6943 Å. This photon can stimulate emission from other excited atoms and the process can build up. To ensure that the amplification continues, mirrors are placed at the ends of the laser tube, as shown in Fig. 30.6. One of the mirrors is only partially silvered, so that some of the light passes out to be used. The rest oscillates back and forth in the system, causing further stimulated emission and thus being amplified in the process. Only photons which travel parallel to the axis of the system give rise to the light beam which is being magnified, because off-axis rays after several reflections tend to be absorbed in the sides of the system. The laser beam output is thus inherently parallel and will diverge little over distances of several miles if properly collimated.

In such a laser the output must be pulsed, and at not too rapid a rate, otherwise the crystal will overheat. Gas lasers which give a continuous output are, however, available if a pulsed output is undesirable. Semiconductor lasers using *p-n* junctions (cf. Section 33.3) have now been developed and these are of much smaller size than the other types.

The coherence of the output beam, and the fact that it is narrow and parallel, have many advantages. Direct measurements of distance by sending out a pulse of light, reflecting it, and measuring the transit time have been achieved even in daylight. It has been reported that the variations caused to a laser beam reflected from a window by the movements of the glass pane due to the sound waves from the people inside the room are sufficient to allow the conversation to be reproduced at the point where the laser beam is received. Because the beam is essentially a sequence of continuous sine waves, it can be focused by a lens to almost a mathematical point. This means that the energy per unit area in the focal spot can be made enormous, and small regions can be vaporized without harming the surroundings.

It is by the use of this last property that lasers have until now found most of their applications in biology and medicine, although the use of lasers is so recent that new fields are likely to be discovered in the near future. In biology the role of the various portions of cells and the function of various parts of the genetic material occupy a good proportion of current research. One line of study takes the form of damaging a particular cell component and finding out the effect this produces on the functioning of an organism or the characteristics of progeny. Until recently this has been achieved by inserting microprobes though a cell wall, and the damage done to other parts of the cell has been difficult to estimate. The ability to reproduce the same damage on subsequent occasions has been strictly limited. It is now possible by employing a focused laser beam to confine the damage to the particular component required. This technique has also been employed on developing embryos, specific cells being destroyed at a stage at which division has proceeded so far that the function of the cells has become specialized. In this way it may be possible to discover how defective offspring occur and perhaps relate this to defects and diseases in the mother.

In medicine one of the most spectacularly successful uses of lasers has been in ophthalmic surgery. In the eye the retina may become detached from the choroid owing to disease, injury, or degenerative changes. Unless treatment is prompt, blindness results, because the alteration in nutrition and metabolism of the retina causes rapid degeneration. It is necessary to attempt to tie the retina back to the choroid by damaging small portions of each in the hope that the resultant scar tissue will bond the two together at that point. Before the advent of the laser this was difficult and not always successful. It has been found that a 1 ms flash of light from a laser focused on to the retina is highly efficient in welding the retina to the choroid. Further, the patient feels no pain and an anaesthetic is not required. The exposure time is so short that the patient does not react to the light and move his eye, so that no mechanical fixing of the eye is required. The operation does not require an operating theatre and many cases can be dealt with as outpatients.

Although it is not possible to treat deep-seated tumors by use of the laser, skin tumor therapy has been attempted by using high-energy focused laser beams. The results have proved very encouraging. Although not all results agree on this point, there is evidence that not only the treated cells but others in the tumor and also adjacent secondaries regress and heal over a period of weeks. It is believed that this may be due either to thermal denaturization of key enzymes in the structure or to disruption of the molecular structure of the tumor cells by the electric field of the beam, which results in rejection of the cells by the host body.

Continuous lasers of high power are also being used in surgery. In vascular organs such as the liver, normal surgery produces a large amount of bleeding, which may obscure the operative field. The focused beam of light from a laser tends to seal off the vessels as they are incised and very much less bleeding occurs. Laser surgery is also being used in the removal of cancerous growths. Since there is no physical contact with the growth or surrounding tissue, the risk of spreading the growth during surgery is eliminated.

Lasers combined with instruments using fiber optics are now employed for internal treatment in the human body. For example, gastric bleeding is often difficult to stop and surgery may be required to save a patient's life. If a fiber optics gastroscope is inserted and the seat of the trouble found, power of the order of 12 W cm^{-2} can be directed on to the region in the form of light from a gas laser. This has proved very successful in the cases so far tried.

30.14 FLUORESCENCE ASSAY

In many molecules the absorption of energy results in a change, not only in the electronic energy levels, but also in rotational or vibrational energy states as well. If the molecule dissipates all available vibrational or rotational energy before it drops back to the ground state, it emits radiation of lower energy and longer wavelength than it absorbed. Very often the emitted radiation is in the visible region if the absorbed radiation is in the ultraviolet. The phenomenon is known as fluorescence.

The fluorescence of many plant materials such as quinone has been known for over a century. The presence of trace elements, vitamins and drugs has been checked by fluorometric methods for many years. Because of the sensitivity of fluorescence detection and the fact that almost all molecules are capable of fluorescing, if necessary after suitable chemical treatment, it is surprising that it is only recently that it has been used extensively for identification and location studies. It affords many of the possibilities offered by radioactive techniques without any of the consequent hazards. For *in vivo* work many substances to be used in radioactive assays must be laboriously synthesized. The same job can often be done more simply and with equal accuracy by fluorescence assay. Combined with various separation techniques, it can detect impurities at concentrations of less than 1 in 10^{10}.

If the substance to be investigated is in solution in a container, the container, the solvent and other substances present may fluoresce also. The light used for stimulation of the emission must be made strictly monochromatic, by filters if necessary, and no solvent should be used which fluoresces in the same general region. The presence of fluorescence may be all that is required in routine work. In more elaborate investigations, the fluorescent light may have to be examined by spectrometers (cf. Chapter 21) and the wavelengths of the various components measured before identification is possible.

Routinely, fluorescence can be used to measure protein content in milk, to assay vitamin contents of food after various types of preparation and processing, to follow flow rates and pool formation in sewerage systems and to follow metabolic activity by the excretion rates of substances in urine. Because of its sensitivity it is much in use in biochemistry for such studies as the location of enzymes and co-enzymes in organisms, the following of a drug and its metabolic products (which may be very similar chemically), and determining the amino acid sequence in proteins. It is now being used for the detection of drug abuse. The absorption and excretion of quite small quantities of LSD, for instance, can be detected by fluorescence techniques.

Example 1. Protons are accelerated through a potential difference of 1.0 MV and strike a lead target. What is the closest distance of approach of a proton to a lead atom?

Solution. The kinetic energy acquired by a proton is 1.0 MeV $= 10^6 \times 1.60 \times$

10^{-19} J = 1.6 × 10^{-13} J. The closest distance of approach r will occur when the proton travels towards the lead atom (Z = 82) along their line of centers and all the available kinetic energy is turned into potential energy. When this happens

$$\frac{Q_1Q_2}{4\pi\varepsilon_0 r} = 1.60 \times 10^{-13}\ \text{J}.$$

Therefore

$$r = \frac{Q_1Q_2}{4\pi\varepsilon_0 \times 1.60 \times 10^{-13}\ \text{J}}$$

$$= \frac{9.0 \times 10^9\ \text{N m}^2\ \text{C}^{-2} \times 1.60 \times 10^{-19}\ \text{C} \times 82 \times 1.60 \times 10^{-19}\ \text{C}}{1.60 \times 10^{-13}\ \text{J}}$$

$$= 1.2 \times 10^{-13}\ \text{m}.$$

Example 2. What energy is possessed by one quantum of infra-red radiation of wavelength 1000 nm?

Solution. The energy is given by Planck's formula

$$E = hf.$$

But $f\lambda = c$, the velocity of light. Therefore

$$E = hf = hc/\lambda$$

$$= \frac{6.63 \times 10^{-34}\ \text{J s} \times 3.00 \times 10^8\ \text{m s}^{-1}}{1000 \times 10^{-9}\ \text{m} \times 1.60 \times 10^{-19}\ \text{J eV}^{-1}}$$

$$= 1.24\ \text{eV}.$$

Example 3. The maximum energy of the electrons emitted from a photoelectric surface of work function 2.2 eV illuminated by monochromatic radiation is 0.8 eV. What is the wavelength of the light incident on the surface?

Solution. Einstein's equation is

$$\tfrac{1}{2}mv^2 = hf - \mathscr{W}.$$

Therefore

$$hc/\lambda = \tfrac{1}{2}mv^2 + \mathscr{W} = E + \mathscr{W}$$

$$\therefore\ \lambda = \frac{hc}{E + \mathscr{W}}$$

$$= \frac{6.63 \times 10^{-34}\ \text{J s} \times 3.00 \times 10^8\ \text{m s}^{-1}}{(2.2 + 0.8)\ \text{eV} \times 1.60 \times 10^{-19}\ \text{J eV}^{-1}}$$

$$= 414\ \text{nm}.$$

Example 4. Calculate the speed and the de Broglie wavelength of an electron which has fallen through a potential difference of 100 V.

Solution. We know that the energy acquired is 100 eV and this appears as kinetic energy. Thus

$$\tfrac{1}{2}mv^2 = 100 \text{ eV},$$

therefore

$$v^2 = \frac{2 \times 100 \text{ eV} \times 1.60 \times 10^{-19} \text{ J eV}^{-1}}{9.1 \times 10^{-31} \text{ kg}},$$

$$v = 5.9 \times 10^6 \text{ m s}^{-1}.$$

The de Broglie wavelength is

$$\lambda = h/mv$$

$$= \frac{6.63 \times 10^{-34} \text{ J s}}{9.1 \times 10^{-31} \text{ kg} \times 5.9 \times 10^6 \text{ m s}^{-1}}$$

$$= 0.124 \text{ nm}.$$

Example 5. What is the energy of the quantum of radiation emitted when an electron goes from the ($m = 2$)-level in hydrogen to the ($n = 1$)-level?

Solution. From Eq. (30.11) we have

$$\frac{1}{\lambda} = R\left(\frac{1}{n^2} - \frac{1}{m^2}\right).$$

But the energy of one quantum is

$$E = hf = hc/\lambda = hcR\left(\frac{1}{1^2} - \frac{1}{2^2}\right) = \frac{3}{4}hcR$$

$$= \frac{\frac{3}{4} \times 6.63 \times 10^{-34} \text{ J s} \times 3.00 \times 10^8 \text{ m s}^{-1} \times 1.10 \times 10^7 \text{ m}^{-1}}{1.60 \times 10^{-19} \text{ J eV}^{-1}}$$

$$= 10.02 \text{ eV}.$$

PROBLEMS

30.1 The density of liquid hydrogen is 68 kg m^{-3}. Estimate the maximum possible diameter of a hydrogen atom.

30.2 Doubly charged α-particles of energy 7.33 MeV are emitted by one isotope of thorium. What is the closest distance of approach of such an α-particle to a gold nucleus? The atomic number of gold is 79 and the mass of an α-particle is 6.70×10^{-27} kg.

30.3 A blue lamp rated at 100 W emits light of mean wavelength 4.5×10^{-5} cm. If 12% of the energy appears as light, how many quanta are emitted per second?

30.4 What energy is carried by one quantum of sodium light of wavelength 5893 Å?

30.5 When the wavelength of the incident light exceeds 5.6×10^{-7} m, the emission of photoelectrons from a surface ceases. What is the maximum energy in electron-volts of the electrons emitted when the surface is irradiated with light of wavelength 4.2×10^{-7} m?

30.6 When light of wavelength 4.0×10^{-7} m falls on a sodium surface, the maximum energy of the emitted electrons is 0.8 eV. What is the work function of sodium?

30.7 A photon is scattered through 60° in collision with an electron which acquires an energy of 12.4 keV. What are the initial and final wavelengths of the photon?

30.8 What are the de Broglie wavelengths of (a) a bullet weighing 4 g and having a speed of 500 m s^{-1} and (b) a proton with a speed of 1.25×10^6 m s^{-1}?

30.9 What is the wavelength associated with thermal neutrons which have an energy of 0.02 eV?

30.10 What is the smallest amount of energy that can be absorbed by a hydrogen atom when it is in its normal state?

30.11 Electrons of energy 12.75 eV cause radiation to be emitted from hydrogen atoms. Calculate the quantum number of the orbit to which the electron in the atom has been raised by the incoming electrons.

30.12 What energy would be required to ionize all the atoms in one mole of hydrogen?

30.13 What is the minimum value for the quantum number of the circular orbit in a hydrogen atom which would be of such a size as to be above the limit of resolution of an optical microscope?

CHAPTER 31

X-RAYS

31.1 INTRODUCTION

At about the same time that radioactivity was observed came the discovery of X-rays by Roentgen. He found that when electrons were accelerated to high speeds in a discharge tube and allowed to strike the glass of the envelope or an anode, some kind of radiation or particle, which he called X-rays, and which could affect a photographic plate, was being given off. It has already been mentioned in Section 29.2 that particles and radiation are absorbed in a different manner in passage through matter. A study of the absorption of the X-rays appeared to indicate that they were radiation. Von Laue suggested that crystals had the correct order of magnitude for their molecular spacings in order that diffraction effects might be produced on X-rays impinging on them. Friedrich and Knipping performed the experiment and showed that the resultant diffracted beams occurred in discrete directions only. The wave nature of X-rays was established.

The usefulness of X-ray crystallography has already been described (cf. Section 20.8). The use of X-rays in medicine for diagnostic and therapeutic purposes is well known, and biological research employing X-rays will be described later in the chapter. For a proper understanding of these, a few sections will now be taken up in describing the production and properties of this radiation.

31.2 THE PRODUCTION OF X-RAYS

X-rays are usually produced at the present time by employing a thermionic tube in which the electrons are obtained from a heated filament in a glass envelope evacuated as fully as possible (as shown in Fig. 31.1). The current for the filament is normally obtained from the electrical supply with a step-down transformer, and the electrons produced are accelerated from cathode to anode by a large difference of potential, which might be of the order of 50 kV. This is supplied either from a step-up transformer if the current is low, since the tube provides its own rectification and a pulsed current may be no problem, or by a rectified high-voltage unit (cf. Section 33.3). The electrons are focused toward the anode by a cup round the filament which is kept at negative potential.

Since there is virtually no gas in the tube, the supply of electrons depends solely on the temperature of the filament and, hence, on the filament current. It follows that the current flowing through the X-ray tube depends solely on the filament current, whatever the potential difference across the tube; and the current through the tube controls the amount of X-rays produced. A current of perhaps 20 mA might be considered typical.

Only a small fraction of the energy of the electrons, less than 1%, goes into the production of X-rays, the rest appearing as heat in the target anode, which must therefore be cooled, normally by passing a current of water through it. In addition, the anode may be rotated in order to bring different parts of the target into the electron beam and thus spread the source of the heat. The substance of the anode and the potential difference across the tube control the type of X-ray which is being produced, i.e. control the wavelength and the frequency. For medical purposes the target is often made of tungsten, which has a very high melting point. In research work the target is often demountable, so that it can be changed and therefore different types of X-rays produced.

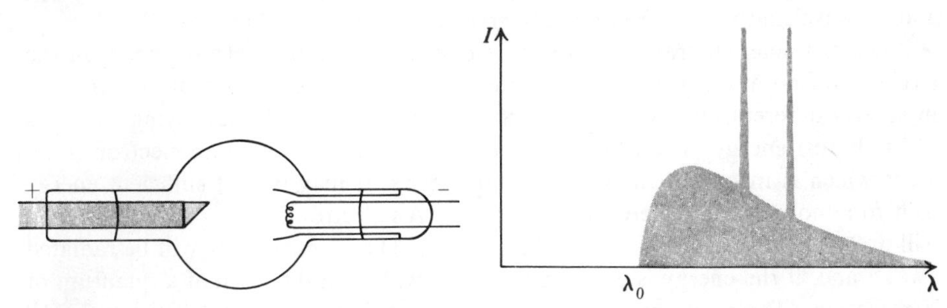

Fig. 31.1. A thermionic X-ray tube.

Fig. 31.2. A typical emission spectrum of an X-ray tube.

31.3 X-RAY SPECTRA

If the spectrum emitted from any X-ray tube is investigated, it looks somewhat similar to the one shown in Fig. 31.2. It starts abruptly at a minimum wavelength λ_0, rises rapidly to a maximum, and then drops off gradually. This describes the general or white X-radiation, and on top of this will be noted sharply peaked maxima. The number of these present depends on the potential difference across the X-ray tube, and the wavelength at which they occur depends on the material of the anode. Since the latter is the case, they are called characteristic X-rays, being characteristic of the substance used as target material.

The heat produced in the target arises from electrons losing their energies gradually in collisions with the target atoms. In addition to this, if the electron is suddenly brought to rest, it radiates the energy remaining in the form of X-rays. Any rapidly accelerating or decelerating particle emits radiation. The German name of this is braking radiation, or Bremsstrahlung, which is very descriptive. In the case of the X-ray tube, the radiation emitted is in the wavelength range of X-rays. As was seen in Section 30.4, the radiation is given off in the form of quanta, each of which has an energy hf. The reason for the sharp cut-off at λ_0 immediately becomes clear. If an electron arrives at the target having passed through a potential difference V, it has energy eV, which is the maximum energy any electron can possess. If this is immedately converted into a quantum of X-radiation, it follows that

$$hf_{\text{max}} = eV, \qquad \text{or} \qquad \frac{hc}{\lambda_0} = eV. \tag{31.1}$$

Therefore

$$\lambda_0 = \frac{hc}{eV}.$$

If, however, the electron loses a certain amount of energy by collision in the target before producing X-radiation, its energy when it radiates is less than the maximum possible energy. Hence, the frequency of the emitted radiation is less than f_{max}, and its wavelength is correspondingly greater than λ_0.

The characteristic radiation must be due to an internal rearrangement in the structure of an atom of the target material. The electrons in the atom can exist in several different energy states (cf. Section 30.8), normally occupying the ones of the lowest energy. If an incoming fast electron collides with an electron in an atom which is in one of the lowest energy levels, it may impart sufficient energy to it to remove it completely from the atom. An electron in a higher energy level will drop back to occupy the now vacant level. The excess energy will be radiated away; and, if the energy is great enough, this will be in the form of a quantum of X-radiation. The radiation emitted will have a definite energy, and thus a fixed f and λ, corresponding to the difference between the fixed energies of the two levels involved in the electronic jump. Since there are obviously several such possible jumps available, there will be several characteristic lines in the X-ray spectrum. The number in any particular case will depend on the energy of the incident electron. The lower the energy state from which an electron is removed the more energy the incident electron requires; and the lower the vacated energy level the greater is the number of electrons of higher energy which will be available to jump to the vacant level.

Moseley investigated the characteristic X-ray spectra of all elements obtainable which were suitable for use as the target in an X-ray tube. He found that particular characteristic lines appeared in all the spectra but at slightly differing wavelengths. Each type of line obeyed a specific equation. For instance, the lines known as the $K\alpha$-lines of X-ray spectra obey the law

$$\frac{1}{\lambda} = \frac{3}{4}R(Z - a)^2, \tag{31.2}$$

where R and a are constants, R being the Rydberg constant, and Z is the atomic number of the target material. The equation may be alternatively written as

$$\frac{1}{\lambda} = R\left(\frac{1}{1^2} - \frac{1}{2^2}\right)(Z - a)^2. \tag{31.3}$$

Other lines appeared to obey equations of the type

$$\frac{1}{\lambda} = R\left(\frac{1}{n^2} - \frac{1}{m^2}\right)(Z - b)^2, \tag{31.4}$$

n and m being integers, and b being a constant.

The similarity of the equations quoted above to those which govern the spectral series emitted by hydrogen (cf. Section 30.8) will not have gone unnoticed. If, in Section 30.8, an atom of nuclear charge Ze and only one electron had been considered in place of hydrogen, the analysis would have been the same everywhere except that Ze would have replaced e in the expression for the centripetal force. The net result would have been that the equation giving the wavelength of a particular emitted line would have changed to

$$\frac{1}{\lambda} = RZ^2\left(\frac{1}{n^2} - \frac{1}{m^2}\right). \tag{31.5}$$

This is indeed the correct form of equation to describe the spectrum of ionized helium or doubly ionized lithium.

The last equation is very similar to those describing X-ray emission. One need only note one fact to explain the final difference. In the X-ray case an electron about to jump back to a vacated level is not only attracted by the nucleus, it is repelled by the electrons in the shells closer to the nucleus than itself. In effect, it is traveling not in the field of a positive charge Ze but in the field of a positive charge $(Z - c)e$, $-ce$ representing the masking or screening effect of the shell electrons. If an electron is ejected from the $(n = 1)$ shell, only one electron is left. An electron in the $(n = 2)$ shell which drops back is under the influence of the nuclear charge Ze and a further charge $-e$ due to the solitary electron in the $(n = 1)$ shell. The $K\alpha$-lines are indeed produced when an electron drops from the $(n = 2)$ shell to the $(n = 1)$ shell. Referring to Eq. (31.2), one might deduce that the value of the constant a would be in the region of 1, and this is in fact correct. The constants a, b, etc., are interpreted as screening constants, and, with this proviso, Bohr's theory explains the nature and form of X-ray spectra.

Moseley, by plotting $\sqrt{f}$ or $1/\sqrt{\lambda}$ against Z for any of the types of X-radiation, obtained a straight-line graph as in Fig. 31.3. At that time, as has already been mentioned, elements were assigned an atomic number according to their atomic mass. Moseley found that he could only get all points to fall on a straight-line graph if the value of Z was reassigned in certain cases. For instance, the atomic numbers of nickel and cobalt, two neighboring elements in the series, had to be reversed in order to obtain values which would obey the correct law. Further, he had to assign all elements which had been given a value of Z of 43 or above a new value one greater in magnitude than had been assigned on the basis of the atomic masses. He therefore predicted that $Z = 43$ had been wrongly assigned and that there must exist an unknown element with this atomic number. This was subsequently discovered and called technicium. He also predicted missing elements at $Z = 61$, 72 and 75.

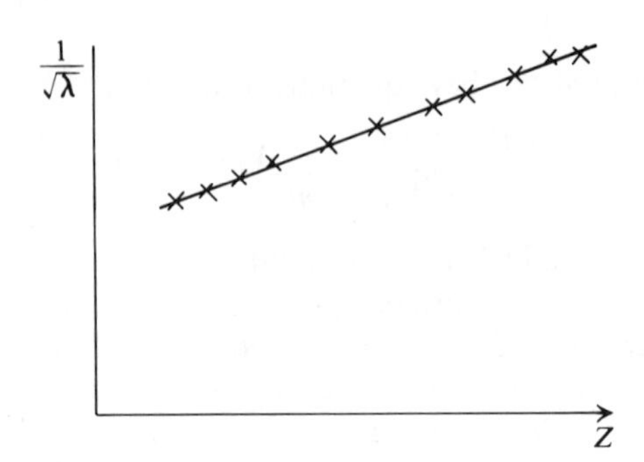

Fig. 31.3. A Moseley plot of the reciprocal of the square root of characteristic X-ray wavelength against atomic number for the $K\alpha$-lines.

The X-ray spectrum emitted by a substance is as characteristic of that substance as is the optical emission spectrum. It follows that an unknown substance, trace impurities in a sample, and many other things, may be identified from the X-ray spectra they emit when bombarded by electrons, or by photons energetic enough to eject electrons from low levels in the atoms.

31.4 THE ABSORPTION OF X-RAYS

The photons of a beam of radiation are removed from that beam by either absorption or scattering, according to a random law. As was seen in Section 21.5, this means that the form of the law is

$$I = I_0 e^{-\mu x}, \tag{31.6}$$

where I_0 and I are the initial and final intensities of a beam of radiation which passes through a thickness x of absorber, and μ is the absorption coefficient. This law is as true for X- and γ-rays as it is for light, although, of course, μ depends on somewhat different factors.

The main processes by which X- and γ-rays are absorbed are the photoelectric effect and the Compton effect (cf. Sections 30.5 and 30.6). In the first of these processes a photon is completely annihilated, part of its energy being used to free an electron from an atom or molecule, the rest being given to the freed electron in the form of kinetic energy. In the second process a photon is scattered by an effectively free electron, continuing with less energy, the difference having been given to the electron in the form of energy of recoil.

The photoelectric effect is greatest for low-energy photons and increases very rapidly with atomic number. The Compton effect, on the other hand, does not vary much with the energy of the photons and is only directly proportional to the atomic number. In biological materials, for photon energies of 0.3 MeV and over, the Compton effect is the predominant effect. It should of course be pointed out that the degraded radiation scattered in this effect tends to be absorbed fairly rapidly owing to the photoelectric effect.

For photon energies in excess of 1 MeV a third effect becomes important—pair production. The photon is annihilated and its energy is used in creating a pair of positive and negative electrons. Part of the energy is used to create the two particles, the rest being distributed between them in the form of kinetic energy. In biological materials, where the atomic number is generally low, the effect is not very marked.

31.5 MEDICAL AND BIOLOGICAL EFFECTS OF RADIATION

Each quantum of X-rays (or γ-rays or high-frequency ultraviolet light) carries energy of 1000 eV or more. If this energy is imparted to a cell, biochemical effects are produced similar to those produced by radioactive particles. All the applications mentioned in Section 29.13 therefore apply to X-rays also.

The old unit of dosage of X-rays, now superseded but still widely used, is the roentgen (R). It was defined in terms of the number of ion-pairs produced in air, but may be stated as that quantity of X-rays which loses 83.4×10^{-7} joules of energy per gram in air. The unit of dosage now agreed upon is the rad. One rad of X-rays is that quantity which produces energy losses of 10^{-5} joules per gram of absorbing material. Unfortunately, other radiations and particles in passing through tissues produce differing biological effects for the same amount of energy absorbed. Dosages are therefore often measured in terms of the rem or the reb. One rem (rad equivalent man) produces an effect in man equivalent to 1 rad of X-rays, and 1 reb (rad equivalent biological) a similar effect in some biological material. These two units depend on the specific effect chosen as the criterion, since different effects vary with energy absorbed in quite distinct ways from one organism to another.

The ratio of the rem to the rad is called the relative biological effectiveness (RBE), the value depending on the criterion used for the rem. Fast neutrons have an RBE of around 10 and 1-MeV α-particles an RBE of around 25.

For therapy X-rays have the advantage over radioactive isotopes in that no implant operation is necessary before their use. A collimated beam of X-rays can be directed to the target volume, machines operating under 250 kV being used for tumors on or near the skin, others operating at around 4 MV for deep-seated tumors. Some of the X-rays pass right through the body, so that all cells along the line of the beam are irradiated. The direction of the beam relative to the body is often altered during exposure or for treatment on different occasions, the beam always of course passing through the volume to be treated. In this way no cells other than those of the tumor receive large dosages. The use of machines operating at high voltage is comparatively recent. The ouput of these machines is high and exposure times are correspondingly short. The higher-energy X-rays produced penetrate soft tissue better and are less absorbed in bone.

The use of X-rays for diagnostic purposes is so common that it hardly needs mention. But it is not well understood that the marked increase of absorption of X-rays with atomic number is a function of low-energy X-rays only. Hence, diagnostic X-ray machines always operate at voltages of less than 150 kV. Since many organs show no greater absorption than the surrounding tissue, it is often necessary to insert into the organ to be studied material which has a much greater absorption. Barium salts are often fed to patients and iodine compounds are injected intravenously for this purpose.

In biology X-rays and other ionizing radiations are used for various research purposes. Apart from the obvious study of the damage caused to cell and organism function by the radiations and of the mutations which may be induced, one interesting line of research may be mentioned in somewhat simplified form. We have already seen that the absorption of one quantum of radiation can destroy a cell completely. This result can be used to determine which parts of an organism are concerned with a specific function.

If a solution of viruses, for example, is irradiated, certain of the viruses will be unable to reproduce. Since it is possible that part of this effect is due to free radicals, as described in Section 29.13, this secondary effect must be suppressed. The addition of substances such as glycerol or cysteine is found to do this, the radicals in general producing effects in the glycerol or cysteine rather than the virus. In such a solution the loss of reproductive function in the viruses is thus due to ionization by the radiation only.

The radiation will be absorbed randomly throughout the solution. The probability of the sensitive region of a virus absorbing a photon of the radiation, and thus being destroyed, depends directly on the volume of the region. If the dosage is such that one photon is absorbed on average by a volume of the solution equal in magnitude to the sensitive region in the virus, then by pure chance some regions will absorb no photons, some one photon, some two photons, and so on. If we were able to count the numbers of regions in each of these categories, the values would follow a Poisson distribution with a mean of unity (cf. Section 2.13). The fraction of regions which would not have absorbed a photon would therefore be $e^{-1} = 0.37$.

The experiment therefore consists of irradiating treated solutions of viruses and determining what dosage is required to produce a situation in which only 37% of the viruses will be able to reproduce. From this dosage one may work out the total number of photons absorbed. Dividing the volume of the solution by this number must give, from the above analysis, the volume of the region in a virus which controls the reproductive function. It is found in experiments of this type that the volume calculated is equal to that occupied by the nucleic acid in the virus. This technique, or a similar one, may obviously be employed with any function of an organism.

31.6 RADIATION HAZARDS AND PROTECTION

It is clear that exposure to ionizing particles or radiation is very harmful to all living organisms and, in particular, to human beings. Many of the earliest pioneers in these fields suffered serious injury to their skins and this drew attention to the problem. It was soon found that other cells of the body could be destroyed and that gene mutations could be produced. High doses of radiation can produce cancer with latent period of up to 20 years, though the most serious risk appears to be leukemia, which normally appears within a few years after irradiation. The gene mutations produced are generally detrimental and, since most of them are recessive, the effect may not be observed until the mutant genes have been distributed throughout a large population.

The population is subjected to radiation from naturally occurring sources, the main agents being cosmic rays, natural radioactive materials in the soil and rocks, and small amounts of radioisotopes, principally ^{40}K, found in the human body. The dosages from these sources are approximately 0.05, 0.05, and 0.025 rem per year. The total natural dosage per year is thus 0.125 and in an average lifetime around 9 rem. The lethal dose of radiation is 400 rem, 50% of people who receive such a dose over a short period dying. The body has, or course, considerable recuperative ability, and 400 rem spread over several years will not be likely to cause death, although health may be seriously impaired. A dose of 200 rem over a short period is likely to lead to leukemia.

The International Commission on Radiological Protection has laid down safety standards for the protection of radiological workers and for the population as a whole. If a person is exposed in his occupation to radiation hazards, he must receive a dosage of no more than 5 rem per year (40 times the natural dosage), and no more than 3 rem in any period of 13 weeks. Anyone working in the vicinity of a radioactive area must receive no more than 1.5 rem per year; and the population as a whole must not receive a dosage of more than 0.5 rem per individual per year.

The increasing use of X-rays in diagnostic and therapeutic medicine, and the testing of atomic weapons with consequent fall-out, represent the greatest hazards to the population as a whole. The dosage per individual from these sources is still small in comparison with that from natural sources. Nonetheless, this is no cause for complacency. All radiation causes biological damage and, in particular, gene mutation. From naturally occurring sources the mutation rate is already very high, resulting at the moment in gross abnormality in around 3% of all births. Any increase in the dosage rate of the population will increase the number of mutations, the most noticeable likely result being an increase in mental diseases.

For the protection of radiation workers all radioactive materials must be shielded in store, normally by being surrounded by a lead container thick enough to absorb all the particles and most of the γ-radiation. In use, radioactive samples must be handled either by remote control or at a safe distance (the dosage falling off at least as rapidly as the inverse square of the distance from the source). All equipment producing radiation must be adequately shielded and the operators must be protected from scattered radiation. All workers are required to wear film badges or pocket ionization chambers which are checked regularly to calculate dosages received, and must adhere strictly to codes of practice which have been laid down by legislation. If it is found that any worker has received more than the permitted dose, he is immediately removed from radiation work for a stipulated period. Since genetic effects are the most serious for the population as a whole, it is recommended that, where appropriate, workers wear lead-rubber aprons to protect the gonads or ovaries.

31.7 RADIATION HAZARDS IN SPACE FLIGHT

Astronauts must pass through the Van Allen belts on their way to outer space. These are regions in which charged particles are trapped in the magnetic field of the earth, and in passing through them the astronaut is exposed to an appreciable dose of radiation. In space he is subjected to further radiation—while no longer shielded by the earth's atmosphere. Nonetheless, under normal circumstances he will receive dosages not much higher than those experienced by radiation workers and only for relatively short periods, with plenty of recuperation time between flights. Lengthy space flights may, however, require to be limited to once only in any astronaut's lifetime.

The real danger arises if a solar flare occurs during a journey to the moon or the other planets. Vast quantities of electromagnetic and particulate radiation are emitted from the sun during these events, and it is impossible to shield the astronaut from them; nor indeed, even if it were a practical maneuvre, could he hope to take evasive action in his space-ship. It is most unfortunate that the maximum of sun-spot activity, which occurs on an 11-year cycle, should coincide with the culmination of the Apollo space program. Any astronaut passing through the output from a solar flare will undoubtedly receive radiation dosages much higher than the permissible level and may require medical treatment at the end of his trip.

Example 1. An X-ray tube operates at a potential difference of 150 kV. What is the short wavelength limit of the X-radiation emitted? What energy does one quantum of this radiation carry?

Solution. The energy of the electron on striking the target is 150 keV. All this energy is turned into one quantum of the shortest wavelength emitted. This therefore also carries energy of $150 \text{ keV} \times 1.60 \times 10^{-19} \text{ J eV}^{-1} = 2.4 \times 10^{-14} \text{ J}$. Further

$$2.4 \times 10^{-14} \text{ J} = hf_{\max} = hc/\lambda_0,$$

therefore

$$\lambda_0 = \frac{6.63 \times 10^{-34} \text{ J s} \times 3.00 \times 10^{8} \text{ m s}^{-1}}{2.4 \times 10^{-14} \text{ J}}$$

$$= 8.29 \times 10^{-12} \text{ m} = 0.0829 \text{ Å}.$$

Example 2. The linear absorption coefficients for the $K\alpha$- and $K\beta$-radiations of silver are 155 cm^{-1} and 661 cm^{-1} when palladium is used as an absorber. What thickness of palladium foil reduces the intensity of the $K\alpha$-radiation to one-tenth of its incident value? What is then the percentage reduction in the intensity of the $K\beta$-radiation?

Solution. If the $K\alpha$-radiation is to be reduced to one-tenth of its incident value, the thickness required is

$$x = \frac{1}{\mu}\ln\left(\frac{I_0}{I}\right) = \frac{2.303}{\mu}\log\left(\frac{I_0}{I}\right)$$

$$= \frac{2.303}{155\ \text{cm}^{-1}}\log(10)$$

$$= 0.0149\ \text{cm}.$$

For the $K\beta$-radiation for this thickness of absorber,

$$\ln(I_0'/I') = \mu x = 661\ \text{cm}^{-1} \times 0.0149\ \text{cm} = 9.82,$$

therefore

$$\log(I_0'/I') = 9.82/2.303 = 4.264,$$

therefore

$$(I_0'/I') = 1.84 \times 10^4,$$

therefore

$$\frac{I_0' - I'}{I_0'} \times 100 = 99.995.$$

The percentage reduction in the intensity of the $K\beta$-radiation is 99.995%.

Example 3. A source emits γ-radiation and even when shielded the dose rate is 0.15 rad h^{-1} at a distance of 1 m. If the maximum permissible dose rate is 6.25 millirad h^{-1}, how close to the shielded source may a scientific worker approach?

Solution. The intensity of the radiation falls off at least as rapidly as the square of the distance from the source because of the normal geometrical factors involved. We wish to find the distance r at which the dose rate will have fallen to the permissible value, and we can therefore say that, at the worst,

$$0.15\ \text{rad h}^{-1} \propto 1/(1\ \text{m})^2$$

$$6.25 \times 10^{-3}\ \text{rad h}^{-1} \propto 1/r^2.$$

Therefore

$$\frac{r^2}{(1\ \text{m})^2} = \frac{0.15}{6.25 \times 10^{-3}} = 24,$$

therefore

$$r = 4.9\ \text{m}.$$

Example 4. Polio virus is inactivated if it absorbs one quantum of radiation. A dose of 125 rad of γ-rays produces a 37% reproduction rate in a solution of polio virus. If 1 rad produces 6.1×10^{11} ionizations per cubic centimeter, estimate the size of the sensitive volume of the virus.

Solution. 125 rad of γ-rays produce $125 \times 6.1 \times 10^{11}$ ionizations per cubic centimeter $= 7.6 \times 10^{13}$ ionizations per cubic centimeter.

A 37% survival rate implies that on average each ionization is being absorbed in a volume equal to the size of the active volume of the virus. This volume is thus

$$V = 1\ \text{cm}^3/7.6 \times 10^{13}$$
$$= 1.32 \times 10^{-20}\ \text{m}^3.$$

PROBLEMS

31.1 An X-ray tube is operating at 150 kV and 20 mA. If 1% of the energy appears in the form of X-rays, at what rate is heat produced in the target? If no attempt were made to cool the target, how quickly would a copper target of mass 15 g and specific heat 0.395 J g $^{-1}$ deg $^{-1}$ begin to melt? The melting point of copper is 1083° C.

31.2 An X-ray tube is operated at 100 kV. What is the short-wavelength limit of the radiation emitted?

31.3 What are the short-wavelength limits of the radiation emitted when an electron is brought to rest on a TV screen after passing through potential differences of 1000 V and 10,000 V?

31.4 An X-ray tube with a copper target is found to be emitting lines other than those due to copper. The $K\alpha$-line of copper has a wavelength of 1.54 Å, and the other two $K\alpha$-lines present have wavelengths of 0.71 Å and 1.66 Å. Identify the impurities.

31.5 The linear absorption coefficients for 0.71-Å X-rays in aluminum, nickel, and lead are 14.3 cm $^{-1}$, 421.9 cm $^{-1}$, and 1593 cm $^{-1}$, respectively. What thicknesses of each absorber are necessary to reduce the intensity of a beam of 0.71-Å X-rays to one-tenth of its incident value?

31.6 The copper $K\alpha$-line and $K\beta$-line have wavelengths of 1.54 Å and 1.39 Å, respectively. The corresponding linear absorption coefficients in nickel are 439 cm $^{-1}$ and 2546 cm $^{-1}$. What thickness of nickel is necessary to reduce the intensity of the $K\beta$-radiation in a beam to 0.1% of its incident value, and what is the percentage transmission of $K\alpha$-radiation through this thickness of nickel?

31.7 Copper $K\alpha$-X radiation of wavelength 1.5405 Å undergoes Compton scattering in a carbon block. What is the wavelength of X-radiation scattered at 90° to the incident direction, and what is the energy of recoil of the electron?

31.8 If 34 eV is necessary to produce an ion-pair in standard air, how many ion-pairs would be produced per gram in air by the complete absorption of 1 R of X-radiation?

31.9 The maximum permissible dosage for scientific workers using X-radiation is 6.25 millirads per hour. What is the safe working distance from a shielded source of cobalt 60 which produces a dose rate of 0.256 rad per hour at a distance of 1 m?

31.10 A spherical virus of radius 150 Å is inactivated by a single-hit process. If 1 rad is equivalent to 6.1×10^{11} ionizations per cubic centimeter, what is the 37% dose for a solution containing the virus?

31.11 An influenza virus is inactivated by a single-hit process. One hundred rads of X-radiation produce a 37% survival rate. Using the data of Problem 31.10, estimate the size of the virus.

CHAPTER 32

THE NUCLEUS

32.1 INTRODUCTION

Bohr theory and its subsequent refinements have permitted an understanding of the way in which the electronic portion of the atom is built up. The nucleus must have structure also, since the phenomenon of radioactivity shows that one type of atom can change into another type of atom merely by emitting a simple particle. It would seem, in addition, that the nucleus can exist in various energy levels, like the atom, since γ-radiation appears when an emitted particle has taken away less than the maximum energy available. It is assumed that the nucleus is left in an excited state, dropping back to the normal energy state and emitting the excess energy in the form of radiation.

One method of attempting to discover of what basic units an atom is made is to break the atom into pieces. Rutherford's scattering experiments mentioned in an earlier chapter were the starting point of many investigations of this type. He discovered, while firing α-particles into nitrogen, that protons, the nuclei of hydrogen atoms, were emerging. He was only able to explain this on the assumption that a nuclear reaction had taken place according to the equation

$$^{14}_{7}\mathrm{N} + {}^{4}_{2}\mathrm{He} = {}^{1}_{1}\mathrm{H} + {}^{17}_{8}\mathrm{O}\,.$$

Note that in such an equation the sums of the atomic numbers and of the atomic mass numbers on each side of the equation must separately be equal.

This discovery by Rutherford, that an atom of one species could be changed into the atom of another species artificially, opened up a whole new field of investigation. The α-particles from radioactive materials were not energetic enough to take such investigations very far, and since that time atom-smashing machines of greater and greater energy have been continually developed. Thousands of types of nuclear disintegrations under fast particle and photon bombardment are now known, and knowledge of the structure of the nucleus has increased in consequence.

32.2 THE DISCOVERY OF THE NEUTRON

In certain of the early bombardment experiments, radiation and not particles appeared to be emitted. This apparent radiation did not seem to be absorbed according to the laws obeyed by γ-rays; and, if directed on to a hydrogen-containing material, knocked the protons out. Chadwick showed that the observed effects could only be explained on the assumption that the radiation was indeed uncharged particles, which he called neutrons. Thus, for instance, beryllium under bombardment by α-particles must undergo the reaction

$${}^{9}_{4}\text{Be} + {}^{4}_{2}\text{He} = {}^{1}_{0}\text{n} + {}^{12}_{6}\text{C}.$$

With the discovery of the neutron it was at last possible to obtain a coherent idea of how nuclei were built up. The neutron has a mass almost the same as the proton but, of course, no charge.

Earlier attempts to explain the formation of nuclei, using as the building bricks involved protons and electrons, had foundered on various scores, principally the ones listed below.

a) It is reasonable to assume that nuclei will only form a stable arrangement if the electrons in them circle, probably in complicated orbits, around the protons. This requires orbits of linear dimensions of around 10^{-15} m, with the electrons traveling round them at very high speeds, approaching that of light. At such high speeds the mass increases markedly owing to relativistic effects, and, with large numbers of electrons involved, it is not clear why isotopic masses do not therefore markedly depart from integral values.

b) Several isobars occur (nuclei with the same A, but different Z) among the stable elements. Emission of electrons from one member of such a group can produce another member with lower energy. It is difficult to see why this does not occur if there are free electrons in the nucleus.

c) The electron, the proton, and some nuclei appear to spin about an axis. If nuclei are built up from protons and electrons, it should be possible to calculate the spin of any nucleus from the spins of the component particles. This attempt often gives the wrong result.

32.3 NUCLEAR BINDING ENERGY

The arrival of the neutron on the scene was therefore greeted with some relief. If nuclei are made up of protons and neutrons, then Z represents the number of protons, $A - Z$ the number of neutrons, and A the total number of particles in the nucleus. The nuclei of different elements are formed from one another by the addition or subtraction of a number of protons and neutrons. The mass of the proton is measured as 1.00813 a.m.u. and the mass of the neutron as 1.00893 a.m.u. The mass of any nucleus is always less than the sum of the masses of the component protons and neutrons, the difference being a measure of the potential energy of the arrangement and thus the stability of the resultant nucleus.

If M is the mass of any nucleus and Σm the sum of the masses of the constituent particles, then $\Delta m = \Sigma m - M$ is called the mass defect, $c^2 \Delta m$ the binding energy, $\Delta m/A$ the packing fraction, and $c^2 \Delta m/A$ the binding energy per particle. The atomic mass number A is the total number of particles in the nucleus and c is the velocity of light. Any mass m represents one form of energy of amount mc^2, according to Einstein's law. The binding energy per particle gives a better idea of the stability of the nucleus than does the total binding energy.

The type of force which holds the nuclear particles together is an unusual one, called an exchange force. It is possible to explain the inverse square law of force between charged particles by assuming that they exchange photons. The net result of doing so is to produce a force between the particles of the expected form. The forces in the nucleus, which act over very short ranges, result from the component particles in the nucleus exchanging not photons but particles with a finite though small mass. These exchange particles were predicted theoretically by Yukawa, and subsequently found experimentally. They are mesons, and there are now several different types of these known.

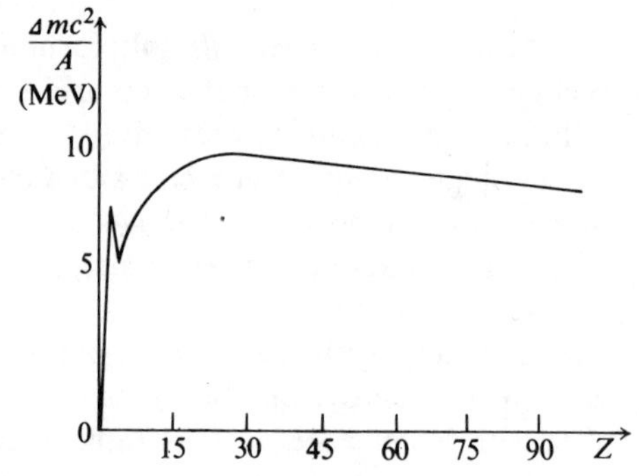

Fig. 32.1. The binding energy per particle of nuclei as a function of atomic number.

Although there are attractive forces holding the nucleus together, the electrostatic repulsion forces between the protons must be attempting to disrupt the nucleus. The attractive force between two protons is greater than the repulsive force between them but, unfortunately, the attractive force is one which saturates. As soon as the proton is exchanging mesons with a small number of particles, it is unable to do so with any others. The electrostatic force, on the other hand, acts between every pair of charged particles in the nucleus. Thus, as more and more particles are added to form heavier nuclei, the disruptive forces increase much more rapidly than the attractive ones. Light atoms contain in their nuclei roughly equal numbers of protons and neutrons, but after about $Z = 10$ the nuclei tend to add more than one neutron for every proton added, in order to increase the stability. It is also, for a given Z, possible to add differing numbers of neutrons and still get stability, which explains why several different stable isotopes of the one element may exist. Above lead, the stability has become so uncertain that all heavy elements are radioactive, having no completely stable isotopes; and it is clear that if one formed elements of around $Z = 120$ or above, they would immediately disrupt spontaneously.

This is fairly clearly seen in Fig. 32.1. The binding energy per particle is greatest for elements with medium Z, which are therefore the most stable elements.

Note that there is a sharp maximum at $Z = 2$, i.e. at helium, the nucleus of this element being particularly stable, which is why it tends to be emitted from radioactive elements. Since electrons are not present as free particles in the nucleus, they must be emitted during some change inside the nucleus. It is believed that in an attempt to get rid of instability a neutron may turn into a proton, emitting an electron in the process.

Just as with the electrons in the outer orbits, the protons and neutrons are occupying energy levels in the nucleus, and in general all the lowest-energy states are filled. Exchange forces act only between particles in the same energy state, which explains the saturation effect. If particles are raised to higher energy levels, the nucleus is said to be in an excited state; and when it drops back to the normal energy state, it emits γ-radiation. If an element has several stable isotopes, these isotopes have atoms with differing numbers of neutrons in the lowest available energy level. If neutrons are added to a higher energy level after the lowest available one is filled, the resultant isotope is normally radioactive, which explains the occurrence of radioactive isotopes. Note that these, being often very much less stable than the naturally occurring radioactive elements, can emit other types of particles besides α's and β's. In particular, positive electrons are often emitted, particles with the electronic mass but a single positive charge.

32.4 NUCLEAR FISSION AND FUSION

The fact that the lighter and the heavier elements are less stable than the intermediate ones leads to the possibility of obtaining energy by, in the one case, combining low mass elements and, in the other, splitting elements of high mass. The latter process is the easier one to achieve. If uranium is struck under the right conditions by a neutron and absorbs it, it can become so unstable that the easiest way to lose the instability is to split in half. In this process two or three neutrons are released to cause further fissions, and energy is made available, roughly of an amount equal to the difference in the binding energies per particle of the uranium and the products formed, multiplied by the number of particles involved. In the fission of uranium about 200 MeV of energy is released each time. Unfortunately, this only happens in $^{238}_{92}U$, the normal isotope, if the neutron energy is high. For slow neutrons the effect is very marked in $^{235}_{92}U$, an isotope only present in natural uranium in the ratio of about 1:160. For many purposes the $^{238}_{92}U$ isotope is therefore to be regarded as an impurity, and in an atomic bomb it is removed by physical means, only the other isotope being used. In an atomic reactor more elaborate methods are used to ensure the capture of most of the neutrons by the $^{235}_{92}U$ isotope and cause fission, although in some types of reactor the $^{238}_{92}U$ isotope is still employed; for when it captures a slow neutron, it changes into a transuranic element $^{238}_{94}Pu$, unknown in nature, which is a good fissionable element.

Fusion, the welding together of light atoms, is much more difficult, although weight for weight it produces more energy than fission. Here the difficulty is that to make, say, four atoms of hydrogen fuse together to form an atom of helium, it is necessary to force them together against their mutual electrostatic repulsion. This can only be done if the kinetic energies of the hydrogen ions are greater than their mutual potential energy. This is normally only true at very high temperatures. At the temperature, of the order of 10^7K, which exists in the sun or other stars, this process goes on, and it is from this process, and from others of a similar nature, that the energy radiated from the stars comes. On the earth this sort of temperature is only achieved under exceptional circumstances, such as during the explosion of a fission bomb, when if hydrogen is included inside the bomb case, it turns into helium, releasing large amounts of energy. This is the basic idea behind the hydrogen bomb.

The controlled use of fusion energy is very much under investigation at the moment. Passing very large discharge currents, of the order of 10^6 A, through hydrogen or heavy hydrogen gas causes the kinetic temperature of the molecules to achieve the sort of value at which fusion can take place. The control and extraction of the energy released in such a process is, however, very difficult, and the commercial use of fusion energy is still very much in the future.

Example 1. When $^{27}_{13}\text{Al}$ is bombarded by deuterons, α-particles are emitted. Identify the atom remaining. How many protons and neutrons does its nucleus contain?

Solution. A deuteron is the nucleus of the isotope of hydrogen of mass number 2. Thus the equation of the reaction taking place is

$$^{27}_{13}\text{Al} + {}^{2}_{1}\text{H} = {}^{4}_{2}\text{He} + {}^{A}_{Z}\text{X},$$

where X stands for the unknown atom. It follows that

$$A + 4 = 29$$

$$Z + 2 = 14,$$

therefore

$$A = 25, \qquad Z = 12.$$

The atom is therefore $^{25}_{12}\text{Mg}$.

The number of protons is given by Z and is therefore 12. The total number of nucleons is A and is 25. The number of neutrons is thus 13.

Example 2. Determine the mass defect and binding energy per particle of $^{40}_{18}\text{Ar}$. The masses of the proton, the neutron and the $^{40}_{18}\text{Ar}$ nucleus are 1.00813 a.m.u., 1.00893 a.m.u., and 39.97505 a.m.u. respectively. 1 a.m.u. is 1.6598×10^{-27} kg.

Solution. The total masses of the constituents of the $^{40}_{18}\text{Ar}$ nucleus are

$$18 \times 1.00813 \text{ a.m.u.} = 18.14634 \text{ a.m.u.}$$

$$22 \times 1.00893 \text{ a.m.u.} = \underline{22.19646 \text{ a.m.u.}}$$

$$40.34280 \text{ a.m.u.}$$

from which we subtract

$$\frac{39.97505 \text{ a.m.u.}}{0.36775 \text{ a.m.u.}}$$

The mass defect is therefore 0.36775 a.m.u. × 1.6598 × 10^{-27} kg (a.m.u.)$^{-1}$

$$= 6.1039 \times 10^{-28} \text{ kg}.$$

The binding energy per particle is $c^2 \, \Delta m/A$

$$= \frac{(2.9979 \times 10^8 \text{ m s}^{-1})^2 \times 6.1039 \times 10^{-28} \text{ kg}}{40 \times 1.6021 \times 10^{-13} \text{ J (MeV)}^{-1}}$$

$$= 8.56 \text{ MeV}.$$

Example 3. The explosive TNT releases energy of 3.8 × 10^9 J per ton. How much ^{235}U is used up in the one-megaton explosion of an atomic bomb?

Solution. A one-megaton explosion means that the energy released is equivalent to that from one million tons of TNT. That energy is 3.8 × 10^{15} J. A quantity x of ^{235}U contains $x/(235 \text{ g})$ moles and thus $xN_0/(235 \text{ g})$ atoms, where N_0 is Avogadro's constant. Each fission releases 180 MeV of energy. Thus

$$\frac{xN_0}{235 \text{ g}} \times 180 \text{ MeV} = 3.8 \times 10^{15} \text{ J},$$

therefore

$$x = \frac{3.8 \times 10^{15} \text{ J} \times 235 \times 10^{-3} \text{ kg}}{6.02 \times 10^{23} \times 180 \text{ MeV} \times 1.60 \times 10^{-13} \text{ J (MeV)}^{-1}}$$

$$= 51.6 \text{ kg}.$$

PROBLEMS

32.1 If a $^9_4$Be atom is struck by an α-particle and a neutron is emitted, what is the atom which remains?

32.2 Complete the equation $^{23}_{11}$Na + $^4_2$He → $^1_1$H +

32.3 How many neutrons and protons are there in

$$^{14}_{7}\text{N},\ ^{39}_{19}\text{K},\ ^{96}_{42}\text{Mo},\ ^{197}_{79}\text{Au, and } ^{238}_{92}\text{U?}$$

32.4 If the masses of the proton, the neutron and a $^7_3$Li nucleus are 1.00813 a.m.u., 1.00893 a.m.u., and 7.01600 a.m.u., respectively, what is the binding energy per particle of $^7_3$Li? One atomic mass unit is equivalent to 932 MeV.

32.5 When a proton strikes a $^7_3$Li nucleus, two α-particles are formed. Calculate the amount of energy released in the process using data from Problem 32.4 and given that the mass of the α-particle is 4.00386 a.m.u.

32.6 If the masses of $^3_1$H and $^3_2$He are 3.01700 a.m.u. and 3.01695 a.m.u., respectively, calculate their binding energies. The difference in these values may be regarded as being due to the repulsion energy between the protons in the helium nucleus. Estimate the separation of these protons.

32.7 The energy released in the fission of one ^{235}U atom is 180 MeV. Express this in J mol $^{-1}$.

32.8 Calculate the energy release in J mol $^{-1}$ of deuterium when deuterium atoms fuse to form helium. The masses of a deuterium atom and of a helium atom are 2.01472 a.m.u. and 4.00386 a.m.u., respectively.

32.9 How much electrical energy is obtained from a nuclear power station when 5 g of ^{239}Pu are used up? The conversion efficiency is 18% and the fission of each ^{239}Pu atom releases 180 MeV of energy.

32.10 A space rocket of mass 10^5 kg and powered by atomic motors can reach the escape velocity from the earth of 11.3×10^3 ms $^{-1}$ in 3 min under constant acceleration. If air resistance and the change in gravitational potential energy during acceleration are neglected, calculate how much ^{235}U is used up by the rocket in escaping from the earth, if the motor has an efficiency of 10% and the fission of one ^{235}U atom releases 180 MeV of energy.

32.11 The solar constant measures the rate at which energy arrives at the earth's surface from the sun and has a value of 1.35 kW m $^{-2}$. The distance of the earth from the sun is 1.50×10^8 km. How much hydrogen is being converted to helium in the sun per second if the masses of hydrogen and helium atoms are 1.00813 a.m.u. and 4.00386 a.m.u., respectively?

CHAPTER 33

ELECTRONICS

33.1 INTRODUCTION

The term electronics is used to cover phenomena associated with the movement of electrons in, and their emission from, substances, with the control of their behavior, often in electric and magnetic fields, and, in short, with any physical phenomena in which electrons are involved. However, in popular parlance, and in the context of this chapter, we are chiefly concerned with those aspects of the subject which are pertinent to an understanding of practical devices. Daily, electronics plays an ever more important role in everyday life, as more and more electronic gadgets invade the home and the office. No form of science or technology can afford to be without its backing of electronic devices, and measuring apparatus of all types has become increasingly electronic.

Biology and medicine are no exceptions. Electronic apparatus is being employed routinely in the laboratory and in the hospital. Some background is thus obviously necessary for aspiring biologists and physicians. At the heart of electronic instruments are various semiconductor devices but, before these can be understood, it is necessary to look more closely at the structure of solids, in particular the differences between conductors, semiconductors, and insulators, and this will be the subject matter of the next section.

33.2 THE ELECTRONIC STRUCTURE OF SOLIDS

As we have seen, the atoms in solids are situated very close together, and although this has only a slight affect on the tightly bound core electrons, it can alter the behavior of the valence electrons completely. In a simple solid the outermost electronic orbits no longer belong to individual atoms but are shared by the lattice as a whole and play a part in holding the atoms together in the crystal. Because of this, what were discrete energy levels in the isolated atom become broadened out into bands of several permitted energy levels very close together, as shown in Fig. 33.1, the broadening being greatest for the most loosely bound levels. All levels in any band can, in general, be occupied by electrons, but between the bands there are gaps containing no energy levels. Electrons can get from a level in one band to a level in a higher one only by acquiring sufficient energy to jump the gap.

This is the situation in an insulator. All the levels of the valence band B_1 are filled with valence electrons and all the levels of B_2 are empty. The electrons in the material are therefore firmly held at a particular site and take place in local bonding. The only way in which they could migrate throughout the whole solid would be if they acquired sufficient energy to transfer to the higher band B_2, called the conduction band. The energy gap between B_1 and B_2 is, however, too large for this to happen.

In a conductor there is no gap between B_1 and B_2, they may indeed overlap, and not all the levels of B_1 are filled. The next unoccupied level is only very slightly above the last occupied one and an electron needs only a very small amount of energy, which may be supplied by the thermal energy available in the lattice, to lift it to a free level and allow it to migrate throughout the solid.

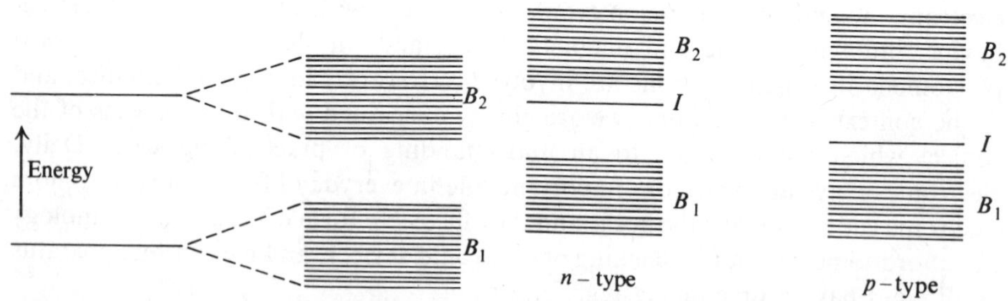

Fig. 33.1. The discrete energy levels in an atom become bands of levels in a solid.

Fig. 33.2. The positioning of the impurity levels in *n*-type and *p*-type semiconductors.

In an intrinsic semiconductor the gap between B_1 and B_2 is small, so that electrons may cross the gap if sufficient thermal energy is acquired. Such semiconductors are insulators at low temperatures but become increasingly conducting as the temperature is raised (cf. Section 23.8). But the most important type of semiconductor is the impurity semiconductor, which is produced by doping an insulator with atoms of another element.

In an impurity semiconductor the atoms of the impurity provide additional energy levels between B_1 and B_2 (Fig. 33.2). In an *n*-type semiconductor, the impurity level I lies just below the empty B_2-band and electrons in I can be raised to levels in B_2 by thermal energy. These electrons then take part in conduction, and such a semiconductor is called *n*-type because negative charge carriers are involved. The impurity levels are called donor levels, since they supply electrons to the conduction band.

In a *p*-type semiconductor the impurity levels lie just above B_1 and are acceptor levels. An electron in B_1 can acquire sufficient energy to move to an impurity level. A hole is left, as has already been mentioned in connection with the Hall effect. The empty hole behaves like a positive charge when an emf is applied and the conduction process is therefore achieved by the movement of the positive charge carriers (hence *p*-type).

Germanium doped with arsenic (to around 1 part in 10^6) is an example of an n-type semiconductor; if it is doped with gallium in the same proportion, a p-type semiconductor is produced. Note that there will always be a few holes in an n-type semiconductor, and conduction electrons in a p-type semiconductor, but conduction in either case is predominantly by the mechanism implied by the name.

Conduction electrons are fairly loosely bound to the lattice and under certain conditions can be given sufficient energy to detach them from the lattice altogether. The simplest way of doing this is by passing a current through the substance, often in the form of a filament, or otherwise conveniently heating it to a high temperature. The electrons acquire sufficient kinetic energy to overcome their potential energy in the lattice and they are emitted from the filament. Such a process is known as thermionic emission.

33.3 THE SEMICONDUCTOR RECTIFIER

When two semiconductors, one n-type and one p-type, are joined together, the resulting device has the properties of a rectifier. On one side of the junction is a substance in which conduction takes place by the movement of electrons, and on the other side a substance in which the conduction results from a movement of holes. Both portions of the device in isolation are electrically neutral.

When the junction is formed as in Fig. 33.3, a diffusion of electrons to the right takes place across the junction and of holes to the left. This continues until A has become positively charged and B negatively charged to such an extent as to set up a potential difference across the junction sufficient to prevent further diffusion. The electrons and holes which have diffused across the junction find carriers of the other sort awaiting them and tend to be neutralized by combination. A small region on either side of the junction has thus been denuded of charge carriers and is called the

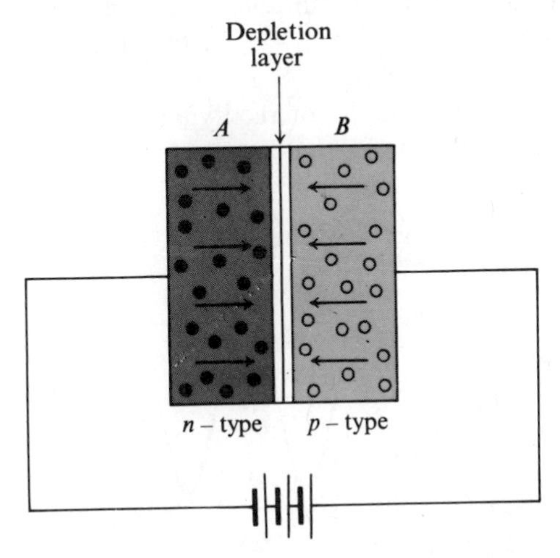

Fig. 33.3. The semiconductor rectifier.

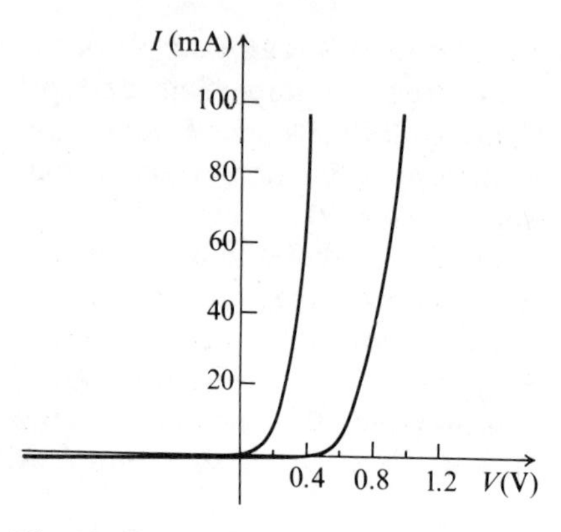

Fig. 33.4. Typical characteristic curves of a semiconductor rectifier.

Fig. 33.5. Half-wave rectification.

depletion layer. It is over this region that the established potential difference acts.

If an external potential difference is set up as shown in Fig. 33.3, this opposes the potential difference across the junction. When the applied voltage is great enough (of the order of 0.5 V for silicon; rather smaller for germanium), electrons will once more move to the right across the junction and holes to the left. A current passes through the junction and around the circuit and keeps on flowing as new electrons and holes are formed. The current increases rapidly with an increase in the applied voltage. If the external potential difference is reversed, it reinforces the barrier across the junction and no current flows. In fact this explanation is over-simplified; there is always a slight leakage current in the reverse direction as shown in Fig. 33.4. But it is clear that the device acts as a rectifier. Current can pass easily in one direction through the junction but not appreciably in the other. This device is therefore suitable for converting a.c. to d.c. A rectifier is denoted in diagrams by the symbol —|◁—, the arrowhead indicating the easy direction of flow of positive current.

Fig. 33.5 shows a circuit suitable for rectifying the a.c. output of a generator. The emf is applied across the rectifier and a resistor in series and the voltage drop

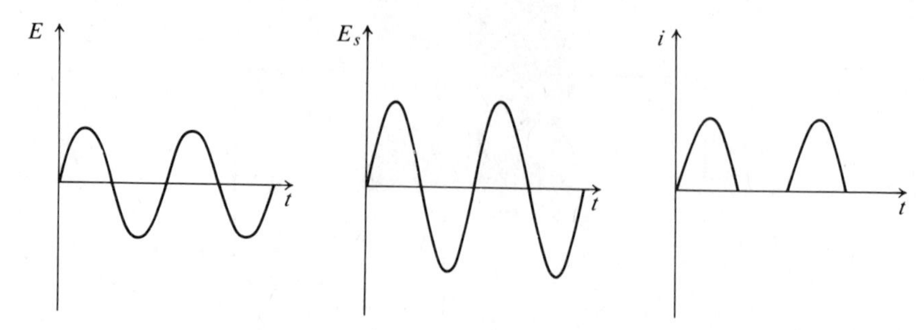

Fig. 33.6. The input and output curves of a half-wave rectifier.

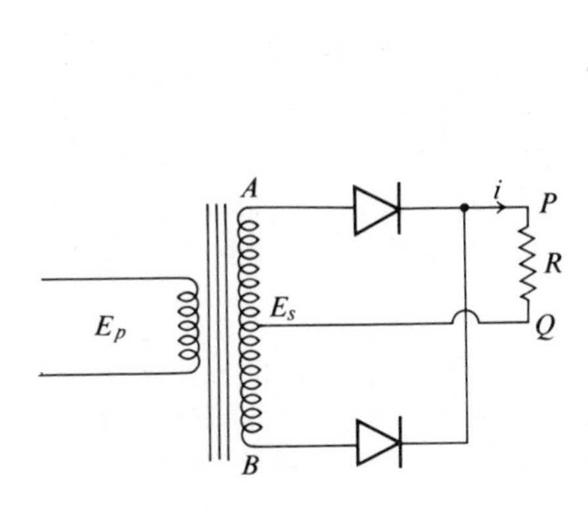

Fig. 33.7. Full-wave rectification.

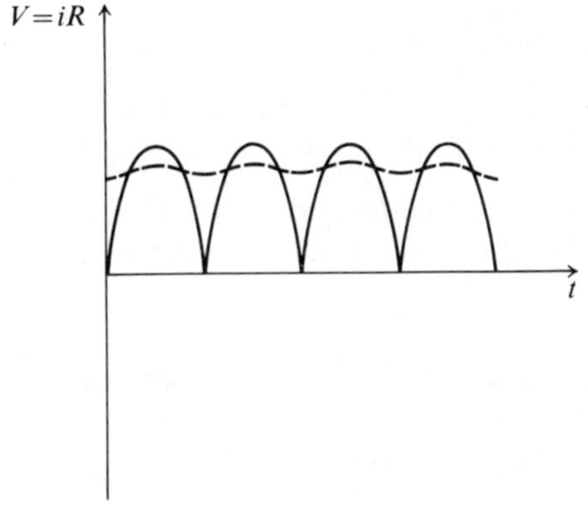

Fig. 33.8. The output voltage from full-wave rectification.

across the resistor acts as the source of rectified voltage. Only the positive portions of the input signal can produce an appreciable current in the circuit, and the output current flowing through the resistor produces an almost unidirectional voltage. This circuit is called a half-wave rectifier and has the disadvantage that the output voltage is a varying one and only has an appreciable value for half the time.

If two rectifiers are employed, better results are obtained in a full-wave rectifier as in Fig. 33.7. When A is positive with respect to B, the upper rectifier is forward biased and the lower rectifier reverse biased. An appreciable current passes only through the upper rectifier and the output resistor, the current going from P to Q. When A is negative with respect to B, the lower rectifier is forward biased and the upper reverse biased. The current now passes through the lower rectifier and resistor but still goes from P to Q. The form of the output voltage across the resistor is shown by the full line in Fig. 33.8.

This is still not a constant d.c. voltage. However, if at PQ a smoothing circuit is inserted, as in Fig. 33.9, the output voltage has the form of the dashed rather than the full line in Fig. 33.8. The capacitors have reactances which are small compared with the values of the high resistance r and load R. The capacitors charge quickly up to the peak value of the voltage and discharge only slowly. The voltage has therefore hardly dropped appreciably before the capacitors charge up again. In this way the voltage is kept at an almost constant value.

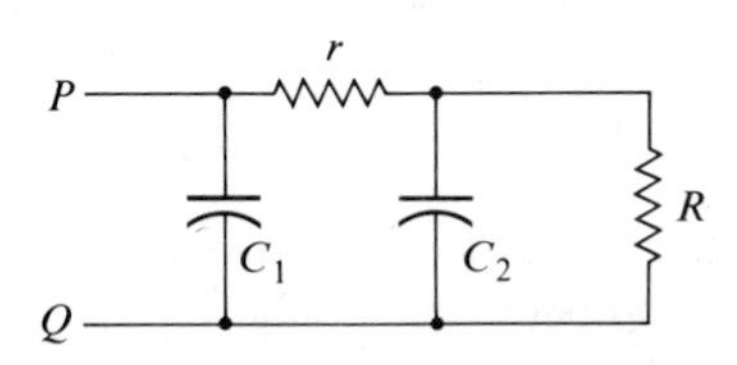

Fig. 33.9. A smoothing circuit for use with a full-wave rectifier.

Example. A semiconductor rectifier has the characteristics given in the table:

V(V)	0	0.40	0.50	0.525	0.55	0.575	0.60	0.625	0.65	0.675	0.70
I(mA)	0	0	0.4	1.0	2.0	3.5	5.4	7.9	11.2	15.0	20.0

When it is connected in series with a resistor, an ammeter, and a 6-V d.c. supply, 5.4 mA is recorded. If a current of 20 mA is required, what should be the supply voltage?

Solution. When 5.4 mA passes through the rectifier, the potential difference across the junction according to the table is 0.6 V. It follows that a potential difference of 5.4 V is dropped across the resistor. The resistance of the resistor is thus

$$R = 5.4 \text{ V}/5.4 \text{ mA} = 1 \text{ k}\Omega.$$

If 20 mA is required, the potential difference across the rectifier according to the table is 0.7 V. Further the drop across the resistor will be

$$V = 1 \text{ k}\Omega \times 20 \text{ mA}$$
$$= 20 \text{ V}.$$

Hence the supply voltage must be 0.7 V + 20 V = 20.7 V.

33.4 THE TRANSISTOR

The transistor is a semiconductor device for the amplification of small signals. It takes many forms, but one of the simplest consists of a sandwich of a *p*-type semiconductor between two layers of *n*-type semiconductor, or alternatively a *p-n-p* sandwich. Figure 33.10 shows in a simplified manner what is happening. The emitter is heavily doped, the base lightly doped. The emitter-base portion is biased as in the rectifier with a fraction of a volt potential difference. Because of the heavy doping, the emitter is rich in conduction electrons and these flow into the base. The base is poorly supplied with holes and is only 10^{-4} to 10^{-3} cm in thickness. Most of the electrons flowing in thus do not combine with holes but pass on to the collector under the attraction of the potential difference of perhaps 10 V set up from base to collector. The current in the main emitter-collector circuit is therefore very large.

There is also a small current flowing from emitter to base in the external circuit. Perhaps two per cent of the electrons from the emitter combine with holes in the base material. To keep the base in electrical equilibrium electrons must leave the base and return to the emitter through the cell. The current I_b is of the order of 50 times less than I_2.

If V_2 is disconnected, the current I_b still flows. If V_1 is disconnected, the current I_2 ceases. If both V_1 and V_2 are connected, altering the current I_b produces corresponding but much larger variations in I_2. A large current is thus being controlled by a small current, and this is what makes the transistor useful as an amplifying

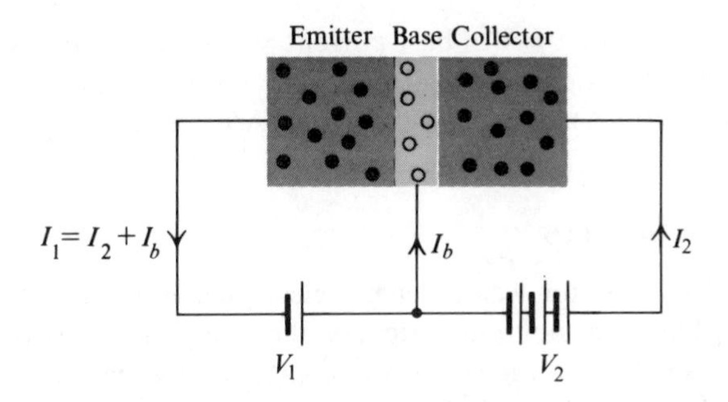

Fig. 33.10. An *n-p-n* transistor with correct bias voltages.

device. Because it appears to be acting as an instrument in which resistance is transferred from one circuit to another the device is known as a *trans*fer res*istor* or *transistor*. In diagrams a transistor is shown by the symbol

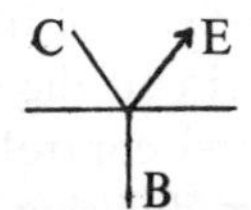

Figure 33.11 shows the circuit diagram of a typical amplifier. A small alternating signal is fed into the base-emitter circuit and varies the current there. Much larger variations in current occur in resistance R and the potential difference across it varies by a large amount. The small signal has thus been amplified by the action of the transistor.

A transistor has the advantage over the tubes that were used until recently that it requires far less power, is much smaller and lighter, is more robust and has a far longer useful life. It does suffer from being somewhat sensitive to temperature changes, and can not handle very large voltages. These problems can be solved by proper circuit design. In all new equipment vacuum tubes have been replaced by semiconductor devices, except for special purposes.

Example. If the value of R in Fig. 33.11 is 5 kΩ and the base-emitter current is 2% of the current in the collector circuit, determine the voltage variation across R for a 1 μA variation in base current.

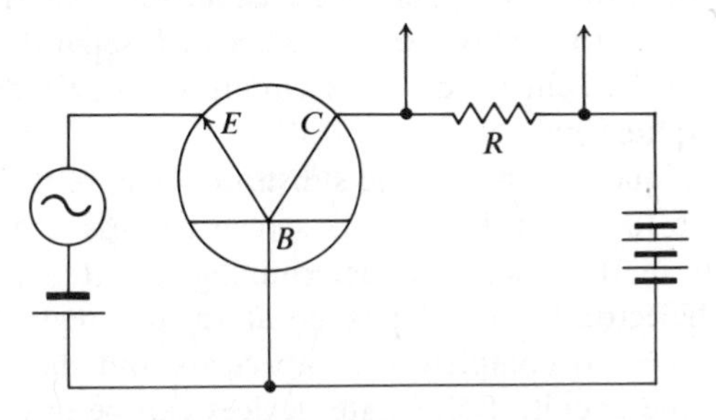

Fig. 33.11. A typical amplifier circuit.

Solution. A small change of 1 μA in base current produces fifty times the change, i.e. a variation of 50 μA, in collector current. The voltage variation produced is between IR and $(I + 50\ \mu\text{A})R$. It follows that the variation is $50\ \mu\text{A} \times 5\ \text{k}\Omega =$ 0.25 V.

33.5 INTEGRATED CIRCUITS

Until 1948, when the transistor was invented, electronic equipment tended to be bulky and heavy. This was not only because the components used were large but also because power for the tubes had to be supplied by massive transformers and rectifying circuits. The advent of solid-state devices, allied to the ability at that time to deposit layers of metal on glass or ceramic, opened up possibilities of drastically reducing the size and weight of electronic apparatus. Integrated circuits were used where all the components for a particular circuit were inseparably joined on a single small module. Over the intervening years many methods of producing integrated circuits have been investigated and employed. We shall discuss the two main methods of doing so in the rest of this section.

In a semiconductor integrated circuit the substrate is made of semiconductor material. The surface is masked, and controlled diffusion of different impurities through the various holes produces differing characteristics at different regions. Thus at one region a transistor may be formed by producing successive layers of *p*-, *n*-, and *p*-material; at another region a track of known conducting properties may be formed to produce a resistor of carefully calculated resistance; and a capacitor may be formed across a *p-n* junction. Connections are finally made by depositing thin layers of aluminum between the components. A circuit formed in this way may be only a fraction of a square millimeter in size.

In integrated circuits which employ films deposited on a substrate the latter is of glass or some other insulating material. In thin-film deposition a layer of nichrome is evaporated on to the surface and a layer of gold is deposited on top, the total thickness being less than 10^{-6} m. The nichrome layer has a resistance of around 3000 times that of the gold. Selected areas are etched away to form suitable conducting and resistive tracks on the base. At other points silicon oxide can be deposited after the etching process and gold can be deposited on top. One conducting path will lead to the base of the oxide and another will be connected to the top. What has been formed is therefore a capacitor. Alternatively, and more economically, the capacitors may be manufactured separately, and they and solid-state devices can be connected to the circuit and covered with resin. Once again, the circuit is quite tiny.

A different technique uses a ceramic substrate. The resistive and conductive tracks are printed on to the base by a silk-screen process, the quantity of metal and other materials in the inks used determining the resistive properties of a particular track. Dielectric is then deposited at various regions through a mask and printing is repeated to complete the capacitors and the connections of the latter to the rest of the circuit. Solid-state devices can be inserted appropriately, as before.

Circuits of these types are always employed nowadays in implants such as for artificial pacemakers. They require less space and are much more reliable. Theater equipment has been drastically reduced in size and weight, and many instruments using transducers with an electrical output can now be made quite tiny. In patient-monitoring the leads to bulky equipment have always presented a problem, since they are liable to come loose owing to patient-movement and detract from comfort. With the miniature circuits now possible, the whole instrument can be strapped to the patient, the output being transmitted to a nearby receiver which gives an audible or visual indication of the patient's condition. This is of particular importance when patients are being transported by stretcher or trolley and immediate emergency treatment may have to be given. Microminiature integrated circuits are also incorporated into monitoring pills fed to patients. The conditions in the tracts of the patient are continuously measured and the information is transmitted to the external receiver. This is of great assistance in the diagnosis of many conditions where information is difficult to obtain by other means.

33.6 THE CATHODE-RAY OSCILLOSCOPE

It is very useful to have a method of visually displaying wave-forms which may be quite complicated. The CRO is an apparatus which does this. It consists of an evacuated glass tube of the shape shown in Fig. 33.12, the larger end being coated with a material which fluoresces when struck by electrons. The electrons are produced in an electron gun, which consists essentially of an emitting filament, a collimating arrangement which confines the electrons to a fine beam of negligible lateral dimensions, and an electric field which accelerates them to a high velocity.

The electron beam, if there were nothing else in the tube, would strike the fluorescent screen and produce a point of light. In fact it passes between two vertical plates X and X' which, in normal operation, are connected to a capacitor which is being charged in the manner prescribed in Section 23.14. The voltage between the plates thus varies, as in Fig. 23.19. The electron beam is deflected

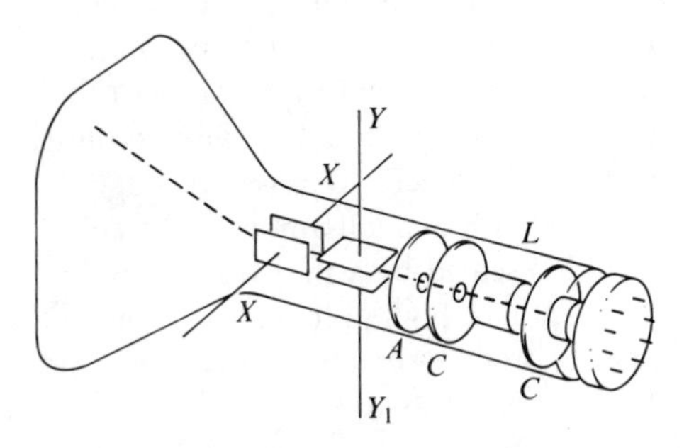

Fig. 33.12. The cathode-ray tube.

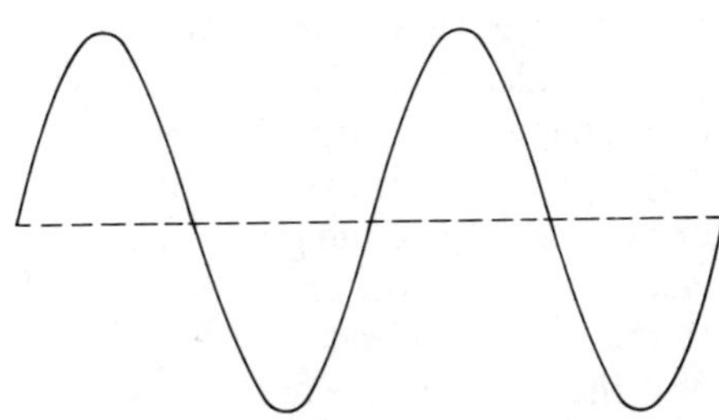

Fig. 33.13. A simple sine-wave signal as seen on a CRO with synchronized timebase.

more and more with time up to a certain value, after which it returns almost instantaneously to the undeflected position and begins the deflection process again. The point of light therefore moves uniformly across the tube, flicks back to the starting position, and begins to move uniformly again. Since the fluorescence lasts for a finite time, the trace on the screen is a line. What has been produced is a time base, since the voltage between the X–X' plates, and thus the deflection, varies linearly with time.

If a varying signal is also applied to a pair of horizontal plates Y and Y', the spot is deflected vertically as well as horizontally. The line traced by the spot of light thus shows the variation of the wave-form against time. If this is sinusoidal, the trace on the CRO screen will be as in Fig. 33.13.

There are of course various controls on the CRO. Some of these are concerned with centering the trace but there are others which alter the time it takes for the trace to move across the screen. Various capacitors may be switched in to vary the time scale coarsely, and the resistance in series with the capacitor can be varied to control any particular time range finely. There are also devices which lock the incoming signal and time base so that a stable pattern appears on the screen, or which start the trace only if a signal arrives. This latter control is useful for studying pulses of irregular repetition rate. More refined CROs have two traces so that two signals may be compared. Other devices will be found on more sophisticated instruments.

The cathode-ray tube is also, of course, employed in TV sets. Here the deflection is often done by magnetic rather than electric fields. The trace runs in a number of lines one below the other, the intensity of the spot at any place being adjusted to produce a brighter or darker spot of light. The combination of these intensities gives the picture of the event which is being televised.

33.7 THE GEIGER-MÜLLER COUNTER

As a bridge between electronics and modern physics, let us discuss counters. Since we have dealt with the measurement and counting of atomic particles, it is important to show how the presence of particles and radiation can be detected, and how the number present in any place can be counted. There are many counters now in use but one of the simplest and most widely used is the Geiger-Müller tube.

Consider, as in Fig. 33.14, a cylindrical metal tube with a fine wire running down the middle of it, but insulated from the tube where it passes through it. The tube is filled with, say, argon at low pressure with a small amount of alcohol or polyatomic organic gas mixed in with it, and is sealed by a thin window at the end. The wire is the anode of the tube, being kept at a high potential with respect to the case or cathode. The potential is high enough to create a very high field near the anode, where the lines of force are very concentrated.

Atomic particles and radiation are allowed to enter the window, which is thin enough for this to happen, causing ionization in the gas, i.e. they knock electrons from the atom producing free electrons, or negative ions, and atoms with a net

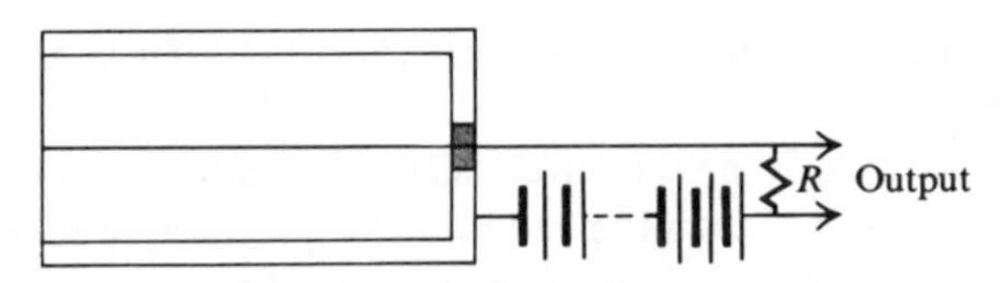

Fig. 33.14. The Geiger-Müller tube.

positive charge, or positive ions. These ion-pairs are attracted in opposite directions in the electric field. The positive ions, being heavier, move slowly toward the cathode, and the light, negatively charged electrons are accelerated more and more rapidly toward the anode. The electrons indeed acquire sufficient energy if the field is high enough to cause ionization themselves. The further electrons produced are accelerated toward the anode, cause further ionization, and so on. In addition, radiation produced in some of the collisions spreads in all directions throughout the tube, causing further electrons to be produced at various points. An avalanche of electrons, triggered by the initial ion-pair or ion-pairs, arrives along the whole length of the tube at the anode, producing an appreciable pulse of current through the high resistance. The change in potential across the resistor can be carried to further amplifying or recording apparatus. The height of the pulse is independent of the number of ion-pairs initially produced in the tube, i.e. is the same for all types and energies of incoming particles. It is thus impossible with this type of counter to tell from the output what type of particle has entered. This can be done with proportional or scintillation counters, in which the pulse height is dependent on the type and energy of the incoming particle. But the Geiger tube has the advantage that it can detect the presence of even weakly ionizing particles which would not produce an appreciable pulse in the other types of counter mentioned. It is therefore one of the most useful, and one of the most widely used, instruments for the detection of atomic particles and radiation.

During the period when the avalanche of electrons is being produced and for a short time thereafter, the current passing through the resistor produces an appreciable potential drop across it, lowering the potential difference across the tube. During that time the tube will not produce the same effect from ions produced inside it. The tube for that period is effectively dead and will not count incoming particles. This lowers the counting rate that can be achieved with this tube, and also requires that a correction be made to the number recorded during any given period to account for the estimated number of particles which would have entered during the dead time.

33.8 THE SCINTILLATION COUNTER AND PHOTOMULTIPLIER TUBE

The geiger counter is most useful for detecting high-energy β-particles. The scintillation counter is more efficient in detecting γ-radiation. In addition, as was mentioned in the last section, its output-pulse height is dependent on the properties of the incoming particle, so that discrimination between several types of simultaneous emission is possible.

The initial portion of the counter consists of a material known as a scintillator or phosphor in which the incoming radiation or particles produce trails of ionization in a time of the order of one nanosecond. In a solid, inorganic phosphor with an impurity, such as thallium-activated sodium iodide, the electrons produced leave holes in various electron levels. Other electrons dropping into these holes come from higher levels and emit energy in the process in the form of photons of light. These are emitted within a time of 1 μs after the initial production of the ionization. In organic scintillators the process taking place is more complicated but similar. The phosphor suffers from no dead time.

The phosphor must be transparent to the photons produced and the latter are generally detected by allowing them to pass through a light-pipe of Perspex on to a photoelectric surface in the second part of the counter, which consists of a photomultiplier tube. If the phosphor is sodium iodide, the most popular and useful solid scintillator, it and the light-pipe must be sealed in an aluminum can, since the phosphor is hygroscopic. The can must have reflecting or diffusing inside walls so that no photons are lost by absorption.

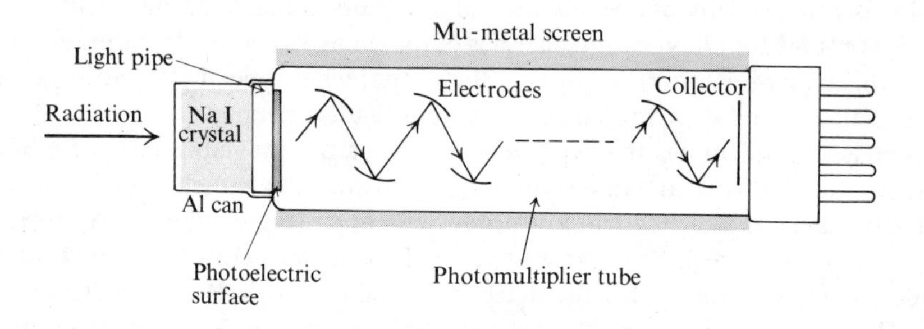

Fig. 33.15. The scintillation counter.

In the photomultiplier tube, electrons are emitted from the photoelectric cathode and accelerated by a potential difference to the first of a series of electrodes. At this electrode the impact of each arriving electron ejects between two and five further electrons in a process known as secondary emission. These are accelerated to the next electrode, where further secondary electrons are produced, and so on. In a ten-stage tube the number of electrons reaching the final collector plate for every single electron emitted from the photo-cathode may be of the order of 10^6. An appreciable pulse of current is thus obtained from the instrument and the height of the pulse is proportional to the initial ionization produced in the phosphor. It should be noted that the photomultiplier tube is used in many other instruments besides the scintillation counter. A diagrammatic representation of the counter is shown in Fig. 33.15.

Because the scintillator is a solid, there is a greater chance of γ-ray absorption than there is in the gas-filled geiger counter: but neither the solid scintillator nor the geiger counter is very efficient at detecting low-energy β-particles. These tend to be absorbed in the end-window of a geiger counter or the aluminum can surrounding a sodium iodide crystal. Unfortunately, ^{3}H and ^{14}C, both of which are immensely useful in investigation of living matter, emit only low-energy β-rays. To overcome this difficulty, liquid scintillators are often used where practicable. These consist of fluorescent materials dissolved in toluene or some other organic solvent. If the sample of radioactive material can be added to the solution, the β-particles produce ionization in the fluorescent materials and consequent emission of light. The detection is still by photomultiplier tube.

33.9 THE SEMICONDUCTOR DETECTOR

More recently semiconductors have been proving of immense value as counting mechanisms in low-energy particle detection. In Section 33.3, it was shown that, when a *p*-type semiconductor and an *n*-type semiconductor are brought together to form a junction, electrons and holes tend to cancel one another out in a region around the junction and a potential difference is set up across it. If the junction is reverse biased so that this potential difference is enhanced, any electrons or holes produced around the junction are rapidly swept away. When a charged particle passes across the junction, electron-hole pairs are produced, and, if these are in sufficient number, a small but detectable pulse passes through the circuit. This can be magnified by a high-gain amplifier and applied to a counter which will total the number of pulses in a given time.

The energy required to produce an electron-hole pair is of the order of 3 eV compared with 30 eV to produce an ion-pair in a gaseous detector. This explains the importance of semiconductor detectors in low energy work. In addition they are very small and can be used in areas where conventional counters cannot operate, or many separate counters can be used to map out spatial distributions. Semiconductor detectors permit very fast counting times, and, since they are fairly insensitive to γ-rays and neutrons, they are very useful when employed as charged particle detectors under conditions of high backgrounds of γ-radiation or in nuclear reactors. They operate at very low voltages and, since their response is related to the energy deposited at the junction, they allow very fine discrimination between different types of charged particles.

33.10 THE IMAGE INTENSIFIER

Another use of photoelectric emission is in the image intensifier often employed with X-rays. The back surface of a fluorescent screen is covered with a thin layer of photoelectric material. When X-rays fall on the screen, the photons of light emitted interact immediately with the corresponding, adjacent portion of the photoelectric surface. The electrons resulting from the interaction are accelerated across an evacuated chamber and focused by electron lenses on to a much smaller fluorescent screen perhaps 1/100 of the size of the first. The picture obtained is increased in brightness not only because its area is smaller and the energy is more concentrated but also because the accelerated electrons increase the amount of light emitted by a factor of around 50.

To enable the internal organs of a patient to be seen clearly, fairly large dosages of X-rays must be passed through him. It is clear therefore that a continuous X-ray picture cannot be taken of a patient during a delicate operation without serious risk of radiation damage. By using an image intensifier a doctor may cut down the dosage to acceptable proportions while still getting a clear picture of what is happening. A cardiologist, for example, may thus follow the passage of a catheter through an artery and into the heart—a procedure which would be dangerous if he were unable to see what was happening.

33.11 FEEDBACK MECHANISMS

Control and regulation are essential parts of most systems. Often the control has to be exercised from outside the system by human agency, but many biological systems have evolved in such a manner that regulating mechanisms have been incorporated. In manufactured devices it is now standard procedure to build in automatic control systems wherever possible.

A simple example of a control system which can be made automatic with reasonable ease occurs in the steering of a boat. Once it is decided what course the boat is to take, a helmsman can be sat at the tiller with a compass and given a simple instruction. He is told that if the course deviates from the fixed compass bearing to the right, he moves the tiller in that direction, and vice versa. He has no thinking to do; he merely responds to that instruction; and it is clear that an automatic device can do the job as efficiently. One can arrange that if the compass swings from the preset bearing to the left an electrical switch is closed and current is supplied to a mechanism which moves the tiller to the right. As the compass swings back to its correct position the switch is released and the correcting movement ceases.

Such a system is fairly crude. It works on a stop-go principle; it has no method of judging whether the deviation is slight or great and the inertia of the system is such that it is likely to overcorrect, pass the desired bearing direction and suffer a correction in the opposite sense. The progress of the vessel would therefore be a zig-zag motion about the correct direction. A more sophisticated device would arrange not only that deviation from the correct course operated a switch but that increasing deviation operated a variable resistance, increasing the current in the circuit so that the response on the tiller was proportional to the angle of displacement from the true direction. One would also arrange that the movement of the tiller followed the variation of current without appreciable time-delay, ensuring that there was no inertial lag in the system as there was in the cruder version.

Such a system is said to be using *negative feedback*. The output, in this case the deviation from the correct course, is being made to provide a signal which is used to correct or reduce the deviation from the desired behavior. The purpose of negative feedback is therefore to stabilize the system at the position of correct operation. The thermostat in a central heating system is one of the most familiar instances of such a feedback system.

In ecological systems, an automatic type of feedback system is found in the stabilization of populations in predator–prey relations. If predators increase, prey are killed off too rapidly and their population diminishes. The food supply becomes too small to maintain the increased predator population, which therefore falls off rapidly either by death or by migration to other areas. The prey population then increases again and the balance of species is maintained.

Many biochemical reactions are controlled by feedback mechanisms. In anaerobic glycolosis, for example, a sequence of chemical reactions takes place, each catalyzed by a specific enzyme. The whole process is regulated by one of these enzymes, phosphofructokinase, since its catalytic activity is very sensitive to the concentrations of ATP and ADP present. Adenosine triphosphate (ATP) is used as an energy currency by cells, the energy being released when ATP is converted to adenosine diphosphate (ADP) in a chemical reaction (cf. Chapter 24). The purpose of the glycolosis is to convert ADP back to ATP. As ATP is produced in the process it increasingly inhibits the continued production by diminishing the activity of the regulatory enzyme. If the cell begins to use energy and ATP is turned into ADP, the glycolosis restarts since the control of the enzyme is relaxed. Very many processes are controlled in this way, the final product of a sequence of reactions regulating the activity of the enzyme involved in one of the reactions of the chain.

In many pieces of electronic equipment, the output may be very sensitive to changes in the operating conditions, such as variations in the electricity supply. In these circumstances, negative feedback is often employed. One of the most sensitive stages is the amplifier, and almost all electronic devices have at least one amplifier stage. This may be a good deal more elaborate and complicated than the simple amplifier circuit described in this chapter. What it actually contains need not concern us too much, the details of the circuitry being a technical matter. But what it is possible to discuss is how the stage can be made relatively insensitive to changes in the working conditions.

If a very high amplification is required, all other considerations may have to be sacrificed to achieve it; but, in general, this is not the problem. If the output of a simple amplifier stage is investigated, several things will be noted. The wave-form of the output will not be exactly the same as the wave-form of the input; there may well be harmonics present (cf. Section 11.4) and there will be a certain amount of background noise mixed in with the signal. In other words, the output suffers from distortion and noise. In addition, changes in the voltages supplied to the circuit may alter the output substantially. From the necessity to keep the output stable against changing electrical supply conditions and also to keep the output as distortion- and noise-free as possible, it is normal to sacrifice some of the gain by using negative feedback.

As is shown in Fig. 33.16, a small portion of the output is fed back to the input. Without the feedback network, an input voltage v_i becomes an output voltage $v_o = Av_i$ after amplification. If $-\beta v_o$ is fed back to the input, the input to the amplifier is now

$$v_i' = v_i - \beta v_o. \tag{33.1}$$

Therefore

$$v_o = Av_i - A\beta v_o. \tag{33.2}$$

Therefore

$$v_o = \frac{Av_i}{1 + A\beta}. \tag{33.3}$$

Thus the gain with negative feedback is

$$A' = \frac{A}{1 + A\beta}. \tag{33.4}$$

It is not difficult to show that

$$\frac{\Delta A'}{A'} = \frac{\Delta A}{A(1 + \beta A)}. \tag{33.5}$$

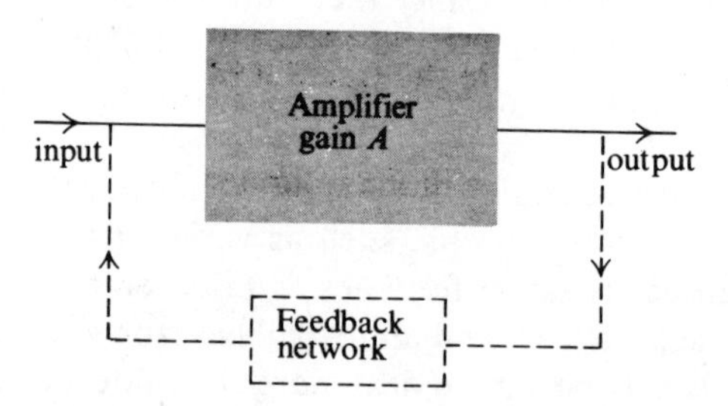

Fig. 33.16. A block diagram of an amplifier with feedback.

If $A = 10^3$ and $\beta = 10^{-2}$, $A' = 1000/11 = 90.9$. But if the supply voltage changes in such a way that A changes by 10%,

$$\frac{\Delta A'}{A'} = \frac{0.1}{(11)} = 0.0091 = 0{\cdot}91\%.$$

Because of the feedback, the change is now less than 1%.

The use of negative feedback in fact assures greater benefits than those mentioned. In addition to almost eliminating the effects of changes caused by power supply variations or by temperature, aging etc., it tends to produce a more constant gain over a wider range of frequencies than would be the case with a simple amplifier, and it allows a greater choice of impedances at input and output without the necessity for complicated matching techniques.

Experiments on conscious and not automatic feedback control are proceeding in a number of centers in the U.S. It has been believed for centuries that most of the organs of the body are outside the range of conscious control, but this has been demonstrated to be quite untrue. In Baltimore City Hospital patients have been trained to control their heart rhythms. The beat is measured by an electrocardiograph and a red light is flashed when the rhythm slows down and a yellow light flashes when it speeds up. Eventually patients can alter the frequency of the heart beat by 2 or 3 per minute at will, and this allows some control over abnormal rhythms in patients suffering from heart abnormalities. Similar techniques have been applied in other centers to the control of blood pressure, but the results are not consistent.

One of the most spectacular uses of conscious feedback control is in the regulation of brain waves (cf. Section 27.12). In the Sepulveda Hospital, California, a brain rhythm associated with motionless posture in cats has been discovered, and patients suffering from epilepsy have been taught to produce such rhythms at will. They now appear to have fewer fits and more normal brain waves.

Brain waves have also been used to train children in the art of concentration. Alpha rhythms are not present as long as concentration is maintained. A child is allowed to watch TV, but as soon as he produces α-waves the set switches off and will not turn on again until the waves disappear. It is hoped that this will increase a child's ability to maintain concentration.

33.12 ELECTRONIC COMPUTERS

One great revolution in man's condition arose from his introduction of machinery. This represented an unprecedented extension to his hands and muscles and allowed him to produce goods and food in such quantities that the work of only a small number of people could support a whole population. At present a second revolution is in progress with the development of the computer. This represents an extension to the mind. A computer can now far outstrip a human being in the speed with which it can absorb, process and communicate information and in its capacity to store the information and retrieve it in a short time. It is unrivalled in the speed and reliability with which it can carry out repetitive operations, and it can even be programmed to take logical decisions based on the results it obtains. Although a computer cannot think for itself and must be carefully programmed before it can do any job whatsoever, it is increasingly taking over jobs from men, leaving them free to do more creative tasks.

In business stock inventories, personnel information and payrolls are all areas where the computer can increase efficiency, and banks and income tax offices have now removed much of the drudgery involved in keeping records by using machines for the purpose. In medicine not only can patient records be maintained and retrieved readily, but computers have even been used for aid in diagnosis. Symp-

Fig. 33.17. A modern computer system including printers, tape and disc storage, card reader and punch and a communications controller. (Photograph by courtesy of I.B.M. Inc.)

toms can be fed in to a properly programmed machine and all the possible illnesses which fit the symptoms are immediately available to the doctor. When the doctor has made a decision among the alternatives, possibly by asking supplementary questions or performing tests, the computer will also suggest the various treatments

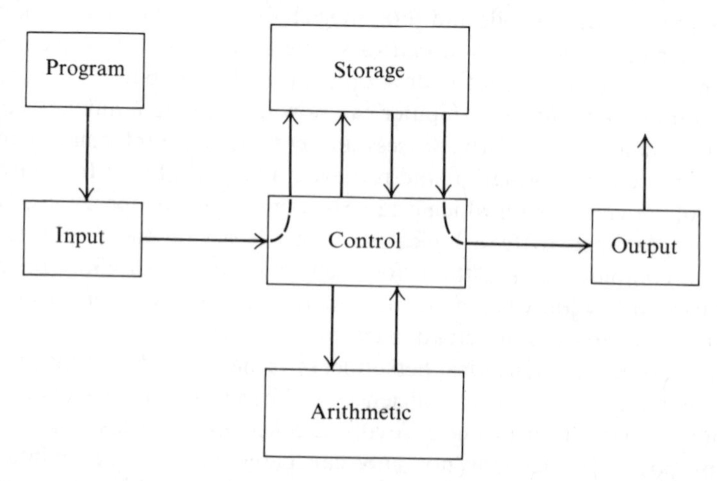

Fig. 33.18. A schematic diagram of a computer showing the function of each part.

possible. Small computers are also in use in intensive care units to control patient monitoring.

In science as a whole the computer's greatest impact has been on the computational side. Lengthy and complex calculations which a generation ago would either never have been attempted or would have constituted a life's work can now be done in a matter of days. Problems once thought impossible of solution are tackled routinely; the impact on biological and biomedical subjects has been immense. It is now easy to simulate complex situations and to find out the effect of changing some or all of the conditions. The revolution in theoretical biology over the last two decades is very much tied to the availability of high speed computers.

A digital computer deals essentially only with numbers and this in a special way. It works using binary arithmetic which is simple to reproduce on the machine. The numbers we normally deal with are to the base of ten. Thus 64312 means, reading from the right, $2 \times 10^0 + 1 \times 10^1 + 3 \times 10^2 + 4 \times 10^3 + 6 \times 10^4$ and 10^0 is, of course, 1. Using a base of ten, ten different symbols are required, the numbers from 0 to 9. There is no reason why a base ten should be employed; any number at all would serve as well. If one adopts a base of two, only two symbols 0 and 1 are required. A number like 1 0 1 1 1 means $1 \times 2^0 + 1 \times 2^1 + 1 \times 2^2 + 0 \times 2^3 + 1 \times 2^4$ and would therefore be written as $1 + 2 + 4 + 16 = 23$ in the base of ten. Any number to the base of ten can be expressed as a number to the base of two and then only contains 0's and 1's. In a computer a closed switch which allows current to pass can represent 1 and an open switch can represent 0. In practice, switches take too long to operate, and integrated circuits which do the same thing are substituted. To add two numbers, one applies the output from the closed or open switches to further integrated circuits which are programmed to respond in the following way. If no current reaches the circuit, there is no output so $0 + 0 = 0$. If a current enters from one terminal but not the other, the output is a current; $0 + 1 = 1$ or $1 + 0 = 1$. If a current enters from both terminals no current output occurs but the next circuit has a current passed to it: $1 + 1 = 10$. Subtraction is merely negative addition, multiplication is successive addition and division successive subtraction.

A computer also requires a memory or store so that numbers can be held until required, used and returned to store for later use. Magnetic core memories provide rapid-access storage. Leads run to each core (there may be of the order of a million of these) and current passed along one wire magnetizes the material in one direction, representing 1; current passed along the other magnetizes in the other direction, representing 0. A third wire is used to test the state of magnetization and report to the main computer whether the core holds 1 or 0. This is frequently the mechanism which transfers from store to computer when calculations are to be performed with the numbers. More storage space can be provided by magnetic disks, drums or tapes but access to the information is not then so rapid.

Information is fed in to the machine in the form of holes in cards or paper tape or as magnetic tape. The information will contain data to be stored and handled and also a program of instructions as to how the calculations are to proceed. A control system sees that the instructions are obeyed and does checks on the tasks performed. The output is usually in the form of a print-out of the answers on a teleprinter, but may take other forms.

An analog computer represents the values fed to it in some physical form such as the voltages set up between two particular points of a circuit. The simple arithmetical operations and more complicated algebraic ones are accomplished by applying the inputs to electronic circuits specially designed to simulate all the operations. The great advantage of analog devices is that the inputs can be continuously varied to simulate a changing physical situation and give the results of these changes almost instantaneously. Analog machines are not as accurate as digital computers, but they are of great use in dealing with the effects on problems of changing variables and are often used for automatic control systems.

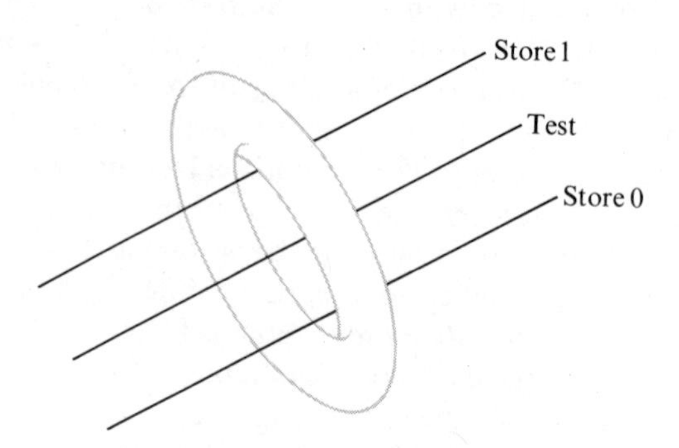

Fig. 33.19. Magnetic core storage.

Combinations of digital and analog computers can be very useful. The great speed and accuracy of the digital portion is used for calculations but special routines and complex operations can be performed on some of the data using the analog portion. This type of use is likely to increase.

Example 1. In a particular use of an intrinsic semiconductor, it is found that three-quarters of the current is being carried by electrons and one-quarter by holes. If at that temperature the drift speed of the electrons is two and one-half times that of holes, determine the ratio of electrons to holes present.

Solution. If the subscript 1 refers to electrons and the subscript 2 to holes, then the current passing is

$$I = n_1 e v_1 + n_2 e v_2,$$

where n is the number of charge carriers per unit length and v is the drift speed. But we know that

$$\frac{n_1 e v_1}{I} = \frac{n_1 e v_1}{n_1 e v_1 + n_2 e v_2} = \frac{3}{4}.$$

Therefore

$$\frac{1}{1 + \frac{n_2}{n_1}\frac{v_2}{v_1}} = \frac{3}{4},$$

therefore

$$3\frac{n_2}{n_1}\frac{v_2}{v_1} = 1,$$

therefore

$$\frac{n_1}{n_2} = 3 \times \frac{1}{2.5} = \frac{6}{5}.$$

Example 2. A Geiger-Müller tube has a dead time of t. Find the relation between the actual counting rate R and the correct counting rate R_0.

Solution. The counter records R particles per second, but it is dead for a period t after each count. It thus recorded the R particles in a period $(1 - Rt)$s while it was capable of counting. The correct counting rate is thus

$$R_0 = R/(1 - Rt).$$

PROBLEMS

33.1 In an intrinsic semiconductor the current may be carried both by free electrons and by holes. At a particular temperature it is found that there are twice as many free electrons as holes present and that the drift speed of the electrons is three times that of the holes. Calculate the fraction of the current flowing which is due to the electrons.

33.2 The current I through a rectifier when a potential difference V is applied across it is given in the following table:

V(V)	0	0.4	0.5	0.525	0.55	0.575	0.60	0.625	0.65	0.675	0.70
I(mA)	0	0	0.05	0.13	0.28	0.50	0.75	1.05	1.40	2.50	4.00

Sketch the characteristic curve.

The rectifier is connected in series with a resistor, an ammeter and a 5-V d.c. supply and 1.40 mA is recorded. If a current of 4.00 mA is required, what supply voltage is necessary?

33.3 The rectifier of Problem 33.2, a 3-kΩ resistor and a 2-V a.c. supply of negligible internal resistance are joined in series. What is the maximum instantaneous potential difference set up across the resistor?

33.4 In the amplifier circuit of Fig. 33.11 it is required to have a variation of 0.1 V over the resistance R for a change of 1 μA in the base current. If the base current is 5% of the collector current, what value must R have?

33.5 A Geiger-Müller tube has a dead time of 200 μs. In a particular experiment

it registers 5000 counts per minute. What would be the correct counting rate if there were no dead time?

33.6 A Geiger-Müller tube in a particular experiment registers 500 counts per minute after dead-time corrections have been made. The background count is 35 per minute. How long should the count be continued in the experiment in order to obtain 1% accuracy?

33.7 An input signal of 1 mV is converted by an amplifier to an output signal of 0.5 V. What is the amplification produced? If a negative-feedback circuit returns 2% of the output to the input to what figure is the amplification reduced?

33.8 If the amplifier of Problem 33.7 suffers a drop in supply voltage which reduces the amplification achieved by 0.1%, what would have been the reduction which would have resulted if no negative feedback had been used?

33.9 Express 148 to the base of 10 as a number to the base of 2.

33.10 Give the binary number 1 0 1 1 1 0 as a number to the base of 10.

APPENDIX A

MATHEMATICS THE READER SHOULD KNOW

A.1 THE LOGARITHMIC AND EXPONENTIAL FUNCTIONS

The laws of indices

The expression a^n, where a is a number and n is an integer, is the shorthand way of writing the product $a \times a \times a \times \ldots$ to n factors. By writing the expression out in full, the following three results are easily proved, n being considered greater than m.

$$a^n \times a^m = a^{n+m}, \tag{A.1}$$

$$a^n \div a^m = a^{n-m}, \tag{A.2}$$

$$(a^m)^n = a^{mn}. \tag{A.3}$$

These results follow from a definition which is intelligible only on the supposition that the indices m and n are positive and integral. But it is very convenient to use the expression a^x, where x can be any number at all—fractional, negative, or even transcendental (i.e. a number such as π that cannot be written as a fraction), and this expression can be given a definite meaning in any case by assuming that the above three laws are always obeyed, and by accepting the conclusions they lead to. Thus the meaning of a^0 can be found by considering the product $a^0 \times a^n$. From Eq. (A.1)

$$a^n \times a^0 = a^n,$$

and therefore

$$a^0 = \frac{a^n}{a^n} = 1. \tag{A.4}$$

So any quantity with zero index is equivalent to 1. In the same way, since $a^n \times a^{-n} = a^0$,

$$a^{-n} = \frac{1}{a^n}. \tag{A.5}$$

From this it follows that any factor may be transferred from the numerator to the denominator of an expression, or vice versa, by merely changing the sign of the index.

The exponential function

The function $y = a^x$, where x can take any value at all, is called an exponential function. The value of this function increases very rapidly with x. The graphs of two exponential functions, $y = 10^x$ and $y = 2^x$, are shown in Fig. A.1.

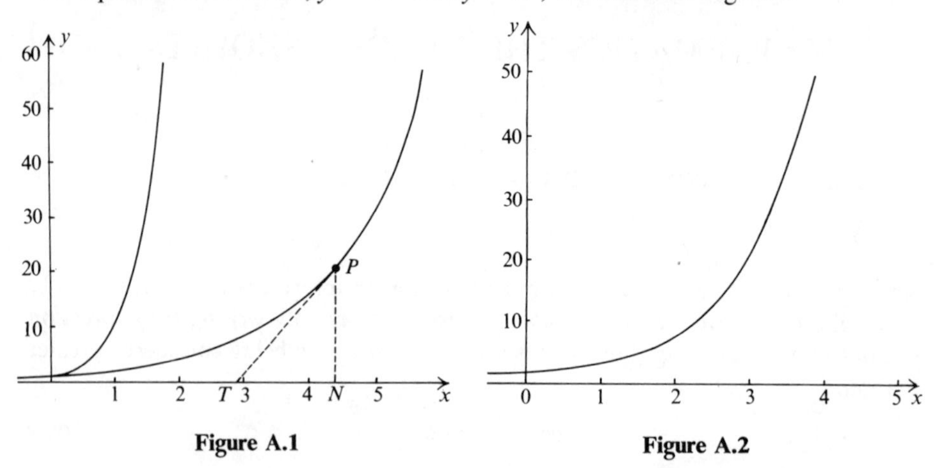

Figure A.1 **Figure A.2**

These curves are of great importance in all branches of science. The behavior exhibited by them—viz. a very rapid increase of the function $y = a^x$ with increase of x—has given rise to the name "the law of gangster growth". Such a law is characteristic of uninhibited growth processes. In the commerical field it is represented by the law of compound interest. To elucidate the nature of the process, it is very instructive to consider the slope of such a curve at any point P. This is illustrated for the curve $y = 2^x$, where the tangent to the curve at P cuts the horizontal axis in T and the vertical line through P cuts it in N. The slope of the curve at P is the ratio PN/TN, measured in their appropriate units, $PN = y$.

It is found that the length of TN is constant, no matter where the point P is chosen on the curve. In other words, the slope m is such that

$$m \propto y = ky,$$

where k is a constant equal to $1/TN$. The numerical value of k depends on which exponential curve we are considering. Thus for $y = 2^x$, $k_2 = 0.6931$, and for $y = 10^x$, $k_{10} = 2.3026$. Clearly, there must exist some number e between 2 and 10 such that k_e is exactly 1. This very special number has the magnitude 2.71828. The corresponding exponential curve $y = e^x$ is called *the* exponential curve, and for it $m = y$. The graph of the function is shown in Fig. A.2.

In many physical situations the manner in which a quantity varies with respect to another one involves exponential functions. The curve of $y = e^{-x}$ is shown in Fig. A.3 and is typical of the decay of many physical quantities, such as radioactivity, with time. It will be easily seen that this is merely the portion of the curve of Fig. A.2. for which x has negative values drawn on an enlarged scale.

Few physical situations grow as rapidly as the exponential function, but many grow exponentially to some definite final value. Figure A.4 shows the curve of $y = 1 - e^{-x}$ which is the curve of Fig. A.3 reflected across the line $y = 0.5$. Many examples of this type of growth could be cited. One described in this book is the voltage on a capacitor as it charges through a resistor.

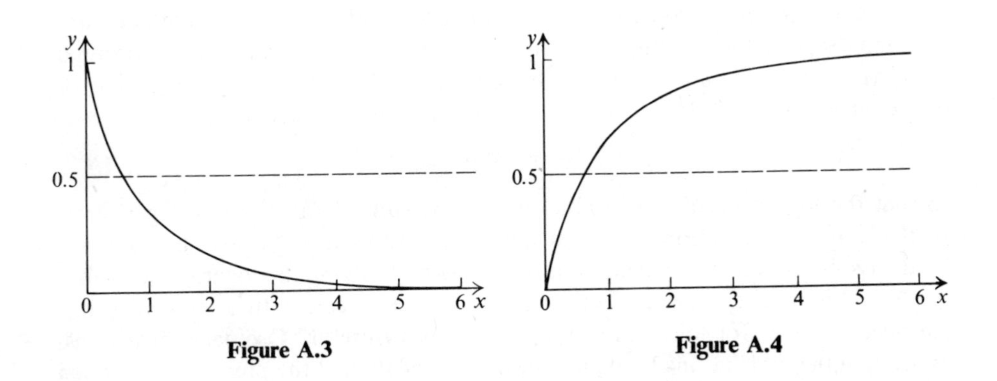

Figure A.3

Figure A.4

The curve $y = e^{-x^2}$ is shown in Fig. A.5. Experimental measurements are found to cluster about the mean value according to a curve of this type.

Finally, Fig. A.6 is a graph of the function $y = x^2e^{-x^2}$. A curve of this form is found to represent the distribution of velocities among the atoms or molecules of a gas.

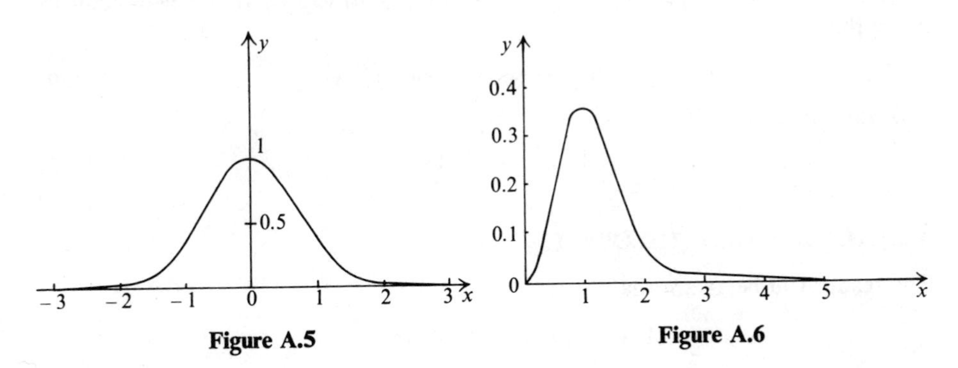

Figure A.5

Figure A.6

Values of e^x, e^{-x} and e^{-x^2} for any value of x are to be found in tables.

Since it is inconvenient to write e^{-x^2} and even more inconvenient to print it in that form, it is more usual to write it as $\exp(-x^2)$. With an even more complicated expression such as $\exp[-(x - \bar{x})^2/2s^2]$, no other form would even be considered.

Logarithms

For the exponential function $y = a^x$, the value of y is determined for every value of x. Conversely, the value of x is determined for every value of y, and this value is called the logarithm of y to the base a. In symbols,

$$x = \log_a y. \tag{A.6}$$

In practice, logarithms to the base 10 or the base e are the only ones ever employed. The curve of $x = \log y$ is simply $y = e^x$.

Logarithms were introduced as a computational aid in the seventeenth century by John Napier, the Scottish mathematician, who also introduced the decimal point as we now know it. Their usefulness derives from the fundamental law of indices given in Eq. (A.1). If we wish to multiply two numbers y_1 and y_2, then

$$y_1 \times y_2 = a^{x_1} \times a^{x_2} = a^{x_1 + x_2}, \tag{A.7}$$

so that the logarithm of the product is just the sum of the separate logarithms. In this way multiplication is carried out by the addition of logarithms, division by their subtraction, and raising to a power by multiplying the logarithms by that power. Although the advent of electronic pocket calculating machines has curtailed the need for logarithms, they are still very useful. Ordinary logarithms, as given in the table in this book, are to the base of 10, and the student is assumed to be familiar with handling them. Napier's original logarithms were to the base e, and these are called natural logarithms. They are written in two alternative ways,

$$x = \log_e y \quad \text{or} \quad x = \ln y. \tag{A.8}$$

The latter convention is adopted throughout this book. It follows that $\log y$ means $\log_{10} y$ wherever it appears.

It is sometimes necessary to change from $\ln y$ to $\log y$. It is not difficult to prove that

$$\ln y = (\ln 10)(\log y), \tag{A.9}$$

and, since $\ln 10 = 2.303$,

$$\ln y = 2.303 \log y. \tag{A.10}$$

A.2 THE BINOMIAL THEOREM

The reader will be aware that

$$(1 + x)^2 \equiv 1 + 2x + x^2,$$

$$(1 + x)^3 \equiv 1 + 3x + 3x^2 + x^3,$$

and so on. In general, if n is an integer

$$(1 + x)^n \equiv 1 + nx + \frac{n(n-1)}{2} x^2 + \ldots + nx^{n-1} + x^n, \tag{A.11}$$

the coefficient of x^r in the expression being

$$\frac{n(n-1)(n-2)\ldots(n-r+1)}{r(r-1)\ldots 2.1}.$$

This expansion is known as the binomial expansion.

If x is small in comparison with 1, it is found that a very good approximation to $(1+x)^n$ is given by taking only the first two terms of the expansion. To this degree of accuracy therefore

$$(1+x)^n = 1 + nx \quad \text{(for } x \text{ small).} \tag{A.12}$$

Further, it is found that Eq. (A.12) is true not only for integral values of n, but for negative and fractional values as well.

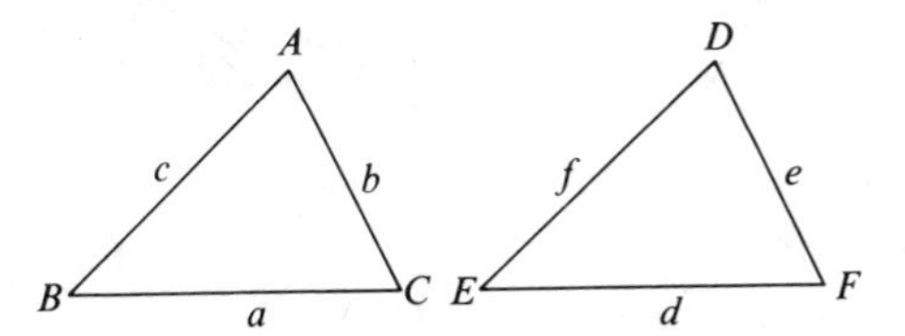

Figure A.7

Equation (A.12) is used throughout this book. Examples of its application in numerical work are given below.

$$(1.01)^3 = (1+0.01)^3 = 1 + 0.01 \times 3 = 1.03,$$

$$\frac{1}{1{\cdot}01} = (1+0.01)^{-1} = 1 + 0.01 \times (-1) = 0.99,$$

$$\frac{1}{0{\cdot}99} = (1-0.01)^{-1} = 1 + (-0.01) \times (-1) = 1.01,$$

$$\sqrt{1.01} = (1+0.01)^{\frac{1}{2}} = 1 + 0.01 \times \tfrac{1}{2} = 1.005.$$

A.3 CONGRUENT AND SIMILAR TRIANGLES

Two triangles are said to be congruent if all the sides and angles of one are equal in magnitude to the corresponding sides and angles of the other. In Fig. A.7, triangles *ABC* and *DEF* are congruent and one could therefore be lifted up and placed on top of the other so that they fitted perfectly. It is customary to label the side opposite angle *A* as side *a*, and so on, as shown in Fig. A.7.

In practice it is not necessary to show that all sides and all angles are equal to prove congruence, since the 6 quantities for any triangle are not independent, as will be shown in a later section of this appendix. Only three of the quantities need to be shown to be the same in the two triangles. The three must be carefully chosen, since not any three will do. The only criterion for congruence used in this book is that $a = d$, $B = E$ and $C = F$.

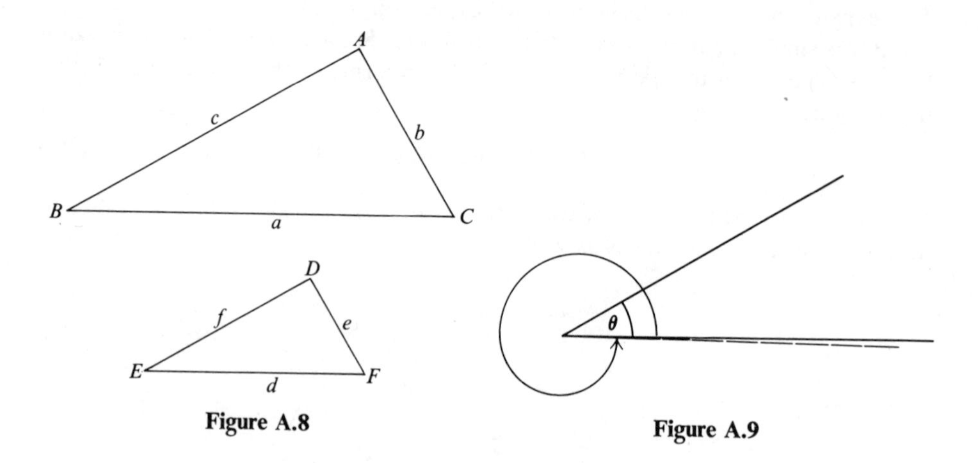

Figure A.8

Figure A.9

Two triangles are said to be similar if, as in Fig. A.8, they have all corresponding angles but not all corresponding sides, equal. If the angles can be proved equal, the triangles are similar and it follows that

$$\frac{a}{d} = \frac{b}{e} = \frac{c}{f}.$$

A.4 TRIGONOMETRY

The angle between any two lines which pass through a common point may have any value from zero, through the value θ shown in Fig. A.9, up to a maximum value when the two lines are almost coincident once more. In everyday life this maximum angle is divided into 360 equal parts, called degrees, with further subdivision into minutes and seconds of arc if necessary. It is very inconvenient to work with such an arbitrarily chosen unit of angle in scientific work, and the unit of the radian is therefore adopted.

If a circle of any radius r and with its centre at the common point of the two lines is drawn as in Fig. A.10, the circle cuts the two lines; an arc of the circle of length s is included between the two lines and the angle between the lines is defined as $\theta = s/r$ radians. Since for the whole circle $s = 2\pi r$, the maximum angle possible between the lines is 2π radians. This is true whatever the size of the circle, and it is therefore clear that the definition of the radian does not depend on the radius chosen. The angles 0°, 90°, 180°, 270°, and 360° are therefore 0, $\pi/2$, π, $3\pi/2$, and 2π radians respectively.

If θ is the angle between the two lines in Fig. A.11, and if a perpendicular is dropped from the point P on the one line to the point Q on the other, then

$$\frac{PQ}{OP} = \sin\theta, \quad \frac{OQ}{OP} = \cos\theta, \quad \frac{PQ}{OQ} = \tan\theta, \quad \frac{\text{arc } PR}{OP} = \theta.$$

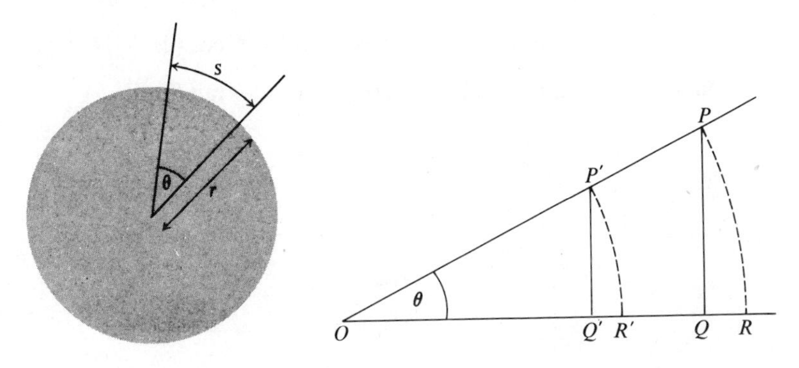

Figure A.10 **Figure A.11**

It should be clear that the trigonometrical functions $\sin\theta$, $\cos\theta$, and $\tan\theta$ do not depend on which point P is chosen. If P' had been selected, the ratios would have been

$$\frac{P'Q'}{OP'} = \sin\theta, \quad \frac{OQ'}{OP'} = \cos\theta, \quad \frac{P'Q'}{OQ'} = \tan\theta.$$

But triangles OPQ and $O'P'Q'$ are similar. From the preceding section it follows that

$$\frac{PQ}{P'Q'} = \frac{OP}{OP'} = \frac{OQ}{OQ'}.$$

The two different ratios for $\sin\theta$ (and $\cos\theta$ and $\tan\theta$) are consequently equal.

If θ becomes very small, the differences in length between the arc PR and PQ and between OR and OQ become smaller and smaller. Thus good approximations, valid for θ less than about 0.1 rad, are

$$\begin{aligned} \sin\theta = \tan\theta &= \theta. \\ \cos\theta &= 1 \end{aligned} \tag{A.13}$$

It is customary to label one of the lines forming the angle as the x-axis, to take the origin at the junction of the two lines, and to erect a y-axis as is shown in Fig. A.12. It follows that for the angle θ in Fig. A.12, PQ is a positive quantity, being above the x-axis and therefore in the region where y is positive, but OQ is a negative quantity, being to the left of the y-axis and therefore in the region of negative x. The radial line OP is always taken as positive. Therefore $\sin\theta$ has a positive value, cos θ a negative value, and tan θ a negative value, when θ has a value between $\pi/2$ and π. Similarly, when θ lies between π and $3\pi/2$, sin θ and cos θ are negative but tan θ is positive; and when θ lies between $3\pi/2$ and 2π, sin θ and tan θ are negative but cos θ is positive. Indeed

$$\sin \theta = \sin (\pi - \theta) = - \sin (\pi + \theta) = - \sin (2\pi - \theta), \tag{A.14}$$

$$\cos \theta = - \cos (\pi - \theta) = - \cos (\pi + \theta) = \cos (2\pi - \theta), \tag{A.15}$$

$$\tan \theta = - \tan (\pi - \theta) = \tan (\pi + \theta) = - \tan (2\pi - \theta). \tag{A.16}$$

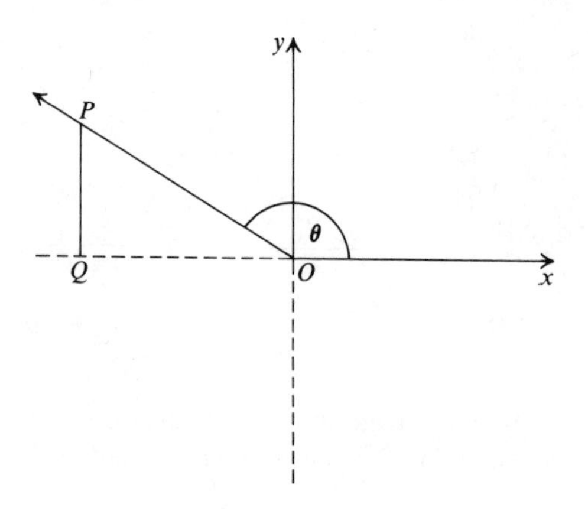

Figure A.12

The specific values sin θ, cos θ, and tan θ when θ has the values of $\pi/2$, π, and $3\pi/2$ are

$$\sin \pi/2 = 1, \cos \pi/2 = 0, \tan \pi/2 = \infty, \tag{A.17}$$

$$\sin \pi = 0, \cos \pi = - 1, \tan \pi = 0, \tag{A.18}$$

$$\sin 3\pi/2 = - 1, \cos 3\pi/2 = 0, \tan 3\pi/2 = - \infty. \tag{A.19}$$

It is necessary that the reader should know a few simple relations linking trigonometrical quantities. These are listed below without proof.

$$\frac{\sin \theta}{\cos \theta} = \tan \theta, \tag{A.20}$$

$$\sin^2 \theta + \cos^2 \theta = 1, \tag{A.21}$$

$$\sin 2\theta = 2 \sin \theta \cos \theta, \tag{A.22}$$

$$\cos 2\theta = \cos^2 \theta - \sin^2 \theta = 2 \cos^2 \theta - 1 = 1 - 2 \sin^2 \theta, \tag{A.23}$$

$$\sin (\theta + \phi) = \sin \theta \cos \phi + \cos \theta \sin \phi, \tag{A.24}$$

$$\sin (\theta - \phi) = \sin \theta \cos \phi - \cos \theta \sin \phi, \tag{A.25}$$

$$\sin(\theta + \pi/2) = \sin\theta\cos\pi/2 + \cos\theta\sin\pi/2 = \cos\theta, \tag{A.26}$$

from Eq. (A.17), and, similarly,

$$\sin(\theta - \pi/2) = -\cos\theta, \tag{A.27}$$

$$\sin(\theta + \phi) + \sin(\theta - \phi) = 2\sin\theta\cos\phi, \tag{A.28}$$

$$\sin(\theta + \phi) - \sin(\theta - \phi) = 2\cos\theta\sin\phi. \tag{A.29}$$

From Eqs. (A.28) and (A.29), it follows that

$$\sin\theta + \sin\phi = 2\sin\frac{\theta+\phi}{2}\cos\frac{\theta-\phi}{2}, \tag{A.30}$$

$$\sin\theta - \sin\phi = 2\cos\frac{\theta+\phi}{2}\sin\frac{\theta-\phi}{2}. \tag{A.31}$$

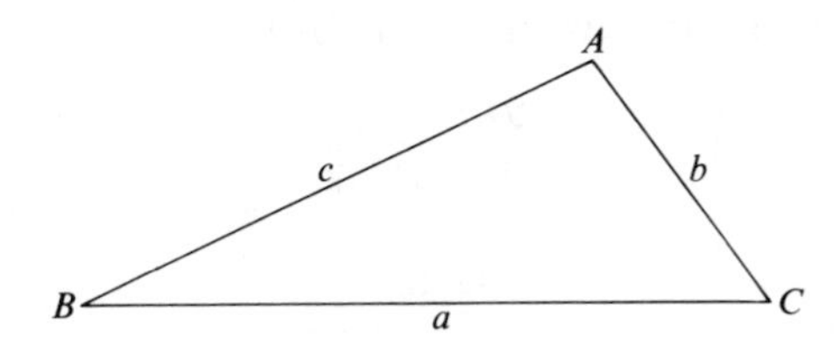

Figure A.13

In the triangle shown in Fig. A.13, it should be known that

$$c^2 = a^2 + b^2 - 2ab\cos C, \tag{A.32}$$

with similar expressions for b and a. It follows that once a, b, and C are known, the value of c is immediately calculable. The magnitude of the three sides now being known allows the determination of A and B from the two equations analogous to Eq. (A.32). This justifies the statement made earlier in this appendix that, of the 6 quantities a, b, c, A, B, and C, only three are independent.

A.5 RATIO

It eases the difficulty of algebraic manipulation, as will be found, for example, in the chapters on electricity when parallel circuits are being considered, if a little is known about the handling of ratios. It can easily be shown that if

$$\frac{a}{b} = \frac{c}{d}$$

then

$$\frac{ma + nb}{pa + qb} = \frac{mc + nd}{pc + qd},$$

where m, n, p, and q are any numbers. In particular,

$$\frac{a}{a+b} = \frac{c}{c+d} \quad \text{and} \quad \frac{a}{a-b} = \frac{c}{c-d}.$$

This knowledge is often helpful also in solving equations. A student faced with the equation

$$\frac{54.8}{48.4} = \frac{x}{x-3.2}$$

will probably groan and begin the tedious expansion. But the arithmetic he goes through is unnecessary, for the equation may be rewritten as

$$\frac{54.8}{54.8-6.4} = \frac{x}{x-3.2}.$$

Application of the ratio rules given above leads to

$$\frac{54.8}{6.4} = \frac{x}{3.2}.$$

Therefore

$$x = \frac{54.8}{2} = 27.4.$$

A.6 THE SOLUTION OF A QUADRATIC EQUATION

The general quadratic equation may be written as

$$ax^2 + bx + c = 0. \tag{A.33}$$

The solution of this equation should be known. It is

$$x = \frac{-b \pm \sqrt{b^2 - 4ac}}{2a}. \tag{A.34}$$

If $b^2 > 4ac$, the equation has two roots (or solutions) for x, corresponding to taking $\sqrt{b^2 - 4ac}$ as positive or negative. If $b^2 = 4ac$, there is only one solution for x, which is $-b/2a$. No equation with $b^2 < 4ac$ will be found in this text.

A.7 USEFUL SYMBOLS

In a preceding section we wrote "if x is small in comparison with 1". It is inconvenient to write frequently recurring phrases in full and shorthand notations exist for this and other phrases. A few are listed below:

x is smaller than 1	$x < 1$
x is very much smaller than 1	$x \ll 1$

x is less than or equal to 1	$x \leqslant 1$
x is greater than 1	$x > 1$
x is very much greater than 1	$x \gg 1$
x is greater than or equal to 1	$x \geqslant 1$
x is greater than a but smaller than or equal to b	$a < x \leqslant b$
the magnitude of the quantity x if x is positive, but minus the magnitude of the quantity x if x is negative (known as the modulus of x)	$\lvert x \rvert$
the angle whose sin is x	$\arcsin x$
the angle whose cos is x	$\arccos x$
the angle whose tan is x	$\arctan x$
x is approximately equal to y	$x \approx y$

A.8 SOME MATHEMATICAL CONSTANTS

The following constants may prove useful when attempting the problems in this book:

$\pi = 3.1416$	$\log \pi = 0.4971$
$4\pi/3 = 4.1888$	$\log(4\pi/3) = 0.6221$
$\pi/180 = 0.01745$	$\log(\pi/180) = \bar{2}.2419$
$e = 2.7183$	$\log e = 0.4343$
$1/e = 0.3679$	$\log(1/e) = \bar{1}.5657$
$\log(e^x) = 0.4343\,x$	$\log(e^{-x}) = -0.4343\,x = x \times \bar{1}.5657$

APPENDIX B

DIMENSIONAL ANALYSIS

B.1 THE DIMENSIONS OF PHYSICAL QUANTITIES

In Section 1.4 the concepts of units and dimensions were discussed. The dimensions of a physical quantity determine what kind of quantity it is. Thus a velocity is not the same kind of quantity as a mass, and a focal length is not the same kind of quantity as an electric charge. This basic distinction between different kinds of quantity has nothing whatever to do with the *magnitudes* of the quantities. Thus 5 m s $^{-1}$ and 5 kg are no more comparable than 5 m s $^{-1}$ and 8 kg; the identity of magnitude is irrelevant. Also, the units of two quantities of the same kind need not be the same: a velocity of 5 m s $^{-1}$ and one of 20 knots are comparable in the basic sense, since they are physical entities of the same kind. But they have neither magnitude nor units in common.

The dimensions of one quantity need not be independent of the dimensions of another quantity of a different kind. For example, area and volume are two physical quantities, different in kind from each other, and both different in kind from length. Yet it is clear that their respective dimensions are L^2 and L^3, where L stands for the basic dimensionality of length. In the same way the basic dimension of length may be combined with the basic dimension of time, T, to give the dimensions of velocity and acceleration as LT^{-1} and LT^{-2}, respectively. All geometric quantities will have dimensions that are powers of L; all kinematic quantities will involve combinations of powers of L and T; to introduce dynamical quantities, a third basic dimension, mass, must be introduced. Thus force, being mass times acceleration, has dimensions MLT^{-2}, and so on. It is found that the whole of mechanics can be dealt with using only the three basic dimensions, M, L, and T. The dimensions of any physical quantity can be deduced once the quantity has been properly defined in terms of other quantities. For example,

$$[\text{torque}] = [\text{force} \times \text{distance}] = MLT^{-2} \times L = ML^2T^{-2}.$$

Here a square bracket has been used as a shorthand to mean "the dimensions of . . .". In the same way,

$$[\text{work}] = [\text{energy}] = [\text{force} \times \text{distance}] = ML^2T^{-2},$$

and work and energy are seen to have the same dimensions as torque. A more complicated example is viscosity, which was defined in Section 9.5 as a shear stress per unit velocity gradient. The dimensions of viscosity can be worked out step by step as follows:

$$[\text{viscosity}] = \left[\frac{\text{shear stress}}{\text{velocity gradient}}\right] = \frac{[\text{force/area}]}{[\text{velocity/length}]}$$

$$= \frac{(MLT^{-2})(L^2)^{-1}}{(LT^{-1})(L)^{-1}} = ML^{-1}T^{-1}.$$

B.2 THE PRINCIPLE OF DIMENSIONAL HOMOGENEITY

This principle states that in any physical equation the dimensions of the left-hand side must be the same as the dimensions of the right-hand side. This principle is perhaps rather obvious, but is of very great practical utility. Thus any equations derived after a long analysis should *always* be dimensionally checked to see if mistakes have been made. If the equations are not dimensionally homogeneous, then they are wrong. If they are found to be dimensionally homogeneous, although this does not necessarily guarantee their correctness, nevertheless it will inspire some confidence in the results. As a trivial example, suppose one could not remember which of the two expressions, v^2/r or v/r^2, for the centripetal acceleration in uniform circular motion, was the correct one. A dimensional check (which in this simple example could be done mentally) shows up the wrong expression at once. For [acceleration] $= LT^{-2}$, and only an expression with these dimensions can possibly be correct. Now, $[v^2/r] = (LT^{-1})^2L^{-1} = LT^{-2}$, and $[v/r^2] = (LT^{-1})L^{-2} = L^{-1}T^{-1}$. It is therefore clear that the second form cannot be the correct one.

But the principle can be used much more directly than this. It can, in fact, be used to *predict* the form of the equation in certain cases without the need for detailed mathematical analysis. All that it is necessary to know beforehand is the physical quantities on which the answer is going to depend. If there are no more than three of these, then the form of the physical equation can be completely determined in all those cases where the equation is purely algebraic. Numerical constants, such as π, or 6, can never be determined by the use of dimensional analysis, and such constants must be determined by an exact mathematical analysis, or by performing certain experiments if this is easier. For the same reason, where the physical equation contains trigonometric, or logarithmic, or exponential terms, the method of dimensions cannot completely succeed, and recourse has to be made again either to a complete analysis or to experiment. It is important to remember that trigonometric, logarithmic, and exponential functions are dimensionless, and also that the arguments of such functions must be dimensionless. Thus in the expression for the acceleration of a simple harmonic oscillator, $-\omega^2A\sin(\omega t + \gamma)$, we may check that the algebraic expression ω^2A indeed has the dimensions LT^{-2} of acceleration, and that the argument $(\omega t + \gamma)$ is dimensionless, but there is no way to check the validity of the trigonometrical function. This requires other techniques.

The method is best understood by seeing it in action, and several examples will now be worked out.

B.3 EXAMPLES OF DIMENSIONAL ANALYSIS

1) To find an equation for the frequency of a simple pendulum.

The first step is to decide on the relevant variables. It is easy in this simple example but often presents great difficulties. Here let us assume that the frequency depends on the mass of the bob (m), the length of the string (l), and the acceleration of gravity (g). Then we assume an algebraic equation

$$f = C.\, m^a.\, l^b.\, g^c,$$

where a, b, and c are the unknown powers of the three variables and C is a pure number. Using the principle of dimensional homogeneity, we may write

$$[f] = [m]^a.\, [l]^b.\, [g]^c$$

or

$$T^{-1} = (M)^a\, (L)^b\, (LT^{-2})^c = M^a L^{b+c} T^{-2c}.$$

It will be noticed that the pure number C does not figure in the dimensional equation and that consequently it cannot be determined. This is an important limitation on the dimensional method. The dimensions of the left-hand side may alternatively be written as $M^0L^0T^{-1}$, since the dimensions of mass and length are not involved. Then mass occurs to the power zero on the left-hand side and to the power a on the right-hand side, so that the principle of dimensional homogeneity demands that $a = 0$. Systematically equating in this way the exponents of M, L, and T gives

$$\begin{array}{rcrcl} \text{exponent of } M & : & 0 & = & a, \\ L & : & 0 & = & b + c, \\ T & : & -1 & = & -2c. \end{array}$$

These three equation are easily solved to give $a = 0$, $b = -\frac{1}{2}$, $c = +\frac{1}{2}$. This determines the algebraic form of the relation

$$f = C\sqrt{g/l}.$$

This example has been explained because the result is very well known, and because it illustrates most of the important points. Notice that if we asked for the equation connecting the *angular* frequency, ω, with the same variables we should have obtained the same answer, since f and ω are simply proportional. The connecting factor of 2π will never show up in the analysis. For the same reason, the numerical constant C is left undetermined. More important, the choice of mass as a relevant variable has been shown to be wrong, and the analysis indicates quite unequivocally that the frequency of a pendulum does not depend on the mass of the bob.

2) To find the volume rate of flow of a viscous liquid through a tube of circular cross-section.

Again the first step is to decide on the relevant physical variables. Clearly, the radius of the tube and the viscosity are involved, and it would seem that the difference of pressure between the ends of the tube and the length of the tube would also be involved. But this would make four unknown exponents, to be determined by only three equations—these being obtained by equating the exponents of M, L, and T, respectively. A way out of the dilemma is to use physical intuition and recognize that only the pressure gradient down the tube can possibly be involved. In this way the problem becomes explicitly solvable. We assume an algebraic equation of the form

$$\dot{V} = Cr^a\eta^b \left\{\frac{p_1 - p_2}{l}\right\}^c,$$

where r is the radius of the tube, with $[r] = L$; η is the viscosity with $[\eta] = ML^{-1}T^{-1}$ as discussed above; and $\{p_1 - p_2/l\}$ is the drop in pressure per unit length of the tube. Remembering that pressure is a force per unit area, the dimensions of this pressure gradient are easily found to be $ML^{-2}T^{-2}$. The left-hand side is a volume per unit time with dimensions L^3T^{-1}. Hence, equating exponents of M, L, and T in the dimensional equation

$$L^3T^{-1} = (L)^a(ML^{-1}T^{-1})^b(ML^{-2}T^{-2})^c = M^{b+c}L^{a-b-2c}T^{-b-2c}$$

gives

$$\begin{aligned} \text{exponent of } M \quad &: \quad 0 = b + c, \\ L \quad &: \quad 3 = a - b - 2c, \\ T \quad &: \quad -1 = -b - 2c. \end{aligned}$$

Solving the first and third of these simultaneously gives $b = -1$, $c = +1$. Substituting these values in the other equation gives $a = +4$. So, finally, we get the Poiseuille formula

$$\dot{V} = C\left\{\frac{p_1 - p_2}{l}\right\}\frac{r^4}{\eta}.$$

Exact analysis gives the undeterminable constant to be $C = \pi/8$.

3) To find an expression for the torsional constant of a wire.

The torsional constant is the torque required to twist one end of the wire through unit angle (1 rad) relative to the other end. Since twisting involves a shearing of the wire, the rigidity modulus, n, of the material of the wire must be involved, with $[n] = [\text{force/area}] = ML^{-1}T^{-2}$. Clearly, both the radius, r, of the wire and the length, l, will be involved, both of dimensions L simply. Again an algebraic equation is assumed of the form

$$\tau = C\,n^a r^b l^c,$$

which gives the dimensional equation

$$ML^2T^{-2} = (ML^{-1}T^{-2})^a (L)^b (L)^c = M^a L^{-a+b+c} T^{-2a}.$$

Proceeding to the determination of the exponents in the usual way gives

$$\begin{array}{rcrcl} \text{exponent of } M & : & 1 & = & a, \\ L & : & 2 & = & -a+b+c, \\ T & : & -2 & = & -2a, \end{array}$$

and scrutiny of these three equations reveals something unusual. The first and third of them give the same result, namely $a = 1$, and the other gives $(b + c) = +3$. In other words, the radius and the length of the wire have not been disentangled. This is a very useful warning that the dimensional method is not by any means an automatic process guaranteed to give results even when, as here, there are only three variables. However, in this case, as in the last, physical intuition can complete the solution, since it is fairly obvious that doubling the length of wire will halve the torsional constant, and so $c = -1$. Consequently $b = +4$, and the final formula for the torsional constant is

$$\tau = C\frac{nr^4}{l}.$$

Conventional analysis of an advanced sort gives the constant as $C = \frac{1}{2}\pi$.

4) To find the retarding drag on a small sphere falling with its terminal velocity through a viscous liquid.

It seems reasonable to assume that the viscous drag will depend on the viscosity η, the terminal velocity v, and the radius of the sphere r. The assumed algebraic equation is then

$$F = C\eta^a r^b v^c,$$

and the dimensional equation is

$$MLT^{-2} = (ML^{-1}T^{-1})^a (L)^b (LT^{-1})^c = M^a L^{-a+b+c} T^{-a-c}.$$

The remainder of the derivation is left as an exercise, but the final result is obtained in the usual way as

$$F = C\eta r v,$$

the constant being given by more exact analysis as 6π. This is Stokes' law.

B.4 CONCLUSION

The four examples given in the previous section illustrate in a very direct way the power of dimensional analysis. It is seen that under certain conditions, if one knows the variables that must be involved in a problem, then the algebraic form of the solution is determined without recourse to detailed mathematical analysis. It is therefore not surprising that much philosophical speculation has gone into the exact status of dimensional analysis. The prudent and practical attitude is the best one here: to regard the method of dimensions as a powerful aid, to be used with discretion and know-how to give valuable information in cases where exact analysis is either impossible or at least very difficult. The undetermined constants can in such cases be determined experimentally. In addition its value as an aid to memory, and as a check on the accuracy of more normal analysis, should not be underestimated.

It is possible to extend the method to deal with problems in heat (which requires the introduction of a fourth basic dimension, temperature) and in electricity (which requires the introduction of current), but these applications have never been as successful as those in the simpler field of mechanics. In addition, the method can be made more powerful by splitting up the basic length dimension into component lengths along three mutually perpendicular directions. In this way the obviously distinct concepts of energy and torque, which on the simple system have the same dimensions, become distinct, and the range of problems that can successfully be tackled is substantially increased. For information on these and other topics the reader is referred to more specialized texts.*

---

* H. E. Huntley, *Dimensional Analysis*, Macdonald and Company, London (1952).

APPENDIX C

CONVERSION FACTORS

Time

1 year	$= 3.156 \times 10^7$ s
1 day	$= 8.640 \times 10^4$ s
1 hour	= 3600 s
1 minute	= 60 s

Length

1 mile	$= 1.609 \times 10^3$ m
1 ft	= 0.3048 m
1 in	$= 2.540 \times 10^{-2}$ m
1 Å	$= 10^{-10}$ m
1 light-year	$= 9.46 \times 10^{15}$ m

Angle

1°	$= 1.745 \times 10^{-2}$ rad
1′	$= 2.909 \times 10^{-4}$ rad
1″	$= 4.848 \times 10^{-6}$ rad

Mass and Density

1 lb	= 0.4536 kg
1 slug	= 14.59 kg
1 tonne	= 1000 kg
1 ton (U.S.)	= 907 kg
1 ton (U.K.)	= 1016 kg
1 a.m.u.	$= 1.66024 \times 10^{-27}$ kg
1 lb ft $^{-3}$	= 16.02 kg m $^{-3}$

Velocity and Acceleration

1 m.p.h.	= 0.4470 m s^{-1}
1 ft s^{-1}	= 0.3048 m s^{-1}
1 ft s^{-2}	= 0.3048 m s^{-2}

Force and Pressure

1 dyne	$= 10^{-5}$ N
1 lbf*	= 4.448 N
1 lbf in^{-2}	$= 6.895 \times 10^3$ N m^{-2}
1 torr (= 1 mmHg)	= 1.333 N m $^{-2}$
1 atm (= 760 torr)	$= 1.013 \times 10^5$ N m^{-2}
1 bar	$= 10^5$ N m $^{-2}$

Energy, Work and Power

1 erg	$= 10^{-7}$ J
1 cal	= 4.186 J
1 ft lbf	= 1.356 J
1 kWh	$= 3.6 \times 10^6$ J
1 eV	$= 1.602 \times 10^{-19}$ J
1 horse-power	= 746 W

* 1 lbf is the weight of a mass of 1 lb.

APPENDIX D

IMPORTANT PHYSICAL CONSTANTS

Constant	*Symbol*	*Value*
Velocity of light	c	2.9979×10^{8} m s $^{-1}$
Elementary charge	e	1.6021×10^{-19} C
Mass of the electron	m	9.1083×10^{-31} kg
Mass of the proton	M	1.6724×10^{-27} kg
Planck's constant	h	6.6256×10^{-34} J s
Charge-to-mass ratio of the electron	e/m	1.7589×10^{11} C kg $^{-1}$
Rydberg constant	R	1.097×10^{7} m $^{-1}$
Avogadro's constant	N_0	6.0249×10^{23} mol $^{-1}$
Boltzmann's constant	k	1.3804×10^{-23} J K $^{-1}$
Universal gas constant	R	8.3170 J K $^{-1}$ mol $^{-1}$
Faraday	F	9.6522×10^{4} C mol $^{-1}$
Permittivity of free space	ε_0	8.8544×10^{-12} C^2 N $^{-1}$ m $^{-2}$
Permeability of free space	μ_0	1.2566×10^{-6} Wb m $^{-1}$ A $^{-1}$
Gravitational constant	G	6.670×10^{-11} N m^2 kg $^{-2}$
Gravitational acceleration (at Washington, D.C.)	g	9.801 m s $^{-2}$
Mass of the sun		1.99×10^{30} kg
Mass of the earth		5.98×10^{24} kg
Mass of the moon		7.34×10^{22} kg
Radius of the sun		6.96×10^{8} m
Radius of the earth (mean)		6.37×10^{6} m
Radius of the moon		1.74×10^{6} m
Mean earth-sun distance		1.50×10^{11} m
Mean earth-moon distance		3.84×10^{8} m

Periodic Table of the Elements

The atomic masses, based on the exact number 12.00000 as the assigned atomic mass of the principal isotope of carbon, ^{12}C, are the most recent (1961) values adopted by the International Union of Pure and Applied Chemistry. The unit of mass used in this table is called *atomic mass unit* (amu): 1 amu = 1.6604×10^{-27} kg. The atomic mass of carbon is 12.01115 on this scale because it is the average of the different isotopes naturally present in carbon. (For artificially produced elements, the approximate atomic mass of the most stable isotope is given in brackets.)

Group→		I	II	III	IV	V	VI	VII	VIII			0
Period	Series											
1	1	1 H 1.00797										2 He 4.0026
2	2	3 Li 6.939	4 Be 9.0122	5 B 10.811	6 C 12.01115	7 N 14.0067	8 O 15.9994	9 F 18.9984				10 Ne 20.183
3	3	11 Na 22.9898	12 Mg 24.312	13 Al 26.9815	14 Si 28.086	15 P 30.9738	16 S 32.064	17 Cl 35.453				18 Ar 39.948
4	4	19 K 39.102	20 Ca 40.08	21 Sc 44.956	22 Ti 47.90	23 V 50.942	24 Cr 51.996	25 Mn 54.9380	26 Fe 55.847	27 Co 58.9332	28 Ni 58.71	
	5	29 Cu 63.54	30 Zn 65.37	31 Ga 69.72	32 Ge 72.59	33 As 74.9216	34 Se 78.96	35 Br 79.909				36 Kr 83.80
5	6	37 Rb 85.47	38 Sr 87.62	39 Y 88.905	40 Zr 91.22	41 Nb 92.906	42 Mo 95.94	43 Tc [99]	44 Ru 101.07	45 Rh 102.905	46 Pd 106.4	
	7	47 Ag 107.870	48 Cd 112.40	49 In 114.82	50 Sn 118.69	51 Sb 121.75	52 Te 127.60	53 I 126.9044				54 Xe 131.30
6	8	55 Cs 132.905	56 Ba 137.34	57–71 Lanthanide series*	72 Hf 178.49	73 Ta 180.948	74 W 183.85	75 Re 186.2	76 Os 190.2	77 Ir 192.2	78 Pt 195.09	
	9	79 Au 196.967	80 Hg 200.59	81 Tl 204.37	82 Pb 207.19	83 Bi 208.980	84 Po [210]	85 At [210]				86 Rn [222]
7	10	87 Fr [223]	88 Ra [226.05]	89–Actinide series**								

* Lanthanide series:	57 La 138.91	58 Ce 140.12	59 Pr 140.907	60 Nd 144.24	61 Pm [147]	62 Sm 150.35	63 Eu 151.96	64 Gd 157.25	65 Tb 158.924	66 Dy 162.50	67 Ho 164.930	68 Er 167.26	69 Tm 168.934	70 Yb 173.04	71 Lu 174.97
** Actinide series:	89 Ac [227]	90 Th 232.038	91 Pa [231]	92 U 238.03	93 Np [237]	94 Pu [242]	95 Am [243]	96 Cm [245]	97 Bk [249]	98 Cf [249]	99 Es [253]	100 Fm [255]	101 Md [256]	102 No	103

FOUR-PLACE LOGARITHMS OF NUMBERS

N	0	1	2	3	4	5	6	7	8	9
10	0000	0043	0086	0128	0170	0212	0253	0294	0334	0374
11	0414	0453	0492	0531	0569	0607	0645	0682	0719	0755
12	0792	0828	0864	0899	0934	0969	1004	1038	1072	1106
13	1139	1173	1206	1239	1271	1303	1335	1367	1399	1430
14	1461	1492	1523	1553	1584	1614	1644	1673	1703	1732
15	1761	1790	1818	1847	1875	1903	1931	1959	1987	2014
16	2041	2068	2095	2122	2148	2175	2201	2227	2253	2279
17	2304	2330	2355	2380	2405	2430	2455	2480	2504	2529
18	2553	2577	2601	2625	2648	2672	2695	2718	2742	2765
19	2788	2810	2833	2856	2878	2900	2923	2945	2967	2989
20	3010	3032	3054	3075	3096	3118	3139	3160	3181	3201
21	3222	3243	3263	3284	3304	3324	3345	3365	3385	3404
22	3424	3444	3464	3483	3502	3522	3541	3560	3579	3598
23	3617	3636	3655	3674	3692	3711	3729	3747	3766	3784
24	3802	3820	3838	3856	3874	3892	3909	3927	3945	3962
25	3979	3997	4014	4031	4048	4065	4082	4099	4116	4133
26	4150	4166	4183	4200	4216	4232	4249	4265	4281	4298
27	4314	4330	4346	4362	4378	4393	4409	4425	4440	4456
28	4472	4487	4502	4518	4533	4548	4564	4579	4594	4609
29	4624	4639	4654	4669	4683	4698	4713	4728	4742	4757
30	4771	4786	4800	4814	4829	4843	4857	4871	4886	4900
31	4914	4928	4942	4955	4969	4983	4997	5011	5024	5038
32	5051	5065	5079	5092	5105	5119	5132	5145	5159	5172
33	5185	5198	5211	5224	5237	5250	5263	5276	5289	5302
34	5315	5328	5340	5353	5366	5378	5391	5403	5416	5428
35	5441	5453	5465	5478	5490	5502	5514	5527	5539	5551
36	5563	5575	5587	5599	5611	5623	5635	5647	5658	5670
37	5682	5694	5705	5717	5729	5740	5752	5763	5775	5786
38	5798	5809	5821	5832	5843	5855	5866	5877	5888	5899
39	5911	5922	5933	5944	5955	5966	5977	5988	5999	6010
40	6021	6031	6042	6053	6064	6075	6085	6096	6107	6117
41	6128	6138	6149	6160	6170	6180	6191	6201	6212	6222
42	6232	6243	6253	6263	6274	6284	6294	6304	6314	6325
43	6335	6345	6355	6365	6375	6385	6395	6405	6415	6425
44	6435	6444	6454	6464	6474	6484	6493	6503	6513	6522
45	6532	6542	6551	6561	6571	6580	6590	6599	6609	6618
46	6628	6637	6646	6656	6665	6675	6684	6693	6702	6712
47	6721	6730	6739	6749	6758	6767	6776	6785	6794	6803
48	6812	6821	6830	6839	6848	6857	6866	6875	6884	6893
49	6902	6911	6920	6928	6937	6946	6955	6964	6972	6981
50	6990	6998	7007	7016	7024	7033	7042	7050	7059	7067
51	7076	7084	7093	7101	7110	7118	7126	7135	7143	7152
52	7160	7168	7177	7185	7193	7202	7210	7218	7226	7235
53	7243	7251	7259	7267	7275	7284	7292	7300	7308	7316
54	7324	7332	7340	7348	7356	7364	7372	7380	7388	7396

N	0	1	2	3	4	5	6	7	8	9
55	7404	7412	7419	7427	7435	7443	7451	7459	7466	7474
56	7482	7490	7497	7505	7513	7520	7528	7536	7543	7551
57	7559	7566	7574	7582	7589	7597	7604	7612	7619	7627
58	7634	7642	7649	7657	7664	7672	7679	7686	7694	7701
59	7709	7716	7723	7731	7738	7745	7752	7760	7767	7774
60	7782	7789	7796	7803	7810	7818	7825	7832	7839	7846
61	7853	7860	7868	7875	7882	7889	7896	7903	7910	7917
62	7924	7931	7938	7945	7952	7959	7966	7973	7980	7987
63	7993	8000	8007	8014	8021	8028	8035	8041	8048	8055
64	8062	8069	8075	8082	8089	8096	8102	8109	8116	8122
65	8129	8136	8142	8149	8156	8162	8169	8176	8182	8189
66	8195	8202	8209	8215	8222	8228	8235	8241	8248	8254
67	8261	8267	8274	8280	8287	8293	8299	8306	8312	8319
68	8325	8331	8338	8344	8351	8357	8363	8370	8376	8382
69	8388	8395	8401	8407	8414	8420	8426	8432	8439	8445
70	8451	8457	8463	8470	8476	8482	8488	8494	8500	8506
71	8513	8519	8525	8531	8537	8543	8549	8555	8561	8567
72	8573	8579	8585	8591	8597	8603	8609	8615	8621	8627
73	8633	8639	8645	8651	8657	8663	8669	8675	8681	8686
74	8692	8698	8704	8710	8716	8722	8727	8733	8739	8745
75	8751	8756	8762	8768	8774	8779	8785	8791	8797	8802
76	8808	8814	8820	8825	8831	8837	8842	8848	8854	8859
77	8865	8871	8876	8882	8887	8893	8899	8904	8910	8915
78	8921	8927	8932	8938	8943	8949	8954	8960	8965	8971
79	8976	8982	8987	8993	8998	9004	9009	9015	9020	9025
80	9031	9036	9042	9047	9053	9058	9063	9069	9074	9079
81	9085	9090	9096	9101	9106	9112	9117	9122	9128	9133
82	9138	9143	9149	9154	9159	9165	9170	9175	9180	9186
83	9191	9196	9201	9206	9212	9217	9222	9227	9232	9238
84	9243	9248	9253	9258	9263	9269	9274	9279	9284	9289
85	9294	9299	9304	9309	9315	9320	9325	9330	9335	9340
86	9345	9350	9355	9360	9365	9370	9375	9380	9385	9390
87	9395	9400	9405	9410	9415	9420	9425	9430	9435	9440
88	9445	9450	9455	9460	9465	9469	9474	9479	9484	9489
89	9494	9499	9504	9509	9513	9518	9523	9528	9533	9538
90	9542	9547	9552	9557	9652	9566	9571	9576	9581	9586
91	9590	9595	9600	9605	9609	9614	9619	9624	9628	9633
92	9638	9643	9647	9652	9657	9661	9666	9671	9675	9680
93	9685	9689	9694	9699	9703	9708	9713	9717	9722	9727
94	9731	9736	9741	9745	9750	9754	9759	9763	9768	9773
95	9777	9782	9786	9791	9795	9800	9805	9809	9814	9818
96	9823	9827	9832	9836	9841	9845	9850	9854	9859	9863
97	9868	9872	9877	9881	9886	9890	9894	9899	9903	9908
98	9912	9917	9921	9926	9930	9934	9939	9943	9948	9952
99	9956	9961	9965	9969	9974	9978	9983	9987	9991	9996

FOUR-PLACE TRIGONOMETRICAL TABLES

Degrees	Radians	Sin	Cos	Tan	Cot	Sec	Csc		
0° 00′	.0000	.0000	1.0000	.0000	——	1.000	——	1.5708	**90° 00′**
10	029	029	000	029	343.8	000	343.8	679	50
20	058	058	000	058	171.9	000	171.9	650	40
30	.0087	.0087	1.0000	.0087	114.6	1.000	114.6	1.5621	30
40	116	116	.9999	116	85.94	000	85.95	592	20
50	145	145	999	145	68.75	000	68.76	563	10
1° 00′	.0175	.0175	.9998	.0175	57.29	1.000	57.30	1.5533	**89° 00′**
10	204	204	998	204	49.10	000	49.11	504	50
20	233	233	997	233	42.96	000	42.98	475	40
30	.0262	.0262	.9997	.0262	38.19	1.000	38.20	1.5446	30
40	291	291	996	291	34.37	000	34.38	417	20
50	320	320	995	320	31.24	001	31.26	388	10
2° 00′	.0349	.0349	.9994	.0349	28.64	1.001	28.65	1.5359	**88° 00′**
10	378	378	993	378	26.43	001	26.45	330	50
20	407	407	992	407	24.54	001	24.56	301	40
30	.0436	.0436	.9990	.0437	22.90	1.001	22.93	1.5272	30
40	465	465	989	466	21.47	001	21.49	243	20
50	495	494	988	495	20.21	001	20.23	213	10
3° 00′	.0524	.0523	.9986	.0524	19.08	1.001	19.11	1.5184	**87° 00′**
10	553	552	985	553	18.07	002	18.10	155	50
20	582	581	983	582	17.17	002	17.20	126	40
30	.0611	.0610	.9981	.0612	16.35	1.002	16.38	1.5097	30
40	640	640	980	641	15.60	002	15.64	068	20
50	669	669	978	670	14.92	002	14.96	039	10
4° 00′	.0698	.0698	.9976	.0699	14.30	1.002	14.34	1.5010	**86° 00′**
10	727	727	974	729	13.73	003	13.76	981	50
20	756	756	971	758	13.20	003	13.23	952	40
30	.0785	.0785	.9969	.0787	12.71	1.003	12.75	1.4923	30
40	814	814	967	816	12.25	003	12.29	893	20
50	844	843	964	846	11.83	004	11.87	864	10
5° 00′	.0873	.0872	.9962	.0875	11.43	1.004	11.47	1.4835	**85° 00′**
10	902	901	959	904	11.06	004	11.10	806	50
20	931	929	957	934	10.71	004	10.76	777	40
30	.0960	.0958	.9954	.0963	10.39	1.005	10.43	1.4748	30
40	989	987	951	992	10.08	005	10.13	719	20
50	.1018	.1016	948	.1022	9.788	005	9.839	690	10
6° 00′	.1047	.1045	.9945	.1051	9.514	1.006	9.567	1.4661	**84° 00′**
10	076	074	942	080	9.255	006	9.309	632	50
20	105	103	939	110	9.010	006	9.065	603	40
30	.1134	.1132	.9936	.1139	8.777	1.006	8.834	1.4573	30
40	164	161	932	169	8.556	007	8.614	544	20
50	193	190	929	198	8.345	007	8.405	515	10
7° 00′	.1222	.1219	.9925	.1228	8.144	1.008	8.206	1.4486	**83° 00′**
10	251	248	922	257	7.953	008	8.016	457	50
20	280	276	918	287	7.770	008	7.834	428	40
30	.1309	.1305	.9914	.1317	7.596	1.009	7.661	1.4399	30
40	338	334	911	346	7.429	009	7.496	370	20
50	367	363	907	376	7.269	009	7.337	341	10
8° 00′	.1396	.1392	.9903	.1405	7.115	1.010	7.185	1.4312	**82° 00′**
10	425	421	899	435	6.968	010	7.040	283	50
20	454	449	894	465	6.827	011	6.900	254	40
30	.1484	.1478	.9890	.1495	6.691	1.011	6.765	1.4224	30
40	513	507	886	524	6.561	012	6.636	195	20
50	542	536	881	554	6.435	012	6.512	166	10
9° 00′	.1571	.1564	.9877	.1584	6.314	1.012	6.392	1.4137	**81° 00′**
		Cos	Sin	Cot	Tan	Csc	Sec	Radians	Degrees

Degrees	Radians	Sin	Cos	Tan	Cot	Sec	Csc		
9° 00′	.1571	.1564	.9877	.1584	6.314	1.012	6.392	1.4137	**81° 00′**
10	600	593	872	614	197	013	277	108	50
20	629	622	868	644	084	013	166	079	40
30	.1658	.1650	.9863	.1673	5.976	1.014	6.059	1.4050	30
40	687	679	858	703	871	014	5.955	1.4021	20
50	716	708	853	733	769	015	855	992	10
10° 00′	.1745	.1736	.9848	.1763	5.671	1.015	5.759	1.3963	**80° 00′**
10	774	765	843	793	576	016	665	934	50
20	804	794	838	823	485	016	575	904	40
30	.1833	.1822	.9833	.1853	5.396	1.017	5.487	1.3875	30
40	862	851	827	883	309	018	403	846	20
50	891	880	822	914	226	018	320	817	10
11° 00′	.1920	.1908	.9816	.1944	5.145	1.019	5.241	1.3788	**79° 00′**
10	949	937	811	974	066	019	164	759	50
20	978	965	805	.2004	4.989	020	089	730	40
30	.2007	.1994	.9799	.2035	4.915	1.020	5.016	1.3701	30
40	036	.2022	793	065	843	021	4.945	672	20
50	065	051	787	095	773	022	876	643	10
12° 00′	.2094	.2079	.9781	.2126	4.705	1.022	4.810	1.3614	**78° 00′**
10	123	108	775	156	638	023	745	584	50
20	153	136	769	186	574	024	682	555	40
30	.2182	.2164	.9763	.2217	4.511	1.024	4.620	1.3526	30
40	211	193	757	247	449	025	560	497	20
50	240	221	750	278	390	026	502	468	10
13° 00′	.2269	.2250	.9744	.2309	4.331	1.026	4.445	1.3439	**77° 00′**
10	298	278	737	339	275	027	390	410	50
20	327	306	730	370	219	028	336	381	40
30	.2356	.2334	.9724	.2401	4.165	1.028	4.284	1.3352	30
40	385	363	717	432	113	029	232	323	20
50	414	391	710	462	061	030	182	294	10
14° 00′	.2443	.2419	.9703	.2493	4.011	1.031	4.134	1.3265	**76° 00′**
10	473	447	696	524	3.962	031	086	235	50
20	502	476	689	555	914	032	039	206	40
30	.2531	.2504	.9681	.2586	3.867	1.033	3.994	1.3177	30
40	560	532	674	617	821	034	950	148	20
50	589	560	667	648	776	034	906	119	10
15° 00′	.2618	.2588	.9659	.2679	3.732	1.035	3.864	1.3090	**75° 00′**
10	647	616	652	711	689	036	822	061	50
20	676	644	644	742	647	037	782	032	40
30	.2705	.2672	.9636	.2773	3.606	1.038	3.742	1.3003	30
40	734	700	628	805	566	039	703	974	20
50	763	728	621	836	526	039	665	945	10
16° 00′	.2793	.2756	.9613	.2867	3.487	1.040	3.628	1.2915	**74° 00′**
10	822	784	605	899	450	041	592	886	50
20	851	812	596	931	412	042	556	857	40
30	.2880	.2840	.9588	.2962	3.376	1.043	3.521	1.2828	30
40	909	868	580	994	340	044	487	799	20
50	938	896	572	.3026	305	045	453	770	10
17° 00′	.2967	.2924	.9563	.3057	3.271	1.046	3.420	1.2741	**73° 00′**
10	996	952	555	089	237	047	388	712	50
20	.3025	979	546	121	204	048	356	683	40
30	.3054	.3007	.9537	.3153	3.172	1.049	3.326	1.2654	30
40	083	035	528	185	140	049	295	625	20
50	113	062	520	217	108	050	265	595	10
18° 00′	.3142	.3090	.9511	.3249	3.078	1.051	3.236	1.2566	**72° 00′**
		Cos	Sin	Cot	Tan	Csc	Sec	Radians	Degrees

Degrees	Radians	Sin	Cos	Tan	Cot	Sec	Csc		
18° 00′	.3142	.3090	.9511	.3249	3.078	1.051	3.236	1.2566	**72° 00′**
10	171	118	502	281	047	052	207	537	50
20	200	145	492	314	018	053	179	508	40
30	.3229	.3173	.9483	.3346	2.989	1.054	3.152	1.2479	30
40	258	201	474	378	960	056	124	450	20
50	287	228	465	411	932	057	098	421	10
19° 00′	.3316	.3256	.9455	.3443	2.904	1.058	3.072	1.2392	**71° 00′**
10	345	283	446	476	877	059	046	363	50
20	374	311	436	508	850	060	021	334	40
30	.3403	.3338	.9426	.3541	2.824	1.061	2.996	1.2305	30
40	432	365	417	574	798	062	971	275	20
50	462	393	407	607	773	063	947	246	10
20° 00′	.3491	.3420	.9397	.3640	2.747	1.064	2.924	1.2217	**70° 00′**
10	520	448	387	673	723	065	901	188	50
20	549	475	377	706	699	066	878	159	40
30	.3578	.3502	.9367	.3739	2.675	1.068	2.855	1.2130	30
40	607	529	356	772	651	069	833	101	20
50	636	557	346	805	628	070	812	072	10
21° 00′	.3665	.3584	.9336	.3839	2.605	1.071	2.790	1.2043	**69° 00′**
10	694	611	325	872	583	072	769	1.2014	50
20	723	638	315	906	560	074	749	985	40
30	.3752	.3665	.9304	.3939	2.539	1.075	2.729	1.1956	30
40	782	692	293	973	517	076	709	926	20
50	811	719	283	.4006	496	077	689	897	10
22° 00′	.3840	.3746	.9272	.4040	2.475	1.079	2.669	1.1868	**68° 00′**
10	869	773	261	074	455	080	650	839	50
20	898	800	250	108	434	081	632	810	40
30	.3927	.3827	.9239	.4142	2.414	1.082	2.613	1.1781	30
40	956	854	228	176	394	084	595	752	20
50	985	881	216	210	375	085	577	723	10
23° 00′	.4014	.3907	.9205	.4245	2.356	1.086	2.559	1.1694	**67° 00′**
10	043	934	194	279	337	088	542	665	50
20	072	961	182	314	318	089	525	636	40
30	.4102	.3987	.9171	.4348	2.300	1.090	2.508	1.1606	30
40	131	.4014	159	383	282	092	491	577	20
50	160	041	147	417	264	093	475	548	10
24° 00′	.4189	.4067	.9135	.4452	2.246	1.095	2.459	1.1519	**66° 00′**
10	218	094	124	487	229	096	443	490	50
20	247	120	112	522	211	097	427	461	40
30	.4276	.4147	.9100	.4557	2.194	1.099	2.411	1.1432	30
40	305	173	088	592	177	100	396	403	20
50	334	200	075	628	161	102	381	374	10
25° 00′	.4363	.4226	.9063	.4663	2.145	1.103	2.366	1.1345	**65° 00′**
10	392	253	051	699	128	105	352	316	50
20	422	279	038	734	112	106	337	286	40
30	.4451	.4305	.9026	.4770	2.097	1.108	2.323	1.1257	30
40	480	331	013	806	081	109	309	228	20
50	509	358	001	841	066	111	295	199	10
26° 00′	.4538	.4384	.8988	.4877	2.050	1.113	2.281	1.1170	**64° 00′**
10	567	410	975	913	035	114	268	141	50
20	596	436	962	950	020	116	254	112	40
30	.4625	.4462	.8949	.4986	2.006	1.117	2.241	1.1083	30
40	654	488	936	.5022	1.991	119	228	054	20
50	683	514	923	059	977	121	215	1.1025	10
27° 00′	.4712	.4540	.8910	.5095	1.963	1.122	2.203	1.0996	**63° 00′**
		Cos	Sin	Cot	Tan	Csc	Sec	Radians	Degrees

Degrees	Radians	Sin	Cos	Tan	Cot	Sec	Csc		
27° 00′	.4712	.4540	.8910	.5095	1.963	1.122	2.203	1.0996	**63° 00′**
10	741	566	897	132	949	124	190	966	50
20	771	592	884	169	935	126	178	937	40
30	.4800	.4617	.8870	.5206	1.921	1.127	2.166	1.0908	30
40	829	643	857	243	907	129	154	879	20
50	858	669	843	280	894	131	142	850	10
28° 00′	.4887	.4695	.8829	.5317	1.881	1.133	2.130	1.0821	**62° 00′**
10	916	720	816	354	868	134	118	792	50
20	945	746	802	392	855	136	107	763	40
30	.4974	.4772	.8788	.5430	1.842	1.138	2.096	1.0734	30
40	.5003	797	774	467	829	140	085	705	20
50	032	823	760	505	816	142	074	676	10
29° 00′	.5061	.4848	.8746	.5543	1.804	1.143	2.063	1.0647	**61° 00′**
10	091	874	732	581	792	145	052	617	50
20	120	899	718	619	780	147	041	588	40
30	.5149	.4924	.8704	.5658	1.767	1.149	2.031	1.0559	30
40	178	950	689	696	756	151	020	530	20
50	207	975	675	735	744	153	010	501	10
30° 00′	.5236	.5000	.8660	.5774	1.732	1.155	2.000	1.0472	**60° 00′**
10	265	025	646	812	720	157	1.990	443	50
20	294	050	631	851	709	159	980	414	40
30	.5323	.5075	.8616	.5890	1.698	1.161	1.970	1.0385	30
40	352	100	601	930	686	163	961	356	20
50	381	125	587	969	675	165	951	327	10
31° 00′	.5411	.5150	.8572	.6009	1.664	1.167	1.942	1.0297	**59° 00′**
10	440	175	557	048	653	169	932	268	50
20	469	200	542	088	643	171	923	239	40
30	.5498	.5225	.8526	.6128	1.632	1.173	1.914	1.0210	30
40	527	250	511	168	621	175	905	181	20
50	556	275	496	208	611	177	896	152	10
32° 00′	.5585	.5299	.8480	.6249	1.600	1.179	1.887	1.0123	**58° 00′**
10	614	324	465	289	590	181	878	094	50
20	643	348	450	330	580	184	870	065	40
30	.5672	.5373	.8434	.6371	1.570	1.186	1.861	1.0036	30
40	701	398	418	412	560	188	853	1.0007	20
50	730	422	403	453	550	190	844	977	10
33° 00′	.5760	.5446	.8387	.6494	1.540	1.192	1.836	.9948	**57° 00′**
10	789	471	371	536	530	195	828	919	50
20	818	495	355	577	520	197	820	890	40
30	.5847	.5519	.8339	.6619	1.511	1.199	1.812	.9861	30
40	876	544	323	661	501	202	804	832	20
50	905	568	307	703	1.492	204	796	803	10
34° 00′	.5934	.5592	.8290	.6745	1.483	1.206	1.788	.9774	**56° 00′**
10	963	616	274	787	473	209	781	745	50
20	992	640	258	830	464	211	773	716	40
30	.6021	.5664	.8241	.6873	1.455	1.213	1.766	.9687	30
40	050	688	225	916	446	216	758	657	20
50	080	712	208	959	437	218	751	628	10
35° 00′	.6109	.5736	.8192	.7002	1.428	1.221	1.743	.9599	**55° 00′**
10	138	760	175	046	419	223	736	570	50
20	167	783	158	089	411	226	729	541	40
30	.6196	.5807	.8141	.7133	1.402	1.228	1.722	.9512	30
40	225	831	124	177	393	231	715	483	20
50	254	854	107	221	385	233	708	454	10
36° 00′	.6283	.5878	.8090	.7265	1.376	1.236	1.701	.9425	**54° 00′**
		Cos	Sin	Cot	Tan	Csc	Sec	Radians	Degrees

Degrees	Radians	Sin	Cos	Tan	Cot	Sec	Csc		
36° 00′	.6283	.5878	.8090	.7265	1.376	1.236	1.701	.9425	**54° 00′**
10	312	901	073	310	368	239	695	396	50
20	341	925	056	355	360	241	688	367	40
30	.6370	.5948	.8039	.7400	1.351	1.244	1.681	.9338	30
40	400	972	021	445	343	247	675	308	20
50	429	995	004	490	335	249	668	279	10
37° 00′	.6458	.6018	.7986	.7536	1.327	1.252	1.662	.9250	**53° 00′**
10	487	041	969	581	319	255	655	221	50
20	516	065	951	627	311	258	649	192	40
30	.6545	.6088	.7934	.7673	1.303	1.260	1.643	.9163	30
40	574	111	916	720	295	263	636	134	20
50	603	134	898	766	288	266	630	105	10
38° 00′	.6632	.6157	.7880	.7813	1.280	1.269	1.624	.9076	**52° 00′**
10	661	180	862	860	272	272	618	047	50
20	690	202	844	907	265	275	612	.9018	40
30	.6720	.6225	.7826	.7954	1.257	1.278	1.606	.8988	30
40	749	248	808	.8002	250	281	601	959	20
50	778	271	790	050	242	284	595	930	10
39° 00′	.6807	.6293	.7771	.8098	1.235	1.287	1.589	.8901	**51° 00′**
10	836	316	753	146	228	290	583	872	50
20	865	338	735	195	220	293	578	843	40
30	.6894	.6361	.7716	.8243	1.213	1.296	1.572	.8814	30
40	923	383	698	292	206	299	567	785	20
50	952	406	679	342	199	302	561	756	10
40° 00′	.6981	.6428	.7660	.8391	1.192	1.305	1.556	.8727	**50° 00′**
10	.7010	450	642	441	185	309	550	698	50
20	039	472	623	491	178	312	545	668	40
30	.7069	.6494	.7604	.8541	1.171	1.315	1.540	.8639	30
40	098	517	585	591	164	318	535	610	20
50	127	539	566	642	157	322	529	581	10
41° 00′	.7156	.6561	.7547	.8693	1.150	1.325	1.524	.8552	**49° 00′**
10	185	583	528	744	144	328	519	523	50
20	214	604	509	796	137	332	514	494	40
30	.7243	.6626	.7490	.8847	1.130	1.335	1.509	.8465	30
40	272	648	470	899	124	339	504	436	20
50	301	670	451	952	117	342	499	407	10
42° 00′	.7330	.6691	.7431	.9004	1.111	1.346	1.494	.8378	**48° 00′**
10	359	713	412	057	104	349	490	348	50
20	389	734	392	110	098	353	485	319	40
30	.7418	.6756	.7373	.9163	1.091	1.356	1.480	.8290	30
40	447	777	353	217	085	360	476	261	20
50	476	799	333	271	079	364	471	232	10
43° 00′	.7505	.6820	.7314	.9325	1.072	1.367	1.466	.8203	**47° 00′**
10	534	841	294	380	066	371	462	174	50
20	563	862	274	435	060	375	457	145	40
30	.7592	.6884	.7254	.9490	1.054	1.379	1.453	.8116	30
40	621	905	234	545	048	382	448	087	20
50	650	926	214	601	042	386	444	058	10
44° 00′	.7679	.6947	.7193	.9657	1.036	1.390	1.440	.8029	**46° 00′**
10	709	967	173	713	030	394	435	999	50
20	738	988	153	770	024	398	431	970	40
30	.7767	.7009	.7133	.9827	1.018	1.402	1.427	.7941	30
40	796	030	112	884	012	406	423	912	20
50	825	050	092	942	006	410	418	883	10
45° 00′	.7854	.7071	.7071	1.000	1.000	1.414	1.414	.7854	**45° 00′**
		Cos	Sin	Cot	Tan	Csc	Sec	Radians	Degrees

ANSWERS TO PROBLEMS

CHAPTER 1

1.1 ML^2T^{-2} and $ML^2T^{-3}I^{-1}$
1.2 2.56×10^6 eV, 2.56 MeV
1.3 6.67×10^{-11} N m^2 kg^{-2}, 0.667 pN m^2 kg^{-2}
1.4 17.9 m s^{-1}
1.5 16 kg m^{-3}
1.6 kg m^{-3}
1.7 N s m^{-2} or kg m^{-1} s^{-1}

CHAPTER 2

2.1 2/47
2.2 8/47
2.3 (a) 125/512; (b) 5/28
2.4 1/6
2.5 5/72
2.6 151,200
2.7 210
2.8 2,944,656
2.9 All black; 3 to 1
2.10 0.316, 0.0625
2.11 BRPS and RRSS
2.12 0.35 and 3/8; 0.5 $\pm$ 0.03
2.13 9.80; $\pm$0.12
2.14 340 Å
2.16 Highly
2.17 No
2.18 6
2.19 11
2.20 No
2.21 No

CHAPTER 3

3.1 (a) 7.5 m s^{-1}; 7.2 m s^{-1}; 5.0 m s^{-1}
(b) 9.2 m s^{-1}; 9.8 m s^{-1}
(c) 32.4 m; 42.3 m; 9.9 m s^{-1}
(d) 10.0 m s^{-1}
(e) 4.4 m s^{-1}; -3.6 m s^{-1}
(f) 57.9 m; 8.3 s
3.2 (a) 6.86 m s^{-1}; 9.64 m s^{-1}; 11.0 m s^{-1}
(b) 11.93 m s^{-1}; 12.05 m s^{-1}
(c) 28.46 m; 40.56 m; 12.10 m s^{-1}
(d) 12.11 m s^{-1}
(e) 3.46 m s^{-1}; 16.51 m s^{-1}
3.3 m s^{-3}; $3At^2$
3.4 m; m s^{-1}; m s^{-2}; $B + 2Ct$
3.5 $\omega A \cos \omega t - \omega B \sin \omega t$; m; s^{-1}
3.6 kXe^{-kt}; m; s^{-1}
3.7 (a) 4 m s^{-2}; 0.4 m s^{-2}; -2 m s^{-2}
(b) 2.8 m s^{-2}; 2.8 m s^{-2}
(c) 17.5 m s^{-1}; 20.3 m s^{-1}; 2.8 m s^{-2}
(d) 2.8 m s^{-2}
(e) $a = (10 \text{ m s}^{-2}) - (2.4 \text{ m s}^{-3})t$
(f) 115.2 m; 58.2 m

3.8

t (s)	x_T (m)	x_S (m)
1	4.4	—
2	16.4	16.8
3	33.6	—
4	53.6	54.4
5	74.0	—
6	92.4	93.6
7	106.4	—
8	113.6	115.2
9	111.6	—
10	98.0	100.0

3.9 236 m

3.10 (a) $dA/dr = kr$; (b) $dx/dt = kx$; (c) $dH/dt = -kH^{\frac{1}{2}}$; (d) $dv/dt = kv^2$; (e) $\dot{N} = kN$; (f) $\dot{T} = -k(T - T_0)$

3.12 24.8 m s^{-1}

3.13 1.5 s; -1.5 m s^{-1}

3.14 -8 m s^{-1}

3.15 [*Note.* An arbitrary constant may be added to all these solutions]
(i) $\frac{1}{3}x^3 + 3x$; (ii) $\frac{8}{3}x^3$; (iii) $\frac{3}{2}x^4$; (iv) $\frac{1}{3}x^6$; (v) $\frac{3}{2}x^{12}$; (vi) $x^4 + \frac{5}{3}x^3 - \frac{1}{2}x^2 + 7x$; (vii) $\frac{1}{3}x^3 + 2x^2 + 4x$; (viii) $\frac{2}{3}x^3 - \frac{3}{2}x^2 - 2x$; (ix) $-\frac{1}{2}x^{-2}$; (x) $-\frac{9}{7}x^{-7}$; (xi) $\frac{1}{3}x^3 - x^{-1}$; (xii) $2x^{\frac{1}{2}}$; (xiii) $\frac{5}{3}x^{0.6}$; (xiv) $-3x^{-1} - \frac{8}{3}x^{3/2}$; (xv) $-2x^{-1} + \frac{1}{2}x^{-2}$; (xvi) $\frac{1}{5}x^5 - x^4 + 2x^3 - 2x^2 + x$.

3.16 (i) 9; (ii) 15; (iii) 15

3.17 20 m s^{-1}; 26 m s^{-1}; 3 s

3.18 10 m s^{-1}; 1 m s^{-2}; 50 m

CHAPTER 4

4.1 10.4 m at 35°13′ N of E

4.2 0

4.3 36 ms^{-1} at 56°18′ N of E

4.4 0.4 ms^{-1}; 0.3 m s^{-1}

4.5 5.0 ms^{-1}; 8.66 m s^{-1}

4.6 4 m s^{-1}; 8 m s^{-1}; 6.3 m s^{-1} at 71°34′ to bank.

4.7 1 m s^{-1}; 5 m s^{-1}

4.8 5°43′ E of S; 249 m s^{-1}

4.9 10 m s^{-1}; 8° W of S; 10 m s^{-1}; 8° E of N; 1.4 km

4.10 6.6 m at 80°24′ to x-axis

4.11 14.8 m s^{-1} at 168°52′

4.12 22.8 m; 2.28 m s^{-1}

4.13 16.9 m and 1.69 m s^{-1}, both at 71°21′ to the x-axis

4.15

t (s)	$\dot{x}$ (m s^{-1})	$\dot{y}$ (m s^{-1})	v (m s^{-1})
0.5	2.17	1.15	2.45
1.5	2.13	1.25	2.47
2.5	1.97	1.35	2.38
3.5	1.69	1.45	2.22
4.5	1.29	1.55	2.01
5.5	0.77	1.65	1.82
6.5	0.13	1.75	1.75
7.5	-0.64	1.85	1.96
8.5	-1.52	1.95	2.47
9.5	-2.52	2.05	3.25

4.17 (-1.0 m s^{-2}, $+0.1$ m s^{-2})

4.18 1000 m

4.19 20.4 s; 500 m s^{-1}; 9.8 m s^{-2}; 40.8 s; 20,400 m

4.20 10 m; 0.74 s

4.21 8414 m; 7907 m; 8165 m

4.23 19.5 m

4.24 783 m s^{-1}

4.25 2.7 m s^{-2}; centripetal

4.26 29,850 ms^{-1}; 5.94×10^{-3} m s^{-2}

CHAPTER 5

5.1 3.75 kg m s^{-1}; 53° S of E

5.2 1875 N; 53° S of E

5.3 1875 N and 3.75 kg m s^{-1}; 53° N of W; 0.625 m s^{-1}

5.4 31.6 s

5.5 44 7 s

5.6 1 m s^{-2}; 37° E of N

5.7 3.84×10^8 m

5.9 2.0×10^{30} kg
5.11 2.7 m s^{-2}
5.12 2.95×10^4 N; 1.95×10^4 N
5.14 0.087 m s^{-2}; East–West is the greater.
5.15 902 J; 823 J
5.16 250 N
5.17 −0.25 MJ
5.18 260 J; −200 J; 30 N; −60 J
5.19 115 J
5.20 42 J
5.23 Zero; zero
5.25 3.83 m s^{-1}
5.26 0.31 m s^{-1}
5.27 2.96×10^{14} J
5.28 2 N; 0.41
5.30 1.4 m s^{-1}; 3.3 m s^{-1}
5.31 2500 J; 442 J; 588 J; 1470 J

CHAPTER 6

6.1 133 N; 167 N at 37° to horizontal
6.4 118 N; 335 N at 73°41′ to ground; 4.2 m
6.5 512.5 g; 337.5 g
6.6 277 N
6.7 16.75 m s^{-1}
6.8 4 rad s^{-1}; 4 rad s^{-2}
6.10 1500 kg m^2 s^{-1}; 5 m
6.11 4500 J
6.12 3.075×10^{-2} kg m^2; 5.6 cm
6.13 3.122×10^{-2} kg m^2; friction
6.14 6.2 rad s^{-2}; 9.68 N; 59.9 J
6.15 180°
6.16 10 rad s^{-1}
6.17 32° 24′
6.18 1.87 mm
6.19 4.2×10^7 N m^{-2}
6.20 2.1×10^5 N m^{-2}
6.21 1.4×10^8 N m^{-2}
6.22 8.25×10^4 N
6.23 1750 N m rad^{-1}; 153 N m
6.24 156 N m

CHAPTER 7

7.1 310.5K
7.2 20°C
7.3 $a = +4.122 \times 10^{-3}$ deg^{-1}
$b = -1.72 \times 10^{-6}$ deg^{-2}
7.4 2400 J
7.5 1500 J; 15 kW
7.6 22.1°C
7.7 0.18 J g^{-1} deg^{-1}
7.8 308 deg
7.9 1 deg
7.11 14.9 g; 61,960 J; 8.9°C
7.12 188 kg
7.13 2.5 g
7.14 2500 A g deg^4
7.15 11.95 g s^{-1}
7.16 133 W m^{-2}
7.17 3.84 cm
7.18 1.59 cm
7.19 (a) 75.0°C; (b) 67.2°C
7.20 0.115 mm
7.21 0.164 m
7.22 5760K

CHAPTER 8

8.1 74 cm Hg
8.2 1.2 atm
8.3 7.5 liter
8.4 5.77 cm
8.5 3000 J
8.6 1.73 kg m^{-3}
8.7 0.083 lit-atm $mol^{-1}K^{-1}$
8.8 7.72×10^{-21} J; 345 m s^{-1}
8.9 0.75 atm; 1.25 atm; 25 liter
8.10 1.87×10^4 J
8.12 209°C
8.13 60%; 40%

CHAPTER 9

9.1 2.15×10^4 N m^{-2}; 9.69×10^4 N
9.2 1.03×10^7 N m^{-2}
9.3 1.19
9.4 1.17×10^3 kg m^{-3}; 1.17
9.5 1.43×10^3 kg
9.6 2.74 m^3
9.7 1.05
9.8 2.15; 0.72
9.9 1.08
9.10 1.04 m s^{-1}
9.11 6.63×10^{-3} m^3 s^{-1}
9.12 25.5 cm s^{-1}
9.13 2.4×10^{-3} kg m^{-1} s^{-1}
9.14 1 atm
9.15 26.5 cm s^{-1}
9.16 6.5×10^{-13} kg
9.17 4.2×10^{-7} m
9.18 3.8×10^4 N m^{-2}
9.19 6.7×10^4 N m^{-2}
9.20 5.96 cm
9.21 0.30 μm

CHAPTER 10

10.1 0.49 s
10.2 6.4 kg
10.3 8.7×10^3 kg m^{-3}
10.4 0.45 s
10.5 1%
10.6 $2mr^2$
10.7 2.6 beats
10.8 0.375 m
10.9 3.14 s
10.10 2.5 s; 6.25 m s^{-2}
10.11 10 cm; 3.14 s
10.12 $x = (-20 \text{ cm}) \sin (10\pi \text{ s}^{-1})t$
$\dot{x} = (-200\pi \text{ cm s}^{-1}) \cos (10\pi \text{ s}^{-1})t$
$\ddot{x} = (+2000\pi^2 \text{ cm s}^{-2}) \sin (10\pi \text{ s}^{-1})t$
10.13 31.58 m s^{-2}; 2.51 m s^{-1}; 15.79 m s^{-2}; 2.18 m s^{-1}; 0.0675 s
10.14 0 h 53 min 4 s
10.15 0.075
10.16 19.6 cm; 0.63 s
10.17 $\left(-\frac{8}{\pi} \text{ m}\right) \cos \left\{\frac{\pi t}{2 \text{ s}} + \frac{\pi}{3}\right\}$; 0.746 J
10.20 0.17 s; 6.7 J
10.21 3.16 cm
10.22 5 cm
10.23 17.4 cm

CHAPTER 11

11.1 6 cm; $\frac{2\pi}{3}$ m; $\frac{50}{\pi}$ s^{-1}; 33.3 m s^{-1}
11.2 0.5 m; 12 s^{-1}; 3 m; 36 m s^{-1}; left to right
11.3 11.2 m; 0.5 s^{-1}; 16 m; 8 m s^{-1}; right to left
11.4 2 kg
11.5 20 m s^{-1}
11.6 (a) 323 m s^{-1}; (b) 1320 m s^{-1}
11.7 (a) 0.85%; (b) 0.99%
11.8 399 c/s
11.9 288 c/s
11.10 5
11.11 1%
11.12 340 m s^{-1}
11.13 257.6 c/s
11.14 253 c/s; 259 c/s
11.15 3.62×10^{-12} m
11.16 3.62×10^{-6} m
11.17 74 dB
11.18 10^6
11.19 12 dB

CHAPTER 13

13.2 12.5 cm
13.3 6 m
13.4 0.88 m; 0.78 m
13.5 0.25 m × 0.17 m; 1.92 m
13.6 2.26×10^8 m s^{-1}
13.8 1.65
13.9 32°
13.10 1.71 m
13.12 28°; 32°; 28°
13.14 33.7°
13.16 37°
13.17 2.41
13.18 62.75°
13.20 1.33
13.22 28°

CHAPTER 14

14.2 0.33 cm
14.3 18 cm, real, −2
14.4 24 cm
14.5 −120 cm
14.6 1.82 cm
14.7 24 cm
14.8 −2 cm; 12 cm
14.9 40 cm
14.10 Real, 16 cm from concave mirror, −1/9
14.11 Plane, −3
14.12 −4.8 cm, virtual, 0.4
14.14 4.03 cm, 4 cm
14.15 5 m, 5.05 cm
14.16 8 cm
14.17 1 ft, 12.5 in
14.18 12 cm; 36 cm; −3, $-\frac{1}{3}$
14.20 60 cm; 3
14.21 3.5 cm
14.22 20 cm

CHAPTER 15

15.1 6; 6
15.2 5
15.3 3.33
15.4 From 4 to 5, from 3.33 to 3.13
15.5 −28.75 cm; 12.5, 10.7
15.6 (a) 6.25 (b) 4.4 (c) 3.4
15.7 11.7 cm
15.8 0.65 cm
15.10 200; 200 cm
15.11 26°7′
15.12 13.96 cm; −650
15.13 1.08 cm; −162.5
15.14 3.06 mm; 842; 789
15.15 3.06 mm; 789; 789
15.16 1.33 cm
15.17 0.11 cm

CHAPTER 16

16.1 (a) 50 diopters (b) 50.2 diopters (c) 54 diopters
16.2 4 diopters
16.3 3.75 diopters
16.4 $f = -400$ cm; 26.7 cm
16.5 42.9 cm focal length
16.6 37.5 cm
16.7 22.2 cm, 200 cm
16.8 40 cm; 13.3 cm; ordinary glass, 150 cm.
16.9 Bifocals of 37.5 cm and −250 cm; 32.6 cm to 107.1 cm
16.10 66.7 cm; 300 cm
16.11 4
16.12 (a) 86%; (b) 95%
16.13 1.74×10^{-4} rad

CHAPTER 17

17.1 $\sqrt{7}\,A$
17.2 $\sqrt{2}\,A$
17.3 6.5 mm
17.4 1.18 mm
17.5 None
17.6 2.94 cm
17.7 It contracts
17.8 0.628 mm
17.9 10 m
17.10 1.31×10^4
17.11 54.54π
17.12 1/16
17.14 $0.125\ \mathrm{g\ cm^{-3}}$
17.15 36°56′
17.16 63°25′
17.17 $2.28 \times 10^8\ \mathrm{m\ s^{-1}}$, 3.24×10^{-7} m
$2.25 \times 10^8\ \mathrm{m\ s^{-1}}$, 3.20×10^{-7} m
17.18 0.236 rad
17.19 3.27×10^{-3} cm

CHAPTER 18

18.2 5.5×10^{-3} rad
18.3 6.71×10^{-3} rad
18.4 0.25 mm
18.5 1.31×10^{-3} cm
18.6 7.47 km
18.8 6.25×10^9 m
18.9 1.21 cm
18.10 20
18.11 5×10^{-5} cm
18.12 2.69 μm
18.13 29°, 75°30′
18.14 6250 Å, 5000 Å, 4167 Å; 6250 Å
18.15 5.89 Å
18.16 7°30′; 3.44 cm

CHAPTER 19

19.1 From +7 to −7
19.2 0.93
19.3 1.1
19.4 1.58×10^{-4} cm; 2.76×10^{-5} cm
19.5 63; 362
19.6 3.4×10^{-5} cm
19.7 12 cm; 1.5

CHAPTER 20

20.1 0.582 rad
20.2 10.5 rad
20.3 3.0056 mm; 3.0656 mm
20.4 2.285 rad
20.5 (a) 2.65×10^{-34} m;
(b) 0.8×10^{-10} m
20.6 2.43 Å; 243 Å
20.7 (a) 2°52′ (b) 17°28′ (c) 36°52′

CHAPTER 21

21.2 48.2°, 46.6°
21.3 0.82 cm
21.4 9°30′; 22°10′
21.5 5.4 cm; 17.2 cm
21.7 Up to 525 nm
21.8 $0.101\ \mathrm{cm^{-1}}$
21.10 $1.38\ \mathrm{cm^{-1}}$

CHAPTER 22

22.2 2.37×10^{-43} to 1
22.4 6.25×10^{10}
22.6 4.9×10^{-9} C
22.7 6.25 to 1; 9×10^{-19} N
22.8 5.38×10^{-8} C; yes
22.9 4.3×10^{-7} N from center
22.10 860 N C^{-1} from center; 121.8 V
22.11 1 m from + charge: 40 cm and 15.4 cm from + charge are two points on a surface in the line of the charges
22.12 8.85×10^{-10} C m^{-2}; 927 N; 1.96×10^{20} N
22.13 27.7 V
22.14 81 μJ; 9.2×10^{-4} g
22.15 10.9 eV
22.16 4×10^{10} J; 29.5° C
22.18 2.69×10^{-7} m
22.19 13
22.20 (a) 5.93×10^{6} m s^{-1}
(b) 9.79×10^{6} m s^{-1}
(c) 7.73×10^{6} m s^{-1}
22.21 8.34×10^{-7} C; 1.5×10^{5} V
22.22 (a) All in series
(b) 3 μF in parallel with the other two in series
(c) 10 μF in parallel with the other two in series
22.23 200 μC on all; 66.7 V, 20 V, 13.3 V
22.24 100 V on all; 300 μC; 1000 μC, 1500 μC
22.25 214.3 V, 428.6 μC; 214.3 V, 1071.4 μC; 0.114 J
22.26 71.4 V, 142.9 μC; 71.4 V, 357.1 μC; 0.257 J
22.27 1.33 μF
22.28 1000 V; 2.375 mJ
22.29 4.25 μF

CHAPTER 23

23.1 12 in series
23.2 12 in parallel
23.3 24 V, 12 V
23.4 124 mA, 15 mA, 139 mA; 1.39 V
23.6 0.37 A
23.7 35 Ω, 28.6 Ω
23.8 105 Ω, 4.76 Ω
23.9 60 V, 1 Ω
23.10 9.95×10^{-3} cm s^{-1}
23.11 0.478 Ω, 0.030 Ω
23.12 1.08 mm
23.13 2 A, 14 V; $\frac{2}{3}$ A, 4 V; $\frac{4}{3}$ A, 4 V; 1 A, 18 V
23.14 427°C
23.15 750 cm
23.16 1.54 V, 7.69 V, 10 V
23.17 2500 Ω; 5000 Ω
23.18 12.2 Ω
23.19 867 deg
23.20 45 Ω, 1:1.1
23.21 49.5
23.22 2 Ω
23.24 9×10^{-8} Ω m
23.25 0.96 kW, 1.92 kW, 3.84 kW
23.26 20 Ω
23.27 480 W
23.28 8.33 V; $\frac{2}{9}$ W, $\frac{1}{4}$ W, $\frac{1}{12}$ W
23.29 163 cm^3 s^{-1}
23.30 50 Ω; 900 W, 100 W, 200 W
23.31 223 Ω
23.32 5 ms; 0.01 A; 5 ms

CHAPTER 24

24.2 11.6 deg
24.3 3.05×10^{-2} mm
24.4 1.576 g; 7.49×10^{21}
24.5 91.1 per minute
24.6 9.1 μm
24.7 46 min 52 s
24.8 $1.2 \times 10^{-3}\ \Omega^{-1}\ m^{-1}$
24.9 3.93 millimoles per liter.
24.10 59.5 Ω.
24.13 12.58 mV
24.14 64 m s^{-1}
24.15 22.5 μm s^{-1}
24.16 1.8×10^{7} V m^{-1}
24.17 0.26 MΩ

CHAPTER 25

25.1 3.14×10^{-2} Wb m^{-2}; 94.25 Wb m^{-2}
25.2 1.03 cm
25.3 14.4 cm
25.4 0.656 μs
25.5 9.8 A
25.6 (a) 0.1005 Ω in parallel
(b) 19,980 Ω in series
25.7 0.0105 Ω, 0.0948 Ω, 0.9477 Ω
25.8 180 Ω, 1800 Ω, 18000 Ω
25.9 1800 Ω
25.10 0.15 μV, 3.84 A
25.11 2.5×10^{22} m^{-1}

CHAPTER 26

26.1 0.1 V
26.2 0.6 Wb m^{-2}
26.4 0.25 V, 0.177 V
26.5 0.83 A, 6.94 W
26.6 0.625 V
26.7 4 mH
26.8 0.542 V
26.9 1 m s^{-1}

CHAPTER 27

27.2 2 A, 480 W, 339.4 V
27.3 55.54 Ω, 0.413 H
27.4 796 c/s, 31.8 c/s
27.5 (a) 1.67 A
(b) 167 V, 50.1 V, 222 V
(c) 0.6955, −45°56′
(d) 278 W
(e) 100.6 c/s
(f) 0.63
27.6 33.3 Ω, 35.7 μF
27.7 20.1 μF
27.8 10 Ω, 21.1 mH
27.10 25 mA, 50 mA, 571 mA
27.11 2.53 μF; 125.7 V, 126.0 V
27.12 8.37 pF to 0.99 pF
27.13 398 c/s; 388 c/s, 408 c/s; 20
27.14 40 W; 39.981 W
27.15 1320 V; 1 to 12
27.16 r; $5r/4$ and $5r$
27.17 5 to 1

CHAPTER 28

28.1 3.06×10^{24}
28.2 2.58×10^{-10} m
28.3 2.81×10^{25}
28.4 4.897×10^{5} C kg^{-1}
28.5 5.987×10^{6} C kg^{-1}; oxygen
28.6 19.37 cm, 1370 V
28.7 10.44 cm
28.8 (a) 0.993 MeV
(b) 0.497 MeV

CHAPTER 29

29.1 82, 206, Pb
29.2 30 per minute
29.3 0.554 μg
29.4 28 days
29.5 11.5 mCi
29.6 3.627×10^{10} s^{-1}
29.7 6 h 14 min
29.8 1.21 mg, 4.06×10^{-10} g, 5.64×10^{-17} g
29.9 2236 year
29.10 4.05×10^{9} year
29.11 3.63×10^{-3} m^3
29.12 73.2%

CHAPTER 30

30.1 2.91×10^{-10} m
30.2 3.10×10^{-14} m
30.3 2.72×10^{19} s^{-1}
30.4 3.36×10^{-19} J
30.5 0.74 eV
30.6 2.3 eV
30.7 0.104 Å, 0.116 Å
30.8 3.30×10^{-34} m; 3.16×10^{-13} m
30.9 2.02×10^{-10} m
30.10 10.2 eV
30.11 $n = 4$
30.12 1.31×10^{6} J
30.13 $n = 62$

CHAPTER 31

31.1 2.1 s
31.2 0.124 Å
31.3 12.4 Å, 1.24 Å
31.4 Molybdenum and nickel
31.5 0.161 cm, 5.46×10^{-3} cm, 1.45×10^{-3} cm
31.6 2.71×10^{-3} cm; 30.4%
31.7 1.5648 Å, 124 eV
31.8 1.53×10^{12}
31.9 6.4 m
31.10 1.16×10^{5} rad
31.11 1.64×10^{-14} cm^3

CHAPTER 32

32.1 $^{12}_{6}C$
32.2 $^{26}_{12}Mg$
32.3 7,7; 20,19; 54,42; 118,79; 146,92
32.4 5.87 MeV
32.5 15.29 MeV
32.6 8.379 MeV, 7.680 MeV; 2.06×10^{-15} m
32.7 1.74×10^{13} J mol^{-1}
32.8 1.15×10^{12} J mol^{-1}
32.9 1.82×10^{4} kW h
32.10 865 g
32.11 5.98×10^{11} kg s^{-1}

CHAPTER 33

33.1 6/7
33.2 13.1 V
33.3 2.25 V
33.4 5 kΩ
33.5 5,085 per minute
33.6 21.5 min
33.7 500, 45.5
33.8 5.5
33.9 10010100
33.10 46

INDEX